Immune Recognition

ACADEMIC PRESS RAPID MANUSCRIPT REPRODUCTION

PROCEEDINGS OF THE
NINTH LEUKOCYTE CULTURE CONFERENCE
HELD IN WILLIAMSBURG, VIRGINIA
DECEMBER 1-5, 1974

Immune Recognition

EDITED BY

Alan S. Rosenthal

Laboratory of Clinical Investigation
National Institute of Allergy and Infectious Diseases
National Institutes of Health
Bethesda, Maryland

Academic Press, Inc.

NEW YORK SAN FRANCISCO LONDON 1975

A SUBSIDIARY OF HARCOURT BRACE JOVANOVICH, PUBLISHERS

ACADEMIC PRESS, INC.
111 Fifth Avenue, New York, New York 10003

United Kingdom Edition published by
ACADEMIC PRESS, INC. (LONDON) LTD.
24/28 Oval Road, London NW1

LIBRARY OF CONGRESS CATALOG CARD NUMBER: 75-10504

ISBN 0-12-597850-2

PRINTED IN THE UNITED STATES OF AMERICA

CONTENTS

SECTION IV. EARLY EVENTS IN LYMPHOCYTE ACTIVATION II: Membrane and Biochemical Events

SECTION V. COUPLING OF CYTOPLASMIC AND NUCLEAR EVENTS TO MEMBRANE SIGNALS

SECTION VI. CELLULAR AND MOLECULAR BASIS OF CELL RECOGNITION AND INTERACTION IN NONLYMPHOCYTE SYSTEMS

SECTION VII. THE ROLE OF THE MACROPHAGE IN THE INITIATION AND REGULATION OF THE IMMUNE RESPONSE

SECTION VIII. GENETIC CONTROL OF IMMUNOCOMPETENT CELL INTERACTIONS

SECTION IX. REGULATION OF THE IMMUNE RESPONSE: CELLULAR MECHANISMS

PARTICIPANTS

G. Ashwell, Laboratory of Biochemistry and Metabolism, National Institute of Arthritis, Metabolism and Digestive Diseases, National Institutes of Health, Bethesda, Maryland 20014

F. H. Bach, Immunobiology Research Center, University of Wisconsin–Madison, Madison, Wisconsin 53706

J. A. Bash, Department of Pathology, Yale University School of Medicine, New Haven, Connecticut 06510

R. D. Berlin, Department of Physiology, University of Connecticut Health Center, Farmington, Connecticut 06032

I. Betel, Radiobiological Institute TNO, Rijswijk (Z.H), The Netherlands

H. G. Bluestein, University of California, San Diego, School of Medicine, La Jolla, California 92037

G. D. Bonnard, Laboratory of Immunodiagnosis, National Cancer Institute, National Institutes of Health, Bethesda, Maryland 20014

I. R. Cohen, Department of Cell Biology, The Weizmann Institute of Science, Rehovot, Israel

Z. A. Cohn, The Rockefeller University, New York, New York 10021

H. L. Cooper, Laboratory of Pathophysiology, National Cancer Institute, National Institutes of Health, Bethesda, Maryland 20014

B. A. Cunningham, The Rockefeller University, New York, New York 10021

J. M. Davie, Department of Pathology, Washington University School of Medicine, St. Louis, Missouri 63110

G. M. de Oliveira-Castro, Instituto de Biofísica, Universidade Federal de Rio de Janeiro, Ilha do Fundão, Rio de Janeiro, Brasil

M. Edidin, Mergenthaler Laboratory for Biology, Johns Hopkins University, Baltimore, Maryland 21218

V. P. Eijsvoogel, Central Laboratory, Blood Transfusion Service, The Netherlands Red Cross, Amsterdam, The Netherlands

R. Elie, Department of Physiology, McGill University, Montreal, Quebec H3G 1Y6, Canada

M. Feldmann, I. C. R. F. Tumour Immunology Unit, Department of Zoology, University College London, London WC1E 6BT, England

V. Ginsburg, Laboratory of Biochemical Pharmacology, National Institute of Arthritis, Metabolism and Digestive Diseases, National Institutes of Health, Bethesda, Maryland 20014

L. Glaser, Department of Biological Chemistry, Washington University School of Medicine, St. Louis, Missouri 63110

J. Gordon, Division of Surgical Research, Department of Surgery, McGill University, Montreal 110, Quebec Canada

S. Gordon, The Rockefeller University, New York, New York 10021

M. F. Greaves, I. C. R. F. Tumour Immunology Unit, Department of Zoology, University College London, London WC1E 6BT, England

I. Green, Laboratory of Immunology, National Institute of Allergy and Infectious Diseases, National Institutes of Health, Bethesda, Maryland 20014

J. W. Hadden, Laboratory of Immunopharmacology, Memorial Sloan-Kettering Cancer Center, New York, New York 10021

Charles A. Janeway, Jr., Laboratory of Immunology, National Institute of Allergy and Infectious Diseases, National Institutes of Health, Bethesda, Maryland 20014

J. G. Kaplan, Department of Biology, University of Ottawa, Ottawa, Ontario, Canada

J. A. Kapp, Department of Pathology, Harvard Medical School, Boston, Massachusetts 02115

D. H. Katz, Department of Pathology, Harvard Medical School, Boston, Massachusetts 02115

P. Lonai, Division of Immunology, Department of Medicine, Stanford University School of Medicine, Stanford, California 94305

V. C. Maino, Department of Allergy–Clinical Immunology, National Jewish Hospital and Research Center, Denver, Colorado 80206

H. O. McDevitt, Division of Immunology, Department of Medicine, Stanford University School of Medicine, Stanford, California 94305

H. Metzger, Arthritis and Rheumatism Branch, National Institute of Arthritis, Metabolism and Digestive Diseases, National Institutes of Health, Bethesda, Maryland 20014

G. Möller, The Wallenberg Laboratory, Division of Immunobiology, Karolinska Institute, 104 05 Stockholm 50, Sweden

D. E. Mosier, Laboratory of Immunology, National Institute of Allergy and Infectious Diseases, National Institutes of Health, Bethesda, Maryland 20014

E. F. Neufeld, Section on Human Biochemical Genetics, National Institute of Arthritis, Metabolism and Digestive Diseases, National Institutes of Health, Bethesda, Maryland 20014

A. Novogrodsky, Department of Biophysics, The Weizmann Institute of Science, Rehovot, Israel

J. M. Oliver, Department of Physiology, University of Connecticut Health Center, Farmington, Connecticut 06032

J. J. Oppenheim, Laboratory of Microbiology and Immunology, National Institute of Dental Research, National Institutes of Health, Bethesda, Maryland 20014

C. W. Parker, Department of Medicine, Washington University School of Medicine, St. Louis, Missouri 63110

W. E. Paul, Laboratory of Immunology, National Institute of Allergy and Infectious Diseases, National Institutes of Health, Bethesda, Maryland 20014

C. W. Pierce, Department of Pathology, Harvard Medical School, Boston, Massachusetts 02115

M. R. Quastel, The Soroka Medical Center, Ben Gurion University of the Negev, Beer-Sheba, Israel

K. Resch, Institut für Immunologie der Universität Heidelberg, D-69 Heidelberg, Germany

D. L. Rosenstreich, Laboratory of Microbiology and Immunology, National Institute of Dental Research, National Institutes of Health, Bethesda, Maryland 20014

D. L. Rosenstreich, Laboratory of Microbiology and Immunology, National Institute of Dental Research, National Institutes of Health, Bethesda, Maryland 20014

A. S. Rosenthal, Laboratory of Clinical Investigation, National Institute of Allergy and Infectious Diseases, National Institutes of Health, Bethesda, Maryland 20014

S. Roth, Mergenthaler Laboratory for Biology, Johns Hopkins University, Baltimore, Maryland 21218

D. H. Sachs, Immunology Branch, National Cancer Institute, National Institutes of Health, Bethesda, Maryland 20014

R. E. Scott, Department of Laboratory Medicine and Pathology, University of Minnesota Medical Center, Minneapolis, Minnesota 55455

G. M. Shearer, Immunology Branch, National Cancer Institute, National Institutes of Health, Bethesda, Maryland 20014

E. M. Shevach, Laboratory of Immunology, National Institute of Allergy and Infectious Diseases, National Institutes of Health, Bethesda, Maryland 20014

R. M. Steinman, The Rockefeller University, New York, New York 10021

T. Tada, Department of Pathology, School of Medicine, Chiba University, 280 Chiba, Japan

D. W. Talmage, Department of Microbiology and Medicine, University of Colorado Medical Center, Denver, Colorado 80220

M. J. Taussig, Immunology Division, Department of Pathology, University of Cambridge, Tennis Court Road, Cambridge CB2 1QP, England

E. R. Unanue, Department of Pathology, Harvard Medical School, Boston, Massachusetts 02115

E. Vitetta, Department of Microbiology, Southwestern Medical School, University of Texas Health Science Center, Dallas, Texas 75235

H. von Boehmer, Basel Institute for Immunology, 487 Grenzacherstrasse, Basel, Switzerland

M. J. Waxdal, Laboratory of Immunology, National Institute of Allergy and Infectious Diseases, National Institutes of Health, Bethesda, Maryland 20014

H. J. Wedner, Department of Medicine, Division of Immunology, Washington University School of Medicine, St. Louis, Missouri 63110

H. Wigzell, The Wallenberg Laboratory, University of Uppsala, Uppsala, Sweden

PREFACE

Perhaps by design or by tradition, the preface to an edited volume is a rather dull ritualistic acknowledgment of financial support and the contributions of one's co-workers. Many colleagues have indeed contributed to the success of the Ninth Leukocyte Culture Conference and, sadly, I find that it is impossible dutifully to acknowledge them all if this is to remain within my own page limitations.

This has been an experimental leukocyte culture conference, although we have attempted to maintain the established tradition of maximal opportunity for informal exchange and to bring together, in high density, basic and clinical scientists interested in immunobiology. To better accomplish this goal, the organizing committee modified the format to emphasize a theme, "Immune Recognition." In general, half of each working day was devoted to symposia and the other half to abstract-based presentations. Invited participants conversant with various aspects of macrophage and lymphocyte biology were chosen to form the nucleus of the central symposia. We have chosen to expand some symposia by including in this volume selected abstracts, which appear either to represent a unique point of view, or which represent an approach with potential impact on various facets of immune recognition. One symposium requires special note. Midway throught the Conference, four presentations were given by speakers from areas other than lymphocyte biology. These presentations have been included in abstract only, although their insights into nonlymphocyte cellular recognition phenomenon gave many immunobiologists considerable food for thought.

As in previous years, the preparation of these proceedings has been based on our belief that rapid publication is an extremely important goal in disseminating scientific information. We have attempted to assure publication within months after the conference by making only the necessary editorial revisions and by placing minimum emphasis on uniformity of style.

To what extent did we achieve our goal of summarizing our present understanding of a variety of immune recognition phenomenon? While I must leave that final judgment to each individual reader, I can say with some confidence that we have, at least, identified those areas of greatest controversy and ignorance. Finally, the organizing committee wishes to express its deep appreciation to the nearly five hundred participants whose attendance and enthusiasm was essential. We also wish to give special recognition to our FASEB-based conference organizing staff led by the undaunted Barbara Nichols, who, by right of effort, was the real chairman of the Ninth Leukocyte Culture Conference.

ACKNOWLEDGMENTS

The information upon which this publication is based was collected pursuant to Contract No. NIH-73-C-1178 sponsored by the NATIONAL INSTITUTE OF ALLERGY AND INFECTIOUS DISEASES, NATIONAL INSTITUTE OF DENTAL RESEARCH and the NATIONAL CANCER INSTITUTE.

CONTRIBUTORS

ADAPS, INC.
BURROUGHS WELLCOME COMPANY
CIBA-GEIGY CORPORATION
DIFCO LABORATORIES, INC.
FISHER SCIENTIFIC COMPANY
GRAND ISLAND BIOLOGICAL COMPANY
HOFFMANN-LA ROCHE, INC.
LAB-TEK PRODUCTS
MICROBIOLOGICAL ASSOCIATES
NYEGAARD & COMPANY A/S
PHARMACIA FINE CHEMICALS, INC.
G. D. SEARLE & COMPANY
SKATRON A/S
SMITH KLINE & FRENCH LABORATORIES
THE UPJOHN COMPANY
WARNER-LAMBERT RESEARCH INSTITUTE
CARL ZEISS, INC.

The chairman and the Organizing Committee of the Ninth Conference express their appreciation to the sponsoring government agencies and the industrial organizations listed above for their generous financial support which made possible the arrangements for this conference.

SECTION I

SIGNALS FOR LYMPHOCYTE ACTIVATION I: NONANTIGENIC SIGNALS AND RECEPTORS

Chairman

MELVYN F. GREAVES

University College, London, England

Co-Chairman

DAVID W. TALMAGE

University of Colorado Medical Center, Denver, Colorado

SCRATCHING THE SURFACE

Melvyn F. Greaves
I.C.R.F. Tumour Immunology Unit, Zoology Department,
University College, London, England

The organizers of this conference clearly recognize the very considerable current interest in early events in lymphocyte activation. Although we are ignorant of the precise sequence of molecular events underlying the initiation responses to either immunogens or polyclonal ligands (PCL) this state of affairs has not inhibited the flowering of hypotheses purporting to reveal the secrets of the lymphocytes black box. In introducing this subject I should like to briefly comment on how I visualize the central problems and to touch on one or two key questions. I shall not discuss at all critically the competitive models of lymphocyte activation currently under debate since space is not available and would anyway I suspect obscure the few points I wish to make.

Comparative receptor biology

We have all become used to the idea that lymphocytes are really rather unique as indeed they are, at least with respect to clonal diversity. However, in terms of molecular mechanisms involved in activation and regulation by extracellular signals there is no compelling reason to suppose that lymphocytes have devised ingenious tricks beyond the scope of other cells. There may be much to be gained therefore by considering what is already known of receptor function in other biological systems and in seeking to identify common or ground rules. Even a cursory consideration of the multiplicity of recognition systems in biology leads to an appreciation of the great diversity of chemical signals (hormones, neurotransmitters, pheromones, etc.) and the corresponding variety of complimentary recognition structures of receptors, usually associated with cell surfaces. Multiple signals impinge upon individual cells to elicit or regulate a preprogrammed performance. For example, rat adipocytes are responsive to at least four different hormones (1) and a similar 'sensory spectrum' has been described for an adrenal cortex tumor (2) and a brain astrocytoma cell line (1). It seems highly probable that this arrangement is not unique: indeed we already know that lymphocytes, aside from their receptors for antigen, have cell surface binding sites for histamine, catecholamines and several hormones (3,4). Each specialized cell type, like each animal species, probably has its own sensory world or "merkwelt" (5) as reflected in its diversity of cell surface receptors or sense organs respectively.

This wealth of signal diversity and sensory imput of individual cells needs not however imply that we have to consider multiple mechanisms for the transduction (6) of signals across membranes. It would not be very surprising to find that relatively few coupling devices were used by many different signal response systems. There are already strong hints that this is indeed the case. In the instance of the three cell types described above to have multiple distinct hormone receptors there is reasonably convincing evidence that each receptor type having bound its corresponding ligand initiates the cellular response via the same or shared device - adenyl cyclase. The ubiquitous occurrence of the cyclic nucleotides cyclic AMP and cyclic GMP and the abundant evidence, at least for the former, of second messenger or regulatory qualities (7,8) provides ample justification for the view that these small molecules are likely to play an important role in signal-response coupling in many different systems. Similarly, calcium ions appear to have potential to influence numerous critical metabolic pathways and considerable evidence exists for a general regulatory role for this cation (9,10).

The significance of calcium and the cyclic nucleotides for immune responses is beginning to emerge (reviewed in 10,11, 12). Cyclic AMP and cyclic GMP or drugs/hormones influencing the intracellular levels of these substances clearly can modulate immunological responses (e.g. proliferation, cytoxicity, antibody secretion, mast cell degranulation); usually the two nucleotides operate antagonistically with cAMP depressing and cGMP enhancing the response under consideration which may reflect a basic competitive dualism of this regulatory system to which many cell types are subject (8). A definite second messenger role (i.e. equivalent to cAMP in many hormone responses) for cGMP has yet to be established although the early transient increase cGMP levels inactivated T and B lymphocytes (13,14) encourages one in this belief. cGMP derivatives have recently been reported to directly induce at least a limited degree of lymphocyte proliferation (14,15). In our experience using a wide variety of conditions this has not yet been possible. However, it is worth bearing in mind the general difficulty that exists in demonstrating cGMP effects in vitro even using systems where a messenger role in vivo seems reasonably clear (8). The fibroblast activation system studied by Rudland and colleagues (16,17) provides a very interesting parallel with lymphocytes and it is significant that cGMP initiates division in these cells, and other stimulating ligands (FGF - a pituitary hormone and serum mitogens) rapidly activate guanyl cyclase. Insulin appears to have similar effects on rat adipocytes and liver cells (18). Fertilization of eggs may also elevate cGMP levels (8). It would therefore not be surprising if cGMP was a fairly widespread coupling device involved in the initiation of proliferation in resting or quiescent cells including lymphocytes.

With respect to calcium ions we know most immunological phenomena in vitro to be calcium dependent (9) and recent important experiments with the calcium ionophore A23187 provide evidence for a possible primary role of calcium ions in initiating activation of lymphocytes (19) and mast cells (20). A23187 also has a capacity to mimic epinephrine effects (21) and to induce cell division in eggs of several vertebrate and invertebrate (22) species. It may be very significant that whereas calcium antagonizes most cAMP responses it generally facilitates cGMP associated responses (8).

Another important 'common' aspect of numerous signal response systems is the 'pleiotypic' (23) pattern of early membrane changes (24,25). Within minutes of ligand binding changes can be detected in binding site distribution (i.e. clustering), alterations in membrane fluidity (increased mobility of incorporated spin label, 26), phospholipid turnover, permeability to calcium, potassium, amino acids, nucleosides, phosphate ions and sugars and activation of membrane bound enzymes (e.g. Na^+ K^+ independent ATP'ase, adenyl cyclase, guanyl cyclase). Most of these changes have been demonstrated with lymphocytes (usually T cells with lectins; 27,28,29); however it is very unlikely that anyone is peculiar to lymphocytes. The causal relationships between these different measures of membrane reponse have not been identified for any receptor system.

Lymphocyte activation

The central problem can be expressed as follows. We have a great diversity of cell surface receptor sites for extracellular (soluble or cell bound) signals, a conservative pattern of membrane response and evidence for the utilization of common coupling devices. How then do we go about determining the precise causal sequence of events for any particular system? We clearly need to identify three elements, the receptors involved, the membrane transduction mechanism and the coupling device or 'second messenger' for initiating the intracellular cascade of events.

With respect to activation of T and B lymphocytes the following points might meet with common agreement:

1. Activation is essentially a cell surface initiated event since insolublized lectins (30-32) and antigens (33) in

parallel with other non-steroid ligands (polypeptide and peptide hormones (34,35), catecholamines (36), and perhaps embryonic inducers (37) provide efficacious stimulants.

2. The specificity of B cell response (and that of mast cells and basophils) to antigens resides in cell surface associated immunoglobulins (Ig) (38-40) and that of T cells in as yet uncharacterized receptors possibly products of genes within the I region of the major histocompatability locus of the species (43).

3. Cell interactions between macrophages, B cells and T cells are of considerable importance to the full expression of immune responses (40-42). Several non-antigen specific signals may be involved but interactions between I region coded glycoprotein molecules could well be of central importance (43).

4. The structural attributes of an immunogen over and above the stereochemistry of individual determinants has an important influence on the activation process (44). The same is probably true of polyclonal ligands (PCL) (27). Thus, size, digestibility, the existence of repeating identical determinants and charge of an immunogen profoundly influence the quality of response evoked. These data can be interpreted to suggest that the physical form of antigen presentation and the number of repetitive interactions or persistence of interactions carried out may be crucial. In particular, as argued forcibly by Feldmann (45), the extent of cross-linkage of Ig molecules on B lymphocytes into a two dimensional lattice may be of considerable importance. This might simply reflect the avidity premium of simultaneous multi-point binding; however recent emphasis on membrane fluidity (46) and the capacity of cell surface proteins to be aggregated (47,48) raises more intriguing possibilities.

5. cGMP and calcium ions may be critically involved in the initiation of proliferation and in subsequent regulation or modulation of ongoing responses (see above).

These observations provide some of the 'superstructure' of lymphocyte activation but leave several points unresolved. A key question is undoubtedly the nature of the relationship between the receptor structure itself and the transduction mechanism, i.e., the molecular device for converting receptor site occupation into intracellular response (6). If, for the sake of argument, we accepted that immunoglobulin was the only physiological relevant receptor for antigen and calcium influx and/or guanyl cyclase activation involved with the second signal(s) we are still left with the fundamental problem of linking the two events (i.e., signal 1 with signal 2). The theoretical possibilities for Ig, or in principle any other receptor, are as follows:

1. That it is physically part of, or linked to, the transductional device, i.e., an allosteric enzyme such as a cyclase or a calcium gating protein (c.f. the ionophore hypothesis for the acetylcholine receptor (49)).

2. That it had an affinity (or <u>increased</u> affinity) for the transduction device when its ligand binding sites are occupied. For example, cross-linkage and concomitant aggregation of Ig might facilitate an interaction with the IgFc receptor (of the same cell) which in turn might have transductional qualities (50). This would be an attractive idea since the IgEFc receptor is presumably a crucial component in the activation of mast cells by antigen and anti-Ig (51).

3. Ig molecules, either occupied or unoccupied, have no physical contact with the relevant transductional molecules but influence the latter indirectly (e.g., via changes in the membrane fluidity).

In the past, studies on oligomeric enzymes and membrane cooperativity phenomena have tended to promote explanation 1 (52). Although allosteric or regulating enzymes are very probably involved in membrane associated recognition events the existence of membrane fluidity greatly increases the range of potential interactions and relieves any demand for intimate physical proximity between interacting components.

This set of opportunities will equally well suffice for receptors on B cells, T cells, macrophages and not only for antigens but PCL (e.g., lectins) also. However with respect to cell surface Ig on B cells we have to contend with another possibility. Coutinho and Möller have suggested that Ig may be an 'acceptor' rather than 'receptor' in the sense that it serves <u>solely</u> to concentrate a nonspecific (i.e., potentially polyclonal) activating signal onto the adjacent surface of the same cell (53,54). The nonspecific signal is envisaged as either being part of the antigen itself (e.g., bacteria polysaccharides and most other T cell independent antigens), but distinct from the antigenic determinants, or a PCL derived from T cells and/or macrophages. Whether one then considers Ig as "actively involved in initiating responses" or not is clearly a matter of semantics. If the pharmacological definition of receptors (55) is accepted and these molecules considered as that part of the signal transduction mechanism responsible for discriminatory binding then by all accounts Ig is active on B lymphocytes. The question of whether another nonantigen specific molecule is necessarily involved in B cell activation is however very important not only for an understanding of the 'immunogenicity' of different antigens (e.g., bacterial polysaccharides - versus hapten-protein carriers) but also for the more basic issue of the activating mechanism itself. Despite the attractive simplicity of the Coutinho-Möller hypothesis, the question of the precise role(s) of B cell surface Ig is, in my view, still quite open. Evidence that B lymphocytes can be triggered independently of Ig (i.e. by PCL) in no way denies that they can also be activated via Ig. In fact limited evidence compatible with an active or direct receptor role for Ig is available:

1. Rabbit B cells are activated by anti-Ig (56) and with additional T cell help develop into Ig secreting cells (57, 58).

2. Anti-Ig (and presumably allergens also) activate mast cells and basophils via Ig molecules which are themselves bound to IgE Fc receptors (51, 59).

3. Purified mouse B lymphocytes can be activated by insolubilized haptens which have no convincing polyclonal activating capacity (33,60) and therefore may interact solely with surface Ig.

Although several interpretations of these observations may be possible it seems reasonable to conclude that it is still premature to delegate or deny receptor function to particular surface structures. This in no way negates the heuristic value of models which do just this.

Activating or Regulating Signals?

Considerable debate revolves around the question of how many and which particular signals are involved in B cell activation (Fig. 1). Proffered models vary from the single nonspecific signal (53) to various double signal models (14, 61,62) which portray antigen -Ig interaction as signal one generator and a variety of candidates as initiating signal two (e.g., T cell factors, activated C3). They provide different interpretations of thymus dependency and tolerance and also suggest a different significance for activation of lymphocytes by PCL (60) (Fig. 1). These attractive models (hum hum!) may well be oversimplifications in some respects.

Unlike other rapid signal response systems (e.g., catecholamine responses, mast cell degranulation) lymphocyte activation has two disadvantages as a general model for analyzing receptor mechanisms:

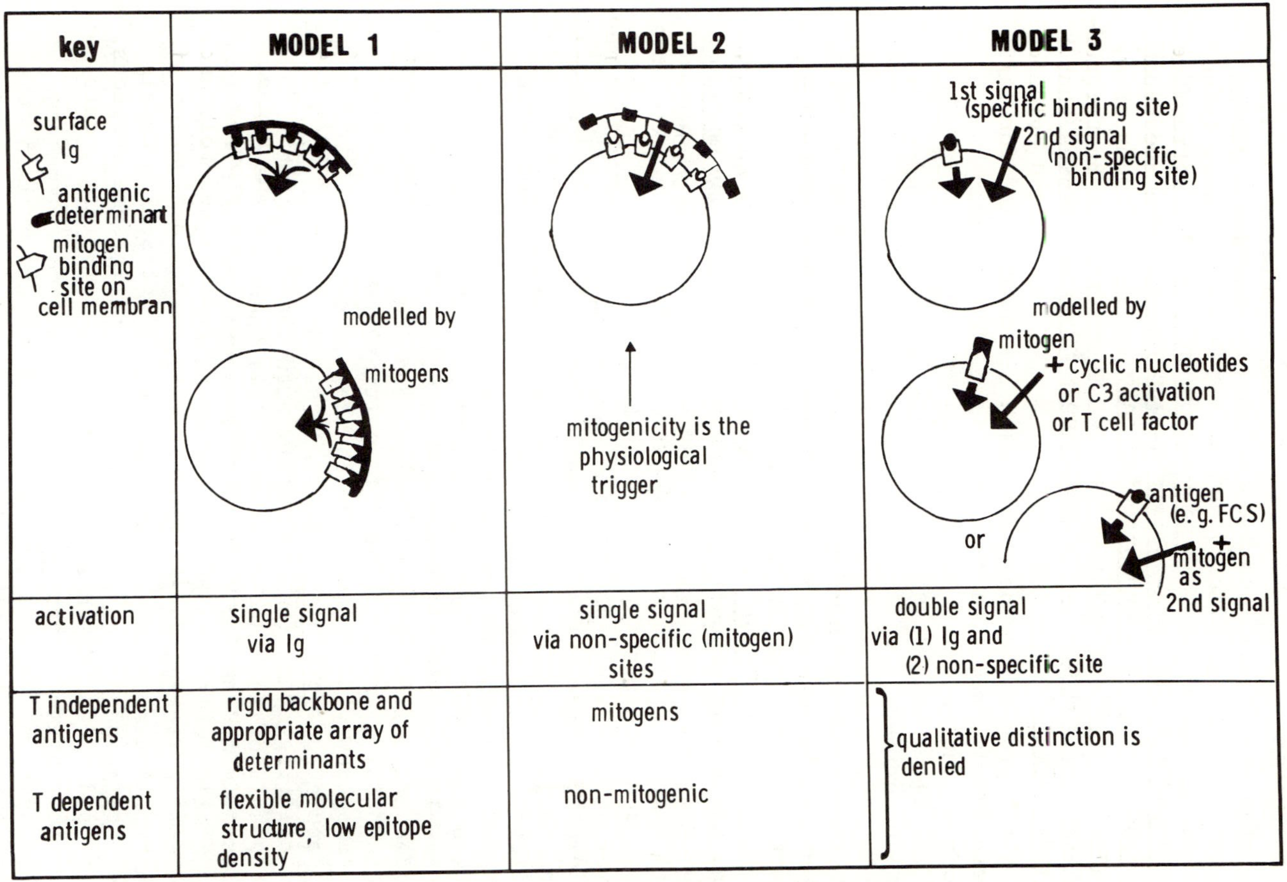

Fig. 1 Models for B cell activation by antigens and poly-clonal ligands.

1. The responses we are usually interested in take at least two to three days to be fully expressed.

2. The cell types involved are very heterogeneous with respect to both specificity and more general but nevertheless relevant, properties. A dissection and eventual description of B and T lymphocyte activation mechanisms might therefore depend upon a greater appreciation of the subset heterogeneity of both B and T cells and the sequential events in a protracted response which might be subject to separate regulatory signals.

This problem has been discussed elsewhere (60) but can be briefly analyzed with reference to a simple diagrammatic representation of the life history of single T or B cell (Fig. 2). Three categories of cell surface reactive signals are of primary significance - inducing, activating and regulating. Whether a particular signal activates a cell can be seen to depend upon what one considers 'activation'. If for example one analyzes B cell activation by IgG antibody secretion in vitro it is difficult to distinguish those signals that initiate the proliferative response upon which the end result depends and other signals which may modulate the response, e.g., by promoting the development of plasma cells from proliferating precursors. When such a dissection has been attempted by subjecting B cells to T cell products over differing periods of time the results (63) support the view that putative second signals have a regulatory rather than primary inductive role - an interpretation which conflicts with the basic tenets of the Coutinho-Möller (single signal) and Cohn-Watson-Bretcher (double signal) models, but which seems very likely to be correct. It also seems quite possible that clarification of the roles of cyclic nucleotides and calcium ions will also depend upon dissection of sequential events. These difficulties are compounded by the heterogeneity of lymphocytes themselves not only into T and B populations but also into subsets. Most current models of B cell triggering (unfortunately with the exception of the proposals of Katz and colleagues (64) there is a veritable dearth in models for T cells) involve the implicit assumption that all B cells can be considered equal. 'Conventional wisdom' reflected in the popular models dictates, for example, that molecular interactions involved in the activation of IgM or IgG secreting precursors by T cell dependent antigens can be deduced from a study of T cell independent responses or alternatively that all responses are T cell dependent. This approach denies, or fails to adequately explain, the qualitative distinction between these two types of antibody

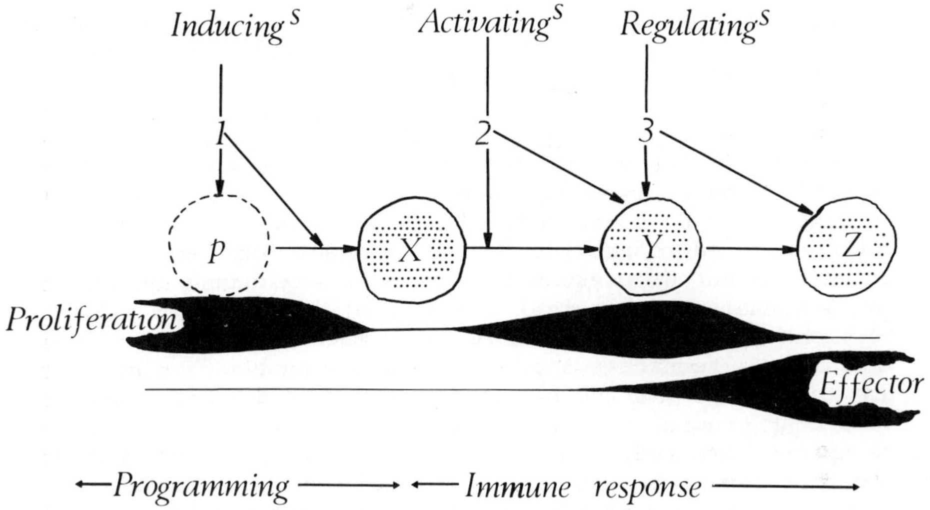

Fig. 2 Signals in the life of lymphocytes.
Inducing^S (signal): Influence of thymic or bursal - equivalent epithelia on precursor (p) differentiation into T or B lymphocytes respectively.
Regulating^S (signal): e.g. antigen-antibody (+C3) complexes, T cell 'factors', hormones.
Activating^S (signal): primarily antigen itself.
(Effector: i.e. expressions of immunological activity such as antibody secretion, MIF release, cytotoxicity.)

responses and ignores the considerable evidence for B cell heterogeneity (e.g., responses to PCL (60,65) and antigens (66,67), adherence and charge properties (68,69), recirculation capacity (70)). We have argued elsewhere (60) that responses of different B cell subsets might well differ in their susceptibility and/or dependence upon various inductive and regulatory signals. Evidence for T cell heterogeneity is equally compelling (71,72) and Katz et al. have discussed (64) how differential expression of cell surface structures (probably I region coded (73)) might critically affect T cell responses. An emphasis on subset heterogeneity may seem an unnecessary complication with respect to the basic issue, however, I would be much happier to see it disposed of satisfactorily rather than ignored.

Polyclonal Ligands versus Antigens

This still leaves us with the question of the significance of activation of lymphocytes by PCL (27,74). This is a crucial question since not only is a lymphocyte activation by PCL considered to provide a general model for analyzing cell triggering but we presume that it also tells us something about normal phyisological activation; indeed, with respect to biochemical events and molecular mechanisms, there is no immediate alternative approach to studying how lymphocytes are activated. At the same time we may well be missing something crucial to antigen induced responses, e.g., actual receptors involved, number of signals, regulatory influences etc. Obviously the nature of the cell surface structure involved in binding of PCL and subsequent activation is of particular significance in this respect. How many different receptors are involved and what is the physiological relevance of these sites with respect to normal antigen initiated activating or regulatory signals? The great chemical and structural diversity of PCL for B cells (e.g. polysaccharides, polymerized proteins, and lipid A) and T cells (e.g. protein lectins and antibodies and sodium periodate) would appear to demand the existence of several different receptors. In the absence of strong evidence to the contrary, I would suggest the general following view:

1. Most individual PCL interact with a number of different surface structures, e.g., charged groups (polyanions), different glycoproteins with similar sugar residues (lectins), glycoproteins with either shared or distinct antigenicity (heterospecific antilymphocyte antibodies). However of these distinct binding sites only a very limited number may have the capacity to initiate a response. Novogrodsky has suggested (75,76) that the same T cell membrane site might be

the 'target' for mitogenic oxidizing agents (periodate, galactose oxidase) and lectin (soybean). Similarly Toyoshima et al. (77) have suggested that mitogens from L. culinaris, W. floribuna, PHA and Con A might all act on a 'common' receptor glycoprotein bearing ($(Man)_3$ - GLcNAc - GLcNAC) residues. It is interesting that unabsorbed antisera to B lymphocytes although binding well to T cells are nonactivating (Greaves and Verbi in unpublished observations, Woody, J. in personal communication) implying that the relevant structure involved in the T cell response to mitogenic antithymocyte/T cell sera might be a T cell specific entity. Could this same molecule be the receptor also for responses induced by lectins and alloantigens, i.e., is the T cell antigen receptor and PCL receptor the same glycoprotein (as indicated in Fig. 3)? We are currently investigating this possibility and preliminary results suggest that mitogenic ALS and lectins may indeed compete for a common receptor. Recent 'blocking' experiments with antisera to B_2 microglobulin can be similarly interpreted (78).

A parallel situation may again be provided by mast cells where the capacity of concanavalin A to degranulate seems to relate to its binding to surface associated Ig - the physiological receptor for antigen (74). With respect to B cell mitogens and with the exception of anti-Ig, there is very little evidence that any PCL interacts directly with Ig; however as Coutinho and Möller have argued, PCL and antigens might both directly or indirectly affect some common non-antigen specific surface structures capable of initiating or modulating B cell responses. If these interpretations and speculations are valid, then we are provided both with a means of isolating physiological receptors and a route to unraveling the triggering sequence.

2. Some PCL may activate lymphocytes by a process which bypasses cell surface receptors. I suspect LPS acts in this way and by integrating into membrane lipids. The corollary of this interpretation would be that experiments with LPS may tell us something about the membrane events involved in physiological activation but they do not necessarily bear on receptors involved, assuming these to be in the main proteins. It is interesting in this respect that LPS does not appear to induce changes in phospholipid turnover (80) in contrast to lectins (on T cells (81)) and anti-immunoglobulin (on pig - B cells? (82)).

3. If any correlation exists at all between ligand-receptor polar redistribution (i.e., capping) and response of lymphocytes (83,84,85) or mast cells (86), it is inverse. However

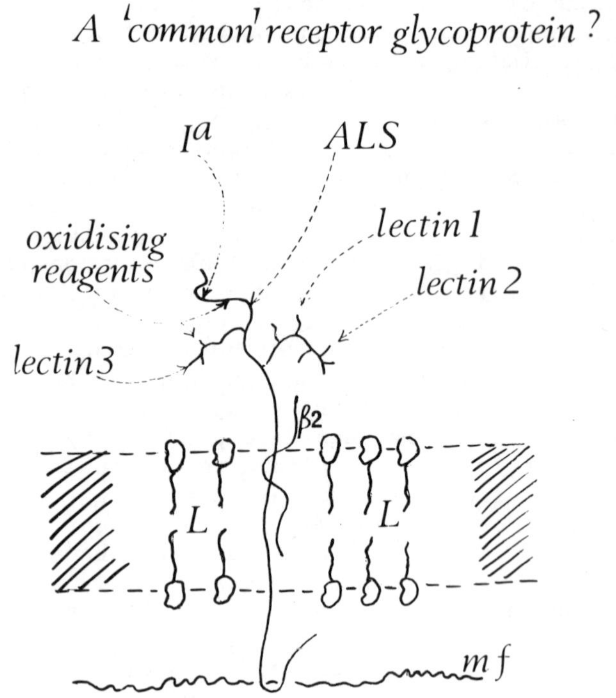

Fig. 3 L Lipid
Mf Microfilaments
ALS Antilymphocyte sera
I^a: I^a antigens of lymphocytes (Potent MLR <u>inducing</u> surface structures)
C.M. Cell membrane
β2 = β2 microglobulin

the demand for at least divalency (27,87) is unlikely to be simply a reflection of avidity requirements and may therefore implicate receptor 'micro-patches' as being of crucial importance.

These points along with other details described above (i.e., involvement of calcium ions and cyclic nucleotides) provide ingredients for further 'model' constructions.

An eminent immunologist recently suggested that the day of the lectin had now passed (for immunologists only presumably!) and that in the future we would attempt to unravel lymphocyte activation mechanisms using more physiological systems - perhaps for example, stimulating purified antigen reactive clones. It seems to me that this is somewhat optimistic with respect to the alternative proposed but more significantly it fails to recognize the important fact that lymphocytes, clever little beasts that they are, are not too smart at distinguishing antigens from PCL. Studies on complex collaborative responses with antigens are essential for constructing a total picture of the cellular physiology and regulatory mechanisms of immune responses, however polyclonal stimulants provide a realistic approach to the analysis of activating mechanisms and in my view will probably provide the crucial answers.

REFERENCES

1. Perkins, J.P., in Advances in Cyclic Nucleotide Research, Vol. 3, Raven Press, 1973.
2. Schorr, L., Ratham, P., Saxena, B.B. and Ney, R.L., J. Biol. Chem. 246: 5806, 1971.
3. Hollenberg, M.D. and Cuatrecasas, P., in Control of Proliferation of Animal Cells, eds. B. Clarkson and R. Baserga, Cold Spring Harbour Laboratory, Cold Spring Harbour, 1974, p. 423.
4. Weinstein, Y., Melmon, K.L., Bourne, H.R. and Sela, M., J. Clin. Invest. 52: 1349, 1973.
5. von Uexküll, quoted in Tinbergen, N., The Study of Instinct, Clarendon Press, Oxford, 1951.
6. Rodbell, M., in Current Topics in Biochemistry, eds. C.B. Anfinsen, R.F. Goldberger and A.N. Schechter, Academic Press, New York, p. 187, 1972.
7. Sutherland, E.W. and Robinson, C.A., Pharmacol. Rev. 18: 145, 1966.
8. Goldberg, N.D., O'Dea, R.F. and Haddox, M.K., Adv. Cyclic Nucleotide Res. 3: 155, 1973.
9. MacManus, J.P., Whitfield, J.F., Boynton, A.L. and Rixon, R.H., Adv. Cyclic Nucleotide Res. 5: in press.

10. Rasmussen, H., Goodman, D.B.P. and Tanenhouse, A., CRC Critical Rev. Biochem. 1: 95, 1972.
11. Lichenstein, L.M., in Progress in Immunology II, eds. L. Brent and E.J. Holborow, ASP Biological and Medical Press B.V. (North Holland Division), Amsterdam, 1974, in press.
12. Kaliner, M. and Austen, K.F., Biochem. Pharmacol. 23: 763, 1974.
13. Hadden, J.W., Hadden, E.M., Haddox, M.K. and Goldberg, N.D., Proc. Nat. Acad. Sci. USA 69: 3024, 1972.
14. Watson, J., in The Immune System: Genes, Receptors and Signals, eds. E. Sercarz, A.R. Williamson and C.F. Fox, Academic Press, New York, 1974, p. 511.
15. Weinstein, Y., Chambers, D.A., Bourne, H.R. and Melmon, K.L., Nature 251: 352, 1974.
16. Rudland, P.S., Gospodarowicz, D. and Seifert, W., Nature 250: 741, 1974.
17. Seifert, W.E. and Rudland, P.S., Nature 248: 138, 1974.
18. Illiano, G., Tell, G.P.E., Seigel, M.I., Cuatrecasas, P., Proc. Nat. Acad. Sci. USA 70: 2443, 1973.
19. Maino, V.C., Green, N.M. and Crumpton, M.J., Nature 251: 324, 1974.
20. Foreman, J.C., Mongar, J.L. and Gomperts, B.D., Nature 245: 249, 1973.
21. Selinger, S., Eimerl, S. and Schram, M., Proc. Nat. Acad. Sci. USA 71: 128, 1974.
22. Steinhardt, R.A., Epel, D., Carroll, E.J. and Yanagimachi, R., Nature 252: 41, 1974.
23. Hershko, A., Maont, P., Shields, R. and Tomkins, G.M., Nature New Biol. 232: 206, 1971.
24. De Asua, L.J. and Rozengurt, E., Nature 251: 624, 1974.
25. Reed, P.W. and Lardy, H.A., in Role of Membranes in Metabolic Regulation, eds. M.A. Melman and R.W. Hanson, Academic Press, New York, 1972, p. 111.
26. Barnett, R.E., Scott, R.E., Furcht, L.T. and Kersey, J. H., Nature 249: 466, 1974.
27. Greaves, M.F. and Janossy, G., Transpl. Rev. 11: 87, 1972.
28. Ferber, E. and Resch, K., The lymphocyte: structure and function, ed. J.J. Marchalonis, M. Dekker, New York, 1974.
29. van den Berg, K.J., The role of amino-acids in the mitogenic activation of lymphocytes, Doctoral Thesis, published by Radiobiological Inst. Organization for Health Research TNO, Rijswijk (2H), The Netherlands.
30. Greaves, M.F. and Bauminger, S., Nature New Biol. 235: 67, 1972.

31. Betel, I. and van den Berg, K.J., Eur. J. Biochem. 30: 571, 1972.
32. Andersson, J., Edelman, G.M., Möller, G. and Sjöberg, O., Eur. J. Immunol. 2: 233, 1972.
33. Feldmann, M., Greaves, M.F., Parker, D.C. and Rittenberg, M., Eur. J. Immunol. 4: 591, 1974.
34. Cuatrecasas, P., Proc. Nat. Acad. Sci, USA 63: 450, 1969.
35. Hecht, J.P., Dellacha, J.M., Jantome, J.A. and Paladini, A.C., FEBS Letters 20: 83, 1972.
36. Venter, J.C., Dixon, J.E., Maroko, P.R. and Kaplan, N.O., Proc. Nat. Acad. Sci, USA 69: 114, 1972.
37. Levine, S., Pictet, R. and Rutter, W.J., Nature New Biol., 246: 49, 1973.
38. Makela, O. and Cross, A.M., Prog. Allergy 14: 154, 1970.
39. Paul, W.E., in 'Defense and Recognition' MTP Int. Rev. Sci. Biochem. Series 10: 329, 1964.
40. Greaves, M.F., Owen, J.T. and Raff, M.C., T and B lymphocytes: their origins, properties and roles in immune responses, Excerpta Medica Amsterdam, 1973.
41. Katz, D.H. and Benacerraf, B., in The Immune System, Genes, Receptors, Signals, eds. E. Sercarz, A.R. Williamson and C.F. Fox, Academic Press, New York, 1974, p. 569.
42. Feldmann, M. and Nossal, G.J.V., Transpl. Rev. 13: 3, 1972.
43. Katz, D.H., Graves, M., Dorf, M.E., DiMuzio, H. and Benacerraf, B., J. Exp. Med., 1974 in press.
44. Sela, M., Harvey Lect. 67: 213, 1972.
45. Feldmann, M., Contemp. Topics in Molec. Immunol. 3: 57, 1974.
46. Singer, S.J. and Nicolson, G.L., Science 175, 720, 1972.
47. Raff, M.C. and de Petris, S., Fed. Proc. 32: 48, 1973.
48. Unanue, E.R., Perkins, W.D. and Karnovsky, M.J., J. Exp. Med. 136: 885, 1972.
49. Karlin, A., Cowburn, D.A. and Reiter, M.J., in Drug Receptors, ed. H.P. Rang, MacMillan, London, 1972, p. 193.
50. Ramasamy, R., Munro, A. and Milstein, C., Nature 249: 573, 1974.
51. Stanworth, D.R., Clin. Exptl. Immunol. 6: 1, 1970.
52. Changeux, J.P., in Nobel Symposium II, eds. A. Engström and B. Strandberg, Wiley Press, New York, 1969, p. 235.
53. Coutinho, A. and Möller, G., Scand. J. Immunol. 3: 133, 1974.
54. Coutinho, A., Gronowicz, E. and Möller, G., in Proceedings of the Ninth Leukocyte Culture Conf., 1975.
55. Fastier, F.N., Pharmacol. Rev. 4: 51, 1964.

56. Shek, P.N., Chou, C.T., Dubiski, S. and Cinader, B., Int. Arch. Allergy 46: 753, 1974.
57. Frensdorff, A., Jones, P.P., Netter, Y.B., Cebra, J.J. and Mage, R., Science 171: 391, 1971.
58. Kishimoto, T. and Ishizaka, K.J., Immunol., in press.
59. Lichtenstein, L.M., in Cellular and Humoral Mechanisms in Anaphylaxis and Allergy, ed. H.Z. Movat, Karger, Basel, Switerland, 1969, p. 176.
60. Greaves, M.F., Janossy, G., Feldmann, M. and Doenhoff, D., in The Immune System, Genes, Receptors and Signals, eds. E. Sercarz, A.R. Williamson and C.F. Fox, Academic Press, New York, 1974, p. 271.
61. Dukor, P. and Hartmann, K.U., Cell Immunol. 7: 349, 1973.
62. Schrader, J.W. and Nossal, G.J.V., J. Exp. Med. 139: 1582, 1974.
63. Dutton, R.W., in The Immune System, Genes, Receptors and Signals, eds. E.E. Sercarz, A.R. Williamson and C.F. Fox, Academic Press, New York, 1974, p. 485.
64. Katz, D.H. and Benacerraf, B., Transpl. Rev. 22: 1974, in press.
65. Gronowicz, E., Coutinho, A. and Möller, G., Scand. J. Immunol. 3: 413, 1974.
66. Katz, D.H., Hamaoka, T., Newburger, P.E. and Benacerraf, B., J. Immunol. 113: 974, 1974.
67. Klinman, N.R., Press, J.L., Pickard, A.R., Woodland, R.T. and Dewey, A.F., in The Immune System, Genes, Receptors and Signals, eds. E.E. Sercarz, A.R. Williamson, C.F. Fox, Academic Press, New York, 1974, p. 357.
68. Schlegel, R.A. and Shortman, K., Cell Immunol. 1974, in press.
69. Schlegel, R.A., von Boehmer, H. and Shortman, K., Cell Immunol. 1974, in press.
70. Stober, S., Nature New Biol. 237: 247, 1972.
71. Cantor, H., T Cells and the Immune Response. Prog. Biophys. Molecular Biol. 25: 71, 1972.
72. Stobo, J.D. and Paul, W.E., J. Immunol. 110: 362, 1973.
73. Shreffler, D.C. and David, C.S., Adv. Immunol. 1974, in press.
74. Andersson, J., Sjöberg, O. and Möller, G., Transplant. Rev. 11: 132, 1972.
75. Novogrodsky, A. and Katchalski, E., Proc. Nat. Acad. Sci., 70: 1824, 1973.
76. Novogrodsky, A., in Proceedings of the Ninth Leukocyte Culture Conf., 1975.
77. Toyoshima, S., Fukuda, M. and Osawa T., Biochem. 11: 4000, 1972.
78. Solheim, B.G., Transpl. Rev. 21: 35, 1974.

79. Magro, A.M., Nature 249: 572, 1974.
80. Betel, I., Martijnse, J. and van den Berg, K.J., Cell Immunol., 1974, in press.
81. Resch, K. and Ferber, E., Eur. J. Biochem. 27: 153, 1972.
82. Maino, V.C., Hayman, M.J. and Crumpton, M.J., Biochem. J., 1974, in press.
83. Greaves, M.F., FEBS 26: 17, 1972.
84. Elson, C.J., Singh, J. and Taylor, R.B., Scand. J. Immunol. 2: 143, 1973.
85. Loor, F., Eur. J. Immunol. 4: 210, 1974.
86. Becker, K.E., Ishizaka, T., Metzger, H., Ishizaka, K. and Grimley, P.M., J. Exp. Med. 138: 394, 1973.
87. Fanger, M.W., Immunol. Communication 1: 315, 1972.

RECOGNITION OF CHEMICALLY MODIFIED AUTOLOGOUS CELLS: IMPORTANCE OF THE MURINE MAJOR HISTOCOMPATIBILITY COMPLEX

Gene M. Shearer
Immunology Branch, National Cancer Institute
Bethesda, Maryland

Recognition of and reactivity against self-antigens is generally considered to be an exceptional immune phenomenon, and has been attributed to the appearance of "forbidden clones" of lymphocytes possessing specificity for autologous antigens (1-3). However, an increasing number of experiments demonstrating immunological auto-reactivity have been reported, particularly those in which the induction phase of the response was performed in vitro (4-6). In such experimental models, reactivity to "self immunogens" can be divided into two broad categories: those involving autologous components which have not been intentionally (or knowlingly) altered, and those involving self entities which have been experimentally modified by agents such as chemicals or viruses. In vitro induction of cell-mediated cytotoxicity against syngeneic murine tumor cell antigens (7-8), as well as against tissue-specific syngeneic cellular antigens in the rat (4-6), mixed lymphocyte reactivity (MLR) between syngeneic cells of certain mouse strains (9-11), and experimental autoimmune reactions such as NZB disease (12) and thyroiditis (13) are examples that fall within the first category.

Chemical modification of syngeneic serum or cellular components also leads to immune reactions, the specificity of which: (a) may be directed against the modifying agents (haptens) or against new antigenic determinants created as a result of the modification; and (b) appears to involve "modified self" and possibly the modifying agent as well. Thus, Rathbun and Hildemann observed that trinitrophenyl-(TNP-) modified mouse serum albumin was a good immunogen for the generation of TNP-specific antibodies in certain inbred mouse strains (14). More recently, Rubin and Wigzell have investigated antibody production to dinitrophenyl-(DNP-) modified mouse serum albumin (MSA) (15-17). It was found that antibodies could be elicited to DNP alone or to DNP as well as to new antigenic determinants presumably involving modified MSA,

---

The author gratefully acknowledges the skilled technical assistance of Carol Garbarino and preparation of the manuscript by Marilyn Schoenfelder.

depending on the extent of DNP-lation. Furthermore, these investigators demonstrated that both B- and T-lymphocytes exist with specificity for the new antigenic determinant and for the hapten (17). MLR and blast transformation have been reported in the guinea pig for modified autologous serum albumins (18,19). An interesting example of viral infection which may involve modification of autologous lymphocytes has been reported by Zinkernagel and Doherty (20,21). These investigators observed that mice injected with lymphocytic choriomeningitis (LCM) virus possessed T-lymphocytes capable of lysing LCM-infected target cells of the same H-2 haplotype, but not LCM-infected allogeneic targets nor syngeneic targets infected with another virus (21).

Susceptibility to LCM infection in mice is a genetically controlled phenomenon which is linked to the major histocompatibility H-2 complex (MHC) (22). Immune responsiveness to many synthetic and natural immunogens has been shown to be regulated by intra-MHC genes (23). The murine MHC has been divided into four regions which include at the extremities of this genetic complex the H-2K (K) and H-2D (D) serological regions, which control the expression of the major transplantation antigens of the mouse. H-2K and H-2D are separated by the I (immune response) and S (serum serological variant) regions. (For review of H-2, see ref. 24.) The I region maps to the left of S with respect to the centramere of the seventeenth chromosome, and is adjacent to K (24). Functions attributed to the I region include the genetic regulation of immune responsiveness to synthetic and natural immunogens, and the control of cell surface antigens on subpopulations of lymphocytes known as Ia (immune response associated) antigens. The I region has been recently divided into three subregions (24). The S region is located to the right of I and to the left of D, and controls the synthesis of serum alpha globulins (24). A number of alleles are known to exist for each of these regions. Congenic resistant and recombinant mouse lines have been bred and selected, which differ only at H-2 or for certain regions within H-2. Such genetically defined animals are extremely useful in mapping new immunological functions found to be associated with the MHC.

The induction of T-cell mediated cytotoxicity against TNP-modified mouse syngeneic lymphocytes has been recently reported (25). Effector cell specificity was such that TNP-modified syngeneic spleen target cells were lysed, but not unmodified targets nor TNP-modified splenic targets which differed from the modified stimulating cells throughout the H-2 complex (25). Furthermore, MLR was obtained by mixing unmodified and irradiated, TNP-modified autologous splenic lymphocytes (Shearer and Lozner, unpublished observations), and TNP-specific hemolytic plaque-forming cells were detected in the cultures which generated cytotoxic effector cells (Shearer, unpublished observations). The present report which summarizes the cytotoxic results collected to date (both published and unpublished), demonstrates: (a) distinct effector cell specificity for TNP-modified H-2K and H-2D serological regions of the MHC; and (b) probably H-2 linked genetic control of responsiveness to a specificity resulting from TNP-modification of the H-2D region of B10.A spleen cells, and speculates on a possible role for syngeneic modified H-2 controlled cell surface components in nature.

MATERIALS AND METHODS

Mice. All mice used in the experiments were males, 7-9 weeks of age. The B10 congenic strains were purchased from the Jackson Laboratory, Bar Harbor, Me. The A.TL and A.TH mice were raised in our mouse colony from breeders obtained from Dr. Donald Shreffler, Univ. of Mich., Ann Arbor, Mich. Other inbred strains were obtained from the NIH Animal Production facility.

Trinitrophenyl modification of spleen cells. Cells to be modified with TNP were prepared from the spleens of young adult mice by gentle washing with blunt forceps and aspiration through a syringe in Hank's balanced salt solution (BSS), and filtered through nylon mesh to remove clumps and debris. Erythrocytes were lysed by a two-minute exposure of the cell suspension to ammonium chloride lysing buffer, followed by washing in BSS. The cell pellet was suspended in a phosphate buffered solution containing 10 mM 2,4,6-trinitrobenzene sulfonate (TNBS) (Pierce Chemical Co., Rockford, Ill.) pH 7.3 at a volume:volume ratio of cells to reagent of 1:4. The cell-reagent mixture was incubated at 37°C for 10 minutes and then washed twice in BSS containing fetal bovine serum (FBS). Spleen cells to be used for sensitization and for targets were modified with TNBS in an identical fashion. Cells to be used as cytotoxic targets were labelled with $^{51}CrCl_3$ (Amersham/Searle Corp., Arlingtion Heights, Ill.) for 40 minutes at 37°C (25) prior to modification with TNBS.

In vitro cell culture and assay for cytotoxic effector cells. Splenic lymphocytes were sensitized in vitro with TNP-modified cells as previously described (25) by the Mishell-Dutton procedure (26). Following a 5-day incubation period, effector cells were harvested and their cytotoxic reactivity was determined by a modification of method of Canty and Wunderlich (27). The mean and standard error of the percentage of specific cytotoxicity was computed as described elsewhere (28).

RESULTS

B10.BR splenic lymphocytes were sensitized against TNP-modified autologous spleen cells. Five days later, the effector cells generated were assayed for cytotoxic activity using unmodified and TNP-modified spleen target cells from the C57BL/10, B10.BR, B10.D2, and B10.A congenic resistant mouse strains. The results are summarized in Table I. Cytoxicity was obtained using TNP-modified B10.BR syngeneic spleen target cells, but not when unmodified targets were used. In order to test whether the cytotoxic reaction was TNP-(hapten-) specific, B10.BR effector lymphocytes were tested on TNP-modified spleen targets from B10.D2 and B10.A donors. No appreciable cytolysis was detected using TNP-modified targets from these two congenic strains which express different H-2 haplotypes than the responding and stimulating cell donor. These results indicate that TNP-modification of the target cell population is a necessary but insufficient requirement for cytotoxicity. Thus, the specificity requirements between stimulating and target cells are such that both TNP and some other component are needed in order to obtain a cytotoxic reaction. Since B10.BR and B10.A mice are similar in two of their H-2 subregions (24), effector cells generated by sensitization of B10.BR lymphocytes with B10.BR-TNP cells were assayed on B10.A-TNP spleen targets to determine if the second specificity component depends on a shared common MHC subregion. As shown in the last line of Table I, cytotoxicity was detected with the B10.A-TNP targets similar to that obtained when TNP-modified syngeneic B10.BR targets were used. This observation raised the possibility that one or more shared H-2 subregions are required between sensitizing and target cells or between responding and target cells in order to obtain cytotoxicity.

Cytotoxicity as measured by ^{51}Cr-release may be a T-cell mediated reaction involving the cytolytic action of specific thymus-derived lymphocytes (7,29-33), or it can be an antibody-mediated phenomenon involving lysis of antibody-coated target cells by nonspecific effectors (K cells) from unimmunized donors (34,35). A number of criteria have been used to establish that the cytotoxicity obtained against TNP-modified spleen target cells was initiated and effected by thymus-derived cells and was not due to the lymphocyte-dependent antibody mediated cytotoxicity (25). These experiments summarized in Table II include: (a) failure to detect lymphocyte-dependent antibody in the effector cell culture media; (b) failure of TNP-lysine to block the effector phase under

TABLE I

IN VITRO INDUCTION OF CYTOTOXICITY OF B10.BR SPLEEN CELLS TO TNP-MODIFIED SYNGENEIC SPLEEN CELLS ASSAYED WITH UNMODIFIED AND TNP-MODIFIED SYNGENEIC AND CONGENIC SPLEEN TARGET CELLS

Responding Cells	Stimulating Cells	Target Cells	% Specific Lysis ± S.E.	Target Cell H-2 Haplotype
B10.BR	B10.BR-TNP	B10.BR	2.4 ± 2.7	k
↓	↓	B10.BR-TNP	27.3 ± 5.6	k
↓	↓	B10.D2	-0.6 ± 8.7	d
↓	↓	B10.D2-TNP	6.1 ± 3.0	d
↓	↓	B10	2.3 ± 3.0	b
↓	↓	B10-TNP	4.1 ± 3.2	b
↓	↓	B10.A	-0.1 ± 1.4	a
↓	↓	B10.A-TNP	20.9 ± 0.6	a

Effector:Target Cell Ratio = 20:1

TABLE II

SUMMARY OF EXPERIMENTS INDICATING THAT THE CYTOTOXICITY OBTAINED TO TNP-MODIFIED SYNGENEIC SPLEEN CELLS IS A T-CELL-MEDIATED PHENOMENON

Parameter Tested in the Induction or Effector Phase	% Specific Lysis ± S.E. Positive Control	Experimental
Lymphocyte-dependent antibody is not detected in culture media.	37.3 ± 1.6	0
TNP-lysine does not block the effector phase.	23.3 ± 1.1	20.0 ± 2.2
Spleen cells from athymic nude donors do not generate effector cells.	13.8 ± 2.1	-0.9 ± 0.4
Effector cells are sensitive to rabbit anti-mouse brain serum.	17.0 ± 1.5	0.2 ± 1.0
Effector cells can be generated by cortisone-resistant thymocytes.	10.4 ± 2.2	10.0 ± 1.1

conditions that abrogate antibody-mediated lysis; (c) failure of spleen cells from athymic nude donors to generate effector cells; (d) abolition of effector cell function by rabbit anti-mouse serum and complement; and (e) the generation of effector cells by cortisone resistant thymocytes.

In order to explore the possibility raised above (last line of Table I) that intra-H-2 homology is required either between modified stimulating and modified target cells or between effector cells and modified target cells, splenic lymphocytes from three C57BL/10 congenic mouse strains were sensitized with TNP-modified autologous cells. The effectors generated were tested on a battery of TNP-modified congenic and allogeneic spleen target cells, which differed from the responding and stimulating cells throughout the MHC or within distinct regions of H-2.

The results of such experiments using B10.BR spleen cells as responders and modified stimulating cells are summarized in Table III. The lower case letters shown beneath the responding and stimulating cells (i.e., kkkk) indicate the H-2 alleles for the H-2K, I, S, and H-2D regions, respectively. Thus, B10.BR, whose H-2 haplotype is k, expresses the k allele for all four regions. For the sake of simplicity, the I region has not been further divided into the IR-1A, Ir-1B, and I-C subregions in this discussion. The K, I, S, and D H-2 alleles of the various target cells are shown in the right column of the table. B10.BR effector lymphocytes sensitized with TNP-modified autologous cells did not lyse unmodified syngeneic target cells, although they did lyse TNP-modified splenic target cells from both B10.BR and B10.A donors. These results verify that homology between modified target cells and either modified stimulating or unmodified responding cells is required and indicate that the homology is associated with the K and/or I regions of H-2. In order to distinguish between homology at K and I, TNP-modified target cells were used from A.TL donors, which express the same I and S regions as B10.BR, but differ at K. No cytotoxicity was obtained, indicating that homology between modified target and either modified stimulating cells or responding cells is associated with the H-2K serological region. Modified target cells from C3H.H-2° donors which share only the k allele of the D region of H-2 with responding and stimulating cells also failed to be lysed by the B10.BR effectors. This indicates that homology at the D serological region was not adequate to detect cytotoxicity. B10.D2-TNP targets which would be identical with B10.BR except at H-2 were not effective targets, whereas C3H/HeJ-TNP targets which are identical with B10.BR at H-2 but differ elsewhere were effective tar-

TABLE III

IN VITRO INDUCTION OF CYTOTOXICITY OF B10.BR SPLEEN CELLS TO TNP-MODIFIED SYNGENEIC SPLEEN CELLS ASSAYED WITH TNP-MODIFIED SYNGENEIC, CONGENIC, AND ALLOGENEIC SPLEEN TARGET CELLS

Responding Cells	Stimulating Cells	Target Cells	% Specific Lysis ± S.E.	Target Cell H-2 Region K	I	S	D
B10.BR	B10.BR-TNP	B10.BR	-4.0 ± 1.9	k	k	k	k
kkkk	kkkk	B10.BR-TNP	26.8 ± 1.9	k	k	k	k
↓	↓	B10.A-TNP	26.5 ± 2.2	k	k	d	d
↓	↓	A.TL-TNP	-2.2 ± 2.2	s	k	k	d
↓	↓	C3H·H-2°-TNP	2.1 ± 2.1	d	d	d	k
↓	↓	B10.D2-TNP	-2.2 ± 2.4	d	d	d	d
↓	↓	C3H/HeJ-TNP	19.4 ± 3.1	k	k	k	k

Effector:Target Cell Ratio = 20:1

gets.

Spleen cells from B10.A donors were then used as a source of responding and TNP-modified stimulating and target cells to further explore the cross-reactivity noted between modified B10.BR and B10.A cells shown in Tables I and III. As shown in Table IV, B10.A responding lymphocytes sensitized to modified autologous spleen cells generated effectors which did not lyse unmodified syngeneic cells, but which did lyse both TNP-modified B10.A and B10.BR targets. This cross-reactivity is similar to that noted above between B10.A and B10.BR when cells from the latter were used as responding and modified stimulating cells, and appears to be associated with the fact that these two strains share the K and I regions of H-2. Modified spleen cells from the other four strains shown in Table IV, i.e., A.TL-TNP, B10.D2-TNP, B10-TNP, and C3H-H-2°-TNP, were not lysed by the B10.A effectors. The A.TL-TNP targets which share the I and D regions with B10.A, are of particular interest here for two reasons: first, substitution of the k allele at the K end of H-2 by the S allele rendered the modified A.TL cells ineffective targets for B10.A effectors which had been sensitized with B10.A-TNP; and second, the modified A.TL targets were not lysed even though they expressed the d allele at H-2D, as do the effector and stimulating cells. Furthermore, B10.D2-TNP targets were also ineffective targets, although they share the S and D regions with the B10.A responding and stimulating cells.

Although TNP-modified cells were ineffective as targets for B10.A lymphocytes, it was worth re-examining these two strains by reversing their functional roles and using B10.D2 cells as the source of responding lymphocytes, since they are similar in H-2 at S and D but differ at K and I. Thus, a strain of mouse expressing different K and I region alleles (i.e., d instead of k) was used for responding cells. Similar to the results obtained with B10.BR and B10.A, B10.D2 responding lymphocytes sensitized with TNP-modified autologous cells lysed B10.D2-TNP targets, but not unmodified B10.D2 targets (Table V). In contrast to B10.BR and to B10.A lymphocytes which did not respond to modified syngeneic S and D alleles when sensitized with modified autologous cells, B10.D2 did respond to the TNP-modified d allele of the D region. (Note the cytolysis obtained with B10.A-TNP and A.TL-TNP targets.) This reactivity by B10.D2 lymphocytes toward modified D end is verified by the fact that B10.D2 lymphocytes responded to B10.A-TNP and lysed B10.D2-TNP targets (last line of Table V). That B10.D2 lymphocytes also respond to TNP-modified intra-H-2 antigenic specificities controlled by loci other than H-2D, is shown by the fact that

TABLE IV

IN VITRO INDUCTION OF CYTOTOXICITY OF B10.A SPLEEN CELLS TO TNP-MODIFIED SYNGENEIC SPLEEN CELLS ASSAYED WITH TNP-MODIFIED SYNGENEIC, CONGENIC AND ALLOGENEIC SPLEEN TARGET CELLS

Responding Cells	Stimulating Cells	Target Cells	% Specific Lysis ± S.E.	Target Cell H-2 Region K	I	S	D
B10.A	B10.A-TNP	B10.A	-3.2 ± 3.1	k	k	d	d
kkdd	kkdd	B10.A-TNP	25.6 ± 3.1	k	k	d	d
↓	↓	B10.BR-TNP	17.9 ± 0.9	k	k	k	k
↓	↓	A.TL-TNP	3.9 ± 4.8	s	k	k	d
↓	↓	B10.D2-TNP	4.3 ± 1.9	d	d	d	d
↓	↓	B10-TNP	0.0 ± 1.7	b	b	b	b
↓	↓	C3H·H-2°-TNP	0.7 ± 1.6	d	d	d	k

Effector:Target Cell Ratio = 20.1

TABLE V

IN VITRO INDUCTION OF CYTOTOXICITY OF B10.D2 SPLEEN CELLS TO TNP-MODIFIED SYNGENEIC SPLEEN CELLS ASSAYED WITH TNP-MODIFIED SYNGENEIC, CONGENIC AND ALLOGENEIC SPLEEN TARGET CELLS

Responding Cells	Stimulating Cells	Target Cells	% Specific Lysis ± S.E.	Target Cell H-2 Region K	I	S	D
B10.D2	B10.D2-TNP	B10.D2	-2.4 ± 2.8	d	d	d	d
dddd	dddd	B10.D2-TNP	19.2 ± 2.4	d	d	d	d
↓	↓	B10.A-TNP	17.6 ± 3.6	k	k	d	d
		A.TL-TNP	15.6 ± 3.7	s	k	k	d
		C3H·H-2°-TNP	14.7 ± 1.2	d	d	d	k
		B10.BR-TNP	2.4 ± 0.6	k	k	k	k
	B10.A-TNP	B10.D2-TNP	12.8 ± 2.0	d	d	d	d
	kkdd						

Effector:Target Cell Ratio = 20:1

C3H·H-2°-TNP spleen cells served as effective targets. B10.BR-TNP cells, which do not share known intra H-2 alleles with B10.D2, were ineffective targets and provide a negative control for the experiment.

Before leaving Tables IV and V, it is worth emphasizing that B10.A-TNP spleen cells were: (a) adequate targets when B10.D2 lymphocytes were sensitized with B10.D2-TNP; (b) adequate stimulating cells for sensitization of B10.D2 lymphocytes subsequently assayed on B10.D2-TNP targets; but (c) poor effectors when stimulated with autologous B10.A-TNP and assayed for the TNP-modified d allele of H-2D by B10.D2-TNP or A.TL-TNP targets. Taken together, this series of experimental results suggest that H-2 linked Ir genes may be involved in the reaction of lymphocytes to modified self (see DISCUSSION).

If TNP is modifying cell surface components such as the H-2 serological antigens, one might expect that TNP-modification of either stimulating or target cells would result in a reduced allogeneic cytotoxic response involving an H-2 difference. To test this possibility, B10.A spleen cells were sensitized against B10.A-TNP, B10.D2, or B10.D2-TNP spleen cells. The effectors generated were assayed on the corresponding unmodified or modified target spleen cells. The results shown in Table VI indicate that TNP-modification of either or both stimulating and target cells had no detectable effect on the cytolysis of B10.A effectors toward the H-2 antigens of B10.D2. Furthermore, the level of cytotoxicity obtained was as great to TNP-modified autologous spleen cells as it was to the alloantigens of B10.D2.

DISCUSSION

Modification of mouse spleen cells with TNBS sufficiently alters surface structure so that modified cells are recognized by autologous lymphocytes. Reaction to the modified surfaces can be demonstrated in vitro both by MLR (Lozner and Shearer, unpublished observations) and CML (ref. 25 in this report). Alteration of the sensitizing cell population alone is not sufficient, however, to detect the cytolytic response. The target cell must also be modified by TNP. Furthermore, as shown in Table I, TNP-lation of the target cell is a necessary but insufficient requirement for cytolysis. Some one or more common denominators associated with the H-2 complex are required either between the modified stimulator and target cells or between the responding lymphocyte population and the modified target cells for the CML response to be effected.

Cell-mediated cytotoxicity is an immunological function attributed to thymus-derived lymphocytes. In the mouse, evi-

TABLE VI

IN VITRO INDUCTION OF CYTOTOXICITY OF B10.A SPLEEN CELLS TO UNMODIFIED OR TNP-MODIFIED B10.D2 SPLEEN CELLS ASSAYED WITH UNMODIFIED OR TNP-MODIFIED B10.D2 TARGETS

Responding Cells	Stimulating Cells	Target Cells	% Specific Lysis ± S.E.
B10.A	B10.A-TNP	B10.A	2.7 ± 2.6
↓	B10.A-TNP	B10.A-TNP	28.4 ± 3.3
↓	B10.D2	B10.D2	25.9 ± 2.1
↓	B10.D2	B10.D2-TNP	21.9 ± 3.8
↓	B10.D2-TNP	B10.D2	26.3 ± 2.6
↓	B10.D2-TNP	B10.D2-TNP	24.7 ± 2.4

Effector:Target Cell Ratio = 20:1

dence that both the induction and effector phases of specific CML involve cells of thymic origin comes from studies demonstrating that these phases are sensitive to anti-θ serum (31,33), that lymphocytes from athymic nude mice do not generate CML (29), and that in vitro induction of CML can be generated by cortisone-resistant thymocytes (7,29,30). Another type of antibody-mediated cytotoxicity has been described which involves the lysis of antibody-coated target cells by nonspecific lymphoid cells (34,35). A number of criteria were used to establish that the cytotoxicity observed to TNP-modified target cells was initiated and effected by T-lymphocytes and was not due to antibody-mediated lysis (ref. Table II).

The H-2 associated specificity of the T-cell mediated cytotoxicity generated by sensitization with TNP-modified syngeneic spleen cells is of particular interest. By sensitizing with TNP-modified autologous splenic lymphocytes and testing for cytolysis on modified splenic targets differing from the effector and sensitizing cells within H-2, it was possible to map the particular MHC region responsible for the specificity to the H-2K alleles of B10.BR and B10.A, and to both the K and D alleles of B10.D2.

B10.BR and B10.A effectors generated by sensitization with the respective TNP-modified autologous cells lysed all modified targets tested which expressed the k allele in the K region. As shown by the failure of B10.BR effectors to lyse A.TL-TNP targets (Table III), the I and S regions do not appear to be important for this specificity. Using identical targets, the k allele of H-2K, but not the d allele of H-2D, was found to be necessary for B10.A effectors to be cytolytic (Table IV). Furthermore, both B10.BR and B10.A effectors were unable to lyse modified target cells, which were homologous with effector and sensitizing cells at the D or D and S regions. Modified C3H·H-2° targets expressing the k allele at H-2D were not adequate targets for B10.BR effectors, nor were B10.D2 targets, which express the same S and D regions as the cells of the sensitizing phase, adequate for B10.A effectors (Table III and IV). From the B10.BR and B10.A examples which are similar at K and I, homology between TNP-target cells and either TNP-sensitizing cells or unmodified responding cells at only the K region (or at K and I) appears to be a prerequisite for detection of cytotoxicity in this system.

This type of homology, however, is not restricted to the K region in all mouse strains. For example, B10.D2 effectors generated by sensitization with modified autologous cells lysed: (a) TNP-targets which share only the D region (i.e., A.TL-TNP, Table IV) with the cells of the sensitizing phase as well as; (b) C3H.H-2°-TNP which differ from B10.D2 at D but are similar at K, I, and S. Based on the mapping to the H-2K region in the B10.BR and B10.A examples, the C3H.H-2°-TNP targets are presumed to be lysed by virtue of their K region as well, but this was not proven. Thus, in contrast to B10.BR and B10.A, B10.D2 can be shown to have reactivity for two TNP-modified syngeneic H-2 regions, are involving the d allele of H-2D and the other somewhere to the left of the D region, possibly involving the d allele of the K region.

What is the nature of this specificity? Does the specificity depend on intra-H-2 compatibility between effector and modified target cell at only the K serological region for the B10.BR and B10.A strains, but at the D serological region as well as some other locus, possibly K, for the B10.D2 strain? This type of compatibility might take the form of some kind of H-2-dependent effector-target cell interaction resembling the H-2 dependence reported for optimal T- and B-cell cooperation leading to antibody production (36, 37). If homology is required between effector and target cell in order to obtain lysis, then why should it be necessary to modify the target cell with TNP? It is possible to envision a complex model in which the antigenic specificity involves TNP, whereas the specificity for interaction leading to cytolysis is H-2 dependent. However, no such H-2 compatibility for cytotoxic effector cell killing is known in allogeneic models. In fact, the best cytolysis is obtained between H-2 incompatible effector and targets (32). Furthermore, one would have to propose that in two of the strains shown (B10.BR and B10.A), one H-2 region is exclusively important for interaction, whereas in the other strain (B10.D2), either of two regions is sufficient.

A less complicated, and more plausible alternative, is that the H-2-dependent specificity actually depends on intra-H-2 homology between TNP-modified sensitizing cells, i.e., "immunogen", and TNP-modified target cells, i.e., the assaying antigen. According to this interpretation, cell surface components controlled by the K serological region in the case of B10.BR and B10.A and by the D serological region, as well as presumably K, in the case of B10.D2 can be modified by TNP, resulting in conformationally altered "self antigens", which are recognized as foreign by syngeneic lymphocytes. In vitro sensitization to these new cellular antigens would lead to proliferation and differentiation, resulting in a population of cytotoxic effector cells which are capable of lysing only targets bearing the identical specificity, which includes some modified H-2 controlled cell surface components and possibly TNP as well. As shown in Table I, TNP is necessary in order to obtain cytolysis, but it has not yet been established whether the modifying agent is an integral part essential for antigenic recognition, or whether it simply serves as a modifier. Experimental proof has not yet been obtained to distinguish between the two alternative models outlined above, although the latter less complicated interpretation involving intra-H-2 homology between modified stimulating and modified target cells is presently favored. Experiments are currently in progress to answer this question.

The immunogenic entities resulting from TNP-modification could be described as "modified self" and may represent altered histocompatibility antigens. If new cellular antigens are being created by modification of the H-2 serological specificities, the strict specificity observed raises the possibility that the new determinants are restricted to private specificities, since $H\text{-}2^d$ and $H\text{-}2^k$ share at least two public specificities (24). Alternatively, it is possible that new antigens are formed by both modified private and public specificities, but that the mouse strains investigated are responsive to the new cellular antigens resulting from TNP-lation of private specificities, but not to those involving modification of public specificities, possibly due to lack of expression of the proper Ir (immune response) genes in the responding cells. Whatever the cell surface changes resulting from TNP modification are, they do not detectably affect expression of the H-2 antigens involved in either the sensitization or effector phases of an in vitro allogeneic cytotoxic response (see Table VI).

Zinkernagel and Doherty have recently reported H-2 associated specificity for T-cell mediated cytotoxicity to LCM-infected target cells which is similar to the TNP results described here (20). Specific cytotoxicity was not detected above control levels between $H\text{-}2^k$ effector cells from LCM-infected mice and $H\text{-}2^d$ LCM-infected targets and vice versa. Two possible mechanisms were mentioned for the H-2-associated homology: (a) H-2 dependent interaction between effector and target; and (b) the creation of altered self due to LCM infection (20). If the former possibility is correct, one has to assume that a type of H-2 homology which is not essential for allogeneic cytotoxicity is required for effectors to lyse LCM-target cells which are identical at H-2. If the latter possibility is correct, then injection of LCM into mice results in cell surface modification creating new antigenic specificities involving H-2 controlled products. These modified cells would then be the stimulating population for in vivo sensitization of the relevant lymphocytes. The progeny of such lymphocytes would then lyse only targets expressing the new antigenic specificity, which includes some modified H-2 dependent component(s) as well as LCM (21). It should be possible to differentiate between these two mechanisms by in vitro sensitization of LCM-infected cells in which the genotypes of the responding, sensitizing and target cells can be varied. Such studies are currently in progress.

It is noteworthy that the specificity of the B10.A effector cells appeared to be directed exclusively toward the modified H-2K serological region, but not to the modified D region (Table IV), whereas B10.D2 effectors sensitized against TNP-modified autologous cells lysed modified targets expressing homology at H-2D only (Table V). The observation that the TNP-modified d allele of H-2D served both as adequate stimulator and target cells for B10.D2 effectors indicates that TNP modification of the D region created a new antigenic structure, which was detected only if the proper responding antigen-sensitive lymphocytes were present, i.e., from B10.D2 donors. These observations indicate that formation of a new antigenic specificity and intra-H-2 homology between either target and effector or target and stimulator are not sufficient in all instances to generate cytotoxicity in this system. By replacing the k alleles at the K and I regions with the corresponding d alleles and therefore at H-2 regions not detectably involved in this particular specificity, a cytotoxicity response was obtained. It is noteworthy that the intra-H-2 regions changed: (a) are not associated with the TNP-modified H-2D region; and (b) the regions changed include the I region, which controls immune response potential to many immunogens (23). Thus, it is probable, although not yet proven, that the failure of B10.A responding lymphocytes to generate an in vitro cytotoxic response to the TNP-modified d allele of the D region is due to the lack of expression of the proper H-2 linked Ir gene(s). The gene(s), however, is apparently expressed in the B10.D2 strain, which differs from B10.A at K and I. Further studies using mouse strain exhibiting intra-H-2 recombination between the K and I regions will hopefully clarify whether this difference in responsiveness is associated with genes mapping in the K or I regions. If H-2-linked Ir genes are involved in sensitization to TNP-modified autologous cells, then this model offers a number of unique properties which include one of the first demonstrations: (a) of an Ir-gene-controlled primary immune response in vitro; (b) of an Ir-controlled T-cell mediated cytotoxicity, and therefore probably indicating an Ir gene defect attributable to a functional thymus-derived cell population; and (c) that the H-2 complex can play an important role <u>both</u> in the response potential to <u>and</u> in the formation of new cellular antigenic determinants. This model may also be useful in elucidating the lymphocyte defined and serologically detectable regions of the MHC (38,39). Although the major H-2-dependent specificity components detected thus far appear to be associated with the K and D serological regions, reactivity against modified I region components (Ir gene products or Ia antigens) have not been excluded. Preliminary results from this laboratory suggest that there may be a weaker but detectable CML to TNP-modified components that map in the Ir-1B and/or I-C regions of H-2. Finally, this approach may be useful in demonstrating the existence of Ir genes in humans, since a primary response can be generated in vitro by lymphocytes from unsensitized donors.

The TNP-modification of cells in these experiments is likely to induce many conformational cell surface alterations which need not necessarily be associated with the MHC. Are such non-H-2-associated conformational changes not immunogenic? Are they immunogenic but undetected due to an absence of the proper Ir genes necessary for responsiveness? What is the significance of control by MHC of both the formation of new antigenic specificities and the response potential to these specificities? Is it possible that the H-2K and H-2D serological regions (and possibly others) of the MHC control the synthesis of cell surface glycoproteins whose function in the syngeneic system is to serve as readily modifiable autologous cell surface components, which in turn, are immunogenic -- providing that the proper Ir genes, controlled by the <u>same</u> genetic complex, are expressed? Such H-2-controlled surface components might combine with chemical (TNP) or biological (viruses and autoimmune complexes) entities to form new immunogens -- which can be realized and identified only if the proper Ir genes are expressed. If H-2 linked Ir genes are demonstrable for "altered self immunogens" which in turn are themselves controlled by the MHC, then perhaps the absence of expression of some of these Ir genes can be important for failure to react against self. Elimination of self clones in utero or during neonatal life is one possibility (1,2). An alternative and not necessarily mutually exclusive possibility is that the Ir genes necessary to recognize and react against self are not usually expressed (3). In this case, autoimmunity could result from a minor but significant alteration in self antigens which would be immunogenic, providing that Ir genes that regulate responsiveness to altered self are expressed. If H-2 linked Ir genes exist for modified self, then it is also possible that minor but significant somatic Ir gene mutations (3) could occur, leading to the development of antigen-reactive cell clones directed against unmodified self antigens. Experiments are in progress in this laboratory to establish if the intra-H-2 associated specificities demonstrated for TNP-modified autologous cells, as well as possible Ir genes, can be demonstrated for other chemical modifying agents, viruses, and in experimental autoimmunity.

SUMMARY

Splenic lymphocytes from C57BL/10 congenic resistant mouse strains were sensitized in vitro to trinitrophenyl (TNP)-modified syngeneic spleen cells. The effector cells generated were incubated with ^{51}Cr-labelled unmodified or TNP-modified target spleen cells and the percentage of specific lysis determined. The results obtained using syngeneic, congenic, and allogeneic modified target cells indicated that TNP-modification of the target cells was a necessary but insufficient requirement in order to obtain cytolysis. Intra-H-2 homology either between modified stimulating cells and modified target cells or between responding lymphocytes and modified target cells was equally important in the specificity for cytolysis. Homology at the K serological region in the B10.BR and B10.A strains, and at the D serological region as well as some other region (possibly K) in the B10.D2 strain, were necessary in order to detect cytotoxicity. Two possible mechanisms are considered for interpreting this type of homology. The first involves homology between intra-H-2 components of the responding and TNP-modified target cells, in which the cytotoxic reaction might be hapten-specific, in addition to which some form of H-2 dependent effector-target cell interaction not known in allogeneic cytotoxic models is required for lysis. The second involves the creation of new H-2-controlled antigenic specificities resulting from TNP-modification of the stimulating cells, which in turn, require the same H-2 dependent modified components in order to be lysed by the specifically sensitized effector cell population. Furthermore, evidence is presented suggesting that H-2 linked Ir genes are expressed in the TNP-modified autologous cytotoxic system. These findings may indicate that the major histocompatibility complex can play a role in the response potential to and in the formation of new antigenic specificities involving modified self components. The possible relevance of the major histocompatibility complex in self-recognition and autoimmunity is discussed.

REFERENCES

1. Burnett, M., "The Colenal Selection Theory of Acquired Immunity", Vanderbilt University Press, Nashville, Tennessee, 1959.
2. Burnett, M., "The Integrity of the Body", Harvard University Press, Cambridge, Massachusetts, 1963.
3. Jerne, N.K., Eur. J. Immunol. 1: 1, 1971.
4. Cohen, I.R., Globerson, A. and Feldman, M., J. Expt. Med. 133: 834, 1971.
5. Cohen, I.R. and Wekerle, H., Science 176: 1324, 1972.
6. Orgad, S. and Cohen, I.R., Science 183: 1038, 1974.
7. Wagner, H. and Rollinghoff, M., Nature-New Biol. 241: 53, 1973.
8. Lundak, R.L. and Raidt, D.L., Cell. Immunol. 9: 60, 1973.
9. Howe, M.L., Goldstein, A.L. and Battisto, J.R., Proc. Nat. Acad. Sci., U.S. 67: 613, 1970.
10. von Boehmer, H. and Byrd, W.J., Nature-New Biol. 235: 50 1972.
11. von Boehmer, H. and Adams, P.B., J. Immunol. 110: 376, 1973.
12. Talal, N. and Steinberg, A.D., in "Current Topics in Microbiology and Immunology", Springer-Verlag, New York, in press.
13. Rose, N.R., Kite, J.H., Flanagan, T.D. and Witebsky, E., in "Cellular Interactions in the Immune Response" (ed. by S. Cohen, G. Cudkowicz and R.T. McCluskey), S. Karger, New York, 1971.
14. Rathburn, W.E. and Hildemann, W.H., J. Immunol. 105: 98, 1970.
15. Rubin, B., Eur. J. Immunol. 3: 26, 1973.
16. Rubin, B., Schinmacher, V. and Wigzell, H., Scan. J. Immunol. 2: 189, 1973.
17. Rubin, B. and Wigzell, H.J., Exptl. Med. 137: 911, 1973.
18. Geczy, A.F. and Baumgarten, A., Immunology 19: 189, 1970.
19. Baumgarten, A. and Geczy, A.F., Immunology 19: 205, 1970.
20. Zinkernagel, R.M. and Doherty, P.C., Nature 248: 701, 1974.
21. Doherty, P.C., Zinkernagel, R.M. and Ramshaw, I.A., J. Immunol. 112: 1548, 1974.
22. Oldstone, M.B.A., Dixon, F.J., Mitchell, G.F. and McDevitt, H.O., J. Exptl. Med. 137: 1201, 1973.
23. Benacerraf, B. and Katz, D.H., Advanc. Cancer Res., in press.
24. Shreffler, D.C. and David, C.S., Advanc. Immunol., in press.
25. Shearer, G.M., Eur. J. Immunol. 4: 527, 1974.

26. Mishell, R.I. and Dutton, R.W., J. Exptl. Med. 126: 423, 1967.
27. Canty, T.G. and Wunderlich, J.R., J. Nat. Cancer Inst. 45: 761, 1970.
28. Martin, W.J., Wunderlich, J.W., Fletcher, F. and Inman, J.K., Proc. Nat. Acad. Sci., U.S. 68: 469, 1971.
29. Wagner, H., J. Immunol. 109: 630, 1972.
30. Wagner, H., Harris, A.W. and Feldman, M.J., Cell. Immunol. 4: 39, 1972.
31. MacDonald, H.R., Phillips, R.A. and Miller, R.G., J. Immunol. 111: 575, 1973.
32. Cerottini, J.-C., Nordin, A.A. and Brunner, K.T., Nature 228: 1308, 1970.
33. Goldstein, P., Schirrmacher, V., Rubin, B. and Wigzell, H., Cell. Immunol. 9: 211, 1973.
34. Möller, E., Science 147: 873, 1965.
35. Perlmann, P. and Hohn, G., Advanc. Immunol. 11: 117, 1969.
36. Katz, D.H., Hamaoka, T., Dorf, M.E. and Benacerraf, B., Proc. Nat. Acad. Sci., U.S. 70: 2624, 1973.
37. Katz, D.H., Dorf, M.E. and Benacerraf, B., J. Exptl. Med. 140: 290, 1974.
38. Bach, F.H., Widmer, M.B., Bach, M.L. and Klein, J., J. Exptl. Med. 136: 1430, 1972.
39. Alter, B.J., Schendel, D.J., Bach, M.F., Bach, F.H., Klein, J. and Stimpfling, J.H., J. Exptl. Med. 137: 1303, 1973.

LYMPHOCYTE ACTIVATION INDUCED BY MODIFICATIONS OF SURFACE MEMBRANE SACCHARIDES

Abraham Novogrodsky
Department of Biophysics, The Weizmann Institute of Science, Rehovot, Israel.

Interaction of phytomitogens with saccharide containing sites in the surface membrane induces lymphocyte transformation (1-3). The nature of the triggering signal is as yet unknown. Our main approach in studying the mechanism of lymphocyte transformation was to search for well defined modifications of the cell membrane, induced chemically or enzymically, which will activate the cell or will alter its response to stimulation by other agents.

Induction of lymphocyte transformation by periodate or by sequential treatment with neuraminidase and galactose oxidase (NAGO)

Lymphocytes isolated from different sources undergo extensive blastogenesis after mild treatment with sodium periodate ($NaIO_4$) (4-6). The following observations lead to the conclusion that the $NaIO_4$ target site is a glycoprotein(s) containing sialic acid that yields on oxidation an aldehyde moiety that is essential for transformation.

1) Incubation of lymphocytes with neuraminidase markedly reduced their response to $NaIO_4$ (5) (Table I, exp. 1). The effect of neuraminidase is specific since it can be reversed by excess of free sialic acid (substrate competition).

2) Blastogenesis induced by $NaIO_4$ is markedly decreased when the $NaIO_4$-treated cells are reacted with reagents that reduce (KBH_4) or combine (NH_2OH) with aldehydes (5,7) (Table I, exp. 2). We have recently found (8) that cysteine, a compound that reacts with an aldehyde through its amino and thiol groups to form a thiazolidine derivative also decreased transformation of $NaIO_4$-treated mouse lymphocytes. Other thiol compounds like reduced glutathione, dithiothreitol and mercaptoethanol did not show this effect.

The above conclusion is further supported by the early observations (9,10) that the two terminal exocyclic carbon atoms of sialic acid glycosides are selectively cleaved upon exposure to $NaIO_4$ without destruction of the ring structure or cleavage of the glycosidic bond. Furthermore, Van Lenten and Ashwell (11) have shown that $NaIO_4$ under mild conditions selectively oxidizes sialyl residues in sialoglycoproteins to yield a 7 carbon sialic acid analog containing an aldehyde moiety. The latter could be reduced by KBH_4. Recently Liao

TABLE I

DECREASE OF $NaIO_4$-INDUCED TRANSFORMATION BY PREINCUBATION OF THE LYMPHOCYTES WITH NEURAMINIDASE AND BY REACTION OF THE $NaIO_4$-TREATED CELLS WITH BOROHYDRIDE OR HYDROXYLAMINE

	Treatment		[^{3}H] Thymidine incorporation (cpm) into cells after additional treatment with			
	I	II	None	Con A	$NaIO_4$	$NaIO_4$+ Con A
Exp. 1	PBS	-	1,086	136,900	130,400	143,900
	Neuraminidase	-	1,156	159,200	27,380	131,100
Exp. 2	PBS	PBS	1,303	151,900	183,600	162,800
	PBS	KBH_4	1,965	150,500	215,500	178,000
	PBS	NH_2OH	1,766	266,400	101,200	194,000
	$NaIO_4$	PBS	172,800	192,400	130,100*	-
	$NaIO_4$	KBH_4	13,480	135,500	15,130	-
	$NaIO_4$	NH_2OH	10,820	255,800	-	-

Exp. 1. Mouse spleen cells (70 x 10^6/ml PBS) were treated with neuraminidase (50 units/ml) for 30 min at 37°C.

Exp. 2. Cells were treated with $NaIO_4$ (1 mM) for 10 min at 23°C or with KBH_4 and NH_2OH (1 mM each) for 30 min at 23°C.

Con A was added at a final concentration of 2 μg/ml. Cells were cultured (5 x 10^6/ml, Dulbecco's modified Eagle's medium, supplemented with heat inactivated human serum (5%) for 72 hr. [^{3}H] Thymidine incorporation during 2 hr was then determined.

*Cells were treated with $NaIO_4$ (0.4 mM).

et al. (12) have shown that the main target of $NaIO_4$ oxidation of human erythrocytes is the sialic acid moieties of surface membrane sialoglycoproteins.

The observation that $NaIO_4$ triggers lymphocytes by the formation of an aldehyde moiety on the cell surface, prompted us to search for an enzymic modification of the cells that will mimic the $NaIO_4$ effect. Sialic acid in glycoproteins always occupies a non-reducing terminal position and is glycosidically linked either to D-galactose or to N-acetyl-D-galactosamine (13). Galactosyl residues exposed by the action of neuraminidase on plasma glycoproteins could be oxidized at the carbon 6 position by galactose oxidase to yield the 6-aldehyde analogs (14,15). We have found (16) that treatment of mouse spleen cells with galactose oxidase after incubation with neuraminidase induced extensive blastogenesis. Treatment of the cells with galactose oxidase before incubation with neuraminidase had very little stimulatory effect. Either of these enzymes alone had practically no effect on the cells (Table II). The mitogenic action of the galactose oxidase preparation used is abolished upon treatment of the enzyme with bis(2-hydroxyethyl) dithiocarbamate, an inhibitor of galactose oxidase, or upon incubation of the cells with the enzyme in the presence of excess of D-galactose (16). These findings indicate that the action of the enzyme is specific and is not due to the activity of contaminating enzymes. Lymphocyte transformation induced by sequential treatment with neuraminidase and galactose oxidase (NAGO) was decreased upon reacting the treated cells with KBH_4 or NH_2OH. It is suggested that galactosyl residues exposed by the action of neuraminidase on the cell membrane are oxidized by galactose oxidase and the aldehyde moiety thus formed is involved in the induction of blastogenesis (Fig. 1). Human peripheral lymphocytes are also stimulated by $NaIO_4$ (6, 17) or by NAGO (18). The blastogenic response of lymphocytes from chronic lymphatic leukemia patients to $NaIO_4$ is markedly reduced (6,17).

Removal of phagocytic cells from human peripheral lymphocytes markedly reduced their blastogenic response to treatment with $NaIO_4$ and NAGO, and had less effect on PHA-induced stimulation (17,18) (Table III). Transformation of lymphocytes by $NaIO_4$ or NAGO provides a suitable system for studying the separate effects of the mitogens on lymphocytes or macrophages alone in the induction of blastogenesis. The blastogenic response of $NaIO_4$ or NAGO-treated purified lymphocytes was markedly enhanced upon incubation on macrophage monolayers. In addition we have found that untreated purified lymphocytes were stimulated upon incubation on $NaIO_4$ or

TABLE II
RESPONSE OF MOUSE SPLEEN CELLS TO TREATMENT WITH NEURAMINIDASE AND GALACTOSE OXIDASE

Treatment I	Treatment II	[^{3}H]Thymidine incorporation (cpm)
-	-	889
Galactose oxidase	-	1,425
Neuraminidase	-	1,059
Neuraminidase	Galactose oxidase	93,995
Galactose oxidase	Neuraminidase	4,380

Mouse spleen cells (20 x 10^6/ml PBS) were incubated with neuraminidase (from Vibrio comma, Behringwerke, AG, Germany) (50 units/ml) for 30 min at 37°C or with galactose oxidase (Worthington Biochemical Corp., U.S.A.) (0.25 units/ml) for 30 min at 23°C. Cells were cultured as in Table I.

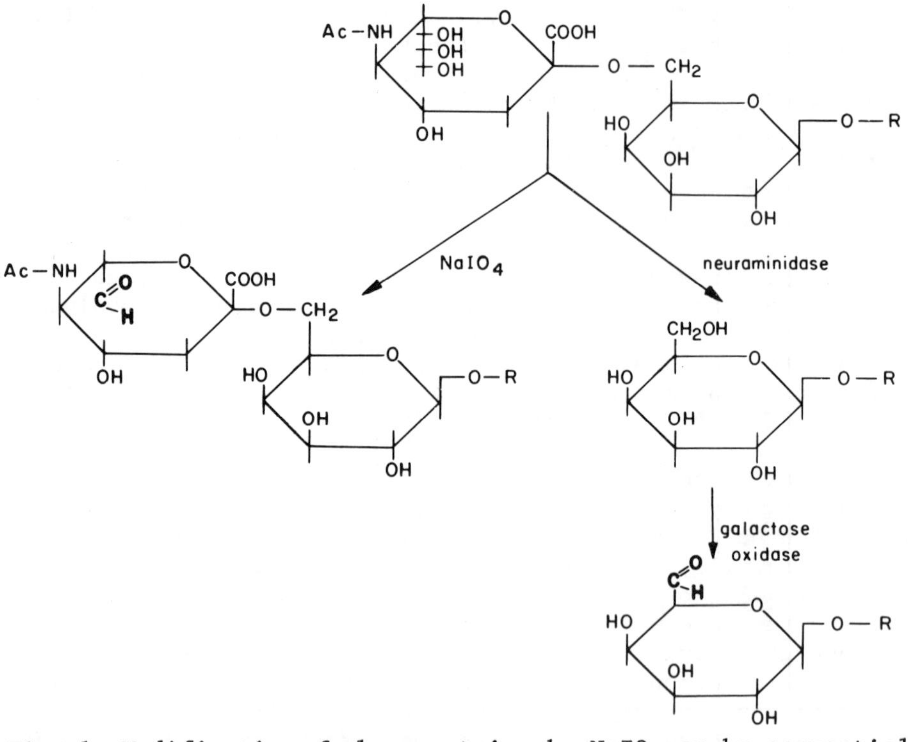

Fig. 1 Modification of glycoproteins by $NaIO_4$ or by sequential treatment with neuraminidase and galactose oxidase.

TABLE III

Effect of macrophages on periodate-induced transformation of normal human lymphocytes

Normal subjects	[^{3}H] Thymidine incorporation (cpm)											
	Mononuclear leucocytes treated with			Purified lymphocytes treated with			Untreated macrophage monolayers incubated with			Periodate-treated macrophage monolayers incubated with		
	None	Periodate	PHA	None	Periodate	PHA	Alone	Untreated purified lymphocytes	Periodate-treated purified lymphocytes	Alone	Untreated purified lymphocytes	Periodate treated purified lymphocytes
1	1600	18500	78300	680	1450	27600	-	-	-	-	-	-
2	730	23500	10450	200	800	11050	-	-	-	-	-	-
3	1250	70100	75600	1330	13250	42100	-	-	-	-	-	-
4	910	4450	-	820	1850	-	1210	1850	7800	-	-	-
5	550	24700	33300	1100	640	-	320	2300	40500	250	13300	36500
6	910	6630	32500	1300	3120	98800	880	1300	55900	730	11700	64500
7	350	9600	92200	530	2650	62400	-	-	34900	-	20700	55600
8	300	14700	-	830	170	-	-	-	15800	1560	16300	44200
9	-	-	-	420	1860	-	490	-	25800	310	-	57100
10	-	-	-	310	480	-	150	860	8950	280	2250	9770

Human peripheral mononuclear leucocytes (containing about 80% lymphocytes and 20% monocytes) were obtained by Ficoll-Hypaque gradients. Purified lymphocytes were obtained by passing of mononuclear leucocytes through nylon-columns. Macrophage monolayers (containing about 2×10^5 cells per tube) were obtained by incubation of mononuclear leucocytes in tissue culture tubes. Lymphocyte suspensions (2×10^6/ml in PBS) were treated with $NaIO_4$ (5×10^{-4} M) for 60 min at 0^0 C, and macrophage monolayers were treated with $NaIO_4$ (5×10^{-4} M) for 10 min at 23^0C. PHA-M (400 μg/ml) was added, as indicated. Cells (10×10^6/ml) were cultured in Dulbecco's modified Eagle's medium, supplemented with heat inactivated autologous serum (10%). Thymidine incorporation within 2 hr was determined after incubation for 48 hr (subject 1-3) and 72 hr (subject 4-10).

NAGO-treated macrophage monolayers (17,18). O'Brien et al. (19) have recently reported that human lymphocytes, inactivated with mitomycin C, ^{60}Co irradiation or UV light and treated with $NaIO_4$ are capable of inducing transformation of untreated autologous lymphocytes. In the above mentioned systems, in which lymphocytes were activated by cell to cell interactions, no stimulating factor could be demonstrated in the cell-free supernatants.

$NaIO_4$ or NAGO stimulates selectively thymus derived lymphocytes (T-cells) (20). This conclusion is based on the following observations:

1) Mouse thymus cells were stimulated after treatment with $NaIO_4$ or with NAGO. The stimulation of these cells was markedly enhanced upon addition of a lymphocyte activating factor which was obtained from human leukocyte cultures.

2) Spleen cells from congenitally athymic mice (nude mice) did not respond to treatment with $NaIO_4$ or with NAGO. Spleen cells from mice deprived of T-cells by thymectomy, irradiation and bone marrow reconstitution, responded poorly to the above mentioned mitogens.

The function of the aldehyde moiety in the induction of blastogenesis is not known. It is possible that the aldehydes react with other functional groups in the cell membrane, and that the cross-linked structure thus formed might play a role in the triggering process. It has already been suggested that cross-linkage and aggregation of specific membrane sites may be involved in the triggering of lymphocytes to undergo blastogenesis (21-24). The functional groups in the cell membrane that might react with the aldehyde group are:

1) Free amino groups (N-terminal amino groups or ϵ-amino group of lysyl residues) to form Schiff bases.

2) Alcoholic groups (saccharide moieties of glycoproteins or glycolipids) to form hemiacetals. The latter could react with additional alcoholic groups to form acetals.

3) Thiol groups (cysteinyl residues) to form hemimercaptals (or mercaptals). The Schiff base is probably the most stable structure among those which were outlined above.

Previous studies have shown (2,3,25,26) that a prolonged contact of phytomitogens with the cells is required for the induction of blastogenesis. We extended these studies to $NaIO_4$-induced lymphocyte transformation (8,27). Blastogenesis of $NaIO_4$-treated lymphocytes (as measured by the induction of DNA synthesis) could be essentially eliminated upon addition of cysteine (0.9 mM) or KBH_4 (0.8 mM) to the cell cultures during the first 5 hr of incubation. Addition of these agents between 5-22 hr after $NaIO_4$ treatment, caused only partial inhibition (Fig. 2,3). Similar results were obtained

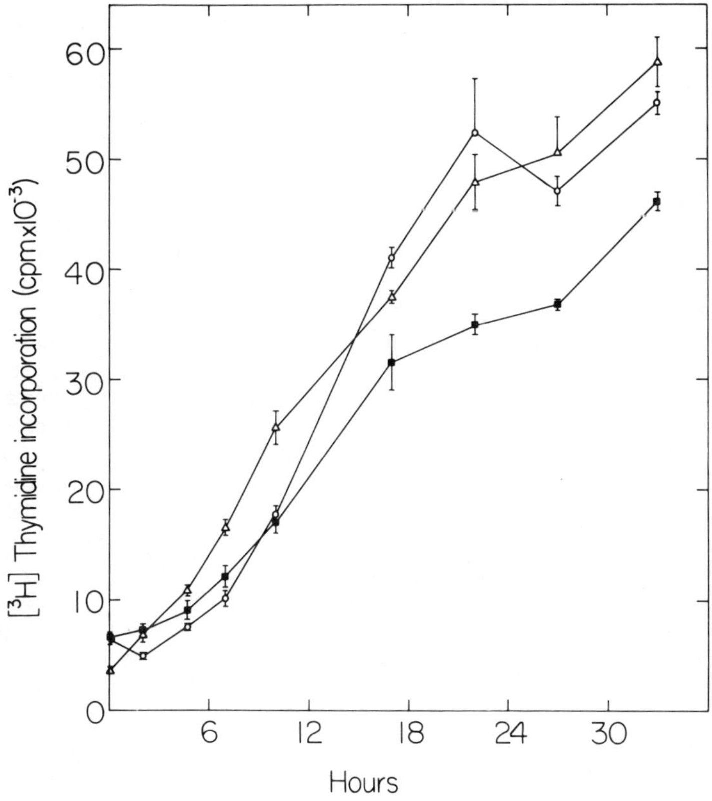

Fig. 2 Effect of KBH_4 and cysteine, added at different times of incubation on $NaIO_4$-induced DNA synthesis. Mouse spleen cells (30 x 10^6)/ml PBS) were treated with $NaIO_4$ (1 mM) for 30 min at 0°C and cultured (5 x 10^6/ml Dulbecco's modified Eagle's medium, containing heat inactivated fetal calf serum (5%) for 48 hr. At different times of incubation, KBH_4 (0.8 mM) (Δ-Δ), or cysteine (0.9 mM) (o --- o) were added. To untreated cell cultures, incubated with Con A (2 μg/ml), α-methyl-mannoside (20 mg/ml) (■-■) was added at different time intervals. After incubation for 48 hr. [^{3}H] thymidine incorporation during 2 hr was determined.

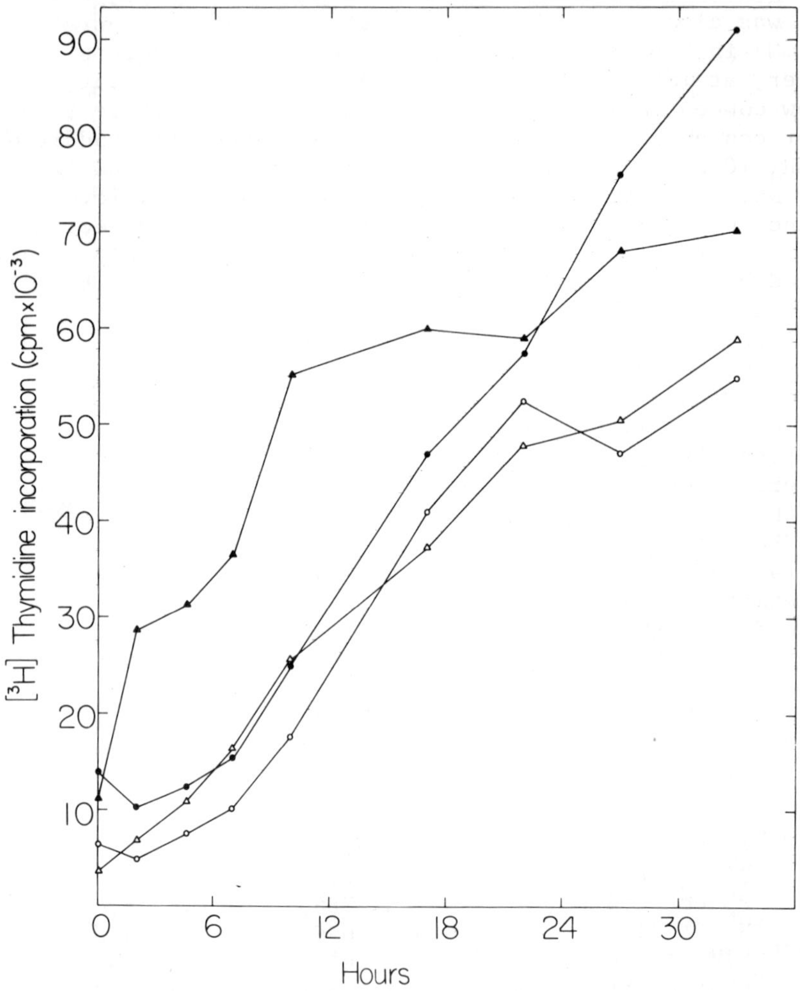

Fig. 3 Effect of different concentrations of KBH_4 and cysteine, added at different times of incubation, on $NaIO_4$-induced DNA-synthesis. KBH_4 (0.2 mM) (▲-▲); KNH_4 (0.8 mM) (Δ-Δ); cysteine (0.6 mM) (●---●); cysteine (0.9 mM) (o---o).

in experiments in which $NaIO_4$-treated cells, at different time intervals, were suspended in PBS and treated for 30 min with KBH_4, cysteine or NH_2OH (1 mM of each). Addition of KBH_4, at low concentration (0.2 mM) soon after $NaIO_4$ treatment, was almost as effective as KBH_4 at high concentration (0.8 mM) in the reversion of $NaIO_4$-induced blastogenesis. However, at later time intervals after $NaIO_4$ treatment, KBH_4 at low concentration (0.2 mM) was much less effective than at higher concentrations (0.8 mM) (Fig. 3). Sequential addition of KBH_4 (0.2 mM) and cysteine (0.9 mM), between 3 and 12 hr after $NaIO_4$ treatment reduced blastogenesis to the levels obtained upon addition of KBH_4 (0.8 mM) alone. The decreased effectiveness of KBH_4 (at low concentration) to inhibit $NAIO_4$-induced blastogenesis might result from structural alterations in the cell membrane, induced by the mitogen, which rendered the aldehydes less accessible to KBH_4 at low concentrations. We have postulated that the aldehyde moiety formed upon treatment of lymphocytes with $NaIO_4$ might interact with an amino group on the cell membrane to form a Schiff base, and that the cross-linked structure thus formed is involved in the triggering process. KBH_4 may reduce aldehyde and Schiff bases with the same efficiency. Therefore, one might have expected that KBH_4, in contrast to cysteine or NH_2OH, will not inhibit the transformation of $NaIO_4$-treated cells when added after the formation of the Schiff base. The experimental findings outlined above showed, however, that KBH_4 (1 mM), cysteine or NH_2OH had a similar inhibitory effect on $NaIO_4$-induced blastogenesis, when added at different time intervals after $NaIO_4$ treatment. The following possibilities should therefore be considered:

1) Schiff base is not involved in the triggering of lymphocytes by $NaIO_4$.

2) Schiff base is involved in the triggering process, however, reduction of the Schiff base results in the "fixation" of the linkage between the aldehyde and the amino group leading to the abolishment of the triggering signal. Cross-linkage between membrane sites of different cells might be essential for blastogenesis. The fixation of the cross-linked structure on single cells might therefore interfere with this process, and thus inhibit blastogenesis.

The kinetics of the reversion of $NaIO_4$-induced transformation by KBH_4, cysteine of NH_2OH (1 mM each) was similar to the reversion of Con A-induced blastogenesis by α-methylmannoside (Fig. 2).

Transformation of neuraminidase-treated lymphocytes by soybean agglutinin

Soybean agglutinin (SBA) is known to agglutinate different cell types (28,29). The agglutination is specifically inhibited by N-acetyl-D-galactosamine and also, but to a lesser degree by D-galactose. SBA agglutinates mouse lymphocytes only at relatively high concentrations and is not mitogenic. The observation that neuraminidase treatment rendered lymphocytes responsive to the mitogenic action of galactose oxidase prompted us to investigate whether blastogenesis can be induced by SBA in lymphocytes after their incubation with neuraminidase. We have found that treatment of mouse lymphocytes with neuraminidase rendered the cells responsive to blastogenic stimulation by SBA (30) (Fig. 4). Incubation of the cells with the enzyme also facilitated their agglutination by the lectin. The SBA-induced agglutination and transformation was inhibited specifically by N-acetyl-D-galactosamine and D-galactose (Table IV). Neuraminidase treatment increased about three-fold the amount of SBA bound to the lymphocytes. It is postulated that the transformation of neuraminidase-treated mouse lymphocytes, induced by SBA, is caused by the binding of the lectin to galactosyl residues exposed by the action of neuraminidase on the cell membrane.

It is possible that the chemical oxidizing agent $NaIO_4$, the enzymic oxidizing agent, galactose oxidase and the chemically inert lectin SBA, trigger lymphocytes by affecting the same glycoprotein containing the sequence sialic acid→ galactose →. The same glycoprotein might also contain the receptor [$(Man)_3$-GlcNAc-GlcNAc] for the mitogens: PHA, Con A, *W. floribunda* mitogen and *L. culinaris* mitogen (31).

Experiments in which the blastogenic effect of SBA on thymus cells, nude mouse spleen cells and T cells-deprived mouse spleen cells was investigated, revealed that SBA stimulates selectively neuraminidase treated-thymus-derived (T-cells) lymphocytes (20).

Potentiation of phytomitogens action by neuraminidase and basic polypeptides

Treatment of lymphocytes with neuraminidase enhances their blastogenic response to Con A and PHA without increasing the extent of lectin binding (24,32). Similar to neuraminidase, basic polypeptides (poly-L-ornithine or poly-D-lysine) also enhanced blastogenesis induced by phytomitogens. The basic polypeptides alone did not stimulate the cells and also did not enhance the response of neuraminidase-treated cells to phytomitogens. The extent of lectin binding to the cells was not affected by the basic polypeptides. It is postulated that

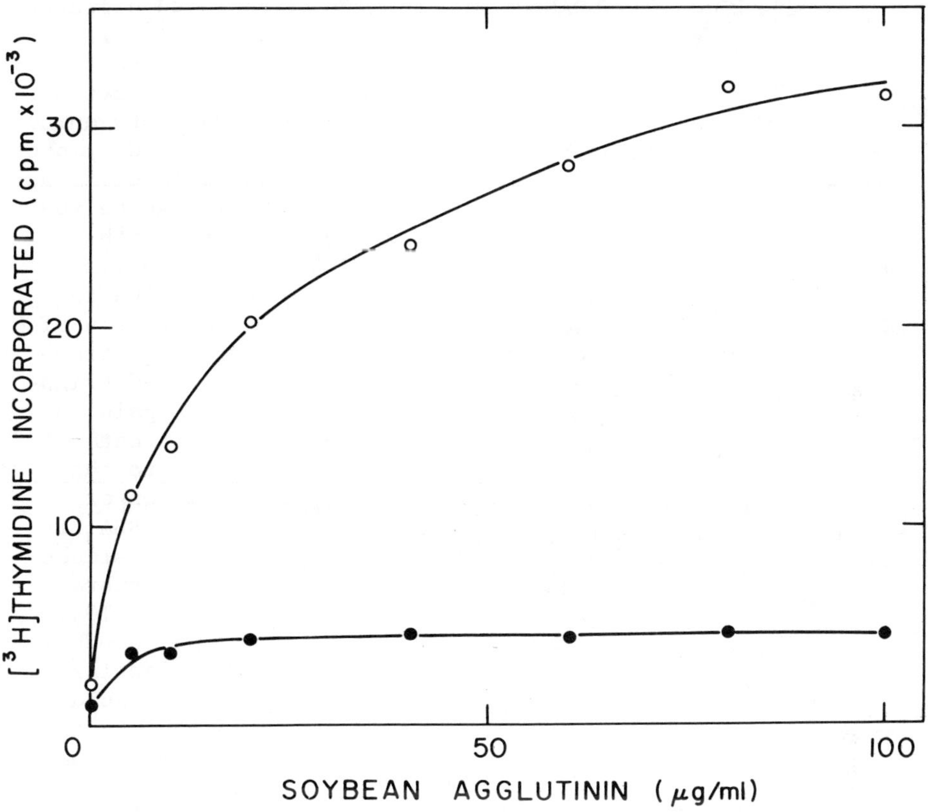

Fig. 4 Stimulation by soybean agglutinin of mouse spleen cells treated with neuraminidase. Mouse spleen cells (80×10^6/ml PBS) were treated with neuraminidase (50 units/ml) for 30 min at 37°C and cultured (5×10^6/ml) in the presence of SBA (as indicated) for 75 hr. [^{3}H] Thymidine incorporation during 2 hr was then determined. Neuraminidase-treated cells (o --- o); untreated cells (● --- ●).

TABLE IV

EFFECT OF DIFFERENT SACCHARIDES ON SBA-INDUCED AGGLUTINATION AND STIMULATION OF NEURAMINIDASE-TREATED MOUSE-SPLEEN CELLS

Additions SBA	Saccharide	[^{3}H] Thymidine incorporated (cpm)	Agglutination
-	-	1,740	-
+	-	21,200	+
+	D-Galactose	4,340	-
+	N-Acetyl-D-galactosamine	4,010	-
+	D-Glucose	22,100	+
+	N-Acetyl-D-glucosamine	23,400	+
+	α-Methyl-D-glucose	16,100	+
+	α-Methyl-D-mannose	21,200	+
+	L-Fucose	24,900	+

Mouse spleen cells were treated with neuraminidase and cultured as outlined in Fig. 2, in the presence of SBA (20 μg/ml). Saccharide was added at a final concentration of 5 mg/ml. Agglutination was recorded after 5 hr of incubation.

neuraminidase treatment or polycations facilitate lectin-induced aggregation of surface membrane glycoproteins by reducing their net charge and that the aggregations formed are involved in the triggering of lymphocytes to undergo blastogenesis (24).

Lymphocyte cytotoxicity induced by modification of effector or target cells with $NaIO_4$ or with neuraminidase and galactose oxidase (NAGO)

Cytotoxic effector cells can be formed in several different ways. Specific effector cells can be obtained from immunized animals or by the exposure of the lymphocyte to antigenically foreign target cells in vitro. Immunologically, non-specific effector lymphocytes can be obtained by incubation of normal lymphocytes with different mitogens (33). The mitogenic agents, $NaIO_4$ or NAGO can be easily removed after modification of the cells, thus it was able to study the effect of modification of the effector or target cells alone on the development of cytotoxicity. We have found that treatment of mouse spleen cells with $NaIO_4$ or with NAGO rendered the cells cytotoxic to mastocytoma (P815) target cells (Table V). Treatment of target cells (P815 cells and turkey erythrocytes) with $NaIO_4$ or with NAGO rendered them susceptible to cytolysis by untreated mouse spleen cells (Table VI). The cytotoxicity induced by $NaIO_4$ was reduced upon reacting the $NaIO_4$-treated, effector or target cells with KBH_4 or NH_2OH (Table VII, VIII). Thus the formation of free surface aldehydes on either the effector or target cell was sufficient to induce a cytotoxic effect. It is postulated that cross-linkage (possibly via a Schiff base) between effector and target cell initiates the cytotoxic effect (Fig. 5).

The cytotoxic potential of the lymphocyte might be an inherent property of the normal, non-activated cell or alternatively, might be induced shortly after cross-linkage between the effector and target cell. It is also possible that the mitogen-induced structural alterations in the target cell surface membrane, are essential for the induction of cytotoxicity. Cytotoxicity induced by $NaIO_4$ or by NAGO is immunologically non-specific and is independent of major antigenic differences between effector and target cells. Phagocytic cells are not involved in $NaIO_4$ or NAGO-induced cytotoxicity towards P815 target cells.

Potentiation of lectins-induced cytotoxicity by treatment of the effector or target cells with neuraminidase

Treatment of different cell types with neuraminidase exposes galactose-like sites on their cell membrane. The

TABLE V

CYTOTOXIC ACTION OF $NaIO_4$-OR NAGO-TREATED MOUSE SPLEEN CELLS ON MASTOCYTOMA (P815) TARGET CELLS

Treatment of spleen cells	[^{51}Cr] release (%) from P815 cells	[^{3}H] Thymidine incorporation (cpm)
None	-1.0 ± 0.02*	7,516 ± 1,220
$NaIO_4$	13.3 ± 0.19	149,615 ± 8,961
Neuraminidase and galactose oxidase	13.9 ± 0.64	177,740 ± 15,602
Neuraminidase	-0.3 ± 0.06	7,897 ± 916
Galactose oxidase	1.5 ± 0.06	8,249 ± 504
Con A	8.7 ± 0.52	149,613 ± 12,279

Mouse spleen cells were treated with $NaIO_4$ (1 mM) for 30 min at 0°C. Cells were treated with neuraminidase (50 units/ml), galactose oxidase (0.25 unit/ml) or with both enzymes for 30 min at 37°C. Con A was added at a final concentration of 2 μg/ml. Cytotoxicity (in the experiments outlined in Tables V - VIII) was measured by [^{51}Cr] release from [^{51}Cr]-labelled target cells (3-5 x 10^5) after incubation with mouse spleen cells (7-10 x 10^6) for 18-22 hr in a total volume of 1 ml Dulbecco's modified Eagle's medium containing heat inactivated fetal calf serum (5%). Thymidine incorporation was measured as outlined in Fig. 3.

* ± Standard error of triplicate cultures.

TABLE VI

CYTOTOXIC ACTION OF UNTREATED MOUSE SPLEEN CELLS ON $NaIO_4$-OR NAGO-TREATED P815 CELLS AND TURKEY ERYTHROCYTES

Treatment of target cells	[^{51}Cr] release (%)	
	P815 cells	Turkey erythrocytes
None	-2.9 ± 0.02	2.0 ± 0.08
$NaIO_4$	10.6 ± 0.13	11.6 ± 0.13
Neuraminidase and galactose oxidase	10.7 ± 0.09	15.9 ± 0.14
Neuraminidase	-3.9 ± 0.13	5.6 ± 0.29
Galactose oxidase	2.1 ± 0.04	2.5 ± 0.04

Treatment of cells and cytotoxicity assay as outlined in Table V.

TABLE VII

DECREASE OF THE CYTOTOXIC ACTION OF $NaIO_4$-TREATED MOUSE SPLEEN CELLS ON P815 TARGET CELLS BY REACTING THE FORMER WITH BOROHYDRIDE OR HYDROXYLAMINE

Treatment of spleen cells		[^{51}Cr] release (%) from P815 cells	
I	II	-	Con A
PBS	PBS	-0.9 ± 0.1	13.8 ± 0.5
PBS	KBH_4	0.1 ± 0.01	12.2 ± 0.16
PBS	NH_2OH	0.3 ± 0.01	12.6 ± 0.06
$NaIO_4$	PBS	12.2 ± 0.61	14.9 ± 0.09
$NaIO_4$	KBH_4	2.8 ± 0.12	15.5 ± 0.34
$NaIO_4$	NH_2OH	5.1 ± 0.24	14.5 ± 0.14

Mouse spleen cells were treated with $NaIO_4$ as outlined in Table V. After washing the cells were suspended (50 x 10^6/ml) in PBS and treated with KBH_4 (1 mM) or NH_2OH (1 mM) for 30 min at 23°C. Cytotoxicity was measured as outlined in Table V.

TABLE VIII

DECREASE OF THE CYTOTOXIC ACTION OF UNTREATED MOUSE SPLEEN CELLS ON $NaIO_4$-TREATED P815 TARGET CELLS BY REACTING THE LATTER WITH BOROHYDRIDE OR HYDROXYLAMINE

Treatment of P815 cells		[^{51}Cr] release (%) from P815 cells	
I	II	-	Con A
PBS	PBS	-1.8 ± 0.09	10.4 ± 0.04
PBS	KBH_4	0.2 ± 0.02	12.2 ± 0.74
PBS	NH_2OH	0.7 ± 0.04	13.2 ± 0.58
$NaIO_4$	PBS	9.5 ± 0.23	11.5 ± 0.14
$NaIO_4$	KBH_4	2.2 ± 0.07	14.1 ± 0.2
$NaIO_4$	NH_2OH	3.0 ± 0.05	12.6 ± 0.15

Treatment of cells and cytotoxicity assay as outlined in Table VII.

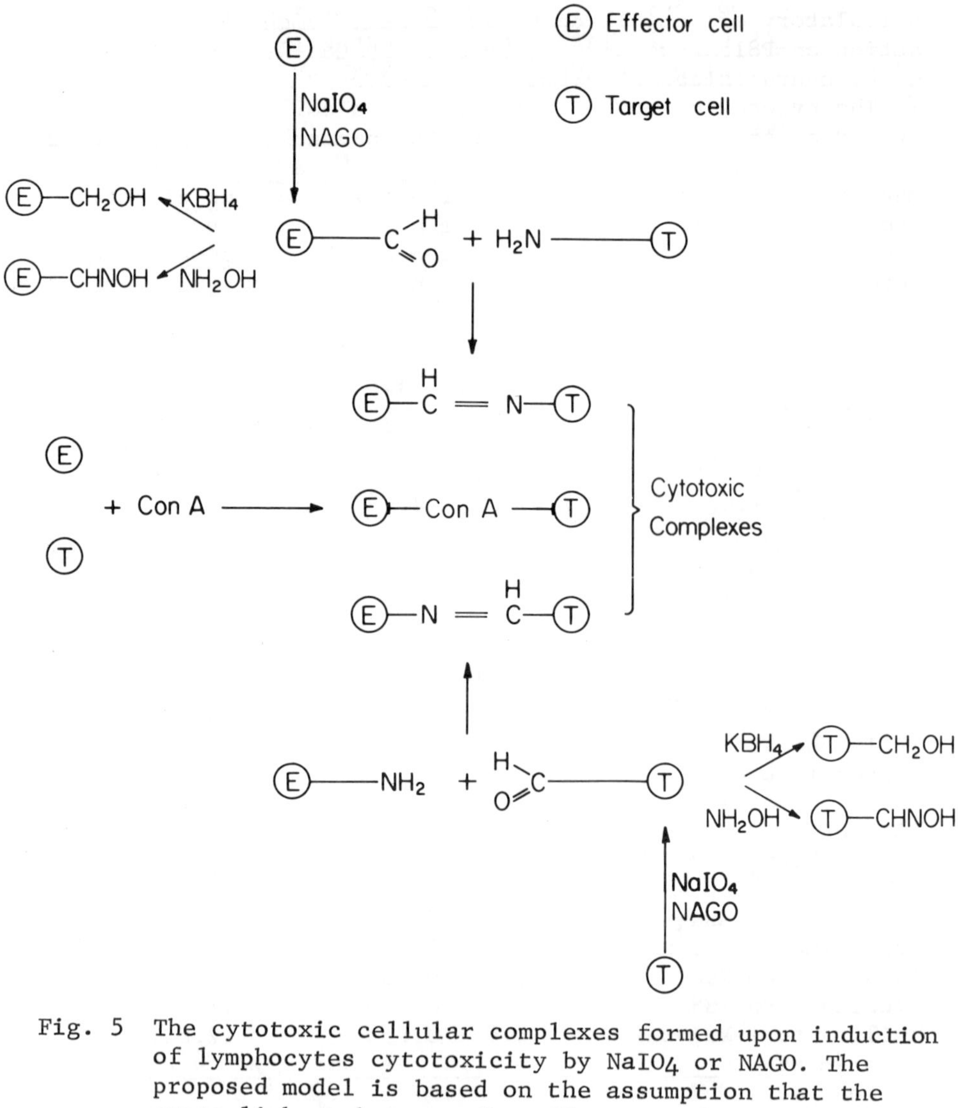

Fig. 5 The cytotoxic cellular complexes formed upon induction of lymphocytes cytotoxicity by $NaIO_4$ or NAGO. The proposed model is based on the assumption that the cross-linkage between the effector and target cells is formed via a Schiff base.

cytotoxic activity of mouse spleen cells on mastocytoma P815 target cells induced by SBA (a lectin which binds specifically galactosyl residues) was markedly enhanced by treatment of both the effector and target cells with neuraminidase. Treatment of either cell type alone with the enzyme had little stimulatory effect. In contrast to SBA, lymphocyte cytotoxic action on P815 target cells induced by Con A, was not affected by neuraminidase treatment.

The cytotoxic action of mouse spleen cells on turkey erythrocytes in the presence of SBA was markedly increased after treatment of the target cells with neuraminidase. Treatment of the effector cells with the enzyme had no effect. A similar but less pronounced effect of neuraminidase on Con A-induced cytotoxicity towards turkey erythrocytes was also noted, although neuraminidase did not expose new binding sites for Con A the target cells. Preliminary experiments have indicated that different effector cell types might be involved in the cytotoxic action of mouse spleen cells on P815 or turkey erythrocytes target cells.

The above findings indicate that cross-linkage between the effector and target cells initiates lymphocyte mediated cytotoxicity. The potentiating effect of neuraminidase on lectin-induced cytotoxicity might be due to:

1) an increase in the lectin binding sites on the effector or target cells;

2) a reduction in the net negative charge of surface membrane glycoproteins.

The latter effect might enhance the aggregation of membrane sites induced by the lectins. Aggregation of membrane sites (on the effector or target cells) in addition to its role in lymphocyte activation, might also render target cells susceptible to lysis by the effector cells.

CONCLUDING REMARKS

Lymphocyte activation induced by $NaIO_4$ or NAGO is a unique system in which a well defined modification of a membrane site induces extensive blastogenesis. The observation that the formation of a free aldehyde moiety on the cell surface is sufficient to induce lymphocyte transformation, might elucidate the nature of the triggering signal. Moreover, the ability to label the modified membrane sites by reduction with $^3H\text{-}KBH_4$ will facilitate their isolation and characterization. Activation of lymphocytes by chemical or enzymic agents permitted the study of the separate effects of the mitogens on different cell types which are involved in blastogenesis (lymphocytes and macrophages) or lymphocyte mediated cytotoxicity (effector and target cells). Lymphocyte activation induced by cell modification is also most suitable for the study of the products of mitogen stimulation (e.g. MIF, blastogenic factor, interferon) since the excess of the mitogen can be readily removed from the cell suspension after its brief action.

REFERENCES

1. Borberg, H., Yesner, I., Gesner, B., and Silber, R., Blood, 31: 747, 1968.
2. Powell, A.E., and Leon, M.A., Exp. Cell. Res., 62: 315, 1970.
3. Novogrodsky, A., and Katchalski, E., Biochem. Biophys. Acta, 228: 579, 1971.
4. Novogrodsky, A., and Katchalski, E., FEBS Lett., 12: 297, 1971.
5. Novogrodsky, A., and Katchalski, E., Proc. Natl. Acad. Sci., 69:3207, 1972.
6. Parker, J.W., O'Brien, R.L., Lukes, R.J., and Steiner, J., Lancet i: 103, 1972.
7. Zatz, M.M., Goldstein, A.L., Blumenfeld, I.O., and White, A., Nature New Biol., 240: 240, 1972.
8. Rosenwasser, A., and Novogrodsky, A., Israel J. Med. Sci., in press, 1974.
9. Yu, R.K., and Ledeen, A.J., J. Biol. Chem., 244: 1306, 1969.
10. Karkas, J.D., and Chargaff, E., J. Biol. Chem., 239: 949, 1964.
11. Van Lenten, L., and Ashwell, G., J. Biol. Chem., 246: 1889, 1971.
12. Liao, T., Gallop, P.M. and Blumenfeld, O.O., J. Biol. Chem., 248: 8247, 1973.
13. Spiro, R.G., Ann. Rev. Biochem., 39: 599, 1970.
14. Robinson, J.C., and Pierce, J.E., Arch. Biochem. Biophys., 106: 348, 1964.
15. Morell, A.G., Van Den Hamer, C.J.A., Scheinberg, I.H., and Ashwell, G., J. Biol. Chem., 241: 3745, 1966.
16. Novogrodsky, A., and Katchalski, E., Proc. Natl. Acad. Sci., 70: 1824, 1973.
17. Biniaminov, M., Ramot, B., and Novogrodsky, A., Clin. Exp. Immunol., 16: 235, 1974.
18. Biniaminov, M., Ramot, B., Rosenthal, E., and Novogrodsky, A., Clin. Exp. Immunol., in press, 1974.
19. O'Brien, R.L., Paolilli, P., Steiner, J., and Parker, J.W., J. Immunol., 112: 1884, 1974.
20. Novogrodsky, A., Eur. J. Immunol., in press 1974.

21. Fanger, M.W., Hart, D.A., Wells, J.V., and Nisonoff, A., J. Immunol., 105: 1484, 1970.
22. Greaves, M., and Janossy, G., Transplant. Rev., 11: 87, 1972.
23. Lotan, R., Lis, H., Rosenwasser, A., Novogrodsky, A., and Sharon, N., Biochem. Biophys. Res. Commun., 55: 1347, 1973.
24. Novogrodsky, A., Nature, 250: 788, 1974.
25. Younkin, L.H., Exp. Cell. Res., 75: 1, 1972.
26. Jones, G., J. Immunol., 110: 1262, 1973.
27. Ravid (Rosenwasser), A., and Novogrodsky, A., unpublished results.
28. Lis, H., Sela, B., Sachs, L., and Sharon, N., Biochem. Biophys. Acta, 211: 582, 1970.
29. Sela, B., Lis, H., Sharon, N., and Sachs, L., J. Membrane Biol., 3: 267, 1970.
30. Novogrodsky, A., and Katchalski, E., Proc. Natl. Acad. Sci., 70: 2515, 1973.
31. Toyoshima, S., Fukuda, M., and Osawa, T., Biochemistry, 11: 4000, 1972.
32. Han, T., Clin. Exp. Immunol., 13: 165, 1973.
33. Perlmann, P.& Holm, G., Adv. Immunol., 11: 117, 1969.

SIGNALS AND RECEPTORS IN B CELL ACTIVATION

Antonio Coutinho, Eva Gronowicz and Göran Möller
Division of Immunobiology, Karolinska Institutet, Wallenberg-laboratory, Lilla Freskati, Stockholm, Sweden

INTRODUCTION

The mechanism by which lymphocytes are activated by antigen constitutes a fundamental problem in immunology. Several lines of evidence suggest that T and B lymphocytes are governed by different rules in this respect. The present paper focuses attention on B cells, since more is experimentally known about these cells and theoretical background concerning their activating mechanism is more advanced.

As will be shortly outlined below, most of the present hypotheses concerning B cell triggering focus attention on the fundamental role of the Ig receptors in delivering at least one necessary signal. However, other membrane structures have also been implicated, such as the Fc and C'3 sites. We will try to show, in this paper that the Ig receptors do not give activating signals to the B cells, but serve as passive focusing devices for the actual triggering signals. Nor do the Fc and C'3 receptors have any active role in activation, but may participate in a role analogous to the Ig receptors, although to a much small extent.

Presently, three conceptually different hypotheses have been advanced to explain antigen-induced activation of B cells. In two of these hypotheses the Ig receptors are considered to be directly responsible for the generation of the triggering signals. The first hypothesis postulates that only the interaction of the antigen with the surface Ig receptors is sufficient for activation to occur, when the pattern of antigen presentation "cross-links" the Ig receptors to a certain, but as yet undefined, extent (1). A second hypothesis claims that B cells are only activated when the Ig receptors of the B cells have first interacted with the antigen (signal one), and thereafter a second signal is delivered (2). The nature of this second signal varies according to different formulations of the hypothesis, from immunologically specific to non-specific. The third hypothesis suggests that only one non-specific signal (which does not involve Ig receptors) is needed to trigger B cells (3).

This work was supported by grants from the Swedish Cancer Society, the Swedish Medical Research Council and the Jeansson Foundation.

For activation to occur, the non-specific signals must be delivered to the cell in a sufficient but not excessive quantity.

It is not our intention to critically analyze the various hypotheses. Rather we want to summarize the experimental evidence for the one non-specific signal hypothesis. However, it is evident that the support for this hypothesis at the same time constitutes arguments against competing concepts.

Direct induction of antibody synthesis in B cells by polyclonal B cell activators

Although it is generally accepted that the Ig receptors are directly responsible for triggering of B cells, the available evidence for this dogma is weak. So far, only the finding that anti-Ig sera can induce DNA synthesis in blood lymphocytes from certain species support this notion (4). However, these findings are restricted to rabbits and chickens in spite of attempts to verify them in other species, and in these species the responding cells have not been characterized with regard to T or B origin. Actually, the evidence is against activation of B cells, since there is no selective increase in immunoglobulin synthesis in the activated cell population (5).

In contrast, it has been clearly shown that resting B cells can be activated to division and/or increased immunoglobulin synthesis by substances which do not react with the Ig receptors of the activated lymphocytes. These are the polyclonal B cell activators (PBA), the first discovered being LPS, but there is now an extensive list of such substances (Table 1) (6,7,8). These substances induce polyclonal immunoglobulin synthesis, morphological transformation and increased DNA synthesis in B, but not T, lymphocytes. Since lymphocytes having Ig receptors of all specificities are activated by different PBA, it is obvious that there is no complementarity between the PBA and the combining site of the Ig receptor. Therefore, the B cells were activated by a non-specific signal. Attempts have been made to explain PBA triggering in terms of a two-signal concept by assuming that the Ig receptors of the activated cells have reacted with cross-reacting antigens in the culture medium or on autologous cells and that the PBA only delivers a second non-specific signal (9). This is unlikely for several reasons such as 1. triggering can be achieved in serum-free medium and is not influenced (enhanced or suppressed) by the addition of free hapten or immunogen (10). Thus, cells that have combined with antigens are triggered to the same extent as

TABLE 1
LIST OF POLYCLONAL B CELL ACTIVATORS

Polyclonal B cell activators	Thymus-independent antigens
Lipopolysaccharides (LPS)	Lipopolysaccharides (LPS)
Pneumococcal polysaccharide SIII	Pneumococcal polysaccharide SIII
Polymerized flagellin (POL)	Polymerized flagellin (POL)
Levan	Levan
Dextran	Dextran
Polyvinyl pyrrolidone (PVP)	Polyvinyl pyrrolidone (PVP)
Pokeweed mitogen (PWM)	N.t.
Purified protein derivative of tuberculin	N.t.
Dextran sulphate	N.t.
Pentosan sulphate	N.t.
Nocardia extracts	N.t.
Fetal calf sera	N.t.
Some KLH preparations	Some KLH preparations
SRC lysates	SRC lysates
Fucose binding protein	N.t.
Soluble factors from accessory cells	N.t.
N.t.	Certain synthetic polypeptides

cells that have not reacted with the specific antigen.
2. If cross-reacting antigens were present on autologous cells, the two signal hypothesis postulates that this would lead to paralysis and such cells could not be induced by a second signal, although this regularly occurs. Furthermore, cells from germ-free and newborn animals can be activated to polyclonal antibody synthesis as well.

It seems inescapable to us that polyclonal B cell activation is caused by one non-specific signal that is not delivered by Ig receptor specificities.

It should be pointed out that B cell activation by PBA represents a direct effect on the B cell and does not require participation of any helper cells such as T cells or adherent cells (unpublished).

Polyclonal B cell activators and thymus-independent antigens

The finding that immunocompetent B cells could be directly activated to antibody synthesis by PBA raised the question whether only polyclonal antibody synthesis could be induced by one non-specific triggering signal, or whether the phenomenon also operated in specific physiological (antigen-mediated) triggering of B cells. There are so far several lines of evidence suggesting that one non-specific signal is the normal mechanism operating also in specific immune activation. Thus, haptens or large protein molecules, when coupled to a TI molecule, also induce TI antibody responses, clearly suggesting that TI independence results from an intrinsic property of the carrier.

The first direct evidence was the finding that all thymus-independent (TI) antigens were also PBA and had the capacity of directly inducing polyclonal antibody synthesis in B cells without participation of helper cells (T cells and macrophages). Thus, all known TI antigens possessed the intrinsic property of activating B cells of all immunological specificities (6).

Based on these findings, it was suggested (3) that the role of Ig receptors in the induction of specific immune responses was only to focus the antigen to the surface of the specific antigen reactive B cells. In this way, B cells which have Ig receptors directly against the antigenic determinants of a TI antigen will selectively bind the antigen to its surface. At low concentrations of antigen only the specific cell will bind a sufficient amount of mitogenic molecules in order to be triggered, but triggering is caused exclusively by the PBA properties of the TI antigen. Non-specific cells - lacking Ig receptors directed against the antigenic determinants of the TI antigen - will at low (antigenic) concentrations not bind a sufficient number of molecules to become triggered. However, when the concentration of TI antigens is raised also the non-specific cell will bind a sufficient number of TI antigen molecules to its surface to become activated (Fig. 1). In this case activation will be polyclonal. However, the specific B cells reacting with the antigenic determinants will now bind so many molecules that they will be specifically turned off (paralyzed). We base our concept of triggering on the findings with polyclonal B cell activators that cells are activated by the quantity, but not the quality of the triggering signals.

This hypothesis was tested experimentally and the results were in agreement with the proposal (10). A typical experiment is presented in Fig. 2. In these experiments, a haptenated PBA (NNP-LPS) was added in different concentrations to spleen cell cultures. Two days later we measured the responses induced in the B cell population which could recognize and bind with high affinity the haptenic determinants on the PBA molecule (detecting anti-NNP PFC) as well as in the B cell population bearing all the other Ig specificities (detecting PFC to an unrelated antigen - SRC, or increased DNA synthesis). As shown in Fig. 2, low NNP-LPS concentration only induced specific high affinity anti-NNP PFC. At increasing concentrations of NNP-LPS the high affinity antibody producing NNP cells disappeared, whereas polyclonal antibody synthesis was induced. When free hapten - in the form of NNP-cap - was added to the cultures followed by NNP-LPS, the specific high affinity immune response to NNP at low concentrations of NNP-LPS was abolished. However, at high concentrations of NNP-LPS, capable by themselves of inducing polyclonal antibody synthesis, also the NNP-reactive cells were activated as all other cells, even in the presence of free hapten. This shows that specific antigen-binding cells can concentrate antigen by means of the Ig receptors and, therefore, be confronted with a sufficient number of PBA molecules to become triggered. This response, although induced by a non-specific PBA appears as a specific thymus-independent response. Free hapten, which binds to the immunoglobulin receptors, prevents the induction of a specific immune response, presumably because the specific cells could not selectively concentrate the PBA, since the combining sites of the Ig receptors were occupied by hapten. Free hapten by itself did not induce antibody synthesis and furthermore failed to induce paralysis in B cells since higher doses of NNP-LPS or LPS triggered also the B lymphocytes that

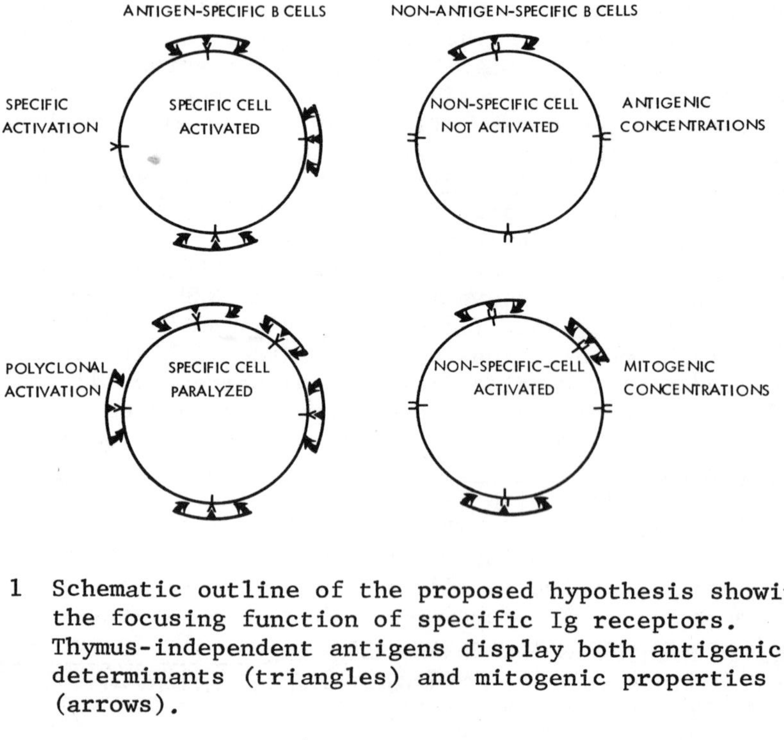

Fig. 1 Schematic outline of the proposed hypothesis showing the focusing function of specific Ig receptors. Thymus-independent antigens display both antigenic determinants (triangles) and mitogenic properties (arrows).

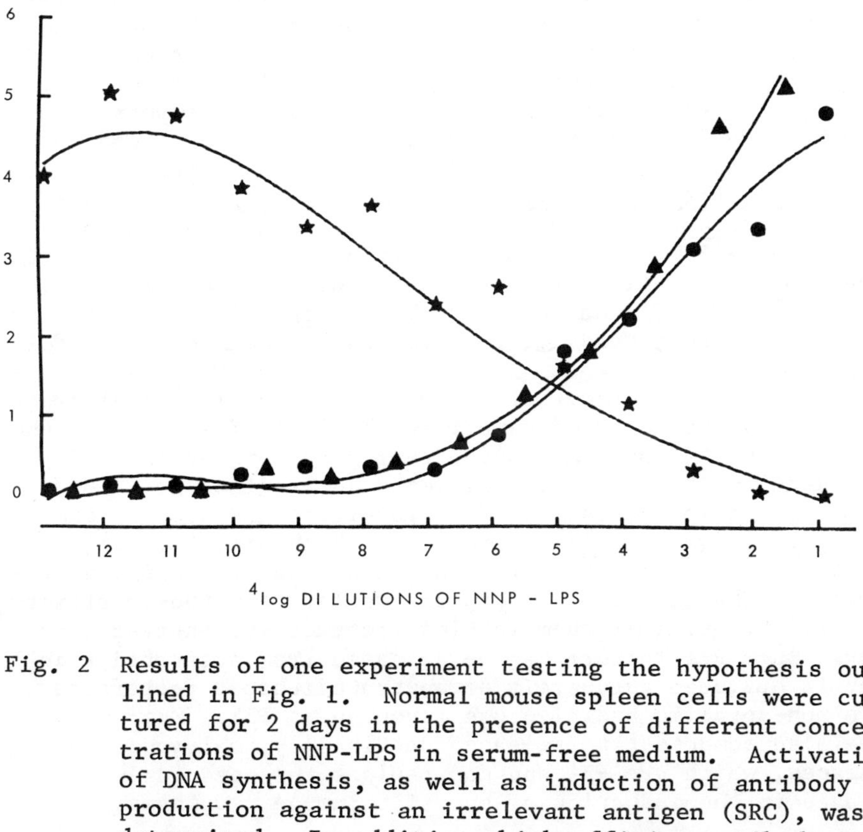

Fig. 2 Results of one experiment testing the hypothesis outlined in Fig. 1. Normal mouse spleen cells were cultured for 2 days in the presence of different concentrations of NNP-LPS in serum-free medium. Activation of DNA synthesis, as well as induction of antibody production against an irrelevant antigen (SRC), was determined. In addition, high affinity antibody synthesis against the specific hapten NNP was measured. ▲——▲ = cpm/culture (Y:0 to 84,000); ●——● = anti-SRC PFC/culture (Y:0 to 300); ★——★ = high-avidity anti-NNP PFC/culture (Y:0 to 1,500).

had reacted with the hapten (10).

Failure of TD antigens to induce B cells in a locally concentrated form

It follows from the one non-specific signal that truly TD molecules would never be competent to induce B cells directly even when presented in a locally concentrated pattern, suitable for cross-linking the surface Ig receptors. To test this prediction, FITC was conjugated to HGG and HSA at various ratios. The same FITC conjugates were also coupled in various concentrations to Sepharose particles. It was consistently found in the same experiments that FITC-LPS induced a primary immune response in cultures, whether it was added in soluble form or bound to Sepharose particles. However, none of the FITC protein conjugates induced a response whether added in soluble or precipitated form or bound to Sepharose particles over a wide range of concentrations (11). Therefore, these experiments failed to support the cross-linking concept, in particular as the conjugation ratio of FITC-LPS (1 FITC per 100,000 molecular weight LPS) was in the range of conjugation ratios of the proteins used (1 per 3,000 to 1 per 150,000), and the density of haptenic determinants on the Sepharose beads in culture was varied over a 10^9 fold range in 10 fold steps.

Several consequences of the one non-specific signal hypothesis have been tested experimentally. If follows from the hypothesis that only molecules having PBA properties (TI antigens) will be able to trigger B cells directly. TD immunogenic molecules lack PBA properties and are not competent to activate B cells directly. Therefore, they require help of other cell types for the induction of the immune response. The surface Ig receptors would focus non-specific B cells by bringing them in close contact via antigen bridging. Such PBA factors can be released from activated T cells and it has been adequately demonstrated that T cell factors are non-specific and also can function as PBA. Macrophages are also competent to release functionally analogous substances, which are PBA and can replace the need for macrophages in the induction of specific immune responses.

PBA synergize with TI but not TD antigens in the induction of specific antibody synthesis

In terms of the one non-specific signal hypothesis it is to be expected that specific immune responses to TI antigens should be additively influenced by the addition of other PBA, since the quantity of the delivered signal determines the magnitude of the B cell response. This should not occur with

TD antigens using purified B cell populations in vitro, since these molecules lack PBA properties and, therefore, the simultaneous addition of TD antigens and PBA would not give a response stronger than that induced by PBA itself. Both these types of experiments have been performed. NNP-LPS conjugate was added in such a low concentration to B cells that high levels of high affinity anti-NNP cells were not induced. In addition, low concentrations of non-haptenated LPS (which by itself gave a very weak polyclonal response) were added simultaneously and it could be shown that the simultaneous addition of both markedly enhanced the specific anti-NNP response, but did not increase the magnitude of the polyclonal antibody response (unpublished). The specific anti-NNP response could be selectively induced in anti-NNP cells, because these cells specifically focused NNP-LPS to its surface and were at the same time confronted with LPS binding non-specifically, and therefore, these cells received more triggering signals than non-specific B cells.

Finally, experiments were performed to study whether a PBA (LPS) would cause an increased response to FITC when added together with the above mentioned FITC protein conjugates or FITC-Sepharose particles or with free hapten (11). These results were consistently negative, one example being shown in Fig. 3. Thus, a TD antigen could not deliver a first signal preparing the B cells for a second signal delivered by LPS, as postulated in the two signal hypothesis.

Hapten PBA molecules suppress specific antibody synthesis

In terms of the one non-specific signal hypothesis a specific immune response, as well as polyclonal induction of immunoglobulin synthesis, could be suppressed by preventing binding of the PBA to the triggering surface structures on the B cells. There exist substances that seem capable of binding to these structures on the B cells without activating them. Thus, the PBA activity of dextrans varies with its molecular weight. Unsubstituted dextrans with a molecular weight of less than 70,000 do not activate B cells. The degree of activation increases linearly with the log of the molecular weight of dextrans with the higher MW (8). Another molecule with analogous properties is polyacrylic acid (PAAC). It could be shown that both PAAC and low MW dextrans could suppress induction of polyclonal antibody synthesis as well as the specific response to TI antigens such as NNP-LPS. This is to be expected from the one non-specific signal hypothesis. It is of great interest that also the response to a thymus-dependent antigen, such as HRC, could be suppressed the same way. As shown in Fig. 4, the addition of PAAC to

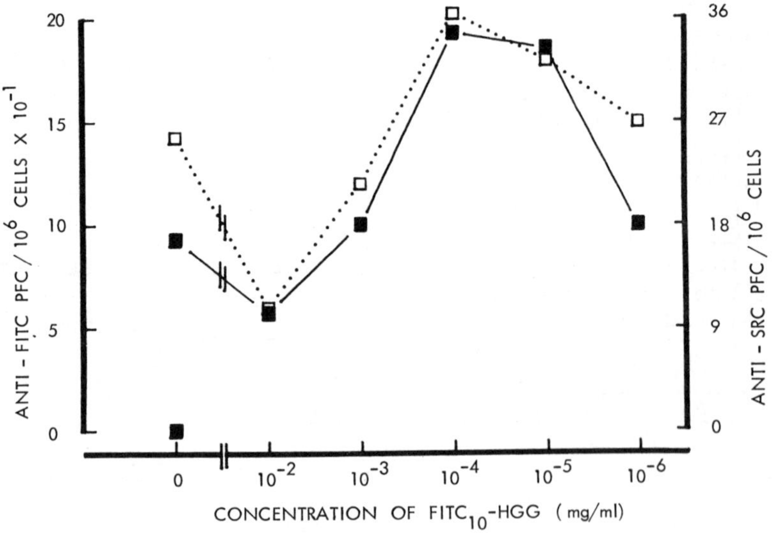

Fig. 3 Attempt to detect synergy between different concentrations of soluble $FITC_{10}$-HGG and 100 μg LPS in serum free cultures tested at day 4. The response was assayed against FITC (□ · · · · · □) and against SRC (■——■). The background against FITC (■) is also indicated.

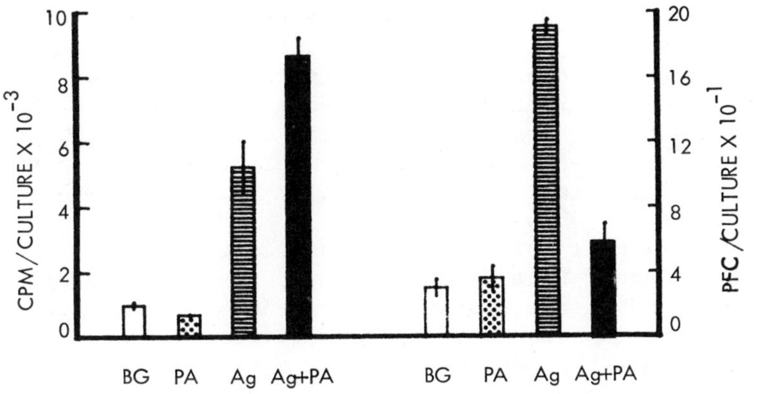

Fig. 4 Spleen cells from (A x 5M)F_1 hybrids primed two weeks before with HRC were cultured in tubes (2 x 10^6 cells/ml) in serum free medium in the presence of the indicated substances. The anti-HRC PFC and the DNA synthetical responses were measured after 4 days in parallel cultures. Bg - no additions; PA - polyacrylic acid 400 µg/ml; Ag - 10^7 HRC.

spleen cell cultures given the antigen HRC caused a marked suppression of the anti-HRC response, whereas it did not suppress the increased DNA synthesis induced by HRC (unpublished). The latter response is probably due to activation of T cells, which should not be expected to be suppressed by the haptenic PBA.

Different PBA act on subpopulations of B cells

It has recently been found that different PBA act on subpopulations of B lymphocytes (12,13). Thus, dextran sulphate acts on rather primitive cells and the response of these cells is mainly increased DNA synthesis with little induction of polyclonal antibody synthesis. Dextran sulphate may actually activate fetal liver and bone marrow cells before they have acquired Ig receptors on their surface. LPS and SIII activate more differentiated cells and induce both to DNA and polyclonal antibody synthesis, whereas other PBA, such as PPD act on very differentiated B cells, where the predominant response is induction of immunoglobulin synthesis with little DNA synthesis.

The experimental evidence for these conclusions is based both on studies of B cells in different stages of ontogenic development and on the results of sequential stimulation of B cells (in vitro and in vivo) with different PBA. The findings support the concept that the various targets for different PBA belong to one single line of differentiating B cells rather than to separate subpopulations. A summary of these findings is shown in Fig. 5.

Evidence from genetic non-responder mice

It is known that certain substrains of C3H do not produce antibodies to LPS, but show a good immune response to thymus-dependent antigens (14). It is also known that the same substrains of C3H do not respond to the endotoxic properties of LPS (15). We have investigated these LPS low responder substrains with regard to their ability to respond to LPS as evidenced by the induction of polyclonal antibody snythesis and to respond specifically to hapten conjugated to LPS. In addition, the response of these substrains to other PBA has been studied. In short, the results show that B cells from certain C3H sublines (C3H/HeJ) are low responders with regard to induction of polyclonal antibody or DNA synthesis by LPS. The low responder mice, however, can respond to LPS added in much higher concentrations. B cells from these strains show a normal or enhanced DNA synthetic and polyclonal antibody response to PPD and a somewhat lower response to dextran sulphate. Therefore, their defect appears to selectively

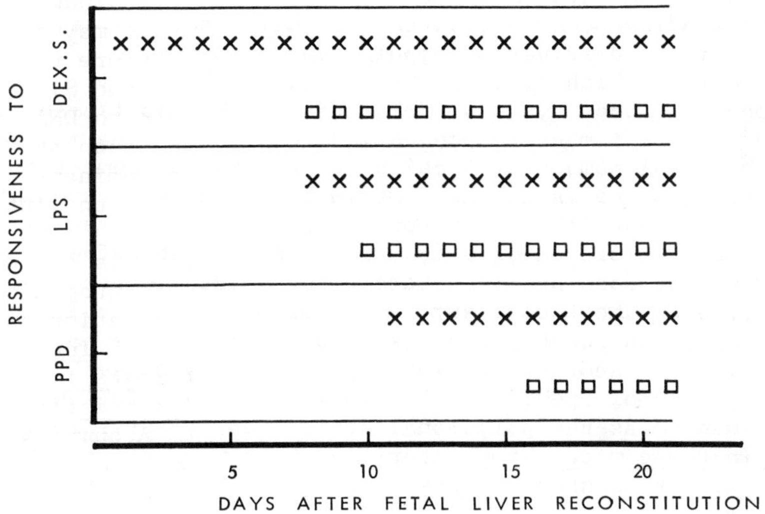

Fig. 5 Summary of results on responsiveness of adoptively transferred fetal liver cells from 15-17 day old embryos into adult lethally irradiated syngeneic hosts. At the indicated days the spleens were removed and tested for responsiveness to dextran sulphate, LPS and PPD with regard to induction of DNA synthesis (X) and of polyclonal antibody synthesis (□).

affect the cells that can be stimulated by LPS. To test the one non-specific signal hypothesis it was studied whether induction of specific anti-hapten responses would be impaired when the hapten was coupled to LPS. As shown in Fig. 6, various concentrations of NNP-LPS were added to cells from responder B10.5M and non-responder C3H/HeJ mice, and the specific anti-NNP response was tested. It can be seen that very low concentrations of NNP-LPS induced high affinity anti-NNP plaques in the responder strain, whereas the low responder started to develop anti-NNP PFC only at very high concentrations of NNP-LPS, concentrations which were paralytogenic for the responder cells.

These findings demonstrate that 1. different PBA selectively activate distinct subpopulations of B cells. 2. a genetic defect which affects the ability of LPS to induce polyclonal antibody synthesis also results in abolishment of the specific TI immune responses induced by the same molecule. Since the total number of B cells as well as the number of hapten specific cells is the same in high and low responder strains, the most likely interpretation for these results would be the existence of a genetic defect which affects triggering but does not interfere with antigen-Ig receptor interaction on the B cell surface. Moreover, as the low responder mice can mount normal responses to the same hapten when coupled to another carrier, this strongly suggests that the triggering of specific cells in TI responses follows the same mechanism as the polyclonal activation. Therefore, the defect in these mice which interferes both with polyclonal and specific responses affects a non-specific triggering mechanism.

The C´3 and the Fc receptors are not triggering receptors

As a first step in elucidating the nature of the relevant triggering receptor we studied whether already known and well characterized cell surface receptors on B cells were responsible for triggering. In addition to the Ig receptors (which have already been excluded) two such receptors have been investigated, namely the one for complement factor 3 and for the Fc part of the IgG molecule. Both these receptors have been implicated as triggering receptors for antibody synthesis.

Many PBA, such as LPS and SIII are known to be complement bypassers, i.e. to be capable of directly binding C´3. Therefore, LPS, SIII as well as other PBA were incubated with fresh normal mouse serum as a source of complement or, in the controls with inactivated mouse serum, or medium only. Thereafter, various concentrations of the PBA were added to

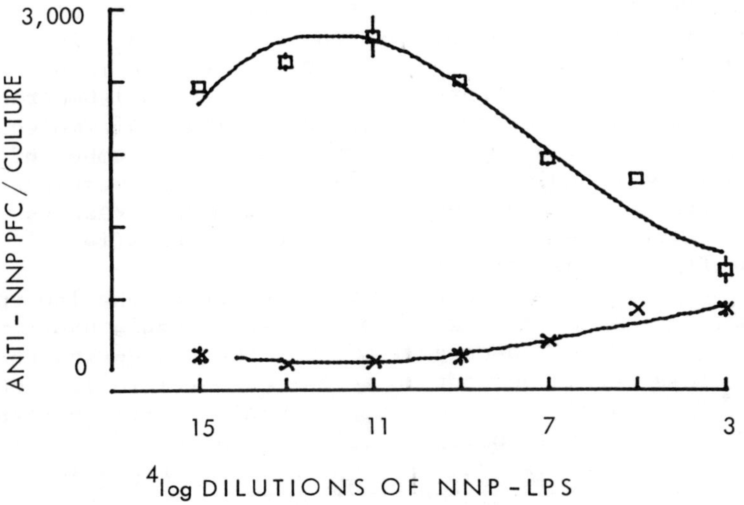

Fig. 6 Spleen cells from LPS high responder B10.5M mice (□——□) or from LPS low responder C3H/HeJ mice (x——x) were cultured in dishes (10^7 cells/ml) in serum free medium, in the presence of the indicated concentrations of NNP-LPS. The high avidity anti-NNP PFC responses were measured on day 3.

B lymphocytes and induction of polyclonal antibody synthesis investigated. As shown in Fig. 7, PBA incubated with fresh mouse serum caused induction of polyclonal antibody synthesis which exhibited a clearly different dose response curve as compared with the same PBA incubated with inactivated serum. Thus, 10 to 100 fold lower concentration of PBA mixed with fresh serum was needed for optimal induction of polyclonal antibody synthesis as compared to PBA incubated with heat-inactivated serum. However, the total number of plaque forming cells was not different in these two groups. Although this could be interpreted to indicate a triggering role for C´3, the following experiments exclude that. Stearoyl dextran has the ability to bind to cell surfaces, such as lymphocytes and red cells. Stearoyl dextran-treated red cells become lysed after addition of guinea pig complement, most likely because the substance dissolves in the cell membrane and is competent of fixing C´3 leading to lysis of red cells. Also lymphocytes lyse in an analogous but less pronounced way after addition of guinea pig complement. Furthermore, stearoyl dextran and complement treated lymphocytes have their C´3 receptors blocked as revealed by inhibition of C´3 (but not Fc) rosette formation.

Consequently, it was investigated whether such cells would by themselves become activated to polyclonal antibody synthesis, as would be expected if C´3 was the triggering receptor. This turned out not to be the case (Fig. 8). The failure of activation was not due to impaired cell survival, because the cells responded normally to LPS and PPD. This also shows that the pretreated lymphocytes with blocked C´3 receptors were fully competent to respond to PBA, thus excluding C´3 as a triggering site.

The shift in the dose response profile observed in Fig. 7 indicates that C´3 may have a helper role in triggering. Most likely PBA which have bound C´3 will now be capable of fixing to B lymphocytes by two sites (the C´3 receptor on the B cell and the receptor for the PBA itself). Thus, the C´3 receptor will function as a passive focusing device in a way analogous to the role of the Ig receptor. However, the difference in doses (10-100-fold) is much less than those found with Ig (10^4 - 10^6 -fold), presumably reflecting the difference in affinity between the two types of receptors.

The role of Fc receptor was investigated in an analogous way. Antibodies to FITC were produced and mixed with $FITC_{10}$-HGG which is a highly thymus-dependent antigen and therefore completely lacking PBA properties. This antigen-antibody complex adhered to B lymphocytes in the same way as heat-aggregated gammaglobulin. However, the complex never induced

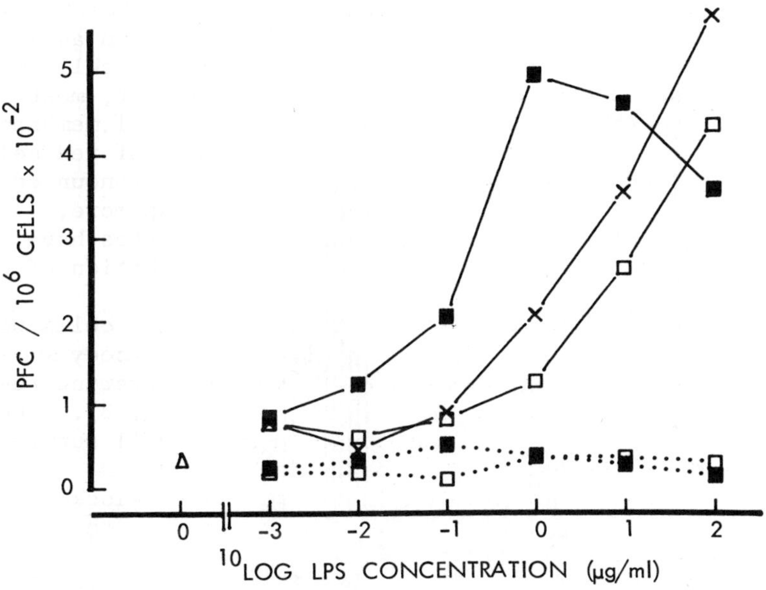

Fig. 7 Effect of LPS activated C´3 on induction of polyclonal antibody synthesis. 1 mg/ml of LPS was reacted for 30 minutes at 37°C with fresh mouse serum as a source of complement (■——■), with heat-inactivated (56°C for 45 minutes) serum (□——□) or with medium (x——x). Thereafter, different concentrations of LPS were added to normal spleen cells in serum free culture and the PFC response against NNP-SRC determined 2 days later. In addition, corresponding concentrations of fresh serum (■···■) or heat-inactivated serum (□···□) in the absence of LPS were added to the cultures. Background response (Δ) is also shown.

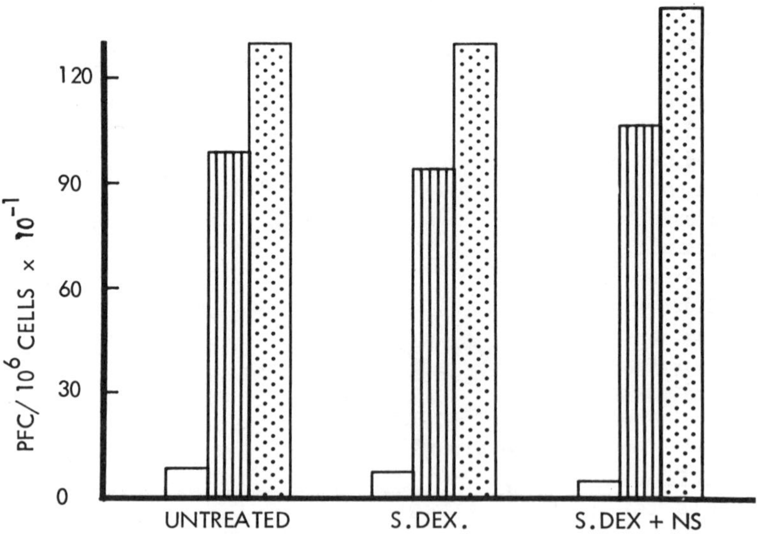

Fig. 8 Failure of activated C´3 to induce or block polyclonal antibody synthesis induced by LPS and PPD. Spleen cells were first treated with medium or with 100 μg/ml of stearoyl dextran. The cells were washed three times and thereafter resuspended. The stearoyl dextran treated cells were either incubated in fresh mouse serum or in medium for 30 minutes at 37°C and thereafter washed three times. The cells were cultivated in serum-free medium alone (▭), or in the presence of a 100 μg/ml of LPS (▥) or PPD (⬚). The antibody response to FITC-SRC was determined at day 2.

polyclonal antibody synthesis in treated lymphocytes irrespective of the concentration of the complex. Furthermore, lymphocytes exposed to antigen-antibody complexes were stimulated by LPS to the same extent as lymphocytes treated with antibody alone, with antigen alone or remained untreated. Thus, interaction of complexes with the Fc receptor on B cells neither triggered nor prevented triggering of polyclonal IgM synthesis by a competent PBA.

Finally, FITC labelled LPS were studied in an analogous way. It was found that the addition of IgG anti-FITC to $FITC_{87}$-LPS reduced the total number of anti-FITC PFC produced, as compared to normal serum-treated FITC-LPS. However, again a shift in the dose response profile was observed analogous to that found with C´3. Presumably the same type of mechanism operates. The antibody bound to the FITC-LPS molecule will expose its Fc part and this will bind to the Fc receptors on the lymphocytes. Therefore, antibody-complexed FITC-LPS will have two combining possibilities with the B cell, and therefore fewer molecules in solution will be needed for triggering concentrations to be reached on the cell membrane.

The results exclude an active triggering role of C´3 and Fc receptors, but suggest that both may participate in PBA activation by stabilizing the bond between a potent PBA and a reactive B cell. Exactly the same mechanisms are likely to operate in TI specific responses in addition to the Ig receptor mediated binding.

SUMMARY

The evidence for the "one non-specific triggering signal" hypothesis for immune activation of B cells has been presented. Polyclonal B cell activators (PBA) directly trigger B cells of all specificities to division and/or selectively increased Ig synthesis, via a mechanism which <u>does not involve their Ig receptors</u>. Different PBA act on distinct subpopulations on B cells most likely belonging to the same cell line in sequential stages of differentiation. The result of activation (division, Ig synthesis or both) depends on the differentiation stage of the cell when activated and it varies, therefore, for each PBA and each B cell population.

B cells can be directly activated to <u>specific</u> antibody synthesis only by substances possessing PBA properties (TI antigens), but not by those lacking PBA properties (TD antigens). Thus, haptens and hapten-TD protein conjugates of various epitope densities could not activate B cells, even when concentrated on Sepharose beads, over a 10^9 fold range of hapten concentrations, whereas specific B cell activation could be readily achieved by coupling the hapten to a PBA.

Therefore, all attempts to substantiate the cross-linking concept were negative. The Ig receptors on the B cells are not responsible for triggering because: a) binding of the antigen to the Ig receptors did not activate the cells nor did it interfere in any detectable way with the PBA-induced activation via other surface sites; b) PBA worked synergistically with hapten-PBA conjugates in the induction of specific anti-hapten responses, whereas this did not occur when the haptens were free or coupled to TD antigens (failure to substantiate the "two-signal" concept for B cell induction); c) specific immune responses could be suppressed by molecules which bind to the PBA receptor structures without triggering the cells, but not by antigen binding to the Ig receptors; d) mice that are genetic low responders to polyclonal induction by LPS were also low responders with regard to specific TI anti-hapten immune responses against a hapten-LPS conjugate, but they responded normally to the same hapten coupled to different carriers. Finally, neither C'3 nor Fc receptors on B cells were the triggering receptors, but they had a passive focusing function analogous to but less efficient than the Ig receptors.

ACKNOWLEDGEMENTS

We want to thank Miss Susanne Bergstedt and Miss Lena Lundin for skilful technical assistance.

REFERENCES

1. Greaves, M., Janossy, G., Feldmann, M. and Doenhoff, M., in The Immune System. Genes, Receptors, Signals, eds. Sercarz, E.E., Williamson, A.R. and Fox, C.P., Academic Press, New York-London, pg. 271-279.
2. Bretscher, P.A. and Cohn, M., Science 169: 1042, 1970.
3. Coutinho, A. and Möller, G., Scand. J. Immunol. 3: 133, 1974.
4. Sell, S., Transpl. Rev. 5: 19, 1972.
5. Sell, S., Personal communication.
6. Coutinho, A. and Möller, G., Nature, New Biol. 245: 12, 1973.
7. Coutinho, A. and Gronowicz, E., Unpublished.
8. Coutinho, A., Möller, G. and Richter, W., Scand. J. Immunol. 3: 321, 1974.
9. Watson, J., Frenkner, E. and Cohn, M., J. Exp. Med. 138: 699, 1973.
10. Coutinho, A., Gronowicz, E., Bullock, W. and Möller, G., J. Exp. Med. 139: 74, 1974.

11. Möller, G., Coutinho, A. and Persson, U., Scand. J. Immunol., in press.
12. Gronowicz, E., Coutinho, A. and Möller, G., Scand. J. Immunol. 3: 413, 1974.
13. Gronowicz, E. and Coutinho, A., Eur. J. Immunol., in press.
14. Amsbaugh, D.F., Hansen, C.T., Prescott, B., Stashak, P.W., Barthold, D.R. and Baker, P.J., J. Exp. Med. 136: 931, 1972.
15. Sultzer, B., Infection and Immunity 5: 107, 1972.

DIFFERENTIAL STIMULATION OF MURINE T AND B CELL POPULATIONS BY PURIFIED MITOGENS FROM POKEWEED

M. J. Waxdal
Laboratory of Immunology, NIAID, National Institutes of Health, Bethesda, Maryland 20014

One of the approaches to the study of the architecture of the cell surface and the control mechanisms of cellular division and differentiation employs certain proteins which bind to cell surface receptors and influence these phenomena. The most prevalent of these are found in plants and the entire class of proteins may be termed lectins. Certain lectins are mitogenic for lymphocytes and have found much use in model studies of the early events of immune stimulation. Mitosis induced by most of these soluble plant lectins requires the presence of thymus dependent lymphocytes (T cells) (1,2) and these lectins are unable to stimulate pure populations of T-independent (B) lymphocytes. There are other mitogens, such as certain bacterial lipopolysaccharides (LPS) which stimulate the thymus independent lymphocytes but fail to cause DNA synthesis by T lymphocytes (3). An exception to the division of specificities is found in the extract of Phytolacca americana or pokeweed which has been reported to stimulate both classes of lymphoid cells (4,5,6,7). It has been reported in humans that the berries and juice of this plant cause mitosis of lymphoid cells in vivo after ingestion or systemic exposure via abrasions or scratches on the hands (8,9), and that extracts from all parts of the plant contain the active components (10).

Extracts of the roots of pokeweed have been fractionated and a single protein was isolated (PWM) which showed both mitogenic and hemagglutinating activity (11,12,13,14). Many recent publications, utilizing the unfractionated extracts of pokeweed or the partially purified protein, have attributed the biological activities in pokeweed extracts to PWM.

We have fractionated extracts of pokeweed by a different procedure than had previously been employed and have found five biologically active proteins (15). They have been designated Pa-1 through Pa-5 for the initials of Phytolacca americana and the order of their purification. These five proteins have been characterized both physicochemically (15) and biologically (16,17) and were found to show different activities in the stimulation of murine lymphoid cells.

Pa-1 through Pa-5 were prepared from saline extracts of ground roots of Phytolacca americana by ion-exchange and gel permeation chromatography (15). The separation of a partially purified sample of the five proteins by gel filtration is shown in Fig. 1. Peak A contains Pa-1, contaminated by several other proteins. The materials contained in peaks B to E are Pa-2, Pa-3, Pa-4 and Pa-5, respectively. Each of these proteins is further purified by additional gel filtration. The materials isolated from each of these peaks showed single bands on polyacrylamide gel electrophoresis in SDS, allowing the calculation of their approximate molecular weights (Table I) and indicating that each of these materials was a single protein. With the exception of Pa-1, the molecular weights were in accordance with the behavior of these proteins during gel filtration. When gel filtered in acidic, neutral, or basic buffers Pa-1 appeared to be a polymer or aggregate of the single subunit observed in SDS gel electrophoresis.

The approximate molecular weights, amino acid compositions, and neutral carbohydrate analyses of these five proteins are presented in Table I. The most striking feature of these analyses is the similarity of composition and the large number of half-cystine residues in Pa-2, Pa-3, Pa-4, and Pa-5. In contrast, Pa-1 has a much lower half-cystine content and has less arginine and carbohydrate. Pa-1 also contains notably more alanine, valine, isoleucine, leucine and lysine than the other four proteins. These compositional differences and its polymeric state are in accord with the biological differences between Pa-1 and the other pokeweed mitogens as described below.

The elution pattern shown in Fig. 1 is found in preparations of roots harvested in late summer or autumn. In winter Pa-5 becomes the predominant protein with an accompanying loss of the other constituents. This observation raised the possibility that the smaller and/or minor components were breakdown products or fragments of the larger proteins, particularly because Pa-5 increased during the dying and rotting of the foliage. However, the amino acid and carbohydrate analyses (Table I) of these proteins do not support this possibility. The large differences between Pa-1 and the other four proteins indicate that it cannot be a fragment of any of them. Although the remaining proteins have similar compositions, there are notable differences. Pa-3 is probably not a fragment of Pa-2 because it contains more carbohydrate that Pa-2. However the amino acid compositions of these two proteins are too similar to further substantiate the point. Pa-4 cannot be a fragment of Pa-2 or Pa-3 because Pa-4 contains methionine, an amino acid lacking in Pa-2 and Pa-3. Furthermore Pa-4 contains more tyrosine than either Pa-2 or Pa-3. Pa-5 cannot be a fragment of Pa-2 nor Pa-3 because it

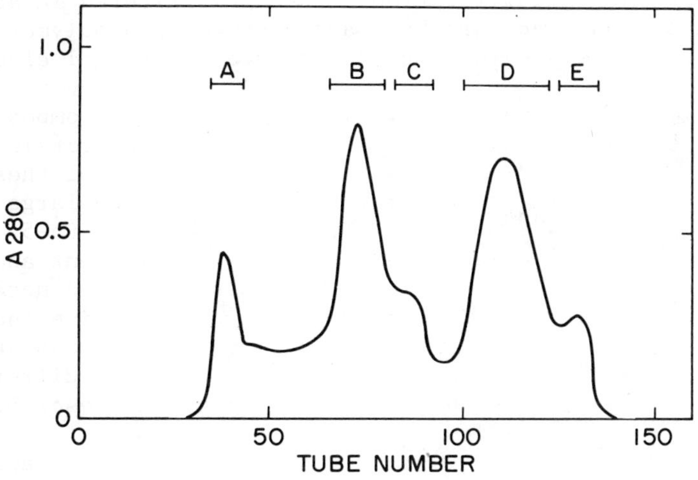

Fig. 1 Fractionation of the partially purified pokeweed mitogens on Sephadex G75 in 1 M propionic acid. The material in cuts A to E was resubjected to gel filtration to yield pure Pa-1 to Pa-5, respectively.

TABLE I

COMPOSITION AND MOLECULAR WEIGHTS OF THE POKEWEED MITOGENS

Amino Acid	Pa-1	Pa-2	Pa-3	Pa-4	Pa-5
Asx	22.4	33.3	31.6	21.7	21.6
Thr	12.1	10.5	12.5	8.0	7.0
Ser	16.9	19.9	15.9	14.3	11.1
Glx	21.6	34.4	32.3	27.3	26.6
Pro	9.6	10.6	6.3	4.8	8.2
Cys/2	7.8	49.7	42.3	38.9	29.9
Gly	21.8	41.9	36.3	35.1	25.6
Ala	15.2	6.9	5.7	5.2	4.6
Val	17.0	11.2	11.6	6.3	2.2
Met	3.0	0	0	1.7	2.1
Ile	9.3	1.1	1.4	0.4	0.4
Leu	14.5	8.5	8.8	7.0	4.6
Tyr	8.0	5.5	5.8	9.2	11.5
Phe	6.8	7.2	6.5	3.1	2.7
His	3.9	7.5	7.1	4.3	3.5
Lys	11.7	7.0	6.9	7.6	8.3
Arg	5.3	12.3	6.7	7.5	10.7
Trp	----	8.1	----	7.8	----
CHO	1.8	8.4	12.5	5.1	2.9
Mol. Wt.	22,000±3300	31,000±4600	25,000±3700	21,000±3200	19,000±2900

Values are expressed as residues/mole (15).

also contains methionine. It cannot be a fragment of Pa-4 because it contains more proline, tyrosine and arginine than Pa-4. Thus it appears that these four similar proteins, and the markedly different Pa-1, are individual proteins and the products of separate genes. The composition of Pa-2, its molecular weight, and its biological activities (see below) indicate that it is the pokeweed mitogen previously isolated (11,12).

Pa-1, Pa-2, Pa-3 and Pa-5 agglutinate human erythrocytes. There is no correlation between the agglutinin titers of these proteins and the ABO nor the Rh blood types. The most potent agglutinin was Pa-1, which caused agglutination at a concentration of 150 nanograms/ml. The weakest agglutinin was Pa-5 (2,500 ng/ml). Pa-4 inhibited agglutination by Pa-2 and Pa-3, suggesting that these 3 proteins bind to the same site on the cell. On the other hand, agglutination by Pa-1 was not inhibited by Pa-4, suggesting that Pa-1 binds to a different site.

To further study the relationships of Pa-2 and Pa-4, binding studies were performed using ^{125}I-labeled proteins. At saturation, both Pa-2 and Pa-4 showed about 2 X 10^5 receptor sites per erythrocyte. Furthermore, at equal concentrations of the ^{125}I-protein, equal numbers of Pa-2 or Pa-4 molecules bound to the cell. This suggests that the two proteins bind with the same avidity and perhaps bind to the same site. Competitive binding studies between ^{125}I-labeled Pa-2 (Pa-4) and unlabeled Pa-4 (Pa-2) indicated that these two proteins competed for the same receptor sites on the erythrocyte.

As expected from the studies with erythrocytes, Pa-2 also aggregated murine splenic lymphocytes, and this aggregation was inhibited by Pa-4. Binding studies with ^{125}I-labeled proteins indicated that both Pa-2 and Pa-4 bind to the same number of receptor sites, and bind with the same avidity. An accurate estimation of the number of receptors per cell was not made because saturation of the binding sites could not be achieved, perhaps due to the heterogeneous nature of this cell population.

The ability of each of these five proteins to stimulate mitosis in different populations of murine lymphoid cells was assayed by incorporation of ^{3}H-thymidine into the newly synthesized DNA. Experiments were conducted to determine the dose-response curves with cells from the spleen and thymus of BALB/c mice, the spleens of adult thymectomized, lethally irradiated and anti-θ treated bone marrow reconstituted (TXBM) BALB/c mice, and the spleens of athymic (nude) mice bearing the <u>nu/nu</u> gene and of their heterozygotic littermates (16). The results showed that all five proteins were mitogenic for thymus cells and spleen cells from normal mice and that Pa-2 was the most potent in this regard. But, only Pa-1 invoked a response of near normal magnitude from TXBM spleen cells, and only Pa-1 was able to stimulate mitosis in spleen cells from nude mice.

These mitogens were used to study the time course of DNA synthesis, protein synthesis and immunoglobulin production in spleen cells from BALB/c and "nude" mice (17). Protein synthesis was measured by the incorporation of ^{3}H-leucine into cellular protein. Immunoglobulin production was determined by lysing the cells in the culture with Nonident P40. The insoluble material was removed by centrifugation and the supernatant dialyzed. The dialysate was first subjected to a nonspecific immune precipitation followed by a specific precipitation of the synthesized immunoglobulin with rabbit anti-mouse kappa chain and goat anti-rabbit immunoglobulin. The radioactivity of this specific precipitate was used as a measure of the amount of immunoglobulin produced.

The results obtained using Pa-1 as the stimulant are presented in Fig. 2. DNA synthesis was elevated on day 1 and reached a maximum on day 2 for both normal (BALB/c) and nude spleen cells. The amount of DNA synthesis, and hence cell proliferation, was about the same for both cell sources. Protein synthesis was also stimulated on day 1 and reached its maxima on day 3 in normal cells and day 2 for nude cells. In contrast, immunoglobulin production was not stimulated on day 1 in either cell population, nor on day 2 in the cells from BALB/c mice. Immunoglobulin production reached its maxima on day 3, only after cell division had occurred, and continued into day 4 in both cell populations. The amount of radioactive immunoglobulin produced was consistently about 2-fold higher in spleen cells from nude mice than from their littermates or BALB/c mice.

To determine the class of immunoglobulin produced and to establish the specificity of the assay system, the immune precipitates were subjected to polyacrylamide gel electrophoresis in SDS (18). The gels were sliced and the radioactivity in each slice was determined. All of the radioactivity applied to the gel was recovered in peaks corresponding to authentic μ and L chain standards (Fig. 3). These results indicate that the assay system used measures only immunoglobulin production and that IgM is the only immunoglobulin synthesized in detectable amounts by murine spleen cells stimulated with Pa-1. The stimulation of IgM production in spleen cells by crude mixtures of the pokeweed mitogens (PWM) has previously been reported (19,20). Our results indicate that of the five known pokeweed mitogens, only Pa-1

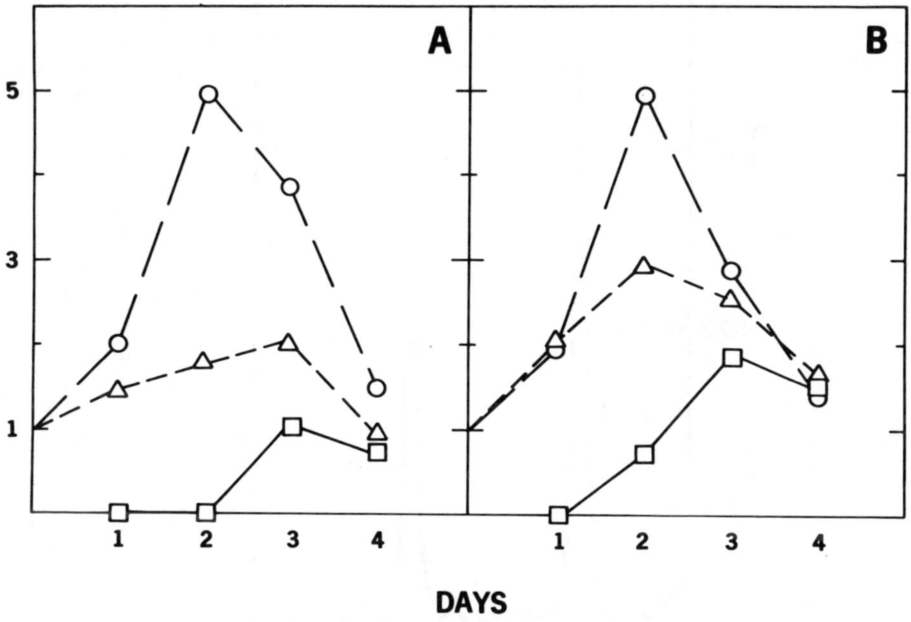

Fig. 2 Time course of the response of spleen cells from BALB/c (A) and nude (B) mice to stimulation by Pa-1. DNA (O) and protein (Δ) synthesis are expressed as experimental cpm/control cpm (E/C). Immunoglobulin production (□) is expressed as 10^3 cpm in the specific immune precipitate.

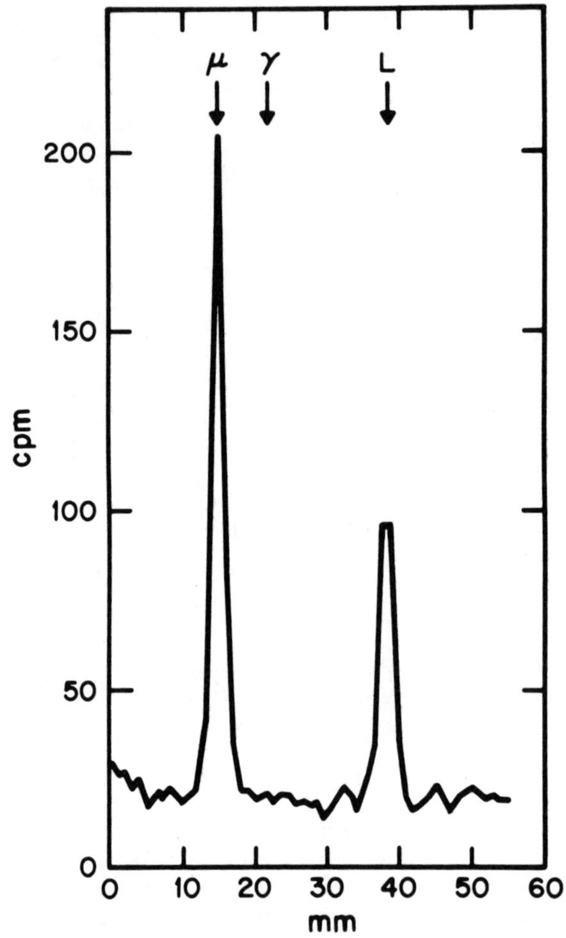

Fig. 3 SDS polyacrylamide gel electrophoresis of the reduced specific immune precipitate. The positions of the marker proteins, μ-chain, γ-chain, and L-chain are indicated.

is able to stimulate immunoglobulin synthesis in murine lymphoid cells. Furthermore, the other four pokeweed mitogens (PA-2 to Pa-5) do not stimulate DNA nor protein synthesis in spleen cells from nude mice, suggesting that they have no direct B cell stimulating activities.

Pa-2 and Pa-3 are the most potent of the pokeweed mitogens for both normal spleen cells and thymus cells. Figure 4 presents the time course of stimulation of spleen cells by Pa-2 and Pa-3. Both DNA and protein synthesis were elevated on day 1 in the Pa-2 stimulated cultures. Similar to the time course of Pa-1 activation, maximum synthesis was reached on day 2. However Pa-3 stimulation produced little synthesis of either DNA or protein on day 1, but also reached its maximum on day 2 and remained at high levels on day 3. There is markedly greater protein synthesis in spleen cells stimulated by Pa-3 than by any of the other pokeweed mitogens.

The time courses of stimulation of spleen cells by Pa-4 and Pa-5 are presented in Fig. 5. Pa-4 is the least potent of the five mitogens, but follows the pattern of reaching maximum DNA synthesis on day 2. Little protein synthesis was stimulated by this mitogen, and it also reached a maximum on day 2. The low activity of Pa-4 may correlate with its inability to cause the agglutination of human erythrocytes or murine splenocytes. In marked contrast to the time course followed by the other four mitogens, Pa-5 stimulation shows very little increase in DNA or protein synthesis on day 2. These maxima occur later in culture and DNA synthesis continued to increase to the end of culture on the fifth day.

In summary, four of the five pokeweed mitogens (Pa-2 to Pa-5) appear to stimulate directly only T cells. They are mitogenic for cells from the thymus and spleens of normal mice, but not for the cells from the spleens of nude mice and have markedly reduced activities for spleen cells from TXBM mice (16). These mitogens do not stimulate immunoglobulin production in any of the lymphocyte preparations tested.

On the other hand, Pa-1 appears to be both a T cell and a B cell mitogen. Pa-1 is mitogenic in all the lymphocyte preparations tested (including thymus cells and cortisone resistant thymus cells) and stimulated immunoglobulin production in both normal and nude mouse spleen cells.

The variations in the magnitude and time course of DNA and protein synthesis in cells stimulated by the T cell mitogens suggest that different subsets of the T cells may be responding to these mitogens, or that these cells are being stimulated along different pathways by the different mitogens.

The response of spleen cells from nude mice to Pa-1 was further analyzed by using the stathmokinetic effect of vin-

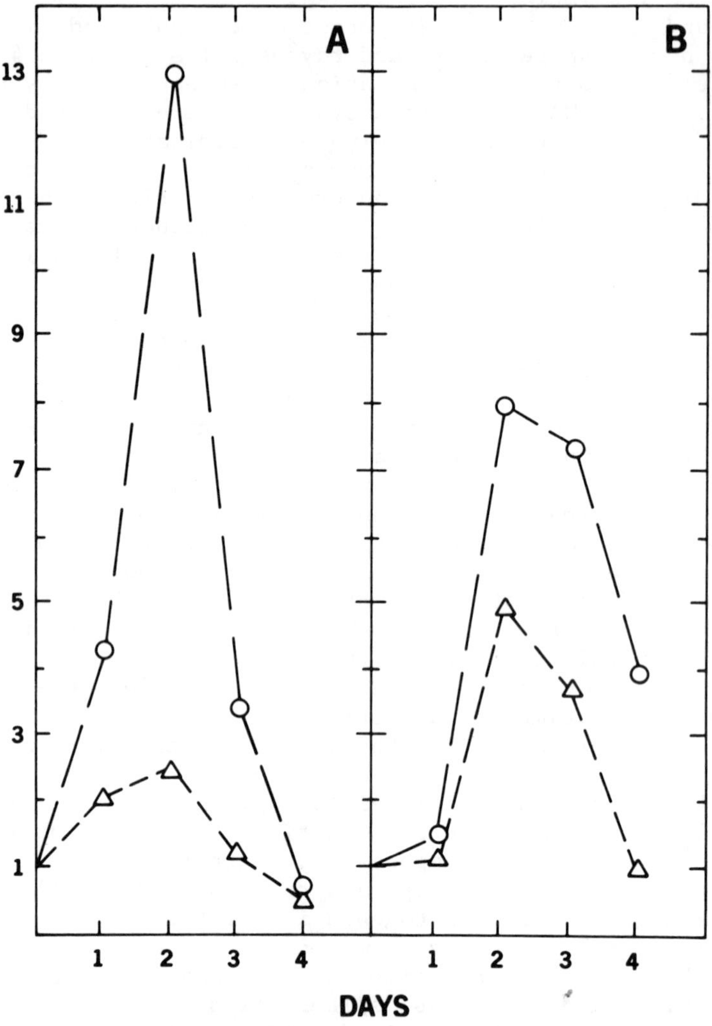

Fig. 4 Time course of the response of BALB/c spleen cells to stimulation by Pa-2 (A) and Pa-3 (B). DNA (O) and protein (Δ) synthesis are expressed as E/C.

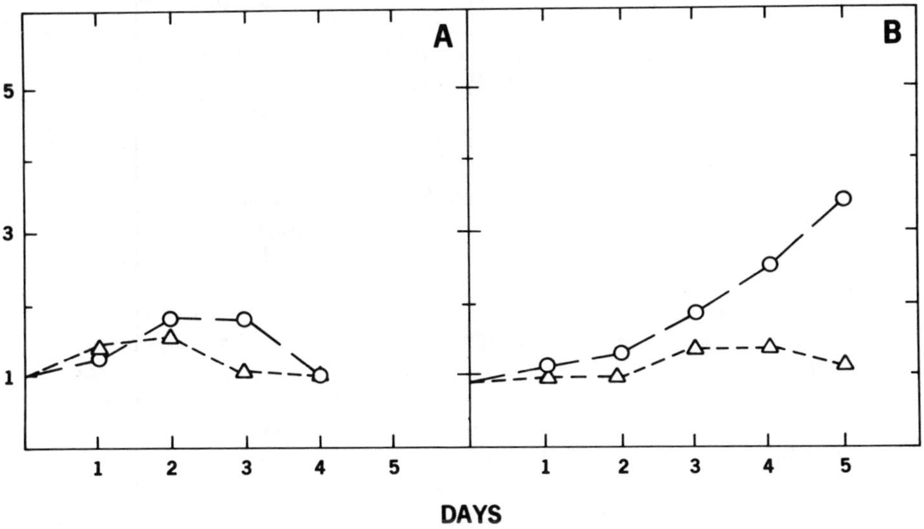

Fig. 5 Time course of the response of BALB/c spleen cells to stimulation by Pa-4 (A) and Pa-5 (B). DNA (O) and protein (Δ) synthesis are expressed as E/C.

blastine (21) to arrest cell division after DNA synthesis. Preliminary experiments indicated that 10^{-8} M vinblastine had little observable effect on DNA synthesis, protein synthesis or immunoglobulin production, and concentrations of 10^{-6} M or greater severely depressed both the responses to Pa-1 and the control values. In most spleen cell preparations, a concentration of 10^{-7} M vinblastine was found to be optimal for blocking cell division without destroying other cellular functions, such as protein synthesis. This concentration was used in the following experiments. In the initial experiments both Pa-1 and vinblastine were added to the cultures at 0 time and ^{3}H-thymidine was added 24 hours prior to harvest on days 1-4. The data are presented in Fig. 6, where the heights of the bars are a measure of the amount of DNA synthesized each day, by all cells (not blocked), and by cells entering mitosis for the first time (vinblastine blocked). We have found that the ^{3}H-thymidine incorporation assay correlates well with the number of blast cells determined by autoradiography and is much easier to perform. Hence, vinblastine-blocked thymidine incorporation analyses provide a usable measure of the relative number of cells dividing on each day.

The data from the vinblastine blocked cultures (Fig. 6) allow the cells which respond to Pa-1 by mitosis to be divided into two groups: about one-third of all the cells, which enter initial mitosis on day 1, and almost all of the remainder which enter mitosis on day 2. A comparison of the responses of these blocked cultures to the normal stimulated cultures indicates that about three-fourths of the daughter cells that resulted from initial mitosis on day 1 continued to proliferate on day 2. From these data it cannot be determined whether the cells dividing on days 3 and 4 are the progeny of cells which entered mitosis for the first time on day 1 or on day 2. However, from the total amount of DNA synthesized in both sets of cultures, it can be estimated that there were about 1.5 cell divisions for each "clone" of stimulated cells. This number must be used with caution because deteriorating culture conditions may have diminished or stopped continued mitosis.

The effect of vinblastine on immunoglobulin production is presented in Fig. 7. The addition of vinblastine at 0 time to Pa-1 stimulated cultures totally abolishes subsequent immunoglobulin production, indicating that cell division is necessary for increased immunoglobulin production. But, if the vinblastine is added at 24 hours, after the first group of cells have entered mitosis, there is less than a 30% reduction in immunoglobulin synthesis. If the vinblastine is not added until 48 hours or later, there is essentially no

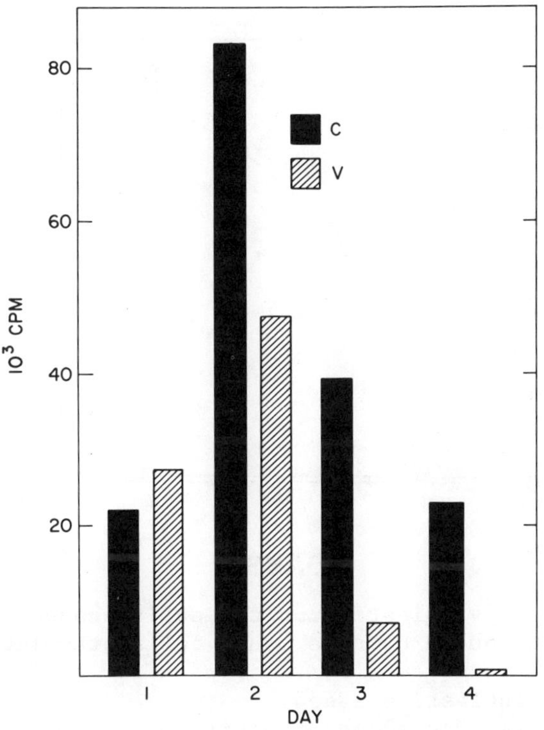

Fig. 6 Daily synthesis of DNA by nude mouse spleen cells stimulated by Pa-1, in the absence (solid bars) and presence (cross-hatched bars) of vinblastine. Ordinate is 10^3 cpm above the unstimulated control.

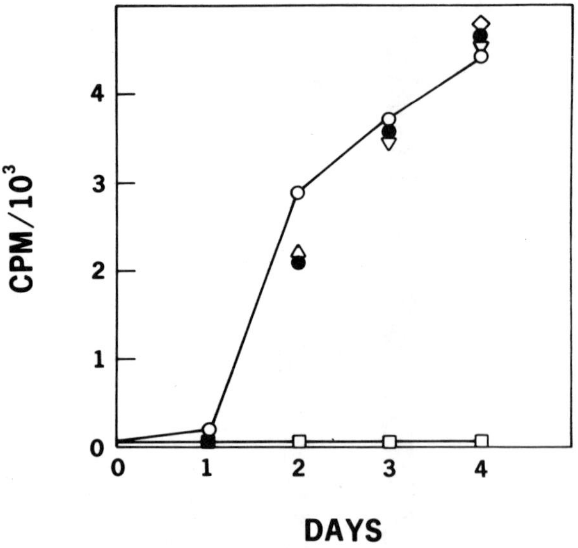

Fig. 7 Effect of vinblastine on the daily production of immunoglobulin by nude mouse cells stimulated by Pa-1.
0 No vinblastine added.
□ Vinblastine added at 0 time.
Δ Vinblastine added at 24 hours.
∇ Vinblastine added at 48 hours.
◊ Vinblastine added at 72 hours.
• Vinblastine added at 24 hours prior to harvesting.

effect on immunoglobulin production.

Preliminary experiments with mitomycin C added at 0 time also allowed a response to Pa-1 by increased DNA and protein synthesis but prevented subsequent immunoglobulin production.

These findings on Pa-1 stimulation of immunoglobulin production, considered in conjunction with the data on cell division (Fig. 6), suggest the following interpretations: 1) Almost all of the cells which will differentiate and produce immunoglobulin respond initially to Pa-1 on day 1, ostensibly by mitosis. The remaining potential immunoglobulin producing cells either respond initially on day 2 or are the progeny of cells which entered mitosis on day 1 and have continued to divide on day 2. 2) Only a portion (a maximum of about one-third) of the cells in the nude mouse spleen which respond to Pa-1 differentiate into immunoglobulin producing cells. 3) A single cell division is necessary for Pa-1 stimulated cells to differentiate into immunoglobulin producing cells. Or, during the first 24 hours of response to Pa-1, vinblastine and mitomycin C inhibit differentiation by a mechanism other than the blocking of cell division, and that the cells are not susceptible to this mechanism at later stages of differentiation.

The results of our studies on the stimulation of immunoglobulin production by Pa-1 are somewhat at variance with the findings of Melchers and Andersson (22,23), who have used hydroxyurea to block cell division in cultures of small resting lymphocytes stimulated by LPS. These authors report an earlier onset of IgM synthesis and did not find the immunoglobulin producing response susceptible to hydroxyurea blockage of cell division. The differences in our observations on the time of onset of production may be due to the cell populations studied or to the different stimulants used. The differences observed by the blocking of cell division may again be due to the stimulants used, or to the different blocking agents.

The pokeweed mitogens have proven to be a useful system for the study of lymphocyte stimulation, providing several agents with varying activities. The unusual biological properties (e.g., mitogenicity for both T and B lymphocytes) of Pa-1 may be due to its polymeric state. We have induced B cell stimulating activity into Pa-2 and Pa-4 by attaching these T cell mitogens to agarose beads. The same type of result has been observed with other T cell mitogens (5,24). Furthermore, we have treated Pa-2 and Pa-4 with a difunctional reagent (dimethyl subermidate) to form molecular aggregates, and have found B cell stimulating activities in these products (25). Hence, it would seem that the interaction of a multimeric mitogen with appropriate receptors on the B cell surface is necessary for stimulation. The similar necessity for multiple interactions by antigen (26) and antibody (27, 28) with immunoglobulin receptors to stimulate B cells is in accord with this view.

REFERENCES

1. Greaves, M.F. and Roitt, I.M., Clin. Exp. Immunol. 3: 393, 1968.
2. Janossy, G. and Greaves, M.F., Clin. Exp. Immunol. 9: 483, 1971.
3. Peavy, D.L., Adler, W.H. and Smith, R.T., J. Immunol. 105: 1453, 1970.
4. Stockman, G., Gallagher, M., Heim, L. and Trenton, J., Proc. Soc. Exp. Biol. Med. 136: 980, 1971.
5. Greaves, M.F. and Bauminger, S., Nature New Biology 235: 67, 1972.
6. Janossy, G. and Greaves, M.F., Clin. Exp. Immunol. 10: 525, 1972.
7. Greaves, M.F., Bauminger, S. and Janossy, G., Clin. Exp. Immunol. 10: 537, 1972.
8. Barker, B.E., Farnes, P. and LaMarche, P.H., Pediatrics 38: 490, 1966.
9. Barker, B.E., Farnes, P. and Fanger, H., Lancet 1: 170, 1965.
10. Farnes, P., Barker, B.E., Brownhill, L.E. and Fanger, H., Lancet 2: 1100, 1964.
11. Borjeson, J., Reisfeld, R., Chessin, L.N., Welsh, P.D. and Douglass, S.D., J. Exp. Med. 124: 859, 1966.
12. Reisfeld, R.A., Borjeson, J., Chessin, L.N. and Small, P.A., Jr., Proc. Nat. Acad. Sci. USA 58: 2020, 1967.
13. Chessin, L.N., Borjeson, J., Welsh, P.D., Douglas, S.D. and Cooper, H.L., J. Exp. Med. 124: 873, 1966.
14. Douglas, S.D., Hoffman, P.F., Borjeson, J. and Chessin, L.N., J. Immunol. 98: 17, 1967.
15. Waxdal, M.J., Biochemistry 13: 3671, 1974.
16. Waxdal, M.J. and Basham, T.Y., Nature 251: 163, 1974.
17. Basham, T.Y. and Waxdal, M.J., (Submitted J. Immunol.).
18. Laemmli, U.K., Nature 227: 680, 1970.
19. Parkhouse, R.M.E., Janossy, G. and Greaves, M.F., Nature New Biol. 235: 21, 1972.
20. Parkhouse, R.M.E., Transplant Rev. 14: 131, 1973.
21. Frei, E. III, Whang, J., Scoggins, R.B., Van Scott, E.J., Rall, D.P. and Ben, M., Cancer Res. 24: 1918, 1964.
22. Melchers, F. and Andersson, Jr., in Cellular Selection and Regulation in the Immune Response: 217. Ed. by

G.M. Edelman, Raven Press, N.Y., 1974.
23. Melchers, F. and Andersson, J., Biochemistry 13: 4645, 1974.
24. Andersson, J., Edelman, G.M., Möller, G. and Sjöberg, O., Eur. J. Immunol. 2: 233, 1972.
25. Basham, T.Y. and Waxdal, M.J., Unpublished results.
26. Feldmann, M. and Basten, A., J. Exp. Med. 134: 103, 1971.
27. Woodruff, M.F.A., Reid, B. and James, K., Nature 215: 591, 1967.
28. Fanger, M.W., Hart, D.A., Wells, J.V. and Nisonoff, A., J. Immunol. 105: 1484, 1970.

DISCUSSION FOLLOWING SYMPOSIUM
SIGNALS FOR LYMPHOCYTE ACTIVIATION 1: NONANTIGENIC SIGNALS AND RECEPTORS

Unidentified Questioner: Dr. Möller, do you have any evidence that LPS does not react with immunoglobulin, that is, that it is really nonspecific?

Möller - Stockholm: Yes. First, LPS activates production of all specificities. If you postulate that LPS reacts with immunoglobulins you must say that it cross reacts with all the antigens in the world. I'm open for that, although I think it is unlikely. Secondly, LPS is as good a mitogen or polyclonal cell B activator in mice which are completely tolerant to it as in normal mice. Thirdly, you can block Ig receptors with thymus dependent antigens or haptens and still induce an immune response against the blocked antigens with LPS.

Greaves - London: Isn't it possible that LPS might be reacting with immunoglobulin by some site other than the antigen specific binding site?

Möller - Stockholm: Yes, than I would call that the one nonspecific site.

Oppenheim - Bethesda: I think that the model should be modified to some extent because there is considerable evidence in the rabbit and the chicken that anti-immunoglobulin (anti-Ig) can activate lymphocytes. Furthermore in mice, Iverson showed that there were some anti-Ig preparations which produced a low degree of transformation and some plaque forming cells against a variety of stimuli.

Möller - Stockholm: The only careful study of which I am aware on the effect of anti-Ig showed that it was T cells which responded to anti-Ig. To be convincing that anti-Ig activates B cells directly this kind of experiment must be done in nude mice and this has not yet been done.

Greaves - London: Perhaps I can make a comment that the work by Shek and Cinader showed very clearly that in the rabbit it is probably B cells that are activated, so I really wouldn't accept that point.

Möller - Stockholm: In the mouse it is pretty solid that anti-Ig does not stimulate.

Greaves - London: How do you account for the specificity of the response to thymus dependent antigens?

Möller - Stockholm: I don't think that is difficult at all. We know from work in vivo that the condition for triggering B cells in a thymus dependent response is that the T and B cells can be linked up by antigen. That could have a variety of explanations, the most simple being that the T cell factor is effective at only a very short distance. That is the only condition required to obtain a high degree of specificity. The one B cell which will be activated is the one which is linked to the secreting T cell.

Sage - Durham: I would like to make a comment on the model Dr. Greaves presented in his second slide and for which Dr. Novogrodsky gave some evidence. I wish to argue against the assumption that because a polysaccharide exists in a molecule like thyroglobulin or on a red cell membrane that it is legitimate to assume that the same molecule or something like it exists on the lymphocyte. Cornfeld described an oligosaccharide isolated from red cells which binds several lectins simultaneously. One of these lectins is from the mushroom and is a nonstimulant and one is from the lentil and is a stimulant. If you make the same assumption that Dr. Novogrodsky made, then a similar oligosaccharide should exist on the lymphocyte, and both of these lectins should be mitogenic. On the other hand we have given some evidence in the past that different mitogens may react with different molecules, because not only do they not compete with one another but the molecules to which they are attached are capable of independent motion on the membrane. If they are capable of independent motion, then it is hard to see how they are on the same polypeptide molecule.

Greaves - London: Do you want to comment on that, Abraham?

Novogrodsky - Rehovot: No doubt there are other possibilities. Unless you isolate in the test tube a well defined structure,

I think it will be very difficult to draw a general conclusion. The possibility exists that there will be several unique triggering sites but I think our results support the notion that a variety of mitogens might effect a single site.

Sage - Durham: I don't think the evidence for that exists! Until you isolate structures from lymphocyte membrane demonstrating that there is only a single site that binds lectins and that this indeed represents the mitogenic sige, then it is a leap in faith to make such an assumption. You could just as easily have obtained the same effect from a mixture of the simple sugars that are present in that structure. If you had taken many other sugars, they might have done the same thing as your polysaccharide.

Greaves - London: But I think the structure has been isolated. The Japanese workers have isolated it from lymphocytes. Crumpton's group has shown that the PHA receptors block the mitogenic response to lentil and concanavalin.

Sage - Durham: That is not enough. You can do that with any glycopeptide which will bind with lectin. That does not prove that the receptor you have isolated is the mitogenic receptor. You have isolated a receptor from the membrane which binds with a lectin. Anything that binds with a lectin will interfere with mitogenic stimulation. I don't think that constitutes proof.

Novogrodsky - Rehovot: I would share some of that view and I don't think there is definite support for a single site. There are no doubt experiments which are now in progress which might advance our knowledge. For example, one of them is to label periodate treated sites and then to see whether you can pick up those labelled sites with immobilized lectin. But I am not going to comment more, since I agree with you that a definite answer is still open to further experiment.

Mosier - Bethesda: I would like to address a question to Dr. Möller. It seems to me that your hypothesis has the advantage that it explains everything but it has the great disadvantage that it is extremely difficult to disprove. I would like to quote an example. It seems to me that the easiest way to disprove the hypothesis is to find a T independent antigen which is not a polyclonal activator. If I could pin you down to a definition of polyclonal activator, that might help. Let me say that DNP-Ficoll which is a thymus independent antigen by all criteria fails to show characteristics of a polyclonal activator. It does not induce antibody to a number of antigens at any of the wide range of concentrations up to 10 mg/ml in vitro.

Möller - Stockholm: I anticipated the question. I had hoped to have the answer but unfortunately I don't. I can say that if Ficoll is an antigen, it is fantastically weak. What I tested was fluorescein labelled Ficoll and that had strong effects on inducing polyclonal antibody synthesis. But there are two things of which I am uncertain. One is whether there is LPS contamination, the second is whether the fluorescein acts on T cells. But even if it would turn out that Ficoll doesn't induce antibody synthesis to whatever antigen you pick, I am not at all disturbed, because in the final analysis and with a weak polyclonal B cell activator there is only one property to define its B cell activation namely thymus independence. This is the most stringent test for polyclonal B cell activation.

Greaves - London: Thank you for presenting a beautiful circular argument there. Perfect circle.

Handwerger - Minneapolis: Dr. Moller, I think your and everyone else's use of the term "thymus independent" has to be changed somewhat. All of these antigens and polyclonal mitogens are thymic independent only with respect to helper cell function. I think that there is now evidence accumulating that indeed thymic independent antigens may be under thymic cell influence with respect to suppressor cell function. With that statement in mind, how does your model system fit with thymic suppressor cell regulation of B cell responses?

Möller - Stockholm: When we talk about thymus independence I think we should be clear that this is a matter of degree and there are antigens, for instance LPS, which are thymus independent, but if there is a protein attached to it, this will give a nice IgG response. That only shows that T helper activity can also help a thymus independent antigen. You can fit suppressor T cells into this thymus independent antigen easily, I think irrespective of your theoretical framework.

Handwerger - Minneapolis: How would this fit in with a one signal model? Here you have one signal being a T independent antigen and then you have suppression of that response by another signal. How is that suppression occurring and does that make one invoke a two signal hypothesis?

Möller - Stockholm: No, it is too much of the same argument. It is like when you make a dose response curve to LPS you always get peak and then a drop and that can drop to zero. Now the model implies that the T cell and macrophage factors are having the same effect as the mitogens. So therefore they work synergistically or at least their function is synergistic. So if you have an optimum response to a thymus independent antigen the addition of T cell help may result in a depressed response.

SUMMARY OF SYMPOSIUM
LYMPHOCYTE ACTIVATION: THE STATE OF OUR IGNORANCE
A REPORT TO IMMUNOLOGY-INDEPENDENT RESEARCHERS (IIR)

David W. Talmage
University of Colorado Medical Center, Denver, Colorado

Two things seem clear from the papers presented at this symposium and discussion which followed. We immunologists have collected a large number of diverse and interesting phenomena which describe the circumstances under which lymphocytes are activated, but we still know almost nothing about the underlying mechanisms.

Lymphocytes comprise a large fraction of the nucleated cells in certain lymphoid organs and in the peripheral blood. They differ from most of the other nucleated cells of the body in several important respects: 1) they move around the body freely, i.e. they are normally metastatic; 2) they generally exist for long periods of time in an inactive or resting state; 3) although monotonously hemogeneous in appearance, they are tremendously heterogeneous in the functions they serve when activated and in the stimuli required to activate them; 4) they respond to stimuli from outside the body (foreign substances) rather than to intrinsic physiological controls.

It is in this last area that we have collected our most impressive array of interesting phenomena and maintained our total ignorance of what is going on. Foreign substances are called antigens if they activate a very small fraction of lymphocytes and produce thereby a persisting memory which alters a subsequent response to the same substance. Other substances are called mitogens if they activate a larger subset of lymphocytes and produce no memory.

Immunologists recognize two major classes of lymphocytes: B cells and T cells. B cells when activated synthesize and secrete a large and heterogeneous set of globulin molecules called Ig (for immunoglobulin), but all the Ig are made on a similar pattern with interchangeable parts and each B cell makes only one of the perhaps 10^6 or more different Ig made by a single animal. Most antigens activate only a very small fraction of B cells, just those few B cells whose Ig product (antibody) reacts specifically with the antigen that induced their production. Some mitogens, on the other hand, may activate a large fraction of B cells, most of which make an Ig which fails to react with the mitogen inducer.

How do the antigens do their work so selectively and why are the mitogens so non-specific? It is impossible to discuss this question without reference to another strange phenomenon. Some antigens activate their lymphocytes only in the presence of T cells and a third kind of cell called a "macrophage" (because it ingests and digests foreign particulate matter). Such antigens are called thymus dependent (TD). Other antigens are thymus independent (TI).

We don't know how either kind of antigen works. We have observed that all TI antigens are polymers, but not all polymers are TI. Most but not all TI antigens are non-specific B cell mitogens at higher concentrations than that required for specific antibody induction.

The function and activation of T cells is even more obscure than for B cells. Some T cells form rosettes with macrophages by binding to their surface and require the presence of macrophages for activation by antigen or non-specific mitogens. But T cells do not bind to the antigen to which they respond specifically, and it is not certain that they make a specific product. It is not known how they recognize the presence of antigen. T cells make a variety of non-specific substances called lymphokines, but it is not known whether these are made by the same or different T cells.

T cells are activated by macrophages from other animals of the same species but not by macrophages from animals of a different species or from animals of the same species with an identical genetic make-up such as inbred animals have. T cells may be activated by macrophages from the same animal if the latter have been modified chemically.

Macrophages ingest numerous antigens, but it is not known whether this process has anything to do with activating lymphocytes. B cells bind antigens, and the surface Ig receptors of B cells are clumped within a few minutes of the binding; but this process seems to be neither necessary nor sufficient for lymphocyte activation. T cells don't bind antigen but respond as specifically as B cells which do. How T cells synergize with antigen to activate TD antigens is a total mystery.

Within a few years we should have all these phenomena straightened out. We will be coming to you for help with the elucidation of the underlying mechanisms.

SECTION II

SIGNALS FOR LYMPHOCYTE ACTIVATION II: ANTIGENIC SIGNALS AND RECEPTORS

Chairman

VINCENT P. EIJSVOOGEL

Central Laboratory, Blood Transfusion Service
The Netherlands Red Cross, Amsterdam, The Netherlands

Co-Chairman

HANS WIGZELL

University of Uppsala, Uppsala, Sweden

THE MECHANISM OF ACTION OF MACROPHAGES IN THE ACTIVATION OF T-LYMPHOCYTES *IN VITRO* BY ANTIGENS AND MITOGENS

David L. Rosenstreich and J. Mark Wilton*
Laboratory of Microbiology and Immunology, National Institute of Dental Research, National Institutes of Health, Bethesda, Maryland 20014

INTRODUCTION

It has been known for some time that macrophages are required for the activation of lymphocytes by antigens *in vitro* (1,2). Recent work has shed light on the means by which they exert their helper function. There appear to be two distinct mechanisms by which macrophages act. The first is through a process of binding, processing and "presenting" antigens to lymphocytes in a form that is stimulatory (3,4). This of course requires direct macrophage-lymphocyte contact. The second mechanism of macrophage action is through the elaboration of soluble factors that enhance immune reactivity in some fashion (5,6).

However, great gaps remain in our understanding of how macrophages function *in vitro*. In terms of T-lymphocyte activation, while it is well accepted that antigen-induced activation is macrophage-dependent (7,8), it is still controversial if this is the case for mitogen-induced activation as well. Thus, while some workers contend that mitogen-induced T-cell activation is macrophage independent (8), others claim that it is absolutely macrophage-dependent (9). Furthermore, it is not known how the different modes of macrophage function i.e. "presentation" vs. soluble factor, interact to produce lymphocyte activation.

In order to further study the role of macrophages in the activation of T-lymphocytes *in vitro*, we prepared a highly purified, macrophage-depleted T-lymphocyte population in the guinea pig. Using this T-cell population we have investigated the macrophage dependency of a number of different types of T-cell stimulants and analyzed the nature of macrophage function for each. Our data demonstrate that T-lymphocyte activation by all the stimulants we have tested, is absolutely macrophage-dependent.

* Present address: Department of Oral Microbiology and Immunology
Guy's Hospital Medical School
London Bridge, SE1 9RT, England

MATERIALS AND METHODS

Animals and immunization. Female strain 13 guinea pigs weighing 350 to 450 g (Division of Research Services, National Institutes of Health) were immunized in each footpad with 0.1 ml of an emulsion of dintrophenylated ovalbumin (DNP_{14}-OVA, 125 μg/ml) and complete Freund's adjuvant (CFA, *Mycobacterium tuberculosis* H_{37} Ra, Difco Laboratories, Detroit, Mich.).

Reagents and media. All tissue cultures were grown in RPMI 1640 (Grand Island Biological Co., Grand Island, N.Y.) containing added glutamine (2 mM/ml), penicillin-G (100 μg/ml) streptomycin (100 μg/ml) and were buffered with Hepes (15 mM). Fetal calf serum (FCS) was obtained from the Reheis Chemical Co., (Rehatiun, Phoenix, Ariz.) and heat-inactivated prior to use. Antigens used in these experiments included DNP-OA and tuberculin purified protein derivative (PPD) (Connaught Medical Research Labs, Toronto, Canada) which were added at a final concentration of 10 or 100 μg/ml. Mitogens included Concanavalin A (Con A, Calbiochem, San Diego, Calif.), phytohemagglutinin (PHA, Burroughs Wellcome, Research Triangle Park, N.C.), 2-mercaptoethanol (2-ME, Eastman Kodak Corp., Rochester, N.Y.).

Two to four weeks after immunization, draining lymph nodes were removed aseptically, freed of fat, minced with scissors and gently pressed through a 60-mesh screen to produce a single cell suspension. Both cell types were strained through a sterile gauze to remove clumps and washed twice with medium.

Peritoneal exudate cells (PEC) were obtained after intraperitoneal injection of sterile mineral oil (Drakeol Penreco, Butler, Pa.) and collected after 3 days as previously described (10). These cells, which contained 70-75% macrophages, were x-irradiated (2500 R) prior to use in culture. Where necessary, red cells were lysed with isotonic ammonium chloride (11).

Preparation of highly purified, macrophage depleted T-lymphocytes by passage over two types of adherence column. Macrophage-depleted T-lymphocytes were prepared by use of two adherence columns (Fig. 1). Lymph node cells were first partially purified by passing them over an adherence column of nylon wool and fine glass beads as previously described (7). These partially purified lymph node lymphocytes (LNL) were then further purified on an adherence column of packed nylon wool, as originally described by Julius et al. (12). 100 x 10^6 LNL were resuspended in 2 ml of RPMI 1640 with 10% FCS and added to a washed, equilibrated and prewarmed nylon wool column, and rinsed into the column with an additional 1 ml of warm medium. The column was sealed and incubated at 37°C for

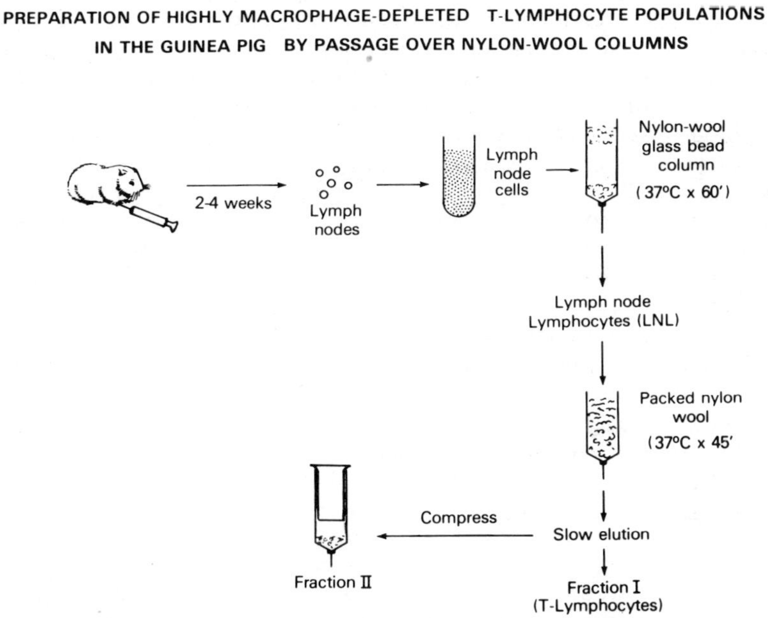

Fig. 1 <u>T-lymphocyte preparation in the guinea pig</u>. The technical details are outlined in the materials and methods.

45 min. T-lymphocytes (Fraction I) were then eluted by the dropwise addition of 20 ml of warm medium and 10% FCS. A second lymphocyte population (Fraction II) could be obtained by using the modification of Handwerger and Schwartz (13). The columns were rinsed with 50 ml of warm medium. Then 10 ml of warm medium was added to the column, the nylon wool compressed with a sterile syringe barrel and the eluted cells collected. This procedure was repeated twice more.

Identification of B and T lymphocytes and macrophages using various marker techniques. All marker techniques were performed using previously described reagents and methods (14): Surface immunoglobulin was stained with fluoresceinated anti-GP Fab; the Fc receptor was detected using aggregated guinea pig γ-globulin and the C3 receptor was detected by using EAC. Viability was analysed using trypan blue dye. Macrophages were detected by either using phagocytosis of latex beads (15) or staining for a cytoplasmic esterase (16).

Determination of lymphocyte proliferation in vitro. All cells were cultured in RPMI 1640 containing additives and 5% FCS, at a concentration of 1×10^6 cells/ml, (2×10^5 cells/well) in round bottom plastic microtrays (Cooke Engineering, Alexandria, Va.). Cells were cultured for 72 hours at 37°C in a humidified atmosphere of 5% CO_2 and air. At the end of the culture period they were pulsed with 0.5 μCi of ^{3}H-TdR for 4 hours (6.0 Ci/mMol.; Schwarz/Mann, Orangeburg, N.Y.), harvested using a multiple automated sample harvester (Mash II, Microbiological Associates, Bethesda, Md.) and counted in a Packard Tri-Carb Scintillation Counter.

Technique of Brief Exposure of Macrophages or Lymphocytes to PHA. Column purified T-lymphocytes or x-irradiated PEC at a concentration of 10×10^6 cells/ml in medium containing 5% FCS were incubated at 37°C for 60 min with PHA (10 μg/ml) and agitated intermittently. They were then washed four times with fifty ml aliquots of medium until free of unbound PHA, restored to the appropriate cell concentration based on their percent of viable cells and used in culture.

Preparation of macrophage supernatants with lymphocyte activating factor (LAF) activity. LAF was produced by a modification of the method of Gery and Waksman (6). Nonirradiated PEC were resuspended at a concentration of 2×10^6 cells/ml in RPMI 1640 containing 5% FCS and cultured in sterile glass petri dishes. After 24 hours, the supernatants were harvested, spun twice at 2,000 RPM x 10' to remove cells and then dialysed for another 24 hours against two changes of RPMI 1640. These LAF-rich solutions were used immediately or frozen at -70°C and used within one week of preparation. Cells were cultured in LAF either diluted 1:1 or 1:2 with medium.

RESULTS

Characterization of nylon-column passaged lymph node lymphocyte sub-populations in the guinea pig. The cellular characteristics of the purified T-lymphocyte population was studied using a variety of cell marker techniques and compared to the partially purified (LNL) population (Table I). In terms of cell recovery and frequency, 60-70% of the cells placed on the column were recovered in the effluent and of these eluted cells 69% were recovered in the T-lymphocyte population. Both populations were 95% viable as judged by exclusion of trypan blue dye. Macrophage contamination was measured using two different parameters. There were less than 0.1% phagocytic cells in both populations. However, using a more sensitive macrophage marker, the presence of a cytoplasmic esterase, the T-population contained at most 0.3% macrophages while the LNLs contained more.

B cell contamination was measured using two different parameters. The T population contained less than 2% B cells as judged by the presence of cells with eaily detectable surface immunoglobulin or cells bearing a C3 receptor. In contrast, the LNL population contained 6-7% B cells.

Finally, these populations were examined for the presence of an Fc receptor. This receptor is present on B cells (18) and is also present on "activated" T cells (14). The T population contained only 1.5% Fc receptor-bearing lymphocytes compared to 10% in the LNL. Therefore this T-lymphocyte population contained less than 0.3% macrophages; 2% B cells and in addition, seemed to be depleted of "activated" T cells as well.

Absolute macrophage dependency of highly purified guinea pig T-lymphocytes for in vitro proliferation induced by antigens as well as mitogens. The ability of these two populations to proliferate *in vitro* in response to antigens and mitogens was next examined (Table II).

It is clear that the T-lymphocyte population no longer responded to the mitogens Con A or PHA, or to the antigen DNP-OA to which the guinea pig cell donor had been immunized. In contrast, LNLs still responded well to the T-cell mitogens.

These populations were then reconstituted with 20%, syngenic, x-irradiated peritoneal exudate cells (MØ). In the presence of MØ, the T-lymphocytes now exhibited a normal proliferative response to both mitogens as well as to DNP-OA. Therefore, these data demonstrate that T-cell activation by mitogens as well as antigens is absolutely macrophage-dependent.

Ability of 2-mercaptoethanol (2-ME) to restore the proliferative response of purified T lymphocytes. In some immune

TABLE I

CHARACTERIZATION OF NYLON-COLUMN PASSAGED LYMPH NODE LYMPHOCYTE SUBPOPULATIONS

Parameter *	LNL	Nylon Fraction I (T)
		(% Positive)
Frequency	---	69 ± 4
Viability	95%	95%
Phagocytosis (Latex)	0.1	0.1
Esterase	0.7	0.3
Surface IG	6.3 ± 0.7	0.25 ± 0.1
C_3 Receptor (EAC)	7.3 ± 0.3	1.8 ± 0.3
Fc Receptor (AGG-1G)	10.3 ± 0.8	1.5 ± 0.2

* See materials and methods for technical details.

TABLE II

ABSOLUTE MACROPHAGE DEPENDENCY OF T-LYMPHOCYTE ACTIVATION BY MITOGENS AND ANTIGENS

Cell Population	Mean Lymphocyte Proliferative Response* (cpm ^{3}H-TdR incorporated/2 x 10^5 cells) 0	Con A	PHA	DNP-OA
LNL	327 ± 187	5734 ± 1317	20,699 ± 6383	698 ± 301
LNL + MØ	294 ± 125	42,841 ± 4584	69,173 ± 12,177	1936 ± 26
T (Fr I)	69 ± 23	155 ± 32	769 ± 286	136 ± 39
T + MØ	104 ± 14	10,835 ± 3999	35,227 ± 9287	1235 ± 285

* Results represent the arithmetic means of 9 separate experiments ± S.E.M.

systems, macrophages can be replaced by the reducing agent, 2-ME (19). 2-ME has also been found to enhance the response of T-cells to PHA (20) and to allogeneic cells in the MLR (21). However, it is not clear whether 2-ME will enhance T-cell responses in the total absence of macrophages in these systems. The effect of 2-ME on the proliferative response of the purified T-lymphocytes to PHA was first examined (Fig. 2). T-lymphocytes alone responded poorly to PHA as expected with only a slight proliferative response at the higher concentrations of PHA. The addition of 2-ME (5×10^{-5}M) produced some enhancement of this response that was most marked at higher PHA concentrations. In contrast, macrophages enhanced proliferation approximately 30-fold over the maximum seen with 2-ME. Therefore, it is clear that 2-ME by itself will not restore the response of pure T-lymphocytes to PHA. Finally, the effect of a mixture of 2-ME and macrophages on the PHA response was investigated. At lower PHA concentrations there was definite synergy between 2-ME and macrophages, suggesting that 2-ME may in fact enhance T-cell activation only in the presence of macrophages.

The effect of 2-ME on T-cell activation by other stimulants was also tested (Table III). As previously observed, 2-ME was only able to slightly enhance the response to Con A and PHA, as well as to DNP-OA. However, this enhancement was always much less than that seen with macrophages.

<u>The role of macrophage "presentation" of mitogens in activation of T-lymphocytes</u>. Macrophages are required to "present" soluble protein antigens to T-cells in a stimulatory form (9). Therefore, it was not surprising that 2-ME alone could not restore the response of these T-lymphocytes to antigen. However, since it could not replace macrophages completely for mitogen responses, it was possible that macrophages were required to "present" mitogens to T-lymphocytes as well. We therefore investigated the ability of macrophages to "present" antigens or mitogens to T-lymphocytes and produce activation. X-irradiated PEC were incubated with PHA (10 μg/ml) for 60 min at 37°C, washed free of unbound mitogen, added back at a 20% concentration to T-lymphocytes and the induced proliferative response measured (Table IV). Whereas the T-cells alone did not respond to antigen or mitogen, they demonstrated a good proliferative response to macrophage-bound PHA (40-fold), thus demonstrating that macrophages could present PHA in a manner analogous to the way they present antigens.

It has been found, that histocompatibility between lymphocyte and macrophage is essential for antigen activation (22). We therefore examined the ability of allogeneic (strain 2)

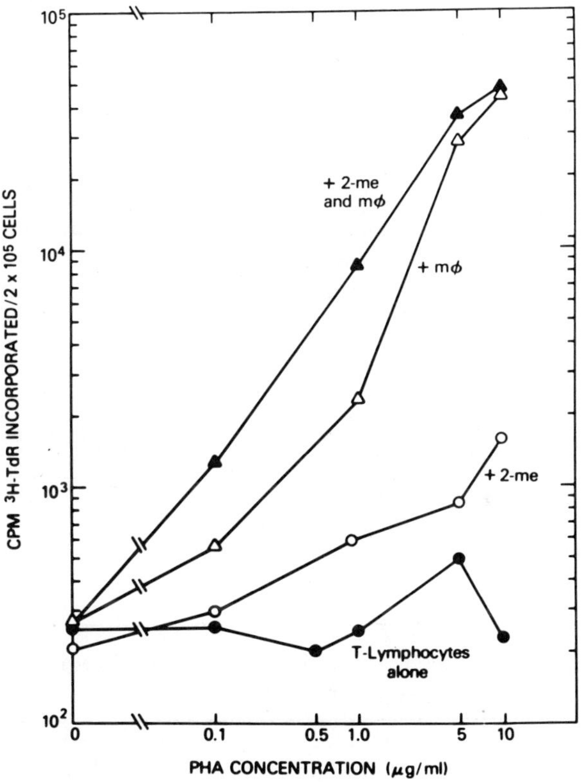

Fig. 2 Comparison of the ability of 2-ME or macrophages to restore the PHA response of purified T-lymphocytes.

TABLE III

COMPARISON OF THE EFFECT OF 2-MERCAPTOETHANOL OR MACROPHAGES ON T-LYMPHOCYTE ACTIVATION BY MITOGENS AND ANTIGENS

	Mean Lymphocyte Proliferative Response (^{3}H-TdR incorporated/2×10^{5} cells)			
Cell Population	0	Con A	PHA	DNP-OA
T-Lymphocytes	110	162	785	145
T-Lymphocytes + 2-ME *	195	1,033	3,460	453
T-Lymphocytes + MØ **	115	30,450	78,088	2,813

* 2-ME = 5×10^{5} M final concentration

** MØ = 20% syngeneic, x-irradiated PEC

TABLE IV

EFFECT OF SYNGENEIC OR ALLOGENEIC MACROPHAGES ON T-LYMPHOCYTE RESPONSE TO MITOGENS AND ANTIGENS

	Mean Lymphocyte Proliferative Response (E/C)		
Stimulant	T-Lymphocytes alone	+ STR 2 MØ	+ STR 13 MØ
Con A	1.7	48.2	37.5
PHA	3.5	49.4	172.7
DNP-OA	1.0	1.7	7.9
STR. 2 MØ-PHA *	28.6	----	----
STR. 13 MØ-PHA	39.8	----	----

* MØ = 20% x-irradiated PEC. For details of PHA pulse technique, see text.

macrophages to "present" PHA to T-cells. In contrast to what is found with antigen, allogeneic macrophages activate T-cells as well as did syngeneic. This observation was further confirmed when we examined the ability of non-pulsed syngeneic or allogeneic macrophages to restore T-cell responsiveness to antigens and mitogens. Thus, while the mitogen responses were restored by both allogeneic and syngeneic macrophages, the response to DNP-OA was restored only by the syngeneic macrophages.

These data demonstrate two points. First, macrophages can "present" mitogens to T-cells, as they do with antigens. However, there must also be different requirements for T-cell activation by antigens and mitogens since only antigens require histocompatibility between macrophages and lymphocytes.

Macrophage requirement for activation of PHA-pulsed T-lymphocytes. It is presumed that T-cells require macrophages to "present" antigen because their receptors are either weak or labile (23,24,25). Since macrophages could present mitogens, and 2-ME did not completely restore mitogen responsiveness, it was possible that T-cells were only activated by macrophage-bound mitogen, and that the interaction of mitogen directly with the T-cell, might be a non-stimulatory or tolerogenic signal. In order to evaluate the effect of mitogen directly bound to the T-cell surface, T-lymphocytes were incubated with PHA for 60 min, washed free of unbound PHA, and then cultured either alone, with additional macrophages or with 2-ME (Table V).

Non-pulsed T-cells, as expected, did not respond significantly to PHA and 2-ME enhanced this response only slightly. On the other hand, both untreated and PHA-pulsed macrophages stimulated significant T-cell proliferation. PHA-pulsed T-lymphocytes by themselves did not respond, nor was their response significantly enhanced by 2-ME. However, in the presence of normal macrophages, these PHA-pulsed T-cells demonstrated significant proliferation. Thus, it is clear that PHA binds to T-cells. However, T-cell bound PHA is stimulatory only in the presence of macrophages. These experiments demonstrate that there are two distinct pathways by which T-cells can be activated by PHA. PHA can bind directly to T-cells or it can be presented by a macrophage. However, no matter which way PHA is presented to the T-cells, macrophages are still essential.

The role of a soluble factor from macrophages (LAF) in antigen and mitogen induced proliferation. Gery and Waksman have reported that a soluble factor from macrophages (LAF) would restore the response of thymocytes or macrophage-depleted spleen cells (6). We therefore investigated whether

TABLE V

MACROPHAGE-DEPENDENCY OF RESPONSE WHEN PHA IS DIRECTLY BOUND TO SURFACE OF T-LYMPHOCYTE

			Mean Lymphocyte Proliferative Response (cpm 3H-TdR incorporated/$2x10^5$ cells)	
Cell Population	Treatment	Add Back:	+ 0	+ PHA
T-Lymphocytes	None	0	142	509
		2-ME	119	2,663
		MØ	256	104,624
		MØ-PHA	36,251	125,672
T-Lymphocytes	PHA-Pulsed*	0	607	630
		2-ME	519	1,666
		MØ	32,091	24,563

* For details of culture technique and method of PHA exposure see text.

a similar factor was responsible for T-cell activation where PHA already was bound to the lymphocyte. The supernatant from cultured macrophages (see Materials and Methods for details of preparation) was tested for its ability to restore the response of purified T-lymphocytes to antigens and mitogens (Table VI). The activity of this supernatant factor (1:1 dilution) was the equivalent of intact macrophages in its ability to restore mitogen responsiveness of T-lymphocytes. It also produced a significant enhancement of the response to antigen.

The effect of LAF on PHA-pulsed T-cells was also investigated (Table VII). As in previous experiments, T-lymphocytes alone did not respond to PHA but their responsiveness was restored by either intact macrophages or by LAF. PHA-pulsed T-cells by themselves did not proliferate. However, either intact macrophages or LAF restored their response in the absence of any additional PHA except for that bound to the T-cell surface.

Thus, these data demonstrate that there are two distinct pathways by which macrophage function in the activation of T-lymphocytes by mitogens. The first pathway involves macrophage binding of PHA and "presentation" to the lymphocyte. The second pathway does not require macrophage "presentation" of mitogen at all. In this mechanism, PHA binds directly to the T-cell and macrophages furnish a necessary second signal in the form of a soluble factor (LAF). These two mechanisms are illustrated in Figure 3.

DISCUSSION

Our data clearly demonstrate that T-cell activation by mitogen is absolutely macrophage dependent. There are several possible explanations for the inability of previous workers to demonstrate this macrophage dependence (8). First, it may be that mitogen stimulation requires very few macrophages so that it is only after extensive purification that this macrophage dependency is revealed. An alternative explanation is that there are subpopulations of T cells. The T-cell subpopulation that we have worked with may be entirely macrophage-dependent for mitogen stimulation, while others are not. It is clear that our T-cells did not contain significant numbers of "activated" T-lymphocytes as judged by the presence of Fc receptor-bearing cells. Therefore, it is possible that these "activated" T cells may be much less macrophage dependent than our "inactive" T-cells. It is noteworthy that the fraction II population which contained an increased number of "activated" T-cells was less macrophage dependent. Whether this was due to its residual macrophage

TABLE VI

RECONSTITUTUION OF ANTIGEN AND MITOGEN INDUCED PROLIFERATION OF T-LYMPHOCYTES BY A MACROPHAGE SOLUBLE FACTOR (LAF)

		Mean Lymphocyte Proliferative Response (cpm ^{3}H-TdR Incorporation/2 x 10^{5} cells)			
Cell Population	Add:	0	Con A	PHA	DNP-OA
T-Lymphocytes	0	113 ± 17	361 ± 10	275 ± 68	371 ± 31
T-Lymphocytes	MØ*	387 ± 67	12,956 ± 92	10,674 ± 1898	1759 ± 242
T-Lymphocytes	LAF**	119 ± 52	17,470 ± 2795	11,486 ± 2505	1494 ± 162

* MØ = 20% syngeneic, x-irradiated PEC
** LAF = 1:1 dilution. See text for details of preparation.

TABLE VII

RECONSTITUTION OF THE PROLIFERATIVE RESPONSE OF PHA PULSED T-LYMPHOCYTES BY A MACROPHAGE SOLUBLE FACTOR (LAF)

		Mean Lymphocyte Proliferative Response* (cpm 3H-TdR incorporated/2 x 10^5 cells)	
Cell Population	Add:	+0	+PHA
T-Lymphocytes	0	96 ± 14	342 ± 61
	MØ	320 ± 13	13,572 ± 938
	LAF	170 ± 35	15,990 ± 156
PHA-Pulsed **	0	915 ± 138	928 ± 70
T-Lymphocytes	MØ	21,330 ± 1256	28,177 ± 1788
	LAF	7,468 ± 787	7,853 ± 660

* Results represent the arithmetic means ± S.E.M. of two experiments.

** Lymphocytes were cultured with PHA (10 μg/ml) for 60' at 37°C, then washed x 4 and cultured. MØ are 20% irradiated, syngeneic peritoneal exudate cells.

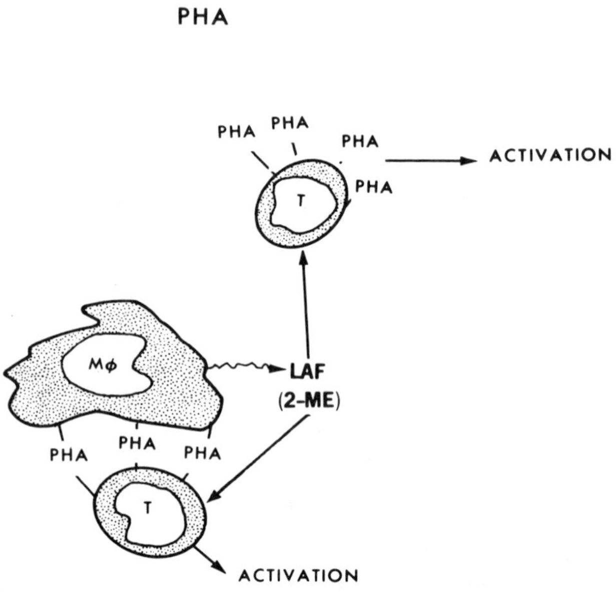

Fig. 3 Two pathways of macrophage action in the activation of T-lymphocytes by mitogens.
There are two pathways by which T-lymphocytes are activated by mitogens, both of which require macrophages. The first pathway involves binding of PHA directly to the T-lymphocyte. Activation occurs only after the interaction of a macrophage-derived soluble factor (LAF). In the second pathway, macrophages first bind PHA, and "present" it to the T-cell. This pathway probably also depends on a soluble factor from macrophages (LAF) to activate the T-cell.

content or B-lymphocyte population is now known.

2-mercaptoethanol has been used to replace macrophages in the response to SRBC (19) and to enhance T-cell responses to PHA (20) or allogeneic cells (21). It has been postulated that 2-ME is active as an analog for LAF in these responses. However, our data on the restoration of T-cell responses to Con A, PHA, and antigen suggest that this is not the case. Thus, in none of these systems will 2-ME replace macrophages whereas LAF will. This suggests that macrophage soluble factors possess more than one activity (or more than one factor) and that 2-ME may be able to exert only one of these functions.

Our data also strongly suggest that there are two distinct pathways for macrophage function in T-lymphocyte activation. By the first pathway two distinct signals are required for T-cell activation. The first signal is given to the T-cell by the binding of the mitogen to the cell surface. The second signal is then delivered by the soluble factor from macrophages (LAF). The interaction of these two signals then results in activation. One would predict, therefore, that any stimulatory agent capable of binding directly to a T cell would activate via this pathway. An example of this is sodium periodate. Periodate is a non-lectin, T-cell mitogen (26), that activates cells by oxidizing surface glycoproteins (17). Furthermore, it has been shown that stimulation of lymphocytes by periodate is macrophage-dependent (31). Since periodate appears to act directly on the T-lymphocyte, one would predict that macrophage "presentation" would not be necessary. While this has not yet been proven, Novogrodsky and Gery have shown that periodate stimulation of mouse thymus cells was greatly enhanced by LAF (27).

Allogeneic cells are also able to bind to T-cells (28). Therefore, it is not surprising that the macrophage requirement of this response can also be satisfied by a soluble factor (5). On the other hand, binding of soluble protein antigens by T cells appears to be weak (23,24) so that antigen stimulation should not be mediated via this pathway. However, our data suggest that there is some enhancement of antigen-responsiveness by LAF; therefore, it is possible that some T cells may bind sufficient antigen to be activated via this pathway. This phenomenon has also been observed with another antigen, PPD (29).

The second pathway of activation involves macrophage presentation of mitogens to T-lymphocytes. Presumably, this requires direct contact between the lymphocyte and the macrophage. This pathway is analogous to the action of macrophages in activation by soluble protein antigens. However, it is clear that this parallel is not exact since mitogen stimulation does not require histocompatible lymphocytes and macrophages. This observation suggests that the requirement for histocompatibility is probably not involved with the actual cell contact mechanism or by the triggering mechanism but rather that surface IR gene products are important for the recognition of the antigen by the T cell.

While it is easy to see that T-cell activation by the first pathway requires two distinct signals, this is not the case for activation via the second pathway. Thus, macrophage-bound mitogen may trigger T-cells because the stimulant is in a polymeric form with multiple repeating epitopes. This form of determinant presentation certainly seems capable of directly activating B cells (30), but whether the same is true for T-cells is not known. Alternatively, T-cell activation by macrophage-bound stimulants may also require 2 distinct signals. Whether the second signal is mediated by a locally derived soluble factor like LAF or by some other mechanism is not known.

Stimulants other than mitogens and antigens are also able to be presented by macrophages. Noteworthy and especially interesting in this regard is sodium periodate. Biniaminov et al., first reported that periodate treated macrophages would activate untreated lymphocytes (31). However, it is very difficult to imagine how this is being mediated. Possibly, the macrophage surface glycoproteins are being oxidized and these in turn may then oxidize T-cell surface glycoproteins resulting in activation. Alternatively, periodate treatment of macrophages may result in the uncovering of surface moieties that are themselves stimulatory for lymphocytes.

It seems clear, that macrophages function to help activate T-lymphocytes by two mechanisms, by "presentation" of stimulants and by the elaboration of certain soluble factors. It is logical to assume that these two functions occur simultaneously with stimulants such as PHA, Con A and periodate which can activate lymphocytes via either macrophage dependent pathway. However, the exact sequence of events involved in either mode of macrophage action remain to be elucidated. Macrophage "presentation", especially of soluble protein antigens, appear to involve a complex interaction of antigen and several different types of receptors and surface moieties on both the T-cell and macrophage. The action of the soluble factors from macrophages may be equally complex and is even less well understood. These factors may act on the lymphocyte surface to expose certain receptors or they may act on the stimulant to produce a necessary configurational change such as cross-linking. Alternatively, these soluble factors may act independently, as true "signals" perhaps via the cyclic nucleotide pathway. Certainly, the more we seem to learn about how macrophages function, the more complex this system becomes. Hopefully, further experimentation will help unravel some of the mysteries of the mechanisms by which macrophages act in the immune response in vitro as well as in vivo.

SUMMARY

A population of highly purified, macrophage depleted T-lymphocytes was prepared from guinea pig lymph node lymphocytes by passage over two types of adherence columns. In vitro proliferation of this population induced by Con A and PHA as well as by antigens was absolutely macrophage dependent. Analysis of the mechanism of action of macrophages using these T-cells, demonstrated that there were two separate pathways involved. For T-cell activation via the first pathway, 2 distinct signals are required. The first signal is given by the interaction of the stimulant directly with the T-cell surface and the second signal is supplied by a soluble factor from macrophages. The second pathway, involves macrophage "presentation", of stimulant to the T cell. T cell activation by all stimulants appears to be mediated via one or both of these pathways.

ACKNOWLEDGEMENTS

We would like to thank: our technicians Lynda Weedon and Sue Dougherty who performed many of the experiments reported here; Dorothy Earman and Carol Oesch for their ever-excellent secretarial assistance; and our wives for their patience, cooperation andencouragement in this and many other projects.

REFERENCES

1. Oppenheim, J.J., Leventhal, B.G. and Hersh, E.M., J. Immunol. 101: 262, 1968.
2. Mosier, D.E., Science 158: 1573, 1967.
3. Seeger, R.C. and Oppenheim, J.J., J. Exp. Med. 132: 44, 1970.
4. Cline, M.J. and Swett, V.C., J. Exp. Med. 128: 1309, 1968.
5. Bach, F.H., Alter, B.J., Solliday, S., Zoschke, D.C. and Janis, M., Cell Immunol. 1: 219, 1970.
6. Gery, I. and Waksman, B.H., J. Exp. Med. 136: 143, 1972.
7. Rosenstreich, D.L. and Rosenthal, A.S., J. Immunol. 112: 1085, 1974.
8. Waldron, J.A., Jr., Horn, R.G. and Rosenthal, A.S., J. Immunol. 111: 58, 1973.
9. Levis, W.R. and Robbins, J.H., Exp. Cell. Res. 61: 153, 1970.

10. Rosenstreich, D.L. and Rosenthal, A.S., J. Immunol. 110: 934, 1973.
11. Roos, D. and Loos, J.A., Biochim. Biophys. Acta 222: 565, 1970.
12. Julius, M.H., Simpson, E. and Herzenberg, L.A., Eur. J. Immunol. 13: 645, 1973.
13. Handwerger, B. and Schwartz, R., Transplantation (in press).
14. Van Boxel, J.A. and Rosenstreich, D.L., J. Exp. Med. 139: 1002, 1974.
15. Rosenstreich, D.L., Blake, J.T. and Rosenthal, A.S., J. Exp. Med. 134: 1170, 1971.
16. Yam, L.T., Li, C.Y. and Crosby, W.H., Amer. J. Clin. Path. 55: 283, 1971.
17. Novogrodsky, A. and Katchalski, E., Fed. European Biochem. Soc. Letter 12: 297, 1971.
18. Dickler, H.B. and Kunkel, H.G., J. Exp. Med. 136: 191, 1972.
19. Chen, C. and Hirsch, J.G., Science 176: 60, 1972.
20. Fanger, M.W., Hart, D.A., Wells, J.V. and Nisonoff, A., J. Immunol. 105: 1043, 1970.
21. Heber-Katz, E. and Click, R.E., Cell Immunol. 5: 410, 1972.
22. Rosenthal, A.S. and Shevach, E.M., J. Exp. Med. 138: 1194, 1973.
23. Hämmerling, G.J. and McDevitt, H.O., J. Immunol. 112: 1726, 1974.
24. Engers, H.D. and Unanue, E.R., J. Immunol. 112: 293, 1974.
25. Ramseier, H., J. Exp. Med. 140: 603, 1974.
26. Thurman, G.B., Giovanella, B. and Goldstein, A.L., J. Immunol. 113: 810, 1974.
27. Novogrodsky, A. and Gery, I., J. Immunol. 109: 1278, 1972.
28. Goldstein, P., Svedmyr, E.A.J. and Wigzell, H., J. Exp. Med. 134: 1385, 1971.
29. Havemann, K. and Schmidt, W., in "Proc. 8th Leukocyte Culture Conference" (ed. by K. Lindahl-Kiessling, Acad. Press, N.Y.), p. 181, 1973.
30. Coutinho, A. and Möller, G., Nature (NB) 245: 12, 1973.
31. Biniaminov, M., Ramot, B. and Novogrodsky, A., Clin. Exp. Immunol. 16: 235, 1974.

THE ROLE OF ANTIGEN PRESENTATION IN B CELL ACTIVATION: ANALYSIS WITH TWO DNP-POLYMER CONJUGATES

Donald E. Mosier and William E. Paul

Laboratory of Immunology, National Institute of Allergy and Infectious Diseases, National Institutes of Health, Bethesda, Maryland

The interaction of antigen with immunoglobulin receptors on the surface of precursors of antibody secreting cells can lead to three possible outcomes: 1) the proliferation and differentiation of the precursors to yield antibody-secreting cells and memory cells; 2) the induction of specific unresponsiveness in the precursor cell; or 3) failure to stimulate either a positive immune response or tolerance induction. How B cells discriminate between surface signals leading to or preventing antibody synthesis has been subject of much experimentation and speculation (1,2). At present it is not known whether one, two, or more discrete signals are required to activate B cells, nor is the nature of these second signals understood. In the case of antibody responses for which T lymphocytes are required, T cells either directly or indirectly provide a needed stimulus for B cell differentiation. Other agents which have been proposed to provide inducing signals for B lymphocytes include activated C3 and nonspecific mitogens (3,4). It has been suggested that tolerance occurs when antigen, in some critical concentration range, binds to surface receptors of B cells which do not receive an appropriate "second" signal. Alternatively, tolerance induction may be the consequence of the topological nature of the binding of antigen to B cell surfaces.

In this study we will examine the relative requirements to activate or inactivate T-independent antibody responses. It is possible, and indeed even likely, that mechanisms of activation and tolerance induction for B cells which respond to T-independent antigens are distinct from those involved in similar phenomena in B cells responding to T-dependent antigens.

We have used dinitrophenyl (DNP) haptenic determinants coupled to two widely different synthetic carriers to investigate the requirements for activating or inactivating DNP-specific B cells _in vitro_. The first antigen is DNP coupled to a linear copolymer of D-glutamic acid and D-lysine (DNP-D-GL), a molecule previously shown to be a strong inducer of DNP-specific unresponsiveness at the B cell level (5,6). The second antigen is DNP covalently bound to Ficoll, a synthetic polysucrose cross-linked with epichlorhydrin (7). Both antigens have the advantage of inducing a primary in vitro anti-DNP antibody response in the absence of T lymphocytes, so they are ideally suited for studying direct activation of B lymphocytes. In the course of these studies it has become apparent that the form of antigen presentation is crucial in determining the outcome of antigen-B cell interaction. Interaction of soluble DNP-D-GL directly with B cells seems invariably to lead to DNP-specific tolerance, even at extremely low concentrations (e.g., 1 picogram/ml) of DNP-D-GL. In contrast, DNP-D-GL bound to macrophages or covalently linked to agarose beads is a potent immunogen. DNP-Ficoll is neither tolerogenic nor macrophage-dependent in vitro at any concentration tested (up to 1 mg/ml), apparently because Ficoll presents DNP to the B cell in a large, rigid, two-dimensional array which favors immunogenicity.

MATERIALS AND METHODS

Animals. BALB/c (NIH Animals Production Section) or $(C3HxDBA/2)F_1$ (Cumberland Farms, Clinton, Tenn.) female mice 8-12 weeks old were used in all experiments. No strain dependence or H-2 linkage of the in vitro response to either DNP-D-GL or DNP-Ficoll has been found (D.E. Mosier and P.L. Cohen, unpublished observations).

Cell Culture. Spleen cells were cultured with antigen for 4 days and the IgM and IgG anti-DNP hemolytic plaque-forming cells assayed against trinitrophenyl-conjugated sheep erythrocytes, as previously described (7).

Antigens. DNP_{75}-D-GL contained 75 moles DNP per mole of D-GL (M.W. = 104,000 Daltons) and DNP_{32}-Ficoll had a substitution ratio of 32 moles of DNP per mole of Ficoll (mean M.W. = 400,000 Daltons). The preparation of these compounds has been described (8,7). DNP-D-GL was covalently bound to agarose beads (Affigel-10, Biorad Laboratories) according to the manufacturer's instructions and extensively washed to remove unbound DNP-D-GL. The beads were washed just prior to each experiment to reduce the amount of antigen spontaneously released in culture.

Cell Fractionation. Spleen cells were separated into adherent and nonadherent fractions by incubation in plastic culture dishes (9). Adherent cells were washed extensively and consisted of $>90\%$ macrophages. Antigen-pulsed adherent cells were prepared by incubating normal spleen cells with DNP-D-GL at $37^{o}C$ for 1 hour, washing the cells three times, then preparing adherent cells, and washing the attached adherent cells twice more. This procedure carried over no more than 0.3 percent of the original antigen concentration with the macrophages.

Cells bearing the θ alloantigen were depleted by lysis with AKR anti-θ-C3H serum and complement as previously detailed (7).

RESULTS

1. The in vitro response to DNP-Ficoll.

The characteristics of the primary in vitro antibody response to DNP-Ficoll have been described (7). In brief, DNP_{32}-Ficoll induces both an IgM and IgG anti-DNP PFC response of considerable magnitude. Removal of adherent cells and/or T lymphocytes does not decrease the magnitude of the response; rather, these procedures lead to an increased and prolonged response, as shown in Figure 1. DNP-Ficoll, therefore, seems to be capable of directly stimulating B lymphocytes to produce antibody.

The effect of the substitution ratio of DNP-Ficoll on the immunogenicity has been examined (7). The optimal ratio for stimulating an anti-DNP response appears to be 30-40 moles of DNP per mole of Ficoll. DNP-Ficoll containing fewer than 10 DNP groups per Ficoll molecule are essentially nonimmunogenic and nontolerogenic, whereas DNP_{85}-Ficoll, the most highly substituted preparation we have tested, is still moderately immunogenic.

DNP-Ficoll has not proved to be tolerogenic, in a formal sense, under any of the conditions we have tested thus far. For example, exposure of spleen cells to 100 μg/ml DNP_{32}-Ficoll (10,000 x the optimal immunogenic dose) for 24 hours failed to inhibit the subsequent response to 10 nanograms DNP_{32}-Ficoll, as shown in Table 1. Nonetheless, 100 μg of DNP-Ficoll elicits a clearly smaller response than does 10 ng, indicating some suppressive action of a high concentration present continuously.

Despite the high degree of immunogenicity of DNP-Ficoll, Ficoll itself has no detectable capacity to activate B cells, suggesting that it is not providing a "second" signal to the cell. For example, when spleen cells are cultured with 1 nanogram of DNP_{32}-Ficoll and various amounts (up to 10 milligrams/ml) of unsubstituted Ficoll, the anti-DNP response is neither augmented nor impaired by Ficoll, even at a millionfold excess. In addition, we have detected no endotoxin activity in DNP-Ficoll, we have failed directly to activate C3 with it and have found it to be nonmitogenic over a wide range of concentrations up to 1 milligram/ml.

B cell activation by DNP-Ficoll appears to be mediated by immunoglobulin receptors, as would be anticipated. This is indicated by the observation that anti-mouse u chain antibody

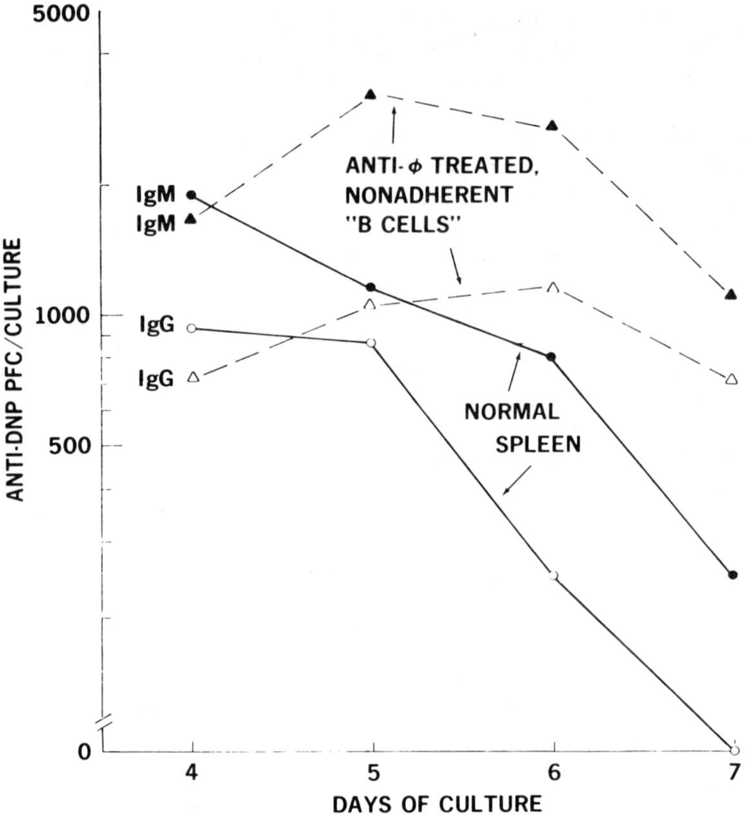

Fig. 1 The *in vitro* anti-DNP response of normal BALB/c spleen cells or anti-θ and complement-treated, non-adherent spleen cells (T cell and macrophage-depleted) to 10^{-2} μg/ml DNP_{32}-Ficoll. Numbers are mean PFC/culture of three replicate cultures. Anti-θ treated nonadherent cells failed to respond to SRBC *in vitro*.

TABLE 1

HIGH DOSES OF DNP-FICOLL FAIL TO INDUCE DNP-SPECIFIC TOLERANCE IN VITRO

24 hr. Preincubation with:*	Cells Washed, cultured with:	4 d DNP-Specific IgM PFC/culture
0	0	440 ± 55
	10^{-2} μg/ml DNP_{32}-Ficoll	2,545 ± 260
	10^{+2} μg/ml DNP_{32}-Ficoll	510 ± 45
	10^{+1} μg/ml DNP_{11}-KLH**	3,435 ± 385
10^{-2} μg/ml DNP_{32}-Ficoll	0	1,540 ± 120
	10^{-2} μg/ml DNP_{32}-Ficoll	3,800 ± 245
	10^{+2} μg/ml DNP_{32}-Ficoll	955 ± 85
	10^{+1} μg/ml DNP_{11}-KLH	5,720 ± 460
10^{+2} μg/ml DNP_{32}-Ficoll	0	810 ± 80
	10^{-2} μg/ml DNP_{32}-Ficoll	3,635 ± 320
	10^{+2} μg/ml DNP_{32}-Ficoll	480 ± 60
	10^{+1} μg/ml DNP_{11}-KLH	5,640 ± 610

* BALB/c spleen cells (10^7/ml) were incubated at 37° for 24 hours with or without DNP-Ficoll at the indicated dose, washed twice, and recultured for 4 days at 10^7/ml before PFC/culture were measured.

** Cultures stimulated with DNP-KLH (keyhole limpet hemocyanin) were supplemented with 1×10^6 irradiated spleen cells from mice primed 7 days previously with 10 μg KLH+2×10^9 Bordetella pertussis organisms. Such "carrier primed" cells did not increase the anti-DNP background response in the absence of antigen.

TABLE 2

FICOLL FAILS TO INHIBIT THE RESPONSE TO DNP-FICOLL IN VITRO

Ficoll concentration in culture ($\mu g/ml$)	DNP_{32}-Ficoll ($\mu g/ml$)	4 d IgM DNP-specific PFC/culture $\pm$ SEM*
0	0	960 $\pm$ 40
0	10^{-2}	6,160 $\pm$ 320
10^{+4}	"	4,870 $\pm$ 440
10^{+3}	"	4,840 $\pm$ 320
10^{+2}	"	4,680 $\pm$ 180
10^{+1}	"	4,470 $\pm$ 830
10^{0}	"	4,700 $\pm$ 200
10^{-1}	"	4,280 $\pm$ 240
10^{-2}	"	6,060 $\pm$ 400
10^{-3}	"	7,080 $\pm$ 620
10^{-4}	"	5,570 $\pm$ 510
10^{-5}	"	5,730 $\pm$ 340

* Mean $\pm$ S.E.M. of three replicate cultures containing 10^7 BALB/c spleen cells, $10^{-2}\mu g/ml$ DNP_{32}-Ficoll, and various concentrations of Ficoll continuously for the 4 day culture interval.

effectively blocks the induction of the anti-DNP response to DNP-Ficoll. The results of one such experiment are shown in Table 3. The suppressive effects of anti-μ serum could be reversed by prior absorption of the antiserum with immobilized MOPC-104E μ chains.

Furthermore, mice rendered tolerant to DNP by DNP-D-GL treatment *in vivo* are incapable of responding to DNP-Ficoll *in vitro* (7), reinforcing the concept that DNP-specific B cells are required for the DNP-Ficoll induced anti-DNP response.

To summarize, the conjugate of DNP and Ficoll is a potent "T-independent" antigen which is capable of directly activating B lymphocytes. This activation has been shown to involve DNP-specific μ-bearing receptors. Ficoll does not nonspecifically activate B cells by any known pathway and the immunogenicity of DNP-Ficoll appears to be due, in large part, to the tertiary structure of the antigen molecule.

2. *The in vitro response to DNP-D-GL.*

The DNP conjugate of the linear copolymer of D-glutamic acid and D-lysine is a potent tolerogen both *in vivo* (5) and *in vitro* (6). In the course of experiments designed to determine whether DNP-D-GL had to interact directly with B cells to induce DNP-specific tolerance (as assessed by *in vitro* challenge with DNP-Ficoll), we observed that mouse splenic adherent cells exposed to DNP-D-GL for 1 hour and recombined with nonadherent cells not only failed to induce tolerance to DNP-Ficoll but stimulated a marked anti-DNP response in the absence of additional antigen. A series of experiments was initiated, therefore, to examine the possible immunogenicity of DNP-D-GL added directly to cultures of normal or nonadherent spleen cells, or added as a macrophage-associated antigen. The pooled results of five such experiments are shown in Figure 2. The effect, initiation of antibody production or induction of tolerance, of DNP-D-GL in culture is related to the effective concentration of the antigen and the cell populations present at the time of initial antigen exposure. Adherent cells prepared from spleen cells pulsed with DNP_{75}-D-GL, and carrying less than 1 percent of the initial antigen dose, were able to stimulate a significant anti-DNP response when recmobined with normal nonadherent cells. The optimal initial concentration of DNP-D-GL during the pulse period was 10^{-2} to 10^{0} μg/ml, which suggests that the amount of macrophage-associated antigen added to the cultures was approximately 10^{-5} to 10^{-2} μg. The peak magnitude of the anti-DNP response induced by DNP-D-GL pulsed adherent cells averaged 2,432 $\pm$ 465 PFC/culture for the five experiments. Direct continuous exposure of normal spleen

TABLE 3

ANTI-μ SERUM INHIBITS THE DNP-FICOLL RESPONSE IN VITRO

Antiserum Dilution	4 d IgM Anti-DNP PFC/Culture ± S.E.M.*
no antiserum	4,260 ± 340
1:500 anti-μ	5 ± 2.5
1:1000 anti-μ	55 ± 15
1:2000 anti-μ	1,510 ± 140
1:500 anti-μ absorbed with insolubilized MOPC 104E**	3,330 ± 460

* All cultures were stimulated with 10^{-2} μg/ml DNP_{32}-Ficoll.

** Purified μ chain from MOPC 104E was bound to cyanogen bromide-activated agarose beads and mixed 1:1 with 1:20 goat anti-μ serum at 4°C for 1 hour. Both reagents were supplied by Dr. R. Asofsky.

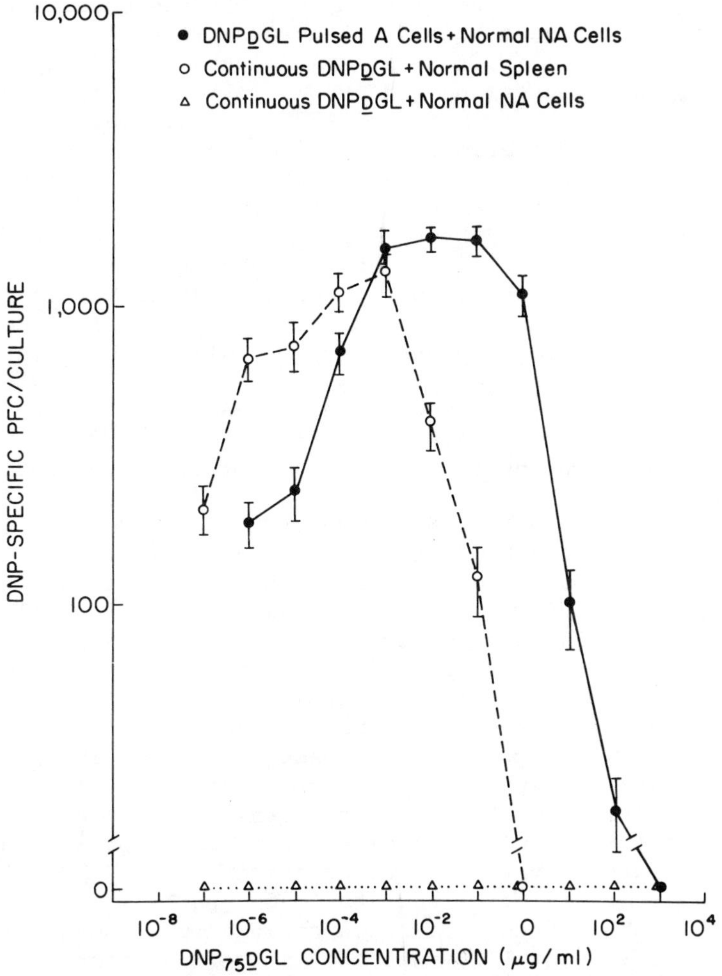

Fig. 2 The *in vitro* PFC response of BALB/c spleen cells to DNP-D-GL presented either as macrophage-associated antigen (DNP-D-GL pulsed adherent cells) or continuously. Responding cells were either normal or macrophage-depleted nonadherent (NA).

cells to DNP_{75}-D-GL also induced a significant anti-DNP PFC response. The optimum antigen concentration for normal spleen cells was 10^{-4} to 10^{-3} μg/ml. Higher concentrations of DNP-D-GL not only failed to stimulate a response, but prevented DNP-Ficoll from inducing an anti-DNP response as well (as will be illustrated below). Concentrations of DNP-D-GL as low as 10^{-8} μg/ml (0.01 picogram) induced a response significantly above background ($P < 0.01$). The magnitude of the mean peak response was 1,760 ± 325 PFC/culture. Nonadherent macrophage-depleted BALB/c spleen cells failed to respond to DNP_{75}-D-GL at any concentration from 10^{-7} to 10^{+3} μg/ml. Nonadherent cells not only failed to respond but were rendered tolerant by DNP-D-GL, since simultaneous addition of DNP-Ficoll failed to stimulate any anti-DNP PFC response (Table 4). Interestingly, the presence of as little as 10^{-6} μg of DNP-D-GL/ml (and in some experiments 10^{-8} μg/ml) completely inhibited the response to DNP-Ficoll. Adding various amounts of DNP-D-GL plus an optimal concentration of DNP_{32}-Ficoll simultaneously to unseparated spleen cells led to an additive response at low (immunogenic) doses of DNP-D-GL but to inhibition of the anti-DNP response at higher concentrations (Figure 3).

In order to ensure that both DNP-D-GL immunogenicity and tolerogenicity reflected events at the B cell level, the influence of T lymphocytes was ruled out by the following two experiments. First, lethally-irradiated spleen cells were pulsed with DNP-D-GL, washed extensively, and antigen-pulsed adherent cells prepared. These were combined with nonadherent cells which had been treated with anti-θ serum and complement, or with control nonadherent cells. The results are shown in Fig. 4, and demonstrate that T lymphocytes are not required for the in vitro IgM response to DNP-D-GL. In the second experiment, nonadherent spleen cells were either treated with anti-θ serum and complement or left untreated, exposed to various concentrations of DNP_{75}-D-GL (10^{-6} to 10^{+1} μg/ml) for 1 hour to induce partial tolerance, thoroughly washed, then cultured with 10^{-2} μg of DNP_{32}-Ficoll to assess the role of T lymphocytes in the induction of unresponsiveness to DNP-Ficoll caused by DNP-D-GL. The results are shown in Table 5. Depletion of T lymphocytes seems to have no significant effect on DNP-D-GL tolerance induction in vitro. Thus, both the DNP-D-GL induced antibody response and tolerance should reflect events at the B cell level, with prior interaction of antigen with macrophages being crucial for B cell activation.

To further investigate the role of antigen presentation in the in vitro response to DNP-D-GL, the antigen was covalently

TABLE 4

EXPOSURE OF NONADHERENT CELLS TO DNP-D-GL MAKES THEM UNRESPONSIVE TO DNP-FICOLL OR DNP-D-GL ON ADHERENT CELLS

Experiment	DNP_{75}-D-GL Concentration (μg/ml)	Antigen Concentration (μg/ml)	4 d IgM DNP-specific PFC/culture ± S.E.M.
1	0	0	470 ± 60
	0	10^{-2} DNP_{32}-Ficoll	5,240 ± 480
	10^{-6}	0	320 ± 30
	10^{-6}	10^{-2} DNP_{32}-Ficoll	375 ± 40
	10^{-4}	0	335 ± 60
	10^{-4}	10^{-2} DNP_{32}-Ficoll	340 ± 25
	10^{-2}	0	150 ± 30
	10^{-2}	10^{-2} DNP_{32}-Ficoll	230 ± 35
	10^{0}	0	110 ± 15
	10^{0}	10^{-2} DNP_{32}-Ficoll	90 ± 20
2	0	0	725 ± 65
	0	DNP-D-GL pulsed-A	1,925 ± 225
	10^{-6}	0	560 ± 100
	10^{-6}	DNP-D-GL pulsed-A	820 ± 40
	10^{-4}	0	760 ± 80
	10^{-4}	DNP-D-GL pulsed-A	725 ± 55
	10^{-2}	0	610 ± 30
	10^{-2}	DNP-D-GL pulsed-A	690 ± 75

BALB/c spleen cells were separated into nonadherent and adherent cells as described. In Experiment 1, nonadherent cells were cultured ± DNP-Ficoll in the presence of varying concentrations of DNP_{75}-D-GL. In Experiment 2, the same conditions obtained except the source of antigen was adherent cells exposed to 10^{-2} μg/ml DNP-D-GL for 1 hour and washed three times before recombining with nonadherent cells.

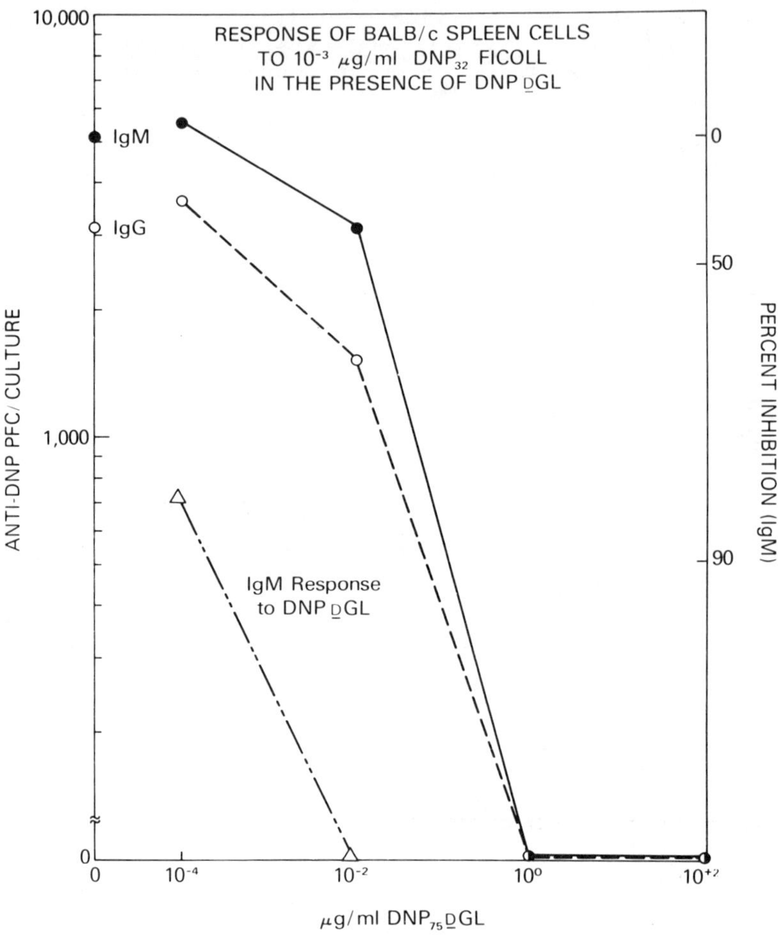

Fig. 3 Dose-dependence of DNP-D-GL induced inhibition of the *in vitro* anti-DNP response induced by DNP-Ficoll. Both DNP-D-GL and DNP-Ficoll were present continuously during the 4-day culture period.

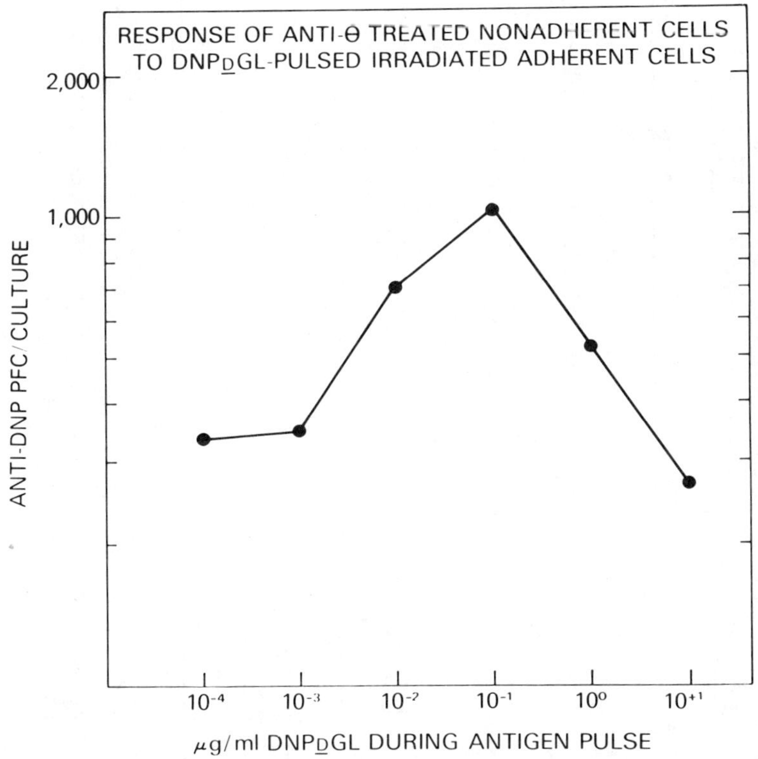

Fig. 4 The anti-DNP response induced by DNP-D-GL-pulsed, irradiated macrophages mixed with anti-Θ treated lymphocytes from BALB/c spleen. Control nonadherent cells treated with brain-absorbed anti-Θ serum responded no better than anti-Θ treated cells (controls not shown). Anti-Θ treatment abrogated the response to SRBC-pulsed adherent cells.

TABLE 5

ANTI-θ PRETREATMENT OF NONADHERENT SPLEEN CELLS FAILS TO ALTER SUSCEPTIBILITY TO DNP-D-GL-INDUCED TOLERANCE

DNP-D-GL Concentration (μg/ml)	4 d IgM DNP-Specific PFC/Culture in response to DNP_{32}-Ficoll of: Response of Anti-θ-Treated Nonadherent Cells	Control Nonadherent Cells
0	2,420	3,260
10^{-6}	225	235
10^{-4}	0	0
10^{-2}	0	0
10^{0}	0	0
10^{+1}	0	0

BALB/c spleen cells (10^7/ml) were separated into adherent and nonadherent cells as described (9) and the nonadherent cells treated with anti-θ serum and complement or brain-absorbed anti-θ serum as a control. The resulting cell populations were cultured at 10^7/ml with 10^{-2} μg/ml DNP_{32}-Ficoll and various concentrations of DNP_{75}-D-GL.

linked to agarose beads (Affigel-10). DNP_{75}-D-GL-Affigel-10 stimulated a significant anti-DNP PFC response in normal spleen cells (data not shown) and, more importantly, in nonadherent spleen cells (Figure 5). Approximately the same concentration of DNP-D-GL presented to the same nonadherent cells in soluble form resulted in no response whatsoever. This experiment suggests that one role of macrophage-bound antigen is to present DNP determinants in the form of highly multivalent complexes. At the very low doses that are immunogenic for normal spleen cells in vitro, DNP-D-GL apparently is concentrated by macrophages and presented in an immunogenic form which competes with soluble DNP-D-GL for interaction with DNP-specific B cells. One paradoxical set of findings which is not yet resolved is the capacity of 10^{-6} μg/ml of DNP-D-GL to induce tolerance in nonadherent cells challenged with DNP-Ficoll and the ability of normal spleen cells to respond to DNP-D-GL under conditions in which the free (non-macrophage associated) concentration of DNP-D-GL is clearly in excess of 10^{-6} μg/ml.

The interaction of macrophages and B lymphocytes with DNP-D-GL or DNP-Ficoll was examined in more detail by separating normal BALB/c spleen cells into adherent or nonadherent fractions, exposing the two populations to antigen separately for 1 hour at 37°C, washing the cells extensively, and recombining adherent and nonadherent cells. The recombined cells were then cultured with or without additional continuous antigen. The general protocol for this type of experiment is shown in Figure 6 and a summary of results is shown in Table 6. Because of the large number of permutations, Table 6 represents pooled data from several different experiments and the relative rather than the absolute magnitude of the anti-DNP response is shown. In each experiment the positive control (4+ response) represents untreated adherent and nonadherent cells recombined and cultured with continuous DNP-Ficoll. The absolute magnitude of this positive control varied between 2200 and 7400 anti-DNP PFC/culture in this series of experiments. These data make several points: first, exposure of nonadherent cells to DNP_{75}-D-GL at 10 μg/ml for 1 hour renders them unresponsive to DNP-Ficoll or DNP-D-GL whatever the form of antigen presentation, and second, addition of DNP-D-GL in concentrations greater than 10^{-1} μg/ml to recombined cell cultures prevented further development of the anti-DNP response. In separate controls (not shown), higher concentrations of DNP-D-GL were shown not to inhibit an in vitro response to sheep erythrocytes, ruling out nonspecific toxicity as a mechanism for inducing this unresponsiveness. DNP-D-GL pulsed adherent cells did not inhibit the

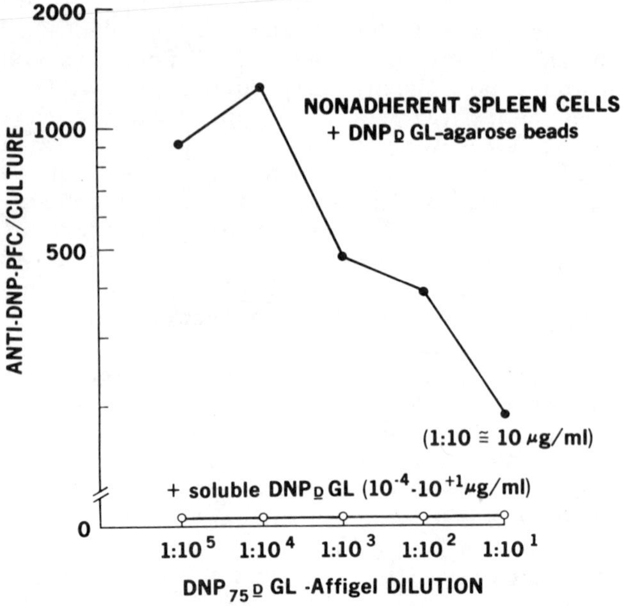

Fig. 5 The response of BALB/c nonadherent (macrophage-depleted) spleen cells to soluble DNP-D-GL or DNP-D-GL insolubilized by covalent linkage to Affigel (agarose) beads. The starting suspension of DNP-D-GL-Affigel was shown to contain about 100 μg/ml DNP-D-GL by dissolving the beads and measuring released DNP. The soluble and insoluble DNP-D-GL concentrations, therefore, are roughly equivalent.

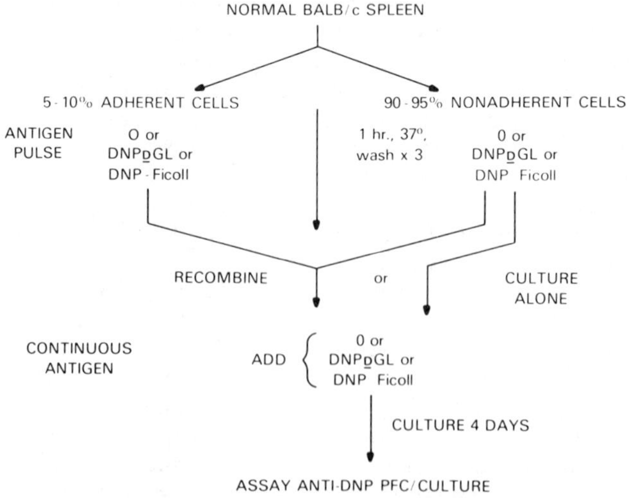

Fig. 6 Protocol for antigen pulse experiments.

TABLE 6

The response of BALB/c spleen cells exposed to antigen either as a macrophage-enriched adherent fraction or as a macrophage-depleted nonadherent fraction, then cultured with or without additional antigen as shown in Figure 6. The IgM anti-DNP PFC response after 4 days in culture is normalized with the reference ++++ number being the response to DNP-Ficoll of recombined adherent and nonadherent cells not previously exposed to antigen. This data summary represents pooled results from nine separate experiments.

Antigen Pulse		Continuous Antigen	Anti-DNP Response
Adherent	Nonadherent		
O	O	O	O
O	O	DNP D GL	O – ++ (dose dependent)
O	O	DNP - Ficoll	++++
DNP D GL	O	O	++++
DNP D GL	O	DNP D GL	O – +++ (dose dependent)
DNP D GL	O	DNP - Ficoll	++++
DNP - Ficoll	O	O	++++
DNP - Ficoll	O	DNP D GL	O – +++ (dose dependent)
DNP - Ficoll	O	DNP - Ficoll	++++
O	DNP D GL	O	O
O	DNP D GL	DNP D GL	O
O	DNP D GL	DNP - Ficoll	O
O	DNP - Ficoll	O	+
O	DNP - Ficoll	DNP D GL	O – ++ (dose dependent)
O	DNP - Ficoll	DNP - Ficoll	++++
DNP D GL	DNP D GL	O	O
DNP D GL	DNP D GL	DNP D GL	O
DNP D GL	DNP D GL	DNP - Ficoll	O
DNP D GL	DNP - Ficoll	O	+++
DNP D GL	DNP - Ficoll	DNP D GL	O – +++ (dose dependent)
DNP D GL	DNP - Ficoll	DNP - Ficoll	++++
DNP - Ficoll	DNP D GL	O	O
DNP - Ficoll	DNP D GL	DNP D GL	O
DNP - Ficoll	DNP D GL	DNP - Ficoll	O
DNP - Ficoll	DNP - Ficoll	O	+++
DNP - Ficoll	DNP - Ficoll	DNP D GL	O – +++ (dose dependent)
DNP - Ficoll	DNP - Ficoll	DNP - Ficoll	++++

response to DNP-Ficoll, again suggesting that such cell-bound antigen is immunogenic but not tolerogenic. The order in which cells were exposed to DNP-D-GL or DNP-Ficoll did not seem to alter the response produced *in vitro*; in particular, prior exposure of nonadherent cells to DNP-Ficoll did not prevent subsequent inhibition of the anti-DNP response by DNP-D-GL. If DNP-D-GL and DNP-Ficoll bind to the same receptors on B cells, then DNP-D-GL either must be able to displace DNP-Ficoll or to reverse DNP-Ficoll induced activation by binding to residual DNP receptors or to receptors which reappear on B cell surfaces after an initial round of endocytosis.

DISCUSSION

The two synthetic antigens studied in these experiments, the DNP derivatives of Ficoll and D-GL, both appear to stimulate substantial *in vitro* anti-DNP responses in the absence of detectable numbers of T lymphocytes. Both antigens appear to activate B lymphocytes, but DNP-D-GL must be cell-bound or insolubilized to do so, while the intrinsic structure of Ficoll suffices for self-presentation. Soluble DNP-D-GL binding to B lymphocytes seems to lead invariably to DNP-specific tolerance. These observations taken together clearly emphasize the importance of the topology of determinant presentation for activation or inactivation of B cells, but do not rule out other surface signals modifying cell differentiation. One advantage of these two antigens is that they can be utilized in situations in which T lymphocytes can be ruled out as regulators of B cell activation.

Of the proposed theories to explain the mechanisms of B cell stimulation by so-called "T-independent" antigens, three have received the most attention and experimental support. The first postulates that all T-independent antigens are intrinsically stimulatory and that binding of specific determinants by immunoglobulin receptors only serves to focus the antigen on the cell surface where the crucial activating interaction takes place between the antigen and "non-specific" non-immunoglobulin receptors (4). A corollary of this theory is that high concentrations of T-independent antigens should lead to the activation of all B cells because sufficient amounts of these antigens should bind to the "non-specific" receptor of all such cells. This has been referred to as polyclonal activation. A second theory is that B cell activation requires two signals, both critical for B cell activation although one signal might suffice to induce tolerance (2). One signal is proposed to be provided by the binding of the antigenic determinant to immunoglobulin on the surface of the B cell, but the nature of the second signal is unspecified. Although portions of the antigen molecule other than the determinant might provide this signal, another possibility is that the second signal is mediated by a different moiety. One version of this two signal theory postulates that T-independent antigens activate C3 which binds to the complement receptor of B lymphocytes thus providing the second signal (3). A third theory to explain B cell activation suggests that T-independent antigens share the ability to present antigenic determinants in an appropriate array or matrix. The carrier portion of the molecule would not interact directly with B cell receptors, but would present determinants spaced in a way to lead to efficient cross-linking of receptors. Corollaries of this theory are that activation or inactivation of B lymphocytes should depend upon the number of surface immunoglobulin receptors bound to antigenic determinants, the spatial array of bound receptors, and the rigidity with which the receptors are cross-linked by the antigen.

The behavior of DNP-Ficoll and DNP-D-GL fails to conform with the predictions of the first two theories, as has been reported (7,10). For example, neither DNP-Ficoll nor DNP-D-GL leads to polyclonal activation of B cells. DNP-Ficoll is not mitogenic over a wide range of doses in a variety of culture media with or without serum supplementation. DNP-D-GL produces little, if any, proliferation of B cells. Neither antigen will activate C3 nor contains contaminating endotoxin.

We, therefore, favor the interpretation that the physical nature of the antigen is instrumental in inducing the differentiation of B cells into antibody-forming cells. In the case of DNP-Ficoll, the rigidly cross-linked sucrose subunits when substituted with the DNP groups at an appropriate epitope density should provide multipoint binding sites capable of cross-linking a number of immunoglobulin receptors into a stable array. This effective immobilization of receptors in a fluid membrane is suggested as the signal for cell activation. The evidence that this signal is mediated by immunoglobulin binding is that cell activation is inhibited by anti-μ reagents. That activation is not blocked by excess Ficoll suggests that only DNP determinants on the antigen are involved in binding and signal generation.

The evidence that the physical nature of DNP-D-GL determines its immunogenicity is even stronger. On the basis of these experiments, soluble DNP-D-GL seems to be tolerogenic and insoluble or cell-bound DNP-D-GL immunogenic. The postulated mechanism of cell activation by macrophage-DNP-D-GL or DNP-D-GL-Affigel would be similar to that of DNP-Ficoll, i.e., the macrophage surface or the surface of agarose beads serves as an inert carrier to present DNP groups in a highly multivalent and relatively fixed array. Tolerance, in contrast, would be induced by binding of anti-DNP-receptors to a smaller non-rigid molecule which would fail to adequately cross-link immunoglobulin receptors to an extent necessary to activate the cell. Subsequent activation of a DNP-D-GL exposed B cell by DNP-Ficoll may be blocked by persistence of D-GL on the cell surface, or, alternatively, by some internal cellular event leading to unresponsiveness (5). In this respect, it is clear that differences between highly tolerogenic DNP-D-GL and more weakly tolerogenic DNP-L-GL exist, probably reflecting the capacity of the former to persist in some critical cellular compartment. This model of B cell activation places great emphasis on the physical nature of the antigen and the presumed role of macrophages and/or T cells in modifying the structure of antigens that may be intrinsically tolerogenic. The observations that deaggregated proteins may be tolerogenic (11) and that the substitution ratio of hapten to carrier may influence immunogenicity or tolerogenicity (12) lend support to the concept that the way in which antigen binds to the surface of the B cell (and not a variety of "second signals") is the determinant for B cell differentiation.

REFERENCES

1. Bretscher, P.A. and Cohn, M., Science 169: 1042, 1970.
2. Lederberg, J., Science 129: 1649, 1959.
3. Dukor, P. and Hartmann, K., Cell. Immunol. 7: 349, 1973.
4. Coutinho, A., Gronowicz, E., Bullock, W.W. and Möller, G., J. Exp. Med. 139: 74, 1974.
5. Katz, D.H., Hamaoka, T. and Benacerraf, B., J. Exp. Med. 136: 1404, 1972.
6. Nossal, G.J.V., Pike, B.L. and Katz, D.H., J. Exp. Med. 138: 312, 1973.
7. Mosier, D.E., Johnson, B.M., Paul, W.E. and McMaster, P.R.B., J. Exp. Med. 139: 1354, 1974.
8. Katz, D.H., Davie, J.M., Paul, W.E. and Benacerraf, B., J. Exp. Med. 134: 201, 1974.
9. Mosier, D.E., Science 158: 1575, 1967.
10. Paul, W.E., Karpf, M. and Mosier, D.E., Proc. Brook Lodge Symposium (in press), 1974.
11. Dresser, D.W., Nature (London) 191: 1169, 1961.
12. Feldmann, M., J. Exp. Med. 135: 735, 1972.

IMMUNE RESPONSE TO PHOSPHORYLCHOLINE
A Model System for the Study of Antibody Diversity*

J. Latham Claflin, Joyce A. Schroer and Joseph M. Davie
Departments of Pathology and Microbiology, Washington Univ. School of Medicine, St. Louis, Missouri

The capacity of the immune system to respond to an enormous array of antigenic materials with remarkable specificity has been well known since Landsteiner (1). Furthermore, the antibodies to any single determinant are generally highly heterogenous so that estimates of the number of different antibody molecules which each individual can produce are very large (10^5 - 10^7) (2). Indeed, in a recent study in mice it was concluded that over 10^4 different antibodies existed with specificity for a single, simple haptenic determinant (3). The genetic problems inherent in accounting for antibody diversity are still very much unresolved but have been focused into two general theories. The first proposes that all genetic information for antibody production is carried in the germ line so that each different immunoglobulin is represented in the genome. The other proposes that only a few genes are carried in the germ line and that mutation of these genes in the soma leads to the generation of new antibody specificities which are therefore not inheritable. The arguments for these theories have been presented in detail by others and need not be repeated here (4-7).

We will review here recent evidence from our laboratory concerning antibody diversity, which, we feel, provides new constraints to those theories of the generation of diversity which rely on hypermutation.

PHOSPHORYLCHOLINE-FINDING MOUSE MYELOMA PROTEINS

The major impetus to our studies came from work by Potter and Lieberman on mouse myeloma proteins which bind phosphorylcholine (PC). They found that 5 of 8 PC-specific proteins

*This work was supported by Public Health Service grants AI-11635, CA-16032 and ES-00128 and a grant from the following companies: Brown & Williamson Tobacco Corporation; Larus and Brother Company, Inc.; Liggett & Myers, Incorporated; Lorillard, a Division of Loews Theatres, Incorporated; Philip Morris, Incorporated; R.J. Reynolds Tobacco Company; United States Tobacco Company; and Tobacco Associates, Inc.

shared variable region (idiotypic) antigenic markers (8). Leon and Young provided support for the similarity of these proteins by showing that the specificity and affinity of these proteins for PC and its analogues were identical (9). The fact that no other myeloma proteins had shown antigenic similarity before, let alone had similar antigen-binding specificities, made this a unique finding.

The implication from these studies was that either the BALB/c mouse possessed a PC-specific clone which, for some reason, had a propensity to develop neoplasms or that the BALB/c mouse possessed a limited repertoire of PC-specific precursor cells. This was resolved by Cosenza and Köhler who found that the majority of anti-PC molecules in immunized BALB/c mice possessed the tumor-specific idiotype (10) and that pretreatment of BALB/c mice with antisera directed to the variable regions of one of the "common" PC-specific myelomas rendered the animals incapable of producing anti-PC antibodies after appropriate immunization (11). With this indication, then, that the PC-specific myelomas probably represent normal products and can be considered prototype antibodies, it is worth considering them in more detail before describing the normal anti-PC immune response.

The PC-specific myeloma proteins which are known at present are listed in Table I along with the known antigenic characteristics of these proteins. It should be noted that all of these proteins possess kappa light chains and, with the exception of one protein, were derived from the BALB/c mouse. Protein CBPC 2 was derived from the CB-20 strain of mice, which is congenic to BALB/c but differs from it by carrying the A^- determinant of C57BL/ka (12). There now exist 3 known variable region markers for some of the myeloma proteins. The first marker which was demonstrated (8), T15, is now known to lie in the variable regions which lie outside the binding site of the immunoglobulin molecule (13). This determinant is detected by antisera raised in mice (A/J or CE) which have been immunized to myeloma T15 or H8. The marker $H8_s$ was shown to lie in the binding region of the immunoglobulin molecules by several means (13). Interaction between anti-$H8_s$ antiserum and TEPC 15 is inhibitable by PC and its analogues, whereas anti-T15 is not. In addition, the $H8_s$ determinant(s) is found on neither heavy nor light chains, but can be reconstituted when isolated heavy and light chains are allowed to reassociate. $H8_s$ is shared by 8 out of the 11 PC-specific myeloma proteins. T15 is found on all $H8_s$ positive proteins of BALB/c origin. A third marker has recently been described which is located on the heavy chain of the myeloma protein and is shared by all members of the PC group

TABLE I
ANTIGENIC MARKERS ON PC-SPECIFIC MYELOMA PROTEINS

Myeloma Protein	Variable Region Markers T15[a]	$H8_s$[b]	H_V-PC[c]	Constant Region Markers[d] a1	a-	K
T15	+	+	+	+	-	+
H8	+	+	+	+	-	+
M299	+	?	?	+	-	+
S63	+	?	?	+	-	+
S107	+	?	?	+	-	+
Y5170	+	?	?	+	-	+
Y5236	+	?	?	+	-	+
CBPC 2	-	+	+	-	+	+
M511	-	-	+	+	-	+
M603	-	-	+	+	-	+
M167	-	-	+	+	-	+

[a]Non-binding site determinant detected by precipitation with mouse (A or CE) anti-T15 (24).

[b]Binding site determinant detected by inhibition of precipitation of ^{125}I-H8 with rabbit anti-$H8_s$ (24).

[c]Heavy chain determinant detected by inhibition of precipitation of ^{125}I-M167 with rabbit anti-H8 (14).

[d]A determinants are allotypes on the heavy chain. K is the light chain type.

(14). While it is not indicated on the table, the variable region markers listed here are found on no other myeloma protein. Thus, 7 of the 11 PC-specific myelomas share all variable region markers, including those in the binding site, and an additional protein differs only in the possession of a different non-binding site marker and allotypic marker. Proteins M511, M603 and M167 each have unique variable region determinants which are not described in this table, but which clearly distinguish these proteins from all other proteins.

The similarities of the T15 group extend to the binding characteristics of these proteins (9). No differences have been found either with respect to the affinity of binding PC determinants or in the cross-reactivity of these binding regions for analogues of PC, glycerophosphorylcholine (GPC) or choline (C). However, totally different binding profiles are found with the non-T15 myeloma proteins, M511, M603 and M167 (Figure 1). Thus, there is a clear association between the presence of a particular antigenic marker(s) within the binding site and the binding specificity of the myeloma protein (15).

Striking confirmation of the structural similarity between the T15 group of myelomas and the unique characteristics of the myeloma proteins in the non-T15 subgroups has come through the sequence analysis of the amino-terminal regions of light and heavy chains from these proteins. Thus, the amino-terminal regions of H8, T15 and S107 are identical through position 36 of both the heavy and light chains (16). In addition, CBPC 2, which differs in allotype, but which possesses the variable region markers of the T15 group, has only 2 substitutions in the first 36 positions on the heavy chain and has identical light chains through this comparable region (12). On the other hand, M603 and M167 have distinctly different amino-terminal sequences in the light chains, but, interestingly, have similar sequences in the heavy chains, in spite of the fact that their binding-site variable region determinants and their binding specificities are clearly different. Examples of these similarities are shown in Figure 2 where the amino acid sequences in the first hypervariable regions of the light chains of T15, H8, S107, CBPC 2 and M603 are listed. In addition, the isoelectric focusing pattern of light chains of these and other PC-specific proteins are shown.

Recent studies of other groups of myeloma proteins with different specificities have also shown striking correlation between binding specificity, cross-idiotypic determinants, as well as amino-terminal sequences of heavy chains, light chains or both. This has been demonstrated among myeloma

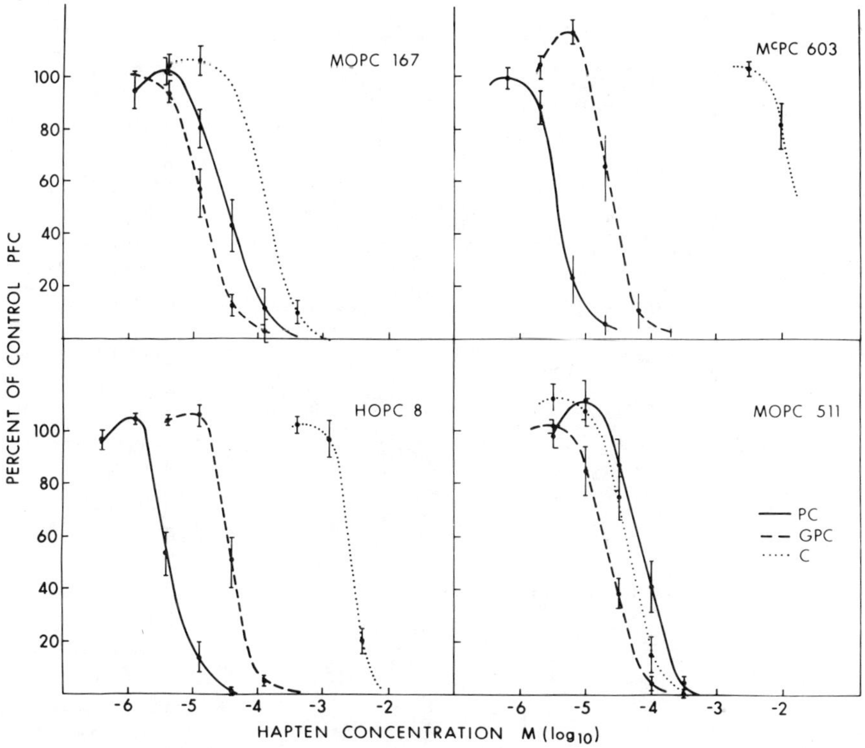

Fig. 1 Hapten inhibition profiles of PC-specific myelomas. The relative specificities of the myeloma proteins secreted by MOPC 167, McPC 603, HOPC 8 and MOPC 511 were determined by inhibition of PC-specific plaques, by phosphorylcholine (——), glycerophosphorylcholine (--) and choline (···) (15).

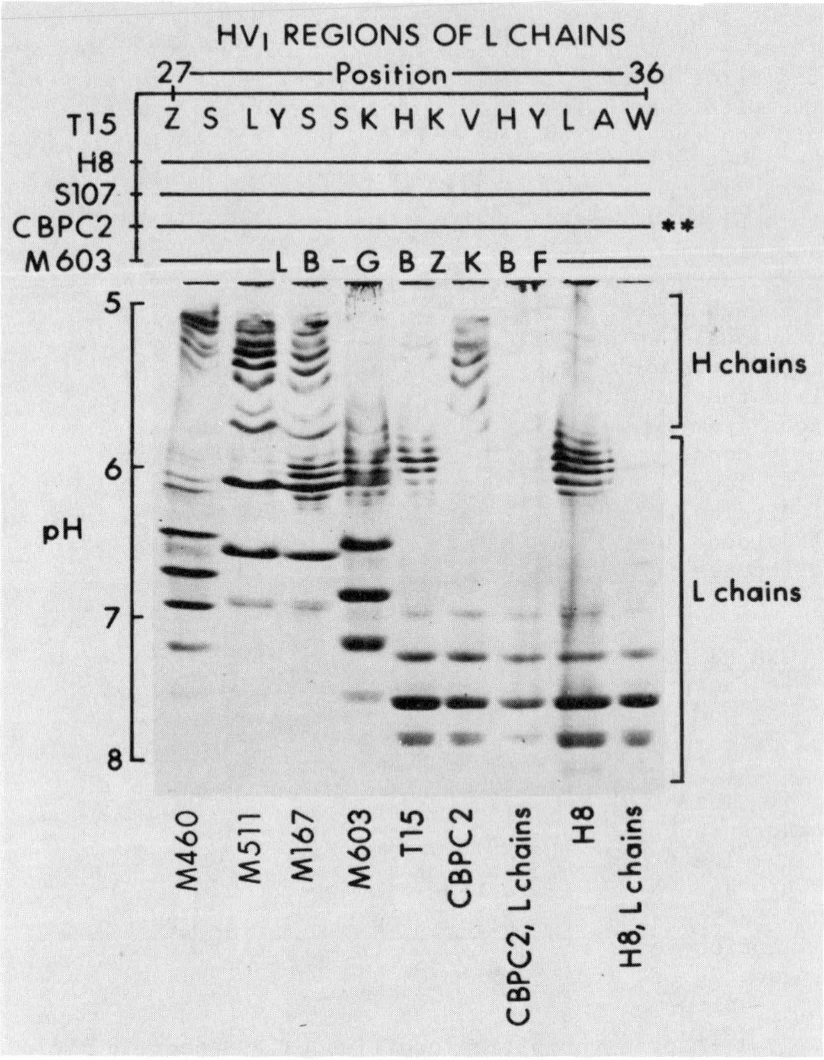

Fig. 2 Amino acid sequence of the first hypervariable region and isoelectric focus profile of light chains from PC-specific myeloma proteins. The amino acid sequences of the HV_1 regions of light chains of several members of the T15 group (T15, H8, S107 and CBPC 2) and a PC-specific protein not in the group (M603) are shown (16,12). In addition, the isoelectric focus patterns of light chains from PC-specific proteins of the T15 group (T15, H8 and CBPC 2), other groups M603, M167, and M511 and DNP-specific protein M460 are shown.

proteins which bind various dextrans (17) and other carbohydrates (18). Particularly relevant to the PC system are studies of human cold agglutinins (19) and anti-gamma globulins with shared idiotypic determinants (20), where amino acid sequence analysis has shown similarities among light chains as well. However, in no other myeloma system of either mouse or human origin have two proteins appeared to be completely identical in either structure or antigenic specificity. Thus, the PC-binding myeloma proteins represent an extraordinarily unusual finding.

Though additional structural differences may exist between individuals in the TEPC 15 group, these differences will probably be minor and may not exist within the binding site, since the binding site function seems uniform. The implications from this are that PC binding sites in mice are strikingly conserved, not only between individual BALB/c mice (i.e. sources of these tumors), but also within other strains of mice as witnessed by the CBPC 2 myeloma protein. It should be evident that the analysis of normally induced anti-PC antibodies in a variety of mouse strains becomes especially pertinent in view of the findings with these myeloma proteins.

IMMUNE RESPONSE TO PHOSPHORYLCHOLINE

If the process which leads to myeloma tumor formation is random and depends only on the relative frequency of individual precursor cells, then certain predictions can be made concerning the PC-specific immune system in mice.

1. The BALB/c mouse should have a limited number of precursors specific for PC.

2. The binding site of anti-PC antibodies of at least BALB/c and C57BL/6 anti-PC antibodies should be similar. If the binding site is highly conserved, it is conceivable that the antibodies directed toward PC in all mice, regardless of genetic background, will be similar.

3. Since non-binding site variants exist in both constant region determinants (CBPC 2) as well as at least one of the variable region determinants, T15, it would be expected that additional non-binding site variants may be found among mice.

In this section we will compare the characteristics of two types of cells involved with the humoral immune response to PC; the precursors of antibody-secreting cells (ABC), namely those small lymphocytes with the capacity to bind antigenic determinants to their surface by means of cell-bound immunoglobulin molecules, and antibody-secreting cells (PFC), the mature progeny of activated precursor cells. Immunization with PC-containing pneumococcal antigens results in stimulation of both of these types of cells as shown in Figure 3 (21). Here the PFC are of the IgM class; repeated attempts to induce an IgG antibody response in mice have failed. It can be seen that appreciable numbers of PFC and ABC are induced upon immunization, but that second immunization results in little, if any, increase over that seen in the primary response. It was shown that the responses of the ABC and PFC were independent of the presence of thymus-derived lymphocytes, and the lack of "memory" is consistent with this feature.

The lymphocyte receptors and secreted antibodies were screened for the variable region marker T15 by measuring the inhibition of antigen binding or antibody secretion by anti-T15 antiserum. In Tables II and III it is seen that the BALB/c strain possesses the T15 variable region idiotypic marker on both the receptors of the antigen-binding cells as well as the antibodies secreted by their progeny. It is evident that the vast majority of the cells involved with the response possess the same variable region determinants and that repeated immunization does not influence the clonal predominance. In addition, this idiotypic determinant was not seen on the receptors or antibody of anti-PC specific cells of the A strain mouse, and was not seen on the DNP-specific cells or antibodies of the BALB/c mouse. The similarities between the receptor molecules and the antibodies secreted by the mature progeny is strong evidence for the identity between the variable regions of these two types of molecules (21).

Support for the monoclonal nature of the PC immune system in BALB/c mice came from work by Köhler and Cosenza. They showed that pretreatment of BALB/c mice in vivo (11), or treatment of their spleen cells in vitro, with anti-T15 antiserum, but not anti-M167 or anti-M603, rendered the animals or their cells incapable of responding to PC-containing bacterial vaccines or PC linked to proteins (22). Thus, removal of T15-bearing lymphocytes rendered the animal incapable of responding to PC.

It is well known that idiotypic antisera may vary with respect to specificity since multiple "idiotypic" determinants may exist on any single molecule (see Table I), and that cross reactions among families of antibody molecules have been known to occur. This is clearly shown in the PC-specific myelomas. It is therefore crucial to this study to provide additional evidence for the restricted nature of the antibody response to PC before conclusions concerning the heterogeneity of the response can be made. Therefore, an analysis was made of the antibodies of BALB/c mice immunized

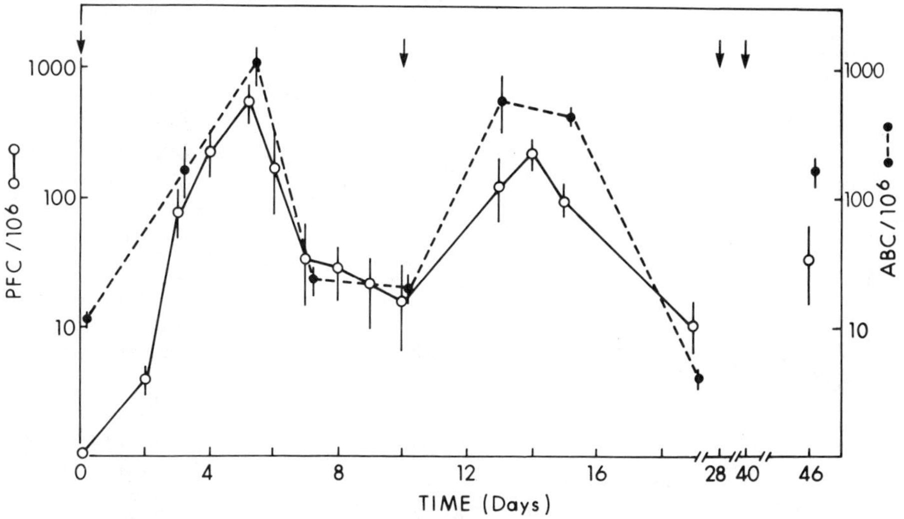

Fig. 3 Change in frequency of PC-specific ABC and PFC after immunization with pneumococci. Mice were immunized i.p. with 10^8 pneumococci (indicated by arrows) and tested at intervals for splenic PC-specific ABC and PFC (21).

TABLE II
IDIOTYPIC DETERMINANTS ON PRECURSOR CELLS[a]

Strain	ABC	Time After Immunization[b]	PC- or DNP-specific ABC/10^5 Cells[c]	Idiotype[d] T15	M603	M167
				% of total ABC		
BALB/c	PC	OT	0.7	N.D.	N.D.	N.D.
		1°, day 6	17	90	-	-
		2°, day 7	7	86	-	-
A	PC	1°, day 7-8	9	-	-	-
BALB/c	DNP	2°, day 8	852	-	-	-

[a]Modified from (21).

[b]Spleens were tested after 1 or 2 injections (days 0 and 10) of 10^8 pneumococci (PC) or after 2 injections (days 0 and 10) of 500 µg of DNP-KLH i.p.

[c]The frequency of small lymphocytes bearing receptors with specificity for PC or DNP determinants was measured by exposing spleen cell suspensions to ~1 µg/ml of ^{125}I-PC-BSA or ^{125}I-DNP-BSA in fetal calf serum and detecting antigen binding to cell surfaces by autoradiography.

[d]The presence of idiotypic determinants T15, M603 and M167 on the surface of ABC was measured by the inhibition of ^{125}I-PC- or DNP-BSA binding to spleen cells by 10% anti-idiotype sera. N.D. signifies not done. - indicates less than 15% inhibition.

TABLE III
IDIOTYPIC DETERMINANTS ON IgM ANTI-PC ANTIBODY[a]

Strain	PFC	Time After Immunization[b]	IgM PFC/10^6 Cells	Idiotype[c] T15	M603	M167
				% of total PFC		
BALB/c	PC	peak 1°	650	85	-	-
		peak 2°	150	72	-	-
A	PC	peak 1°	484	-	-	-
		peak 2°	192	-	-	-
BALB/c	TNP	peak 1°	365	-	-	-

[a]Modified from (15).

[b]Animals were immunized i.p. with 1 or 2 injections (days 0 and 10) of 10^8 pneumococci or TNP-coated duck erythrocytes. The peak of the IgM PFC response varied from day 5 to day 8.

[c]Percent inhibition of PFC by anti-idiotypic antisera diluted 1/80. - signifies less than 15% inhibition.

to PC by measuring the inhibition of PC-specific plaques by soluble PC and its analogues. Inhibition of plaque formation by free hapten is an established procedure for determining the avidity and specificity of antibodies and is particularly useful for the analysis of IgM antibodies where conventional serologic techniques are more difficult.

Figure 4 demonstrates the specificity of the PC-specific PFC at various times after intravenous injection of pneumococci (15). It can be seen that at several time periods of the primary and secondary immune responses the relative specificities of the antibodies vary only slightly. Furthermore, the relative inhibitory concentrations of PC and its analogues GPC and C of BALB/c anti-PC antibodies resembles the inhibitory patterns seen with HOPC 8 (Figure 1). The differences in absolute concentrations of these choline analogues which effectively inhibit the anti-PC antibody and HOPC 8 are probably related to the different valences of the IgA myeloma compared to the IgM anti-PC PFC, rather than to different affinities of binding sites. It can also be seen that the BALB/c induced antibody differs markedly from the specificities of the other myeloma tumors. Thus, BALB/c anti-PC antibodies resemble the T15 group of myeloma proteins both antigenically and functionally. However, also shown in Figure 4 is the specificity of the anti-PC antibody induced in A strain mice. It can be seen that no significant differences can be detected between the specificities of the antibodies of A compared to BALB/c mice, in spite of the fact that the A mouse lacks the T15 variable region marker and therefore must have some variable region differences. More extensive analysis of the specificity of anti-PC antibodies in a variety of strains of mice have supported the general similarities in binding specificities. Thus, hapten inhibition studies of PC-specific antibodies by PC analogues were performed in 17 different strains, including representatives of all histocompatibility types, as well as heavy chain linkage groups. This study showed no significant differences in binding specificity of these antibodies (23). It was this observation, in fact, which stimulated our examination of idiotypic antisera in terms of fine specificity and led to the discovery that conventionally produced mouse anti-mouse idiotypic antisera, such as the anti-T15 reagents used in this study, did not recognize binding site determinants. Thus, heterologous antisera directed toward binding site regions were prepared and used to analyze the antigenic similarities among the binding site regions of anti-PC antibodies of mice (13).

A summary of idiotypic marker distribution on anti-PC

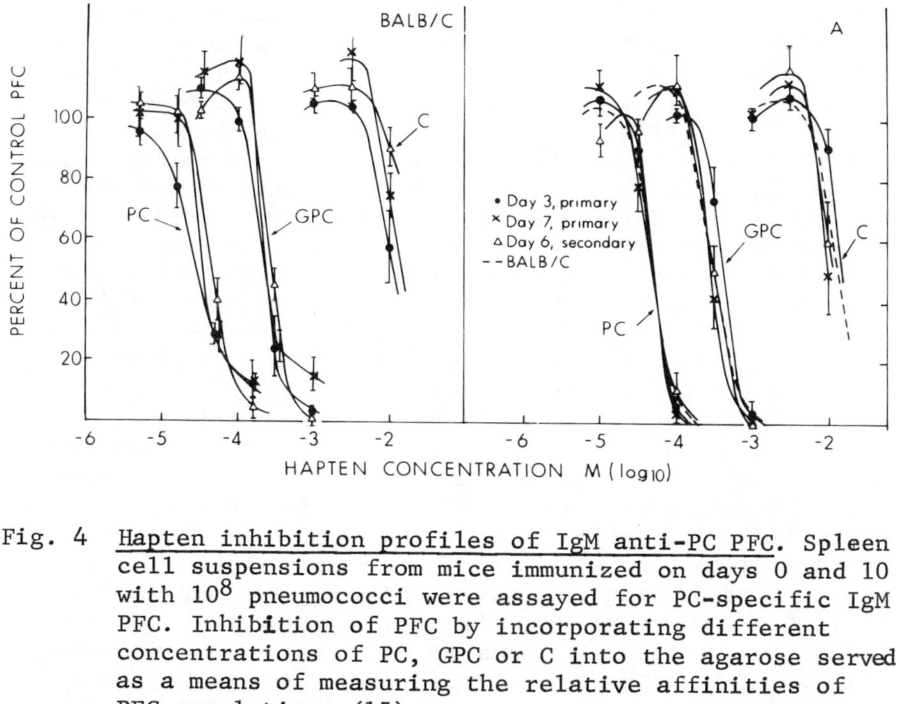

Fig. 4 Hapten inhibition profiles of IgM anti-PC PFC. Spleen cell suspensions from mice immunized on days 0 and 10 with 10^8 pneumococci were assayed for PC-specific IgM PFC. Inhibition of PFC by incorporating different concentrations of PC, GPC or C into the agarose served as a means of measuring the relative affinities of PFC populations (15).

antibodies in some representative strains of mice is shown in Table IV. It can be seen that the T15 variable region marker is limited to those strains of mice of the IgC_H a^1 type and is not found in strains of the other four heavy chain linkage groups (24). However, the binding site-specific variable region marker, $H8_s$, is found on all anti-PC antibodies regardless of the IgC_H type (24). In addition, the $H8_s$ marker is present on antibodies raised in wild Mus musculus. Thus, it appears that binding regions of mouse PC-specific antibodies have been strongly conserved while variations have occurred in non-binding site regions.

Our data support the following model (Figure 5). It is probable that all mice possess PC-specific clones which are derived from a single clone which existed in the distant past (prototype mouse). Variants of this clone have appeared and can be identified by the presence of particular variable region markers. Through inbreeding, the variants may become segregated into different strains of mice. We know of one such variant, namely the T15 positive strains, BALB/c and C58. It is probable that additional variants exist and experiments are in progress to define the antigenic characteristics of such variants. However, the variations which are presently known have not resulted in the change in binding site determinants, at least as detected by the binding specificity and by the presence of antigenic markers within the binding region. One could argue that variation which occurs in the binding site results in clones which no longer are stimulated by PC antigens and therefore escape detection. However, as yet, no such clones are known. This extreme conservation of binding characteristics is most easily explained by germ line or multi-gene theories of immunity, and would be difficult to explain in terms of those theories where hypermutation of the hypervariable regions is responsible for antibody diversity, unless one proposed that individual germ line genes differed with respect to the amount of hypermutation which is possible. Thus, the anti-PC response in mice could be an exceptional case and not typical of the kind of responses normally seen. We, therefore, extended our approach to an examination of the anti-PC responses in other species of animals.

A variety of species of rodents were immunized with pneumococci and the anti-PC antibody responses measured. All rodents tested gave substantial antibody responses, and at least in one species, the guinea pig, IgG, in addition to IgM, antibodies were elicited (23). Several interesting features were found in the immune responses of other species to PC. Like the responses in mice, variations in terms of

TABLE IV
STRAIN DISTRIBUTION OF V-REGION MARKERS ON ANTI-PC ANTIBODIES

Strain	IgC_H Type	V-region Markers		
		T15	$H8_s$	H_V-PC
BALB/c	1	+	+	+
C58	1	+	+	+
C57BL/6	2	-	+	+
DBA/2	3	-	+	+
A	4	-	+	+
CE	5	-	+	+
wild *M. musculus*	-	-	+	+

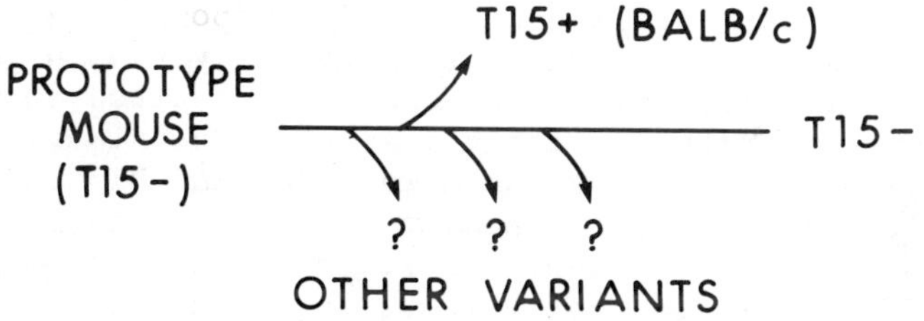

Fig. 5 Generation of variants of PC-specific clones in mice.

the hapten inhibition profile did not occur either as a time-dependent change within an individual (i.e. maturation) or between the responses of individual outbred animals. Secondly, pronounced species-specific differences were found between the fine specificity pattern of the anti-PC antibodies. This is demonstrated in Figure 6. It can be seen clearly that the relative inhibitory capacities of the various PC analogues differ between the species but are quite constant within a species. In view of the PC response in mice, this is an important finding for several reasons. It was conceivable, for example, that the similarities achieved in the binding specificity by the plaque assay in mice could have been due to a lack of sensitivity of the plaque assay. The fact that different myeloma tumors in mice and antibodies induced in different species of rodents produce different inhibition profiles speaks against this possibility. Secondly, it was possible that the similarity of binding profiles in mice was a reflection of the immunochemical simplicity of the PC determinant and that only limited possibilities with respect to PC recognition were possible. The varying species profiles again argue against this.

The finding that a variety of different rodent species respond to PC with restricted antibody responses is intriguing. While idiotypic determinants have not yet been generated to examine antigenic uniformity among the antibodies produced by different individuals within a species, considerable evidence exists concerning the structural and binding site similarities of anti-PC antibodies produced by outbred guinea pigs. Thus, detailed hapten inhibition studies of both IgM and IgG PFC have shown that both kinds of antibodies are restricted and do not show change in specificity or affinity with time after immunization. In addition, hapten inhibition profiles of different outbred animals are identical by this technique. Support for the structural similarity and restriction of IgG antibody comes from the isoelectric focusing pattern of guinea pig anti-PC sera (Figure 7). In this figure the antibody bands were visualized after isoelectric focusing by overlaying the acrylamide gel with agarose containing PC-coated erythrocytes, and the subsequent lysis of antibody-erythrocyte complexes by complement. By this technique, homogeneous antibodies will form 2-4 bands and heterogeneous antibodies, such as guinea pig anti-erythrocyte antiserum, form a diffuse smear. While the anti-PC bands shown in figure 7 may not be homogeneous, they, in fact, are strikingly restricted and may be derived from 1-3 clones. Experiments are underway to enumerate the number of antibody species to PC in guinea pigs, and to compare the

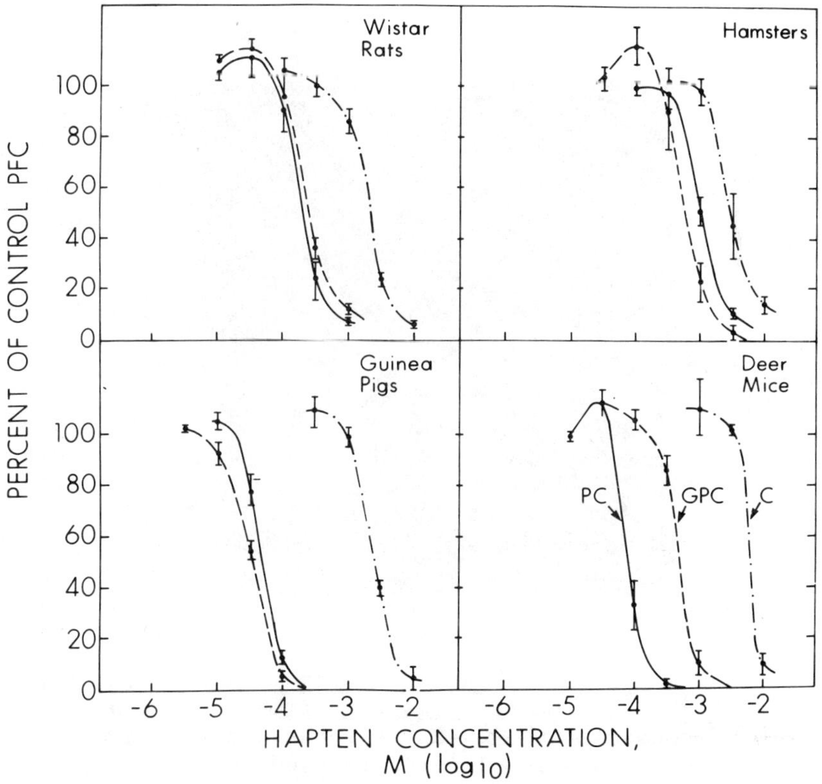

Fig. 6 Hapten inhibition profiles of IgM PC-specific PFC from different rodent species. Splenic IgM PFC were inhibited with PC, GPC or C incorporated into the agarose. The inhibition curves did not vary significantly among individual animals or at various times after immunization (23).

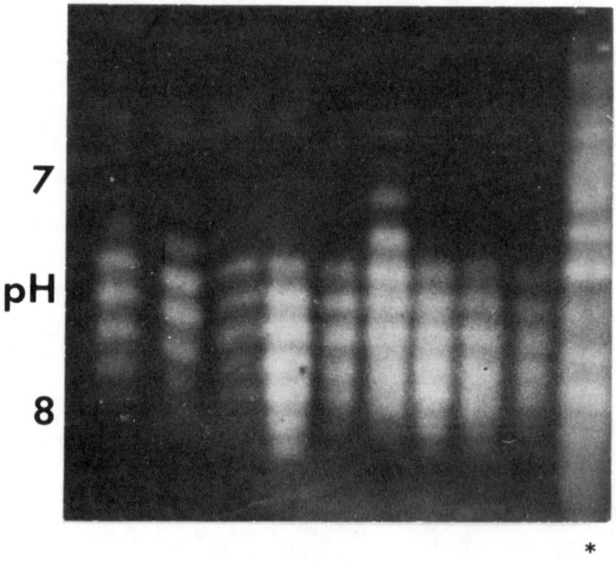

Fig. 7 Isoelectric focus patterns of guinea pig IgG anti-PC antibodies. Guinea pigs were immunized with 10^8 pneumococci in CFA. Three weeks later sera were collected and isoelectric focused in thin slabs of acrylamide. Anti-PC antibodies were detected by overlaying the acrylamide gel with agarose which contained PC-coated erythrocytes and allowing antibody to diffuse to the erythrocytes. The addition of complement caused lysis of antibody-coated erythrocytes which appear here as white bands. Shown here are sera from 9 outbred guinea pigs and a control, heterogeneous guinea pig anti-sheep erythrocyte serum (*).

isolated antibodies for functional, antigenic, and structural similarities.

SUMMARY AND CONCLUSIONS

To non-immunologists it must be difficult to understand how two diametrically opposed theories for the generation of antibody diversity can still be debated. In fact, each theory has strong experimental observations which support it as well as equally strong experimental findings which cannot be explained comfortably. A major fact which has led some to favor the somatic mutation theory has been the finding of "constant" determinants, such as allotypic markers (25) or species-specific residues (26), within the variable region of immunoglobulin molecules. This would suggest that, if multiple V region genes exist in the genome, then each gene must possess the same "constant" region determinants. The somatic mutation proponent would rather argue that only small numbers of variable region genes exist with the allotypic markers or species-specific residues and that mutation within their hypervariable region occurs to account for antibody diversity. These facts are reconciled to the germ line theorist by proposing the additional mechanisms of gene expansion and contraction, whereby massive gene duplication and selection occurs, all within the germ line genome. For that matter, our data concerning non-binding site variable region determinants in the mouse support both somatic mutation and germ line theories. However, it is difficult to reconcile our findings concerning the binding site antigenic determinants with hypermutation theories, since the somatic mutation theory proposes, of course, that most hypervariable regions will not be inherited. In fact, we find very strong inheritance of hypervariable regions both when examined by antigenic determinants as well as binding specificity. It should be pointed out that inheritance of fine specificity antigen binding patterns has been noted by other investigators with other haptens. Very striking inheritance of cross reactivity has been recently described by Mäkelä and his coworkers (27). At the same time, it should be clearly stated that structural comparisons between molecules must ultimately rely on amino acid sequence. We would suggest, however, that the anti-PC immune system will be a particularly important model system for defining the extent of antibody variability and should continue to provide a very interesting tool for the analysis of these genetic questions.

REFERENCES

1. Landsteiner, K., The Specificity of Serological Reactions. (Cambridge, Mass., Harvard Univ. Press, 1945).
2. Jerne, N.K., Proc. Natl. Acad. Sci. 41: 849, 1955.
3. Kreth, H.W. and Williamson, A.R., Eur. J. Immunol., 3: 141, 1973.
4. Hood, L. and Talmage, D.W., Science 168: 325, 1970.
5. Hood, L. and Prahl, J., Adv. Immunol. 14: 291, 1971.
6. Lennox, E.S. and Cohn, M., Ann. Rev. Biochem. 36: 365, 1967.
7. Cohn, M., Ann. N.Y. Acad. Sci., 190: 529, 1971.
8. Potter, M. and Lieberman, R., J. Exp. Med. 132: 737, 1970.
9. Leon, M.A. and Young, N.M., Biochemistry 10: 1424, 1971.
10. Cosenza, H. and Köhler, H., Science 176: 1027, 1972.
11. Cosenza, H. and Köhler, H., Proc. Natl. Acad. Sci. 69: 2701, 1972.
12. Claflin, J.L., Rudikoff, S., Potter, M. and Davie, J.M., (Manuscript submitted for publication)
13. Claflin, J.L. and Davie, J.M., J. Immunol. (In Press).
14. Claflin, J.L. and Davie, J.M. (Manuscript submitted for publication)
15. Claflin, J.L., Lieberman, R. and Davie, J.M., J. Immunol. 112: 1747, 1974.
16. Barstad, P., Rudikoff, S., Potter, M., Cohn, M., Konigsberg, W. and Hood, L., Science 183: 962, 1974.
17. Carson, D. and Weigert, M., Proc. Natl. Acad. Sci. 70: 235, 1973.
18. Weigert, M., Raschke, W.C., Carson, D. and Cohn, M., J. Exp. Med. 139: 137, 1974.
19. Williams, R.C., Kunkel, H.G. and Capra, J.D., Science 161: 379, 1968.
20. Kunkel, H.G., Agnello, V., Joslin, F.G., Winchester, R.J. and Capra, J.D., J. Exp. Med. 137: 331, 1973.
21. Claflin, J.L., Lieberman, R. and Davie, J.M. J. Exp. Med. 139: 58, 1974.
22. Lee, W., Cosenza, H. and Köhler, H. Nature 247: 55, 1974.
23. Claflin, J.L. and Davie, J.M., J. Immunol. (In Press).
24. Claflin, J.L. and Davie, J.M., J. Exp. Med. 140: 673, 1974.
25. Mole, L.E., Jackson, S.A., Porter, R.R. and Williamson, J.M., Biochem. J. 124: 301, 1971.
26. Capra, J.D., Wasserman, R. and Kehoe, J.M., J. Exp. Med. 138: 410, 1973.
27. Imanishi, T. and Mäkelä, O., Eur. J. Immunol. 3: 323, 1973.

RECOGNITION IN MLC AND CML: THE LD-SD DICHOTOMY

Fritz H. Bach, Marilyn L. Bach, Barbara J. Alter, Kirsten F. Lindahl, Dolores J. Schendel and Paul M. Sondel
Departments of Medical Genetics and Surgery, Pediatrics, and the Immunobiology Research Center, The University of Wisconsin, Madison, Wisconsin

The mixed leukocyte culture (MLC) test was described ten years ago by Bain, Vas, and Lowenstein (1) and by Bach and Hirschhorn (2). Since that time it has been used for immunogenetic analysis of histocompatibility antigens [primarily those of the major histocompatibility complex (MHC)] for clinical matching of donor and recipient for transplantation, for studies of lymphocyte cell populations and other facets of cell-mediated immune reactions and for a variety of biochemical studies. Four years ago a second test, the cell-mediated lympholysis (CML) test (3-6), was added to our armamentarium of *in vitro* models of cell-mediated immunity. Whereas the MLC test (usually assayed by studying the incorporation of radioactive thymidine into dividing cells in culture) can be thought of as an assay of recognition by T lymphocytes of alloantigenic differences leading primarily to cell proliferation, the CML test is an assay of cell destruction. We shall use the terms MLC and CML to refer to the proliferative and cytotoxic reactions respectively unless otherwise noted. The "targets" for cytotoxic lymphocytes are recognized by the lymphocytes which will become the cytotoxic cells both during the sensitization phase in MLC and at the time of cell destruction in CML.

Our purpose in this paper is to discuss *recognition* in MLC and CML. This problem can be approached from several perspectives; we shall stress two: the complex genetic control of alloantigenic differences which are recognized and the cellular response that follows recognition. Studies which bear on this topic have been performed primarily in mouse and man; it appears in terms of the broad biological principles which may emerge from these studies that one can freely extrapolate from one species to the other. A major focus of

---

This study was supported by NIH grants GM-15422, CA-14520, AI-08439, AI-11576, NF-MOD grant CRBS 246; MLB is a recipient of the American Cancer Society Faculty Research Award; DJS is an NIH Trainee supported by NIH grant GM-00398. This is paper no. 1814 from the Laboratory of Genetics and paper no. 19 from the Immunobiology Research Center, The University of Wisconsin, Madison, Wisconsin, 53706.

our discussion will be on the genetic control of alloantigens which is more fully understood and more directly analyzed in mouse.

The following terminology will be used. MHC will refer to the major histocompatibility complex, HL-A in man and H-2 in mouse. SD antigens are those antigens detected serologically that are determined by MHC loci (LA, FOUR and AJ in man; H-2K and H-2D in mouse) and which are present on essentially all tissues of the body. LD determinants are those differences of the MHC which lead to proliferation of cells in MLC. Whether the LD differences are currently being recognized serologically will be discussed later in this paper. [The letters Hld (histocompatibility LD) are used for the formal designation of H-2 loci governing the LD antigens, since the letters LD have been used for another locus in mouse.] Finally, we shall refer to the LD factors as "antigens" on the basis that they elicit a clonal cellular response (7-8), and that it has been possible to make animals tolerant to LD (9).

GENETIC CONTROL OF ALLOANTIGENS

Genes Important in MLC Activation

Very early studies in man (10), mouse (11) and rat (12) showed that the genes that led to proliferative events in MLC were to a large extent restricted to the MHC, although it must be noted that in mouse two other genetic systems can, if different in two strains, lead to proliferation in MLC (13,14). Since MHC antigens were first described by serological techniques in both man and mouse, it was assumed that these serologically defined (SD) antigens were responsible for activation in MLC. It is now well established that there are genes of the MHC, other than the SD genes, which are of prime importance in leading to proliferation of lymphocytes in MLC, the evidence for this in humans evolving over several years (15-19). These are the MHC LD genes.

Studies by Widmer (20-22), using mouse strains developed by Snell, Stimpfling and others, led to the mapping of the LD loci in the mouse. Figure 1 gives the genetic maps of the MHC in several species including mouse. There is one locus (loci) in the I region of H-2 which is the strong Hld locus. A second relatively weaker Hld locus has been formally mapped between Ss and H-2D (23). It is quite likely, however, that there are still other loci yet to be discovered in the H-2 complex that should be designated as Hld loci on the basis of their being associated with stimulation in MLC. An analysis of the data obtained in these studies shows the following: first, strong stimulation was observed in strain combinations differing for only the I and S regions, identi-

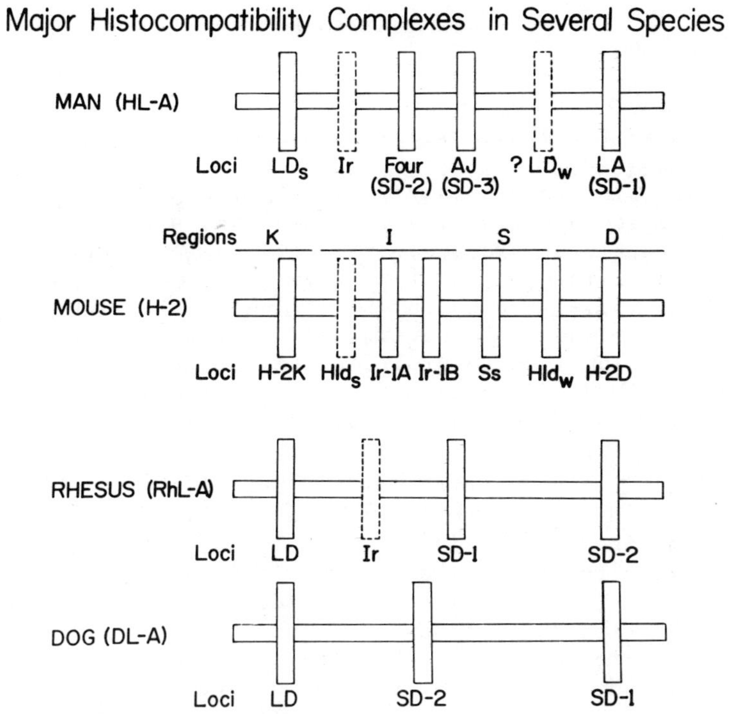

Figure 1. See legend on page 176.

Table 1

Summary of MLC Responses

MHC Regions Which Are Different	Ratio of Stimulation: Range*	Ratio of Stimulation: Average
K, Ir-1, Ss-Slp, D	1.2 - 33.6	7.2
K, Ir-1, Ss-Slp	1.4 - 15.7	6.1
K, Ir-1, D	3.3 - 20.5	7.0
K, Ss-Slp, D	1.5 - 8.6	3.3
Ir-1, Ss-Slp, D	2.7 - 15.1	8.3
K, Ir-1	3.2 - 18.3	6.6
K, Ss-Slp	No experiments	
Ir-1, Ss-Slp	2.7 - 12.8	5.8
K, D	3.0 - 3.8	3.4
Ir-1, D	No experiments	
Ss-Slp, D	0.7 - 4.7	2.0
K	0.8 - 2.2	1.4
Ir-1	No experiments	
D	0.8 - 5.4	1.8
Ss-Slp	0.6 - 4.9	2.0
None	0.6 - 1.9	1.2

*These numbers represent the lowest and highest ratios of stimulation noted.

Fig. 1 A schematic representation of the major histocompatibility complex in four species. In man, there are three loci identified, LA, Four and AJ, alleles of which determine the serologically defined antigens. The locus, differences at which lead to strong activation in MLC, is mapped as LD_S; the LD_W locus, if it does exist, maps between the AJ and LA loci. The Ir locus (loci) are listed in a dotted line since the position of this locus is not known.

The mouse H-2 complex is divided into four regions, K, I, S, and D, each designated by a marker locus, H-2K, Ir-1A, Ss and H-2D. The alleles of the H-2K and H-2D loci determine the classically serologically defined H-2 antigens. The loci differences of which lead to MLC activation are designated as Hld_S and Hld_W respectively; the Hld designation is used since the initials LD have been used for another locus in the mouse genome. The Ir-1A and Ir-1B loci determine immune response of the animal and the Ss locus controls the quantitative levels of a serum protein. Differences of the two LD loci result in activation of lymphocytes in mixed leukocyte culture and are associated with graft versus host reactions in vivo; the alleles of the Hld_S locus are relatively stronger in this regard than alleles of the LD_W locus; differences of LD_S locus are also associated with skin graft rejection without concominant H-2K or H-2D disparity. The Hld_S locus has not been formally separated from the Ir-1A locus.

The order of loci in rhesus and dog are shown. The existence of immune response loci in the RhL-A chromosome of rhesus has been documented; however the position of these loci with respect to the LD and SD loci has not been clearly elucidated. Only in mouse is the position of the centromere known.

cal for the SD antigens; second, relatively weaker stimulation was seen when strains differed for only the SD antigens were identical for the I and S regions (in some of these cases no stimulation was detected); third, when mice differed for the I region plus either K alone or plus other regions, there was on the average stronger stimulation than when there was no I region disparity. These results are shown in Table 1.

These studies, which were confirmed by Meo, Shreffler, et al. (24,25), in aggregate suggested that LD differences in the I region were of the greatest import in leading to MLC activation and that the SD antigens were either not stimulatory at all or only weakly so. More recent work using heat treatment of the stimulating cells (to be discussed later) demonstrates that it is possible to differentially inactivate the stimulating cell so that it no longer induces a proliferative response but the SD antigens are still expressed. We would not want to conclude from these data that the SD antigens are, in the absence of concurrent LD differences, unable to lead to a proliferative response in MLC; however, it would seem that if they can do so, they are certainly weaker in this regard than the LD differences.

Genes Important in CML

At first examination, in culture systems which were less sensitive than those used now, H-2 LD differences in the presence of SD identity led to excellent stimulation in MLC but no cytotoxicity in CML. More recently, using a more sensitive MLC technique (26), Peck, Schendel and Alter (27-30) have demonstrated low level CML in some of these combinations. We shall return to a discussion of this later.

Rather than using LD as the prime target in CML, lymphocytes activated in MLC to both LD and SD differences appeared to recognize primarily the SD antigens themselves or the products of genes very closely linked to those determining the SD antigens as the targets in CML. Because of this genetic inseparability, we shall refer to the CML targets as the SD antigens. Data from an experiment demonstrating this phenomenon are shown in Table 2, confirming similar findings in a human family. Cells activated to LD plus SD differences become cytotoxic to those target cells carrying the SD antigens but not to those carrying LD.

Recent experiments of K.F. Lindahl (30) using xenogeneic combinations are an example of differential cytotoxicity on LD and SD which dramatically demonstrates the difference between these two determinants leading to CML. In these experiments human lymphocytes sensitized to mouse cells are tested for their ability to lyse different mouse target cells.

TABLE 2

	MLC Sensitization			CML Assay		
	MLC (mean cpm ± SD)	Responding Cell (Effector)	Stimulating Cell (Sensitizing)	Target Cell	^{51}Cr released (mean cpm ± SD)	% CML
A[1]	14,819 ± 1406	B10.A (KKDD)	AQR (QKDD)	B10.A (KKDD)	469 ± 11	- 4.6
				AQR (QKDD)	372 ± 36	- 3.6
	44,777 ± 5237	B10.A (KKDD)	B10.T(6R) (QQQD)	B10.A (KKDD)	484 ± 21	- 1.4
				B10.T(6R) (QQQD)	534 ± 19	37.5
				AQR (QKDD)	623 ± 19	42.6
				C57BL/10 (BBBB)	383 ± 33	13.8
	14,487 ± 846	B10.A (KKDD)	B10.A (KKDD)	B10.A (KKDD)	417 ± 54	-15.4
B[2]	13,725 ± 2236	B10.A(2R) (KKDB)	B10.A (KKDD)	B10.A(2R) (KKDB)	532 ± 142	- 4.9
				B10.A (KKDD)	552 ± 13	5.3
	79,236 ± 6902	B10.A(2R) (KKDB)	B10.D2 (DDDD)	B10.A(2R) (KKDB)	889 ± 67	24.9
				B10.D2 (DDDD)	1541 ± 140	72.3
				B10.A (KKDD)	1041 ± 32	78.5
				C57BL/10 (BBBB)	1232 ± 43	40.6
	10,423 ± 623	B10.A(2R) (KKDB)	B10.A(2R) (KKDB)	B10.A(2R) (KKDB)	434 ± 7	-13.1

The % CML is based on the following spontaneous release (SR) and maximum release (MR) values (mean of triplicates SD) for each target cell:

[1] B10.A SR = 491 ± 20 MR = 974 ± 61; AQR SR = 391 ± 24 MR = 935 ± 45; B10.T(6R) SR = 328 ± 25 MR = 877 ± 66; C57BL/10 SR = 721 ± 73 MR = 1978 ± 30.

[2] B10.A(2R) SR = 591 ± 23 MR = 1786 ± 60; B10.A SR = 516 ± 14 MR = 1184 ± 59; B10.D2 SR = 684 ± 47 MR = 1869 ± 34; C57BL/10 SR = 721 ± 73 MR = 1978 ± 30.

Combinations were selected to analyze whether both LD and SD antigens could function as targets or whether one was predominant in this regard. The results show that the SD antigens serve as excellent xenogeneic CML targets; in contrast there was no evidence that LD antigens were recognized by the xenogeneic effector lymphocytes. An example of such an experiment is given in Table 3. Human lymphocytes were sensitized to each of four different mouse strains, B10.A(4R), B10.A(1R), AQR and B10.T(6R). Since the prime targets in such xenogeneic CML tests are H-2 determined (32) we can focus on the H-2 phenotypes of these four strains. The two combinations of 4R - 1R and AQR - 6R are SD identical but differ for LD. AQR and 1R are I and S region identical and SD different. Human effector cells sensitized to AQR or 6R, for instance, are highly cytotoxic to AQR. These two strains are SD identical. On the other hand, the same human lymphocytes sensitized to 1R do not lyse AQR target cells, demonstrating the I and S region components of 1R which are shared by AQR are not recognized as targets. Results on the three other target cells are consistent with those on AQR.

LD - SD COLLABORATION IN CML

While LD differences by themselves or SD differences by themselves on the stimulating cells in MLC do not lead to strong CML on any target cell, the presence of both LD and SD does allow the effective generation of cytotoxic cells. This has been referred to as LD - SD collaboration. This collaboration is most easily studied in three-cell experiments where the determinants are presented on separate stimulating cells. An example of such a three-cell experiment is presented in Table 4. B10.T(6R) is the responding population and LD sensitization is provided by AQR_m cells. $B10.G_m$ cells present SD determinants; in this case SD disparity is associated with H-2D locus differences. Sensitization of B10.T(6R) to only LD (AQR_m) produces strong proliferation but generates no CML. SD sensitization by $B10.G_m$ alone shows minimal proliferation and cytotoxicity. In three-cell culture where both LD and SD determinants are used for simultaneous sensitization, one observes both MLC proliferation and CML. Following the consistently seen pattern, cytotoxic activity is only detected when effector cells are incubated with B10.G (SD-different) targets.

This experiment provides an example of a situation in which sensitization by either LD or SD antigens alone does not stimulate strong cytotoxic potential, but together they collaborate to generate specific CML which is directed towards SD targets. This experiment, together with others previously

TABLE 3

SD SPECIFICITY OF HUMAN EFFECTOR CELLS

Human lymphocytes sensitized to	H-2 haplotype K	I	S	D	% CML on targets from 4R	1R	AQR	6R
B10.A(4R)	k	k/b	b	b	43.4 ± 6.9	29.4 ± 10.0	8.8 ± 7.3	6.5 ± 12.4
B10.A(1R)	k	k	d	b	46.3 ± 6.2	23.5 ± 6.3	- 6.8 ± 11.5	- 1.4 ± 14.2
AQR	q	k	d	d	32.5 ± 6.1	16.2 ± 8.9	43.8 ± 11.3	31.3 ± 11.5
B10.T(6R)	q	q	q	d	24.9 ± 4.4	13.5 ± 7.2	36.6 ± 11.2	34.6 ± 12.4

CML assay was carried out for 3 hours with 70 effector cells:1 target cell. All assays were done in triplicate.

$$\% \text{ CML} = \frac{\text{Exp. release - control release}}{\text{Max. release - control release}} \times 100$$

Control release is the amount of ^{51}Cr released with unstimulated (HH_m) effector cells.

4R: M.R. - 944 ± 50 cpm; C.R. - 500 ± 18 cpm;
1R: M.R. - 1265 ± 28 cpm; C.R. - 556 ± 43 cpm;
AQR: M.R. - 1133 ± 48 cpm; C.R. - 571 ± 60 cpm;
6R: M.R. - 1309 ± 26 cpm; C.R. - 621 ± 89 cpm.

TABLE 4

A THREE-CELL EXPERIMENT

Effector combination	Genetic stimulation	MLC (mean cpm ± SD)	% CML (mean ± SD) AQR	B10.T(6R)	B10.G
B10.T(6R) (qqqd) + AQR_m (qkdd)	LD	16,817 ± 763	-3.1 ± 2.2	-3.9 ± 0.91	
B10.T(6R) + $B10.G_m$ (qqqq)	SD	5,852 ± 480	-1.4 ± 2.0		4.3 ± 1.5
B10.T(6R) + AQR_m + $B10.G_m$	LD + SD	24,334 ± 508	-0.48 ± 2.6	-1.9 ± 1.9	21.4 ± 2.2
B10.T(6R) + $B10.T(6R)_m$	--	2,473 ± 152		-2.2 ± 1.4	

described (33) demonstrates LD - SD collaboration with SD antigens of either H-2K or H-2D stimulating cytotoxic effector cells.

Initial studies with many SD combinations differing only at K or D failed to generate cytotoxic lymphocytes. Using altered culture techniques it is now possible to obtain CML effectors in most SD combinations. Even in situations where stimulation by SD alone can lead to CML, a collaborative effect with stimulation by both LD and SD during the MLC sensitization phase produces an enhanced CML response compared to that detected when only SD differences are used for stimulation. Figure 2 illustrates the cytotoxic potential of cultures stimulated by SD region differences alone [AQR + B10.T(6R)] and LD plus SD differences [AQR + B10.T(6R)$_m$ + B10.A$_m$)]. Data from AQR + B10.T(6R)$_m$ cultures (LD alone) are not presented, as no significant CML response is detected.

At any effector-to-target-cell ratio, stimulation with (LD + SD) differences cause approximately 20% more CML of SD targets. The number of effectors required to cause 50% lysis of a given number of targets is 4.2-fold less in LD enhanced cultures compared to the cultures sensitized to SD alone. Not only is the specific activity of effector populations different in the two cultures, but there is also a two-fold difference in the number of effector cells recovered per culture. Using the preceding two values, approximately eight-fold more effector cells are generated in this experiment by (LD + SD) stimulation compared to SD stimulation alone.

Is collaboration by LD an essential event or only a helper effect? Examples have been cited where neither LD nor SD alone stimulate CML but combined sensitization does; such results would indicate an obligate need for LD. However, stimulation by SD region differences alone can also produce CML responses under some conditions (Fig. 2). It must be emphasized again that when we speak of two strains having SD differences we are speaking of strains that differ for entire SD regions, and certainly only part of the genetic material contained in such a region is coding for SD antigens. Clearly, SD region differences alone are not totally incapable of activating MLC proliferation (34). The SD regions themselves may determine quantitatively weaker LD differences that can both stimulate proliferation and collaborate with SD antigens of the same region to produce CML. We have obtained evidence from one series of experiments which provide data consistent with the hypothesis that there are in fact LD-like determinants in the K region of B10.A which are recognized by AQR and collaborate with SD antigens to produce CML (28) as described

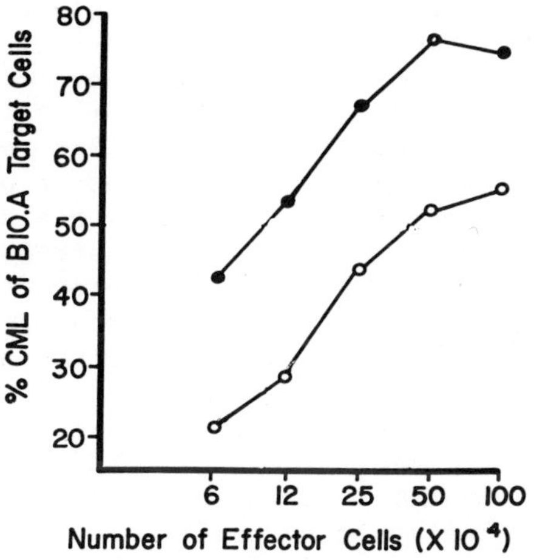

Fig. 2 The cytotoxic activity of effector cells stimulated by an SD region alone (o) and stimulated by LD + SD (●). Different numbers of effector cells are combined with a constant number of target cells (1 x 10^4).

below.

In human CML studies Eijsvoogel et al. (34) demonstrated that heat treatment of an allogeneic stimulating cell destroyed its ability to cause strong MLC proliferation but left its ability to sensitize for CML, provided an LD difference was supplied by another non-heat-treated cell. In addition, heat treatment has been shown to have no effect upon the ability of cells to absorb anti-SD antibody. These results indicate that heat treatment may inhibit or eliminate LD stimulatory potential but leaves SD relatively unaffected. We wished to focus particularly upon the question of whether AQR and B10.A (SD region different) have weak LD differences determined by the SD region which collaborate to produce CML effectors.

The rationale of the experiment was to use heat treatment to damage any LD differences on the B10.A stimulating cell while leaving SD antigens intact. One could then determine whether SD antigens themselves produce CML effectors. Table 5 shows results of such a heat treatment experiment. The first set of data shows control values for the normal experimental system using mitomycin C to inactivate stimulating cells. The SD combination under these conditions gives positive proliferation and cytotoxicity, and CML is enhanced in the three-cell combination where both LD and SD are used as stimuli. When heat treatment is substituted for mitomycin C inactivation essentially all CML activity is lost. The last two lines show responses when both treatments are combined. When the LD stimulating cell is heat treated and the SD stimulating cell is inactivated with mitomycin C, CML is not significantly different from that seen with SD stimulation alone in the normal system (25% vs 22%). More interestingly, though, when the SD stimulating cell is heat treated and the LD stimulating cell is mitomycin C inactivated, CML is as high as that seen in normal three-cell cultures (43% vs 40%). Cell populations stimulated by heat treated lymphocytes differing by SD alone or by both LD and SD are not cytotoxic, but CML can be restored by adding back a normal LD stimulus on a mitomycin C treated cell. This indicates that in AQR + B10.A the SD region difference is still expressed in such a manner that it can sensitize for target recognition, but it is critically affected in its ability to produce cytotoxic effector cells without an added normal LD stimulus. Thus heat treatment may actually be damaging an LD difference which is controlled by a locus closely linked to the H-2K locus of B10.A and located within the SD (K) region. If this interpretation is correct it would suggest that collaboration between LD and SD is essential in generation of CML. Alternatively one

TABLE 5

THE EFFECT OF HEAT-TREATMENT OF STIMULATING CELLS ON THE GENERATION OF CYTOTOXIC RESPONSES

Effector combination	MLC activation* (mean cpm ± SD)	% CML** (mean ± SD)
AQR + B10.A_m	4113 ± 190	21.8 ± 2.2
AQR + B10.A_m + B10.T(6R)$_m$	8292 ± 309	40.4 ± 2.8
AQR + B10.A_Δ***	4332 ± 89	3.1 ± 1.9
AQR + B10.A_Δ + B10.T(6R)$_\Delta$	5771 ± 301	-3.2 ± 2.6
AQR + B10.A_m + B10.T(6R)$_\Delta$	8771 ± 287	25.3 ± 3.5
AQR + B10.A_Δ + B10.T(6R)$_m$	17375 ± 426	42.7 ± 2.9

* The AQR + AQR_m cultures gave 2926 ± 254 cpm and AQR + AQR_Δ gave 2316 ± 107 cpm.

** The % CML of the B10.A target cell is presented.

*** Δ = 45°C for 60 mins.

could argue that heat treatment affects only one molecule (i.e. that carrying the SD determinants). Under this model heat treatment would cause relatively little change of the SD antigens which are important in generation of cytotoxic cells but would inactivate LD-like determinants on the same molecule. Lastly one must rule out that the SD determinants per se are responsible for both reactions. Nevertheless, heat treatment allows operational separation of LD and SD-type activity.

THE H-2 MUTANTS

The most perplexing findings in relation to the genetic control of MLC and CML come from studies with two mutants. One mutant, discovered by Dr. Donald Bailey, carries a spontaneous mutation in the $H\text{-}2^b$ chromosome of the C57BL/6By mouse which leads to reciprocal skin graft rejection with its parent strain; nevertheless, upon reciprocal immunization no antisera defining SD-type antigens had been evoked. Recent evidence suggests that the amount of the H-2K SD antigen in the two strains is different (35) or there may even be a qualitative difference (36). It is difficult to account for the two-way MLC stimulation and CML on the basis of quantitative differences in the amount of the H-2K SD antigen, however. This mutant could be interpreted as showing that LD differences lead to strong cytotoxicity. Similarly the mutant of Egorov, a spontaneous mutation apparently affecting the H-2D locus product of the $H\text{-}2^d$ chromosome, evokes both MLC reactivity and skin graft rejection with the parental strain (37). This might be taken as pure SD-type differences causing strong proliferation. One must be cautious in interpreting these mutations as simple events affecting only LD or SD. It is conceivable that these mutations, while spontaneous, are not point mutations but represent something more complex which affects multiple closely-linked loci. Alternatively LD and SD phenotypic products may interact at the molecular level. Such an interaction could lead to allosteric phenomena whereby a change in configuration of one molecule could lead to a configurational change in the second molecule. Based on the above arguments the mutants may well fit into the LD-SD model we have presented in this paper; by themselves they do not help us critically dissect the relative roles of LD and SD.

CELLULAR BASIS OF THE MLC & CML RESPONSES

The LD - SD dichotomy discussed above posed a problem regarding the cell populations responding to the two types of stimuli. It is known that both the response to LD stimuli as measured by proliferation in MLC and the cytotoxic response in CML are clonally distributed. LD - SD cooperation may involve two subpopulations of lymphocytes. Given two separate responding subpopulations, we might postulate that one population, a proliferating helper cell, responds primarily to LD-type stimuli; this response may then enable a second population, the eventual cytotoxic lymphocyte, to react to the appropriate target (38,39,40). Whether it is proliferation of the first population or a separate response (which might be distinct from the proliferative events themselves) which is essential to allow the second population to respond is a subject for further experimentation.

The cooperative response to LD and SD could be based on a single responding cell population, since the two-cell model has not been critically proven (40). One could either hypothesize that a single responding cell has two types of receptors, one recognizing LD differences and the other SD differences, or that a single receptor recognizes both LD and SD. LD recognition and response might be essential before the cell can recognize SD differences and develop into a cytotoxic lymphocyte. Given this model adsorption studies suggest (39) that within the responding population to any one allogeneic stimulus there are many cells which respond to LD differences, only a fraction of which differentiate to become cytotoxic lymphocytes. Alternatively one might hypothesize, as M. Cohn has done for antibody production, that an associative recognition takes place. The LD different stimulating cell may recognize the eventual cytotoxic effector cell and give it a signal which permits it to recognize the SD difference.

Since MHC LD differences appeared to function as a less effective target for cytotoxic lymphocytes, the question arose whether this was true because LD did not effectively stimulate lymphocytes to become cytotoxic effector cells or whether the cytotoxic cells did develop, but the LD antigens did not serve as a good "target". Alternatively, the LD antigens may be present on so few cells that the destruction of these cells is not detected in CML; this will be discussed later. Alter (41) has used the system developed by Forman and Möller (42) in an attempt to answer this question. These authors demonstrated that lymphocytes sensitized to alloantigens in MLC are cytotoxic not only to target cells carrying those same antigens, but even to isogeneic (self) targets if phytohemagglutinin (PHA) is added to the mixture of sensitized cells and target cells during the cytotoxicity assay (PHA-dependent cytotoxicity). This autokilling is presumably due to agglutination of the killer cell with the target and not to recognition of self-antigens. To the extent that PHA-dependent cytotoxicity can be used as a model for assessing the presence of cytotoxic lymphocytes in CML, we have used this system to determine whether proliferative responses to LD antigens alone can "generate" cytotoxic cells.

Results of two of the six experiments performed are shown in Tables 6 and 7. In Table 6 are given results with PHA stimulated blast cells as target cells; in Table 7, fresh lymphocytes are used. In each table MLC results are given using cells from strains differing by H-2 LD and SD factors and MLC mixtures differing by LD factors alone. The cytotoxic assay is carried out either as in normal CML tests or in the presence of PHA (columns 1 and 2, respectively, under "percent cytotoxicity").

Cultures activated with cells which differ for both LD and SD components generate cytotoxic lymphocytes active against the specific target. In spite of significant and similar MLC activation, combinations which differ only by LD show little if any lysis of specific "targets" [AQR or B10.T(6R)] or of unrelated targets bearing LD and SD differences from the sensitizing cells in CML.

The level of self killing when PHA is added to the cultures during the three-hour incubation depends upon the genetic differences existing during the MLC activation phase in the experiment: LD + SD activated cultures generate greater PHA-dependent cytotoxic activity than when the MLC is activated by LD differences alone, despite equivalent levels of proliferative response.

These data demonstrate that if both SD and LD differences are present during the MLC sensitization phase, the effector cells generated are cytotoxic not only to sepcific target cells in CML but also to isogeneic and other target cells in PHA-dependent cytotoxicity. However, with only LD differences present during the MLC sensitization procedure (identity for the SD antigens) despite an approximately equivalent proliferative response in MLC, the cells are only minimally cytotoxic to any target cells, even in the presence of PHA. This is true both with target cells stimulated three days previously with PHA and with fresh target cells. To the extent that PHA-dependent cytotoxicity can be used to argue about cytotoxic lymphocyte formation in general these results suggest strongly that LD differences present during MLC sensitization do not alone lead to the effective generation of effector cells.

SEROLOGICAL DEFINITION OF THE LD ANTIGENS

In both mouse (43-45) and man (46) there is evidence that products of the MHC genes other than the SD antigens can

Table 6. PHA Dependent Cytotoxicity with PHA-Blast Lymphocytes as Targets

MLC[a]	Sensitization of Effectors[b]	Target	% Cytotoxicity ± SD[d]	
			No PHA Added During CML Assay	PHA Added During CML Assay
	LD + SD			
19819 ± 439	B10.T(6R) + $C57BL/10_m$[c]	AQR	2.3 ± 2.4	36.3 ± 3.6
		B10.T(6R)	-8.0 ± 1.7	31.8 ± 4.7
		C57BL/10	68.2 ± 4.1	77.9 ± 3.5
13303 ± 1075	AQR + $C57BL/10_m$	AQR	0.6 ± 0.8	40.1 ± 0.9
		B10.T(6R)	-7.1 ± 1.7	36.8 ± 2.6
		C57BL/10	71.8 ± 4.4	68.9 ± 4.0
	LD			
12205 ± 604	B10.T(6R) + AQR_m	AQR	-4.4 ± 1.6	1.7 ± 1.2
		B10.T(6R)	-14.7 ± 1.7	-3.1 ± 1.7
		C57BL/10	-11.4 ± 3.6	-1.6 ± 5.4
14771 ± 756	AQR + $B10.T(6R)_m$	AQR	-4.9 ± 1.8	5.4 ± 2.4
		B10.T(6R)	-9.5 ± 1.7	2.8 ± 2.5
		C57BL/10	-11.2 ± 3.8	1.4 ± 5.1

legend for Table 6

[a]Mean cpm of triplicate values ± SD; control values are B10.T(6R) + $B10.T(6R)_m$ = 761 ± 86; AQR + AQR_m = 1577 ± 47.

[b]We refer to the four regions of the MHC as K, I, S, D. The loci, alleles of which determine the SD antigens are in the K and D regions; the strong LD locus appears to be in the I region. AQR = qkdd; B10.T(6R) = qqqd; C57BL/10 = bbbb.

[c]Reciprocal combinations (not shown) were also done with a comparable pattern of results.

[d]Percent cytotoxicity is calculated

$$\frac{ER - SSR}{MR - SSR} \times 100$$

where experimental release (ER) represents mean counts per minute (cpm) released from ^{51}Cr labeled target cells incubated with effectors sensitized to either LD or LD + SD differences; specific spontaneous release (SSR) is the mean cpm released from the target when incubated with cultured cells syngeneic to the target, e.g. AQR + AQR_m/AQR target; maximum release (MR) represents the mean cpm released from target cells after rapid freeze-thaw treatment.

Target AQR, MR = 2063 ± 5.6, SSR = 570 ± 9.5, SSR (PHA added during CML assay) = 355 ± 13.4

Target B10.T(6R), MR = 2967 ± 162.3, SSR = 818 ± 21.0, SSR (PHA added during CML assay) = 571 ± 28.5

Target C57BL/10, MR = 2582 ± 66.0, SSR = 823 ± 52.5, SSR (PHA added during CML assay) = 549 ± 104.9

Table 7. PHA Dependent Cytotoxicity with Fresh Lymphocytes as Targets

MLC[a]	Sensitization of Effectors[c]	Target	% Cytotoxicity ± SD[e] No PHA Added During CML Assay	PHA Added During CML Assay
	LD + SD			
16145 ± 826	B10.T(6R) + $C57BL/10_m$[d]	AQR	2.1 ± 4.6	37.0 ± 5.7
		B10.T(6R)	-1.0 ± 4.5	28.4 ± 4.4
		C57BL/10	55.4 ± 5.0	70.5 ± 2.8
9391 ± 433	AQR + $C57BL/10_m$	AQR	5.3 ± 3.3	40.0 ± 5.1
		B10.T(6R)	3.4 ± 4.4	37.1 ± 2.9
		C57BL/10	52.0 ± 7.1	59.2 ± 5.9
	LD			
10097 ± 274	B10.T(6R) + AQR_m	AQR	1.4 ± 4.7	10.8 ± 6.7
		B10.T(6R)	-4.4 ± 5.6	5.0 ± 3.9[b]
		C57BL/10	3.1 ± 6.2	7.6 ± 5.5
8253 ± 317	AQR + $B10.T(6R)_m$	AQR	0.4 ± 4.6	5.8 ± 4.7
		B10.T(6R)	8.7 ± 4.1	12.4 ± 3.0
		C57BL/10	3.7 ± 4.5	12.6 ± 5.6

legend for Table 7

[a]Mean cpm of triplicate values $\pm$ SD; control values are B10.T(6R) + $B10.T(6R)_m$ = 1128 $\pm$ 88; AQR + AQR_m = 2223 $\pm$ 64.

[b]% cytotoxicity of duplicate values; all other data reflect triplicate values.

[c]We refer to the four regions of the MHC as K, I, S, D. The loci, alleles of which determine the SD antigens, are in the K and D regions; the strong LD locus appears to be in the I region. AQR = qkdd; B10.T(6R) = qqqd; C57BL/10 = bbbb.

[d]Reciprocal combinations (not shown) were also done with a comparable pattern of results.

[e]Percent cytotoxicity is calculated

$$\frac{ER - SSR}{MR - SSR} \times 100$$

where experimental release (ER) represents mean counts per minute (cpm) released from ^{51}Cr labeled target cells incubated with effectors sensitized to either LD or LD + SD differences; specific spontaneous release (SSR) is the mean cpm released from the target when incubated with cultured cells syngeneic to the target, e.g. AQR + AQR_m/AQR target; maximum release (MR) represents the mean cpm released from target cells after rapid freeze-thaw treatment.

Target AQR, MR = 664 $\pm$ 27.1, SSR = 212 $\pm$ 10.1, SSR (PHA added during CML assay) = 215 $\pm$ 19.0

Target B10.T(6R), MR = 689 $\pm$ 18.5, SSR = 239 $\pm$ 18.3, SSR (PHA added during CML assay) = 244 $\pm$ 11.0

Target C57BL/10, MR = 712 $\pm$ 314, SSR = 239 $\pm$ 14.0, SSR (PHA added during CML assay) = 233 $\pm$ 24.1

elicit antibody formation. In mouse, I region products defined with antisera have been called the Ia antigens. It is tempting to hypothesize a priori that these Ia antigens are the stimulatory products in MLC. Furthermore, Meo and his collaborators have recently shown that anti-Ia antisera block the stimulating cells in MLC (47) which is certainly consistent with the Ia antigens being LD. It must be stressed, however, that these studies do not prove that the two types of antigens are one and the same; until definitive proof is available it is probably advisable to maintain them as potentially separate entities. The work in humans is similarly inconclusive; however, if it can be shown that the LD antigens can be defined serologically, this would represent an important and major advance.

IN VIVO RELEVANCE

The relative roles of MHC LD and SD loci in vivo have been recently reviewed (48). In brief, it appears that H-2 LD differences are associated with proliferative graft-versus-host reactions and, in some cases, skin graft rejection; H-2 SD differences are relatively weaker in eliciting a proliferative graft-versus-host reaction and some cases do not do so but lead to skin graft rejection. In both of these analyses it is important to remember that we are speaking of differences for whole regions of H-2 which include either a known SD locus or LD locus; there is always the possibility in the case of the SD region that there are, in addition, LD-type loci or differences included in that region. Results obtained in man are, insofar as testing permits, consistent with the above findings in mouse. There is no evidence yet available which answers the question whether there is LD - SD collaboration in vivo as there is in vitro.

The clinical usefulness of LD matching with MLC and SD typing has been the subject of intense investigation. The results have also recently been reviewed in detail (47).

THE CONCEPTUAL SEPARATION OF LD AND SD?

Do the experimental findings to date justify a conceptual distinction between LD and SD as two functionally distinct MHC antigenic systems, or might these all be H antigens which function similarly in MLC and CML? The findings discussed in previous sections of this paper suggest to us that there are differences between LD and SD components of the MHC which support a model which considers each as a separate system within the MHC. This has been challenged (35) by the argument that in fact the differences are only quantitative, suggesting that the distinctions between them are therefore not justified. We would have to take exception to such a view on two grounds. First, to state that differences are quantitative rather than qualitative is only worthwhile when referring to very specific comparisons. Two gene products that differ only quantitatively in one assay could be both structurally and functionally (qualitatively) distinct. Second, it is well established that quantitative differences are often the basis of differential biological function. The validity of the conceptual separation of LD and SD depends on their differential recognition in these two test systems.

This question can be discussed under three headings, basically those which have led to the suggestion of dichotomy. First, is there a difference between LD and SD in leading to lymphocyte proliferation in MLC? Second, do these two types of antigens function differently in leading to the development and expression of positive CML reactions? Third, is there a difference in their function in the LD - SD collaboration?

With respect to stimulation of proliferation in MLC, LD is of course defined on the basis of this parameter. Speaking for the separation are those studies in mouse and man which show that the LD differences are must stronger in leading to proliferation than are the SD differences. This is true in the studies of Widmer (20-22) quoted earlier as well as the data obtained confirming those studies. Perhaps even more impressive are those human families in which recombinational events occurred between the SD loci and the LD locus and where those siblings that are presumably LD identical but differ for one HL-A SD haplotype stimulate very little, if at all, in MLC (18,19,43). The results obtained with <u>in vivo</u> proliferative graft-versus-host reactions parallel the proliferative studies in MLC (50-52). Thus even <u>in vivo</u> at least some K region different mouse strains fail to provide a proliferation stimulus under conditions where LD is very potent.

Arguing against the separation on this basis are the findings with the Egorov mutant (37) already discussed, as well as the finding that mouse strains differing by only the D region do stimulate in MLC. It could be argued that this rather low stimulation (see Table) is due to the SD antigens themselves; likewise one could argue that at least some of these cases are due to concurrent differences for the weak Hld locus located between Ss and H-2D. As already mentioned, given the data we have to date, the Egorov mutant is not uniquely interpretable. Nevertheless, it should serve as a signal that the biological function of the MHC may be far more complex than the rather simplistic LD - SD model currently envisioned.

The difference between LD and SD as a target in CML is also a complicated issue. Supporting the dichotomy are the findings in man and mouse which show that there is a difference in the extent to which these differences function as targets in CML or, on the basis of the PHA-dependent cytotoxicity studies mentioned in this paper, do lead to the generation of cytotoxic lymphocytes. The studies with xenogeneic effector cells further support this concept. On the other hand, the finding of very strong cytotoxicity with the Bailey mutant [H(z1)] is once again not easily understood, as we have already discussed. Also, it has been found in some studies using human cells that an SD antigen apparently does not serve as a good target in CML (53). This would suggest that some SD antigens do not function well as a CML target.

It is somewhat disturbing that the two exceptions which seem to argue against the data obtained with recombinant families in man or recombinant mouse chromosomes are both spontaneous mutations of H-2. Various interpretations of these mutations, as discussed above, are consistent with the LD - SD dichotomy. Clearly, more work must be done with these mutants.

Last, there is the LD - SD interaction which appears to suggest a dichotomy. LD differences in some culture systems, as already reviewed, do not lead to any cytotoxicity on any target cell. Yet, these same genetic differences potentiate the development of cytotoxicity against an SD target. In contrast, however, the opposite is not true.

It would seem to us that the differences summarized above more than justify our working model of LD and SD antigens as separate systems of the MHC. It would be foolhardy to speak of this separation without recognizing the similarities also; the similarities to some extent are not unexpected when considering the similar function of all arms of the immune system, i.e. recognizing and responding to antigenic differences. Rather, we have found it intriguing to note the differences which seem to be present, suggesting to us that for whatever reasons the products of the LD and SD loci can be distinguished by their ability to stimulate distinct phases of cell mediated responses *in vitro*. A number of reasons can be suggested to explain the dichotomy, both at the level of antigenic expression and cellular recognition, including the molecular configuration, the rate of turnover, the density or topographical location of the antigens on the cell, the nature of specific responding subpopulations, or other factors.

Why does LD not function as a strong target? Any of the reasons just given could be the basis for the inability. In addition, there is the possibility that LD is expressed on only a very small subpopulation of all the lymphocytes and that even though these are killed we do not detect it in the assay. This, of course, could be one basis for the LD - SD difference in at least some tissues. If this is true, then we would stress even more the strong nature of LD as a stimulus to MLC proliferation as compared with SD, which is present on all lymphocytes.

Why does SD not function as an efficient stimulant to lymphocyte proliferation? One possibility would be that the frequency of cells responding to SD is so small that we do not detect a large increase in thymidine incorporation. If this small population of SD-reponsive cells contains the cytotoxic cell precursors, then the LD stimulus is having a marked effect in collaboration. Alternatively, this hypothetical SD-responsive population may not consist of cytotoxic cell precursors but rather of helper cells similar to those responding to LD; the small size of this SD-responsive population may not be able to provide the level of collaboration needed for the development of cytotoxic lymphocytes.

PROBLEMS FOR THE FUTURE

It would be presumptuous to give a list of problems which most need to be investigated in this very large area. Our purpose here is rather to pin-point some of the areas which seem to us to be of great import.

First, we would have to list an understanding of the mouse mutant strains, one of which was in fact the basis on which many of the studies referred to here were started (22). Second, there is the need to determine whether the LD antigens are the same as the Ia antigens, and whether the target for the cytotoxic lymphocytes in CML are the same SD antigens recognized serologically.

More work on the cell populations involved in both the stimulatory events and in the response is clearly needed. We would, for instance, like to study in CML a population of target cells which are all known to carry LD determinants. Likewise it is important to resolve whether there are two completely separable T cell populations which are involved in MLC and CML.

The _in vivo_ relevance of the various reactions tested _in vitro_ must be studied further, most of all the question whether the LD - SD cooperation _in vitro_ also exists _in vivo_. However, questions such as the _in vivo_ relevance of the CML test in various allograft systems must also be further evaluated.

Last, we might try to incorporate the findings with these

in vitro models of the allograft reaction which have been related to the MHC with the many other systems now related to H-2, including the immune response genes and the I region products which are important in T - B cell cooperation (perhaps a model of cell interactions in some respects not different from MLC). Even the further development of the cellular techniques and new protocols using them will, it seems to us, not provide satisfying answers until some of the MHC products involved in this wide spectrum of reactions are isolated in a functional state.

REFERENCES

1. Bain, B., Vas, M.R., and Lowenstein, L., Blood 23: 108, 1964.
2. Bach, F.H. and Hirschhorn, K., Science 143: 813, 1964.
3. Hayry, P. and Defendi, V., Science 168: 133, 1970.
4. Hodes, R.J. and Svedmyr, E.A.J., Transplantation 9: 470, 1970.
5. Solliday, S. and Bach, F.H., Science 170: 1406, 1970.
6. Lightbody, J.J., Bernaco, O., Miggiano, V.C., and Ceppellini, R.G., J. Bact. Virol. Immunol. 64: 273, 1971.
7. Zoschke, D.C. and Bach, F.H., Science 170: 1404, 1970.
8. Thorsby, E., Hirschberg, H., and Helgesen, A., Transpl. Proc. V: 1523, 1973.
9. Wilson, D.B., Blyth, J.L., and Nowell, P.C., J. Exp. Med. 128: 1157, 1968.
10. Bach, F.H. and Amos, D.B., Science 156: 1506, 1967.
11. Dutton, R.W., J. Exp. Med. 123: 665, 1966.
12. Silvers, W.K., Wilson, D.B., and Palm, J., Science, 155: 703, 1967.
13. Festenstein, H., Abbasi, K., Sachs, J.A., and Oliver, R.T.D., Transpl. Proc. II: 219, 1972.
14. Peck, A.B., Bach, F.H., and Boyse, E.A., Transpl. Proc. V: 1611, 1973.
15. Amos, D.B. and Bach, F.H., J. Exp. Med. 128: 623, 1968.
16. Bach, F.H., Albertini, R.J., Amos, D.B., Ceppellini, R., Mattiuz, P.L., and Miggiano, V.C., Transpl. Proc. I: 339, 1969.
17. Plate, J.M., Ward, F.E., and Amos, D.B., in "Histocompatibility Testing 1970" (ed. by P.I. Terasaki), p. 531, 1970.
18. Yunis, E.J. and Amos, D.B., P.N.A.S. (U.S.) 68: 3031, 1971.
19. Eijsvoogel, V.P., du Bois, M.J.G.J., Melief, C.J.M., de Groot-Kooy, M.L., Koning, C., van Rood, J.J., van Leeuwen, A., de Toit, E., and Schellekens, P.Th.A., in

"Histocompatibility Testing 1972" (ed. by J. Dausset and J. Colombani), p. 501, 1972.
20. Bach, F.H., Widmer, M.B., Segall, M., Bach, M.L., and Klein, J., Science 176: 1024, 1972.
21. Bach, F.H., Widmer, M.B., Bach, M.L., and Klein, J., J. Exp. Med. 136: 1430, 1972.
22. Widmer, M.B., Alter, B.J., Bach, F.H., Bach, M.L., and Bailey, D.W., Nature New Biology 242: 239, 1973.
23. Widmer, M.B., Peck, A.B., and Bach, F.H., Transpl. Proc. V: 1501, 1973.
24. Meo, T., Vives, J., Miggiano, V., and Shreffler, D., Transpl. Proc. V: 377, 1973.
25. Meo, T., David, C.S., Nabholz, M., Miggiano, V., and Shreffler, D.C., Transpl. Proc. V: 1507, 1973.
26. Peck, A.B. and Bach, F.H., J. Immunolog. Methods 3: 147, 1973.
27. Alter, B.J., Schendel, D.J., Bach, F.H., Bach, M.L., Klein, J., and Stimpfling, J., J. Exp. Med. 137: 1303, 1973.
28. Schendel, D.J. and Bach, F.H., J. Exp. Med., in press.
29. Peck, A.B. and Bach, F.H., Eur. J. of Immunol., in press.
30. Schendel, D.J. and Bach, F.H., Eur. J. of Immunol., submitted for publication.
31. Eijsvoogel, V.P., du Bois, M.J.G.J., Melief, C.J.M., et al., Transpl. Proc. V: 1301, 1973.
32. Lindahl, K.F. and Bach, F.H., Nature, submitted for publication.
33. Schendel, D.J., Alter, B.J., and Bach. F.H., Transpl. Proc. V: 1651, 1973.
34. Eijsvoogel, V.P., du Bois, M.J.G.J., Meinesz, A., Bierhorst-Eijlander, A., Zeylemaker, W.P., and Schellekens, P.Th.A., Transpl. Proc. V: 1675, 1973.
35. Klein, J., Hauptfeld, M., and Hauptfeld, V., J. Exp. Med. 140: 1127, 1974.
36. Apt, A.S., Blandova, Z., Dishkant, I., Shumova, T., Vedernikov, A.A., and Egorov, I.K., Immunogenetics, in press, 1974.
37. Egorov, I.K., Immunogenetics 1: 97, 1974.
38. Bach, F.H., Transpl. Proc. V: 23, 1973.
39. Bach, F.H., Segell, M., Zier, K.S., Sondel, P.M., Alter, B.J., and Bach, M.L., Science 180: 403, 1973.
40. Bach, F.H., Zier, K.S., and Sondel, P.M., Transpl. Proc. V: 1717, 1973.
41. Alter, B.J. and Bach, F.H., J. Exp. Med. 140: 1410, 1974.
42. Forman, J. and Möller, G., J. Exp. Med. 138: 672, 1973.
43. David, C.S., Shreffler, D.C., and Frelinger, J.A., P.N.A.S. 70: 2509, 1973.

44. Götze, D., Reisfeld, R.A., and Klein, J., J. Exp. Med. 138: 1003, 1973.
45. Sachs, D.H. and Cone, J.L., J. Exp. Med., in press, 1973.
46. van Leeuwen, A., Schuit, H.R.C., and van Rood, J.J., Transpl. Proc. V: 1539, 1973.
47. Meo, T., presented at International Transplantation Conference, Jerusalem, 1974.
48. Bach, F.H. and van Rood, J.J., N. Engl. J. Med., in preparation.
49. Bonnard, G.D., Chappuis, M., Glauser, A., Mempel, W., Baumann, P., Grosse-Wilde, H., and Albert, E.D., Transpl. Proc. V: 1679, 1973.
50. Livnat, S., Klein, J., and Bach, F.H., Nature New Biology 243: 42, 1973.
51. Klein, J. and Park, J.M., J. Exp. Med. 137: 1213, 1973.
52. Elkins, W.L., Kavathas, P., and Bach, F.H., Transpl. Proc. V: 1759, 1973.

MOLECULAR EVENTS IN THE INDUCTION OF AUTOSENSITIZATION OF INITIATOR T LYMPHOCYTES[1]

Irun R. Cohen, Shmuel Livnat, Daniella Livnat, Eugen Steiner and Harlan Waksal
Department of Cell Biology, The Weizmann Institute of Science, Rehovot, Israel

Immune responses involve the differentiation of small resting lymphocytes into effector lymphocytes which mediate an immunospecific reaction. To facilitate study, immune differentiation of T lymphocytes may be divided into three phases: the afferent or sensitization phase, the central phase, and the efferent or effector phase (1).

Sensitization of circulating T lymphocytes is triggered by contact between receptors on the lymphocyte membrane and immunogenic structures. The sensitized lymphocytes appear to journey to the lymph nodes which drain the area in which sensitization was induced (1). The central phase of the response takes place within the lymph nodes and involves further differentiation of lymphocytes into effector cells. The effector lymphocytes, upon leaving the lymph nodes, are capable of recognizing with their receptors the specific immunogen. Contact with the immunogen induces the effector lymphocytes to produce a measurable immune effect.

In this laboratory we have been studying the induction and expression of autosensitization both as a problem in autoimmune disease, and as a particular example of the differentiation of T lymphocytes (2).

In this paper we shall review our previous findings and describe the results of experiments designed to study early molecular events in the induction of autosensitization. Related studies of allosensitization will also be described.

SELF-RECOGNITION AND ITS CONTROL

We first began to suspect the existence of self-recognizing lymphocytes in normal animals when we found that lymph node cells of inbred rats could be induced to transform into large lymphocytes by contact in vitro with syngeneic embry-

---

[1]This research was supported in part by the National Institutes of Health, under agreement No. NCI-G-72-3890, and by the Talisman Foundation, Inc., New York.

[2]Abbreviations used in this paper: CA, cytosine arabinoside; EM, Eagle's medium; PBS, phosphate buffered saline: CH, cycloheximide.

onic fibroblasts. Such lymphocytes were able to specifically kill symgeneic, but not foreign target fibroblasts and produced splenomegaly upon injection into syngeneic animals (3). Similar observations were made using mouse spleen cells. In additional studies we found that autosensitization could also be induced in vitro by thymus reticulum cells from adult syngeneic (4) or autochthonous rats (5).

These findings raised the question of the origin of the T lymphocytes which were triggered in autosensitization in vitro. Were such lymphocytes merely an artifact of cell culture, or did a population of lymphocytes with receptors programmed to recognize self antigens actually exist in intact animals? To answer this question, we modified a system of immunoadsorption which had been developed to characterize normal, unsensitized lymphocytes by their receptors to foreign antigens (6). Monolayers of syngeneic fibroblasts or thymus reticulum cells were used to adsorb lymphocytes before testing their capacity for sensitization in vitro against syngeneic or foreign cells. We found that adsorption on syngeneic monolayers for several hours specifically depleted suspensions of lymphocytes of those cells capable of undergoing autosensitization (5, 7). Lymphocytes which could react against allogeneic fibroblasts were not removed by syngeneic monolayers, but could be removed by specific allogeneic monolayers. Hence, lymphocytes with receptors specific for self-antigens exist among the diversity of T lymphocytes, and autosensitization in vitro seems to be a property of such lymphocytes. Thus, the sensitization phase of autosensitization, similar to sensitization against foreign substances, is triggered by contact between specific immunogen and preformed lymphocyte receptors.

How, then, is the body protected against immune activation of self-recognizing lymphocytes and the development of autoimmunity in vivo? By using the same system of immunoadsorption described above, and by studying the kinetics of self-recognition in vitro, we found that fresh autologous serum contained factors that specifically inhibited the recognition of self-antigens. Autologous serum did not inhibit recognition of foreign antigens and foreign serum did not inhibit the recognition of self-antigens. This inhibitory activity was demonstrable by treating the lymphocytes, but not by treating the immunoadsorbent monolayers (5, 7). Hence, autosensitization in vivo appeared to be prevented by factors behaving like nonimmunogenic soluble self-antigens which blocked self-recognition. In the absence of serum blocking factors in vitro, potentially self-recognizing T lymphocytes could interact with immunogenic self-antigens and undergo immune differentiation leading to the development of specific autoreactivity in vivo as well as in vitro (3, 5, 7).

The relationship of autosensitization in vitro to autoimmune disease was studied by sensitizing rat thymus lymphocytes against extracts of syngeneic brains in vitro. We found that injection of such lymphocytes intravenously into syngeneic rats led to the appearance of brain lesions indicative of autoimmune encephalitis (8). Injection of the brain antigens alone, or mixed with unsensitized lymphocytes, did not produce any lesions. Another model of local autoimmune disease has recently been developed by H. Wekerle and M. Begemann of the Max Planck Institut für Immunbiologie in Freiburg (H. Wekerle and M. Begemann, in preparation). They found that a progressive orchitis could be produced by injecting rats with lymphocytes sensitized against syngeneic testis cells in vitro. Autosensitization in vitro was prevented by the presence of autologous serum (1%) in the culture medium. Thus, it appears that the induction of specific autosensitization in vitro may be associated with local autoimmunity in intact animals. Furthermore, control of autosensitization against certain self-antigens seems to occur at the sensitization phase and is regulated by extracellular soluble factors, probably non-immunogenic, blocking self-antigens. Preliminary studies suggest that the blocking factors have a molecular weight of less than 50,000.

INITIATOR T LYMPHOCYTES RECRUIT EFFECTOR T LYMPHOCYTES

What is the origin of the autoreactive effector lymphocytes that produced the immune effects? Burnet proposed that effector lymphocytes were the clonal descendants of the original lymphocytes that interacted with the inducing immunogen (9). The question was, therefore, whether the cells which differentiated into autoreactive effector lymphocytes in the central phase of the response were the progeny of the lymphocytes that first recognized self-antigens.

We were able to investigate this question by developing a system which made it possible to physically separate the sensitization from the central phases of the response (10, 11). The induction of autosensitization was initiated during several hours of culture in vitro, and the lymphocytes were then separated from the sensitizing cells and injected into the footpads of syngeneic rats or mice. The development of specific effector lymphocytes was found to occur during the next 5 or 6 days within the draining popliteal lymph nodes. The syngeneic lymph nodes enlarged and suspensions of lymphocytes from these nodes were cytotoxic to target cells in vitro. Cytotoxicity was found to be immunospecifically directed against the same immunogens that triggered the induction of sensitization in vitro. Autosensitization led to a specifically autoreactive response, while allosensitization produced an alloreactive response (11).

The magnitude of the central phase was measured by the ratio of the enlargement of the test to the contralateral control popliteal lymph nodes in the same mouse or rat (popliteal lymph node index). An index of 2 or greater was considered as a significant degree of enlargement. It was also possible to assay the degree of cytotoxicity produced by suspensions of lymphocytes from test or control lymph nodes.

A third method used to quantitate the differentiation of effector cells was developed for an allosensitization system (12). Thymus lymphocytes of C3H/eb ($H\text{-}2^k$) mice were sensitized in vitro for 18-20 h against fibroblasts of C57BL/6 ($H\text{-}2^b$) origin. The C3H lymphocytes were then separated from the sensitizing fibroblasts and injected into the footpads of syngeneic C3H mice. Six days later the mice were challenged with 3LL tumor cells of C57BL/6 origin. This tumor progressively grows and kills C3H mice unless they have been specifically sensitized against C57BL/6 alloantigens. Hence, rejection of the 3LL tumor allograft served as a measure of the development of an allospecific effector response.

The above systems were studied to learn the origin of the effector lymphocytes which differentiated during the central phase of the allo- or autosensitization reactions. We used irradiation of the injected sensitized lymphocytes, or of the recipient animals to block the proliferation of cells of either source. In addition, in mouse systems we were able to analyze the origin and type of reactive lymphocytes by employing anti-θ anti-serum and complement, as well as B mice (adult thymectomy, irradiation and injection of bone marrow cells).

The results of these studies (10-12) led to the conclusion that in both autosensitization and allosensitization reactions, the receptor-bearing T lymphocytes that were triggered in vitro by immunogens did not themselves proliferate and differentiate into effector lymphocytes. Rather, the initiator T lymphocytes recruited other T lymphocytes in the central phase of the response to act as effector cells. Hence, the development of T effector lymphocytes can involve the cooperation of two types of T cells. Initiator T lymphocytes are triggered by immunogen and become capable of recruiting potential effector T lymphocytes. Preliminary evidence indicates that the thymus and spleen contain a relatively higher concentration of initiator lymphocytes than do the lymph nodes. Potential effector T cells, on the other hand, are recruited mostly from circulating lymph node-seeking lymphocytes and are deficient in the thymus.

The mechanism by which recruitment is mediated is unknown. However, the specificity of the immune reaction indicates that specific information must be transferred from the initiator to the potential effector lymphocytes. We recently carried out experiments to learn whether initiator lymphocytes sensitized in vitro against alloantigens recruit effector lymphocytes by means of alloantigens passively adsorbed on their surface (13). We used trypsin to digest protein antigens possibly present on the initiator cell membranes. Thymus lymphocytes of C3H/eb mice were sensitized against C57BL/6 fibroblasts in vitro and the initiator lymphocytes were treated with trypsin (0.3%) for 15 min before injecting them into C3H/eb mice. The mice were challenged 6 days later with the 3LL tumor. We found that treatment with trypsin abolished the ability of the initiator lymphocytes to recruit an anti-3LL reaction. However, recruiting ability was spontaneously recovered and the 3LL tumor allograft was rejected after incubating the initiator lymphocytes in vitro for 4 h in the absence of further contact with alloantigens. These findings indicate that recruitment depends on intrinsic properties of the initiator lymphocytes induced by sensitization in vitro. Furthermore, trypsin sensitive surface components synthesized by the initiator lymphocytes appear to be involved in the reaction. These results argue against the possibility that passively adsorbed antigens were a major factor in recruitment. In addition they illustrate the importance of the differentiation of initiator T lymphocytes in the development of an effective allograft response. In the studies to be reported below, we were interested in investigating molecular events underlying the induction of autosensitization. Our general approach was to assay the effects on the triggering of initiator thymus lymphocytes of inhibitors of DNA or protein synthesis.

AUTOSENSITIZATION IS NOT PREVENTED BY INHIBITION OF DNA SYNTHESIS

The processes by which eukaryotic cells differentiate to express limited and specialized functions are unknown. It has been proposed as a general theory (14) and shown in a number of systems such as the development of mammary gland cells (15, 16), or the fusion of myoblasts into muscle fibers (17), that at least one cycle of DNA replication is critical to cell differentiation. A similar conclusion was reached by Nakamura and his colleagues for the induction of a primary antibody response (18). We previously found, however, that irradiation (1000-2000R) of initiator lymphocytes after autosensitization on fibroblasts for 16 to 20 h did not block their ability to recruit autoreactive effector lymphocytes (10, 11). This suggested that a cycle of cell division might not be necessary for the differentiation of immune function in these cells (11). However, it was possible that DNA replication had occurred during contact with the sensitizing fibroblasts before inhibition of DNA synthesis by irradiation. To investigate this possibility, we induced autosensitization in the presence of cytosine arabinoside (CA),[2] an inhibitor of DNA synthesis.

Figure 1 shows that concentrations of CA of 2×10^{-4}, 2×10^{-5} and 2×10^{-6} M inhibited the incorporation of ^{3}H-thymidine by thymus lymphocytes by greater than 95%. The effects of CA in these concentrations on the induction of autosensitized initiator lymphocytes are shown in Table 1. We found that autosensitized lymphocytes caused significant enlargement of the syngeneic popliteal lymph nodes and the index was not at all inhibited by CA. Unsensitized lymphocytes produced no lymph node enlargement. Table 2 shows the results of another experiment in which we measured both the popliteal lymph node index and the autoreactive cytotoxicity of suspensions of lymphocytes from the enlarged nodes. It can be seen that CA during autosensitization failed to depress the ability of initiator lymphocytes to induce cytotoxicity as well as enlargement.

Irradiation (1000 R) of initiator lymphocytes which had been autosensitized for 18 h in the presence of CA (2×10^{-6} M) also failed to depress the popliteal lymph node index. Hence, it appears that a cycle of DNA synthesis is not critical to the immune differentiation of resting thymus lymphocytes into autosensitized initiator lymphocytes. Furthermore, these findings argue against the theory (9) that the induction of autosensitization depends on genetic mutation of lymphocytes into "forbidden clones." Indeed, they complement the conclusions of our earlier work that some T lymphocytes in normal animals are programmed with receptors that recognize certain self-antigens (2-5, 7, 8).

INDUCTION OF AUTOSENSITIZATION PREVENTED BY AN INHIBITOR OF PROTEIN SYNTHESIS

To investigate requirements for protein synthesis in the induction of autosensitized initiator lymphocytes, we added cycloheximide (CH) to the autosensitizing cultures. Figure 2 shows the effects of various concentrations of CH on the incorporation of ^{14}C-labeled amino acids into TCA precipitable material by thymus lymphocytes. Concentrations of 4 μg/ml and above were found to inhibit incorporation by greater than

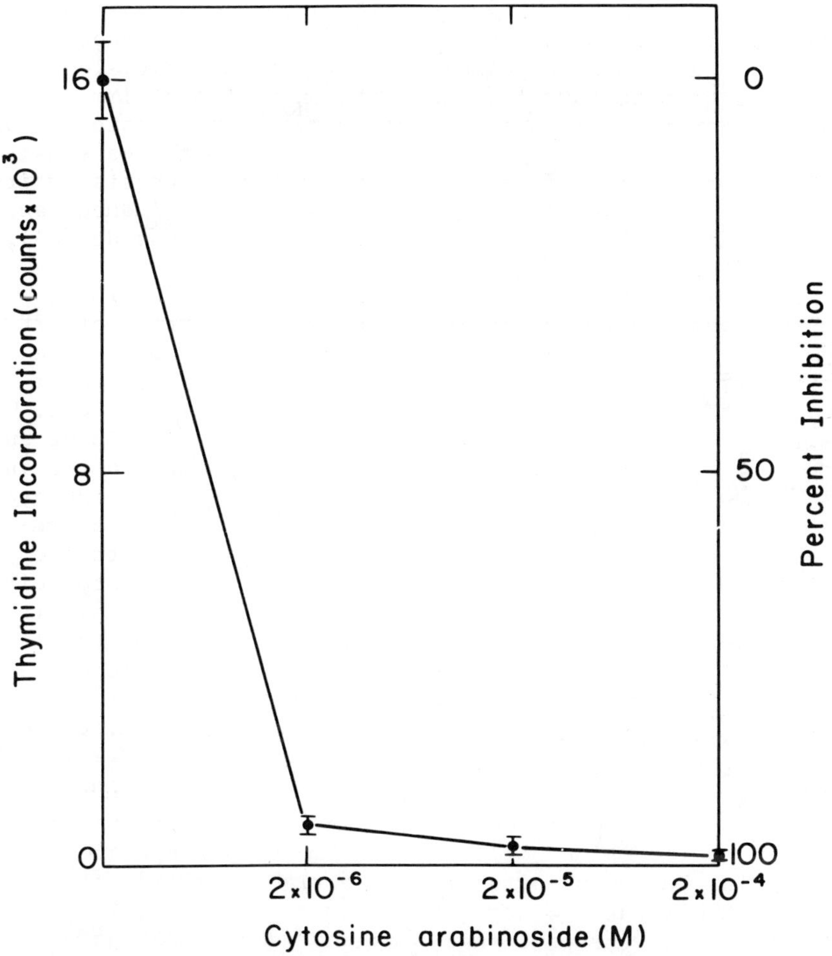

Fig. 1 Effect of cytosine arabinoside (CA) on the incorporation of ^{3}H-thymidine into acid precipitable material of lymphocytes. Thymus lymphocytes of Lewis rats (2 x 10^7) were cultured in quadruplicate in tissue culture tubes for 18 h in 2 ml of Dulbecco's modification of Eagle's medium (EM) containing 10% horse serum, ^{3}H-thymidine (2μCi/ml), and concentrations of CA. The lymphocytes were then washed twice in phosphate buffered saline (PBS), suspended in NaOH (0.3 M) and heated to 80° C for 10 min. The mixture was cooled in ice water, neutralized by HCl (0.3 M), and TCA(10%) was added for 10 min. The precipitate was filtered, washed with TCA, ethanol, and dried. Incorporation of label was measured as counts per min.

TABLE 1
EFFECT OF CYTOSINE ARABINOSIDE (CA) ON THE AUTOSENSITIZATION OF INITIATOR LYMPHOCYTES

Autosensitization culture[a]		Response of syngeneic recipient
Lymphocytes CA (in M)	% inhibition of ^{3}H-thymidine incorporation	Popliteal lymph node index[b]
Autosensitized		
0	0	2.2 ± 0.1
2×10^{-6}	96	3.1 ± 0.5
2×10^{-5}	98	3.6 ± 1.2
2×10^{-4}	99	2.2 ± 0.9
Unsensitized		
0	0	0.9 ± 0.7

[a]Lewis rat thymus lymphocytes (30×10^6) were incubated for 18 h alone (unsensitized) or together with monolayers of Lewis fibroblasts (2×10^6) in 60-mm plastic petri dishes in EM without added serum, containing concentrations of CA. The lymphocytes were separated from the fibroblasts by washing. Induction of initiator function was tested by injecting 10^7 lymphocytes into the right footpads of syngeneic Lewis rats (4 per group).

[b]After 6 days, the popliteal lymph nodes of each rat were removed and weighed.

$$\text{Index} = \frac{\text{weight of right (experimental) node}}{\text{weight of left (control) node}}.$$

TABLE 2
EFFECT OF CYTOSINE ARABINOSIDE (CA) ON AUTOSENSITIZATION OF INITIATOR LYMPHOCYTES AND DEVELOPMENT OF AUTOREACTIVE EFFECTOR LYMPHOCYTES[a]

Autosensitization culture		Response of syngeneic recipient	
Lymphocytes	CA (2×10^{-6}M)	Popliteal lymph node index	% cytotoxicity[b] of Lewis fibroblasts
Unsensitized	+	1.3 ± 0.5	8 ± 2
Autosensitized	-	4.8 ± 0.9	29 ± 3
Autosensitized	+	5.0 ± 1.6	31 ± 1

[a]The procedure was as described in Table 1.

[b]Suspensions of Lewis rat lymph node cells containing 5×10^6 cells in 1.5 ml EM + 15% horse serum were incubated with monolayers of Lewis fibroblasts (10^6 cells) in 35-mm petri dishes for 65 h. The fibroblasts were labeled with ^{51}Cr and the percentage cytotoxicity was computed as described (11).

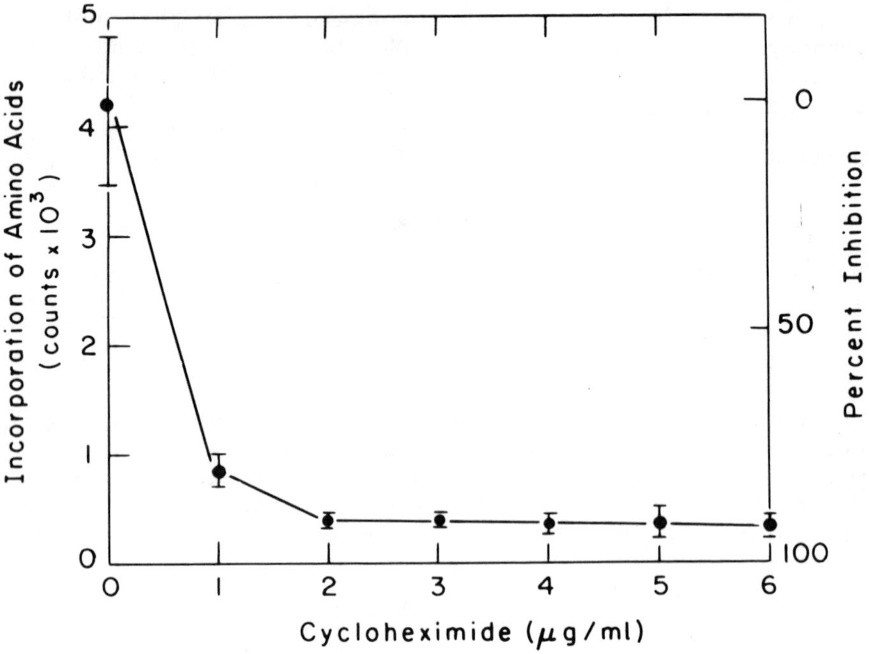

Fig. 2 Effect of cycloheximide on the incorporation of ^{14}C-labeled amino acids into acid precipitable material of lymphocytes. The experiment was performed as described in Figure 1, except that a mixture of ^{14}C-labeled amino acids (2µCi/ml) was added to the cultures in place of ^{3}H-thymidine. Incorporation of label was measured as counts per min.

93%. We therefore studied the effects of CH at a concentration of 5 μg/ml. To decrease the general toxicity of CH for the lymphocytes, we limited the duration of in vitro induction to 8 h, a period of time sufficient for autosensitization (5). Table 3 shows that application of CH during induction inhibited the ability of initiator lymphocytes to produce enlargement of syngeneic popliteal lymph nodes. This could result from a specific requirement for protein synthesis during induction that was blocked by CH. Alternatively, exposure to CH for 8 h may have nonspecifically poisoned the lymphocytes so that they could not produce enlargement of the popliteal lymph nodes after injection. In earlier experiments we found that disruption of autosensitized initiator lymphocytes prevented the popliteal lymph node response (11).

To investigate the nonspecific toxicity of CH on initiator lymphocytes, we carried out experiments in which initiator lymphocytes were first incubated with sensitizing fibroblasts for 8 h to induce autosensitization, and then CH was added to the cultures for another 8 h. The results are shown in Table 4. Incubation with CH subsequent to autosensitization failed to suppress the initiator lymphocytes. As before, we found that the presence of CH during induction prevented the ability of the lymphocytes to produce a lymph node response. This indicates that incubation for 8 h with CH at a concentration of 5 μg/ml inhibits the induction of autosensitization. However, after their induction, initiator lymphocytes are not affected by CH. Treatment of sensitizing fibroblasts with CH failed to affect their ability to sensitize. Hence, induction appears to depend on specific lymphocyte processes which are sensitive to CH, probably protein synthesis. This conclusion is supported by an earlier finding that the induction of initiator lymphocytes depended on time and temperature (11).

THE EFFECTS OF TRYPSIN TREATMENT ON INITIATOR LYMPHOCYTE FUNCTION

In addition to playing a role in induction, protein synthesis also appears to be important in the in vivo activity of the initiator lymphocytes. We found that treatment of autosensitized mouse initiator lymphocytes with trypsin inhibited their ability to produce enlargement of syngeneic lymph nodes (Table 5). However, this function was spontaneously recovered by removing the trypsin and incubating the lymphocytes for 4 h in vitro at 37°C in the absence of further contact with sensitizing fibroblasts. Recovery was prevented by the presence of CH during the recovery period. Hence, trypsin-sensitive surface proteins, resulting from the

TABLE 3
CYCLOHEXIMIDE (CH) DURING INDUCTION INHIBITS AUTOSENSITIZATION OF INITIATOR LYMPHOCYTES[a]

Lymphocytes	Autosensitization Culture CH (5μg/ml)	Response of Syngeneic recipeint Popliteal lymph node index
Autosensitized	-	2.3 ± 0.2
	+	1.4 ± 0.3
Unsensitized	-	1.1 ± 0.4
	+	1.2 ± 0.3

[a]The procedure was as described in Table 1, except that CH was substituted for CA and the duration of autosensitization or unsensitized cultures was 8 h.

TABLE 4
CYCLOHEXIMIDE (CH) FAILS TO INHIBIT INITIATOR LYMPHOCYTES AFTER THE INDUCTION OF AUTOSENSITIZATION[a]

Lymphocytes	Autosensitization Culture CH (5 μg/ml) 0-8 h	 8-16 h	Response of syngeneic recipient Popliteal lymph node index
Autosensitized			
	+		1.0 ± 0.3
	-	+	2.4 ± 0.5
Unsensitized			
	+		0.9 ± 0.4
	-		1.1 ± 0.5

[a]The procedure was the same as described in Table 1, except that in place of CA,CH was added to some of the autosensitization or unsensitized cultures for 8 h and then the lymphocytes were injected into the recipients. Other cultures were autosensitized for 8 h and then received CH for an additional 8 h before injection.

TABLE 5
SPONTANEOUS RECOVERY IN VITRO FROM TREATMENT WITH TRYPSIN OF AUTOSENSITIZED MOUSE INITIATOR LYMPHOCYTES[a]

Autosensitization culture		Response of syngeneic recipient
Trypsin treatment[b]	Recovery	Popliteal lymph node index
Autosensitized lymphocytes		
-	-	2.1 ± 0.3
+	-	1.0 ± 0.1
+	+	2.1 ± 0.3
+	+, CH	1.1 ± 0.2
Unsensitized lymphocytes		
-	-	1.1 ± 0.2

[a]Thymus lymphocytes of C57BL/6 mice were autosensitized against syngeneic fibroblasts by the method described in Table 1.

[b]Initiator lymphocytes were incubated for 15 min at 37° C with a 0.3% trypsin solution. After trypsin was removed by washing, some lymphocytes were allowed to recover by incubating them for 4 h in vitro in the absence of sensitizing fibroblasts. Cycloheximide (CH) at a concentration of 10 μg/ml was present during the recovery period of one group of lymphocytes.

intrinsic metabolic activity of the initiator lymphocytes, appear to be important in carrying out recruitment. These findings argue against the possibility that syngeneic lymph node enlargement was brought about by protein antigens which passively adsorbed in vitro to the surfaces of the injected initiator lymphocytes (13).

SUMMARY AND CONCLUSIONS

In previous studies in this laboratory, it was observed that normal rats and mice possess T lymphocytes with receptors that can recognize self-antigens (5, 7). Incubation of thymus, spleen or lymph node lymphocytes with syngeneic or autochthonous cells led to self-recognition and induced autosensitization. The autosensitized lymphocytes could produce immunospecific autoaggressive reactions either in vitro or in vivo (2-5, 7, 8, 10, 11). Study of the regulation of autosensitization led to the conclusion that self-recognition may be prevented in vivo by soluble factors present in the serum which behave like nonimmunogenic self-antigens and specifically block lymphocyte receptors for such antigens (5, 7).

Further studies of the process of autoreactivity indicated that specific cytotoxic effector T lymphocytes were not the clonal descendants of the original self-recognizing lymphocytes, but were recruited by them from among other lymphocytes during the central phase of the immune response (10, 11). We observed a similar process of recruitment of effector T lymphocytes by initiator T lymphocytes in an allograft reaction (12).

It is noteworthy that extensive proliferation of the initiator lymphocytes did not seem to be required for recruitment to occur. Hence, the recruitment phenomenon provided a means for investigating the process of immune differentiation of initiator T lymphocytes as distinct from the process of cell proliferation. Our general approach was to study the induction of autosensitization of initiator lymphocytes by assaying their ability to produce the central lymph node phase of the reaction. We injected the lymphocytes into the footpads of syngeneic animals and measured their ability to cause enlargement of the draining popliteal lymph nodes. Autoreactive cytoxicity was tested in some cases. This paper reports the results of experiments designed to investigate the need for synthesis of DNA and proteins during induction of autosensitization.

(1) We found that inhibition of DNA synthesis by CA did not affect the induction of autosensitized initiator thymus lymphocytes. Thus, unlike differentiation of mammary gland

cells (15, 16), fusion of myoblasts (17), or induction of antibody production by B lymphocytes (18), the induction of initiator T lymphocytes does not appear to require a critical cycle of DNA replication (14).

It may be claimed that proliferation of lymphocytes is a characteristic feature of immune responses in general. However, the physiologic role of such cell division is unknown. It is likely that the ability to detect and measure immune responses depends on the presence of a critical mass of reacting cells. Augmentation of some immune reactions above the threshold of detection could result from lymphocyte proliferation. In such cases, proliferation may be only ancillary to the fundamental processes of immune differentiation. Important aspects of the immune response itself might develop independently of cell proliferation, although they are accompanied by it. For example, it has been suggested that the lymphocytes which proliferate are not the same lymphocytes which differentiate into cytotoxic effector cells during the mixed lymphocyte reaction (19, 20). An advantage of the recruitment phenomenon is that it can be measured in the absence of proliferation of initiator lymphocytes.

(2) Inhibition of protein synthesis by CH during induction of autosensitization was associated with a decrease in the function of initiator lymphocytes. However, incubation of initiator lymphocytes with CH subsequent to induction did not block their ability to trigger enlargement of syngeneic popliteal lymph nodes in vivo. Although we must interpret with caution the results of experiments of this kind, these findings suggest that the synthesis of proteins is important for differentiation of initiator cell function.

(3) Treatment of autosensitized mouse initiator lymphocytes with trypsin inhibited their ability to produce enlargement of syngeneic popliteal lymph nodes. Spontaneous recovery of this function, however, could be achieved by removing the trypsin and incubating the lymphocytes for 4 h in vitro in the absence of further contact with sensitizing fibroblasts. Recovery was suppressed by the presence of CH in the culture medium. These findings suggest that trypsin-sensitive surface proteins synthesized by the lymphocytes are important for their function as initiators. In addition, the results argue against the possibility that the central lymph node response was mediated by protein antigens passively adsorbed to the initiator lymphocytes.

The mechanism of recruitment must involve the transfer of specific information between initiator and effector cells since the response appears to be immunospecific. The nature of this information is unknown, although it is conceivable that lymphocyte factors such as RNA (21), transfer factor (22) or Ig T (23) may be involved. Trypsin-sensitive surface proteins could be related to the role of some such factors. Basic to these problems is the fundamental question of the role of recruitment in the physiology of the immune response. Is recruitment of effector lymphocytes by initiator lymphocytes only a rare event in nature, or is it a major process in the generation of clinically important cell-mediated immune reactions?

ACKNOWLEDGMENT

We thank Professor Michael Feldman for his support and inspiration.

REFERENCES

1. Elves, M.W., in Immunological Tolerance to Tissue Antigens (ed. by N.W. Nisbet and M.W. Elves), p. 1, Orthopedic Hospital, Oswestry, England, 1971.
2. Cohen, I.R. and Wekerle, H., in Proceedings of the Second International Congress of Immunology, North-Holland Press, Amsterdam, in press.
3. Cohen, I.R., Globerson, A. and Feldman, M., J. Exp. Med. 133: 834, 1971.
4. Cohen, I.R. and Wekerle, H., Science 176: 1324, 1972.
5. Cohen, I.R. and Wekerle, H., J. Exp. Med. 137: 224, 1973.
6. Wekerle, H., Lonai, P. and Feldman, M., Proc. Nat. Acad. Sci. USA 69: 1620, 1972.
7. Wekerle, H., Cohen, I.R. and Feldman, M., Nature New Biology 241: 25, 1973.
8. Orgad, S. and Cohen, I.R., Science 183: 1083, 1974.
9. Burnet, F.M., The Clonal Selection Theory of Acquired Immunity, Vanderbilt University Press, Nashville, 1959.
10. Cohen, I.R., Nature New Biology 242: 60, 1973.
11. Cohen, I.R., Cell. Immunol. 8: 209, 1973.
12. Treves, A.J. and Cohen, I.R., J. Natl. Cancer Inst. 51: 1919, 1973.
13. Livnat, S. and Cohen, I.R., Transplant. Proc., in press.
14. Holtzer, H., in General Physiology of Cell Specialization (ed. by D. Mazia and A. Tyler), p. 80, McGraw-Hill, New York, 1963.
15. Turkington, R.W., Current Topics Dev. Biol. 3: 199, 1968.
16. Topper, Y.J. and Vonderhaar, B.K., in Control of Proliferation in Animal Cells (ed. by B. Clarkson and R. Baserga), p. 843, Cold Spring Harbor Laboratory, Cold Spring Harbor, Massachusetts, 1974.
17. Bischoff, R. and Holtzer, H., J. Cell Biol. 41: 188, 1969.

18. Nakamura, I., Segal, S., Globerson, A. and Feldman, M., Cell. Immunol. 4: 354, 1972.
19. Cohen, L. and Howe, M.L., Proc. Natl. Acad. Sci. USA 70: 2707, 1973.
20. Bach, F.H., Segall, M., Zier, K.S., Sondal, P.M., Alter, B.J., and Bach, M.L., Science 180: 403, 1973.
21. Pilch, Y.H. and Ramming, K.P., Cancer 26: 630, 1970.
22. Lawrence, H.S., Transplant. Proc. 5: 49, 1973.
23. Feldmann, M., Transplant. Proc. 5: 43, 1973.

GENERATION OF CYTOTOXIC LYMPHOCYTES AND INHIBITION OF CELL-MEDIATED LYMPHOLYSIS BY SOLUBLE HL-A ANTIGENS [1,2]

Julius Gordon[3] and Harold N. Rode
Division of Surgical Research, Department of Surgery, McGill University, Montreal, Canada

Recent evidence suggests that transplantation (H)antigens may not be responsible for the induction of the Mixed Leukocyte Culture (MLC) reaction. Thus in some instances, responses occur between cells of HL-A identical individuals as well as between cells of strains of mice without serologically detectable antigenic differences (reviewed in ref 1). Furthermore purified H antigens (2-4) leukocyte extracts (5-7) and other cells rich in transplantation antigens (reviewed in ref 1) fail to provoke a reaction. On the other hand, transplantation antigens have been implicated in the reactivity of effector cells. Thus it has been shown that cytotoxic lymphocytes (CL) do not develop in MLC reactions between cells of HL-A identical, MLC reactive individuals (8,9) or H-2 identical mice (10-12). In addition only target cells which share HL-A antigens with the stimulating cells used in MLC are destroyed in the ensuing CML, regardless whether they themselves can elicit an MLC reaction (8).

In the present investigation, solubilized transplantation antigens were used to establish the role of H antigens in both the inductive and effector phases of the MLC reaction.

MATERIALS AND METHODS

Soluble transplantation antigens were prepared from cultured lymphoblasts, No. RPMI 8068 and 5287, initially obtained from Dr. G. Moore from Roswell Park Memorial Institute, Buffalo. The antigens solubilized by autolysis (13) or by hypertonic KCl (14) were either used as such, or were partially purified by chromatography on Sephadex G-200. The antigenic potency of each preparation was determined by the

---

[1]Supported by a grant from the Medical Research Council of Canada.

[2]Some of the results described have been presented at the Conference of the Federation of American Societies for Experimental Biology, Atlantic City, April, 1974.

[3]Research Associate of the Medical Research Council of Canada.

inhibition of ^{51}Cr release from the parent labelled lymphoblasts in the presence of rabbit complement, and a multivalent antiserum from a kidney transplant recipient. Antigen concentration was expressed in cell equivalents, i.e., relative to that present on the corresponding intact lymphoblasts. Other sources of transplantation antigens used were platelets and neutrophils, purified from fresh blood by differential centrifugation as described elsewhere (1). The CML reaction was carried out by incubating $1x10^5$ ^{51}Cr-labelled target cells (RPMI 8068, 5287, or lymphoblasts raised with PHA) with various numbers of aggressor cells in 2 ml medium RPMI 1640 - containing 15% human serum for 4-5 hours. At the end of the incubation period the ^{51}Cr liberated into the supernatant was counted and the percentage isotope release was calculated according to the formula:

$$\% \text{ release} = \frac{\text{CPM experimental-CPM spontaneous release}}{\text{CPM maximal release-CPM spontaneous release}} \times 100$$

Spontaneous release was established in cultures of target cells incubated alone; maximal release was measured following three cycles of freezing and thawing of the target cells.

RESULTS

The effect of soluble H antigens on the blastogenic response in MLC: induction of the MLC reaction.

The effect of soluble H antigens on the blastogenic response was evaluated in two test systems, one designed to detect stimulation, the other, inhibition. In the former, H antigens were cultured, over a wide range of concentrations, with allogeneic responding cells. In the latter, antigen in the same concentrations was added to the same responding cells exposed simultaneously to stimulating cells. The H antigens and stimulating cells used in any one experiment were derived from the same source; they were added to the responding cells at the same time, or the antigen was added 24 hours earlier. Antigen added to responding cells stimulated with PHA served to reveal non-specific inhibition.

The results obtained are illustrated in Fig. 1. In no instance did stimulation by soluble antigens exceed 10% of that obtained by the corresponding stimulating cells. Inhibition of MLC by soluble antigens, added to responding cells 24 hours before, or together with the stimulating cells was not seen either, except in some experiments at very high antigen concentration. At these high concentrations the response to PHA was also inhibited. Similar results were obtained with antigens prepared from another lymphoblast cell line, and from 18 human spleens. According to these results,

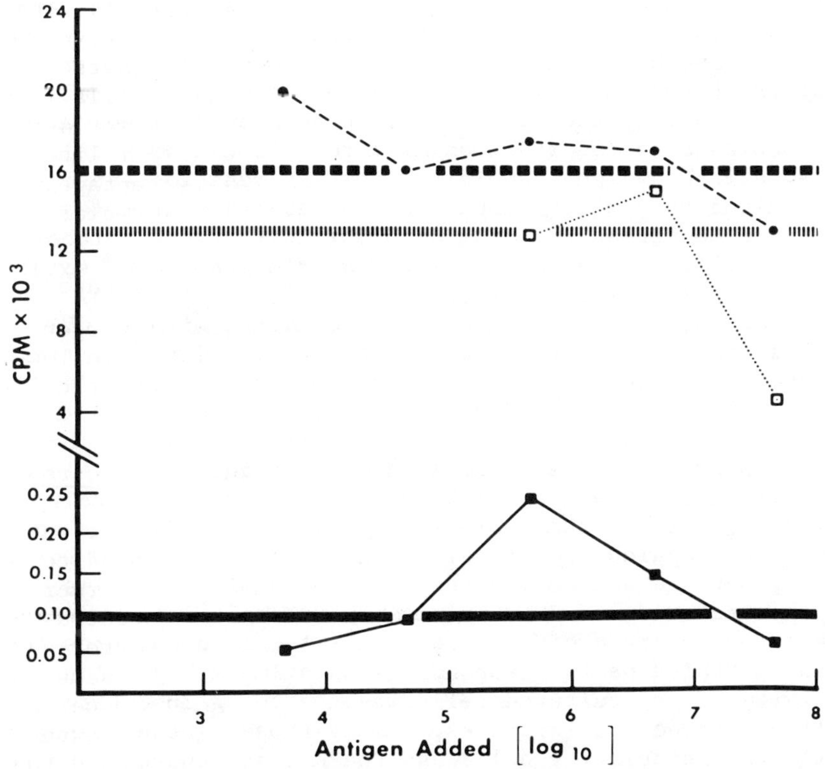

Fig. 1 Elicitation and inhibition of MLC by soluble transplantation antigens. Responding cells were cultured alone (————), with PHA (········) or with mitomycin treated stimulating cells (--------) with (light line) or without (heavy line) added antigen. The concentration of PHA or stimulating cells was the same in all cultures. The amount of soluble antigen added is expressed as the number of lymphoblasts to which the antigen is equivalent.

participation of transplantation antigens was not demonstrated in the proliferative response in MLC either as stimulation, or inhibition.

The effect of H antigens on CML.

Transplantation antigens were assayed for their ability to inhibit and to induce CML. Inhibition assays were carried out by adding soluble H antigens to reaction mixtures consisting of labelled lymphoblast target cells and lymphocytes activated to the lymphoblasts in MLC. The potency of each inhibitor was related to the decrease in ^{51}Cr release brought about by the addition of intact, unlabelled target lymphoblasts to the reaction mixture. As can be seen in Fig. 2, soluble H antigens inhibited the CML reaction: antigen equivalent to 3 x 10^6 lymphoblasts, reduced ^{51}Cr release to the same degree as addition of 0.5 x 10^6 intact, unlabelled target cells. A summary of five experiments, using soluble antigen as inhibitor is shown in Fig. 3. Analysis of variance indicates that the inhibition obtained with soluble antigen is significant at the 97.5% level. The specificity of the inhibition of CML by soluble H antigens is illustrated in Table 1. For this experiment two target cells (M and P) were selected which did not cross-react in CML. Thus addition of unlabelled M cells did not reduce ^{51}Cr release from labelled P target cells by lymphocytes activated to P cells, nor vice versa. Similarly soluble transplantation antigens prepared from M cells inhibited CML using M target cells, but did not modify the reaction against P cells.

The induction of CML activity by transplantation antigens was attempted by adding the antigens to mixed cultures containing "third party" responding and stimulating cells. CML activity of the cultured cells was tested against target cells of three specificities, namely those of the responding cell, the stimulating cell, and the antigen added. The antigenic preparations used were in the form of soluble extracts, platelets or neutrophils. Results of 1 of 4 experiments are shown in Table 2.

The MLC reactions obtained, recorded in the first column gave the expected results: both stimulating cells used alone provoked a reaction, essentially not augmented when both were added together to responding cells. The antigenic preparations used did not elicit more than a marginal reaction, and depressed slightly the stimulation by mitomycin-treated cells. The CML reactivity generated against responding cells (second column) in cultures of responding cells alone was approximately 10% release of ^{51}Cr; this reactivity was lower in some, but not significantly higher in any of the mixed

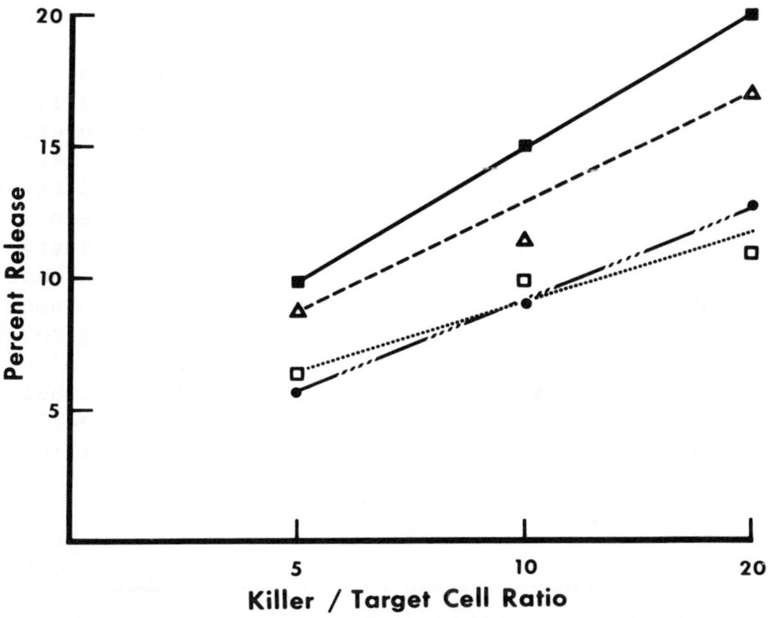

Fig. 2 Inhibition of CML by soluble transplantation antigens. Increasing numbers of killer cells generated in MLC were added to 10^5 target cells labelled with ^{51}Cr (■——■). For inhibition, the reaction was carried out in the presence of soluble transplantation antigen. The amount of antigen added was equivalent to 0.3 x 10^6 (Δ-----Δ) and to 3 x 10^6 (□ .... □) intact cells. The degree of inhibition obtained by soluble antigen was compared to the reduction in the release of ^{51}Cr by the addition of 0.5 x 10^6 intact, unlabelled target cells to the reaction mixture (●.·—·.·—·.●). The lymphoblasts from the same source were used to generate the killer cells in MLC and in the CML reaction as target, and source of soluble antigen. The maximal release and spontaneous release of ^{51}Cr from target cells was 4576 and 526 CPM respectively. Each point on the graph represents a mean from duplicate determinations; the maximal variation from the mean was 10%.

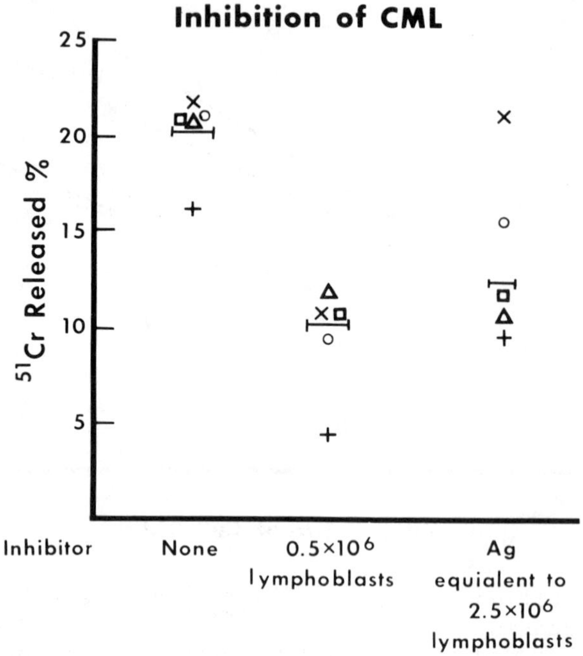

Fig. 3 Inhibition of CML by soluble transplantation antigens (Ag). The results of 5 experiments designated by different symbols are shown. In each experiment the degree of inhibition obtained by soluble antigen was compared to the reduction in the release of ^{51}Cr by the addition of unlabelled lymphoblasts: in each experiment, the lymphoblasts from the same source were used to generate the killer cells in MLC and in the CML reaction as target, and source of soluble antigen.

TABLE 1

INHIBITION OF CML BY SOLUBLE TRANSPLANTATION ANTIGENS

Inhibitor added		^{51}Cr release (%) M	P
---		20.5[b]	21.1
0.5×10^6 M cells		12.9	N.D.
2×10^6 M cells		N.D.	20.0
0.5×10^6 P cells		N.D.	10.9
M soluble antigen[a]		11.0	22.5
Spontaneous release	CPM	526	892
Maximal release	CPM	4576	5504

Responding cells were raised in MLC against mitomycin-treated M cells or P cells; the killer cells raised were used against the appropriate homologous target M or P. Ratio of killer cells to target cells was 10/1. For inhibition, unlabelled M or P cells or soluble transplantation antigens derived from M were used. N.D. stands for experiment not done.

a) The amount of soluble antigen added was equivalent to 3×10^6 cells.

b) Each figure represents the mean from duplicate samples; the duplicates varied from the mean by less than 10%.

TABLE 2

THE INDUCTION OF CML ACTIVITY BY TRANSPLANTATION ANTIGENS

Culture Composition	MLC CPM ^{3}HT	CML % ^{51}Cr Release Target cells A	B	C
A+Cag(8 μg)[a]	556	6.0	6.4	10.5
A+Cag (40μg)	507	4.1	7.4	14.4
A+Bm+Cag (8 μg)	4812	9.3	21.6	24.5
A+Bm+Cag (40 μg)	4902	9.2	27.8	27.7
A	481	9.9	6.1	11.0
A+Bm	5567	4.3	11.7	11.3
A+Cm	7046	8.4	5.8	46.1
A+Bm+Cm	7323	8.8	13.7	68.9
A+B plat ($6.5x10^6$)	585	4.2	2.1	7.6
A+B neut ($1.0x10^6$)	1612	0.9	3.0	6.9
A+Cm+B plat ($6.5x10^6$)	6809	10.3	9.9	52.9
A+Cm+B neut ($1.0x10^6$)	5088	7.2	8.1	41.2
Spontaneous release	CPM	206	181	266
Maximum release	CPM	1376	1282	3155

Killer cells were raised in MLC containing responding cells (A), mitomycin treated stimulating cells (Bm or Cm) and transplantation antigens of B or C origin. The former were in the form of platelets or neutrophils, the latter in the form of soluble preparations. The CML reactions were carried out at a killer cell to target cell ratio of 16; the percentage release was calculated on the basis of spontaneous release and maximal release recorded on the last two lines. All reactions were done in duplicate: the duplicates varied from the mean by less than 10%.

a) 1 mg of soluble antigen was equivalent to 8×10^6 lymphoblasts.

cultures. CML activity was generated against B and C target cells (columns 3 and 4) by only the respective stimulating cells, however CML activity raised in cultures containing both B_m and C_m cells was substantially increased against both targets, suggesting a synergistic action in the generation of CML cells. None of the antigenic preparations cultured alone with the responding cells induced CML activity. However, B antigenic preparations, in the form of platelets and neutrophils when cultured with responding cells in the presence of C mitomycin-treated cells caused a modest increase in the Cr release from B target cells, suggesting that they were instrumental in the generation of anti B killer cells. Similarly soluble H antigens derived from C cells gave rise to anti C killer cells in cultures containing B_m cells; neither the antigens alone, nor B_m cells alone, cultured with responding cells could induce anti C killer cells. However, the CML reactivity induced by the antigen and B_m cells together did not appear to discriminate between B and C target cells. In a subsequent experiment, MLC reactions were generated using cells from 2 HL-A identical individuals, mother and child, possessing HL-A antigens, 1, 2, 8, 12 (Table 3). The cultured cells tested against two unrelated target cells (L and C) gave essentially no release of Cr; however, addition of C mitomycin-treated cells, or C soluble antigens to the MLC reaction did induce CML activity specifically against C, but not against L target cells. Thus, it would appear that solubilized preparations, which do not elicit an MLC response, can nevertheless induce the generation of specific CL in vitro.

DISCUSSION

Blastogenesis by solubilized histocompatibility antigens has been described but the extent of stimulation reported was in fact very low and/or it has been only evaluated by morphological criteria which do not readily lend themselves to quantitation (16,17). Failure to confirm these results by others (2-4) and by the present communication suggest to us that elicitation of the MLC reaction is not a function of HL-A antigens. However, it may be argued that solubilized preparations, although possessing serological activity, might represent incomplete molecules sufficiently altered by the extraction procedures to have lost their biological activity. This latter interpretation is not likely in view of the observation that platelets and neutrophils, not subjected to extraction procedures fail to elicit MLC reactions despite the fact that they are rich in histocompatibility antigens (18,19). Furthermore, antigenic preparations, even if

TABLE 3

THE INDUCTION OF CML ACTIVITY BY TRANSPLANTATION ANTIGENS

Culture Composition	MLC CPM ^{3}HT	CML % ^{51}Cr Release Target cells L	C
H + J	8,443	1.5	6.2
H + J + L_m	17,335	14.0	21.8
H + J + C_m	21,195	2.0	44.2
H + J + Cag (25 μg)[a]	9,130	2.5	13.5
Spontaneous release	CPM	1,139	570
Maximum release		5,130	3,491

Killer cells were raised by the addition of L mitomycin-treated cells, or C mitomycin-treated cells, or solubilized antigens to an MLC reaction generated between leukocytes of two HL-A identical donors, H and J. CML reactions were carried out as indicated in Table 2.

a) 1 mg of soluble antigen was equivalent to 8 x 10^6 lymphoblasts.

incomplete, might have been expected to inhibit stimulation by the corresponding intact leukocytes, however, such inhibition was not detected in the experiments described.

Finally, the most compelling argument that the lack of MLC activation by solubilized antigens is not due to a loss of biological activity is our demonstration of their ability to inhibit CML and generate CL during the MLC reaction. Inhibition of the CML reaction by HL-A antigenic preparations in the present investigation and by H-2 antigenic extracts as recently reported (20) suggest that H antigen is the actual target of cytotoxic lymphocytes in the CML reaction. This conclusion is further strengthened by the demonstration that solubilized antigens added to an MLC reaction can generate cytotoxic activity against cells bearing the same antigens indicating that they may not only determine the specificity of CML once induced, but may also be responsible for its elicitation. This conclusion must be tempered inasmuch that the antigenic preparations used in the present investigation contained substances other than HL-A antigens, however, the demonstrated specificity of both the inhibition of CML and generation of CL lends strong support to this interpretation.

The demonstration that LD antigens give rise to the proliferative reaction in MLC, and SD antigens are responsible for the generation of effector cells, suggests that the MLC-reactive cell and the precursor of CL may not be one and the same cell. According to such a model the precursor of CL, upon interaction with SD antigens may differentiate into mature effector cell under the influence of a mediator derived from the activated MLC-cell. Such a mediator may be blastogenic factor (5). Indeed in preliminary experiments cytotoxic lymphocytes could be generated when responding cells were cultured with solubilized HL-A antigens and culture supernatants derived from mixed leukocyte cultures containing blastogenic factor activity.

REFERENCES

1. Rode, H.N. and Gordon, J., Cellular Immunology 13: 87, 1974.
2. Leventhal, B.G., Mann, D.L. and Rogentine, G.N., Jr., Transplant. Proc. 3: 243, 1971.
3. Ranney, D.F., Gordon, R.O., Pincus, J.H., and Oppenheim, J.J., Transplantation 16: 558, 1973.
4. Ling, N.R., Hardy, D.A., and Steel, C.M. in Proceedings of Eighth Leukocyte Culture Conference, in press.
5. Gordon, J., and MacLean, L.D., Nature (London) 208: 795 1965.

6. Main, R.K., Cole, L.J., Jones, M.J., and Haire, H.M., J. Immunol. 98: 417, 1967.
7. Hardy, D.A., and Ling, N.R., Nature (London) 221: 545-548, 1969.
8. Eijsvoogel, V.P., Du Bois, M.J.G.J., Melief, C.J.M., de Groot-Kooy, M.L., Koning, C., van Rood, J.J., van Leeuwen, A., du Toit, E. and Schellekens, P.Th.A., in "Histocompatibility Testing" (ed. by Dausset, J. and Colombani, J.), p. 501, Munksgaard, Copenhagen, 1973.
9. Bach, F.H., Segall, M., Stauber-Zier, K., Sondel, P.M., Alter, B.J., Science 180: 403, 1973.
10. Alter, B.J., Schendel, D.J., Bach, M.L., Bach, F.H., Klein, J. and Stimpfling, J.H., J. Exp. Med. 137: 1303, 1973.
11. Abbasi, K. and Festenstein, H., Europ. J. Immunol., 3: 435, 1973.
12. Howe, M. and Cohen, L., in Proceedings of the Eighth Leucocyte Culture Conference, in press.
13. Davies, D.A.L., Manstone, A.J., Viza, D.C., Colombani, J. and Dausset, J., Transplantation 6: 571, 1968.
14. Reisfeld, R.A., Pellegrino, M.A. and Kahan, B.D., Science 172: 1134, 1971.
15. Rode, H.N. and Gordon, J., J. Immunol. 104: 1453, 1970.
16. Viza, D.C., Degani, O., Dausset, J. and Davies, D.A.L., Nature 219: 704, 1968.
17. Adler, W.H., Takiguchi, B., Marsh, B. and Smith, R.T., J. Immunol. 105: 984, 1970.
18. Fellous, M. and Dausset, J., Nature (London) 225: 191, 1970.
19. Lucas, Z.J., Transplant. Proc. 3: 240, 1971.
20. Bonavida, B., J. Immunol. 112: 926, 1974.

THE SPECIFICITY OF GUINEA PIG T LYMPHOCYTE RESPONSES TO CHEMICALLY DEFINED ANTIGENS

Charles A. Janeway, Jr. and William E. Paul
Laboratory of Immunology, National Institute of Allergy and Infectious Diseases, National Institutes of Health, Bethesda, Maryland

While the receptor for antigen on the surface of the B lymphocyte is generally accepted to be immunoglobulin, and to have the specificity characteristics of the antibody produced by such cells, the nature and specificity of the receptor for antigen on T lymphocytes is a matter of controversy (1). One of the strongest arguments against a basic similarity for T and B cell receptors has been the failure, in most instances, to observe T cell responses specific for simple chemical haptens, which are readily recognized by B cells, and which have been shown to bind specifically to B cell receptors. Although T cells capable of interacting with such haptens have been described (2,3), careful study of such T cells reveals a recognition of carrier determinants as well (3).

We have approached the problem of T cell receptor specificity by exploiting an apparently hapten specific T cell response comparable to that first described by Benacerraf and Gell (4). In our experiments, guinea pigs which were immunized with the hapten 2,4-dinitrophenyl (DNP) coupled directly to *Mycobacterium tuberculosis* of strain H37Ra (H37) showed delayed hypersensitivity (DH) to many DNP-carrier conjugates, and their peritoneal exudate lymphocytes (PELs) produced migration inhibition factor (MIF) and proliferated *in vitro* on contact with the same DNP conjugates. Preliminary analysis of this system has shown that the *in vitro* proliferation of PELs is a T cell response, and, while the DNP group is required for the elicitation of all the T cell responses in this system, carrier determinants contiguous with the hapten are also clearly involved in functional antigenic recognition by these T cells (5). In this report, these results are confirmed, and further evidence is presented which strengthens the contention that such apparently hapten specific T cells in fact manifest a striking degree of conjugate specificity. In particular, several clones of hapten-reactive T cells have been demonstrated in PELs from DNP-H37 primed guinea pigs, the clones being distinguished by the carrier rather than the hapten, which in all instances is DNP.

All the methods and antigen preparations used here have been described previously (5), except for the copolymers of L-glutamic acid, L-lysine and L-alanine (GLA), in which the subscript denotes the percent alanine residues, and which were kind gifts of Dr. Paul H. Maurer.

The experiment shown in Table I illustrates the basic system employed, as well as several of the antigens used in these studies. Strain 13 guinea pigs were immunized with H37 or DNP-H37, and tested for DH, MIF and in vitro proliferation with several DNP conjugates 3 weeks later. H37 primed animals show little reaction with any of the DNP conjugates, but react strongly with PPD, as expected. Animals primed with DNP-H37 respond well to PPD, but they also show DH, MIF and in vitro proliferative responses to DNP-ovalbumin (OVA), DNP-guinea pig albumin (GPA) and DNP coupled to a polymer of L-glutamic acid, L-alanine and L-tyrosine (L-GAT). These carriers are unrelated chemically, and one might therefore conclude that these T cell responses are specific for the hapten DNP exclusively. However, simply interposing the flexible tripeptide spacer β-alanyl-glycyl-glycyl between the DNP group and the carrier (DAG-carrier) abolished the T cell responses, while such compounds interact readily with anti-DNP antibody and DNP-specific B cells (6). In addition, DNP conjugates of a copolymer of L-glutamic acid and L-lysine (L-GL) do not elicit T cell responses in either strain 2 or 13 guinea pigs which have been immunized with DNP-H37. Thus, while many DNP-carrier conjugates will stimulate a variety of T cell responses in guinea pigs primed with DNP-H37, carrier alone, or DNP on certain carriers, will not elicit such responses.

The role of the hapten and its structure in the activation of T cells and in binding to anti-DNP antibody from the same guinea pigs immunized with DNP-H37 is illustrated in Table II. It can be seen that DNP, paranitrophenyl (pNP) and trinitrophenyl (TNP) conjugates stimulate such T cells and bind efficiently to anti-DNP antibody, as shown by the low I_{50} values. By contrast, orthonitrophenyl (oNP) and 2,6-dinitrophenyl (2,6-DNP) conjugates stimulate T cells only very weakly, and also do not bind anti-DNP antibody strongly. While nitrophenyl haptens that bind effectively to anti-DNP may also elicit such T cell responses, the ability to bind anti-DNP is not a sufficient condition for T cell activation. Thus ε-DNP-L-lysine and DAG-OVA, both of which bind anti-DNP antibody avidly, fail to stimulate T cell responses in this system.

As a result of such experiments, it was postulated that the receptor for antigen on T cells from guinea pigs primed with DNP-H37 recognizes DNP in a fashion similar to conventional anti-DNP antibody, in which the para nitro group plays a dominant role, but that T cell activation also requires recognition of further determinants contributed by the car-

TABLE I

IN VITRO PROLIFERATION, INHIBITION OF MACROPHAGE MIGRATION, AND DELAYED HYPERSENSITIVITY IN STRAIN 13 GUINEA PIGS IMMUNIZED WITH H37 OR DNP-H37

Responses in Strain 13 Guinea Pigs Immunized with:

Stimulating Antigen	H37			DNP-H37		
	Proliferation[a]	D.H.[b]	M.I.F.[c]	Proliferation[a]	D.H.[b]	M.I.F.[c]
0	(1121)	---	100 ± 5	(1778)	--	100 ± 7
PPD[d]	35180	14	4 ± 1	51691	14	65 ± 5
OVA	0	0	-----	0	0	90 ± 12
DNP_6-OVA	673	0	110 ± 6	20345	6	78 ± 3
DAG_7-OVA	201	0	132 ± 15	746	6[e]	94 ± 5
DNP_{12}-GPA	---	0	-----	18297	13	---
DAG_{14}-GPA	---	0	-----	1373	3[e]	---
DNP-L-GAT	173	0	154 ± 20	29209	23	50 ± 7
DNP-L-GL	0	0	-----	14	5[e]	---

a) *In vitro* proliferative response of PELs with antigen at 100 μg/ml expressed as CPM, E-C.

b) Average diameter of induration and erythema of 24 hour skin reactions to 50 μg antigen per site.

c) Inhibition of migration of peritoneal exudate cells by antigen at 100 μg/ml as a percent of control migration.

d) 1 μg/ml for in vitro proliferation and 5 μg per skin test site.

e) Strong, hemorrhagic 4 hour skin reactions which were fading at 24 hours.

TABLE II

ROLE OF HAPTEN AND LINK TO CARRIER ON *IN VITRO* PROLIFERATION OF DNP-H37 IMMUNE STRAIN 13 PELS AND ON BINDING BY ANTI-DNP ANTIBODY FROM THE SAME ANIMAL

Stimulating Antigen (100 μg/ml)	Stimulation (CPM, E-C) After Priming with:		Inhibition of Antibody Binding to DNP-Lysine†
	H37	DNP-H37	(I_{50},‡ Hapten x 10^{-8} M)
0	(1216)	(2520)	–
PPD*	28819	22086	–
OVA	0	0	–
DNP_6-OVA	144	14331	105
DAG_7-OVA	0	0	18.6
$oNP_{0.5}$-OVA	607	0	> 3300
pNP_1-OVA	1723	8263	36.4
TNP_{14}-OVA	1767	6347	310
2,6-DNP_4-OVA	531	2613	25000
ε-DNP-L-Lysine	–	–	10.7

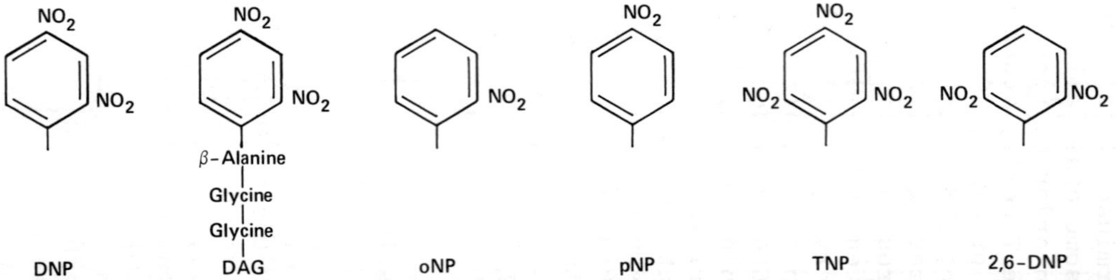

*1 μg/ml
†0.33 x 10^{-8} M ^{3}H-ε-DNP-L-Lysine binding by anti-DNP antibody
‡I_{50} Concentration of hapten required to cause 50% inhibition of binding of ^{3}H-DNP-Lysine by anti-DNP (x 10^{-8} M)

rier. Such determinants were felt to be contiguous with the hapten moiety, as the insertion of small, flexible spacer molecules between the hapten and the carrier can prevent T cell activation by DNP-proteins. On the basis of the strong stimulation obtained with DNP-L-GAT, in which most of the DNP groups are coupled to the hydroxyl of tyrosine, one such determinant is thought to comprise DNP-O-tyrosine, and possibly some other amino acid residues on the carrier as well.

In order to further explore the detailed specificity of these T cells, other DNP-synthetic polypeptide conjugates were prepared. Of particular interest are DNP-GLA_{30}, DNP-GLA_{40}, DNP-GLA_{60} and DNP-GLAT, a conjugate of a random linear polymer of L-glutamic acid, L-lysine, L-alanine (35%) and L-tyrosine (5%). As previously described, DNP-GLAT could be treated with 2-mercaptoethanol to remove DNP coupled to the hydroxyl group of tyrosine, leaving intact the DNP coupled to amino groups, particularly those of lysine. DNP-GLA and DNP-GLAT were used to stimulate PELs from strain 2 and strain 13 guinea pigs primed with DNP-H37. In the representative experiment shown in Table III, DNP-OVA, DNP-GPA, DNP-keyhole limpet hemocyanin (KLH) and DNP-L-GAT all stimulate strain 2 and strain 13 PELs to a similar extent, while DNP-L-GL fails to stimulate PELs from either strain. DNP-GLA stimulates strain 2 PELs strongly, while stimulating strain 13 PELs only weakly. This has been a consistent finding in this system, and it may be related to the fact that responsiveness to both DNP-L-GL and GA is controlled by immune response genes linked to the strain 2 major histocompatibility complex. The response of PELs from these guinea pigs to DNP-GLAT is also interesting in this respect. Both strains respond well to DNP-GLAT, but thiolysis with 2-mercaptoethanol lowers the response of strain 13 PELs far more than that of strain 2 PELs. It seems likely that strain 13 PELs respond primarily to the DNP-O-tyrosyl residues in DNP-GLAT, while strain 2 PELs respond strongly to DNP-GLA-like determinants in DNP-GLAT. Finally, in this experiment PELs from strain 2 guinea pigs responded very poorly to DNP coupled to a copolymer of L-glutamic acid and L-tyrosine (GT), while PELs from strain 13 animals responded well to DNP-GT. This result is interesting in light of the fact that immune responses of guinea pigs to GT are controlled by an Ir gene linked to the strain 13 major histocompatibility complex.

In order to examine whether only one, or alternatively several distinct clones of PELs from strain 2 guinea pigs primed with DNP-H37 account for the responses to these diverse DNP-carrier conjugates, an attempt to destroy cells proliferating to one DNP-conjugate and then to restimulate

TABLE III

THE IN VITRO PROLIFERATIVE RESPONSE OF PELS FROM STRAIN 2 AND STRAIN 13 GUINEA PIGS IMMUNIZED WITH DNP-H37 TO VARIOUS DNP CONJUGATES

Stimulating Antigen	Stimulation (CPM, E-C) of PELs from	
(100 μg/ml)	Strain 2	Strain 13
0	(3884)	(2604)
PPD[a]	62766	68639
DNP-KLH	27548	29573
DNP-OVA	12811	24248
DNP-L-GL	0	0
DNP-L-GLA_{40}	11868	3472
DNP-L-GLA_{60}	18486	4186
DNP-L-GLAT	14369	17962
Thiolysed DNP-L-GLAT[b]	10717	7845
Re-DNP-L-GLAT[c]	13788	19155
DNP-L-GT	848	7914
DNP-L-GAT	10496	18061

a) 1 μg/ml.
b) Treatment with 2.5M 2-mercaptoethanol to remove DNP coupled to tyrosine.
c) Reconjugation of thiolysed compound with DNP.

cells with a second DNP-conjugate was made. To do so, PELs were stimulated with antigen *in vitro*, 5-bromodeoxyuridine (BUdR) being added after 2 days. The cells were exposed to light one day later, washed thoroughly, counted, and recultured with antigen in the presence of an equal number of fresh, Mitomycin C inactivated peritoneal exudate cells as a source of macrophages. The results of such an experiment are given in Table IV. It can be seen, in each instance, that preculturing PELs with a particular DNP conjugate under these conditions specifically reduces the subsequent response of the cells to the same DNP-conjugate, but has only a limited effect on the response of the cells to other DNP conjugates. This finding has been reproduced several times. It strongly suggests that DNP-KLH, DNP-GLA_{30}, and DNP-L-GAT interact with different clones of T cells. The data are not yet sufficiently quantitative to determine whether these clones overlap and, if so, to what extent they do so. Further experiments using this technique have suggested that there may be several other readily defined clones of DNP-reactive T cells in this system. These studies have also shown that structurally related antigens interact with the same clone (i.e. all the DNP-GLA compounds interact with cells which respond to DNP-GLA_{30}). The finding of distinct, DNP-reactive clones of T cells, in which each clone is delineated by a given conjugate or family of structurally related conjugates, is very strong evidence for conjugate specificity in the responses of these T cells. Thus, while the specificity of such T cells is superficially similar to that of DNP-specific B cells, in that these T cells can respond to DNP on a variety of unrelated carriers, the precise specificity of such T cells can be shown to involve the hapten plus contiguous regions of that carrier, almost certainly including the amino acid to which the hapten is coupled and an adjacent amino acid residue. In general, anti-DNP antibody and DNP-specific B cells do not so distinguish between carriers. Therefore, it seems likely that T cells and B cells have distinctive libraries of specificities, with those of T cells being at least as precise as those of B cells (7).

An alternative possibility is that each of these T cell clones is in fact DNP-specific but differs in regard to carrier recognition mediated by an independent recognition unit, such as an Ir gene product. This would require that T cells be specialized, at least to some extent, not only in terms of the expression of the principal antigen binding receptor but also (and independently) in terms of the expression of the Ir gene product. This proposal has the disadvantage that it adds another order of complexity to the problem

TABLE IV

EFFECT OF PRECULTURE WITH ANTIGEN FOLLOWED BY BUdR AND LIGHT ON THE ABILITY OF PELS FROM STRAIN 2 GUINEA PIGS IMMUNE TO DNP-H37 TO RESPOND TO ANTIGEN

Stimulating Antigen	Stimulation (CPM, E-C) of PELs Following Preculture with:			
(100 μg/ml)	0	DNP-KLH	DNP-GLA$_{30}$	DNP-L-GAT
0	(937)	(2052)	(1964)	(3427)
DNP-KLH	5997	1296[a]	3622	3474
DNP-GLA$_{30}$	6125	4855	2329	5717
DNP-L-GAT	5082	4674	5490	370

a) The lowest stimulation given by each antigen is underlined.

of T cell specificity. Nevertheless, the finding that the responses of PELs from guinea pigs immunized with DNP-H37 to certain defined antigens appear to be related to the Ir gene status of the cell donor suggests that Ir gene products play some role in antigen recognition and cell activation by these T cells, either as receptor or as an ancillary specific recognition unit.

REFERENCES

1. Paul, W.E., Transplant. Rev. 5: 130, 1970.
2. Rutishauser, U. and Edelman, G.M., Proc. Nat. Acad. Sci. 69: 3774, 1972.
3. Möller, E., Scand. J. Immunol. 3: 339, 1974.
4. Benacerraf, B. and Gell, P.G.H., Immunology 2: 219, 1959.
5. Janeway, C.A., Jr., Cohen, B.E., Ben-Sasson, S.Z. and Paul, W.E., J. Exp. Med., in press.
6. Janeway, C.A., Jr., Submitted for publication.
7. Janeway, C.A., Jr., Sakato, N. and Eisen, H.N., Submitted for publication.

DISCUSSION FOLLOWING SYMPOSIUM SIGNALS FOR LYMPHOCYTE ACTIVATION II: ANTIGENIC SIGNALS AND RECEPTORS

Cooper - Bethesda: Dave, you have shown clearly that DNA synthesis in T lymphocytes depends on macrophages, but is it certain that all lymphocyte activation proceeds through DNA synthesis? Do you have any evidence that some of the earlier functions of T lymphocyte activation are dependent on macrophages?

Rosenstreich - Bethesda: We have looked for early events in activation. In our hands we can measure chemotactic factor at least within four hours of the initiation of our experiment using absolutely macrophage dependent antigens.

Rosenthal - Bethesda: I disagree with your major thesis that the mechanisms of activation of lymphocytes by mitogens and by antigens are identical and only different in minor insignificant details. How do you reconcile the observation that one can put PHA or Con A on an allogeneic macrophage and initiate the proliferation of a T lymphocyte whereas with antigen, there is restriction to syngeneic macrophages in terms of the triggering of lymphocyte proliferation?

Rosenstreich - Bethesda: I did not mean to reject a requirement for syngeneic macrophages and lymphocytes in antigen induced proliferation. However, that mechanism may not be absolute either; and it may depend on the specific antigen. Exactly what the requirement of the syngeneic macrophage is and exactly what the function of the Ir gene product systemis, is not clear. It may be related to the actual recognition by the T cell of the antigen coupled to the Ir gene product.

Siegel - New York: For antigen stimulation you need syngeneic macrophages. Does stimulation to LAF also require that it comes from syngeneic macrophages?

Rosenstreich - Bethesda: I don't know.

Rubin - New York: I would like you to comment on other species. For instance, very clean preparations of human T cells respond very nicely to PHA and Con A. Is there some problem here with your general thesis?

Rosenstreich - Bethesda: I think it is a problem of how you

define macrophage depletion or lymphocyte purity. I think it is clear from the data of Levis and Robbins that when they performed an extensive purification on human lymphocytes, they were able to completely obliterate PHA response. We used to assume that our lymphocytes were macrophage depleted, and in fact when you look for macrophage, you see very few. Nevertheless, they respond well to mitogens. One possibility is that you need very few macrophages to enhance the response to mitogens. The second possibility which deals directly with your question is that there are T cell subpopulations. We are then dealing with one subpopulation which appears to be absolutely macrophage dependent. I have no doubts that there may be other T cell populations that are not so macrophage dependent. The second fraction that we collect off the nylon wool contains some B cells and a few macrophages, and this may explain why they respond to mitogens directly. On the other hand, this population may actually contain an entirely different T cell population. In mice and rats macrophages exert a real suppressive function. We have done this ourselves and if you use more than 5% macrophages of any type, unactivated or activated, you can suppress responsiveness. I am not sure why the mouse differs, but in the guinea pig you can add 15% or 100% macrophages and still not suppress. Even in the guinea pigs, a suppressive factor may be released. This material is very low molecular weight, and we dialyze our supernatants to get rid of this possible suppressive activity. If, in fact, we take fresh cultured supernatants, they do not enhance as well and we can get a 5 fold greater enhancement after dialysis.

Bach - Madison: You did not tell us much about the macrophage population which you use. Is it possible that it contains enough T lymphocytes so that in fact the cells you are adding to your mitogen stimulated cultures are not only macrophages but also a T lymphocyte subpopulation which can respond and therefore give you a larger response, maybe by secondary factors or something? You may not need a macrophage per se for the mitogen response but rather a different T cell population.

Rosenstreich - Bethesda: It's possible.

Sage - Durham: Have you ever studied the effect of the density of DNP-D-GL molecules per Sepharose bead on stimulation? If your model is right, then theoretically if you attached a very small number of these molecules on a Sepharose bead, the complex should be a very poor stimulator. If you load up the bead, you should get a portionality of stimulation to the density of these molecules on individual beads.

Mosier - Bethesda: We have not done that in a systematic way. The beads I have shown you were of an intermediate substitution ratio and they were the most immunogenic. We had more highly substituted which were less immunogenic and less highly substituted which were poorly immunogenic but that was over a fairly narrow dose range of DNP-D-GL substitutions.

Oppenheim - Bethesda: One could argue at least as far as it goes from your data that in the case of the macrophage B cell interaction is merely a presentation phenomenon. In that light, does the 2ME substitute for the macrophage help for the DNP-D-GL?

Mosier - Bethesda: No, we have not tested that.

Handwerger - Minneapolis: I am interested in the mechanism of the soluble DNP-D-GL inhibition of anti-DNP responses to DNP-Ficoll. All your work that you presented has been looking at populations of nonadherent cells. Have you looked at anti-theta treated nonadherent cells to find out if the DNP-D-GL inhibition is mediated by T cell?

Mosier - Bethesda: Yes, we have. If you deplete T cells from nonadherent cells, the tolerogenic profile of DNP-D-GL is exactly the same. So we feel that T cells do not play a role in tolerance induction.

Unanue - Boston: I would like to raise the point concerning your explanation that DNP-D-GL in soluble forms tolerizes the T cell because it does not cross link. In experiments that we did with Katz, what we found was indeed that DNP-D-GL was highly tolerogenic when you put it on B cells. It did cross link and cap all Ig receptors for D-GL.

Mosier - Bethesda: That may be dependent on antigen dose. At what antigen concentration were these experiments made?

Unanue - Boston: These were done usually at high antigen concentrations so we can detect it by autoradiography. You are at a situation where you are definitely tolerizing with perfectly good cross-linking but you have a blockade somewhere that the cell does not interiorize that complex. As it does not get rid of it from the membrane at the same time the cell is blockading its subsequent re-expression of new surface immunoglobulin.

Katz - Boston: Fritz, I might have missed it, but what were Karen Zier's absorption experiment results?

Bach - Madison: They were published long ago and that is why I did not go into them in detail. The results are that if you take a mixed lymphocyte culture, activate it for six days, say AB_m, you absorb this on a B monolayer for one hour, then you take your nonadherent cells and your adherent cells; essentially what one finds is that in the nonadherent cells is 92% on the average of all the cells that are proliferating; in the adherent cells are virtually all of the cytotoxic potential. One can also do the experiment the other way of taking A cells, absorbing them on a B monolayer and then stimulating with B_m and show an excellent proliferative response and yet very little if any cytotoxic response. Based on the work of Brondz and many others in terms of absorption.

Greaves - London: Am I right in assuming that your target cell is a PHA activated lymphoblast?

Bach - Madison: In most of the experiments I think, which I have shown you, I can tell you that one can use either PHA stimulated lymphoblastic cells or normal lymphocytes obtained from lymph node of mice as targets. Certainly this is true in the PHA dependent cytotoxicity, in the LD different combinations which I showed you. Barbara Alter did these experiments to counter the possible argument that LD is simply not expressed on PHA stimulated cells, that normal lymph node cells behave identically to the PHA stimulators.

Greaves - London: Well that is exactly the point I wanted to get at. The possibility that the PHA blasts lack the LD determinants is a very real one and I wondered if you had tried LPS activated T cells as the target?

Bach - Madison: Yes. In addition to trying the normal lymphocytes which Alter tried, Schendel has used LPS stimulated cells and MLC stimulated cells. The question which we wanted to address ourselves to here and which I suspect is on your mind is what if it is a very small subpopulation of B cells or if it were the Ia antigens or something like this. We have tried to expand this population and we have not been able to. All of those have given the same results in the LD and SD different combinations.

Eijsvoogel - Amsterdam: I am also intrigued by the question of whether you do not induce cytotoxic cells with LD or whether LD is a poor target and now you raise the point of the PHA dependent cytotoxicity. I suppose that you see the PHA in your assay as a linking mechanism which otherwise is the LD. Right? Did you do absorptions with LD identical activated cell and then see if your PHA was any different after absorption on the monolayer?

Bach - Madison: Not yet. That is obviously a very important thing to try to get at. That is to combine the absorption techniques with this kind of studies and we have not done it yet.

SECTION III

EARLY EVENTS IN LYMPHOCYTE ACTIVATION I: MEMBRANE AND BIOCHEMICAL EVENTS

Chairman

RICHARD D. BERLIN

University of Connecticut Medical Health Center
Farmington, Connecticut

Co-Chairman

ELLEN VITETTA

University of Texas Southwestern Medical School
Dallas, Texas

REGULATION OF CELL SURFACE TOPOGRAPHY

Richard D. Berlin, M.D.
Department of Physiology, University of Connecticut Health Center, Farmington, Connecticut.

Many papers of this conference will abundantly document the importance of the cell surface in determining the physiological and immunological functions of leukocytes. We shall hear in this session of specific alterations in membrane lipid and in the behavior of membrane proteins which result from interactions with exogenous molecules. I shall briefly review some evidence which indicates mechanisms by which the membrane response can be modified.

The binding of the same ligand to different cell types produces widely different effects on the distribution of the ligand receptors and in the subsequent physiological response. For example, bound Con A assumes a different topographical distribution in viral transformed as opposed to normal mouse fibroblasts and is considerably more toxic to the transformed cell (1,2).

As we shall discuss, changes in surface topography usually involve the migration of receptors in the plane of the membrane. Hence, many membrane properties have been rationalized in terms of the fluid mosaic model of membrane structure in which proteins are free to move in a sea of lipid (3). The "fluidity" would be dependent essentially on the fatty acid and cholesterol composition, and possibly secondary interactions of the lipid with protein receptors. However, I wish to develop two points which require us to focus on elements extrinsic to the membrane in order to understand the regulation of its function:

(1) Although it is convenient to rationalize the movement of surface receptors in terms of membrane fluidity, it appears that the apparent "fluidity" of the membrane is alterable on a time scale so short with respect to the surface phenomena observed that significant alterations in membrane composition tend to be excluded.

(2) Pharmacologic agents at concentrations which probably do not affect membranes directly - and also on a short time scale - profoundly affect surface phenomena. These include inhibitors of metabolism, which I shall touch on briefly, alkaloids such as colchicine which affect microtubules, and cyclic nucleotides.

The 3T3 and SV3T3 mouse fibroblasts have proved particularly useful in studies of the surface topography of Concanavalin A (Con A). Using hemocyanin to visualize bound Con A on surface replicas (4) it could be shown that bound Con A was dispersed over the surface of the normal fibroblast but aggregated into small clusters in the virus transformed cell (1,5). More importantly, the clustering was induced by the addition of Con A, since Con A topography of cells fixed prior to addition of the Con A was dispersed (5) (Table I). This capacity for aggregation of receptors has been attributed to an increased fluidity of the membrane and by inference to differences in membrane composition between normal and transformed cells. However, treatment of SV40 3T3 with cyclic AMP and theophylline for only one hour changes their response: the Con A topography is disperse (6). Conversely when 3T3 are briefly treated with low activities of trypsin, their behavior becomes like the SV3T3, and cAMP and theophylline treatment of the trypsinized 3T3 reverts their pattern to disperse (7) (Table II). It seems highly unlikely that significant alterations in fatty acid composition (for example) could have occurred during any of these manipulations involving from 10-60 minutes incubation.

The addition of Con A to certain cell types leads to its aggregation over one pole of the cell into a "cap" (8). Cap formation clearly suggests a high mobility of Con A receptors in the plane of the membrane. But this mobility too can be rapidly modified. Thus, the addition of high concentrations of Con A to lymphocytes blocks the mobility of Con A receptors and of other receptors as well (9). This immobilization is relieved by colchicine. Similarly, although SV40 3T3 do not normally show cap formation with the addition of Con A at high or low concentrations, colchicine-treated SV3T3 cap readily (6). Again cAMP with theophylline antagonizes this cap formation (7). It should be emphasized, that these observations were made of fibroblasts on monolayers in the absence of serum - conditions in which there is extremely little cell movement.

It is now becoming apparent that cyclic GMP (cGMP) also modifies the capping response (10). This development will be reviewed in some detail by Dr. Janet Oliver. However, at this time we should point out that cGMP or agents that increase its intracellular concentration (carbachol or phorbol myristate acetate PMA) antagonize the colchicine - associated capping of SV40 3T3 fibroblasts (11). cGMP also antagonizes the capping responses of polymorphonuclear leukocytes (PMN) (10). In PMN microtubules appear to increase when treated with PMA (12). One is sometimes bewildered by large variations in the effects of cyclic nucleotides which seem to depend not only on cell type but on species as well. A thorough comparison between systems should encompass a range of

TABLE I
CON A/H TREATMENT OF 3T3 and SV3Ts CELLS

Cells	Con A/H* labeling °C	Incubation† post Con A/H °C	CABS§
3T3	4	4	Dispersed
	4	37	Dispersed
	37	37	Dispersed
3T3 Prefix‡	37	37	Dispersed
SV3T3	4	4	Dispersed
	4	37	Clustered
	37	37	Clustered
	37	4	Clustered
SV3T3 Prefix‡	37	37	Dispersed

*Cells were treated with Con A (10 min), rinsed three times in PBS, treated with hemocyanin (10 min), and rinsed and †monolayers were then incubated for an additional 10 min. ‡Fixation before Con A/H labeling was in 1% paraformaldehyde for 10 min. §CABS = Con A-binding-site topography after Con A/H labeling.

From Rosenblith et al. (5).

TABLE II
CON A SURFACE TOPOGRAPHY

Cells	Treatment	Topography
3T3	None	Dispersed
	trypsin	Clustered
	trypsin and dbcAMP, theophylline	Dispersed
	colchicine	Dispersed
SV3T3	None	Clustered
	dbcAMP, theophylline	Dispersed
	colchicine	Capped
	colchicine, dbcAMP, theophylline	Dispersed

From Ukena et al. (7). a) Cells incubated with dbcAMP and theophylline, 1 mM each for 1-24 hours. b) Trypsin treatment: 10 μg/ml at 4°C for 10 minutes followed by 4 serial rinses in 100-fold excess (w/w) ovomucoid and 4 rinses in phosphate buffered saline. c) Colchicine, 1 μM, 1 hour at 37°. All cells were then labelled with Con A and hemocyanin as described (6).

drug and Con A concentrations with careful attention to temperature and period of incubation. Our working hypothesis is that Con A and similar multivalent agents that bind to surfaces induce increases in intracellular cGMP which in turn promote microtubule polymerization. The interaction of microtubules with membranes modifies the translational movement of surface receptors.

During phagocytosis there are also translational movements of lectin receptors which are quantitatively as impressive as those seen during cap formation but they occur in the absence of extrinsic multivalent cross-linking agents. In these experiments(13) Con A and Ricin communis agglutinin were used simply to titrate their membrane receptors before and after phagocytosis. The titrations were normalized to the total membrane protein. A sharp decrease in the specific activities of the receptors was observed which paralleled the degree of phagocytosis, signifying their selective inclusion in the membrane interiorized by engulfment of the phagocytized particle (Figure 1). This selective inclusion was abolished by treatment of the phagocytes with colchicine, but not by lumicolchicine, a photochemical derivative of colchicine which does not affect microtubules but which shares with colchicine a capacity to affect the membrane transport of nucleosides. In this instance the role of microtubules appears more dynamic. Con A receptors are not simply anchored by colchicine-sensitive elements. Con A receptors on PMN (as for other cells) are dispersed prior to phagocytosis and appear to migrate into the internalized membrane (Fig. 2).

At the same time other surface proteins appear to migrate out of the membrane which envelops the phagocytized particle. This process also seems to require intact microtubular structures. Some time ago, we showed that several specific membrane transport systems for nonelectrolytes are not reduced (that is, are preserved) during extensive phagocytosis (14). Since we have no means of identifying their topographical locus prior to phagocytosis, we assume the distribution of transport sites also to be disperse. If so, this means that transport proteins are moving out of internalized membrane at the same time lectin receptors are moving into it. As in the latter case, the selective movement of transport proteins is also colchicine dependent (15).

Using certain fluorescent techniques we have recently shown that microtubules can associate with cytoplasmic membranes (16) but we are far from identifying specific membrane protein - microtubule linkages which would seem to be required for the directed movement of membrane-protein-microtubule complexes. The precise mechanism by which microtu-

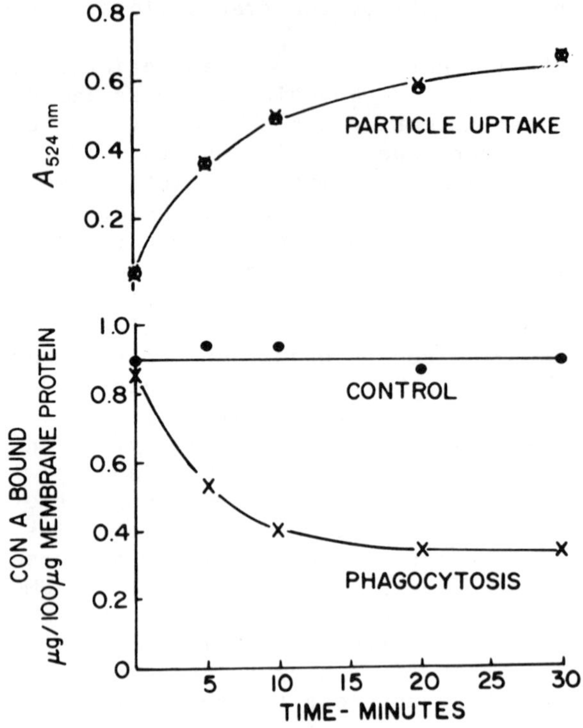

Fig. 1 Top: A plot of the extent of phagocytosis as a function of time with (0-0) or without (X-X) colchicine. Cells were treated with the alkaloid at 5 µM for 30' before initiation of phagocytosis.
Bottom: Effect of phagocytosis on specific activity of CABS of isolated membranes. Cells at 4 x 10^6/ml of medium were incubated without (control) and with (phagocytosis) emulsion for the times indicated. Cells were then centrifuged, washed and resuspended in MH containing 7 µg/ml of labelled Con A and incubated at 4° for 30'. Membranes were then isolated and the specific activity of Con A bound determined. From (13).

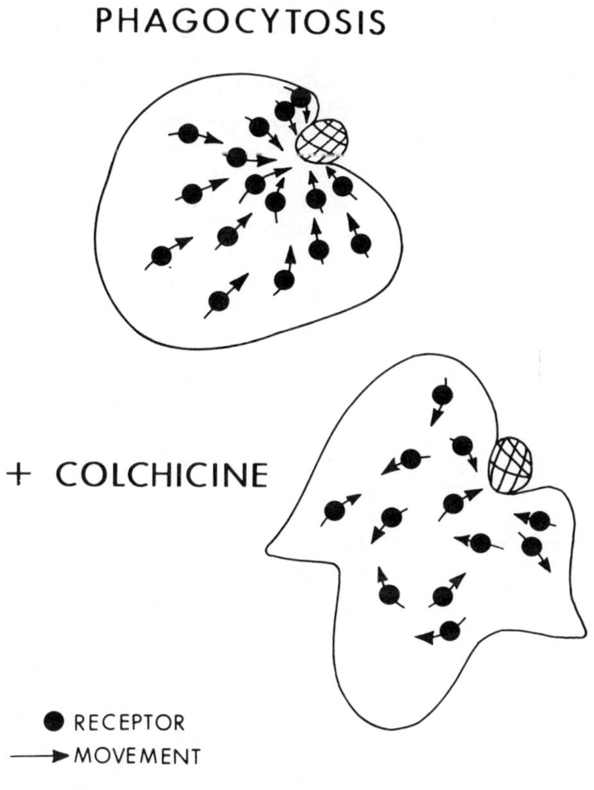

Fig. 2 Diagrams indicating movement of Con A receptors during phagocytosis. Receptors are initially dispersed over the surface. With phagocytosis Con A receptors are swept into the membrane that envelops the particle. In this representation, transport proteins would be moving out of the enveloping membrane. Treatment with colchicine (bottom diagram) abolishes selective movement of Con A receptors and transport proteins (not shown).

bules interact with membranes to alter their surface distribution is unclear. However, we have considered the possibility that local phase changes in the membrane lipid - which could be modified by microtubules - occur during phagocytosis and could result in the partitioning of membrane proteins between internalized and residual membrane.

It is apparent that many membrane proteins are associated with specific phospholipids and depend on them for optimal activity (17). Träuble and Overath (18) have determined that in the bacterial cells they studied, roughly 20% of the phospholipid did not undergo a phase change with the bulk of the lipids, i.e., it appeared segregated presumably with proteins in a way in which chain-chain interactions were disturbed. Looked at from another view, one could say that the immediate lipid environment of a membrane protein might not be equivalent to the bulk lipid. In fact an aggregate of such proteins might create its own unique lipid phase quite distinct from the remaining lipid matrix or from proteins with other dissimilar phospholipid "clouds" about them. The aggregation of proteins with a relatively high proportion of saturated fatty acids (of lower melting point) would in effect create a condensed phase which would tend to freeze-out membrane lipids or proteins of a more fluid nature (i.e., with higher melting-point or saturated fatty acids). Some data of Smolen and Shohet (19) can be interpreted (by us!) to support the hypothesis that during phagocytosis such a partitioning occurs. From their data the fatty acids of phagocytic vesicles are distinctly more saturated than that of the membrane. They suggest that this is due to oxidation of fatty acids within the vesicle, but it appears that the phospholipids internalized are enriched for phosphatidylcholine and this phospholipid in membranes had a relatively high proportion of saturated fatty acids. Our interpretation thus is that during internalization a local aggregation of relatively saturated fatty acids occurs at the sites of internalization which include lectin receptors but freeze out transport proteins which might be associated with phospholipids containing a higher proportion of unsaturated fatty acids. Several experimental tests of this hypothesis are underway in our laboratory, but one rather indirect test has already proven interesting.

If indeed it is the saturated fatty acids which are internalized and others frozen out, then the partitioning of membrane proteins would be influenced by the temperature at which phagocytosis and internalization occurred. Specifically one would predict that at elevated temperatures the specific partitioning would not take place. To test this, we have studied the effect of phagocytosis on the adenine and

TABLE III
EFFECT OF PHAGOCYTOSIS ON ADENOSINE TRANSPORT AT 37° AND 42°Δ

37°			42°		
Time	Ingestion	% Control Transport	Time	Ingestion	% Control Transport
15'	.18	99	8'	.12	69
30'	.26	96	15'	.21	67
45'	.40	88	25'	.34	48

Δ Berlin, R.D., Fera, J., and Oliver, J.M., unpublished observations.

Monolayers of rabbit PMN were prepared on 22 mm glass coverslips as previously described (20). They were then allowed to phagocytize by inverting them over 0.3 ml of a lipid emulsion (21) spread on parafilm. Phagocytosis was quantified from Oil Red O included in the emulsion (13,21). The adenosine transport system has been well characterized (22). Adenosine transport was determined as described and corrected for contamination of the coverslip and non-carrier mediated transport (<10% of the total uptake) as described.

There was no decrease of transport in control (non-phagocytic) cells during incubation at either 37° or 42°. At 37° phagocytosis did not reduce transport for 30' as described for the phagocytosis of polystyrene beads (4) and depressed it only slightly at 45'. At 42° on the other hand, transport is depressed as early as 8' and fell more than 50% at 25'. The periods of phagocytosis at the two temperatures were adjusted so that transport could be compared at intervals during which the magnitude of particle ingestion was greater at the lower temperature. The depression of transport seen at 42°, therefore, could not be due to an increase in the total membrane internalized.

adenosine transport systems in rabbit PMN (Table III). Monolayers of PMN are allowed to phagocytize at several temperatures, rinsed, and their transport then tested at the <u>same</u> temperature (37°). A modest temperature range (32-42°C) was found sufficient to demonstrate the effect. At 32° and 37° phagocytosis did not reduce transport. However, at 42° transport was reduced in parallel with the amount of phagocytosis. The period of phagocytosis was adjusted so that the magnitude of phagocytosis, i.e., membrane internalization, was actually greater at 37° (when there was no reduction of transport) than at 42°. In control experiments it was shown that the transport systems were not affected in the absence of phagocytosis. In these terms the microtubules effect this partition and operationally create a zone enriched for saturated fatty acids (low fluidity). The mechanism(s) by which this could occur are unclear but there are essentially two possibilities: one is that the microtubule-membrane association is primarily with the lipid. This could be direct, or indirect. For example, by mechanically changing the curvature of the membrane, microtubules could alter surface energy and thus the selective movement of component phospholipids. The second is a direct interaction with specific membrane proteins which have in turn specific adsorbed lipid "clouds". The aggregate of these microclouds could create another lipid "environment" for excluded protein elements and thus govern their behavior.

Microtubules can physically associate with membranes. It is clear now that microtubules can modify surface behavior. The mechanism(s) by which they do so are uncertain. However, certain features of the interaction of membranes with microtubules suggest that they are required for the partitioning of specific proteins by mechanisms dependent on local alterations in membrane fluidity. This concept is being developed by analyses of the lipid composition of the phagocytic vesicle and cytoplasmic membrane and by physical studies of their lipid matrices.

REFERENCES

1. Nicolson, G.L., Nature New Biol. 233: 244, 1971.
2. Sharon, N. and Lis, H., Science 177: 949, 1972.
3. Singer, S. and Nicolson, G.L., Science 175: 720, 1972.
4. Smith, S.B. and Revel, J.P., Develop. Biol. 27: 434, 1972. 1972.
5. Rosenblith, J.Z., Ukena, T.E., Yin, H.H., Berlin, R.D. and Karnovsky, M.J., Proc. Nat. Acad. Sci. USA 70: 1625, 1973.

6. Ukena, T.E., Borysenko, J.Z., Karnovsky, M.J. and Berlin, R.D., J. Cell Biol. 61: 70, 1974.
7. Ukena, T.E., Borysenko, J.Z., Black, P.R., Karnovsky, M.J. and Berlin, R.D., submitted for publication.
8. Taylor, R.B., Duffus, W.P.H., Raff, M.C. and de Petris, S., Nature New Biol. 233: 225, 1971.
9. Edelman, G.M., Yahara, I. and Wang, J.L., Proc. Nat. Acad. Sci. USA 70: 1442, 1973.
10. Oliver, J.M., Zurier, R.B. and Berlin, R.D., submitted for publication.
11. Oliver, J.M. and Berlin, R.D., unpublished.
12. Goldstein, I.M., Hoffstein, S.T. and Weissmann, G., Ann. N.Y. Acad. Sci., in press.
13. Oliver, J.M., Ukena, T.E. and Berlin, R.D., Proc. Nat. Acad. Sci. USA 71: 394, 1974.
14. Tsan, M.F. and Berlin, R.D., J. Exp. Med. 134: 1016, 1971.
15. Ukena, T.E. and Berlin, R.D., J. Exp. Med. 136: 1, 1972.
16. Becker, J.S., Oliver, J.M. and Berlin, R.D., submitted for publication.
17. Cronan, J.E. and Vagelos, P.R., Biochim. Biophys. Acta 265: 25, 1972.
18. Träuble, H. and Overath, P., Biochim. Biophys. Acta 307: 491, 1973.
19. Smolen, J.E. and Shohet, S.B., J. Clin. Invest. 53: 726, 1974.
20. Hawkins, R.A. and Berlin, R.D., Biochim. Biophys. Acta 173: 324, 1969.
21. Stossel, T.P., Pollard, T.D., Mason, R.J. and Vaughan, M., J. Clin. Invest. 50: 1745, 1971.
22. Berlin, R.D., J. Biol. Chem. 248: 4724, 1973.

THE MODULATION OF IMMUNOGLOBULIN IN B LYMPHOCYTES AND ITS RELEVANCE TO IMMUNE STIMULATION

Emil R. Unanue and George F. Schreiner
Department of Pathology, Harvard Medical School, Boston, Massachusetts 02115

Interaction of a ligand with the immunoglobulin (Ig) receptor on the B class of lymphocytes produces a series of reactions both at the cell surface and in the cytoplasm, each of which is regulated by complex biochemical events. These cytological reactions are interrelated; and when modulated by appropriate stimuli--such as the helper activity of T cell--result in proliferation and differentiation, the hallmark of a complete immune response. We have been trying to dissect some of the series of events using anti-Ig antibodies as a probe (1). Anti-Ig antibodies have the advantage of producing a limited but well-defined response amenable to analysis because of its rapid development and because technically one can study large numbers of cells. The studies are limited in the sense that they do not lead to the whole process of cell stimulation which, in general terms, takes place upon interaction with the T cell signal. Experiments with labeled antigen have also been made, but the information with these is much more limited because of the difficulties inherent in studying populations containing very small numbers of specific antigen-committed lymphocytes (2). Nonetheless, where they have been compared, the B cell response to anti-Ig and to antigen in the absence of T helper activity is similar, giving us confidence in extrapolating from the experiments with anti-Ig.

The early events that take place in the lymphocyte following interaction of anti-Ig antibodies with surface Ig include: 1) a redistribution of the complexes into aggregates of variable sizes which flow into one pole of the cell--this is the phenomenon termed "capping" first described by Taylor et al. (3); 2) the internalization of the complexes in vesicles into cytoplasm (3,4) with subsequent catabolism (5); and 3) the

---

This study was supported by National Institutes of Health Grants AI-10091 and AI-10677. George Schreiner was supported by a Karin Grunebaum Cancer Research Foundation Fellowship. Emil Unanue is a recipient of a Research Career Award from the National Institutes of Health.

stimulation of translational motility (6,7). Following these processes, the lymphocyte will have lost the bulk of its antigen receptor from the membrane. New Ig molecules are detectable by immunofluorescence by about six to eight hours following this process.

This paper reviews some of the mechanisms involved in the redistribution of surface Ig into caps with emphasis on the process of stimulation of motility, which has been studied at depth more recently. One of the interests in studying capping is that it involves an energy-dependent process modulated by internal events. Hence, it can be used as a tool to learn about cell surface-cytoplasmic interrelationships.

Early events following complexing of anti-Ig with the B cell

The flow of anti-Ig-Ig complexes to one pole of the cell--capping--is a temperature-dependent energy-requiring process (3,9,10). Drugs that affect the cell's oxidative phosphorylation or glycolysis stop capping. In these instances the complexes remain over the entire cell surface forming patches of variable sizes. These kinds of experiments have been extensively discussed in the past and need no reiteration. At the time that the complexes flow into a cap, the cell becomes motile. Indeed, examination of murine splenic lymphocytes at the time of interaction with anti-Ig antibodies demonstrates many peculiar ameboid features (4,6). Most lymphocytes that are planted on the surface of a culture dish attach loosely to it exhibiting a round shape and little displacement; however, a few lymphocytes spontaneously exhibit slow translational motility. The addition of anti-Ig produces a change in the B cell population in that many readily show a slow translational motion adopting pear- or hand-mirror shape forms. These changes in shape, accompanied by true translatory motion, albeit of a limited span in distance, take place in about ten to twenty minutes after which the lymphocyte returns to its round shape. These changes described above take place at the same time that the cell is involved in the process of capping. Indeed, the capped complexes are found at the trailing edge of the cell (6). These changes in motility have had our attention, and experiments have been designed for studying them at more depth and to establish their relationship to capping.

We have used three methods to assess motility induced by anti-Ig antibodies (6,7,11). The first is by direct microscopic examination of cells under phase contrast. A second method consists of counting at a given time after interaction with anti-Ig the percentage of lymphocytes exhibiting ameboid morphology. This is done as follows: cells are placed at a density of 10^6 per ml in a culture well (Linbro Company, 17 mm in diameter) and allowed to settle; anti-Ig is added and the tray incubated at 37°C; at various times fixative is added to the wells to stop the reaction. The motile lymphocytes exhibit the typical morphological changes demonstrated in Figure 1. This method, which is diagrammed in Figure 2, offers us the number of cells that are in motion at a given moment, although it obviously gives no information on rates or characteristics of motility. The method does allow one to determine how a series of experimental variables will influence the number of lymphocytes in motion. The third method that we have employed is to examine the lymphocytes coming through a filter using the double chamber employed in chemotactic assays (11). We have followed the parameters established by Zigmond and Hirsch (12) to determine the extent of directional and random motility. The extent of penetration into the filter of the front of the lymphocytes and the number of cells at different depths of the filter are determined. The three methods have clearly established that anti-Ig stimulates random motility in B cells. For the sake of simplicity, we will report only the results using the second method, outlined in Figure 2.

Relationship between capping and cell motility

It became obvious that addition of anti-Ig stimulated a motile response that took place immediately--within the first ten to twenty minutes. This increase in cell motion was best seen when lymphocytes were placed in a dish rather than when in suspension (6). Presumably, the attachment to a surface favored the motility process. Our first thought was that there was a correlation between capping and motility; that is to say that the membrane activity that stimulated the cell for the motile changes was identical to that involved in carrying the complexes of anti-Ig-Ig to the trailing edge of the cell. However, by various experimental manipulations, both processes could be easily dissociated. For example, the use of the drug cytochalasin B--under appropriate conditions--had little to moderate effect on capping, yet it stopped the motile process completely (6). One experiment using this drug is shown in Figure 3. The exact mode of action by cytochalasin B is not clear, although we assume that directly or indirectly it affects microfilament function. A second drug that reduces motility without affecting the capping process per se is diisopropylfluorophosphate--DFP--an inhibitor of serine esterases (Fig. 4). Lymphocytes incubated with anti-Ig in the presence of DFP show a marked reduction in mobility. In contrast, if the lymphocytes are treated with DFP prior to

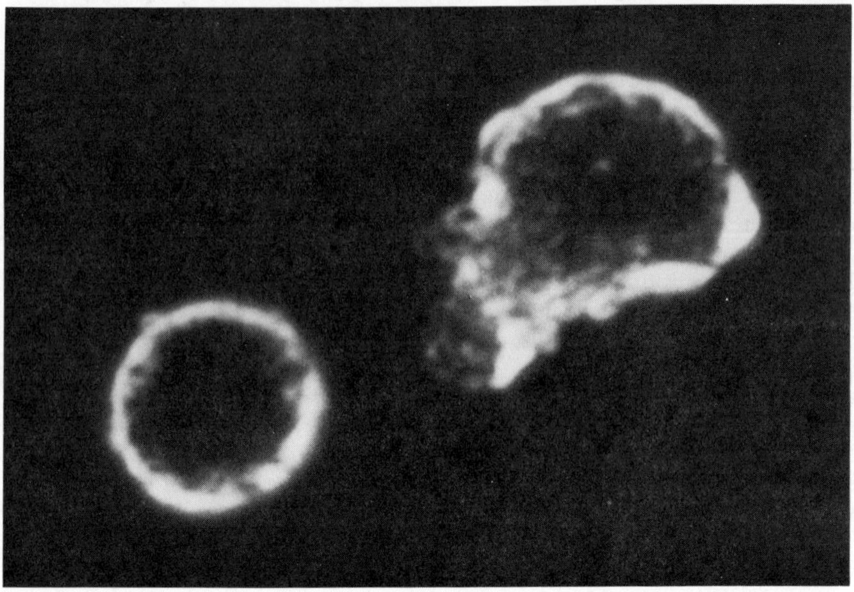

Fig. 1 This is a dark-field photomicrograph of two lymphocytes; the one at the right shows the typical pear-shaped motile cell; the round lymphocyte at the left is stationary. In a normal spleen population, there is a small number of cells that move spontaneously; of these motile cells, about two-thirds are T cells.

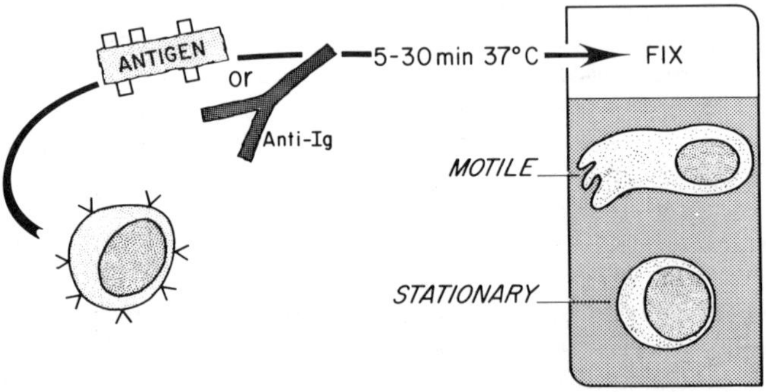

Fig. 2 The method outlined above gives an indication of the number of lymphocytes exhibiting motile forms at a given time. The cells are incubated on the surface of a dish with anti-Ig, then fixed, and the percent of ameboid cells determined microscopically.

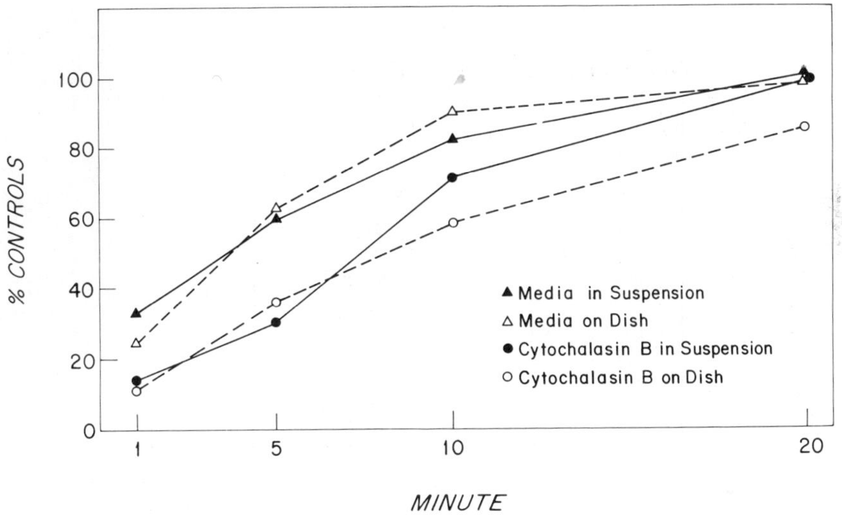

Fig. 3 Lymphocytes were incubated with fluorescent anti-Ig at 4^oC, washed, and portioned as above; in regular media (5 x 10^6 per ml) in test tubes with frequent mixing; in regular media allowing the cells to settle; and in media containing cytochalasin B (10 μg per ml) either in suspension or in a dish. At the different times the cells were examined for percentage exhibiting caps. Mean number of cells with caps when incubated in regular media after 20 minutes was taken as the 100% value. Temperature of the reaction was 37^oC. Note that the tempo of capping is slower in cytochalasin B-treated cells. The caps in cells treated with the drugs occupied a larger area of the cell membrane. Modest decrease in percentage of capped cells with cytochalasin B could be produced by doing the test at room temperature. This figure was published in The Journal of Experimental Medicine 139: 295, 1974, and is reproduced with permission.

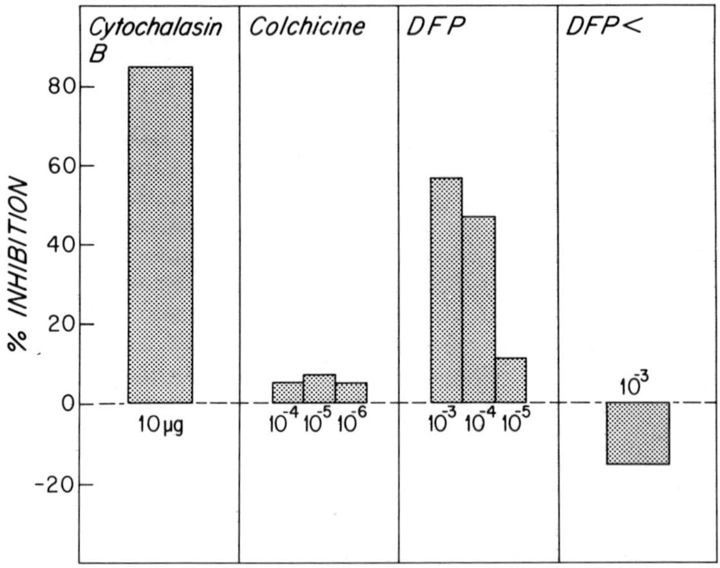

Fig. 4 Lymphocytes were incubated with anti-Ig (100 µg) at 4°C, washed well, placed in a dish, and incubated five minutes at 37°C. The net percentage of motile forms was 24.5%. Similarly treated lymphocytes were also incubated at 37°C in media containing cytochalasin B, colchicine, or DFP. One portion was first incubated with DFP for 15 minutes at 37°C before incubation with anti-Ig at 4°C (DFP <). This experiment was first reported in The Journal of Experimental Medicine <u>139</u>: 295, 1974.

exposure to anti-Ig and the drug is then washed away, the cell responds in a normal fashion. These results are taken to indicate that anti-Ig gives rise to an enzymatic activity sensitive to the inhibitor and which is part of a series of metabolic steps necessary for cell translation. This enzyme activity is not present in (or on) the cell prior to anti-Ig treatment, and hence is not the target of DFP until such a time as anti-Ig binds to the cell. Clearly, the action of DFP suggests that a critical enzymatic step is involved during the motile process. These experimental results with cytochalasin B and DFP suggest that motility in itself involves processes besides the membrane activity involved in capping. It is possible that the surface activity of both events is similar but that the latter process--motion--involves cytoplasmic events which are not required for capping. Conclusions similar to these were reached by Ryan et al. in our department when studying the neutrophil (13).

Regulation of cell motility

The anti-Ig stimulation of lymphocyte motility can be modulated by a variety of drugs, among them those that presumably affect the levels of cyclic AMP or cyclic GMP (7). It is known that movement of cells like fibroblasts (14) or neutrophils (15) can be markedly reduced by drugs that increase the levels of cyclic AMP. Such is also the case with the lymphocyte. Figure 5 shows results of an experiment in which lymphocytes were incubated with dibutyryl cAMP, theophylline (an inhibitor of phosphodiesterases), or both drugs in combination. The combined use of both drugs inhibited completely the anti-Ig-stimulated motility.

Of interest are the observations on the level of spontaneously moving cells, two-thirds of which have been shown to be T lymphocytes. Note in Figure 5 that the level of the spontaneous motile forms was not reduced after treatment with both drugs. However, examination of the live cells (first method described earlier) disclosed that in this instance the lymphocytes exhibiting ameboid or motile morphology were not moving but were fixed in their ameboid shape as if the drugs had "snap frozen" the cells. Indeed, this seemed to be the case. For example, when the preparation of lymphocytes was taken to a lower temperature--from 37°C to room temperature--after a short period most untreated lymphocytes assumed a round shape; however, in the cell preparation tested with cyclic AMP and theophylline, those few exhibiting ameboid morphology retained their shape. In other words, while the spontaneously moving cells had stopped moving at a lower temperature and had retracted their cytoplasm adopting a

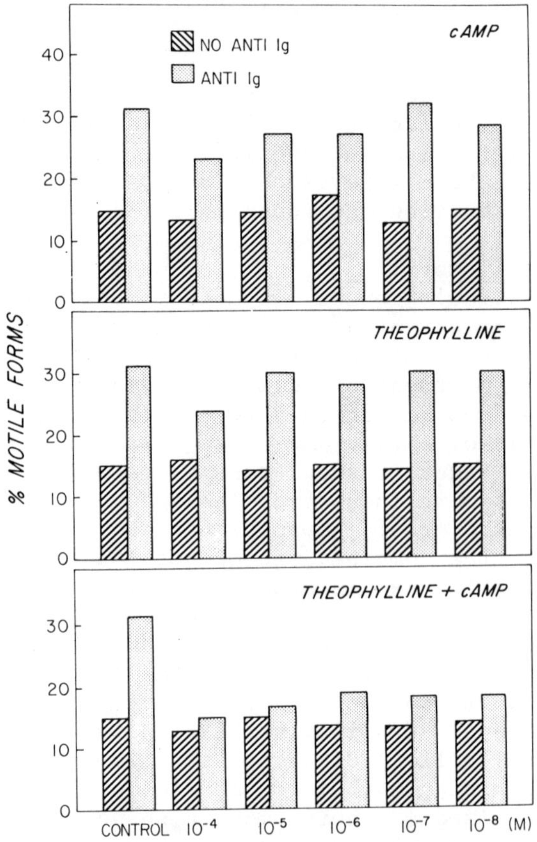

Fig. 5 Lymphocytes were incubated with the drugs (dibutyryl cyclic AMP, theophylline, or both) for ten minutes prior to addition of anti-Ig. The percentage of motile forms was determined as per the test shown in Fig. 2. Note marked inhibition of anti-Ig-induced motility by the combined use of dibutyryl cyclic AMP and theophylline (from Schreiner and Unanue, Journal of Immunology, in press).

round shape, the drug-treated cells were unable to do so. Of interest is the observation that the cyclic AMP-modulating drugs did not affect capping, this then being another example of the dissociation of these two cellular events.

Limited experiments have been made with cholera toxin, an irreversible activator of adenyl cyclase. Cholera toxin at a concentration of 1 μg per ml for 30 minutes reduced significantly the number of motile forms produced by anti-Ig. In the experiment in question the number of spontaneous motile forms was 10%, and this increased after addition of anti-Ig to 26%; the motile response to anti-Ig in cholera toxin-treated cells was only 14%.

Experiments testing other lymphocyte functions have indicated the presence of β-adrenergic receptors. Catecholamines binding to β-adrenergic receptors are known to increase the levels of cAMP (reviewed in 16). In Table I we show an experiment that indicates a reduction in anti-Ig motility by the β-adrenergic agent isoproterenol and its counteraction by its antagonist propanolol. These results do suggest the presence of B cells of β-adrenergic receptors which when combined with their appropriate agonist produce a response similar to that induced by an increase of cyclic AMP.

In summary, the results suggest that the elevation of cyclic AMP restricts the motility of the lymphocytes. In the case of the B cell, it is no longer able to move following anti-Ig treatment; in the example of the T cell, it no longer moves spontaneously. One can predict that those ligands which result in increased cyclic AMP elevation will reduce lymphocyte translatory processes, resulting in impairment of those events requiring the cells to displace themselves. It is possible that the reduction in killing of tumor cells by T cells treated with cyclic AMP-modulating drugs could at least in part be ascribed to the restriction of cell motion. As we discuss below, the restriction in cell motility produced by these drugs could have as a target the microtubular system.

<u>Cyclic GMP</u>

In contrast to the effects seen with cyclic AMP-modulating drugs, addition of cyclic GMP or drugs that presumably elevate this cyclic nucleotide results in increased motility (17). In the experiment shown in Figure 6 we incubated lymphocytes with dibutyryl cyclic GMP and observed a very significant increase in the levels of spontaneously moving cells (7). Addition of anti-Ig resulted in a very slight additive effect. Other experiments indicated that the increase in spontaneous motility was in great part accounted by B cells which increased in number of motile forms by fourfold--T cells increased at

TABLE I
EFFECT OF ISOPROTERENOL ON ANTI-Ig-INDUCED MOTILITY

Treatment	% Motile Forms
No treatment	9
Anti-Ig	23
Anti-Ig + 10^{-5} M Isoproterenol	10
Anti-Ig, 10^{-5} M Isoproterenol, 10^{-5} M Propanolol	25

Explanation is in the text. The increase in motility produced by anti-Ig is antagonized by β-adrenergic stimulation of the cell.

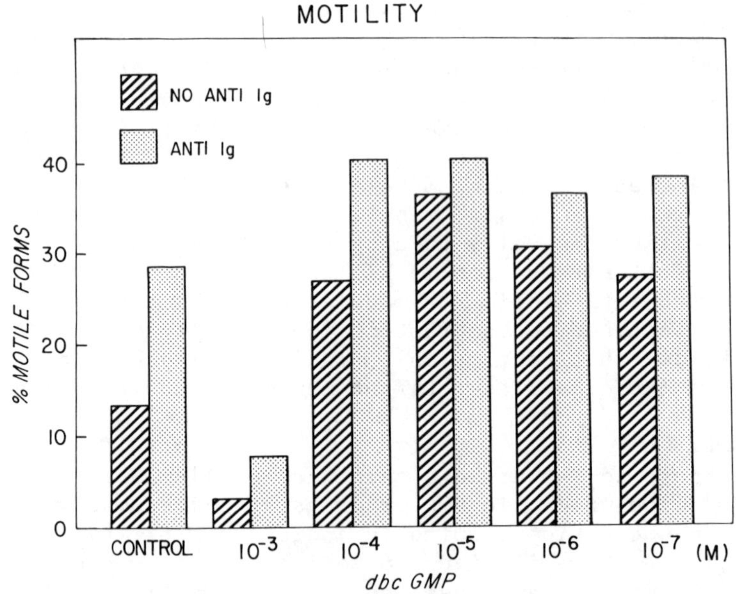

Fig. 6 An experiment identical to that shown in Figure 5 but using dibutyryl cyclic GMP.

the most twofold.

Evidence presented by Strom et al. suggests that lymphocytes have receptors for cholinergic agents. Strom et al. (18) increased the anti-tumor activity of specific T cells by incubation in carbamylcholine. It has been postulated that cholinergic agents such as acetylcholine and carbamylcholine exert their effects by increasing the intracellular levels of cyclic GMP. We decided to test the effects of these two drugs on lymphocyte motility. Indeed, their addition to lymphocytes increased motility dramatically; and, furthermore, the action of carbamylcholine was blocked by atropine. For example, the addition of 10^{-6} M carbamylcholine increased the number of spontaneously moving cells from 12 to 22%, but addition of atropine (10^{-6} M) to the solution of carbamylcholine resulted in only 10% of motile cells.

The experiments have indicated an opposing effect on lymphocyte motility of drugs that are affecting cyclic AMP or cyclic GMP levels, the former blocking cell motion, the latter stimulating it. In this regard these observations support Goldberg's yin-yang hypothesis which indicates different, sometimes opposing effects of these two cyclic nucleotides on a variety of cellular functions (19). The response of lymphocytes to cholinergic agents antagonized by atropine suggests the presence on their surfaces of cholinergic receptors of the muscarinic type. The function of these receptors in the overall response of the lymphocyte to antigen is puzzling at present. Cholinergic effects have also been shown to be exerted on hematopoietic stem cells and, as was mentioned above, have also been seen on T lymphocytes by Strom. In the experiments of Strom et al. the cholinergic response was antagonized by atropine and to lesser extent by curare or bungarotoxin (20). So far, in unpublished experiments with Dr. Fishbock we have been unable to detect binding of radioiodinated α-bungarotoxin to B cells.

<u>The role of microtubules</u>

There is suggestive evidence in favor of microtubules having some control function both in capping and in lymphocyte motility. It was Berlin and coworkers who first pointed to a possible relationship between microtubules and the control of surface macromolecules (21,22). Studying the neutrophil or the macrophage, they were able to discriminate a difference in cell topography between two membrane sites following membrane stimulation--this difference was lost following treatment with colchicine. Yahara and Edelman later reported that the binding of the lectin Con A inhibited the capping of other surface molecules like Ig (23). This inhibition, however, was counteracted by treatment of the cell with colchicine. Yahara and Edelman postulated two states of microtubules in the cell: one fixed to the surface and restricting its motility and the other unfixed. Con A led to the fixation of the cell in its fixed state.

We have evaluated the contribution of microtubular function using colchicine. The following results have been obtained:

1. Treatment of lymphocytes with colchicine does not impair the cell with respect to capping (3,10) or for translational motility following anti-Ig treatment (6); if anything, there is a tendency for lymphocytes treated with anti-Ig and colchicine to be more motile than those cells treated with anti-Ig without the drug. Hence, the absence of assembled microtubules in an otherwise normal lymphocyte is not reflected in a reduction in cell motion and may be necessary.

2. In limited experiments done with Karnovsky, Ault, and Abbas, the topography of surface Ig has not been changed by colchicine treatment.

3. Colchicine and cytochalasin B produce a synergistic negative effect on capping in lymphocytes (24,25). The effects are consistent, albeit not extensive, and are best noticed when cells are incubated at room temperature. A cell treated with cytochalasin B at room temperature is moderately affected in its capacity to cap anti-Ig-Ig complexes. This cell is further impaired if colchicine is added. These effects, however, are dramatic in cells treated with Con A. A representative experiment done in collaboration with Karnovsky is shown in Figure 7. Notice that Con A inhibits capping and that this effect is counteracted by treatment with colchicine; however, treatment with both colchicine and cytochalasin B stops capping altogether in these cells. Unanue and Karnovsky interpreted these experiments as indicating the involvement of two processes in capping: one sensitive to cytochalasin B and one to colchicine. Both processes are integrated in the normal cell, and each can result in capping provided the remaining one is not impaired.

4. This last example provides evidence for the involvement of microtubules in the control of cell motility. As was mentioned before, treatment of lymphocytes with dibutyryl cyclic AMP and theophylline stops cell motility. This inhibition is no longer seen when cells are treated first with colchicine (Figure 8). Our conclusion is that increased cyclic AMP activity produces increased numbers of polymerized microtubules which restrict the translatory motion of lymphocytes. This needs to be confirmed by morphological examination.

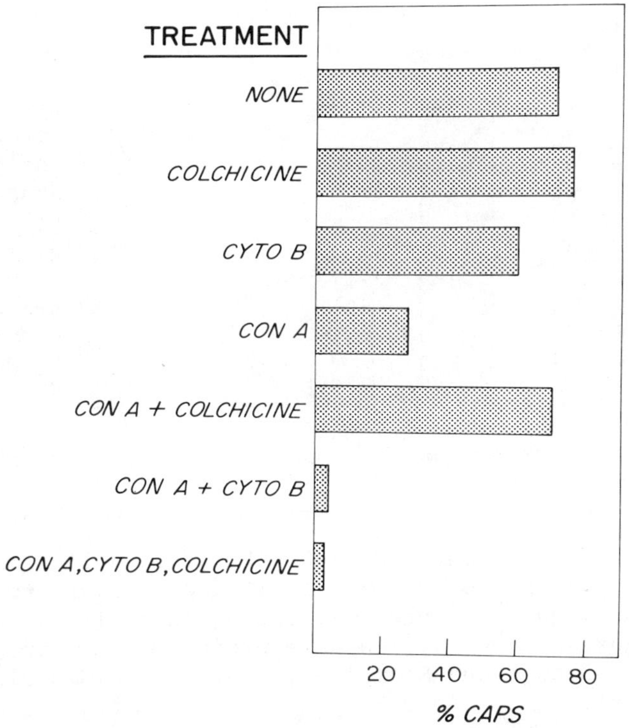

Fig. 7 Lymphocytes were treated with the drugs at room temperature for 30 minutes and anti-Ig was added and the reaction continued for 30 more minutes. Note the inhibition of capping by Con A and the release of this inhibition by addition of colchicine. Cytochalasin B totally blocks capping in the presence of Con A and colchicine and reduces capping of normal cells treated with colchicine (from Unanue and Karnovsky in The Journal of Experimental Medicine, in press).

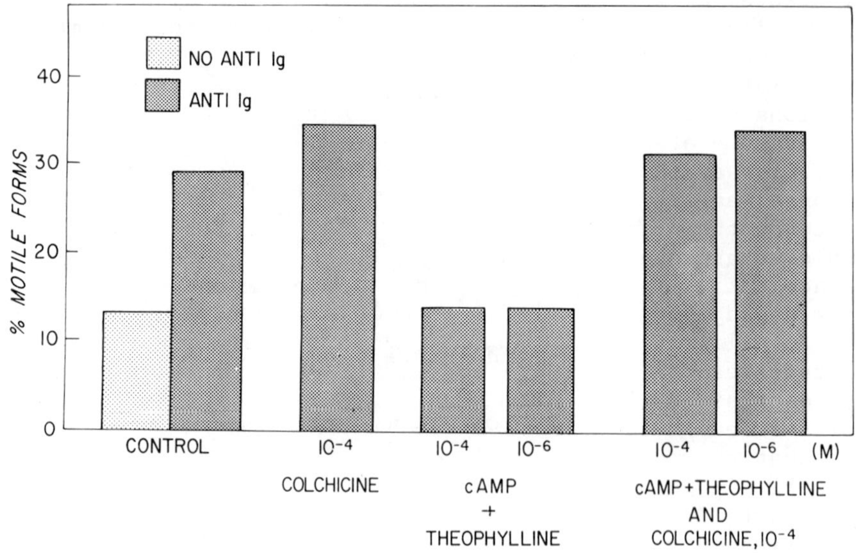

Fig. 8 Lymphocytes incubated with dibutyryl cyclic AMP plus theophylline do not move following exposure to anti-Ig antibodies. This restriction in motility is no longer seen when cells were preincubated with colchicine (Unanue and Schreiner, Journal of Immunology, in press).

It is difficult with the restricted information at hand to formulate a hypothesis that would integrate all the phenomena described here and also studied in many other cells with different results. Despite these difficulties, we believe the following hypothesis may serve to guide us in our future research. When anti-Ig interacts with surface Ig, it leads to a series of biochemical events--perhaps involving the cyclic nucleotides--which are going to exert a rapid influence on the contractile and microtubular systems. These biochemical events, as we detail in the next section, may be unique to some ligands and do not have to be the same for all membrane-ligand complexes. The activation of the microtubular system may involve a rapid series of assembly-disassembly reactions while the contractile activity may involve contraction in some areas and relaxation in others. Both these activities are coordinated so they bring about a harmonious response manifested as a change in cytoplasmic mass accompanied by the surface activity that makes diffuse surface complexes aggregate at one pole of the cell. A coordinated surface activity in which one area of the membrane contracts while an adjacent area is maintained in its normal state may lead to a lifting of the lattice of complexes from the contracted area, the lattice still being anchored to the adjacent stable area. As this process continues from one end of the cell to the other, it results in a wave motion at the same time that the complexes are all pulled together to the end of the cell. Together with the surface activity, which is in great part microfilament dependent, the cell cytoplasmic mass is rearranged and given orientation by some level of microtubular activity. Cytochalasin B affects microfilament function, stopping motion altogether; it is likely, however, that some microfilament activity remains--or that some of it is cytochalasin B insensitive--and that hence in its presence the cell can still cap complexes, albeit in a less efficient way. Microtubules are seen playing a role in the particular circumstances where the cell is also affected by cytochalasin B, indicating that some level of microtubular function is needed for regulating the surface activity of capping. In contrast, an increase in microtubular polymerization as suggested by the experiments of Yahara and Edelman or ours with cyclic AMP could result in various levels of restriction of the surface-cytoplasmic events.

The exact mode by which the cell's structures are activated following the membrane reaction are not known, since precise biochemical studies have not been made. One possibility is that certain surface molecules are in some way directly linked to the cytoplasmic organelles. Alternatively, it is possible that the ligand receptor effects introduce metabolic alterations--such as shift in ions--which are then reflected on the functions of particular organelles. We have been doing preliminary experiments on the role of Ca ions thinking along the lines that Ca fluxes resulting from the membrane receptor-ligand interaction could be instrumental in the process. Our results so far have been surprising because they have been unable to show a requirement for extracellular Ca in capping or motility. Lymphocytes incubated in Ca-deficient media together with a Ca-chelating agent like EGTA are able to cap or move normally. We have incubated these cells for periods as long as four hours thinking that by doing this the cell-associated Ca could be reduced, but the results have not shown an impairment of capping nor of motility. Furthermore, the addition of the Ca ionophore A23187 has consistently produced a complete interruption of capping and motility. The ionophore A23187 produces a flow of extracellular Ca into the cell which we have easily verified in experiments using Ca^{45}. The effects of A23187 in stopping capping are not seen in medium devoid of Ca or in regular medium in the presence of EGTA. Furthermore, experiments with labeled Ca^{45} have shown that anti-Ig produces an efflux of intracellular Ca seen within a few minutes. Thus, we have clearly seen there is no role for extracellular Ca in the necessarily contractile phenomena underlying capping and motility in lymphocytes. The speed and extent of changes in lymphocyte membrane and morphology, however, suggest a plausible role for ions in this ligand-specific transformation. To determine the basis for the apparent interrelationship between cell organelles and membrane-restricted surface reactions, we are currently examining three alternative explanations for the critical step subsequent to binding surface Ig that allows the cell to cap and move: 1) a triggered release of Ca from intracellular stores consistent with, but not proved by, preliminary data; 2) the involvement of ions other than Ca; and 3) a possible non-ionic regulatory factor governing the activity of contractile events.

<u>Effects of other ligands in stimulating motility</u>

We have questioned whether any ligand-membrane reaction stimulates the changes in motility observed with anti-Ig (11). So far, we have tested anti-θ antibodies and polyvalent anti-H2 antibodies without any positive results. Of great interest have been studies with heterologous anti-lymphocyte antibodies (ALS). ALS did not stimulate motility in B or T lymphocytes; and, furthermore, lymphocytes treated with ALS were unable to exhibit a motile response to anti-Ig antibodies. This is to say, a lymphocyte which first binds ALS is now blocked in its response to anti-Ig. The effect of ALS can be totally absorbed by removing its antibodies by spleen cells. This is the situation of two different ligands, each evoking a contrastingly different response in the B cell. The story with colchicine and cyclic AMP-modulating drugs repeats itself in that the ALS-treated lymphocytes exposed to colchicine can now respond perfectly well to anti-Ig antibodies. It is tempting to wonder whether ALS produces an elevation of cyclic AMP and thus polymerizes microtubular subunits with resulting restriction in motility and morphological fluidity. If indeed it does, such an elevation need not be physiological. It may only represent a nonspecific response to "membrane stress", though it is interesting, with respect to this point, that anti-H2 and anti-θ antisera appear not to affect lymphocyte movement, and anti-Ig antiserum has the opposite effect. The precise mechanism by which ALS can affect microtubules remains inexplicable.

We might add at this point that we have certainly not concluded that cyclic nucleotides are effector factors in lymphocyte response to specific ligands studied herein, whether antigens or antigen-analogues. Certainly, we have observed very potent effects at very low concentrations on certain of the early lymphocyte reactions; this is suggestive, not conclusive. Many other investigators have, however, ascribed to cyclic AMP or cyclic GMP central importance in lymphocyte function and transformation (16). Others argue that both nucleotides may play a role, depending on both the signal and the phase of the lymphocyte. If cyclic nucleotides are second messengers in the immune system, in fact antigens are "hormones", the earliest lymphocyte response to the binding of antigen may well be mediated by an increase in cyclic GMP and/or a decrease in cyclic AMP, judging from their effects on the anti-Ig-stimulated events in the lymphocyte. Adding to the "ifs", if ALS does turn out to be stimulating cyclic AMP-inducing membrane molecules, we face the fascinating question of why two ligand-membrane complexes lead to two different metabolic pathways, while other complexes have no effect other than to promote clearing of the complexes from the surface and regeneration of new molecules.

<u>Endocytosis and regeneration of new surface receptors</u>

The complexes of anti-Ig-Ig are rapidly interiorized in vesicles leaving a cell bare of its antigen receptor. Studies of Engers and Unanue using ^{125}I-labeled anti-Ig showed that products of hydrolysis--iodotyrosines and partially digested Ig--appeared in the culture fluid after several hours (5).

All the evidence, then, suggests that the interiorized complexes are digested by intracellular processes involving participation of lysosomes. Evidence was also presented for the release of some complexes from the membrane into the extracellular fluids. Hence, it appears clear that the lymphocyte is perfectly capable of eliminating the complexes from its membrane in a rapid and efficient way. Further proof for this statement was given by studies of Ault with radiolabeled antigens (2). B cells exposed for a brief period of time to radiolabeled DNP compounds (to which no T cell "activity" was present) rapidly eliminated those showing new fresh receptors within hours. In one example, the interruption of this cycle led to marked functional impairment of the cell. Katz and Benacerraf have shown that a polymeric molecule made up of D-glutamic acid and D-lysine (D-GL) is a potent, irreversible tolerogen for B cells (26). Ault et al. exposed DNP-specific B cells to DNP bound covalently to D-GL or L-GL (27). The B cells rapidly eliminated the latter compound but were unable to eliminate the D form which was not metabolizable. Furthermore, the B cell showed a marked impairment of its capacity to interiorize the complexes of DNP-D-GL surface Ig which now remained on the membrane for many hours. Associated with this impairment of internalization, new receptors were never reexpressed. In some way or another, the binding of the D nonmetabolizable compound had reduced both interiorization and reexpression of new receptors on the surface. The blockade in interiorization suggests a surface-specific metabolic step in the process requiring perhaps an enzymatic step which is blocked by the D compound.

Summary

We have described in this paper three early events that take place in the B cell following its interaction with a ligand--redistribution of surface complexes into caps, endocytosis, and cell motility. It is likely that each of these is modulated by a series of coordinated metabolic steps. It is clear, however, that the interaction of a ligand like anti-Ig with the B cell receptor produces a limited response similar to that induced by many garden-variety types of antigens in the absence of thymic helper cells. We view the events described here as part of a continuous series of reactions which, in the complete stimulation involving antigen binding and T cell signal, leads to the cell's proliferation and differentiation. What is the role of receptor aggregation in cell stimulation? Capping per se of Ig receptors is not a significant event for stimulating the B cell to a complete response (28). However, receptor aggregation clearly does play a role in allowing the formation of a more avid receptor-ligand complex. Indeed, binding of labeled antigen best takes place in conditions where receptor aggregation occurs leading to a more efficient immune response. This point has been discussed extensively in reference 1. The second event described today--motility--may play a crucial role in inductive phenomena perhaps by allowing B and T cells to meet as a result of their increased random motility. In lymphoid organs B cells and T cells are compartmentalized--the B cell in the superficial cortex, the T cells in the deep cortex--and apparently separated from each other. If the interaction of antigen with B cells increases their random motility as predicted from our results, it could be possible that this mobilization of B cells would allow them to migrate to the deep cortex and there interact with the T cells. We have contemplated the possibility that a ligand to the Ig receptor could stimulate directional rather than random motility. In this way, perhaps, antigen could focus the lymphocyte to the areas in nodes of high antigen concentration, such as the areas where macrophages are lodged. However, our initial studies using chemotactic chambers and anti-Ig have only shown an increase in random but not directional motility (7). Finally, the process of interiorization and digestion associated with shedding leads to clearance of the complexes allowing the cell to express new receptors and be ready for a new interaction. The experiments with DNP-D-GL described above are very eloquent in this regard, showing how a blockade or dysfunction of a membrane reaction can stop the whole cell cycle.

REFERENCES

1. Unanue, E.R., Am. J. Pathol. 77: 101, 1974.
2. Ault, K.A. and Unanue, E.R., J. Exp. Med. 139: 984, 1974.
3. Taylor, R.B., Duffus, W.P.H., Raff, M.C. and de Petris, S., Nature New Biol. 233: 225: 1971.
4. Unanue, E.R., Perkins, W.D. and Karnovsky, M.J., J. Exp. Med. 136: 885, 1972.
5. Engers, H.D. and Unanue, E.R., J. Immunol. 110: 465, 1973.
6. Unanue, E.R., Ault, K.A. and Karnovsky, M.J., J. Exp. Med. 139: 295, 1974.
7. Schreiner, G.F. and Unanue, E.R., J. Immunol., in press, 1975.
8. Elson, C.J., Singh, J. and Taylor, R.B., Scand. J. Immunol. 2: 143, 1973.

9. Loor, F., Forni, L. and Pernis, B., Eur. J. Immunol. 2: 203, 1972.
10. Unanue, E.R., Karnovsky, M.J. and Engers, H.D., J. Exp. Med. 137: 675, 1973.
11. Schreiner, G.F. and Unanue, E.R., J. Immunol., in press, 1975.
12. Zigmond, S. and Hirsch, J., J. Exp. Med. 137: 387, 1973.
13. Ryan, G.B., Borysenko, J.Z. and Karnovsky, M.J., J. Cell Biol. 62: 351, 1974.
14. Johnson, G., Morgan, W.D. and Pastan, I., Nature 235: 54, 1972.
15. Rivkind, I. and Becker, E.L., Fed. Proc. 31: 657, 1972.
16. Bourne, H.E., Lichtenstein, L.M., Melmon, K.L., Henney, C.S., Weinstein, Y. and Shearer, G.M., Science 184: 19, 1974.
17. Estensen, R.D., Hill, H.R., Quie, P.G. and Hogan, N., Nature 245: 458, 1973.
18. Strom, T.B., Carpenter, C.B., Garovoy, M.R., Austen, K.F., Merrill, J.P. and Kaliner, M.A., J. Exp. Med. 138: 381, 1973.
19. Goldberg, N.D., O'Dea, R.F. and Haddox, M.K., in Advances in Cyclic Nucleotide Research, Vol. 3, edited by P. Greengard and G.A. Robison, pp. 155-223, Raven Press, New York, 1973.
20. Strom, T.B., Sytkowski, A.J., Carpenter, C.B. and Merrill, J.P., Proc. Nat. Acad. Sci. USA 71: 1330, 1974.
21. Berlin, R.D., Oliver, J.M., Ukena, T.E. and Yin, H.H., Nature 247: 45, 1974.
22. Ukena, T.E. and Berlin, R.D., J. Exp. Med. 136: 1, 1972.
23. Yahara, I. and Edelman, G.M., Proc. Nat. Acad. Sci. USA 69: 608, 1972.
24. Unanue, E.R. and Karnovsky, M.J., J. Exp. Med., in press, 1974.
25. de Petris, S., Nature 250: 54, 1974.
26. Katz, D.H., Davie, J.M., Paul, W.E. and Benacerraf, B., J. Exp. Med. 134: 201, 1971.
27. Ault, K.A., Unanue, E.R., Katz, D.H. and Benacerraf, B., Proc. Nat. Acad. Sci. USA 71: 3111, 1974.
28. Katz, D.H. and Unanue, E.R., J. Immunol. 109: 1022, 1972.

THE ROLE OF PHOSPHOLIPIDS IN LYMPHOCYTE ACTIVATION

Klaus Resch and Ernst Ferber
Institut für Immunologie der Universität Heidelberg, Germany
and Max-Planck-Institut für Immunbiologie, Freiburg, Germany

Lymphocytes without being exposed to external stimuli are arrested within the G_1 phase of the cell cycle. They do not express demonstrable function. When antigens or mitogens bind to the surface of the cells, a sequence of metabolic events is triggered leading to proliferation and/or expression of function. Reaction of mitogens or antigens with the plasma membrane alone is sufficient to activate resting lymphocytes (1-3). The activation, therefore, must be initiated by processes which the plasma membrane undergoes interacting with a ligand (stimulant). At least one of these reactions must be responsible for transmitting a "signal" into the interior of the cell leading finally to gene activation.

Several hypotheses have been put forward as to how such a "signal" can be transmitted across the plasma membrane: Thus it has been suggested that a stimulant such as concanavalin A may bind to a receptor glycoprotein which penetrates the outer membrane. Inside of the cell these glycoproteins then are associated to structures of the microtubuli. Conformational changes of the glycoproteins result in changes of the association to the microtubuli and thus induce intracellular alterations (4). This hypothesis barely is able to explain the initiation of lymphocyte activation. As binding sites (receptors) for different mitogens are independent (5,6) sufficient transmembranal proteins do not appear to be present within the plasma membrane. The main objection, however, is that the plasma membrane itself is bypassed by such a mechanism.

A variety of plasma membrane changes have been described to occur very early after a mitogen has bound to lymphocytes, some of them within minutes. Thus the permeability for nucleosides (7), sugars (8), amino acids (9,10) or ions such as K^+ (11) or Ca^{++} (12) is increased. Membrane bound enzymes such as ATPases (11,13) or lipid metabolizing enzymes (14-18)are also activated. The rapidly induced disbalance of intracellular cyclic nucleotides - i.e. increase in cyclic GMP and concomitant decrease in cycle AMP (19,20) - may also be regulated by cyclases located within the plasma membrane (21). In addition, the plasma membrane of activated lympho-

Supported by the Deutsche Forschungsgemeinschaft

cytes exhibits gross changes measured by an increase in membrane fluidity (22,17) or an accelerated turnover of membrane phospholipids (18,23,24).

For some of these early membrane events, e.g. Ca^{++} or K^{+} influx or the changes in cyclic nucleotide levels, a role in the initiation of lymphocyte activation has been postulated. In these cases, too, the way a signal may be transmitted into the interior of the cell was envisaged exclusively and the concomitant membrane changes were neglected (11,12,19).

Thus either one postulates that a ligand primarily has multiple sites of action or one has to look for an amplifying mechanism which underlies the multiple membrane processes. In the following we postulate such a role for the changes of membrane phospholipids which occur immediately when a stimulant binds to the lymphocyte membrane.

Current concepts of membrane structure

To discuss the role of phospholipids in lymphocyte activation some current views of membrane architecture have to be outlined. Though there does not exist a unique structure of all different cellular membranes, such principles will be described which appear to be shared by plasma membranes (25, 26).

The plasma membrane is constituted of a matrix of phospholipids which are organized mainly in a bilayer structure. Integral membrane proteins are imbedded into this matrix tightly bound by hydrophobic interactions with the phospholipid fatty acids. Most of these proteins possess catalytic properties which are regulated by their interaction with different lipids (see conclusions).

The transition temperature of lipids from biological membranes lies between 16°C and 18°C (27,28). Thus, under physiological temperatures the lipid phase appears to be in a liquid state. From many investigations it has become evident that distinct membrane protein units are mobile independent of each other within the phospholipid lattice. This has prompted the deduction of a membrane model (recently reviewed by Singer (26)), describing the protein constituents of the membrane being distributed randomly in the fluid lipid layer (fluid mosaic membrane). This implies that long range cooperative effects - which can change the functions of the entire membrane - can be transmitted solely through the lipid phase. The highly selforganizing structure of phospholipids favours these cooperative interactions (29).

The major components of phospholipids in lymphocytes are lecithin (= phosphatidylcholine) and phosphatidylethanolamine which represent 70% of the total phospholipids (30,31).

(Table 1). Compared to rat liver the content of polyunsaturated fatty acids is lower in the main phospholipid lecithin suggesting relatively rigid membranes occuring in resting lymphocytes. A detailed analysis of the fatty acid composition of the phospholipids of normal and activated lymphocytes is published elsewhere (32,33).

Phospholipid turnover in stimulated intact lymphocytes

Up to some years ago membrane phospholipids have been considered to be relatively inert constituents of cellular membranes. However, as shown in Table 2, the incorporation of some precursors into phospholipids is extremely high, even in resting lymphocytes. As resting lymphocytes do not exhibit a net synthesis of phospholipids (34) these incorporation rates indicate a continuous turnover of the individual phospholipids.

Some characteristic pathway of the turnover of the main phospholipid in lymphocytes, namely lecithin, are depicted in Fig. 1. The whole molecule can be synthesized de novo from glyerol-3-phosphate. During the synthesis choline phosphate is transferred from cytidine diphosphate choline to diacylglycerol. As this appears to be the only de novo pathway in mammalian cells the incorporation of labelled choline closely reflects the de novo synthesis of lecithin.

Besides this, there exists a separate turnover of the fatty acid moieties. A fatty acid residue can be split off by the action of phospholipase A generating the highly surface active compounds lysolecithin and free long chain fatty acids. Lysolecithin can be reacylated to lecithin by lysolecithin acyltransferases.

Activated lymphocytes transform into large blast cells and finally undergo mitotic divisions. This implies that cell constituents are newly synthesized, including cellular membranes. Increased net synthesis of phospholipids becomes apparent after 6 hours of stimulation, at which time morphological changes are still not discernible (34). Increased incorporation of ^{14}C-choline into PHA stimulated lymphocytes starts earlier. Enhanced uptake is discernible one hour after the addition of a mitogen and is two to threefold after four hours activation (34-37). De novo synthesis measured by incorporation of ^{14}C-glycerol or ^{32}P into lecithin follows similar kinetics (36-37).

The incorporation of ^{14}C-oleate into lecithin exceeds, under optimal conditions, the choline uptake by more than hundredfold. This strongly suggests that long chain fatty acids are incorporated into lymphocyte phospholipids by the separate metabolic cycle described above. The capacity of

TABLE 1
LIPID CONTENT AND PHOSPHOLIPID DISTRIBUTION
OF HUMAN BLOOD LYMPHOCYTES AND CALF THYMOCYTES

	Human blood lymphocytes Gottfried (30)	Calf thymocytes Ferber et al. (31)
A Lipid Content moles x 10^{-15}/cell		
total lipid		4.75
total phospholipid	4.73	2.87
cholesterol	3.24	1.20
neutral lipid		0.28
free fatty acids		0.40
B Phospholipid Distribution mole %[a]		
phosphatidyl choline	43.6	56.5
phosphatidyl ethanolamine	28.7	29.1
sphingomyelin	10.1	14.4
phosphatidyl serine +phosphatidyl inositol	13.4	
phosphatidic acid	3.2	
lysolecithin	1.0	

[a]Sum of compounds determined set to be 100%.

TABLE 2

TURNOVER RATES OF PHOSPHOLIPID PRECURSORS IN RESTING LYMPHOCYTES

Precursor	Measured Product	Turnover (p mole/10^7 cells/h)	Reference
^{14}C choline	lecithin	20 20	Resch & Ferber (35) Fisher & Mueller (34)
^{3}H-glycerol	lecithin	56	Fisher & Mueller (18)
	phosphatidyl-inositol	43	
$^{32}P\ O_4^{3-}$	phosphatidyl-inositol	0.05 (<0.5[1)])	Fisher & Mueller (36)
myo-^{3}H-inositol	phosphatidyl-inositol	0.075 (<1.5[a)])	Fisher & Mueller (18)
^{14}C-oleate (free)	lecithin	2 200	Resch & Ferber (35)
(complexed to albumin)	lecithin	300	
^{14}C-acetate	lecithin	10	Resch & Ferber (35)

a) PHA stimulated lymphocytes. As incorporation into resting lymphocytes approaches apparatus background, data of stimulated lymphocytes are included.

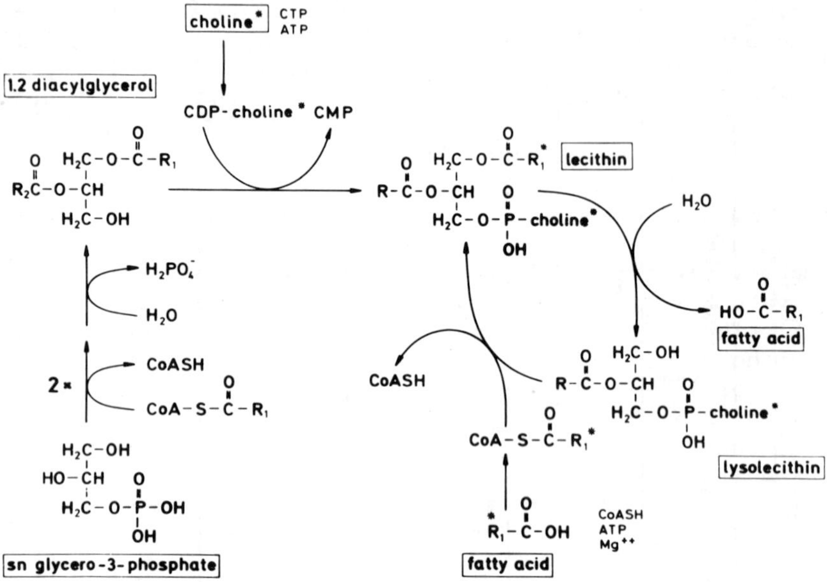

Fig. 1 Pathways of lecithin metabolism

this pathway is sufficient to replace total cell lecithin fatty acids within less than 10 hours. Because of the high turnover rates of this metabolic pathway the cellular steady state concentrations of its components appears to be prone to immediate perturbations. In addition, a high turnover of the phospholipid fatty acids allows to change efficiently (and rapidly) the fatty acid composition of phospholipids. Changed concentrations of intermediate products such as lysophosphatides or free long chain fatty acids as well as altered composition of the fatty acid residues (e.g. increase in unsaturation) are able to change membrane structure and function efficiently. This prompted us to investigate the turnover of the fatty acid moieties of membrane phospholipids in detail.

When lymphocytes were activated by mitogens the incorporation of ^{14}C-oleate into lecithin increased rapidly (Fig. 2). Using PHA or con A the amount of oleate converted into lecithin approximately doubled (35). In rabbit lymphocytes anti-immunoglobulin induced a similar increase in the uptake of ^{14}C-oleate. As in rabbits PHA or con A stimulate T-lymphocytes exclusively, while anti-immunoglobulin can only activate B-lymphocytes (38,39), this suggests that the initiation of lymphocyte activation may be similar in T- or B-lymphocytes. During the first hours of stimulation the uptake of ^{14}C-oleate was rather restricted to lecithin. Between 73 and 80% of the long chain fatty acid were incorporated into this phospholipid (Table 3). As well, only phospholipids showed increased incorporation rates after stimulation, whereas uptake into neutral lipid was enhanced only marginally.

Using long chain fatty acids one can follow the acylation process directly. However, as long chain fatty acids are surface active and nearly insoluble the interpretation of the results may sometimes be difficult. Therefore, the acylation process was also followed using endogenous long chain fatty acids synthesized within the cells from acetate. Such studies were facilitated in lymphocytes since acetate was incorporated exclusively into the fatty acid residues and not into the glycerol backbone (35). The incorporation of acetate was increased in activated lymphocytes, staring within one hour after the addition of a mitogen (Fig. 3). Nearly no accumulation of ^{14}C-acetate was detected in the free fatty acid pool which was increased only marginally after stimulation (Table 4). The incorporation into phospholipids and neutral fats was more than twenty fold higher. However, only the uptake into phospholipids was stimulated several fold. The small pool of free fatty acids (which is not changed markedly even

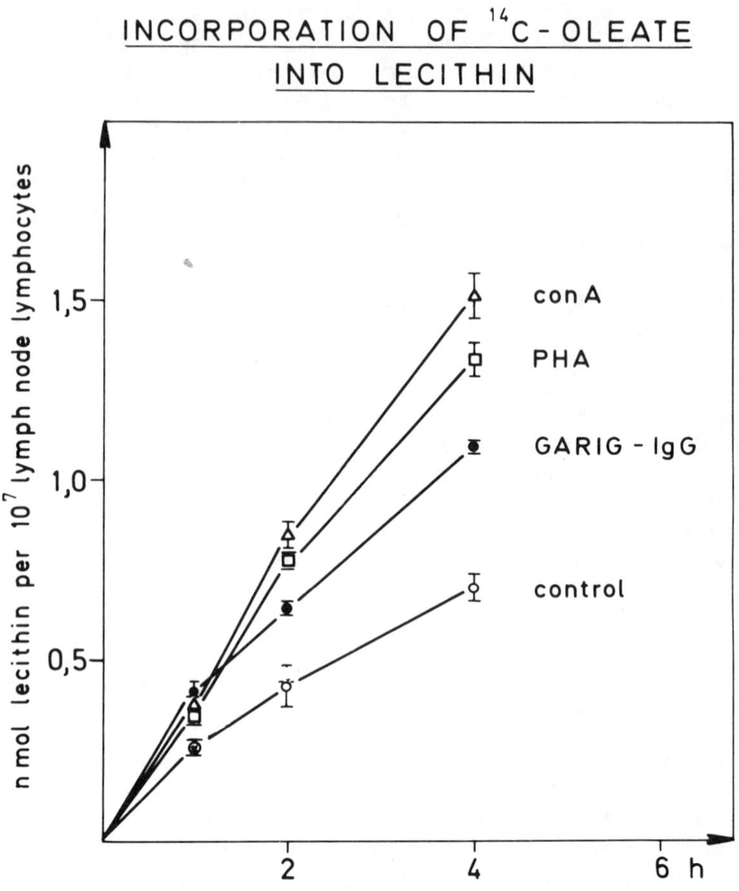

Fig. 2 Incorporation of ^{14}C-oleate into lecithin of activated lymphocytes.
10^7 rabbit lymph node lymphocytes were incubated in 1 ml Dulbecco's modified Eagle's medium containing 10 n moles ^{14}C-oleate and 2.5 mg defatted bovine serum albumin. Stimulants were added to the following concentrations: PHA 20 µg/ml, conA 10 µg/ml, goat-anti-rabbit-immunoglobulin (IgG fraction) 57 ug/ml.
After incubation the lipids were extracted, separated with thin layer chromatography, and their radioactivity determined as described in (35).

TABLE 3

INCORPORATION OF ^{14}C-OLEATE INTO PHOSPHOLIPIDS AND NEUTRAL LIPIDS: EFFECT OF STIMULATION AFTER 4 HOURS CULTURE

n moles / 10^7 lymphocytes [1)]

	phosphatidyl-choline	lysolecithin	phosphatidyl-ethanolamine	sphingomyelin	neutral lipids
Control	0.71 ± 0.04	0.02 ± 0.01	0.02 ± 0.01	0.07 ± 0.01	0.15 ± 0.01
PHA	1.34 ± 0.05	0.04 ± 0	0.13 ± 0.04	0.12 ± 0.04	0.15 ± 0.01
Con A	1.51 ± 0.07	0.02 ± 0	0.05 ± 0.05	0.11 ± 0.02	0.19 ± 0.01
GARIG-IgG [2)]	1.19 ± 0.01	0.02 ± 0.01	0.05 ± 0.01	0.08 ± 0.01	0.18 ± 0.02

1) Values are given with ranges from experiments made in duplicate

2) Goat-anti-rabbit-immunoglobulin (IgG fraction)

Culture conditions and lipid analysis as in Fig. 2

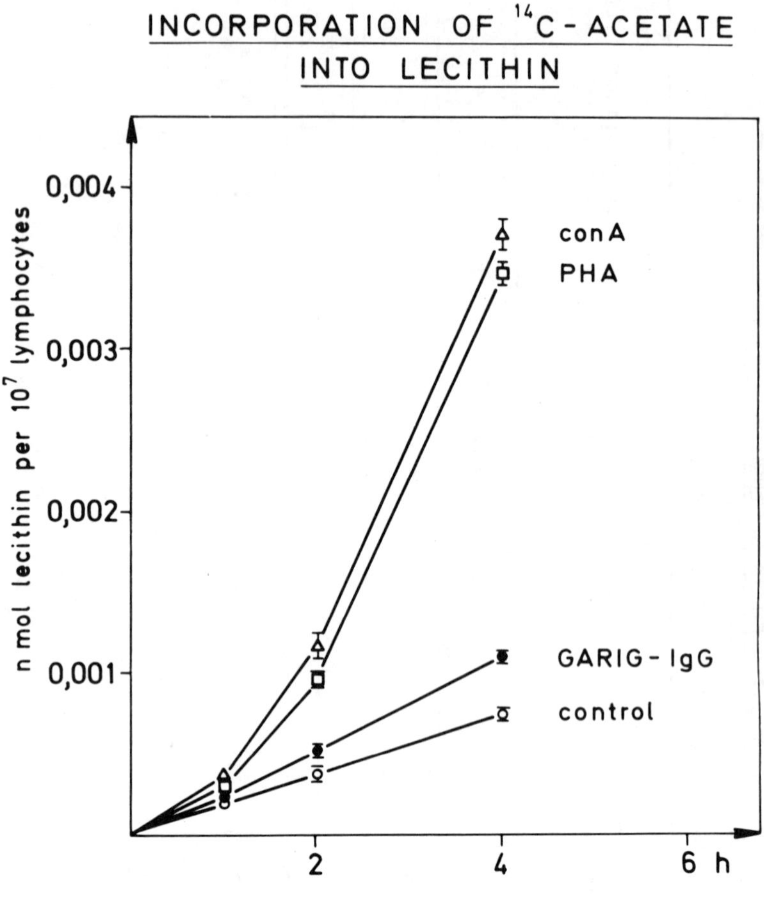

Fig. 3 Incorporation of ^{14}C-acetate into lecithin of activated lymphocytes.
10^7 rabbit lymph node lymphocytes were incubated in 1 ml Dulbecco's modified Eagle's medium containing 25 n moles ^{14}C-acetate.
Culture conditions and lipid analysis as in Fig. 2.

TABLE 4

INCORPORATION OF ACETATE INTO PHOSPHOLIPIDS AND NEUTRAL FATS: AFTER 4 HOURS CULTURE

p moles / 10^7 lymphocytes [1)]

	phosphatidyl-choline	lysolecithin	phosphatidyl-ethanolamine	sphingomyelin	long chain fatty acids	neutral lipids
Control	17 ± 1[1)]	1	4 ± 1	3 ± 1	2 ± 1	16 ± 1
PHA	84 ± 1	2 ± 1	15 ± 1	30 ± 2	3 ± 2	22 ± 2
Con A	90 ± 2	2 ± 1	14 ± 1	30 ± 2	3 ± 2	21 ± 1
GARIG-IgG[2)]	25 ± 2	1	5 ± 1	6 ± 1	2 ± 1	18 ± 2

1) Values are given with ranges from experiments made in duplicate

2) Goat-anti-rabbit immunoglobulin

Culture conditions as in Fig. 3

after a stimulation period of 68 hours (35)) together with an only slightly enhanced incorporation rate into neutral lipids strongly suggest that in activated lymphocytes the turnover of phospholipids is indeed accelerated preferentially.

Isolation and characterization of lymphocyte plasma membranes

To support a role of phospholipids in the initiation of lymphocyte activation changes occurring within the plasma membrane have to be analyzed. To enable this, a method for the isolation of well characterized plasma membranes had to be developed (16) which is briefly outlined: (see also Fig. 4).

Lymphocytes from lymph nodes or thymus were equilibrated with nitrogen at 30 to 50 atmospheres in a "bomb". After equilibration the cells were released to atmospheric pressure. The intracellular development of gaseous nitrogen disrupts the cells very gently, preserving intracellular organelles such as nuclei, mitochondria or lysosomes. The plasma membrane fragments into small pieces which form vesicles (50 to 300 nm diameter). A microsomal fraction containing vesicles derived from the plasma membrane and the endoplasmic reticulum were isolated by differential centrifugation. The plasma membranes were then purified by microsomal subfractionation using dextran gradient ultracentrifugation. The plasma membranes obtained appeared to be highly enriched as assessed by chemical composition and marker enzyme studies (Table 5).

Turnover of the fatty acid moieties of lecithin in plasma membranes of activated lymphocytes

To measure the turnover rates of the fatty acid moieties in the plasma membrane of activated lymphocytes experiments were designed as in Fig. 4 (23,24). Lymphocytes were cultured in the presence of ^{14}C-oleate with or without PHA (or other mitogens, as well as antigens). After varying time intervals the cells were washed with ice-cold buffer and disrupted in the cold as described above. The plasma membrane was isolated and incorporation of ^{14}C-oleate into lecithin was determined. Within 10 minutes after the addition of PHA lymphocytes incorporated approximately three times more oleate into microsomal lecithin than controls (Table 6). Microsomes of lymphocytes consist of more than 60% of plasma membrane into which approximately 90% of the oleate is incorporated (24). Thus uptake into microsomes closely reflects that into the plasma membrane, but can be monitored more easily as during purification of plasma mem-

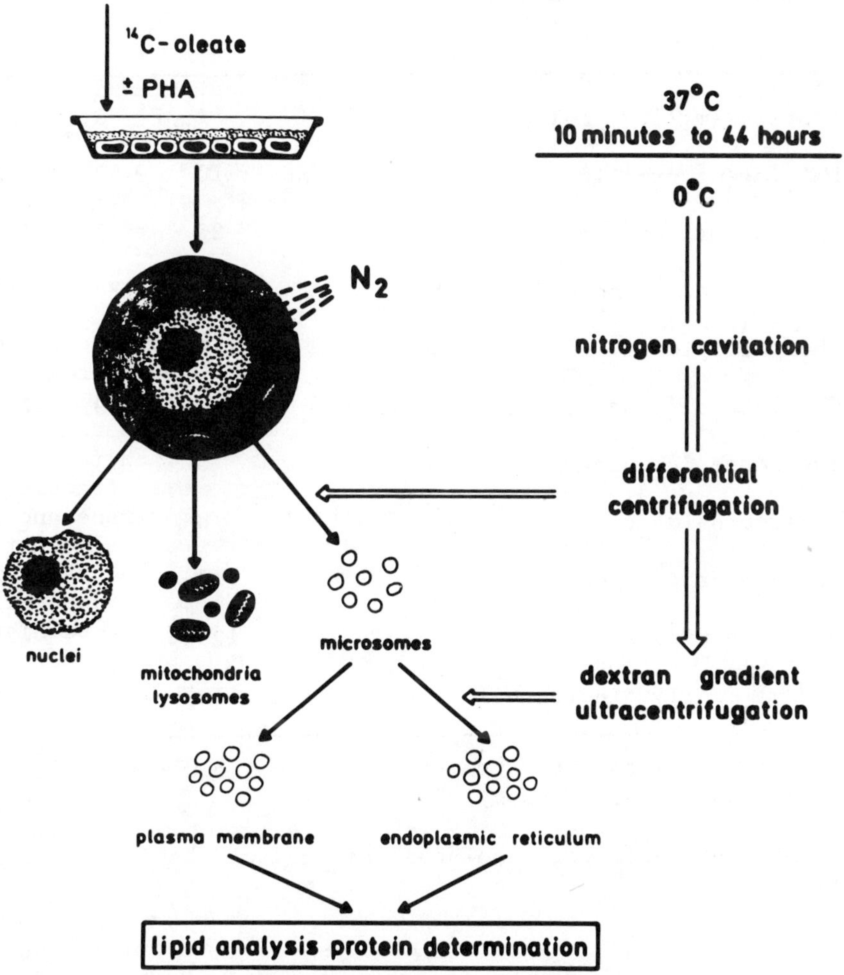

Fig. 4 Incorporation of ^{14}C-oleate into microsomal or plasma membrane phospholipids.

TABLE 5

PROPERTIES OF PIG LYMPHOCYTE PLASMA MEMBRANES

Enzymes activities are expressed as:
n moles x mg protein^{-1} x min^{-1}

Protein (mg/10^{10} cells)	2.87 ± 0.18	(10)
DNA (μg/mg protein)	10 ± 2	(2)
RNA (μg/mg protein)	25 ± 2.1	(2)
Cholesterol (n moles/mg protein)	751 ± 20	(6)
Phospholipid (n moles/mg protein)	727 ± 30	(6)
Cholesterol/phospholipid molar ratio	1.03 ± 0.04	(6)
Succinate dehydrogenase	0	(3)
β-glucuronidase	0	(3)
Lactate dehydrogenase	0	(2)
5'-nucleotidase	125.4 ± 10.5	(8)
Oleoyl-CoA:lysolecithin acyltransferase	10.0 ± 1.5	(8)

Values are given with standard deviation (numbers of determinations in parentheses)

Plasma membranes were isolated and analyzed as described in (16)

TABLE 6

INCORPORATION OF ^{14}C-OLEATE INTO MICROSOMAL LECITHIN

incubation (min)	cmp per 10^9 lymphocytes		
	controls	lymphocytes stimulated by PHA	PPD
10	728 ± 29	2134 ± 24	n.d.
30	2563 ± 71	5540 ± 117	n.d.
60	3098 ± 282	8787 ± 88	3819 ± 152
120	4439 ± 66	15301 ± 556	n.d.
180	5887 ± 251	n.d.	7337 ± 37
240	6583 ± 627	19276 ± 636	8818 ± 282

1 x 10^9 lymphocytes were cultured in 20 ml HEPES-buffered (pH 7.2) Eagle's medium containing 200 nmoles ^{14}C-oleate. PHA was added to a final concentration of 40 μg/ml, PPD to give a final concentration of 50 μg/ml. After incubation the cells were washed twice with PBS, resuspended and disrupted. The microsomes were isolated and washed once with 0.001 M HEPES buffer pH 7.5 to remove cytoplasmic protein.

Incorporation of ^{14}C-oleate into lecithin was measured in microsomes recovered from 10^9 lymphocytes (24). The recovery of microsomal protein from 10^9 cells (± s.d.) was 0.77 ± 0.19 mg in controls and 0.91 ± 0.13 in PHA stimulated lymphocytes. In PPD stimulated lymphocytes the same amount of microsomal protein was recovered as in controls.

n.d. = not done

Data are given with ranges from experiments made in duplicate.

branes material is lost. The degree of the increase in incorporation of ^{14}C-oleate into microsomal lecithin remained constant up to 40 hours culture. This suggested that this enhanced turnover is due to an immediate metabolic switch occurring in the plasma membrane as a consequence of ligand binding.

Mitogens are very potent in stimulating lymphocytes, therefore being used for most of the biochemical studies concerning the mechanism of activation, especially when early events are observed. The changes obtained in the incorporation of oleate into microsomal lecithin after stimulation with mitogens were sufficient to be detected with antigenic stimulation, too. Rabbits were sensitized with BCG. The lymphocytes of the draining lymph nodes were stimulated with PPD and incorporation of ^{14}C-oleate into lecithin was followed. Significant increase of uptake into microsomal lecithin became apparent after one hour stimulation (Table 6). This increase was specific as neither the addition of PPD to unsensitized lymphocytes nor culturing sensitized lymphocytes without antigen could be distinguished from normal lymph node lymphocytes. The degree of antigenic stimulation of the fatty acid turnover of the plasma membrane lecithin was low reflecting the low number of cells triggered to grow and divide.

Phospholipid metabolizing enzymes in the plasma membrane

The turnover of the long chain fatty acids of membrane phospholipids (e.g. lecithin) requires the participation of two specific enzymes: Phospholipase A generates the acceptor lysophosphatid by cleaving an ester bond to release a long chain fatty acid residue. Activated long chain fatty acids are then transfered to this acceptor molecule through lysophosphatid acyltransferase(s), to result in the formation of the fully acylated phospholipid (e.g. lecithin) (see Fig. 1).

Acyl-CoA: lysolecithin acyltransferase(s) are located in the plasma membranes of lymphocytes (16). Phospholipase A (and the enzyme which catalyzes the activation of the fatty acids -acid-coenzyme A ligase) could also be detected in microsomal membranes (14).

The activity of phospholipase A was difficult to determine. Using phospholipid micelles as substrate no activity could be detected. However, when erythrocyte membranes were used, which contained lecithin labeled biosynthetically, small activities became demonstrable (14) (Table 7). After stimulation with PHA, phospholipase A activity was increased by 30%. The method applied was not suitable for detailed

TABLE 7

PHOSPHOLIPASE A IN MICROSOMES OF PHA STIMULATED RABBIT LYMPH NODE LYMPHOCYTES

	p moles x mg^{-1} x min^{-1}
Control	6.0
PHA	7.8

Rabbit lymph node lymphocytes were cultured for 3 hours with or without PHA (200 µg/ml) at cell densities of 10^8 cell /ml Dulbecco's modified Eagle's medium. Microsomes were isolated as described in (16).

Phospholipase assay: 20 µg microsomal protein were incubated for 2 to 4 hours in TRIS buffer (10^{-1}M, pH 7.5) containing 10^{-2}M $CaCl_2$ together with erythrocyte membranes which were labelled in their phospholipids by incorporation of ^{14}C-oleate and lysolecithin (90 percent of the label in lecithin).

Phospholipase A activity is calculated from liberated fatty acid.

TABLE 8

SUBSTRATE SPECIFICITY OF MICROSOMAL ACYL-CoA: LYSOLECITHIN ACYLTRANSFERASE OF CON A STIMULATED CALF THYMOCYTES

Substrate	Oleoyl-CoA		Arachidonoyl-Coa	
	Vmax	Km (M)	Vmax	Km (M)
Control	3.9	1.0×10^{-5}	7.6	6.4×10^{-7}
Con A	7.9	1.2×10^{-5}	31.7	8.5×10^{-7}

Vmax is given as nmoles x mg^{-1} protein x min^{-1}

Culture conditions and enzyme assay as in Fig. 5

enzymatic analysis and the absolute values of enzyme activity, therefore, may be completely irrelevant.

Acyl-CoA: lysolecithin acyltransferase - which incorporates activated long chain fatty acids into lysolecithin - was studied in detail as this enzyme has a crucial role in determining the fatty acid composition of the main phospholipid lecithin (15,17,40). Very shortly after a mitogen had bound to lymphocytes, acyl-CoA: lysolecithin acyltransferase was activated. Maximal activation was obtained within 60 minutes at 37°C or at 0°C after PHA (or con A) was added to lymphocytes (Fig. 5).

Within very short time intervals after the addition of con A (5 to 30 min) the degree of activation depended on the amount of con A bound to the lymphocytes and approximately paralleled binding kinetics (41). The amount of con A bound to intact lymphocytes reached a plateau after 30 to 60 minutes of incubation with a wide range of con A concentrations (42). Without removing the unbound con A, the amount of con A bound to the surface of lymphocytes remained stable within several hours of culture. This allowed to investigate the dependence of the degree of the activation of acyl-CoA: lysolecithin acyltransferase on the concentration of this mitogen. The activation depended quadratically on the concentration of con A (Fig. 6). The data fitted well in a Hill equation with $n = 1.76$ (17). This suggests that the activation was due to allosteric transitions in which two ligand binding sites cooperate at one enzyme entity.

Acyl-CoA: lysolecithin acyltransferases can catalyze the incorporation of activated fatty acids with different chain length or degree of unsaturation into lysolecithin. In lymphocyte plasma membranes there exists a striking selectivity for polyunsaturated long chain fatty acids, predominantly for arachidonic acid (20:4). (Table 8) The affinity for arachidonic acid (k_m) was about 30 times higher than for oleic acid. Correspondingly, the highest specific activities (v_{max}) were observed with arachidonoyl-CoA in membranes from stimulated lymphocytes (17).

<u>Changes in fatty acid composition of phospholipids during stimulation</u>

The high selectivity of acyl-CoA: lysolecithin acyltransferases for polyunsaturated long chain fatty acids suggest that lymphocytes gain a higher degree of unsaturation when the turnover of the phospholipid fatty acids is increased during activation. Therefore, the fatty acid composition of phospholipids was studied in lymphocytes activated with con A (32). Thymus lymphocytes from rabbits were cultured in serum free medium with or without con A added. After 4

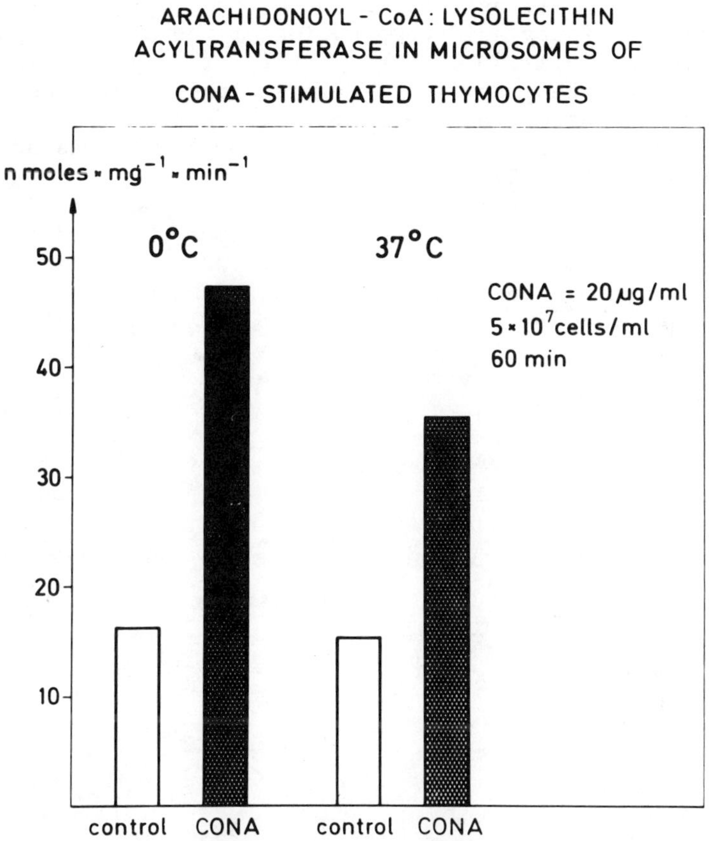

Fig. 5 Acyl-CoA: lysolecithin acyltransferase in microsomes of conA stimulated thymocytes.
10^{10} calf thymocytes were incubated in HEPES buffered Dulbecco's modified Eagle's medium for 60 min. Cell density was 5 x 10^7 cells /ml; conA was added at a concentration of 20 µg/ml.
Microsomal membranes were isolated as described in (16).
Acyl-CoA: lysolecithin acyltransferase was assayed using arachidonoyl-CoA as substrate as described in (15).

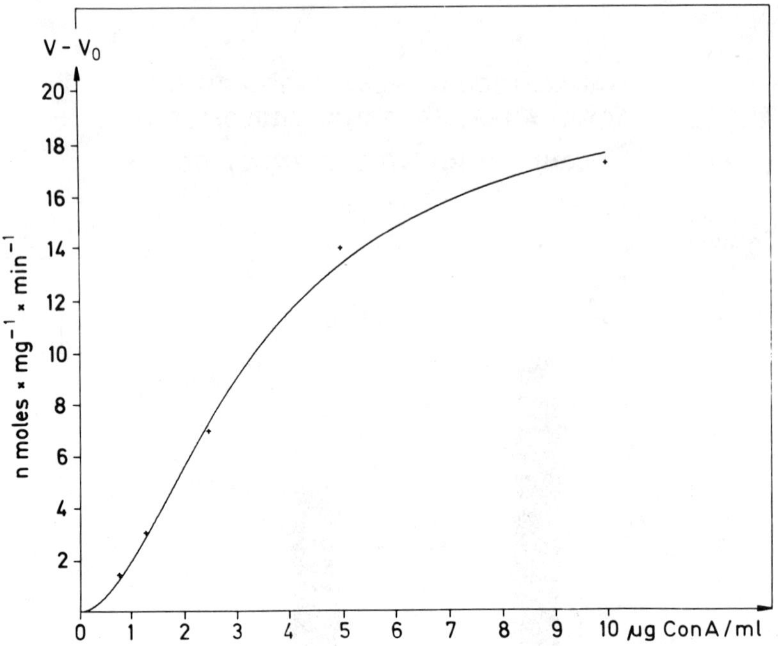
V - V0
n moles × mg⁻¹ × min⁻¹
20
18
16
14
12
10
8
6
4
2
0
1
2
3
4
5
6
7
8
9
10 µg ConA/ml

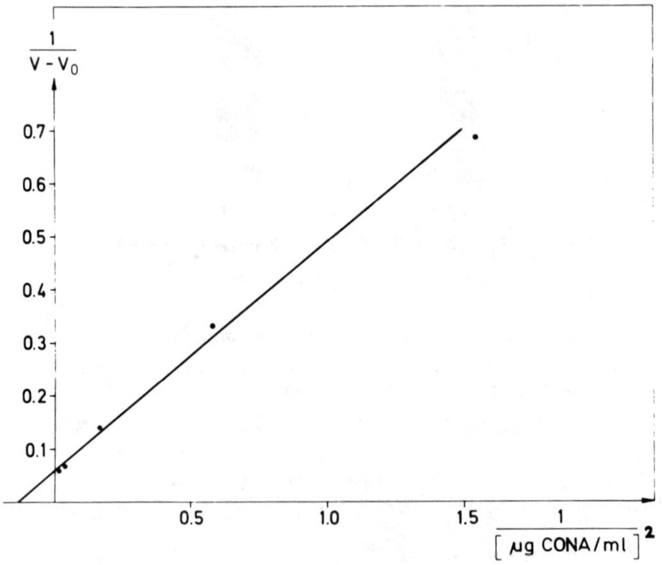
1 / V - V0
0.7
0.6
0.5
0.4
0.3
0.2
0.1
0.5
1.0
1.5
1 / [µg CONA/ml]²

Fig. 6 Activation of microsomal acyl-CoA: lysolecithin acyltransferase. Dependence on the activator (conA) concentration.

v = specific activity with conA

v_o= specific activity without conA

A) measured activities (+) and computed quadratic function according to Cleland (63) with the following constant: $v_{max} = 21$, $k = 3.6$, $b = 1.0$

B) Reciprocal plot of the square of conA dosage to the increase in transferase activity.

hours culture the fatty acids of lecithin (= phosphatidylcholine) were analyzed. Polyunsaturated fatty acids are transfered preferentially into the position 2 of these phospholipid molecules, hence, here the changes in the fatty acid composition could be detected (Table 9). The concentration of the highly unsaturated arachidonic acid was more than doubled in activated lymphocytes. Similar changes were found in phosphatidylethanolamine. As the concentration of monounsaturated fatty acid, oleic acid, did not change, the increase in unsaturation may be evaluated best from the ratio of polyunsaturated fatty acids versus saturated fatty acids, which was enhanced markedly in lecithin or phosphatidylethanolamine extracted from activated lymphocytes.

Increase in membrane fluidity during activation

The architecture and the physicochemical properties of a membrane depend largely on the relative content of saturated or polyunsaturated fatty acids. Higher concentration of polyunsaturated fatty acids in phospholipids are associated with an increase in the membrane fluidity (43). This may be counterbalanced at least partly by modulating the concentrations of the content of cholesterol (44), however, no significant alteration of the cholesterol content or synthesis was detected in activated lymphocytes (unpublished observations).

Increase in membrane fluidity occurring in stimulated lymphocytes has been reported by several authors using different methods of determination (17,22). In our own studies, the fluidity was measured using the fluorescence polarization of perylene (17). As this compound is highly hydrophobic (5 benzene rings without charged groups) it preferentially locates within the fatty acid region of the lipid phase. Fluorescence polarization can only be recorded when the time of emission of light (excitation-relaxation time) is shorter than the time which a molecule requires to rotate to its perpendicular position (rotational relaxation time). Thus polarization increases when the rotational motion is slowed down by increasing the viscosity (or decreasing the fluidity) of the environment. At low temperatures viscosity and therefore the degree of polarization of fluorescence of perylene increases in lipid phases. Relative higher fluidity, therefore, is indicated by a lower temperature at which complete depolarization occurs. In microsomal membranes from stimulated lymphocytes complete depolarization was observed between 20°C and 30°C whereas microsomes from unstimulated lymphocytes showed still fluorescence polarization of perylene even at 40°C. That this higher fluidity of

TABLE 9

CHANGES OF FATTY ACID DISTRIBUTION OF PHOSPHATIDYL CHOLINE AND PHOSPHATIDYL ETHANOLAMINE IN RABBIT THYMOCYTES DURING STIMULATION WITH CONCANAVALIN A

	16:0	18:0	18:1	18:2	20:4	22:6	ratio $\frac{\text{polyenoic}}{\text{saturated}}$
Phosphatidyl choline							
Control Pos.1	50.6	17.9	22.5	4.9	-	-	0.072
Con A Pos.1	58.1	16.0	20.7	5.2	-	-	0.070
Control Pos.2	47.4	3.2	20.0	19.4	7.8	tr.	0.538
Con A Pos.2	39.9	1.6	21.7	20.3	17.2	tr.	0.924

tr. = trace amounts

Rabbit thymocytes were cultured for 4 hours in HEPES buffered Dulbecco's modified Eagle's medium at cell densities of 5×10^7 cells /ml. Con A was added at 5 ug/ml. After incubation the lipids were extracted and separated with thin layer chromatography.

Individual phospholipids were eluted from the silica gel and their fatty acids were cleaved as methyl esters in methanol and sulfuric acid. The fatty acid distribution was determined by gas chromatography. The detailed methods are described in (32).

16:0 palmitic acid
18:0 stearic acid
18:1 oleic acid
18:2 linoleic acid
20:4 arachidonic acid

microsomal membranes was a property of the lipid phase is illustrated in Fig. 7. When fluorescence polarization of perylene was recorded in the extracted microsomal lipids from normal or stimulated lymphocytes, the same differences in the fluidity were found: Whereas lipids from stimulated lymphocytes came to complete depolarization at 24°C, lipids from control cell microsomes did not reach complete depolarization until 40°C.

CONCLUSIONS

The activation of lymphocytes is initiated by interactions of a ligand (mitogen, antigens) with specialized structures of the plasma membrane (receptors) (1-3). Whether these receptors for an individual ligand represent homogeneous entities is unknown at present. Because of its simplicity the assumption that the binding of a ligand to its receptor primarily results in one or only a few membrane changes appears to be most attractive.

As a consequence of activation the plasma membrane undergoes a variety of discernible changes. The permeability for nucleosides (7), amino acids (9,10) or sugars (8) is increased, at the same time the flux rates for Ca^{++} (12) or K^{+} (11) are enhanced. The onset of these changes appears to lie so early that it is difficult to recognize one alteration being the consequence of another.

In addition, very early changes in activated lymphocytes concern the plasma membrane as a whole. As shown recently, the fluidity of the membrane increases within minutes after the addition of a mitogen to lymphocytes (22).

If one now assumes that at most a very restricted number of alterations can result directly from the action of a stimulant the questions which an hypothesis must reconcile to explain the initiation of lymphocyte activation have to be asked more precisely:

1. Which membrane change(s) is (are) caused directly by the binding of a stimulant (ligand) to plasma membrane receptors? (= Primary membrane reactivity)
2. How are changes transmitted laterally within the plasma membrane to induce a variety of membrane alterations such as increase in fluidity, enhanced permeability or the activation of membrane bound enzymes? (= Intramembranal amplification)
3. How is the plasma membrane involved in affecting intracellular processes responsible for cell activation and mitosis? (= "signal" transmission)

The first two questions may be discussed in terms of the changes of the metabolism of membrane phospholipids de-

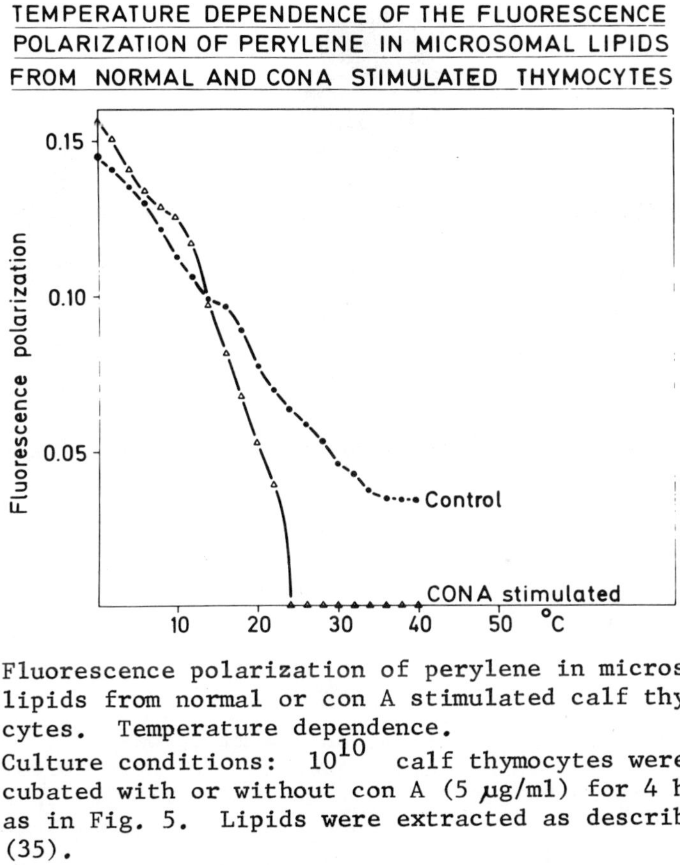

Fig. 7 Fluorescence polarization of perylene in microsomal lipids from normal or con A stimulated calf thymocytes. Temperature dependence.
Culture conditions: 10^{10} calf thymocytes were incubated with or without con A (5 μg/ml) for 4 hours as in Fig. 5. Lipids were extracted as described in (35).
Fluorescence polarization: 5 ml samples containing 300 n moles phospholipid and 0.8 n moles perylene in phosphate buffer (10^{-3}M, pH 7.2) were subjected to fluorescence polarization from 0°C to 40°C. Excitation at 410 nm, emission was recorded at 488 nm.

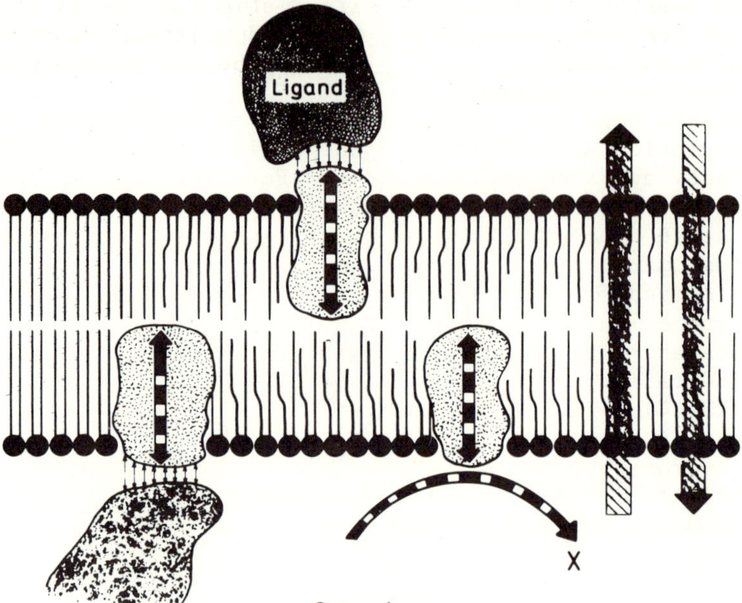

Fig. 8 The role of phospholipids in the initiation of lymphocyte activation.

Binding of a stimulant to receptors of the lymphocyte plasma membrane causes enhanced turnover of the phospholipid fatty acids which results in increase of the degree of unsaturation (primary reactivity). This leads to changed physicochemical properties of the lipid phase which are transmitted immediately through the plasma membrane. As a consequence membrane fluidity increases and plasma membrane function such as permeability or the activity of membrane bound enzymes are altered (intramembranal amplification). A changed intracellular milieu due to influx (e.g. of Ca^{++} or K^{+}) or efflux (e.g. of cyclic AMP) or the generation of effector molecules ("second messengers") within the membrane (e.g. cyclic GMP) may be the signal for cell activation ("signal" transmission).

scribed above. Being aware of the highly superficial understanding of the intracellular controls of function and proliferation, only some ideas will be outlined, how a resting lymphocyte may be triggered to acquire function and/or grow and divide.

1. We suggest that the first reaction which the plasma membrane undergoes interacting with a stimulant is an increase in the turnover of the fatty acid moieties of membrane phospholipids. Membrane alterations may be due to two phases in this metabolic pathway.

Increased turnover rates result from an accelerated de- and reacylation cycle with lysophosphatides and free long chain fatty acids as intermediates. Perturbations of the membrane can be caused by even subtle deviations of the steady state concentrations of these highly surface active compounds.

Within minutes after the administration of a mitogen to lymphocyte cultures long chain fatty acids are incorporated at a constant elevated rate into the plasma membrane lecithin, indicating a new equilibrium of all intermediate products (23,24). Removal of one fatty acid from lecithin (lysolecithin generation) leads to less constraint packing of the hydrocarbon chains in the lipid bilayer, resulting in higher fluidity (45). Increased membrane fluidity has been found to occur within minutes in mitogen activated lymphocytes (22).

The reacylation process is directed in activated lymphocytes: As a consequence of the activation of acyl-CoA: lysolecithin acyltransferases which preferentially transfer polyunsaturated long chain fatty acids (15, 17) the content of linoleic acid and especially arachidonic acid increases in membrane phospholipids of stimulated lymphocytes (32). Polyunsaturated fatty acids exhibit a skewed shape, hence, their presence induces a similar loosely packed lipid membrane as lysolecithin (45). In contrast to the immediate changes of lysophosphatides or free long chain fatty acids, replacement of lecithin fatty acids with polyunsaturated fatty acids requires time. After a first peak of increased fluidity occurring within minutes (22) activated lymphocytes gain a second long lasting enhancement of fluidity (58). It is intriguing to speculate that this second phase of increased fluidity is due to the increased degree of unsaturation of membrane phospholipids.

2. That subtle changes in phospholipids indeed can underlie functional alterations of plasma membranes has been established recently in various experimental approaches.

Phospholipids exhibit a highly self organizing structure

favoring cooperative interactions (29). Due to these cooperative properties, small perturbations at distinct sites of a phospholipid matrix are able to change immediately the whole lipid phase. This has been shown in bacteria auxotrophic for long chain fatty acids (46) or by fusion experiments with mammalian cells (47). Recent experiments with defined artificial membranes allowed a better understanding of the underlying physicochemical principles (45).

One of the intrinsic properties of membrane phospholipids consists in the ability to form a barrier for water and water soluble substances, resulting in compartimentalization and thus in preservation of an internal steady state. The properties of this barrier depend on the fatty acid composition of the membrane phospholipids. Fluxes of glycerol or erythritol are considerably higher in membranes with a higher content of polyunsaturated fatty acids (48,49). Active functions too, e.g. enzyme catalyzed transport are regulated by the degree of unsaturation of the phospholipid fatty acids or the content of surface active constituents such as lysophosphatides or free long chain fatty acids. Some examples of this are the β-galactosid transport system in bacteria (46,50), NADH-cytochrome c reductase in microsomal membranes (51), or ATP ases in the plasma membrane of mammalian cells (52,53). As shown above, membrane phospholipids with a higher degree of unsaturation exhibit increased fluidity. In addition, the incorporation of natural lysolecithin or non-metabolizable synthetic lysolecithin analogs into erythrocyte membranes may also enhance membrane fluidity (54).

3. The replication of DNA and mitosis are very late events in the course of lymphocyte activation. Before lymphocytes enter the S phase they express function and growth.

The central role in the acquisition of function or in cellular growth plays the induction of increased protein synthesis. Whether the early increase in protein synthesis requires the formation of new mRNA is unclear. No new mRNA species have been detected by hybridization studies in mitogen activated lymphocytes (55). Using cell free systems it has been suggested recently that protein synthesis of activated lymphocytes is regulated by translational processes (56,57). The lack of understanding about cellular regulation of growth and function makes it difficult to ascertain where a putative "signal" may have its site of action. Thus the evidence demonstrating such a "signal" is rather circumstantial, derived either from changed intracellular concentrations of the suggested regulator molecule, or from inhibition studies.

All "signals" discussed may be explained on the basis of the properties which the plasma membrane gains immediately when a stimulant binds to its receptors: the increase in fluidity caused by alterations in the phospholipid metabolism.

Several principles can be discussed:

The increase in membrane permeability can cause increased influx of ions (as shown for K^+ (11,13) or Ca^{++} (12) which are required for the activity of regulatory enzymes (e.g. acting on nuclear histones) or affecting the rate of translation.

Elevated concentrations of specific amino acids may be required for the synthesis of regulatory proteins. Thus lowering the exogenous supply of glutamine and cystine under certain levels impairs the initiation of DNA synthesis (59).

Increased membrane permeability can also cause leakage of intracellular substances. Some of these may block cell function or proliferation. In the last years such a control function has been discussed for cyclic AMP (60,61,62). An altered lipid phase of membranes is also able to regulate the activity of membrane bound enzymes, many of which are involved in active transport (see above). A membrane bound enzyme may catalyze the production of a regulator substance ("second messenger") within the membrane which is secreted into the interior of the cell. As a candidate for such a messenger cyclic GMP has been suggested (19,20).

It remains to be established whether the activation of lymphocytes is caused by a single "signal" transduced from the plasma membrane to the interior of the cell. In the view of the multiple changes occurring in the plasma membrane of activated lymphocytes it appears attractive that such a "signal" is the changed plasma membrane itself which determines by a variety of mechanisms the internal milieu which is required for cell activation.

REFERENCES

1. Greaves, M.F. and Bauminger, S., Nature New Biology 235: 67, 1972.
2. Betel, I. and van den Berg, K.J., Eur. J. Biochem. 30: 571, 1972.
3. Andersson, J., Edelman, G.M., Möller, G. and Sjöberg, O., Eur. J. Immunol. 2: 233, 1972.
4. Edelman, G.M., Yahara, I. and Wang, J.L., Proc. Nat. Acad. Sci. 70: 1442, 1973.
5. Taylor, R.B., Duffus, W.P.H., Raff, M.C. and

de Petris, S., Nature New Biology 233: 225, 1971.
6. Loor, F., Forni, L. and Pernis, B., Eur. J. Immunol. 2: 203, 1972.
7. Peters, J.H. and Hausen, P., Eur. J. Biochem. 19: 502, 1971.
8. Peters, J.H. and Hausen, P., Eur. J. Biochem. 19: 509, 1971.
9. Mendelsohn, J., Skinner, S.A. and Kornfeld, S., J. Clin. Invest. 50: 818, 1971.
10. van den Berg, K.J. and Betel, I., Exp. Cell. Res. 76: 63, 1973.
11. Quastel, M.R. and Kaplan, J.G., Exp. Cell. Res. 63: 230, 1971.
12. Whitney, R.B. and Sutherland, R.M., Cellular Immunol. 5: 137, 1972.
13. Averdunk, R., Hoppe-Seyler's Z. Physiol. Chem. 353: 79, 1972.
14. Resch, K., Ferber, E., Odenthal, J. and Fischer, H., Eur. J. Immunol. 1: 162, 1971.
15. Ferber, E. and Resch, K., Biochim. Biophys. Acta 296: 335, 1973.
16. Ferber, E., Resch, K., Wallach, D.F.H. and Imm, W., Biochim. Biophys. Acta 266: 494, 1972.
17. Ferber, E., Reilly, C.E., de Pasquale, G. and Resch, K., in Proceedings of Eighth Leucocyte Culture Conference, eds. Lindahl-Kiessling, K. and Osoba, D., Academic Press, New York and London, 1974, p. 529.
18. Fisher, D.B. and Mueller, G.C., Biochim. Biophys. Acta, 248: 434, 1971.
19. Hadden, J.W., Hadden, E.M., Haddox, M.K. and Goldberg, N.D., Proc. Nat. Acad. Sci. 69: 3024, 1972.
20. Hadden, J.W., in Workshop Early Events in Lymphocyte Activation, 2nd International Congress of Immunology, Brighton, 1974.
21. Smith, J.W., Steiner, A.L., Newberry, W.M., Jr., and Parker, C.W., J. Clin. Invest. 50: 432, 1971.
22. Barnett, R.E., Scott, R.E., Furcht, L.T. and Kersey, J.H., Nature 219: 465, 1974.
23. Resch, K., Ferber, E. and Gelfand, E.W., Proceedings of the Seventh Leucocyte Culture Conference, Academic Press, New York and London, 1973, p. 75.
24. Resch, K., Gelfand, E.W., Hansen, K. and Ferber, E., Eur. J. Immunol. 2: 598, 1972.
25. Singer, S.J. and Nicolson, G.L., Science 175: 720, 1972.
26. Singer, S.J., in Structure and Function of Biological Membranes, ed. Rothfield, L.J., Academic Press, New York and London, 1971, pp. 145-222.

27. Overath, P., Schairer, H.-U. and Stoffel, W., Proc. Nat. Acad. Sci. 67: 606, 1970.
28. Reinert, J.E. and Steim, J.M., Science 168: 1580, 1970.
29. Chapman, D., in Biological Membranes, Vol. 2, eds. Chapman, D. and Wallach, D.F.H., Academic Press, New York and London, 1973, pp. 91-144.
30. Gottfried, E.L., J. Lipid Res., 8: 321, 1967.
31. Ferber, E., de Pasquale, G.G. and Resch, K., unpublished results.
32. Ferber, E., de Pasquale, G.G. and Resch, K., Biochim. Biophys. Acta, in press.
33. Ferber, E. and Resch, K., in The Lymphocyte: Structure and Function, ed. Marchalonis, J.J., M. Dekker, New York, in press.
34. Fisher, D.B. and Mueller, G.C., Biochim. Biophys. Acta 176: 316, 1969.
35. Resch, K. and Ferber, E., Eur. J. Biochem. 27: 153, 1972.
36. Fisher, D.B. and Mueller, G.C., Proc. Nat. Acad. Sci. 60: 1396, 1968.
37. Kay, J.E., Nature 219: 172, 1968.
38. Fanger, M.W., Pelley, R.P. and Reese, A.L., J. Immunol. 109: 294, 1972.
39. Elfenbein, G.J., Harrison, M.R. and Mage, R.G., J. Immunol. 110: 1340, 1973.
40. Ferber, E., in Biological Membranes, Vol. 2, eds. Chapman, D. and Wallach, D.F.H., Academic Press, New York and London, 1973, pp. 221-252.
41. Reilly, C.E., Thesis, Freiburg, Germany, 1974.
42. Huber, A., Thesis, Freiburg, Germany, 1974.
43. Overath, P. and Träuble, H., Biochemistry 12: 2625, 1973.
44. Papahadjopulos, D., Jacobson, K., Nir, S. and Isac, T., Biochim. Biophys. Acta 311: 330, 1973.
45. Träuble, H., in Biomembranes, Vol. 3, eds. Kreuzer, F. and Slegers, J.F.G., Plenum Publishing Corp., New York, 1972, pp. 197-227.
46. Wilson, G. and Fox, C.F., J. Mol. Biol. 55: 49, 1971.
47. Frye, L.D. and Edidin, M., J. Cell Sci. 7: 319, 1970.
48. de Kruyff, B., de Greef, W.J., van Eyck, R.V.W., Demel, R.A. and van Deenen, L.L.M., Biochim. Biophys. Acta 298: 479, 1973.
49. McElhaney, R.N., de Gier, J. and van der Neut-Kok, E.C.M., Biochim. Biophys. Acta 298: 500, 1973.
50. Schairer, H.-U. and Overath, P., J. Mol. Biol. 44: 209, 1969.

51. Jones, P.D. and Wakil, S.J., J. Biol. Chem. 242: 5267, 1967.
52. Bloj, B., Morero, R.D., Farias, R.N. and Trucco, R.E., Biochim. Biophys. Acta 311: 67, 1973.
53. Goldemberg, A.L., Farias, R.N. and Trucco, R.E., Biochim. Biophys. Acta 291: 489, 1973.
54. Weltzien, H.U., Joint meeting of European Societies for Immunology, Strasbourg, France, 1973.
55. Neiman, P.E. and McDonnel, D.M., in Proceedings of the Fifth Leucocyte Culture Conference, ed. Harris, J., Academic Press, New York and London, 1970, p. 61.
56. Ahern, Th. and Kay, J., in Proceedings of the Eighth Leucocyte Culture Conference, ed. Lindahl-Kiessling, K. and Osoba, D., Academic Press, New York and London, 1974, p. 51.
57. Ahern, Th. and Kay, J., Biochim. Biophys. Acta 331: 91, 1973.
58. Kersey, J.H., in Workshop Early Events in Lymphocyte Activation, 2nd International Congress of Immunology, Brighton, 1974.
59. van den Berg, K.J., Publication of the Radiobiological Institute of the Organization for Health Research TNO Rijswijk, The Netherlands, 1974.
60. Bombik, B.M. and Burger, M.M., Exptl. Cell Res. 80: 88, 1973.
61. Bösing-Schneider, R. and Kolb, H., Nature 244: 223, 1973.
62. Henney, C.S., Bourne, H.R. and Lichtenstein, L., J. Immunol. 108: 1526, 1972.
63. Cleland, W.W., Nature 198: 463, 1963.

MODULATION IN THE ORGANIZATION OF
PLASMA MEMBRANE INTRAMEMBRANOUS PARTICLES
IN FIBROBLASTS AND LYMPHOCYTES[1]

Robert E. Scott and Leo T. Furcht
Department of Laboratory Medicine & Pathology, University of Minnesota Medical Center, Minneapolis, Minnesota 55455

Studies on a variety of mammalian cells suggest that modulation in the organization of the plasma membrane influences a variety of cell functions. Cell-to-cell contact inhibits movement (1) and proliferation (2) in certain "normal" cell lines, mitogenic lectins stimulate lymphocyte proliferation (3) and chalones influence cell differentiation by mechanisms mediated at the plasma membrane (4). These effects are mediated at least in part by modulation in the activity of membrane enzymes including transport systems (5) and adenylate (6) and guanylate cyclase (7).

The ultrastructural technique of freeze fracture has served as a useful method to analyze certain changes in plasma membrane organization. With this technique mammalian cells are rapidly frozen and fractured under a high vacuum. The fracture plane has been shown to split biological membranes and to expose two fracture faces, an inner fracture face and an outer fracture face. Both fracture faces are associated with 70 Å particulate structures designated plasma membrane intramembranous particles (PMP).[2] Studies on human red blood cells (8), outer rod segment membranes (9) and sarcoplasmic reticulum membranes (10) have established that PMP represent a complex of intrinsic membrane proteins and membrane lipids. Figure 1 illustrates that initial studies suggested that PMP represented a complex of a single intrinsic membrane glycoprotein, glycophorin, and membrane lipids (8). The studies of Marvin and Garvin (11), however, demonstrated that such glycoproteins exist as dimers in the plasma membrane and

---

[1]This work was supported in part by Grant CA-13458 from the National Institutes of Health, United States Public Health Service.

[2]Abbreviations used in this paper: PMP, plasma membrane intramembranous particle; FF, freeze fracture; IFF, inner fracture face; OFF, outer fracture face; PBS, phosphage buffered saline, pH 7.4; CB, cytochalasin B; Con A, concanavalin A; PHA, phytohemagglutinin.

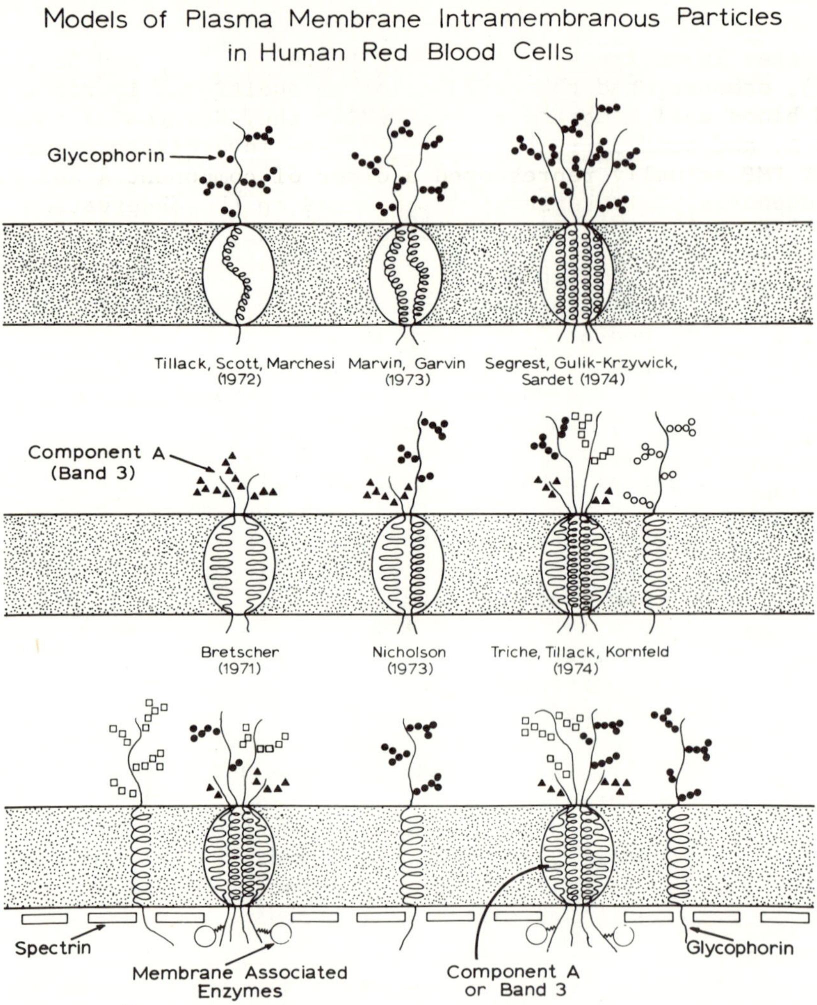

Fig. 1 Model of plasma membrane intramembranous particles in human red blood cells.

raised the possibility that PMP might represent a complex of this dimer and unique membrane lipids. Segrest et al. have also recently presented evidence that PMP in red blood cells may actually represent an aggregate of four or more glycophorin molecules (12).

Other investigators, including Bretscher (13) and Steck (14), demonstrated the presence of an additional intrinsic red blood cell membrane protein which they designated component A, and band 3, respectively. These investigators suggested that PMP actually represented a dimer of component A and not glycophorin. This suggestion was based on the observation that component A is a large molecule with a molecular weight of 90,000, a major portion of which is buried in the membrane lipid bilayer. They calculated that the presumed intramembranous portion of this molecule could account for the mass of a 70 Å particle. Monomeric glycophorin, by contrast, is a smaller molecule with a molecular weight of 30,000 of which only a 23 amino acid region is buried in the lipid bilayer (15). Nicholson attempted to resolve this controversy by proposing that PMPs might represent a complex of glycophorin and component A (16). More recent studies by Trishe et al. suggest a microheterogeneity of glycophorin in human red blood cell membranes and the possibility that PMPs may contain two different glycophorins in addition to component A. It appears that certain membrane glycoproteins are not associated with PMP (17).

These observations support the hypothesis that PMP may represent a complex of unique intrinsic membrane proteins not only in red blood cells but in nucleated mammalian cells. It has been suggested that such protein complexes may be associated with membrane pores, transport sites, and with certain enzymes. Studies on sarcoplasmic reticulum membranes have, for example, demonstrated a correlation between PMP and Ca^{++} -Mg^{++} ATP'ase (10). The functional significance of such protein complexes in most mammalian cells, however, remains to be established. Studies to determine whether changes in the distribution or density of PMP occur as a result of modulation of cell function or by the action of various drugs can nevertheless provide information on basic membrane structure.

For the past two years we have examined the plasma membrane of contact inhibited and Simian Virus 40 transformed 3T3 cells by freeze fracture and electron microscopy and have demonstrated a significant difference in the distribution of PMP. Contact inhibited 3T3 cells contain aggregated PMP at confluence whereas transformed SV3T3 cells showed a random PMP distribution (Fig. 2). Both types contain approximately 600 PMP μ^{-2}. Figure 3 illustrates that at a density below

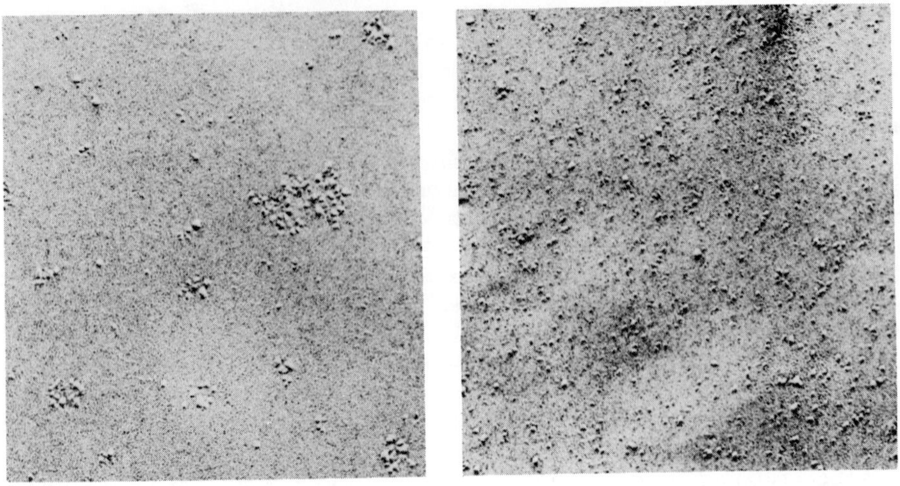

Fig. 2 Differences in the distribution of plasma membrane intramembranous particles in 3T3 (A) and SV3T3 (B) cells. (X 70,000)

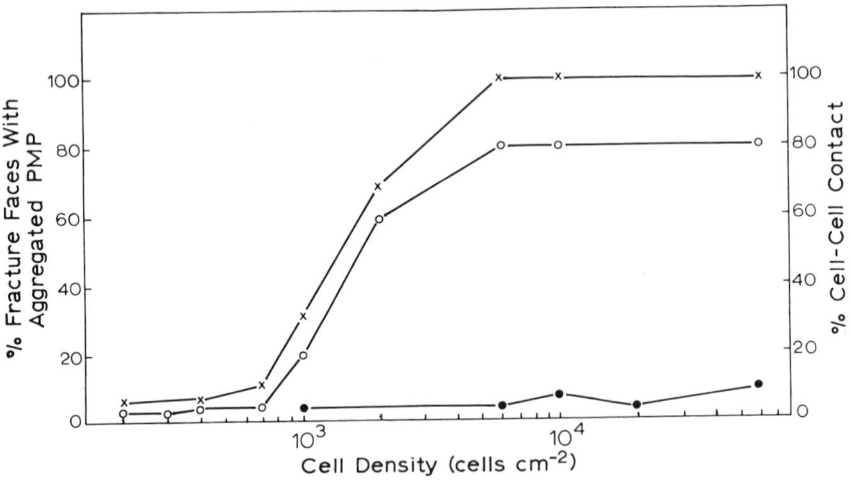

Fig. 3 Association of plasma membrane intramembranous particle aggregation and cell contact in 3T3 cells.

1×10^3 cells cm^{-2} 3T3 cells showed essentially no PMP aggregation and no evidence of cell contact. During the log phase of growth, cell-to-cell contact is established at a density of approximately 1×10^3 cells cm^{-2} and all cells are in contact with adjacent cells at a density of 6×10^3 cells cm^{-2}. During this phase when cell-to-cell contact is established a progressive increase in the percentage of membrane fracture faces showing aggregated PMP is observed. Transformed cells, by contrast, show no evidence of PMP aggregation even when extensive cell contact is apparent. Since PMP aggregation in nonviable trypan blue positive transformed cells is, however, observed at very high cell densities, all experiments have employed only cell populations with greater than 95% viability. Differences in the freeze fracture morphology of normal and transformed cells have also been established with a variety of cell preparation schemes, including freezing cells in media alone, following exposure of cells to 15% glycerol-PBS or 20% sucrose-PBS for fifteen minutes at room temperature or after prefixed in 1% formaldehyde for ten minutes at room temperature prior to glycerination (Fig. 4).

A significant correlation between the appearance of cell contact associated changes in the organization of the plasma membrane freeze fracture morphology of contact inhibited 3T3 cells and increased intracellular levels of cAMP and decreased deoxyglucose transport has been reported (19). However, the physiological significance of this correlation remains to be established.

Attempts to determine the mechanisms that regulate the distribution of PMP in mouse fibroblasts have focused on analysis of the effects of a variety of microtubule and microfilament disruptive drugs. Exposure of SV3T3 cells to colchicine, vinblastine sulfate and lumicolchicine was found to induce a reversible dose and time dependent aggregation of PMP (Fig. 5). For example, incubation of confluent SV3T3 cells with 10^{-6} colchicine at $37^{o}C$ for four hours induced PMP aggregation on 85% of membrane fracture faces. The observation that $10^{-9}M$ colchicine, a dose below that required to disrupt microtubules, can induce PMP aggregation, as can $10^{-5}M$ lumicolchicine, a drug which does not affect microtubular integrity, suggests that PMP aggregation may at least in part be induced by a primary action of colchicine on membrane proteins or lipids or their interaction. These drugs did not influence the freeze fracture morphology of contact inhibited 3T3 cells which show aggregated PMP in the native state.

Cytochalasin B at doses between 0.5 and 0.1 $\mu g\ ml^{-1}$, by

	Approximate % Fracture Faces with Aggregated PMP			
	Medium	Glycerol	Sucrose	Fixed
3T3	80	85	90	80
3T3 + Cytochalasin B	--	20	--	20
SV3T3	20	10	10	10
SV3T3 + Colchicine	80	85	85	85

Fig. 4 Effect of cell preparation on plasma membrane freeze fracture morphology of 3T3 and SV3T3 cells. Specimens prepared in medium were scraped from tissue culture flasks and frozen immediately in Freon 22 cooled by liquid nitrogen. Other specimens were incubated *in situ* in either 15% glycerol-PBS or 20% sucrose-PBS at room temperature for 15 minutes prior to freezing. Fixed specimens were exposed to 1% formaldehyde-PBS for 10 minutes at room temperature and then glycerinated as above prior to freezing. The effect of drugs was assayed following a 4-hour incubation at 37°C in either 0.5 μg ml^{-1} cytochalasin B or 10^{-6}M colchicine. The size of PMP aggregates is slightly reduced in fixed specimens.

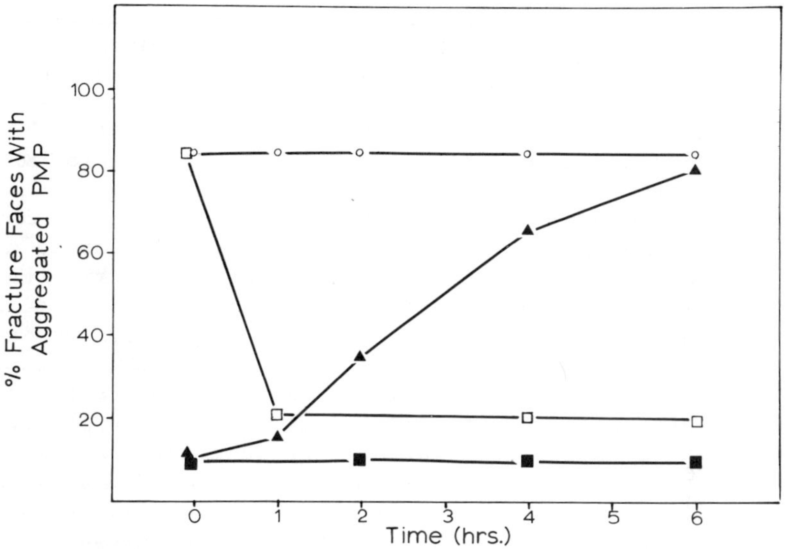

Fig. 5 Effect of cytochalasin B and colchicine on the distribution of plasma membrane intramembranous particles in 3T3 and SV3T3 cells. 3T3 (o——o); 3T3 plus 0.5 μg ml^{-1} cytochalasin B (□——□); SV3T3 (■——■); SV3T3 plus 10^{-6}M colchicine (▲——▲).

contrast, promotes the disaggregation of PMP in 3T3 cells without significantly affecting SV3T3 cells (Fig. 5). Within two hours following exposure of contact inhibited cells to CB at 37°C PMP disaggregation was apparent. This effect could also be reversed within 1 hour following removal of the drug by washing. Although these observations suggest that microfilaments may also be involved in control of the distribution of PMP in the plasma membrane, previous reports suggest that doses of CB required to disrupt microfilaments are up to ten-fold greater than doses that promote PMP disaggregation. We, therefore, interpret these data to suggest that both colchicine and CB may have a primary effect on membrane proteins and their distribution in mouse fibroblasts and that these changes may not be solely mediated by microtubules and microfilaments.

The biochemical characteristics of PMP in most nucleated cells has not been established, but the available data suggest that PMP are not associated with Con A receptors on the cell surface in general. For example, Burger et al. (20) showed that Con A receptors can be aggregated in transformed cells, whereas our data show that PMP are randomly distributed in transformed cells and are aggregated in normal 3T3 cells.

Studies on lymphoid cells also show that there appears to be no association of PMP and Con A or Ig receptors based on studies showing that capping of such receptors can be induced without significantly affecting the distribution of PMP in these cells (21). Further the cellular mechanisms that control the distribution of PMP in lymphoid cells may be unique. Figure 6 illustrates that where colchicine, cytochalasin B and various other treatments affect PMP distribution in mouse fibroblasts, they do not affect the freeze fracture morphology of human RPMI 1788 lymphocytes. These observations further support our hypothesis that PMP represent unique protein complexes in individual cell types, complexes that may serve a variety of physiological functions.

Although the biochemical and physiological characteristics of PMP in "B" and "T" lymphocytes have not been established, and although it has been difficult to modulate PMP distribution in most lymphoid cells (22) artificially, we have established that human peripheral blood lymphocytes and mouse spleen cells do show a significant change in PMP density following stimulation by mitogenic agents. Incubation of human lymphocytes with optimal mitogenic doses of either PHA of Con A induced cell transformation and resulted in a change in plasma membrane organization as demonstrated by the freeze fracture technique (23). During blast transformation human lymphocytes showed a five-fold increase in PMP density

Additives	Duration of Treatment	Conditions for Cell Freezing	PMP Distribution	Approximate PMP Density (PMP/μ^2) OFF	IFF
--------		Medium	Random	800	1650
15% glycerol-PBS	15 min, 24°	15% glycerol	Random	800	1750
15% glycerol-PBS	20 hr, 4°	15% glycerol	Random	800	1750
1% formaldehyde-PBS; 15% glycerol-PBS	10 min, 24° 15 min, 24°	15% glycerol	Random	800	1800
0.5 μg ml^{-1} Cytochalasin B	4 hr, 37°	Medium	Random	750	1550
1 x 10^{-6}M Colchicine	4 hr, 37°	Medium	Random	850	1550
1 x 10^{-6}M Colchicine, 50 μg ml^{-1} Concanavalin A	4 hr, 37° 30 min, 37°	Medium	Random	850	1600
50 μg ml^{-1} Concanavalin A	30 min, 37°	Medium	Random	---	----

Fig. 6 Effect of drugs, glycerol, sucrose and fixation on plasma membrane intramembranous particle density and distribution in human RPMI 1788 lymphocytes.

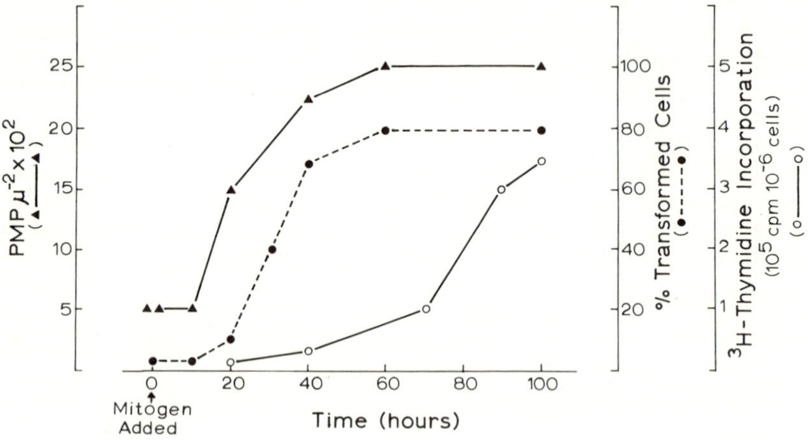

Fig. 7 Modulation in plasma membrane intramembranous particle density in PHA stimulated human lymphocytes. PMP density (▲——▲); percent transformed cells (●——●); ^{3}H-thymidine incorporation (o——o).

(Fig. 7); during the same interval Krug et al. (24) demonstrated the appearance of insulin receptors on lymphocyte plasma membranes. Changes in the density of PMP have also recently been demonstrated to occur in mouse spleen lymphocytes stimulated with sodium periodate. Modulation in the density of PMP in periodate stimulated lymphocytes demonstrated nearly identical kinetics as observed during transformation of human lymphocytes induced by PHA.

On the basis of these data derived from studies on erythrocytes, fibroblasts and lymphocytes we conclude that PMP represent unique complexes of membrane proteins in individual cell types. Changes in PMP density and distribution observed under a variety of conditions and a variety of cell types suggest a real yet presently unresolved physiological significance. Elegant studies demonstrating that intramembranous particles can be produced in artificial liposomes by the incorporation of a variety of intrinsic membrane proteins including glycophorin, rhodopsin and other proteins need to be expanded. We must establish the native composition of PMP in the membranes of nucleated mammalian cells and attempt to determine the functional significance of changes in the organization of these complexes in the plasma membrane.

REFERENCES

1. Abercrombie, M., and Ambrose, E.J., Cancer Res., 22: 525, 1962.
2. Todaro, G.J., Lazar, G.K., and Green, H., J. Cell. Comp. Physiol., 66: 325, 1966.
3. Greaves, M.F., and Bauminger, S., Nature New Biol., 235: 67, 1972.
4. Levine, S., Pictet, R., and Rutter, W.J., Nature New Biol., 246: 49, 1973.
5. Isselbacher, K.J., Proc. Natl. Acad. Sci. USA, 69: 585, 1972.
6. Perry, C.V., Johnson, G.S., and Pastan, I., J. Biol. Chem., 246: 5785, 1971.
7. Hadden, J.W., Hadden, E.M., Haddox, M.K., and Goldberg, N.D., Proc. Natl. Acad. Sci. USA, 69: 3024, 1972.
8. Tillack, T.W., Scott, R.E., and Marchesi, V.T., J. Exp. Med., 135: 1209, 1972.
9. Hong, K., and Hubbell, W.L., Proc. Natl. Acad. Sci. USA, 69: 2618, 1972.
10. Tillack, T.W., Boland, R., and Martonosi, A., J. Biol. Chem., 249: 624, 1974.
11. Marvin, L.S.G., and Garvin, J.E., Biochem. Biophys. Res. Commun., 52: 1457, 1973.

12. Segrest, J.P., Gulik-Krzywicki, T., and Sardet, G., Proc. Natl. Acad. Sci. USA, 71: 3294, 1974.
13. Bretscher, M.S., J. Mol. Biol., 59: 351, 1971.
14. Steck, T.L., J. Cell Biol., 62: 1, 1974.
15. Segrest, J.P., Jackson, R.L., Marchesi, V.T., Guyer, R.B., and Terry, W., Biochem. Biophys. Res. Commun., 49: 964, 1972.
16. Nicholson, G.L., Internatl. Review Cytol., 1974.
17. Triche, T., Tillack, T.W., and Kornfeld, S., Am. J. Path., 74: 86a, 1974.
18. Scott, R.E., Furcht, L.T., and Kersey, J.H., Proc. Natl. Acad. Sci. USA, 73: 3631, 1973.
19. Bannai, S., and Sheppard, J.R., Nature, 250: 62, 1974.
20. Burger, M.M., and Goldberg, A.R., Proc. Natl. Acad. Sci. USA, 57: 359, 1967.
21. Karnovsky, M.J., Unanue, E.R., and Leventhal, M., J. Exp. Med., 136: 907, 1972.
22. Guerin, C., Zachowski, A., Prigent, B., Paraf, A., Dunia, I., Diawara, M.A., and Benedetti, E.L., Proc. Natl. Acad. Sci. USA, 71: 114, 1974.
23. Scott, R.E. and Marchesi, V.T., Cell. Immunol., 3: 301, 1972.
24. Krug, V., Krug, F., and Cuatrecasas, P., Proc. Natl. Acad. Sci. USA, 69: 2604, 1972.

DISCUSSION FOLLOWING SYMPOSIUM EARLY EVENTS IN LYMPHOCYTE ACTIVATION I: MEMBRANE AND BIOCHEMICAL EVENTS

Metzger - Bethesda: A couple of years ago the Ishizakas' and we showed that, at least in the IgE-basophil system, under conditions where the immunoglobulins were capped, one in fact got inhibition of histamine release; only under conditions where one got only microscopic aggregation without gross aggregation could one see histamine release. I wonder whether you might like to summarize for us what in your own mind are the most convincing pieces of evidence which suggest that capping and/or endocytosis and/or motility have in fact anything to do with physiological activation of lymphocytes.

Unanue - Boston: Well David Katz and I published studies two or three years ago in which we capped the lymphocytes with anti-Ig and then asked the question whether they are activated. At least in the mouse we did not induce lymphocyte proliferation. There is no doubt that at one point your receptor aggregation leads to a better binding to antigen. All I can say is that the cell which is able to cap and to clear its membrane totally of receptors do not have helper functions but are able to regenerate their receptors. When you interrupt partially this sequence with D-amino acid polypeptides, the cell leaves it on the membrane and in some way or another distorts the whole sequence. As a result the cell is totally paralyzed although I do not know about cell motility.

Unidentified Questioner: I would like to know if there is a difference between the ability of anti μ and anti kappa to induce motility with B lymphocytes?

Unanue - Boston: I have not tested that.

Paul - Bethesda: What concentrations of anti-Ig are required to induce motility and secondly have you attempted to prepare cross-linked anti-Igs to evaluate whether they may be for example more efficient particularly at low concentrations.

Unanue - Boston: The second part of the question, no I have not. You can induce motility with doses that do not provoke capping. About 10 mg of antibody in a total volume of 0.5 ml containing 2 million cells is generally used. Obviously it is difficult to extrapolate from one to another with different antibody concentration.

Havemann - Marburg: What is the reason for the inhibitory capacity of DFP? Do you inhibit lymphocyte enzymes or are you inhibiting by any chance a protein released from adherent cells?

Unanue - Boston: These have been done on preparations of non-adherent lymphocyte. I cannot rule out that we may have some monocytes in it. They are taken from normal mice but I do not know if they secrete esterases. The effect of DFP is an extremely fast one, you can get inhibition of motility in as little as 3 minutes after addition of DFP. So I think we are inhibiting some kind of cell enzyme on the B cell that is necessary to carry the reaction through, and not as a result of any soluble enzyme. But we really do not have formal proof of this.

Berlin - Farmington: What is the source of unsaturated fatty acids?

Resch - Heidelberg: That is a point we have looked at. In whole cells, there is a pool of fatty acid which is not high, but which astonishingly consists of more than 40% linoleic acid. As we have found that linoleic acid can be easily transformed to arachidonic acid in lymphocytes. This could be the pool for both of them. This pool is high enough to explain all the changes I have described. When we looked whether the lymphocyte can also synthesize double unsaturated fatty acids, it looks like it can a very little bit, but the main source would be the pool of linoleic acid present in the cells.

Gunther - New York: Dr. Adelman, Dr. Wung and I have shown that all lymphocytes do not become committed to stimulation immediately after the addition of a lectin, but rather more and more lymphocytes become committed with longer times of exposure up to about 20 hours. By commitment, I mean the ability to enter DNA synthesis in blast transformation independent of the presence of the lectin. Have you removed Con A from the cells after 4 hours to see if the stimulation of phospholytic turnover now relaxes back to normal levels?

Resch - Heidelberg: Thank you for this question! This is just what is going on now in our laboratory. I do not have the results as yet.

Sage - Durham: A lot of things happen in the first couple of minutes after you add a lectin to a cell. Everybody seems to assume that because one sticks a mitogenic lectin onto the cell and observes metabolic changes that these changes have some significance in mitosis. I think studies from Dr. Edelman's lab and our own have previously shown that there is a mitogenic commitment time that occurs much later than this initial period. But aside from that you can take lectins which don't stimulate cells toward DNA synthesis and mitosis and a great many of the early events you are talking about occur nevertheless. For example, everybody knows about the fact that you get a rise in cyclic AMP when you use Con A or PHA on a cell. You could get more of a cyclic AMP rise if you use a nonstimulating lectin, the mushroom lectin. Furthermore, you can also show that, with this nonstimulating lectin, you can get increased uptake of uridine, and increased protein synthesis and you don't get stimulation. I am just arguing against the a priori assumption that any of these early events is a part of the mitogenic process. That particular simplistic view assumes that the only fate a lymphocyte can possibly have is to either remain in its resting state or go into mitosis. It is very likely that there are other possible fates for lymphocytes, other directions than mitosis.

Edidin - Baltimore: You have been implying I think in the course of your description of what lipids change, that there is a general change in lipid composition and that this is distributed equally over the membrane. Do you have any thoughts about whether or not it might be a very localized lipid pool either surrounding proteins or in another restricted site since the only compartment to increase is arachidonic acid?

Resch - Heidelberg: In that particular slide I showed it was arachidonic acid. In other experiments linoleic acid content also goes up. As we know that linoleic acid can be converted to arachidonic acid, this may just reflect how much this process has been gone, as content of both normally increase. The second part of your question is really an interesting one, that is, of course, we do not know whether the whole membrane changes or just parts of it which one would like to describe as protein associated lipids. We have just succeeded in isolating out plasma membrane areas bearing the receptor, and of course the first thing we will now do is analyze these fragments compared to the bulk membrane. But so far, we do not have the answers.

SECTION IV

EARLY EVENTS IN LYMPHOCYTE ACTIVATION II: MEMBRANE AND BIOCHEMICAL EVENTS

Chairman

CHARLES W. PARKER

Washington University School of Medicine
St. Louis, Missouri

Co-Chairman

IRA GREEN

National Institutes of Health
Bethesda, Maryland

POSSIBLE MECHANISMS OF LYMPHOCYTE ACTIVATION

Charles W. Parker
Department of Medicine, Washington University School of Medicine, St. Louis, Missouri 63110

Lectin induced lymphocyte transformation has been widely used as a model for cell activation. Studies with lectins insolubilized on Sepharose or polyacrylamide beads indicate that the response is initiated at the lymphocyte surface (1, 2,3). In transmitting a signal from a lectin on the cell exterior to the inside of the cell a number of possible mechanisms can be considered. Since it would appear that diverse biochemical pathways in different parts of the cell are affected more or less simultaneously beginning quite early in the response (4,5), the involvement of a diffusable intracellular messenger with broad flexibility as a regulatory agent would logically be required. Tomkins and his colleagues have suggested the use of the term pleiotypic to describe an agent which might fulfill such a role (6). As candidates for pleiotypic agents in the regulation of mitogenic responses in lymphocytes several cyclic nucleotides and a variety of critical intracellular nutrients need to be considered. The possible importance of cyclic nucleotides in lymphocyte activation was studied initially by our group (7,8,9) and by Hirschhorn et al. (10). In many cells, adenylate cyclase, the enzyme that synthesizes cyclic AMP (cAMP), is primarily located in the external plasma membrane, where it modulates the cellular response to polypeptide hormones and other cell surface stimuli (11). It is not unreasonable to assume that cAMP might fulfill a similar role when lectins or antigens bind to lymphocytes, the controlling event being either a rise or fall in cAMP since cAMP has been reported to have both positive and negative effects on cellular proliferation. Cyclic GMP might fulfill a similar role since intracellular cGMP levels are also susceptible to external modulation and cGMP has been reported to be increased in rapidly growing tissues (12). Still another cyclic nucleotide which must be considered is cyclic CMP. It was recently reported that this agent markedly stimulates cell growth in cultured leukemic cells (13). Both cAMP and cGMP undergo rapid increases in lymphocytes in response to PHA and con A consistent with a role for one or both of these

Supported by Allergy Clinical Center Grant AI10405 and Research Grant AI00219

nucleotides in early metabolic events in lectin stimulated cells (4,5,12). Cells stimulated by PHA undergo a later fall in cAMP, suggesting that cAMP may exert different functions in different stages of the cell cycle.

The subject of cyclic nucleotides and their involvement in the modulation of lymphocyte transformation will be dealt with in some detail in this session. The possible role of cAMP as both a positive and negative stimulus to transformation will be considered by Dr. Wedner. He will summarize work conducted in our laboratory over the past 5 years with particular emphasis on recent data indicating an apparent similarity in the effects of cAMP and PHA on protein phosphorylation. Dr. Hadden will summarize the studies he has carried out in collaboration with Dr. Nelson Goldberg in regard to the possible importance of cGMP in lymphocyte activation.

As an alternative mechanism for lymphocyte activation lectin effects on the transport of critical cellular constituents involved in energy utilization or protein, nucleic acid or lipid synthesis need to be considered. This could occur through a direct action on membrane proteins or through an intermediary such as cAMP or cGMP. Nutrients which might be critical in cell growth and differentiation include glucose, K^+, Ca^{++}, $PO_4^=$, nucleosides and amino acids. All of these substances enter lymphocytes more rapidly when PHA or con A is present (14-19); Ca^{++} is of particular interest since it affects many intracellular enzyme systems and has been implicated in cell growth in bone marrow cells, fibroblasts and thymocytes (20,21). The subject of monovalent cations in general and K^+ in particular will be discussed by Dr. Kaplan.

In addition to Dr. Kaplan's presentation it is appropriate to consider some interesting recent data from our laboratory which bears on the possible importance of calcium ion in lymphocyte activation. One approach to defining the role of Ca^{++} in mitogenesis is to see if different classes of mitogenic agents consistently increase Ca^{++} uptake. We have studied this question and found that for 8 different lymphocyte mitogens and 7 related nonmitogens there is a positive correlation between early stimulation of Ca^{++} transport and the ultimate development of a mitogenic response. Human lymphocytes were isolated on a Ficoll-Hypaque gradient, suspended in Eagle's medium in the presence of 2% AB serum and studied for increases in $^{45}Ca^{++}$ uptake at 1 hour. At 1.8 mM Ca^{++} 4 different classes of mitogen-lectins (PHA-P, erythroagglutinating PHA, con A, pokeweed), divalent cations (Hg^{++}, Zn^{++}), oxidizing agents (periodate) and proteolytic enzymes (trypsin) all produced increases in Ca^{++} uptake at one hour. Nonmitogenic divalent cations (Co^{++}, Cd^{++}, Mn^{++}, Cu^{++}) and lectins (wheatgerm and mushroom agglutinins) produced little if any increase in Ca^{++} uptake. The only exception was at highest concentration of mushroom agglutinin (25 μg/ml) where stimulation was obtained but lower concentration of the same agent were inhibitory. The concentrations of heavy metal and periodate which produced maximal stimulation of Ca^{++} uptake were close to the optimal concentrations for mitogenesis. This was not true for the mitogenic lectins where the greatest effects on Ca^{++} transport were obtained at concentrations above the mitogenic optimum. Nor did the degree of increase in Ca^{++} uptake at 1 hour correlate closely with the absolute increase in ^{3}H-thymidine uptake eventually obtained. Thus, it appears that factors other than Ca^{++} uptake are important in determining the magnitude of the mitogenic response. Nonetheless, the overall correlation between mitogenicity and early stimulation of Ca^{++} transport together with the recent demonstration that Ca^{++} ionophores stimulate mitogenesis (22) and the observation by Alford (23) and others (17) that a mitogenic response does not occur if Ca^{++} is omitted from the medium make it likely that Ca^{++} is critically involved in the response.

The mechanism for the increase in ^{45}Ca uptake with the various mitogens remains to be elucidated. In most tissues Ca^{++} influx appears to occur through facilitated diffusion whereas Ca^{++} efflux is an active process (24). The available evidence with P-PHA would suggest that changes are occurring in the Km but not the Vmax for Ca^{++} transport (25). Since lectins and periodate are among the agents that stimulate Ca^{++} accumulation and they appear to act at the cell surface it seems likely that the initial stimulus is at this level. It seems doubtful that cGMP is involved as a secondary messenger since lymphocytes exposed briefly to 8-bromo cGMP do not exhibit increases in Ca^{++} transport. Indeed at 10 and 100 μm 8-bromo cGMP a modest decrease in Ca^{++} uptake appears to be occurring. In view of the ability of Ca^{++} to stimulate guanylate cyclase in some tissues (26) any changes in intracellular cGMP may conceivably be secondary to increases in intracellular Ca^{++}. Whether cAMP is involved in lectin induced increases in Ca^{++} transport remains to be conclusively established. High concentrations (10-1000 μM) of mono and dibutyryl cAMP produce modest increases in Ca^{++} transport in human lymphocytes as early as 2 minutes after the initial exposure (C.W. Parker, unpublished observations) apparently persisting for as long as one hour (25), suggesting that cAMP could be fulfilling the role of secondary messenger. However, the effect is much smaller than that of the lectins themselves and the possible role of other secondary messengers or of simple cross linking of membrane receptors per se, must also be considered.

Just how increases in intracellular Ca^{++} might promote mitogenesis remains to be established. Recent studies from this laboratory indicate that human lymphocyte nuclei contain an adenylate cyclase which is stimulated by Ca^{++} over a broad concentration range (27) (in contrast to an inhibitory effect of Ca^{++} on adenylate cyclase in most tissues). If a similar effect occurrred in intact lymphocytes increases in intracellular Ca^{++} would promote the accumulation of cAMP inside the nucleus, with probable secondary effects on histone or acidic protein phosphorylation. As noted above Ca^{++} also stimulates guanylate cyclase and effects through cGMP dependent protein kinases also are possible.

A final area of interest is the possible importance of microtubules and microfilaments in surface membrane responses to lectins. A variety of multivalent agents which interact with the lymphocyte surface produce changes in membrane distribution culminating in the accumulation of the agent in one pole of the cell (capping). Edelman and his colleagues have reported that mitogenic lectins have a dual effect on the capping response (28). At low concentrations the mitogens themselves cap but at high concentrations they not only fail to cap but inhibit the response to other capping agents. The inhibition appears to be correlated in some way with mitogenesis since nonmitogenic lectins do not inhibit capping. The inhibition is blocked by agents which disrupt microtubules such as vinblastine and colchicine. It has been suggested that receptors for mitogenic lectins are attached to colchicine binding proteins which in turn are attached to other membrane proteins important in modulating the lectin response. We have investigated the possible importance of microtubules in the cAMP response to con A and found that vinblastine and colchicine increase the cAMP response several fold (Greene and Parker, in preparation). By contrast, cytochalasin B, an agent that binds to microfilaments, completely inhibits the cAMP response. These observations suggest a reciprocal relationship between microfilaments and microtubules in modulating the cAMP response. cAMP immunofluorescence measurements indicate that colchicine interferes with the normal localization of cAMP in the plasma membrane in response to con A. Work in progress also indicates important effects of cytochalasin, vinblastine and colchicine on plasma membrane transport in the presence and absence of lectin. By the use of these agents it should be possible to define more precisely the role of cyclic nucleotides, microfilaments and microtubules in early transport events.

In conclusion, it seems likely that changes both in cellular nutrients and cyclic nucleotides are a necessary part of the mitogenic response but their relationship to one another and relative importance have not been delineated. Nor is the role of microtubules and microfilaments clearly defined although it seems likely that one or both of these structures is involved in the maintenance of plasma membrane integrity and quite conceivably in the modulation of plasma membrane transport events as well.

REFERENCES

1. Greaves, M.F. and Bauminger, S., Nature (Lond.) New Biol. 235: 67, 1972.
2. Betel, I. and van den Berg, K.J., Eur. J. Biochem. 30: 571, 1972.
3. Ahmann, G.B. and Sage, H.J., Cellular Immunol. 10: 183, 1974.
4. Parker, C.W., Sullivan, T.J. and Wedner, H.J., in Advances in Cyclic Nucleotide Research, edited by P. Greengard and A.G. Robison, p.1, Raven Press, New York, 1974.
5. Wedner, H.J. and Parker, C.W., in Progress in Allergy, edited by P. Kallós and B.H. Waksman, S. Karger, New York (in press).
6. Mamont, P., Hershko, A., Kram, R., Schacter, L., Lust, J. and Tomkins, G.M., Biochem. Biophys. Res. Communications 48: 1378, 1972.
7. Smith, J.W., Steiner, A.L. and Parker, C.W., J. Clin. Inv. 50: 442, 1971.
8. Smith, J.W., Steiner, A.L., Newberry, W.M., Jr. and Parker, C.W., Fed. Proc. (Abstract) 28: 566, 1969.
9. Smith, J.W., Steiner, A.L., Newberry, W.M., Jr. and Parker, C.W., J. Clin. Inv. 50: 432, 1971.
10. Hirschhorn, R., Grossman, J. and Weissman, G., Proc. Soc. Exp. Biol. 133: 1361, 1970.
11. Robinson, G.A., Butcher, R.W. and Sutherland, E.W., Cyclic AMP, Academic Press, New York, 1971.
12. Hadden, J.W., Hadden, E.M., Haddox, M.K. and Goldberg, N.D., Proc. Nat. Acad. Sci. USA 69: 3024, 1972.
13. Bloch, A., Dutschman, G. and Maue, R., Biochem. Biophys. Res. Communications 59: 955, 1974.
14. Peters, J.H. and Hausen, P., Eur. J. of Biochem. 19: 509, 1971.
15. Averdunk, R., Hoppe-Seyler's Zeitschrift für Physiologische Chemie 353: 79, 1972.

16. Allwood, G., Asherson, G., Davey, M.J. and Goodford, P., Immunology 21: 509, 1970.
17. Whitney, R.B. and Sutherland, R.M., Cell. Immunology, 5: 137, 1972.
18. Peters, J.H. and Hausen, P., Eur. J. Biochem. 19: 502, 1971.
19. Mendelsohn, J., Skinner, A. and Kornfeld, S., J. Clin. Invest. 50: 818, 1971.
20. Gail, M.H., Boone, C.W. and Thompson, C.S., Exp. Cell Res. 79: 386, 1973.
21. Whitfield, J.F., Rixon, R.H., Perris, A.D. and Yondale, T., Exp. Cell Res. 57: 8, 1969.
22. Maino, V.C., Green, N.M. and Crumpton, M.J., Nature 251: 324, 1974.
23. Alford, R.H., J. Immunol. 104: 698, 1970.
24. Borle, A.B., J. General Physiology 53: 43, 1969.
25. Whitney, R.B. and Sutherland, R.M., in Proceedings of the Seventh Leucocyte Culture Conference, edited by F. Daguillard, p. 63, Academic Press, New York, 1973.
26. Schultz, G., Hardman, J.G., Schultz, K., Barid, C.E. and Sutherland, E.W., Proc. Nat. Acad. Sci. USA 70: 3889, 1973.
27. Parker, C.W., Snider, D.E. and Wedner, H.J., in Proceedings of the Second International Congress of Immunology (Brighton, England, July 1974), (in press).
28. Edelman, G.M., Yahara, I. and Wang, J.L., Proc. Nat. Acad. Sci. 70: 1442, 1973.

THE ROLE OF CYCLIC NUCLEOTIDES IN LYMPHOCYTE ACTIVATION*

H. James Wedner+, Floyd E. Bloom, Charles W. Parker+
+Department of Medicine, Division of Immunology, Washington University Medical School, St. Louis, Missouri, and Laboratory of Neuropharmacology, National Institute of Mental Health, St. Elizabeths Hospital, Washington, D.C.

Lymphocyte activation can be defined as the initiation of new DNA synthesis in a resting population of peripheral lymphocytes and the progression of these cells into and through mitosis. The process begins with the surface binding of a variety of specific and nonspecific activating agents to the surface of the lymphocyte. The binding of activating agents induces a wide variety of biochemical alterations within the lymphocyte beginning with changes in the lymphocyte plasma membrane, progressing through changes in cytoplasmic metabolism, and culminating in alterations in intranuclear enzyme activity resulting in new DNA synthesis and ultimately in cell division (for review see 1,2). A critical question in lymphocyte activation has been attempts to find a single, early biochemical alteration, which is common to all lymphocyte activating agents and which can be demonstrated to be responsible for all of the subsequent biochemical alterations which are seen in activated lymphocytes. The cyclic nucleotides, cyclic AMP (cAMP) and cyclic GMP (cGMP) appear to provide the most likely candidates for the biochemical trigger for lymphocyte activation. There are a number of reasons for selecting cyclic nucleotides as the most likely candidate:
1) many of the biochemical changes which can be demonstrated in activated lymphocytes are mediated or modulated by cyclic AMP in nonlymphocytic tissues. These include increases in the transport of glucose (3), amino acids (4,5,6), potassium (7) and calcium (8,9) and changes in phosphatidylinositide turnover (10). Each of these biochemical changes has also been demonstrated to occur in activated lymphocytes (1).
2) Cyclic AMP has been implicated in the induction of DNA synthesis in nonlymphoid tissue. In rat salivary glands isoproterenol produces increases in DNA synthesis in cells which are exposed to this agent (11,12), the dose of isoproterenol which stimulates DNA synthesis increases adenylate cyclase

*Portions of this work were supported by grant #1P01AI12450 and grant #DCG-1281 from the Damon Runyon Cancer Fund

activity (13) and produces 2-4 fold increases in intracellular cAMP (12). Cyclic GMP has also been implicated as a mediator of cell proliferation. 3T3 cells incubated with phorbol acetate show large increases in cGMP and later go on to increased DNA synthesis (14). 3) Alterations in the acetylation and phosphorylation of nuclear histones have been demonstrated in activated lymphocytes, and it has been suggested that these alterations may increase the level of DNA template activity (15,16). Cyclic AMP has been shown to stimulate the activity of protein kinases which phosphorylate selected acidic proteins (17) and lysine rich histones (18,15). It is also possible that cGMP dependent protein kinases (19) may also have this ability although this has not been directly demonstrated. 4) In rat liver slices cAMP has been shown to produce a rapid, dose-related increase in the RNA synthesizing capacity of nuclei isolated at various times after stimulation (20). Substantiation for this action of cAMP comes from the work of Cawthon, Bitte, Krystosek and Cabot (21) that the addition of cAMP to intact rabbit reticulocytes results in the rapid phosphorylation of ribosomal proteins. 5) Finally, cyclic nucleotides appear to be the one class of biochemical agents which is uniquely suited to serve as the second messenger for the surface activation of lymphocytes.

Efforts to implicate cAMP and cGMP as the intracellular "second messenger" have been done using two biochemical approaches; first, measurements of alteration in adenylate cyclase or intracellular cAMP or guanylate cyclase or intracellular cGMP in activated lymphocytes, and second, attempts to mimic the action of activating agents by the incubation of cells with other agents known to increase either cAMP or cGMP or with cAMP or cGMP or their analogs. The most extensive study of the alterations in intracellular cAMP in activated lymphocytes has been carried out in our laboratory (22,23,24, 25,26,27,28,29,30,31). The original studies in this series were done on human peripheral lymphocytes purified by filtration on short nylon columns. Subsequent studies have been done using lymphocytes purified on Ficoll-Hypaque gradients (32) and on populations of lymphocytes enriched in T cells by filtration through long nylon wool columns (32). These studies were done using the plant lectins, phytohemagglutinin (PHA) and concanavalin A (con A) which are mitogenic for a major proportion of human peripheral lymphocytes. In the original studies using PHA-P (a relatively crude preparation of PHA) increases in cAMP ranging from 1.5-8 fold over control were obtained. These increases could be demonstrated as early as 1 minute following the addition of PHA-P with peak values seen at 5-20 minutes. At 6 and 16 hours of incubation cAMP in PHA treated cells was significantly less than control cells. Similar increases in cAMP have also been noted with purer preparations of PHA (E-PHA and L-PHA) and with con A. Adenylate cyclase assayed in crude homogenates of purified peripheral lymphocytes caused a 2 fold increase in the activity of adenylate cyclase indicating that the action of the lectin was probably through direct stimulation of adenylate cyclase. One difficulty with the early studies was that the lectin concentration for maximum increases in cAMP was significantly above the concentration which gave maximum stimulation of the incorporation of tritiated thymidine into newly synthesized DNA. This indicated that the increases in cAMP which were measured might not be occurring in the relevant lymphocyte population. In subsequent studies we have compared rises in cAMP in T cell-B cell mixtures and in purified T cells. In the mixture of T and B cells a biphasic response to PHA was obtained, with the optima seen at the mitogenic range (1-5 μg/ml) and at a lectin concentration greater than 20 ug/ml. When purified T cells were examined a single peak of cAMP stimulation was seen with maximal stimulation in the mitogenic range. Thus the increase in cAMP in T cells, which is the cell believed to be stimulated by PHA and con A, occurs within the mitogenic range. The second peak seen in the T-B cell mixture we believe to be the result of stimulation of cAMP in B cells. In the purified T cell population incubation of the cells with PHA or con A for 2 hours or longer produced levels of cAMP which were significantly below those seen in control cells. This was particularly evident at concentrations of PHA in the mitogenic range. Thus the pattern of cAMP response in T cell populations appears to be an initial rise in cAMP followed by a decrease in cAMP below control levels. Similar increases in cAMP with maximal responses at or near the stimulation maxima have been demonstrated for con A and mercuric chloride, another activating agent.

Increases in cAMP have also been demonstrated in lymphocytes from experimental animals incubated with PHA. Mouse splenic lymphocytes incubated for 1-2 hours with M-PHA demonstrated a 1.5-3.2 fold increase in adenylate cyclase activity (33,34). Mouse lymphocytes preincubated for 3 or 4 hours with 40 μg/ml of PHA (a supramitogenic dose) had significantly depressed adenylate cyclase activity (35). This observation may relate to the observation by our group that intracellular cAMP falls at 2 hours in purified human peripheral T lymphocytes exposed to lectin and does not argue against cAMP as an important mediator early in the lymphocyte activation.

Early increases in intracellular cGMP in activated lymphocytes have also been described. In human peripheral lymphocytes up to 60 fold increases in intracellular cGMP were noted within 30 minutes following the addition of PHA (36). In this study increases in cAMP were obtained in cells incubated with M-PHA (a very crude PHA preparation which has high agglutinating activity), while no increase in cAMP could be demonstrated in peripheral lymphocytes incubated with E-PHA (low but not absent agglutinating ability). In rat peripheral leukocytes large increases in cAMP were again noted in response to PHA and con A (37). However, in this study no attempt was made to obtain a purified population of lymphocytes, and it is impossible to say which population of cells was responding in this system. Early attempts in our laboratory to demonstrate increases in cGMP were unsuccessful. We have re-examined this question and have been able to demonstrate increases in cGMP in some experiments with human peripheral lymphocytes. However, in our hands the increases were usually less than 2.5 fold similar in magnitude to the increases in cAMP which were demonstrated in the same cell populations. We can not explain the difference between the results obtained by others and those obtained by our group. It is not inconceivable that both cAMP and cGMP increase early in the activation of lymphocytes and that both of these cyclic nucleotides may ultimately be demonstrated to be important in lymphocyte activation.

The second method for studying the effect of cAMP and cGMP on lymphocyte metabolism has been attempts to duplicate the action of lectin by agents which increase intracellular cAMP or cGMP or with these agents themselves or attempts to augment the effects of activating agents. In early studies Smith et al. (22,25) evaluated the effect of cAMP or its diacylated analog dibutyryl cAMP on DNA synthesis, RNA synthesis and protein synthesis and demonstrated that high concentrations (10^{-5} to 10^{-3} molar) inhibited all three responses. The inhibition was also obtained with comparable concentrations of other cyclic and noncyclic nucleotides and nucleosides. This was true for both the adenine and guanosine series. This type of study indicated that the cyclic phosphodiester moiety was not responsible for the inhibition seen and suggested that all the nucleotides were being degraded to a common metabolic intermediate. To avoid the possibility of degradation of the products into a common intermediate other agents which are capable of increasing intracellular cyclic nucleotides were studied. These included prostaglandins, catecholamines and methylxanthines which were used in concentrations which had been previously demonstrated to increase intracellular cAMP. All of these agents were able to inhibit parameters of lymphocyte activation similar to that seen in cells incubated with the nucleotides themselves. Since the only common action of these 3 agents appeared to be the elevation of intracellular cAMP it appeared that the inhibition was being exerted through cAMP itself (26). Another agent, glucocorticoids, which are inhibitors of lymphocyte transformation, have also been demonstrated to increase cAMP levels at high concentrations again suggesting that cAMP increased in this manner was an inhibitor of mitogenesis (38,39,40).

Other groups have also examined the effects of cyclic nucleotides and methylxanthines on the response of human peripheral blood lymphocytes. Hirschhorn and her co-workers (41) demonstrated that dibutyryl cyclic AMP and theophylline inhibited the DNA synthetic response to both PHA and con A. They also found that adenosine and other adenine nucleosides were also inhibitors. Results similar to these have also been noted in animal lymphocytes. Novogrodsky and Katchalski (42) demonstrated that dibutyryl cyclic AMP decreased RNA and DNA synthesis in rat lymph node lymphocytes stimulated by lectins. It is interesting that in contrast to the results found in human peripheral lymphocytes PGE_1 did not inhibit the response.

An alternate approach has been to examine parameters of lymphocyte activation in lymphocytes incubated with cAMP or dibutyryl cAMP in the absence of mitogens. Hirschhorn et al. (41) demonstrated a 2 fold increase in the incorporation of tritiated thymidine into DNA or tritiated uridine into RNA in lymphocytes exposed to cAMP. Smith et al. (25) observed similar increases in tritiated thymidine incorporation in lymphocytes incubated with 1-10 μM cyclic AMP or with .01-.02 μM dibutyryl cAMP. The time course of stimulation was similar to that seen with PHA, however, the increase seen was only 2-3 fold (as compared with 50-100 fold with PHA) and might indicate that only a small proportion of the lymphocytes were actually being activated. McCurdy and Rigby (43) also demonstrated modest increases in tritiated thymidine incorporation in the presence of cAMP. In contrast to these results Bourne (44) and Mendelsohn et al. (40) were unable to obtain statistically significant stimulation of blastogenesis with cAMP in human peripheral lymphocytes.

In animal lymphocytes incubation of cells with cAMP or dibutyryl cAMP have also been demonstrated to show increases similar to those seen in lectin treated cells. Cross and Ord (45) demonstrated increases in DNA synthesis in the presence of 0.01 to 0.1 μM dibutyryl cAMP. The changes in DNA synthesis were equal to that seen with PHA. However, it should be pointed out that pig lymphocytes do not respond well to plant lectins which may be due to a relative lack of circulating long-lived lymphocytes in this species (46).

Attempts have also been made to increase DNA synthesis with cGMP or agents which have been demonstrated to increase intracellular cGMP. Hadden, Hadden and Middleton (47) incubated human peripheral lymphocytes with α-adrenergic agents, which increase intracellular cGMP and lower intracellular cAMP in a number of tissues. In the presence of 28 μM hydrocortisone an enhancing effect was noted at low concentrations of norepinephrine. This effect could be blocked with phentolamine which is an α-adrenergic blocking agent. β-adrenergic agents produced no change or were inhibitory (48,49,50). Weistein et al. (51) have demonstrated modest increases in DNA synthesis in mouse lymphocytes incubated with cGMP. We have attempted to duplicate these results in human peripheral lymphocytes and have been unable to do so. Incubating human lymphocytes over a broad range of concentrations of cGMP or 8 bromo cGMP failed to stimulate increases in DNA synthesis and could not augment the increase in DNA synthesis seen in PHA stimulated cells. As discussed above, at high concentrations of cGMP the DNA synthetic response was inhibited whereas there was no change at low concentrations of cGMP or 8 bromo cGMP. In addition incubation of human peripheral lymphocytes with α-adrenergic agents had little effect on the DNA synthetic response. It should be emphasized that our experiments were not carried out in the presence of hydrocortisone which may be important for the response to be seen. It should also be pointed out that in the studies where α-adrenergic agents were effective in augmenting the PHA response intracellular cGMP rises were not measured. We have demonstrated that high concentrations of norepinephrine increase cAMP (24) whereas low concentrations produced little or no change even in the presence of the β-adrenergic blocking agent propanol. Thus there is no evidence at present that norepinephrine has a cAMP lowering effect in these cells.

The result of the studies described above indicate an inability to demonstrate fulfillment of "Sutherland's hypothesis" (52) for lymphocyte activation. Because of this a number of workers have discounted the role of cyclic nucleotides as the mediator in activated lymphocytes. However, an alternate explanation is at hand. If cyclic nucleotides are compartmentalized within the cell with one compartment responsible for activation and other compartments having no effect or being inhibitory this could account for the results of experiments measuring whole cell cyclic AMP or cyclic GMP and attempts to duplicate lectin action with agents which raise intracellular cyclic nucleotides. Other agents would not be expected to increase cyclic nucleotides in the compartment which was responsible for activation. One can take several approaches to examining the question of compartmentalization of cAMP in activated lymphocytes. The first approach is to measure alterations in intracellular cyclic nucleotides when lymphocytes are incubated with combinations of stimulators of adenylate cyclase. If one can be assured that maximal stimulatory concentrations of stimulating agents are being used then compartmentalization would predict that the increase in intracellular cyclic AMP induced by combinations of stimulating agents should be equal to the sum of the increases induced by the agents used singly. Using prostaglandin E_1 (PGE_1), isoproterenol (ISO), and phytohemagglutinin we have examined this question in human peripheral lymphocytes. Figure 1 demonstrates that maximal stimulatory concentrations of isoproterenol and PHA can be obtained. It is not possible to reach a maximum stimulation level with PGE_1 since increases in stimulation are still occurring at 0.3 mM PGE_1 and higher concentrations are precluded because of solubility problems. Nonetheless high levels of PGE_1 in combination with maximal stimulatory concentrations of isoproterenol and PHA should give additive or nearly additive increases in cAMP. In Fig. 2 the results of a series of additivity experiments is presented. Incubation of human peripheral lymphocytes with isoproterenol, PGE_1 and PHA caused statistically significant increases in cAMP when used singly. When the agents were combined the increases in cAMP were equal to or slightly above the predicted increase on the basis of the agents used alone. Similar results have been presented for the combination of PHA and isoproterenol (24). Although this type of experiment does not prove that individual adenylate cyclases are present for each of the stimulating agents used this is the most likely interpretation of the data. A slightly different type of experiment is an attempt to show that the rises in intracellular cAMP induced by one agent do not influence increases in cAMP induced by a second agent added after the first agent has been removed from the cells by washing. The results of this type of experiment are shown in Table 1. In this experiment lymphocytes were incubated in the absence of stimulating agent or with isoproterenol or with PGE_1 for 60 minutes, washed and then a second stimulating agent was added for a 10 minute incubation period. After 60 minutes the response to isoproterenol, following washing, was sustained while the response seen with PGE_1 had returned nearly to control levels. Subsequent addition of PGE_1 to the cells showed no increases, however, addition of isoproterenol to cells previously incubated with PGE_1 had an increase in intracellular cAMP which was similar to that seen in cells incubated for 60 minutes in the absence of stimulating agents. In cells preincubated with

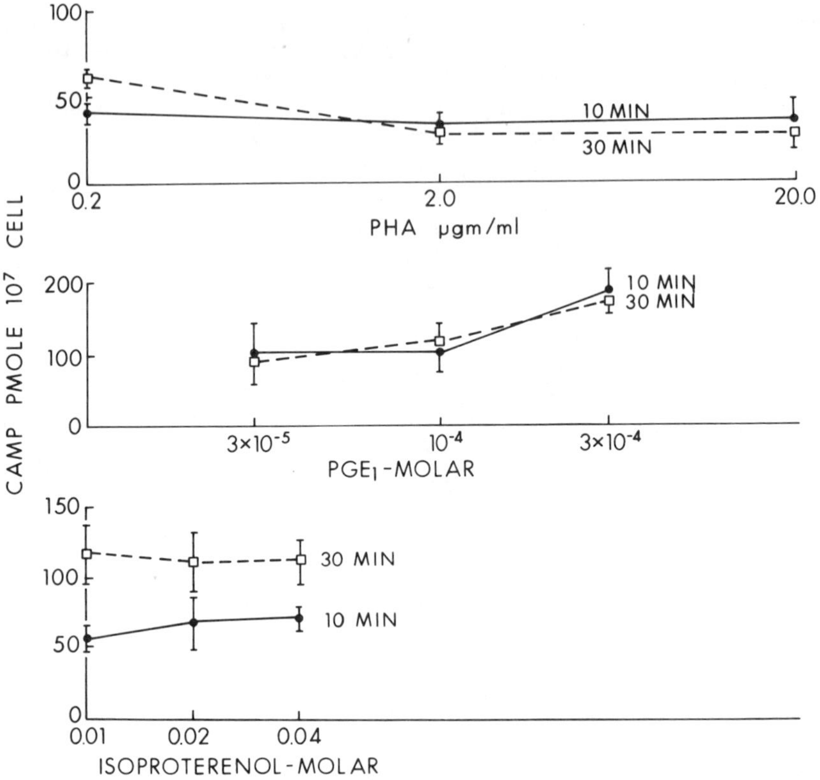

Fig. 1 1.5×10^6 lymphocytes were incubated with PHA, PGE_1 or isoproterenol for 10 minutes (●) or 30 minutes (0) and cAMP was measured by radioimmuno assay. The results are the averages of six experiments. Control value was 18.6 ± 4.3 picomoles/10^7 cells for 30 minute experiments and 22.8 ± 4.7 picomoles/10^7 cells for 10 minute experiments.

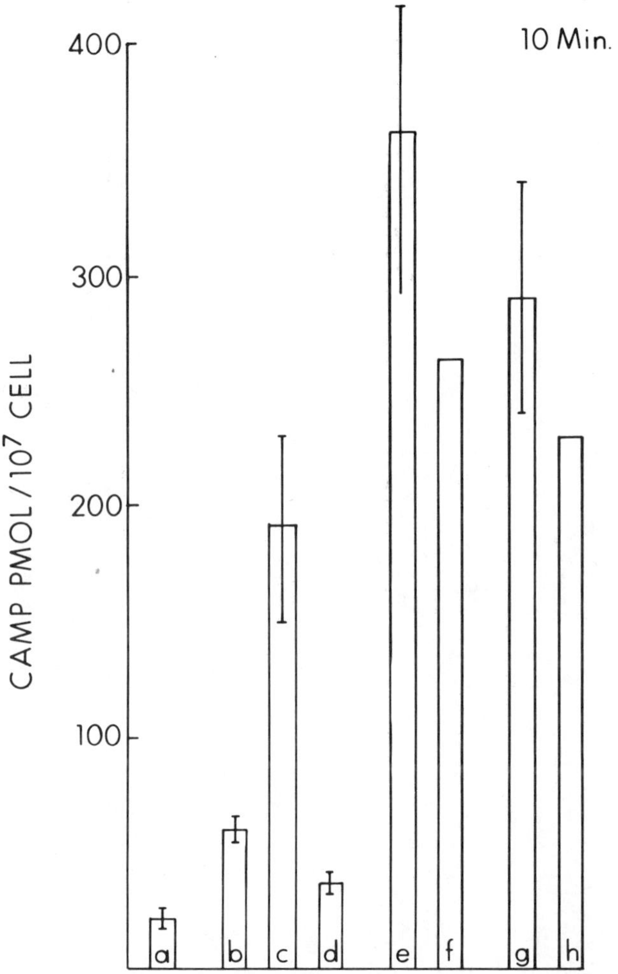

Fig. 2 1.5 x 10^6 lymphocytes were incubated with PHA, isproterenol and PGE_1 alone or in combination for 10 minutes and cAMP was measured by radioimmuno assay. Expected results were calculated from the sum of the agents used singly. A) Control; B) isoproterenol 10 mM; C) PGE_1 1 mM; D) PHA 2 μg/ml; E) PHA + PGE_1 observed; F) PHA + PGE_1 expected; G) Isoproterenol + PGE_1 observed; H) Isoproterenol + PGE_1 expected.

TABLE I

EFFECT OF WASHING ON THE cAMP RESPONSE OF HUMAN LYMPHOCYTES STIMULATED BY ISOPROTERENOL OR PROSTAGLANDIN E_1

First Stimulating Agent	Incubation Time 37°C	Wash	Second Stimulating Agent	Incubation Time 37°C	cAMP picomoles/ 10^7 cells
Control	60 min.	+	Control	10 min.	11.9
Control	60 min.	+	Isoproterenol 10 mM	10 min.	72.6
Control	60 min.	+	PGE_1, 0.3 mM	10 min.	63.4
Isoproterenol 10 mM	60 min.	+	Control	10 min.	79.2
PGE_1, 0.3 mM	60 min.	+	Control	10 min.	16.5
Isoproterenol 10 mM	60 min.	+	PGE_1, 0.3 mM	10 min.	184.8
PGE_1, 0.3 mM	60 min.	+	Isoproterenol 10 mM	10 min.	79.2
PGE_1, 0.3 mM	60 min.	+	PGE_1, 0.3 mM	10 min.	18.2

1.25×10^6 lymphocytes from a single donor were incubated with isoproterenol, PGE_1 or an equal volume of Gey's solution (control) for 60 minutes. The cells were washed, resuspended in the same or a different solution and incubated for an additional 10 minutes. cAMP was measured by radioimmunoassay.

isoproterenol for 60 minutes and then washed the subsequent addition of PGE_1 for 10 minutes gave a rise which was additive or slightly above additive similar to the results described above. Thus the adenylate cyclases which are sensitive to isoproterenol and PGE_1 are apparently unaffected by prior exposure of the lymphocytes to other stimulating agents.

The results of the experiments just described indicate that the adenylate cyclases which are responsive to β-adrenergic agents, PGE_1 and PHA appear to be functionally separated in their activity. The next question which one might ask is: is it possible to demonstrate that these adenylate cyclases are not only functionally but physically separated within the cell? One method of examining this question involves the use of a double antibody technique for the localization of intracellular cAMP (53,54). The first antibody is a purified IgG fraction of rabbit anti-cAMP antisera. This antiserum had been previously demonstrated to be specific for cAMP as measured by radioimmunoassay (55), and the second antibody, a goat anti-rabbit IgG which has been fluorescinated. In this technique cover slip smears of human peripheral lymphocytes incubated with PHA, isoproterenol of PGE_1 are rapidly dried in a stream of filtered dried air. The cover slip smears are immediately covered with the first antibody (anti-cAMP) and incubated at room temperature for 30 minutes. The cover slips are washed extensively and then fixed in a 2% solution of paraformaldehyde for 2 minutes. Following subsequent washing the cover slip smears are covered with the second antibody (goat anti-rabbit IgG) and incubated for a subsequent 30 minute period. The smears are again washed, mounted in 90% buffer glycerol and examined for fluorescence in a Leitz Orthalux fluorescence microscope. The results of these experiments are illustrated in Fig. 3. The patterns of cAMP accumulation seen with the 3 stimulating agents are remarkably different. Cells incubated for 10 minutes without stimulating agents had little or no detectable cAMP by the immunohistochemical method. Cells incubated with PHA demonstrated increases in cAMP which were localized to patchy areas in or near the external plasma membrane. Cells incubated with PGE_1 had increases in cAMP which were localized to the entire cytoplasm of the lymphocyte and cells incubated with isoproterenol had increases in cAMP which were localized throughout the cell. Since the majority of the lymphocyte is occupied by the nucleus much of the fluorescence seen in cells incubated with isoproterenol must be intranuclear. It is interesting that these patterns do not appear to be the result of progression from one pattern to another. We have examined lymphocytes incubated for as short as 1 minute and for as long as 30

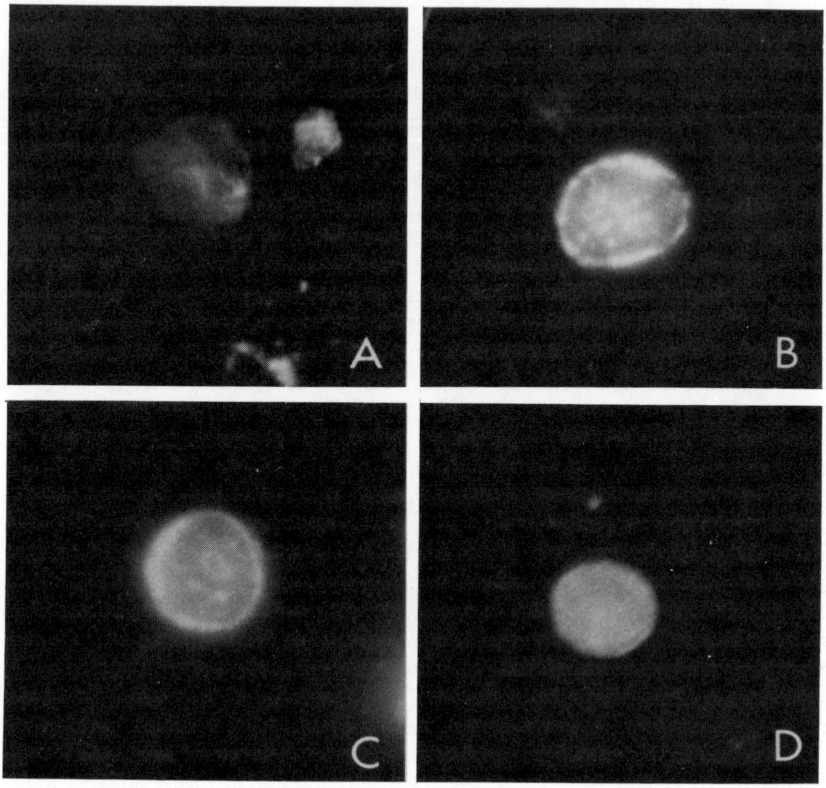

Fig. 3 cAMP was localized in lymphocytes stimulated by PHA, PGE_1 or isoproterenol for 10 minutes by the methods described in the text. A) Control; B) PHA 2 μg/ml; C) PGE_1 1×10^{-4} M; D) isoproterenol 10 mM. Original magnification 1250X (Note: Control cell required increased exposure time to be visible. Other cells were exposed, developed and printed identically.)

minutes. The patterns shown in Figure 1 appear as rapidly as the cells can be processed and are not altered during a 30 minute incubation. Thus the cAMP produced by the 3 agents appears to be the result of adenylate cyclases which are near the areas where cAMP is localized rather than diffusion of cAMP from one area of the cell to another. It should be pointed out that the immunohistochemical technique obviously localizes cAMP which is bound within the cell, most probably to the receptor unit of a cAMP sensitive protein kinase. It is conceivable that the localization pattern is due to alterations in receptor activity rather than a selective activation of different adenylate cyclases. However, this would not invalidate the overall conclusions of the study since the functional consequences as far as the cell is concerned would be identical. The specificity of the immunofluorescence technique for the localization of cAMP can be demonstrated using hapten inhibition. If the cAMP antibody is incubated with 10^{-3} or 10^{-4} molar concentrations of cAMP prior to its use in immunofluoresence experiments the immunofluorescence is blocked. Incubation of the antibody with similar concentrations of other adenosine nucleotides, nucleosides or adenine or with cGMP cyclic IMP, cyclic UMP, succinate or BSA does not alter the fluorescence intensity or fluorescence pattern noted.

The results of the immunohistochemical localization of cAMP suggest that there are different adenylate cyclases in different parts of the cell. It should therefore be possible to fracture the cell, separate different subcellular fractions and compare their adenylate cyclase response. We have carried out experiments in our laboratory on isolated subcellular fractions of human peripheral lymphocytes. Nuclei can be isolated in essentially pure form from human peripheral lymphocytes by incubation of the lymphocytes with low concentrations of triton X 100 followed by purification of the resulting nuclei by sedimentation through sucrose density gradients. The resulting nuclear preparation is pure on the basis of examination under phase contrast microscopy, the absence of 5' nucleotidase, a marker for plasma membranes, and the demonstration that when cells are iodinated by the lactoperoxidase technique prior to the removal of the nuclei the nuclear preparation contains less than 6% of the radioactivity seen in the whole cell. Nuclei isolated in this fashion have an adenylate cyclase which is sensitive to fluoride in the presence of added ATP. In addition incubation of these nuclei with isoproterenol causes a prompt increase in intranuclear cAMP. Interestingly, the increase in intranuclear cAMP is uninfluenced by the addition of ATP suggesting that the reaction utilizes endogenous ATP. Incubation of these cells with PGE_1 or PHA causes no increase in intranuclear cAMP and neither of these agents have any influence on the increase in intranuclear cAMP induced by the incubation of these cells with isoproterenol. The nuclei can also be stimulated with theophylline indicating the presence of an intranuclear phosphodiesterase. Epinephrine at 10^{-6} molar in the presence of theophylline causes a statistically significant increase in intranuclear cAMP. This increase can be partially or entirely blocked by 10^{-4}-10^{-5} M propranolol indicating that the receptor has the characteristics of a β-adrenergic receptor. Thus the results of the experiments using isolated nuclei confirm the immunofluorescent demonstration of intranuclear increases in cAMP seen with the immunohistochemical technique.

In another series of experiments we have examined the adenylate cyclase response in 2 other relatively homogeneous subcellular fractions (Snyder and Parker, in preparation). The first fraction contains plasma membrane and the second fraction microsomes and mitochondria. Using 5' nucleotidase as a marker for plasma membranes and NADH oxidoreductase as a measure for the microsomal fraction we have demonstrated that preparations which contain high levels of 5' nucleotidase contain an adenylate cyclase which is stimulatable by PHA and con A. In contrast fractions which contain high levels of NADH oxidoreductase and low levels of 5' nucleotidase contain an adenylate cyclase which is stimulatable by PGE_1 and not by plant lectins. The results of these experiments again confirm the stimulation results seen with the immunohistochemical localization of cAMP.

These experiments provide 3 lines of evidence for the existence of compartmentalization of cAMP in human peripheral lymphocytes. 1) The adenylate cyclases appear to be functionally distinguishable on the basis of additivity experiments with combinations of stimulatory agents. 2) They appear to be spacially separated within the cell on the basis of an immunohistochemical technique for the localization of cAMP. 3) Spacial separation is also indicated by studies in isolated subcellular fractions. We have not yet had an opportunity to determine whether similar results would be obtained with cGMP although this seems quite possible. Suggestive evidence for compartmentalization of cAMP has also been obtained in non-lymphocytic cells as exemplified by the recent results of Jarett et al. in rat fat cells (56), where different agents which raise cAMP produce quantitatively different metabolic responses. Nor is the concept of functionally and physically separate enzymes with the same catalytic function unique to adenylate cyclase. Neurospora and rat liver both contain 2 carbamyl phosphate synthetase enzymes which are spacially separate in the cell and supply substrate for 2 entirely different metabolic reactions (57). In other words, carbamyl phosphate generated by these enzymes is utilized locally and can not be employed in reactions related to the other enzyme. In neurospora different genes control the synthesis of the 2 enzymes as demonstrated by the isolation of mutant strains which contain one but not both of the enzymes. The carbamyl phosphate synthetase system would appear to be a well validated model for how spacially separated adenylate cyclases might function.

Because of the complexity of the adenylate cyclase cAMP system in lymphocytes and the inability to demonstrate directly that increases in intracellular cAMP or intracellular cGMP can activate lymphocytes we have taken an alternate approach to the problem of cAMP action in human peripheral lymphocytes. Donnelly, et al. (58) have proposed that in eukaryotic cells all cAMP and cGMP function is mediated through the action of protein kinases or phosphoprotein phosphatases. If this is true and if cyclic nucleotides are involved in mitogenic activation one should be able to demonstrate alterations in protein phosphorylation in lymphocytes incubated with activating agents. In the event that increased phosphorylation was observed one could compare the pattern of the response with that of lymphocytes incubated with cAMP, cGMP or analogs of these 2 cyclic nucleotides. In essence this type of analysis emphasizes the known early actions of cAMP rather than the later biochemical alterations seen in activated lymphocytes.

In a series of experiments we have examined protein phosphorylation in purified human peripheral lymphocytes as measured by SDS polyacrylamide gel electrophoresis (Wedner and Parker, manuscript in preparation). Lymphocytes were preincubated in phosphate free medium for 1 hour and subsequently incubated in the same medium containing high levels of ^{32}P orthophosphate for a second 60 minute period to label the intracellular ATP pool. Lymphocytes were then washed and incubated for short periods of time with PHA, 8 bromo cGMP or N_6 monobutyryl cAMP. The cells were then disrupted by Dounce homogenization, the nuclei removed by low speed centrifugation and the resultant supernatant incubated with 1% SDS and 8% 2-mercapto ethanol and subjected to SDS polyacrylamide gel electrophoresis on 5.6% polyacrylamide gels as described by Fairbanks et al. (59). The pattern seen in this technique is demonstrated in Figure 4. Neither PHA, 8 bromo cGMP nor N_6 monobutyryl cAMP altered the gross pattern of protein binding which is demonstrated. Increases in protein phosphorylation could be demonstrated in cells incubated with PHA for 5 minutes and then disrupted as described above. In contrast

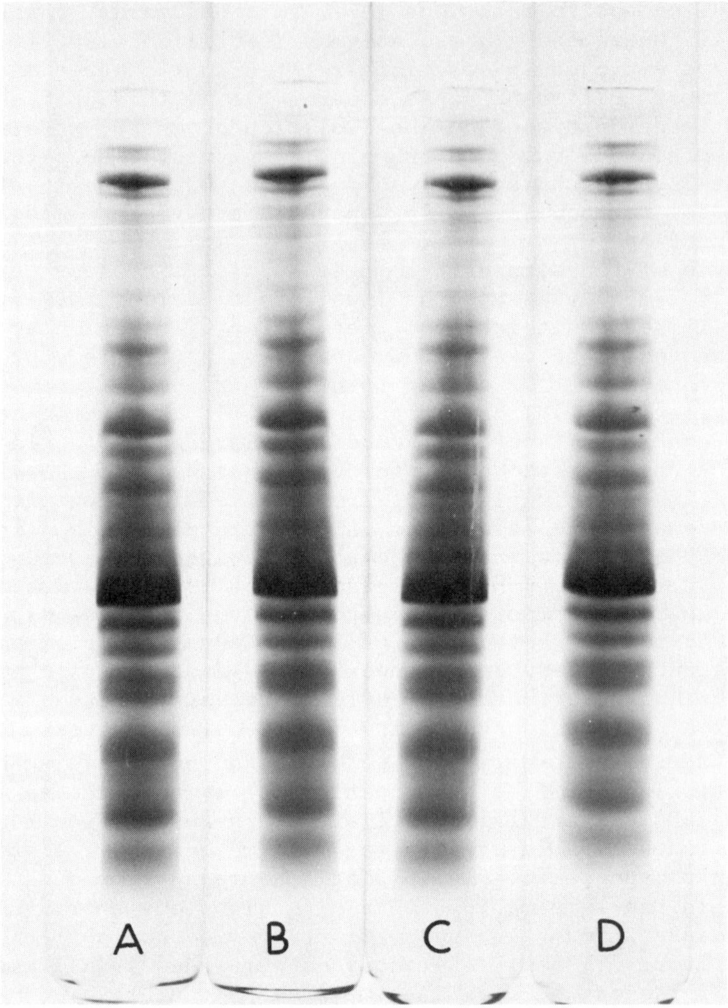

Fig. 4 Human peripheral lymphocytes were incubated with 0.1 M NaCl (control) (a), PHA 20 μg/ml (b), N_6 monobutyryl cAMP 1 x 10^{-5} M (c) or 8 bromo cGMP (d) for 5 minutes. The supernatant following dounce homogenization and centrifugation at 400g for 4 minutes was incubated with 1% SDS and 8% 2-mercaptoethanol for 5 minutes at 100°C. 250 μg of protein was subjected to SDS polyacrylamide gel electrophoresis.

to other studies of protein phosphorylation the increases in phosphorylation were not seen in specific bands within the polyacrylamide gel. Rather there appears to be a general increase in protein phosphorylation in cells incubated with PHA. At 3 minutes the increase in phosphorylation appears to be seen in proteins having molecular weights of 110,000 to 40,000 while at 5 minutes the increase is in those proteins having molecular weights from 70,000 to 10,000. This is similar to alterations in other biochemical parameters in activated lymphocytes where quantitative increases in biochemical parameters are noted with little qualitative changes. When cells were incubated with N_6 monobutyryl cAMP for 3 minutes a pattern which was identical to that seen in cells treated with PHA for 3 minutes emerged. The increase in protein phosphorylation seen in cells incubated with N_6 monobutyryl cAMP was a very transient effect and no increases could be noted at 5 minutes in contrast to those seen at 3 minutes. This may be due to the fact that PHA causes a sustained slow rise in cAMP within the plasma membrane while incubation of the cells with N_6 monobutyryl cAMP caused a rapid burst of cAMP in this area. An alternate explanation may be that N_6 monobutyryl cAMP is more effective in stimulating the action of phosphoprotein phosphatases than is PHA and thus equilibrium between protein phosphorylation and dephosphorylation is reached more rapidly in cells incubated with N_6 monobutyryl cAMP than in cells incubated with PHA. In contrast to the results seen in cells incubated with N_6 monobutyryl cAMP no increases or decreases in protein phosphorylation were noted in lymphocytes incubated with 8 bromo cGMP. These experiments were carried out over time periods from 3 to 30 minutes and at 8 bromo cGMP concentrations ranging from 10^{-3} to 10^{-11} molar. Although this type of experiment seems to rule out a role for cGMP in this particular biochemical parameter, that is protein phosphorylation, and by analogy seems to implicate cAMP rather than cGMP as the mediator of the PHA incubation of lymphocyte activation, it is possible that the conditions used in these experiments were not appropriate for the demonstration of cGMP induced protein phosphorylation. It should be pointed out that in the one instance where cGMP induced protein phosphorylation has been demonstrated (60) the response was a very rapid one and could only be demonstrated in isolated subcellular fractions incubated at 0° for 20 seconds. The experiments as performed in our laboratory do not allow for this type of rapid analysis and thus we may be missing the cGMP response. Nonetheless these experiments do indicate that one of the earliest biochemical events in lymphocyte activation is an increase in protein phosphorylation which would be expected if activating agents operated through cyclic nucleotides. The demonstration that N_6 monobutyryl cAMP mimics almost exactly the PHA response seen at 3 minutes gives credence to the idea that a portion of the modulation or mediation of the PHA response operates through cAMP.

Attempts to demonstrate directly that cyclic nucleotides are involved in the activation of lymphocytes have been disappointing. The bulk of evidence, however, indicates that the majority of activating agents do induce prompt increases in cAMP and cGMP. Although other biochemical or pharmacologic maneuvers which increase intracellular cAMP or cGMP have been unable to mimic the mitogenic response of plant lectins or other activating agents this is most probably due to the inability of these biochemical or pharmacologic agents to alter cAMP or cGMP in the appropriate intracellular compartment so that activation can take place. The results of the experiments done using the immunohistochemical technique and of measurements of adenylate cyclase in isolated subcellular fractions indicate that the increases in cAMP seen in stimulated lymphocytes are limited to specific areas within the external plasma membrane. This is what would be expected since it has been demonstrated that the majority of activating agents do so by binding to the surface of the lymphocyte. The results of the protein phosphorylation experiments indicate that the increases in protein phosphorylation seen in activated lymphocytes appear identical to those seen in lymphocytes with N_6 monobutyryl cAMP and points to a role for cAMP in some of the very early biochemical alterations in activated lymphocytes.

REFERENCES

1. Parker, C.W., Sullivan, T.J. and Wedner, H.F., in Advances in Cyclic Nucleotide Research, edited by P. Greengard and A.G. Robison, p.1, Raven Press, New York, 1974.
2. Wedner, H.J. and Parker, C.W., Progress in Allergy, edited by B.H. Waksman, Karger, New York, (in press).
3. Edelman, P.M. and Schwarts, I.L., Amer. J. Med. 40: 695, 1966.
4. Griffin, D.M. and Szego, C.M., Life Sciences 7: 1017, 1968.
5. Adamson, L.F., Biochim. Biophys. Acta 201: 446, 1970.
6. Weiss, I.W., Morgan, K. and Phang, I.M., J. Biol. Chem. 247: 760, 1972.

7. Riddick, D.H., Kregenow, F.M. and Orloff, J., J. Gen. Physiol. 57: 752, 1971.
8. Entman, M.L., Levey, G.S. and Epstein, S.E., Circulation Res. 25: 429, 1969.
9. Nayler, W.G., McInnes, I., Chipperfield, D., Carson, V. and Doile, P., J. Pharmacology and Experimental Therapeutics 171: 265, 1970.
10. Cunningham, E.B., Biochim. Biophys. Acta 165: 574, 1968.
11. Baserga, R., Cell Tissue Kinet. 1: 167, 1968.
12. Guidotti, A., Weiss, B. and Costa, E., Pharmacologist 13: 256, 1971.
13. Malamud, D., Biochem. Biophys. Res. Comm. 35: 754, 1969.
14. Goldberg, N.D., Haddox, M.K., Estensen, R., Lopez, C. and Hadden, J.W., in Cyclic AMP, Cell Growth and the Immune Response, edited by W. Braun, L.M. Lichtenstein and C.W. Parker, Springer-Verlag Publishers, New York, 1974.
15. Langan, T.A., J. Biolog. Chem. 244: 5763, 1969.
16. Allfrey, V.G., Federation Proceedings 4: 1447, 1970.
17. Johnson, E.M. and Allfrey, V.G., Archives of Biochem. and Biophys. 152: 786, 1972.
18. Langan, T.A., Science 162: 579, 1968.
19. Donnelly, T.E., Kuo, J.F., Reyes, P.L., Liu, Y.P. and Greengard, P., J. Biol. Chem. 248: 190, 1973.
20. Dokas, L.A., Botney, M.D. and Kleinsmith, L.F., Archives of Biochem. and Biophys. 159: 712, 1973.
21. Cawthon, M.L., Bitte, L.F., Krystosek, A. and Kabat, D., J. Biol. Chem. 249: 275, 1974.
22. Smith, J.W., Steiner, A., Newberry, W.M., Jr. and Parker, C.W., Federation Proceedings (Abstract) 28: 566, 1969.
23. Smith, J.W. and Parker, C.W., J. Lab. & Clin. Med. 76: 993, 1970.
24. Smith, J.W., Steiner, A.L., Newberry, W.M., Jr., and Parker, C.W., J. Clin. Inv. 50: 432, 1971.
25. Smith, J.W., Steiner, A.L. and Parker, C.W., J. Clin. Inv. 50: 442, 1971.
26. Parker, C.W., Smith, J.W. and Steiner, A.L., Int. Archives Allergy 41: 40, 1971.
27. Parker, C.W., Bauman, M.L. and Huber, M.F., J. Clin. Invest. 52: 1336, 1973.
28. Parker, C.W., Huber, M.G. and Baumann, M.L., J. Clin. Invest. 52: 1342, 1973.
29. Wedner, H.J., Bloom, F.E. and Parker, C.W., Clinical Research 20: 798, 1972.
30. Wedner, H.J., Hoffer, B.J., Bloom, F.E. and Parker, C.W., Federation Proceedings 32: 744, 1973.
31. Parker, C.W., in Asthma: Physiology, Immunopharmacology,

and Treatment, edited by P.W. Ramwell and B.B. Pharriss, Plenum Press, New York, 1973, p. 185.
32. Eisen, S.A., Wedner, H.J. and Parker, C.W., Immunological Communications 1: 571, 1972.
33. Winchurch, R., Ishizuka, M., Webb, D. and Braun, W., J. Immunology 106: 1399, 1971.
34. Winchurch, R. and Actor, P., J. Immunology 108: 1305, 1972.
35. Makman, M.H., Proc. Nat. Acad. Sci. USA 68: 885, 1971.
36. Hadden, J.W., Hadden, E.M., Haddox, M.K. and Goldberg, N.D., Proc. Nat. Acad. Sci. USA 69: 3024, 1972.
37. Schumm, D.E., Morris, H.P. and Webb, T.E., Europ. J. Cancer 10: 107, 1974.
38. Parker, C.W. and Smith, J.W., J. Clin. Invest. 52: 48, 1973.
39. Parker, J.W., O'Brien, R.L., Steiner, J. and Paolilli, P., Exp. Cell Res. 78: 279, 1973.
40. Mendelsohn, J., Multer, M.M. and Boone, R.F., J. Clin. Invest. 52: 2129, 1973.
41. Hirschhorn, R., Grossman, J. and Weissman, G., Proc. Soc. Exp. Biol. 133: 1361, 1970.
42. Novogrodsky, A. and Katchalski, E., Proc. Nat. Acad. Sci. USA 69: 3207, 1972.
43. McGurdy, J.E. and Rigby, P.G., Proc. Soc. Exp. Biol. and Med. 140: 1456, 1972.
44. Bourne, H.R. and Melmon, K.L., J. Pharmacol. Exp. Ther. 178: 1, 1971.
45. Cross, M.E. and Ord, M.G., Biochem. J. 124: 241, 1971.
46. Binns, R.M., Feinstein, A., Gurnet, B.M. and Coombs, Nature New Biol. 239: 330, 1972.
47. Hadden, J.W., Hadden, E.M. and Middleton, E., Jr., Cellular Immunology, 1: 583, 1970.
48. Hadden, J.W., Hadden, E.M. and Good, R.A., Biochim. Biophys. Acta 237: 339, 1971.
49. Hadden, J.W., Hadden, E.M. and Good, R.A., Exp. Cell Res. 68: 217, 1971.
50. Hadden, J.W., Hadden, E.M., Middleton, E., Jr. and Good, R.A., International Arch. of Allergy and Applied Immunology 40: 526, 1971.
51. Weinstein, V., Chambers, D.A., Bourne, H.R., Melman, K.L. and Segal, S., Proc. 2nd Intl. Conf. on Cyclic AMP, p. 52, 1974.
52. Robison, G.A., Butcher, R.W. and Sutherland, E.W., in Cyclic AMP, Academic Press, New York, 1971.
53. Wedner, H.J., Hoffer, B.W., Battenberg, E., Steiner, A.L., Parker, C.W. and Bloom, F.E., J. Histochem. and Cytochem. 20: 293, 1972.

54. Bloom, F.E.,Wedner, H.J. and Parker, C.W., Pharmacol. Reviews 25: 343, 1973.
55. Steiner, A.L., Parker, C.W. and Kipnis, D.M., J. Biol. Chem. 2117: 1106, 1972.
56. Jarrett, L., Reuther, M., McKeel, D.W. and Smith, R.M., Endocrin. 89: 1186, 1972.
57. Davis, R.H., Science 178: 835, 1972.
58. Donnelly, T.E., Jr., Kuo, J.F., Miyomoto, E. and Greengard, P., J. Biol. Chem. 248: 190, 1973.
59. Fairbanks, G., Steck, T.L. and Wallach, D.F.H., Biochem. 10: 2606, 1971.
60. Casnellie, J.E. and Greengard, P., Proc. Nat. Acad. Sci. 71: 1891, 1971.

CYCLIC GMP AND LYMPHOCYTE ACTIVATION

John W. Hadden, Edward M. Johnson, Elba M. Hadden,
Ronald G. Coffey and Leslie D. Johnson

Laboratory of Immunopharmacology, Memorial Sloan-Kettering Cancer Center, New York, New York

INTRODUCTION

Recent interest in the cyclic nucleotides, cyclic 3',5'-adenosine monophosphate (cyclic AMP) and cyclic 3',5'-guanosine monophosphate (cyclic GMP) as principal intracellular mediators of diverse hormones acting at the cell surface, has found focus in the realm of immunology and in this setting has resulted in a number of important observations concerning the hormonal regulation of immune response and the mechanisms of lymphocyte activation. The observations collected to date suggest that, as appears to be the case in other cells and tissues (1), hormones and other agents which increase lymphocyte levels of cyclic AMP and cyclic GMP promote opposing biological functions.

The present concepts of the mechanisms by which the cyclic nucleotides are generated, find action, and are catabolized, have been summarized extensively (Cf. 2,3). To review them briefly as they apply to the lymphocyte, a number of agents known or suspected to act at the cell surface through discrete receptor interactions activate one or the other membrane bound cyclic nucleotide cyclase (adenylate cyclase or guanylate cyclase) resulting in the generation of the corresponding cyclic nucleotide from its substrate (ATP or GTP). Cellular levels of cyclic AMP are generally 10-100 times those of cyclic GMP. The intracellular actions of cyclic AMP have been in the main attributed to activation of a protein kinase resulting in an enzymatically active subunit acting to phosphorylate and thereby activate various enzymes. The intracellular actions of cyclic GMP remain to be elucidated. The cyclic nucleotides are catabolized to their 5' monophosphates through the action of specific phosphodiesterases. A number of factors like divalent cations are thought to participate in the regulation of cellular cyclic nucleotide levels through influences on generation and catabolism. Further, the two cyclic nucleotide systems appear to interact in such ways that frequently the activation of one results in the suppression of the other. With this short summary of cyclic nucleotide biochemistry as background, we plan in this text to briefly review those events in the lymphocyte which can be attributed to cyclic AMP (Cf. 4,5 for more complete review) and to develop recent observations concerning those attributable to cyclic GMP within the perspective of early events leading to or modifying the activation of lymphocytes.

The earliest studies of cyclic nucleotides in lymphocytes (6-9) utilized agents to increase levels of cyclic AMP and examined their influence on in vitro lymphocyte proliferation induced by phytohemagglutinin (PHA). These studies indicated that cyclic AMP itself in the dibutyryl form, agents which stimulate adenylate cyclase like prostaglandin PGE_1, isoproterenol, and epinephrine, and the phosphodiesterase inhibitor theophylline all inhibit the induction of human lymphocyte proliferation by PHA. Other studies (8,10) on actively proliferating lymphocytes indicated that proliferation once in progress could be augmented by these agents. In addition to bidirectional influences on the process of proliferation, cyclic AMP action has been linked in lymphocytes to the induction of glycogenolysis, the reversal of glycolysis, the inhibition of glucose and potassium transport, and the inhibition of non-histone nuclear protein phosphorylation (11-14). Each of the processes influenced by cyclic AMP is affected in a way which is antagonistic to the effects which PHA exerts on these processes early in activation. These observations underscore the fact that lymphocytes are susceptible to agents which increase their levels of cyclic AMP and that their biological responses can be correspondingly modified; however, they raise two important questions: If cyclic AMP acts to inhibit the induction of proliferation, what naturally occurring event does it promote? If cyclic AMP inhibits, does the other cyclic nucleotide cyclic GMP promote the induction of proliferation?

Insight into the answer to the first question has recently been forthcoming from studies of lymphocyte differentiation, a process which can be envisioned to contrast to the induction of proliferation. These studies indicate that certain putative hormones derived from the thymus are capable of inducing thymocyte precursors derived from bone marrow or spleen to manifest surface markers and functions characteristic of the intrathymic lymphocyte. The actions of these thymic agents have been explored in vitro where such differentiative events can be observed within hours of stimulation (15,16). Indirect evidence based upon pharmacologic manipulations to increase cyclic AMP in these cells suggests that cyclic AMP may mediate processes involved in the induction of thymocyte differentiation (15,17). Our experience in studying the mechanism of one of these thymic hormones thymopoietin (18) in relation to the Komuro-Boyse assay (16) indicates that thymopoietin increases cyclic AMP levels in the precursor population in a manner which directly correlates with its ability to induce differentiative markers (19). In addition, a number of other agents like endotoxin, poly A-poly U, and isoproterenol which act non-specifically to induce differentiation in this system correspondingly appear to do so through mediation by cyclic AMP. This evidence, although preliminary, indicates that cyclic AMP may function in the induction of differentiative events in thymocyte precursors and suggests that other differentiative events in post thymic lymphocytes may be similarly involved.

Insight into the question of cyclic GMP's role in the induction of proliferation began with observations which indicated that the neurotransmitter acetylcholine and its analogs are capable of modifying PHA-induced lymphocyte proliferation in a way opposite to the effects produced by those agents which act through mediation by cyclic AMP.

CYCLIC GMP AND CHOLINERGIC MODULATION OF LYMPHOCYTE PROLIFERATION

In 1970, George and coworkers (20) demonstrated that cholinergic stimulation of the heart is associated with rapid increases in tissue levels of cyclic GMP. This action of acetylcholine has been generalized to brain, lung, uterus, and lymphocyte (Cf. 1). In collaboration with Dr. Nelson Goldberg (Department of Pharmacology, University of Minnesota), we observed that acetylcholine (10^{-6}M) stimulated lymphocyte cyclic GMP levels approximately three-fold at 10 minutes following stimulation (21). This observation raised the possibility of cholinergic modulation of lymphocyte metabolism, and we, therefore, explored the influence of acetylcholine on lymphocyte transformation *in vitro* (22).

These results are depicted in Figure 1. In cultures incubated for 24 hours (calculated to assay the first round of division) acetylcholine stimulates RNA and protein synthesis significantly in non-mitogen stimulated lymphocytes (Fig. 1A). The effects of acetylcholine in the absence of PHA, depicted in Fig. 1A, represent observations made in five earlier experiments. Subsequent experiments have reproduced these observations; however, we have observed considerable variability from donor to donor and day to day in the magnitude of stimulation. The variability has suggested to us that the endogenous antigenic stimulation which is envisioned to produce the background observed in otherwise non-stimulated lymphocytes present in human peripheral blood may be the significant variable in determining the magnitude of cholinergic effects.

In the presence of PHA, acetylcholine has a small, but

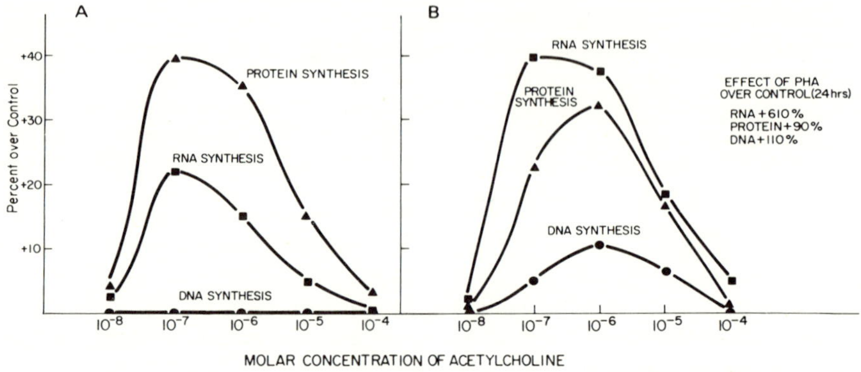

Fig. 1 Effects of Acetylcholine on RNA, Protein & DNA Synthesis of Lymphocytes.

From heparinized human peripheral blood, lymphocytes were purified by Ficoll-Hypaque gradient fractionation yielding preparations of 90-95% pure lymphocytes (<10% monocytes by morphological criteria) virtually free of contaminating granulocytes and erythrocytes. 1×10^6 mononuclear cells were incubated in 1 ml Minimal Essential Media (MEM) containing 10% pooled human AB serum, 1% penicillin-streptomycin solution (GIBCO 514L), 1% L-glutamine and 1 μCi ^3H precursor for 24 hours at 37°C in a 5% CO_2 in air atmosphere. Cultures were terminated by 10% TCA precipitation and incorporation of precursors were assayed by liquid scintillation spectrometry as previously described (7). This figure indicates the effects of addition of acetylcholine 10^{-8} - 10^{-4}M final concentration on incorporation of ^3H-uridine, ^3H-leucine, and ^3H-thymidine in non-mitogen stimulated lymphocytes (A) and in PHA (.033 units/ml PHA-MR69 Burroughs-Wellcome)-stimulated lymphocytes (B). The data are expressed as ratios of experimental CPM incorporation to control. The ratios of B to A are indicated in the inset of B.

significant, effect to stimulate in addition to RNA synthesis, DNA synthesis (Fig. 1B). The influence of PHA to stimulate RNA, protein and DNA synthesis over the first 24 hours of culture is indicated in the inset. The maximal effects of acetylcholine on RNA and protein synthesis are observed at concentrations of 10^{-7} - 10^{-6}M (SEM's .03 - .10%, p's < .01); however, in individual experiments such effects were occasionally observed at concentrations as low as 10^{-9}M. Acetylcholine analogs, carbamylcholine and methacholine also stimulate RNA and protein synthesis and do so at 1-2 logs lower concentrations than acetylcholine. Acetylcholine shows no effect on uridine and leucine uptake measured hourly over the first four hours of culture (data not shown) indicating that the stimulation does not result from a direct influence on transport. The effect of acetylcholine and its analog carbamylcholine on DNA, RNA and protein synthesis are blocked by atropine (10^{-5}M) see Table I) and can, therefore, be interpreted as muscarinic cholinergic effects, that is, characteristic of acetylcholine acting as a post ganglionic neurotransmitter of the parasympathetic nervous system. It is important to note that the influence of acetylcholine and its analogs would logically be restricted to the first minutes of culture as they are rapidly catabolized by acetylcholinesterases present in serum.

These observations provide the basis for the suggestion that the circulating lymphocyte may be cholinergically modified in its response when it finds residence in tissues innervated by the parasympathetic limb of the autonomic nervous system (predominantly the vagus nerve). That functional responses of lymphocytes are modified by cholinergic stimulation has found support in the observations of Strom and coworkers who have shown that cholinergic stimulation enhances the ability of murine T lymphocytes to kill tumor cells to which they are sensitized (23) and graft-versus-host (GVH) induced proliferation of T lymphocytes (24). It is of note that the cholinergic stimulation of murine spleen cells observed in these experiments involved concentrations of agonists 1/100 - 1/1000 those employed in the experiments we have described. Although we have not consistently observed effects of acetylcholine and its analogs at the lower concentrations, the consistency and reproducibility of the two sets of observations suggests important species (man vs. mouse) and source (peripheral blood vs. spleen) differences of the lymphocytes used. We have observed that antigenically-stimulated guinea pig lymphocytes show enhanced production of migration inhibitory factor (MIF) when cholinergically stimulated (22). The stimulation of each of these functional responses has been

TABLE I

EFFECT OF CHOLINERGIC AGENTS AND BLOCKER ON LYMPHOCYTE DNA SYNTHESIS

			+ Atropine
Control with PHA		2874 ± 52 (15)*	2848 ± 75 (15)
Acetylcholine	10^{-6}	3162 ± 174 (6)	2938 ± 271 (6)
	10^{-7}	3063 ± 38 (9)	2286 ± 91 (9)
Carbachol	10^{-7}	3069 ± 122 (6)	2807 ± 81 (6)

* Mean ± standard error of the mean (number of samples) of determinations of CPM ^{3}H-thymidine incorporation in lymphocytes prepared and incubated for 24 hours as described in Figure 1.

observed following prior immunization and each of the responses observed is considered to be a T cell-mediated function (GVH induced proliferation, MIF production and direct lymphocytotoxicity). These observations, while not excluding the B cell, suggest that cholinergic modulation of the lymphocyte finds its in vivo biological perspective in particular relation to antigen-dependent tissue localized T cell responses.

PHARMACOLOGIC MODULATION OF LYMPHOCYTE PROLIFERATION

We have employed several reagents designed to increase cellular levels of cyclic GMP including cyclic GMP, 8-bromocyclic GMP, and five 8-substituted cyclic GMP derivatives kindly provided by ICN (8-mercapto, 8-benzylmercapto, 8-methylmercapto, 8-methylamino, 8-benzylamino) and imidazole, a cyclic GMP phosphodiesterase inhibitor and cyclic AMP phosphodiesterase stimulator (25). In preliminary observations in lymphocytes, we have observed that imidazole lowers the levels of cyclic AMP and increases the levels of cyclic GMP. In a large series of experiments, we have examined the influence of these agents over wide concentration ranges (10^{-12} - 10^{-3}M) on RNA and DNA synthesis of human peripheral blood lymphocytes in the presence and absence of PHA at 24 and 72 hours of culture.

Under the culture conditions employed, we have not observed a consistent effect of cyclic GMP (Fig. 2) or any of its derivatives on the parameters of lymphocyte metabolism studied. These negative findings are in contrast to effects observed by others using murine lymphocytes in which cyclic GMP (10^{-6} - 10^{-7}M) added extracellularly stimulates T cell lymphocytotoxicity and GVH-induced proliferation (23,24). It is of note that using McCoy's medium in the absence of serum, Webb and coworkers (27) have observed effects of cyclic GMP (10^{-9}M) to reverse high dose PHA inhibition of human lymphocyte proliferation.

The effects of imidazole on in vitro lymphocyte responses have been variable. In 16 of 24 experiments, we have observed dose-dependent effects of imidazole to stimulate DNA synthesis of PHA-stimulated lymphocytes at both 24 and 72 hours of culture (Fig. 2). Individual variations in the magnitude and optimal concentrations producing stimulation have been considerable, while high dose inhibition (10 mM) has been consistently observed. Female donors appear to respond more consistently than male donors. The most consistent observation has been stimulation of RNA and DNA synthesis by imidazole in the concentration range of (10^{-5} - 10^{-3}M), the peak concentration being apparently dependent on the individual

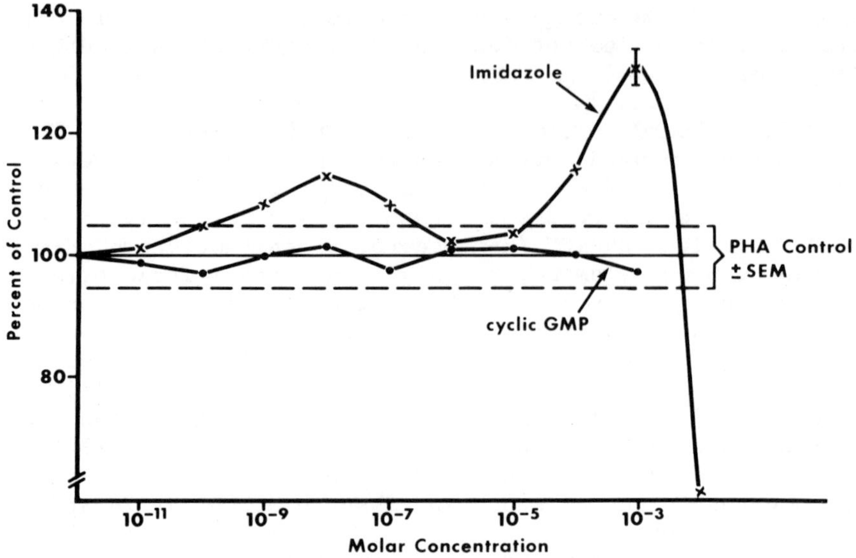

Fig. 2 Effects of Cyclic GMP and Imidazole on PHA-induced DNA Synthesis of Lymphocytes.
Human lymphocytes prepared as described in Figure 1 were incubated for 72 hours and terminated following a four hour ^{3}H-thymidine pulse. Some experiments were performed under the same incubation conditions in microculture as described by Parker & Lukes (26) and terminated with a multiple automatic sample Harvester (MASH-Hiller Co.). The effects of 8-bromo-cyclic GMP and imidazole (10^{-12} - 10^{-3}M) on PHA-induced ^{3}H-thymidine incorporation of lymphocytes are depicted. Data are expressed as ratio of experimental to control. Brackets indicate standard errors of the mean (SEM).

donor. Figure 3 shows the effect of 10^{-3}M imidazole on PHA stimulation at various concentrations of PHA. These results obtained in 8 individuals who responded well to imidazole indicate that imidazole stimulates DNA synthesis over the entire dose response range of PHA. Although it is difficult to ascribe a simple mechanism (i.e. altered levels of cyclic AMP, cyclic GMP or both) for the action of imidazole on human lymphocytes, our studies with mouse lymphocytes indicate that the actions of imidazole to enhance proliferation and T cell mediated lymphocytotoxicity (23) correlate with its ability to increase cyclic GMP levels and not with its ability to lower cyclic AMP levels.

CYCLIC GMP IN LYMPHOCYTE ACTIVATION BY MITOGENS

Despite those studies which have indicated that agents which increased cyclic AMP levels in lymphocytes were antiproliferative, several studies suggested that mitogenic agents like phytohemagglutinin (PHA) could produce small increases in lymphocyte levels of cyclic AMP (8,28). While a number of investigators (29-32), including ourselves, have been unable to show that optimal mitogenic concentrations of purified mitogens consistently produce increases in lymphocyte levels of cyclic AMP, it does appear that mitogens in supramaximal doses are capable of inducing cyclic AMP increases. Although the nature of these increases in relation to patch and cap formation and other membrane phenomena associated with lectin action remains to be clarified, the concentrations which consistently induce cyclic AMP increases are those which exhibit high dose inhibition of proliferation, thus suggesting that cyclic AMP may play a role in the antiproliferative action of supraoptimal mitogen concentrations.

Our interest in cyclic GMP in relation to mitogen action was prompted particularly by the observations which indicated that mitogens bound to agarose beads were capable of inducing proliferation without gaining access to the lymphocyte interior (33) and that mitogens, particularly PHA, need only be present at the surface of the cell for a restricted period of time (34). Such observations suggested a discrete membrane interaction resulting in a "trigger type" of intracellular signal. If cyclic AMP were not involved, based on current information, a restricted number of possibilities existed and cyclic GMP appeared the most likely.

We, therefore, studied the effects of PHA-M, PHA-MR69, and concanavalin A (Con A) on lymphocyte levels of cyclic GMP and cyclic AMP (31). Employing optimal mitogenic concentrations of each, we found that they were comparable in the magnitude of their mitogenic action but different in their capacity to

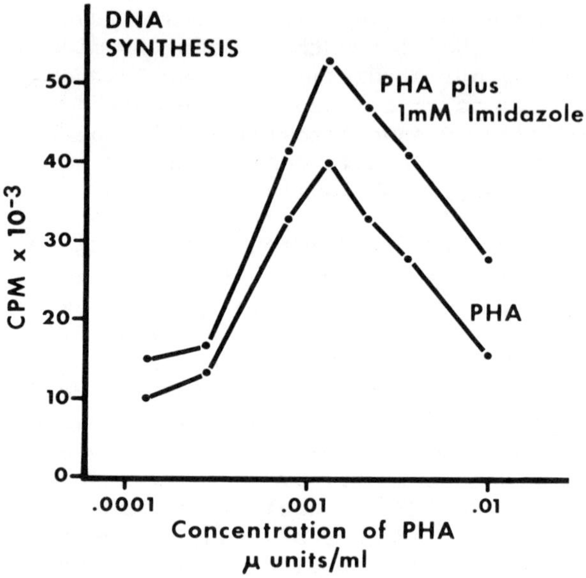

Fig. 3 Effects of Imidazole on PHA-induced DNA Synthesis. The effects of imidazole (1 mM final concentration) on PHA-induced ^3H-thymidine incorporation at 72 hours of culture analyzed over a dose response range of PHA. The data are expressed in CPM ^3H-thymidine incorporation as the mean of triplicated samples of 8 of 24 individuals who responded particularly well to imidazole.

produce agglutination of lymphocytes. The unpurified mitogen PHA-M, in contrast to purified PHA-MR69 and Con A, showed potent agglutinating activity. Correspondingly, under the experimental conditions employed, only PHA-M induced a significant increase in lymphocyte cyclic AMP levels. All three mitogens showed striking increases of 10-50 times in lymphocyte cyclic GMP levels. The increases were progressive to 20 minutes and then declined to near control levels by 30 minutes. These observations prompted us to speculate that increases of cyclic GMP levels might represent an active signal to induce proliferation while increases in cyclic AMP represented a restrictive influence to the induction process. We speculated further that where decreases in cyclic AMP levels had been previously correlated with the induction of proliferation in fibroblasts, increases in cyclic GMP would be observed. To date, the observations of mitogen-induced cyclic GMP increases in human lymphocytes have been confirmed in rat (35) and mouse (36) and the speculations have been confirmed by us and others for a variety of mitogenic influences in fibroblasts (1,37-39), epidermal cells (40), uterus (41), fertilization of sea urchin ova (42), bacteria (1) and plant buds (43).

The measurement of cyclic GMP in lymphocytes has difficulties which are not shared by other cell types. While our previous assays of cyclic GMP levels in lymphocytes were performed using the enzyme cycling assay of Goldberg (44), we have since converted to the simpler radioimmunoassay described by Steiner et al. (45). We have performed a large series of experiments designed to analyze each aspect of the incubation, extraction, purification, and measurement steps of the experiments. These experiments are still in progress; however, we present certain insights and results which should serve to assist others in their cyclic GMP experiments with lymphocytes and other leukocytes. The use of commercial heparin containing benzyl alcohol and of Ficoll-Hypaque gradients to separate lymphocytes appears to decrease basal levels of cyclic GMP and the ability to increase its levels with a variety of agents. Both of these steps in lymphocyte preparations are associated with some impairment of the proliferative response. Sample size is also a factor with 10 x 10^6 lymphocytes/ml being optimal. Larger samples yield progressively lower levels of cyclic GMP/mg protein. Following extraction of cyclic GMP with trichloroacetic acid, either extraction (2 vol x 5) and purification of cyclic GMP is essential for reliable results. Of each of the features examined, the most important concerns the distribution and turnover of cyclic GMP in the cells and the surrounding medium.

Table II shows the results of cyclic GMP and cyclic AMP levels in lymphocytes and their supernatants incubated for 10 minutes under a variety of conditions. In the absence of imidazole, PHA and acetylcholine show negligible effects on both cellular cyclic GMP and cyclic AMP levels. Apparent, however, are significant increases in supernatant cyclic GMP levels with no change in supernatant cyclic AMP levels. The supernatant increases are 6.3 fold for PHA and 4.9 fold for acetylcholine. In the presence of 1 mM imidazole basal cyclic GMP levels are significantly increased in control cells and basal cyclic AMP levels are slightly but significantly increased. The effects of PHA and acetylcholine in the presence of imidazole to increase cyclic GMP levels in the cells are now 3.1 and 2.1 fold, respectively, and the induced accumulations of cyclic GMP in the supernatant are correspondingly greater (see Fig. 4). While cellular levels of cyclic AMP are increased with imidazole, PHA and acetylcholine have no effect to further increase them and have no significant effect to induce accumulation of cyclic AMP in the supernatant. These results indicate several significant points: In these experiments cyclic GMP levels are less than 1/100 those of cyclic AMP in the same lymphocytes and both at rest and during stimulation there occurs significant leakage of cyclic GMP into the medium. Following stimulation significant leakage and accumulation of cyclic GMP occurs in the medium while cellular levels are not increased unless the cyclic GMP phosphodiesterase inhibitor imidazole is added. In the presence of the phosphodiesterase inhibitor cellular levels increase and correspondingly so do levels in the medium. These observations indicate that while acetylcholine and PHA do not influence cyclic AMP levels, they induce significant cellular turnover of cyclic GMP with marked leakage into the medium. The cyclic GMP levels in the supernatant are preserved, spared intracellular phosphodiesterase action. The cellular levels are not preserved unless through inhibition of intracellular phosphodiesterase. These observations are in concert with those in a number of tissues, that cyclic nucleotide turnover during stimulation is perhaps more relevant than cellular accumulation.

Within the perspective of mitogen action PHA in these experiments produced increases in cyclic GMP of cells and medium of maximally 4 fold of those of control at 10 minutes of stimulation. These increases are approximately equivalent to those previously reported at this time period based upon measurements of cells and medium together (31) and are considerably less than the increases which are observable at 20 minutes of stimulation, the optimal time of PHA induced

TABLE II - CYCLIC NUCLEOTIDE LEVELS IN HUMAN LYMPHOCYTES*

		Cyclic GMP pmoles/mg Cellular Protein				Cyclic AMP pmoles/mg Cellular Protein			
	Imidazole 1 mM	Cells	Medium	Total	Ratio to Control	Cells	Medium	Total	Ratio to Control
PHA-MR69 .008 units/ml	-	.043±.012	.177±.070	.220±.070	2.6	9.24±1.8	2.28±1.08	11.52±0.84	1.1
	+	.224±.018	.220±.040	.440±.057	4.2	15.12±3.12	0.6 ±0.2	15.72±3.0	1.1
Acetylcholine (10^{-6}M)	-	.071±.022	.138±.042	.222±.017	2.9	9.48±0.24	1.92±1.32	11.4 ±0.84	1.1
	+	.147±.022	.161±.025	.309±.058	2.9	11.88±1.08	3.12±1.44	15.0 ±1.2	1.0
Control	-	.049±.011	.028±.006	.077±.010		8.52±0.06	2.16±0.12	10.56±0.6	
	+	.071±.091	.035±.011	.106±.017		12.24±0.84	2.16±0.6	14.4 ±0.84	

*Human peripheral blood lymphocytes were prepared as described (7,31). Platelet contamination was reduced to <1/mononuclear cell by differential centrifugation. Incubations of 15×10^6 mononuclear cells in 1 ml Hanks Balanced Salt Solution were performed at 37° without agitation and were terminated by addition of 1 ml 20% Trichloroacetic acid and freezing at -70°C. Purification of the cyclic nucleotides was performed as described (31). Cyclic GMP was assayed by radioimmunoassay (45) (Collaboration Research Inc.) and cyclic AMP by the method of Gilman (46). Recoveries averaged 60% for cyclic GMP and 80% for cyclic AMP and aliquots treated with phosphodiesterase confirmed the purity of the final sample.

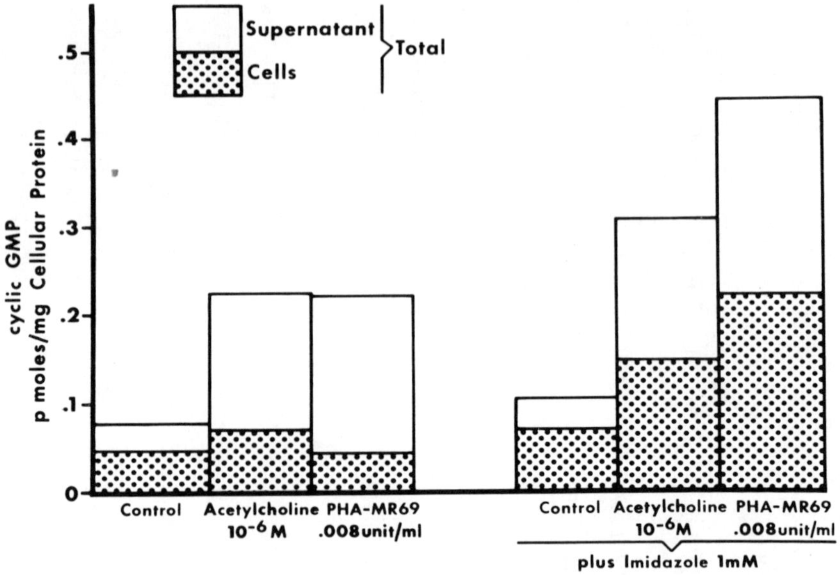

Fig. 4 Cyclic GMP Levels of Lymphocytes and Incubation Medium.
This histogram expresses the cyclic GMP levels of Table II.

increases.

CALCIUM IN LYMPHOCYTE ACTIVATION BY CYCLIC GMP

Calcium has been shown by the studies of Allwood et al. (47) and Whitney and Sutherland (48) to be necessary for mitogen-induced proliferation of lymphocytes and to be involved in a critical way through its entry into the lymphocyte during the early hours following mitogen action. Our interest in calcium in lymphocytes derived particularly from those observations which indicate that calcium participates in the stimulation of guanylate cyclase (49,50) and our observations which indicate the calcium is required for cyclic GMP to stimulate RNA synthesis in isolated nuclei (1). The analysis of calcium flux into lymphocytes has been complicated by the difficulty in separating extracellular calcium rapidly from that associated with the cell and the difficulty in determining whether cell associated calcium represents intracellular calcium. This latter problem is further complicated by the presumed presence of an active calcium extrusion mechanism in the lymphocyte similar to that described in the erythrocyte by Schatzmann (51). We present studies which, while not completely resolving these difficulties, do shed some light on the situation.

We measured calcium (^{45}Ca) influx into lymphocytes using a rapid filter wash technique to separate medium from cell associated ^{45}Ca within 15 seconds. In 10 such experiments we observed that PHA but not acetylcholine (10^{-6}M) stimulated a progressive increase in cell-associated calcium which began at 10 minutes following stimulation (Fig. 5).

We considered the possibility that calcium finding entry to the cell might be actively extruded. By loading lymphocytes with calcium (^{45}Ca) or strontium (^{86}Sr) in isotonic media at 4°C or in hypotonic media at 22°C followed by addition of KCl to restore isotonicity and to reseal the cell, we observed active extrusion. Figure 6 illustrates the influence of the two loading procedures on active extrusion. Lymphocytes loaded in hypotonic media (Fig. 6A (1)) showed rapid extrusion upon resealing. Lymphocytes loaded in isotonic media for 2 hours at 4°C (to inhibit active extrusion) showed less uptake of calcium and a slower extrusion (Fig. 6A (2)). The slower extrusion was not dependent upon ATP or glucose in the loading media and probably resulted from the influence of the low temperature used for loading on the extrusion process. Because of the advantages of the hypotonic loading procedure further experiments employed this method. Extrusion under these circumstances is dependent on the addition of ATP to the loading media (Fig. 6B). Loading is most ef-

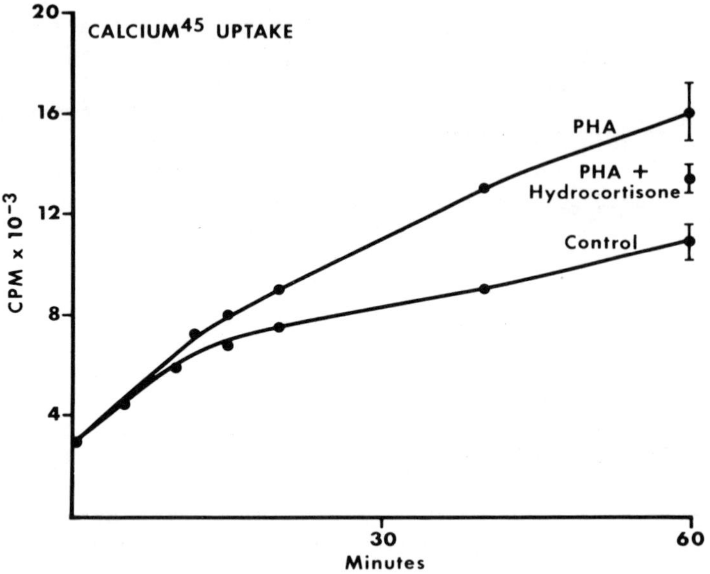

Fig. 5 Effect of PHA on ^{45}Calcium Uptake in Human Lymphocytes. Lymphocytes prepared as described in Figure 1 were incubated at 2 x 10^6 mononuclear cells in 1 ml of Spinner Modified MEM (calcium free) containing 1 μCi $^{45}CaCl_2$ in the absence and presence of PHA-MR69 (.033 units/ml). The incubations were terminated at selected time intervals by pouring the glass tube contents over a glass fiber filter (Reeve Angel 934AH) mounted in a suction apparatus. The tube was washed twice with 10 ml cold isotonic saline and the contents poured over the filter. The filters were dried overnight and assayed by liquid scintillation spectrometry. Uptake is expressed as CPM cell associated ^{45}Ca.

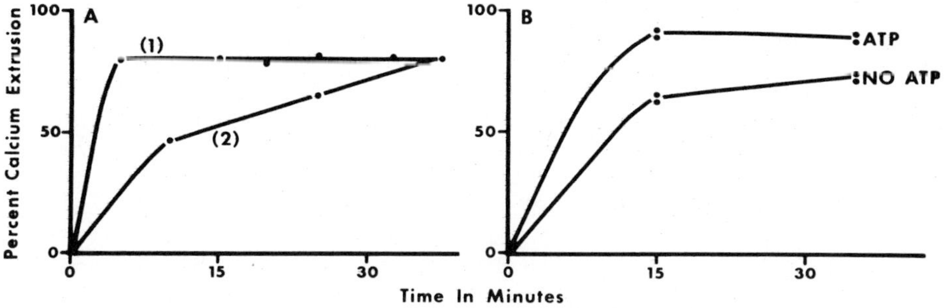

Fig. 6 Calcium Extrusion From Human Lymphocytes.
Human lymphocytes prepared as described were incubated at 5 x 10^6 cells/ml hypotonic media containing 1 μCi ^{45}Ca for periods up to 5 minutes according to the method of Schatzmann (51). Following addition of hypertonic KCl to restore isotonicity incubations were performed at 37° for selected periods and terminated as described in Figure 5. Alternatively (2) the lymphocytes were exposed to isotonic saline containing ^{45}Ca at 4°C for 2 hours. ATP, when added, was at a final concentration of 1.65 mM, ^{86}Sr or Mg, when added, was at 1.5 mM. The extrusion data are presented as percent of the 0 time loaded value.

A-indicates a comparison of the effect of hypotonic loading (1) to 4° loading (2) on subsequent extrusion.

B-indicates the effects of adding ATP during hypotonic loading (1) on subsequent extrusion.

fective and the ATP effect most demonstrable when the ATP is added at the time of resealing. Using this loading procedure, equilibrium is not reached as it is in the erythrocyte so that the response of the calcium extrusion pump to the intracellular concentration of calcium could not be ascertained. Over the time periods of hypotonic loading used (0 to 5 minutes) uptake is linear; however, the rate of extrusion increases progressively with increasing intracellular calcium concentration. We observed that the active extrusion of either ^{45}Ca or ^{85}Sr was competed for by the presence of the other in the loading media but their extrusion was not influenced by the extracellular concentration of either. The presence of magnesium in the loading media or during extrusion did not influence the process. These observations conform to those made in the erythrocyte (51); however, we have not found, as has been shown in the erythrocyte, that this extrusion process in the lymphocyte is inhibited by ethacrynic acid (10^{-4}M). Correspondingly, the effects of PHA to induce calcium influx in the lymphocyte is not affected by 10^{-4}M ethacrynic acid. The presence of an active calcium extrusion pump in the lymphocyte makes more complex the analysis of influx and experiments which employ effective inhibitors of this extrusion mechanism, presumably a Ca^{++}-dependent ATPase, are necessary to control this variable in order to observe unidirectional calcium fluxes induced by PHA.

These experiments do make one effective point that PHA as a mitogen which increases cyclic GMP levels acts to induce calcium influx while acetylcholine, a "hormonal" agent increasing cyclic GMP does not. Notably the action of PHA to induce calcium influx is not instantaneous as implied by the wash-dependent experiments of Whitney & Sutherland (48) and does not occur until 10 minutes following stimulation, a time when cyclic GMP has been observed to be already elevated in lymphocytes. This latter point suggests that the calcium influx may not be acting to stimulate guanylate cyclase to increase cyclic GMP levels but is occurring as an independent event.

Further insight into the relevance of calcium as a participant in mitogenic action has been gained by the use of hydrocortisone, an agent which inhibits DNA synthesis in PHA stimulated lymphocytes. Figure 5 shows the effect of 100 μgm/ml hydrocortisone on PHA induced calcium influx at 1 hour. Hydrocortisone significantly inhibits PHA-induced calcium influx ($p < .01$) and does so to the same extent (i.e. 55%) that this concentration inhibits PHA induced thymidine incorporation at 72 hours of culture (7).

Further insight into the role of calcium in the initiation

of lymphocyte activation has been provided by the development of calcium ionophores which act to alter the membrance permeability of calcium and permit its entry into the cell. Our experiments with one of these ionophores A-23187 (kindly provided by Eli Lilly Company) indicate that this agent induces proliferation in cultured human lymphocytes. Figure 7 shows the effects of A-23187 on RNA or DNA synthesis in human peripheral blood lymphocytes assayed at 24 and 72 hours of culture, respectively. It is of note that the effects of this compound are highly dose dependent with inhibition observed at twice the optimal mitogenic concentration (2 μg/ml). These observations confirm those of others (52,53). Table III shows the effects of this compund to increase cyclic GMP levels of lymphocytes after 10 minutes of stimulation in comparison to those of PHA. As was observed in the experiments depicted in Table II the calcium ionophore promotes medium accumulation of cyclic GMP and does so to an extent comparable to PHA (6.0 vs 6.4 x control). Cellular accumulation of cyclic GMP is not apparent, however, unless imidazole is present to inhibit intracellular cyclic GMP phosphodiesterase activity. In the presence of imidazole the ionophore and PHA induce cellular increases which are comparable (8.2 vs. 8.5 x contol). In contrast to PHA the ionophore shows some effect to increase cellular levels of cyclic AMP (data not shown) which may account for its lesser mitogenicity and its highly restricted dose response curve. This effect, which may result from the DMSO solvent, is currently under study.

EFFECTS OF CYCLIC GMP AND CALCIUM ON DNA-DEPENDENT RNA POLYMERASE ACTIVITY IN ISOLATED NUCLEI

Among the earliest nuclear events associated with mitogen-induced proliferation in lymphocytes are increases in RNA synthesis (54), histone acetylation (55), and phosphorylation of non-histone proteins (56,57). Initial studies were undertaken to correlate the increases in cyclic GMP to the increases in RNA synthesis. The experiments showed that low concentrations of cyclic GMP and calcium increased ^{3}H-uridine uptake into ribonuclease-sensitive material in isolated lymphocyte nuclei (1).

These studies have been extended using the system originally described by Wolff et al. (58) and modified by Pogo (59) in which changes in lymphocyte DNA-dependent RNA polymerases I, II and III are quantitated in nuclear preparations. Pogo (58) has recently shown that nuclei isolated from lymphocytes prestimulated with PHA in culture show a marked increase in RNA polymerase I activity within 10 minutes; RNA polymerase

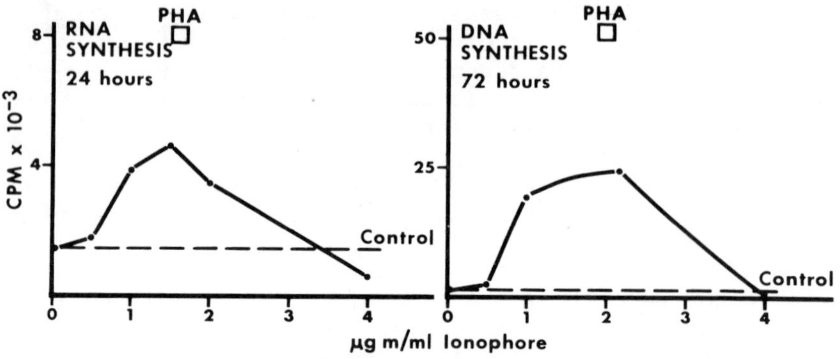

Fig. 7 Effect of a Calcium Ionophore on RNA & DNA Synthesis in Human Lymphocytes.
Lymphocytes were prepared and incubated for 24 hours for RNA and 72 hours for DNA synthesis as described in Figures 1 and 2. The calcium ionophore A-23187 was stored in DMSO and diluted so that DMSO concentrations were < .1%. DMSO at this concentration had no effect on basal RNA and DNA synthesis of lymphocytes. ☐ - indicates the effect of PHA for comparison.

TABLE III

CYCLIC GMP LEVELS AT 10 MINUTES OF INCUBATION IN PMOLES/MG CELLULAR PROTEIN*

	Imidazole 1 mM	Cells	Medium	Total	Ratio to Control
Ionophore 2 μg/ml	-	.052 ± .016	.238 ± .002	.290 ± .018	3.2
	+	.255 ± .039	.271 ± .071	.526 ± .048	6.2
PHA .008 units/ml	-	.043 ± .01	.254 ± .066	.297 ± .048	3.3
	+	.265 ± .005	.305 ± .061	.571 ± .067	6.7
Control	-	.049 ± .005	.040 ± .002	.090 ± .002	
	+	.031 ± .008	.054 ± .018	.085 ± .005	

* See Table II for experimental details.

II activity decreases initially and does not increase until 2 to 4 hours; and RNA polymerase III activity increases between 24 and 48 hours. The increase in RNA polymerase I activity is of particular interest in relating the effects of cyclic GMP to nuclear events. This polymerase is a predominantly nucleolar enzyme, which synthesizes ribosomal RNA requires magnesium and is resistant to α- amanitin.

In an effort to circumvent the PHA-initiated, membrane-restricted events, we isolated nuclei from non-stimulated lymphocytes and examined the activities of RNA polymerase I and II in response to exogenously added cyclic GMP and calcium (60). We observed that cyclic GMP, in concentrations consistent with intracellular levels observed in mitogen-stimulated whole cells, stimulated RNA polymerase I activity in isolated nuclei (see Fig. 8). The concentration of cyclic GMP producing maximal effects depended on the number of nuclei present; however, the median figure of 10^{-9} molar cyclic GMP for 1×10^{6} nuclei was representative. Under these circumstances the magnitude of stimulation of polymerase I activity corresponds to that observed in nuclei of intact lymphocytes prestimulated with PHA. When nuclei of prestimulated lymphocytes were isolated and incubated with cyclic GMP a ten-fold greater concentration was required to achieve maximal stimulation indicating an altered dose-response relationship following mitogen action. The effects of cyclic GMP were only seen at low concentrations and at higher concentrations ($>10^{-7}$M) inhibition was observed.

The effects of calcium on isolated nuclei were variable. Under certain conditions of nuclear isolation (homogenization in hypotonic medium), calcium stimulated RNA synthesis; in detergent-treated nuclei calcium had no observable effect. In the latter circumstance the lack of cyclic GMP and calcium effects when used alone and the stimulation observed when used together is consistent with a co-requirement of the two in the stimulation of polymerase I activity in isolated nuclei. This co-requirement suggests that nuclei isolated under conditions in which soluble intranuclear constituents are less rigorously leached contain sufficient residual cyclic GMP and calcium to allow each when added alone to exert effects. It is of note that if nuclei are treated more rigorously with detergent, all effects of cyclic GMP and calcium alone are lost, possibly implying that the regulation of polymerase activity involves unstable soluble components.

Both cyclic GMP and calcium inhibit RNA polymerase II activity of isolated nuclei while 10X PHA had no effect. The inhibition of RNA polymerase II observed with cyclic GMP and calcium is consistent with the early decreases observed in

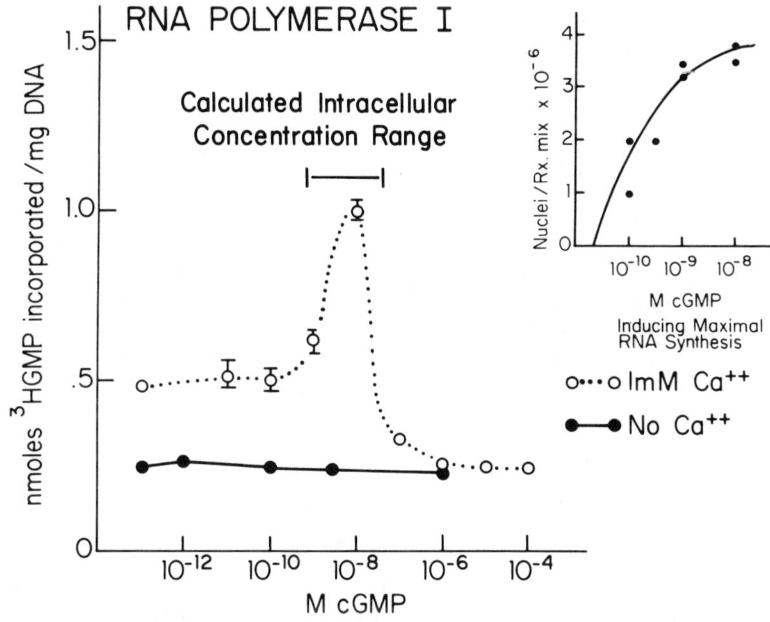

Fig. 8 <u>Effects of Cyclic GMP and Calcium on RNA Polymerase I Activity in Isolated Nuclei</u>.
Isolated peripheral blood lymphocyte nuclei were prepared as described (60) and incubated in Tris-HCl buffer (pH 7.9) containing 1 mM magnesium and ribonucleoside triphosphates one of which was ^{3}H-labeled. RNA polymerase I activity and the increases observed were confirmed to be α-amanatin resistant. The RNA synthesized following incubation at 37°C is precipitated by 10% trichloroacetic acid, collected on Millipore filters and the ^{3}H-labeled RNA quantitated and expressed as nmoles ^{3}H-UMP incorporated into RNA/mg DNA.

the nucleus of the intact lymphocyte stimulated with PHA observed by Pogo (59). Nuclei of lymphocytes pretreated with PHA show increases in RNA polymerase II activity after 2-4 hours, and cyclic GMP and calcium additions to these nuclei inhibit this activity.

Taken collectively the data indicate that cyclic GMP and calcium have a selective ability to reproduce in isolated nuclei the changes observable in RNA polymerase I and II activities of lymphocytes pretreated with PHA and provide the rationale for the hypothesis that cyclic GMP and calcium represent two key components of a membrane to nucleus mitogen signal sequence.

CYCLIC GMP-RELATED ALTERATIONS OF NON-HISTONE CHROMATIN PROTEINS

Phosphorylation of lymphocyte nuclear acidic proteins is an early event which accompanies the extensive stimulation of RNA synthesis induced by the mitogens PHA and Con A (56, 57). In purified equine lymphocytes Con A induces within 15 min a flux of proteins from the cytoplasm into the nucleus and an increased binding of specific non-histone proteins to the chromatin (57). These early changes in nuclear protein distribution are accompanied by a stimulation of phosphorylation of specific phenol-extractable non-histone proteins including a protein of MW = 52,000 bound tightly to the DNA (57). On the basis of the studies on cyclic GMP described, we presumed that some of the early alterations of chromatin structure and activity might be related to cyclic GMP action. We have recently obtained evidence that cyclic GMP can interact directly with lymphocyte proteins which bind to DNA, and that cyclic GMP and agents which selectively raise lymphocyte cyclic GMP levels can stimulate phosphorylation of specific non-histone chromatin proteins (11).

We have observed that acetylcholine selectively raises intracellular cyclic GMP levels in lymphocytes (Table II). Correspondingly, we have found that acetylcholine (10^{-6} - 10^{-8} M), when added to cultured equine lymphocytes increases the rate of phosphorylation of phenol-soluble non-histone chromatin proteins more than 2 fold within 10 min (57). Following this rapid stimulation, the rate of phosphorylation of these nuclear proteins is reduced to control values by 90 minutes after addition of acetylcholine. The acetylcholine analog carbamylcholine (carbachol) stimulates nuclear protein phosphorylation with a time course similar to that obtained with acetylcholine (Fig. 1). Carbachol, however, is active at 1-2 log lower concentrations than acetylcholine, probably reflecting the resistance of carbachol to acetylcholinesterase activity in the culture medium. Stimulation of phosphorylation was observed with carbachol concentrations as low as 10^{-11}M in some experiments. Maximum stimulation was observed at carbachol concentrations ranging from 10^{-7} to 10^{-9}M.

The carbachol-induced stimulation of phosphorylation of equine lymphocyte nuclear acidic proteins is blocked by atropine (Table IV). It can be seen that at 20 min the increased phosphorylation induced by 10^{-7}M carbachol is inhibited about 60% by 10^{-6}M atropine. These results suggest that the nuclear effects of cholinergic agents in lymphocytes are mediated by reaction of these agents with cholinergic receptor sites similar to those found in the parasympathetic nervous system.

It is likely that the effects of cholinergic agents upon lymphocyte nuclear protein phosphorylation reflect the interaction of these agents with specific cholinergic receptors on the lymphocyte cell surface. The existence of lymphocyte cholinergic receptors has been hypothesized on the basis of data indicating that cholinergic agents enhance the cytotoxic T-lymphocyte function of sensitized rat spleen lymphocytes (23).

These findings correlate with results from this laboratory indicating that cholinergic agents enhance labelled uridine and thymidine incorporation into human peripheral blood lymphocytes (see Fig. 1) and, suggest that stimulation of phosphorylation of nuclear acidic proteins by cholinergic agents may be part of the mechanism by which these agents enhance the induction of lymphocyte proliferation. These observations are in accord with the proposal (57) that phosphorylation of specific chromatin proteins is important in the regulation of synthesis of RNA necessary for proliferation.

We have obtained further evidence that cyclic GMP itself modulates phosphorylation of chromatin proteins. Cyclic GMP added exogenously to cultured equine lymphocytes stimulates phosphorylation of a class of phenol-extractable non-histone chromatin proteins (14). Maximum stimulation measured 10 min after addition is obtained with cyclic GMP concentrations in the range of 10^{-6} - 10^{-7}M. The stimulation ranged from 1.5 fold to 3 fold in different experiments. Cyclic AMP added exogenously to lymphocytes has little or no effect upon phosphorylation of these chromatin proteins. 8-Bromo-cyclic GMP also stimulates chromatin protein phosphorylation with maximum stimulation at 10^{-8} - 10^{-9}M. Our experiments performed with butyryl derivatives of the cyclic nucleotides have been difficult to interpret. In contrast to human lymphocytes, it appears that in both murine (23) and equine lymphocytes, effects of agents which elevate intracellular levels

TABLE IV

EFFECTS OF CARBACHOL AND ATROPINE ON INCORPORATION OF [^{32}P] PHOSPHATE INTO LYMPHOCYTE NUCLEAR ACIDIC PROTEIN

Conditions≠	CPM/mg Nuclear Acidic Protein
No additions	8,300
Carbachol (10^{-7})M 10'	22,700
Carbachol 20'	18,700
Carbachol 30'	17,000
Carbachol 20' + Atropine (10^{-6}M)	12,400

≠ Lymphocytes were isolated from equine peripheral blood by Ficoll-Isopaque density centrifugation and incubated as described (56). 10^8 lymphocytes (2 x 10 cells/ml) were incubated for 30 min at 37° with 5 mCi of [^{32}P]NaH_2PO_4 in phosphate-free Eagle's MEM with 10% fetal calf serum. Agents to be tested were incubated in the culture medium for the indicated time periods prior to termination by centrifugation for 2 min at 2000 x g and quick-freezing at -70°C. Phenol-extractable non-histone chromatin proteins were isolated as described (56).

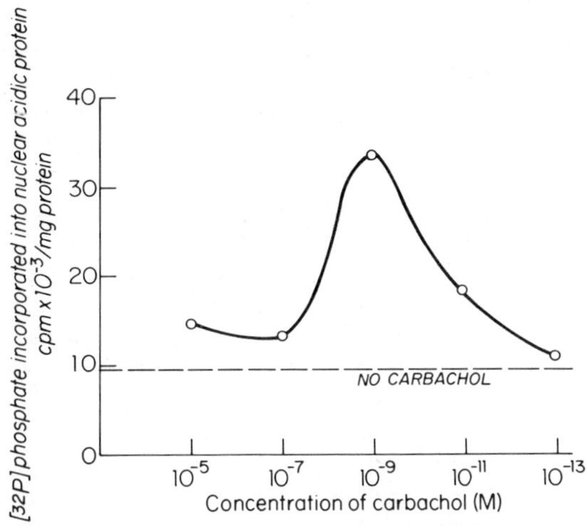

Fig. 9 Stimulation of Equine Lymphocyte Non-histone Chromatin Protein Phosphorylation by Carbachol. Lymphocytes were prepared and incubated with [^{32}P] phosphate as described in the legend to Table IV. Each aliquot of 10^8 cells was incubated with [^{32}P] phosphate for 30 min and with carbachol for 10 min before stopping the reaction. Phenol-extractable non-histone chromatin proteins were isolated and assayed for radioactivity as described (56).

of cyclic nucleotides are accurately mimicked by exogenous addition of the unsubstituted cyclic nucleotides.

In experiments performed in collaboration with Drs. Akira Inoue and Vincent G. Allfrey of the Rockefeller University we have found that calf thymus nuclei contain proteins capable of specifically binding cyclic GMP. When these cyclic GMP-binding proteins are subjected to affinity chromatography on a column of DNA covalently linked to aminoethyl Sepharose 4-B, they are tightly bound and eluted only at high concentrations of salt (0.6M - 1.0M NaCl) (61,62). These experiments suggest the possibility that cyclic GMP may directly affect the activity of DNA-binding nuclear proteins.

The data presented in this report constitute a strong case for a positive regulatory role of cyclic GMP in the induction of lymphocyte proliferation and indicate that a primary site of action of this cyclic nucleotide is in the cell nucleus. In order to ascribe a functional role to an observed increase in cyclic nucleotide levels, it is necessary to show that the cyclic nucleotide regulates specific cellular activities which occur subsequent to its increase. We have presented evidence that the transient increase in cyclic GMP levels observed upon mitogenic stimulation of lymphocytes may serve to stimulate both nuclear protein phosphorylation and RNA polymerase I activity.

ACKNOWLEDGMENTS

The authors appreciate the contributions of Robert A. Good, M.D., Ph.D., Nelson Goldberg, Ph.D., Akira Inoue, Ph.D. and Vincent Allfrey, Ph.D., in making this work possible. This work was supported by grants from the USPHS (NCI-08748), The American Heart Association (AHA 72643), The Jane Coffin Childs Memorial Fund and the American Leukemia Society. J.W. Hadden is an Established Investigator of the American Heart Association.

REFERENCES

1. Goldberg, N.D., Haddox, M.K., Dunham, E., Lopez, C., and Hadden, J.W., in "Control of Proliferation in Animal Cells" (ed. by B. Clarkson and R. Baserga), Cold Spring Harbor Press, New York, p. 609, 1974.
2. Robison, G.A., Butcher, R.W., and Sutherland, E.W., in "Cyclic AMP", Academic Press, New York, 1971.
3. Goldberg, N.D., Haddox, M., and O'Dea, R.F., in "Advances in Cyclic Nucleotide Research" Vol. 3 (ed. by P. Greengard and G.A. Robison), Raven Press, p. 1, 1973.

4. Parker, C.W., Sullivan, W.J., and Wedner, H.J., in "Advances in Cyclic Nucleotide Research" Vol. 4 (ed. by P. Greengard and G.A. Robison), Raven Press, New York, p. 1, 1974.
5. Hadden, J.W., in "Molecular Defect in Combined Immunodeficiency" (Adenosine Deaminase Birth Defect Institute series), in press, 1974.
6. May, C., J. Allergy 43: 163, 1969.
7. Hirschhorn, R., Grossman, J., and Weissman, G., Proc. Soc. Exp. Biol. Med. 133: 1361, 1970.
8. Hadden, J.W., Hadden, E.M., and Middleton, E., Jr., J. Cell. Immunol. 1: 583, 1970.
9. Smith, J.W., Steiner, A.L., and Parker, C.W., J. Clin. Invest. 50: 442, 1971.
10. MacManus, J.P., Whitfield, J.F., and Youdale, T., J. Cell Physiol. 77: 103, 1971.
11. Hadden, J.W., Hadden, E.M., and Good, R.A., Biochem. Biophys. Acta 237: 339, 1971.
12. Coffey, R.G., Hadden, E.M., and Hadden, J.W., Life Sciences, in press, 1974.
13. Hadden, J.W., Hadden, E.M., and Good, R.A., Exp. Cell Res. 68: 217, 1971.
14. Johnson, E.M., Hadden, J.W., Karn, J., and Allfrey, V.G., Fed. Proc. 33: 3030, 1974.
15. Bach, J.F., Dardene, M., and Bach, M., Transplantation Proceedings 5: 99, 1973.
16. Komuro, K. and Boyse, E.A., Lancet 1: 740, 1973.
17. Scheid, M.P., Hoffman, M.K., Komuro, K., Hämmerling, V., Abbott, J., Boyse, E.A., Cohen, G.H., Hooper, J.A., Schulof, R.S., and Goldstein, A.L., J. Exp. Med. 138: 1027, 1973.
18. Goldstein, G., Nature 247: 11, 1974.
19. Hadden, J.W., Sheid, M., Goldstein, G., Boyse, E., and Coffey, R., unpublished observations.
20. George, W.J., Polson, J.B., O'Toole, A.G., and Goldberg, N.D., Proc. Natl. Acad. Sci. U.S.A. 66: 398, 1970.
21. Goldberg, N.D., Haddox, M.K., Hartle, D.K., and Hadden, J.W., in " Proc. V International Congress of Pharmacology" (Karger, Basel, Switzerland), 1973.
22. Hadden, J.W., Hadden, E.M., and Goldberg, N.D., in "Cyclic AMP, Cell Growth and the Immune Response" (ed. by W. Braun, L.M. Lichtenstein, and C.W. Parker, Springer-Verlag, New York, New York), p. 237, 1974.
23. Strom, T.B., Deisseroth, A., Morganroth, J., Carpenter, C.B., and Merrill, J.P., Proc. Natl. Acad. Sci. U.S.A. 69: 2995, 1972.

24. Strom, T.B., Hirsch, M.S., Black, P.H., Carpenter, C.B., and Phillips, S.M., J. Clin. Invest. 53: 79, 1974.
25. O'Dea, R.F., Haddox, M.K., and Goldberg, N.D., Fed. Proc. 29: 473, 1970.
26. Parker, J.W. and Lukes, R.J., Amer. J. Clin. Path. 56: 174, 1971.
27. Webb, D.R., Stites, D.P., Perlman, J.D., Luong, D., and Fudenberg, H.H., Biochem. Biophys. Res. Comm. 53: 1002, 1973.
28. Winchurch, R., Ishizuka, M., Webb, D., and Braun, W., J. Immunol. 106: 1399, 1971.
29. Makman, M.H., Proc. Natl. Acad. Sci. U.S.A. 68: 885, 1971.
30. Novogrodsky, A. and Katchalsky, E., Biochem. Biophys. Acta 215: 291, 1970.
31. Hadden, J.W., Hadden, E.M., and Goldberg, N.D., Proc. Natl. Acad. Sci. 69: 3024, 1972.
32. DeRubertis, F.R., Zenser, T.V., Adler, W.H., and Hudson, T., J. Immunol. 113: 151, 1974.
33. Greaves, M.F. and Bauminger, S., Nature New Biol. 235: 67, 1972.
34. Sören, L., Exp. Cell Res. 78: 201, 1973.
35. Shumm, D.E., Morris, H.P., and Webb, T.E., Eur. J. Cancer 10: 107, 1974.
36. Heidrick, M.L., J. Cell Biol. 59: 139a; and Kersey, J. and Goldberg, N.D., personal communication.
37. Estensen, R.D., Hadden, J.W., Hadden, E.M., Haddox, M.K., and Goldberg, N.D., in "Cold Spring Harbor Symposium, Control of Proliferation in Animal Cells" (ed. by B. Clarkson and R. Baserga), Cold Spring Harbor Press, p. 627, 1974.
38. Armelin, H.A., Nishikawa, K., and Sato, G.H., ibid.
39. Seifert, W. and Rudland, P.S., Nature 248: 138, 1974.
40. Voorhees, J., Stawiski, M., Duell, E., Haddox, M., and Goldberg, N.D., ibid.
41. Kuehl, F.A., Ham, E.A., Zanetti, M.E., Sanford, O.H., Nicol, S.E., and Goldberg, N.D., Proc. Natl. Acad. Sci. U.S.A. 71: 1866, 1974.
42. Bloomquist, C.H., Haddox, M.K., O'Dea, R.F., Hadden, J.W., and Goldberg, N.D., J. Cell Biol. 59: 27a, 1973.
43. Haddox, M.K., Stephenson, J.H., and Goldberg, N.D., Fed. Proc. 33: 522, 1974.
44. Goldberg, N.D. and O'Toole, A.G., in "Method Biochem. Anal." (ed. by D. Glick), John Wiley & Sons, New York, p. 1, 1971.
45. Steiner, A.L., Parker, C.W., and Kipnis, D.M., J. Biol. Chem. 247: 1106, 1972.
46. Gilman, A.G., Proc. Natl. Acad. Sci. U.S.A. 67: 305, 1970.

47. Allwood, G., Asherson, G.L., Davey, M.J., and Goodford, P.J., Immunology 21: 509, 1971.
48. Whitney, R.B. and Sutherland, R.M., J. Cell Physiol. 82: 9, 1973.
49. Schultz, G., Hardman, J.G., Schultz, K., Baird, C.E., and Sutherland, E.W., Proc. Natl. Acad. Sci. U.S.A. 70: 3889, 1973.
50. Hardman, J.G., Beavo, J.A., Gray, J.P., Chrisman, T.D., Patterson, W.D., and Sutherland, E.W., Ann. N.Y. Acad. Sci. 185: 27, 1971.
51. Schatzmann, H.J. and Vincenzi, F.F., J. Physiol. 201: 369, 1969.
52. Kersey, J., in "Advances in Immunology" (II International Congress of Immunology, Brighton, England), in press, personal communication, 1974.
53. Maino, V.C., Green, N.M., and Crumpton, M.J., Nature 251: 324, 1974.
54. Cooper, H.L. and Rubin, A., Blood 25: 1014, 1965.
55. Pogo, B.G.T., Allfrey, V.G., and Mirsky, A.E., Proc. Natl. Acad. Sci. U.S.A. 55: 805, 1966.
56. Kleinsmith, L.J., Allfrey, V.G., and Mirsky, A.E., Science 154: 780, 1966.
57. Johnson, E.M., Karn, J., and Allfrey, V.G., J. Biol. Chem. 249: 4990, 1974.
58. Wolff, J.S., Langstaff, J.A., Weinberg, G., and Abell, C.W., Biochem. Biophys. Res. Comm. 26: 366, 1967.
59. Pogo, B.G.T., Cell Biol. 53: 635, 1972.
60. Johnson, L. and Hadden, J.W., Proc. Natl. Acad. Sci. U.S.A., in press, 1974.
61. Johnson, E.M., Inoue, A., Good, R.A., Allfrey, V.G., and Hadden, J.W., in "Proceedings II International Congress on Cyclic AMP" (Vancouver, Canada), 1974.
62. Allfrey, V.G., Inoue, A., Karn, J., Johnson, E.M., and Hadden, J.W., in "The Ciba Foundation Symposium on 'The Structure and Function of Chromatin'" (Brighton, England), in press, 1974.

LYMPHOCYTE TRANSFORMATION AND CATION TRANSPORT

J. Gordin Kaplan and Michael R. Quastel
Department of Biology, University of Ottawa, Ottawa, Canada, and Ben Gurion University, Beer Sheba, Israel

I. NaK-ATPase

The fundamental discovery by Engelhardt that myosin possessed ATPase activity (1) and could thus assure itself of the metabolic energy required for contraction, together with the subsequent discovery that the energetic requirements for ion pumping after conduction of the nerve impulse were met by hydrolysis of ATP (2) suggested the possibility that nervous tissue might also possess an ATPase at the site where metabolic energy was required: in the neuronal membrane. That precisely was where Skou found a novel membrane-bound ATPase (3); in particulate fractions of crab nerve there was a low basal level of ATPase which was increased about 50-fold if optimal concentrations of Na^+ and K^+ were included in the assay system; this is the NaK-ATPase. Some years previously, Schatzmann showed that the cardiac glycoside ouabain, widely used at one time for its cardiotonic, digitalis-like, action, had a strong inibitory action on the transport of the monovalent cations Na^+ and K^+ (4).

The fundamental discovery was then made that ouabain was a potent and specific inhibitor of that fraction of the ATPase activity resulting from the presence of Na^+ and K^+ (5,6) in the assay system. The evidence is now convincing that the NaK-ATPase is intimately involved in uphill transport of these monovalent cations and that ouabain acts as a competitive inhibitor of K^+ at a site located at the outside surface of the cell membrane; the role of this enzyme system in ion pumping in animals cells (7) and its properties and reaction mechanism (8) have been well described in recent reviews.

The stoichiometric relation between Na^+ and K^+ transport and ATP hydrolysis has been well worked out in the case of the erythrocyte (8):

1. $$ATP + H_2O + 3\ Na^+_{in} + 2\ K^+_{out} \dashrightarrow ADP + P_i + 3\ Na^+_{out} + 2\ K^+_{in}$$

There are several interesting consequences of this stoichiometry. The system is electrogenic, i.e., able to create and maintain a potential difference across the membrane consequent to the relative electronegativity of the cell interior caused by the greater number of Na^+ extruded per mole of ATP.

It is one of the fascinating aspects of this reaction that a scalar process (the hydrolysis of ATP) and a vectorial one (ion transport) are occurring simultaneously; hence the arrow of equation 1 has more than the temporal significance of the usual chemical equations, signifying also in this case a direction of movement of the reactants.

II. Role of the NaK-ATPase in lymphocyte transformation

The fact that phospholipid turnover increased almost immediately after stimulation of lymphocytes with PHA (9) suggested to us that membrane-bound enzymes might be active participants in the very early events linking stimulation with derepression of DNA synthesis and mitosis. Our first experiments (10) established that ouabain, at concentrations higher than 2×10^{-7} M, inhibited virtually completely the incorporation of uridine and thymidine into the RNA and DNA respectively of PHA-stimulated human lymphocytes, a result confirmed by others (11). The effect was perfectly reversible: cells maintained for up to 3 days in the glycoside and then washed responded to subsequent PHA treatment in the normal way (10). Ouabain did not inhibit the initial interaction of PHA with its receptor, since cells treated simultaneously with both PHA and ouabain and then washed twice and re-incubated in fresh medium in the absence of either agent were found to incorporate uridine into RNA at rates characteristic of normal lymphocytes cultured in the presence of PHA for 24 hrs. Furthermore, the inhibitory effect of the glycoside could be completely reversed by including high levels of K^+ in the medium. On the other hand, the characteristic clumping of cells induced by PHA was entirely insensitive to ouabain, even at concentrations higher than 10^{-6} M (10).

Subsequent studies showed that the relation between ouabain and K^+ was that to be expected of a competitive inhibition: 10^5 equivalents of K^+ overcame the inhibition caused per mole ouabain over the entire range of concentrations tested (12). Appreciable inhibition of protein synthesis was noted within one hour of addition of inhibitory concentrations of ouabain whereas the lag period was somewhat longer for inhibition of nucleic acid synthesis (12). Parameters of lymphocyte transformation which are inhibited by ouabain are morphological (blast transformation and mitosis), physiological (cell respiration, and the increased respiration which follows stimulation (13)), and biochemical (synthesis of DNA, RNA and protein).

III. Activation of K^+ transport by PHA

The great sensitivity of such a variety of parameters to ouabain and its reversibility by high K^+ in the medium led us to the hypothesis that a necessary (but not sufficient) condition for lymphocyte transformation was a sustained increased activity of monovalent cation transport via the NaK-ATPase system. We found it difficult to test this hypothesis by direct assay of ATP hydrolysis, since we found that only a very small part of the total ATPase activity of intact cells or of homogenates or particulate fractions was due to a ouabain-inhibitible component; Lichtman et al. (14,15) have reported in fact that only about 5% of the total ATP hydrolysis of lymphocyte homogenates could be ascribed to the NaK-ATPase.

We thus elected to test our hypothesis by studying the effect of PHA stimulation on the rate of uptake of labelled potassium (16). The pooled data from a group of normal human subjects showed that the lymphocytes took up ^{42}K at a mean rate of 2.9 ± 0.4 nmoles/ 10^6 cells/h corresponding at the asymptote to 7-10 nmoles/ 10^6 cells. Within one hour of adding PHA to the cultures, an acceleration of uptake of ^{42}K was observed (16). Subsequent experiments by Averdunk (17) confirmed these results, but his techniques permitted him to observe an increase in rate of uptake of ^{42}K within 30 sec. of addition of PHA; thus activation of monovalent cation transport is one of the most precocious events following stimulation. It is evident that the increase in ion transport occurs much before the increase in volume due to formation of blasts, which suggests that the internal concentration of K^+ increases following stimulation. This was confirmed by the finding that PHA had no effect on the rate of efflux of ^{42}K from lymphocytes which had been pre-loaded with the isotopic cation (16); since rate of influx increased without change in rate of efflux, the concentration of intracellular K^+ must have increased as a result of stimulation. Ouabain strongly inhibited K^+ transport in unstimulated cells and even more strongly that of stimulated cells, reducing the influx rate to that of the ouabain-treated control cells (16).

We next established that the K_M for K^+ transport was 1.5 mM in the unstimulated cells and did not change following stimulation of transport by PHA (13); on the other hand the V_{max} of transport increased after stimulation from 9.6 nmoles to 15.9 nmoles/10^6 cells/ 3 hr. Similarly, the stimulatory effect of PHA on transport of a non-metabolizable amino acid analogue was found to be due to an increase in V_{max} without change in K_M (18,18a) whereas the increased uptake of Ca^{++} following PHA stimulation was on the contrary due to an effect on the K_M (19). The increase in rate of ^{42}K transport in stimulated cells was not affected by including in the medium concentrations of actinomycin-D which inhibited RNA synthesis by more than 95% nor of cycloheximide which inhibited protein synthesis virtually completely, permitting the conclusion that the increased V_{max} of transport was not caused by a PHA-stimulated synthesis of new NaK-ATPase (13).

IV. The model

These data permitted us to make our hypothesis more exigent and heuristic. Our model (20) involves two testable assumptions: 1) one of the reasons for which the resting lymphocyte remains in its metastable state is that the resting internal concentration of K^+ is too low to permit the occurrence of several events which are essential to the active state; and 2) one of the early events following stimulation is exposure of previously synthesized but cryptic, inaccessible cation pumping sites; we suppose the mechanism to be a conformational change in cell surface glycoproteins caused by lectin binding or by other stimulatory agents. These sites become active following the conformational change resulting in increased monovalent cation transport and hence in increase in internal K^+ level which rises above a threshold concentration and thus triggers the train of events which links the initial stimulus to be ultimate DNA synthesis and mitosis.

We prefer to think of the consequences of activation of the NaK-ATPase in terms of a rise in K^+ rather than in terms of a generalized electrogenic effect since, a) it is hard to see why an increase in membrane potential should have a triggering effect, and b) low concentrations of ouabain (4.4×10^{-8} M) were found to be without effect on blastogenesis but to inhibit the mitogenic effect of PHA completely (10). It is easier to account for this by supposing that each process in this train of events has its own K^+ threshold concentration which must be exceeded in order that the process may occur. We can of course not rule out the possibility of Na^+ playing a regulatory role since its intracellular concentration will vary inversely to that of K^+. Nor can we exclude the possibility that ion movement as such is not primarily involved but that the crucial event is rather the activation of the non-vectorial reaction of ATP hydrolysis by the NaK-ATPase, which was found to explain the inhibitory effects of ouabain on glycolysis in ascites tumor cells (21); this seems unlikely in view of the relatively minor proportion of the total lymphocyte ATPase attributable to the NaK-ATPase.

V. Binding of ^{3}H-ouabain

Since K^+ and ouabain compete for the same (or overlapping) sites, our model predicts that stimulation of lymphocytes should expose new binding sites for these substances. In fact, we earlier reported that PHA treatment did result in increased binding of ^{42}K in a labile, rapidly equilibrating compartment, presumably extracellular, from which it could be removed by trypsin treatment or by washing with K^+ or Rb^+ but not with Na^+ or sucrose (20). Preliminary studies on binding of ^3H-ouabain showed that this increased significantly after PHA treatment, and that the increase occurred despite the presence of inhibitors of protein synthesis or of metabolic energy production (22). On the other hand, the increase in binding was completely prevented by incubating the cells at 5^oC; conformational changes of proteins are highly temperature sensitive rate processes (22).

A detailed study of ^3H-ouabain binding (Quastel and Kaplan, submitted for publication (see also ref. 23)) showed that PHA increased the estimated number of binding sites from 1.25×10^5/cell to 2.3×10^5/cell. This compares with values of 5 to 7.5×10^5/cell for HeLa cells reported by Vaughan and Cook (24) and by Baker and Willis (25), and of $1\text{-}3 \times 10^2$/cell for erythrocytes (7) which have both little NaK-ATPase and very low cation fluxes (26).

All of these data accord well with our model. However, we have some preliminary negative data with respect to ConA and pokeweed mitogen (PWM): neither of these agents seemed to increase significantly the binding of ^3H-ouabain. Confirmation of this finding will require us to modify assumption 2 of the above hypothesis and/or our interpretation of the ^3H-ouabain binding data.

VI. Generality of the ouabain effect on stimulated human lymphocytes

Ouabain inhibits DNA synthesis regardless of how the stimulation is achieved; the dose-effect curves are consistently the same as in the case of PHA. Among the agents whose mitogenic action is inhibited by ouabain are:

1. Lectins and chemical agents: highly purified PHA (leukoagglutinin), ConA, PWM, IO_4^- (23,27).
2. Specific antigens, such as streptolysin-O, typhoid-paratyphoid vaccine and diphtheria-tetanus toxoid (27) as well as a crystalline tetanus toxoid preparation (anatoxin) (Kaplan and Bona, see Table 1).
3. The mixed lymphocyte reaction (12).
4. Spontaneously dividing lymphocytes from acute leukemia patients (28).
5. Human lymphoblast cell lines (Lefebvre, Jui, Kaplan and Bona, submitted for publication).

We have more recently been able to generalize the inhibitory effects of ouabain to include B-cell activities. Ouabain was found to inhibit the formation and/or secretion of antibody by pig and rabbit spleen plaque forming cells (29). Furthermore, ouabain inhibits the mitogenic effects of an antiserum to purified human μ chains (Table 1).

VII. Specificity of ouabain action

Could the inhibitory effects of ouabain be due to a nonspecific action of its steroid moiety, or to a non-specific immunosuppressive action like that of hydrocortisone? No; while some other steroids (including hydrocortisone and prednisolone) caused some inhibition in RNA synthesis in PHA-stimulated human lymphocytes, the concentrations required to achieve significant inhibition were some four orders of magnitude higher than in the case of ouabain-induced inhibition (12).

Could the inhibitory effects of ouabain be due to a generalized cytotoxic effect? We have already seen that those effects so far described are freely reversible by simple washing and that the viability of the cells was not impaired. Furthermore, ouabain treatment has no effect on certain of the early events following stimulation, e.g. clumping. Kay has shown that ouabain did not inhibit the early stimulation by PHA of uridine incorporation into RNA even when the ouabain was added to the suspension for from 5 to 21 hrs before PHA stimulation (30). Thus, certain events are more readily inhibited by ouabain than others.

Are all cells in culture inhibited by ouabain? Human lymphocytes and those of the pig (31) are highly sensitive to ouabain as are those of the cat (Stoltz, personal communication). Rabbit lymphocytes are less so (23,31) and lymphocytes of the mouse and rat are not sensitive at least at concentrations up to 10^{-4} M (31). The growth and DNA synthesis of several standard cell lines, including monkey kidney cells, are not inhibited by concentrations of ouabain two orders of magnitude higher than those which cause maximal inhibition of DNA synthesis in stimulated human lymphocytes (32). Furthermore, the uptake and processing of immunologically relevant antigen (33) by guinea pig macrophages and their ability to aggregate with lymphocytes (24) are not inhibited by ouabain.

Is the inhibitory effect of ouabain on lymphocyte transformation a consequence of its primary action on the NaK-ATPase? If so, other agents which inhibit the latter should also inhibit the former. We have shown that the order of effectiveness of five known inhibitors of the NaK-ATPase in

TABLE 1

Effect of 2 x 10^{-7} M ouabain on the incorporation of ^{3}H-thymidine into the DNA of human lymphocytes stimulated by antiserum to human μ chains: comparison with stimulation by mitogen and specific antigen. Mean of three determination and standard error shown.

Category	Ouabain	CPM	%Inhibition
Control	-	1897 ± 121	
Anti-human μ chains	-	7618 ± 150	
	+	148 ± 60	98
Crystalline tetanus toxoid	-	4254 ± 543	
	+	123 ± 14	97
PHA	-	43050 ± 3492	
	+	1365 ± 283	97

vitro is the same for inhibition of the enzyme as it is for inhibition of DNA synthesis in stimulated lymphocytes (13); the major difference was that the lymphocyte system was more sensitive, requiring a concentration of inhibitor generally one order of magnitude less than that required to produce equivalent (50%) inhibition of the in vitro enzyme. Agents such as digoxin whose pharmacological effects are similar to those of ouabain and which inhibit the NaK-ATPase also inhibit the transformation of lymphocytes (35).

VIII. Persistent effects of ouabain treatment

Wright et al. (27) found that if ouabain (2×10^{-7} M) was added to cultures of human lymphocytes stimulated by specific antigens, and if the cells were then washed free of the glycoside they could no longer respond to fresh antigens of any kind, although they could respond to mitogens. They found that antigen did not have to be present during the 48 hrs. incubation with ouabain for the effect to become manifest; they also reported (27,22) that ouabain pretreatment at this concentration and of this duration caused the cells to lose their capacity to incorporate ^{3}H-thymidine in the two-way mixed lymphocyte reaction (MLR).

Recent work on the one-way MLR (Christen, Sasportes, Mawas, Dausset and Kaplan, submitted for publication) has shown that pretreatment of human lymphocytes from both donors (AoBmo) with 2×10^{-6} M ouabain for 48 hours effectively eliminated the one-way MLR. Pretreatment of the responding cells alone (AoBm) caused little or no inhibition; these cells behaved like the controls (ABm). On the other hand, pretreatment of the stimulating cells alone (ABmo) was equivalent to pretreating both stimulating and responding cells: the MLR was abolished.

Ouabain pretreatment at 2×10^{-7} M yielded results which were similar to those at 2×10^{-6} M but much more variable (36). Pretreatment at 10^{-7} M of both stimulating and responding cells caused an astonishing enhancement of incorporation of ^{3}H-TdR in the one-way MLR (AoBmo); incorporation was between 2 to 4 times as great at its peak (at days 5 to 10 of culture) as was incorporation into DNA of the controls (ABm) at its peak at day 5. Pretreatment of responders (AoBm) had no effect, but pretreatment of the stimulators alone (ABmo) caused the same enhancement of incorporation as observed when both stimulators and responders had been pretreated.

We conclude that the persistent effects of ouabain pretreatment on the human MLR are produced at the level of the stimulating cells; these are inactivated at high ouabain concentrations and are activated at low concentrations. The stimulators also can be inactivated simply by prolonged (4-7 days) incubation in medium (37). These results show that stimulation in the MLR is an active and labile process (as Lindahl-Kiessling and others have long insisted (38)) rather than a passive display of appropriate antigens at the cell surface. They also show that prolonged exposure to high concentrations of ouabain does not produce a generalized cytotoxicity, since cells so treated conserve an unimpaired capacity to respond in the MLR, with all the complex biochemical machinery which this involves. It is possible that these persistent effects of ouabain may have significant clinical applications.

IX. Comparison of cation effects

Among the cations whose action is required for lymphocyte transformation are Ca^{++} (19) and Zn^{++} (39), as shown by the inhibition of blastogenesis caused by the presence of chelators with high affinities for these divalent cations; inhibitions caused by either chelator could be overcome by addition of the corresponding cation.

Figure 1 represents schematically a sequence of biochemical events linking the initial activation (step 1) with a round of DNA synthesis and mitosis. Neither Ca^{++} (40), Zn^{++} (39) nor K^+ acts at the level of step 1. Unlike the two other cations, Ca^{++} is required only within the first few hours of stimulation; within some 12 hours of stimulation by PHA, addition to the cultures of the chelating agent EGTA no longer caused significant inhibition of blastogenesis (19); Ca^{++} must thus be required in order for one of the earlier steps in a linear sequence to occur (step 2). Zn^{++} is apparently not required for some of the earlier events in this sequence, since the chelator o-phenanthroline does not inhibit significantly either RNA or protein synthesis, but causes a virtually total block of DNA synthesis (step 4). On the other hand, the pattern of inhibition caused by ouabain is unlike that of the two chelators, since this agent inhibits both at early and late stages of blastogenesis. We thus represent K^+ as being required for all of the steps in the sequence other than the initial interaction of PHA with the lymphocyte membrane (step 1) and with an early RNA synthesis (30) represented as being on a different pathway than the linear sequence leading to DNA synthesis (steps 5 and 6).

X. Conclusion

One of the earliest events following lymphocyte stimulation is an increase in the rate of K^+ influx via the NaK-

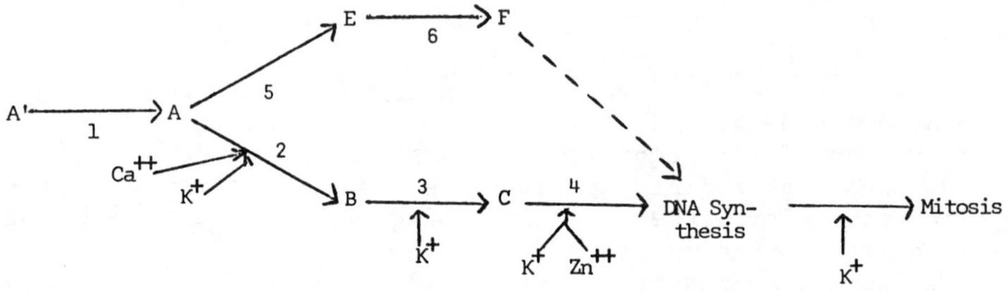

Fig. 1 Temporal sequence of cation action in lymphocyte transformation. Step 1 is a triggering event caused by a mitogen. Steps 5 and 6 represent the ouabain-insensitive RNA synthesis occurring soon after stimulation (30). Steps 2, 3 and 4 are intermediate stages in the train of events linking the activated state, A, and the initiation of DNA synthesis.

ATPase system of the membrane. With the exception of a few early events (development of the 'stickiness' at the membrane which causes clumping; an early synthesis of RNA; interaction of PHA with its receptor) virtually every event, morphological, physiological and biochemical, which characterizes the transformed, activated lymphocyte, has an absolute requirement for the increased level of internal K^+ which follows stimulation; the threshold levels of internal K^+ required for some events (e.g., blastogenesis) are lower than those for others (e.g., DNA synthesis). All activated lymphocytes and lymphoblast lines, whether active spontaneously or as a result of stimulation by some external specific or non-specific agent, require an actively functioning NaK-ATPase system. Most of the effects of the inhibitor of the NaK-ATPase, ouabain, are reversible by washing or by including high K^+ in the medium. Under certain conditions, ouabain pretreatment causes persistent effects which cannot be washed away and which are not prevented by exposure to high K^+ after treatment: such cells have lost the capacity to respond to specific antigens (but not to non-specific mitogens) and to stimulate in the MLR; pretreatment with lower ouabain concentration causes a striking enhancement of the ability of the treated cells to stimulate in the MLR. The temporal pattern of action of the cations essential to transformation, Zn^{++}, Ca^{++} and K^+ is unique for each.

ACKNOWLEDGEMENT

We are grateful to Dr. Mihaescu (Hôpital St. Louis, Paris) for the gift of the antiserum to purified human μ chains, to Dr. Bona for his collaboration in the experiment shown in Table 1, and to the Medical Research Council and National Research Council of Canada.

REFERENCES

1. Engelhardt, V.A., Adv. Enzymol. 6: 147, 1946.
2. Caldwell, P.C., Hodgkin, A.L., Keynes, R.D. and Shaw, T.I., J. Physiol. 152: 561, 1960.
3. Skou, J.C., Biochim. Biophys. Acta 23: 394, 1957.
4. Schatzmann, H.J., Helv. Physiol. Acta 11: 346, 1953.
5. Post, R.L., Merrit, C.R., Kinsolving, C.R. and Albright, C.D., J. Biol. Chem. 235: 1796, 1960.
6. Dunham, E.T. and Glynn, I.M., J. Physiol. 156: 274, 1961.
7. Baker, P.F., in Metabolic Transport, ed. L.E. Hokin, Academic Press, New York, 1972, p. 243.
8. Hokin, L.E. and Dahl, J.L., in Metabolic Transport, ed. L.E. Hokin, Academic Press, New York, 1972, p. 270.
9. Mueller, G.C. and LeMahieu, M., Biochim. Biophys. Acta

114: 100, 1966.
10. Quastel, M.R. and Kaplan, J.G., Nature 219: 198, 1968.
11. Kasakura, S., Proc. 4th Leucocyte Culture Conf., ed. O.R. McIntyre, Meredith, 1969, p. 564.
12. Quastel, M.R. and Kaplan, J.G., Exptl. Cell Res. 62: 407, 1970.
13. Quastel, M.R., Dow, D.S. and Kaplan, J.G., Proc. 5th Leucocyte Culture Conf., ed. J.E. Harris, Academic Press, New York, 1970, p. 97.
14. Lichtman, M.A. and Weed, R.I., Blood 35: 12, 1970.
15. Lichtman, M.A., Jackson, A.H. and Peck, W.A., Cell Physiol. 80: 383, 1972.
16. Quastel, M.R. and Kaplan, J.G., Exptl. Cell Res. 63: 230, 1970.
17. Averdunk, R., Hoppe-Seyler's Z. Physiol. Chem. 353: 79, 1972.
18. Mendelsohn, J., Skinner, A. and Kornfeld, S., J. Clin. Invest. 50: 818, 1971.
18a. van den Berg, K.J., The role of amino acids in the mitogenic activation of lymphocytes. Radiobiological Institute, Organization for Health Research, Rijswijk, The Netherlands, 1974.
19. Whitney, R.B. and Sutherland, R.M., J. Cell Physiol. 82: 9, 1973.
20. Quastel, M.R., Wright, P. and Kaplan, J.G., Proc. 6th Leucocyte Culture Conf., ed. M.R. Schwarz, Academic Press, New York, 1972, p. 185.
21. Scholnick, P., Lang, D. and Racker, E., J. Biol. Chem., 248: 5175, 1973.
22. Wright, P., Quastel, M.R. and Kaplan, J.G., Proc. 7th Leucocyte Culture Conf., ed. F. Daguillard, Academic Press, New York, 1973, p. 87.
23. Quastel, M.R., Milthorp, P., Kaplan, J.G. and Vogelfanger, I., Proc. 8th Leucocyte Culture Conf., eds. K. Lindahl-Kiessling and D. Osoba, Academic Press, New York, 1974, p. 493.
24. Vaughan, G.L. and Cook, J.S., Proc. Nat. Acad. Sci. USA 69: 2627, 1972.
25. Baker, P.F. and Willis, J.S., J. Physiol. 224: 441, 1972.
26. Bonting, S.L. and Caravaggio, L., Arch. Biochem. Biophys. 101: 37, 1963.
27. Wright, P., Quastel, M.R. and Kaplan, J.G., Exptl. Cell Res. 79: 87, 1973.
28. Quastel, M.R., Kaplan, J.G. and Abbatt, J., Lancet 1: 801, 1971.
29. Milthorp, P., Quastel, M.R. and Kaplan, J.G., Cell. Immunol., 14: 128, 1974.

30. Kay, J.E., Exptl. Cell Res. 71: 245, 1972.
31. Quastel, M.R. and Vogelfanger, I.J., Cell Immunol. 2: 504, 1971.
32. Quastel, M.R., PhD. Thesis, University of Ottawa, 1970.
33. Rosenthal, A.S., Rosenstreich, D.L., Blake, J.T., Lipsky, P.E. and Waldron, J.A., Proc. 7th Leucocyte Culture Conf., ed. F. Daguillard, Academic Press, 1973, p. 201.
34. Lipsky, P.E. and Rosenthal, A.S., Proc. 8th Leucocyte Culture Conf., eds. K. Lindahl-Kiessling and D. Osoba, Academic Press, New York, 1974, p. 521.
35. Wright, P., Ph.D. Thesis, University of Ottawa, 1974.
36. Sasportes, M., Christen, Y., Mawas, C., Kaplan, J.G. and Dausset, J., Abstracts, 9th Leucocyte Culture Conf., 1974.
37. Opelz, G. and Terasaki, P.I. Science 184: 464, 1974.
38. Lindahl-Kiessling, K. and Safwenberg, J., Proc. 6th Leucocyte Culture Conf., ed. M.R. Schwarz, Academic Press, New York, 1972, p. 623.
39. Williams, R.O. and Loeb, L.A., J. Cell Biol. 58: 594, 1973.
40. Whitney, R.B. and Sutherland, R.M., Biochim. Biophys. Acta 298: 790, 1973.

DISCUSSION FOLLOWING SYMPOSIUM EARLY EVENTS IN LYMPHOCYTE ACTIVATION II: MEMBRANE AND BIOCHEMICAL EVENTS

Sage - Durham: I would like to echo a little bit of what Dr. Parker had said and some of the warnings too. Like Dr. Parker, we have looked for changes in cyclic AMP in the first 30 minutes of stimulation by lectin. Like Dr. Parker, we find, with certain lectins which are stimulants, you get modest rises, and like Dr. Parker, we have also taken a nonmitogenic lectin, the mushroom lectin, and demonstrated that you can get a greater rise in levels of cyclic AMP in the first 30 minutes if you use enough of it. We also can take another nonmitogenic lectin wheat germ agglutinin and get a decrease in cAMP levels. The point is, if you take a look at enough lectins that are stimulants and nonstimulants you can get almost anything you want in cAMP rises or depressions and also cGMP rises and depressions. Therefore, it is dangerous to generalize the patterns of cAMP distribution with one mitogen or two mitogens to all mitogens or those seen with one nonmitogen to all nonmitogenic lectins. I think it is very important in these cases that whenever you see a situation which apparently distinguishes a mitogen and nonmitogen that you examine a great many other stimulants and nonstimulants. This practice is not observed very often because there are very few people who use lectins other than Con A and PHA. So I think there is warning in there somewhere. I think very similarly the studies on phosphorylation of proteins are subject to the same warning. I would not be surprised to see that if you put mushroom agglutinin on a cell you might also get a change in protein phosphorylation, and the change, if it occurs might be related to mitosis.

Wedner - St. Louis: I agree that the only way that you can show that these are real events is to look at a lot of lectins which are either mitogenic or nonmitogenic. But, the one thing that I do want to point out is that there is compartmentalization and the point is that if you simply measure a rise in something that does not tell you anything because you have to show the rise has some functional significance and that it goes up in the right spot, if it does not go up in the right spot it is insignificant.

Hadden - New York: I would like to ask a technical question concerning your immunofluorescence localization. An antibody-cyclic nucleotide interaction by definition is a hapten interaction unless the cyclic nucleotide is bound. I am sure you make the assumption that you're looking at a membrane bound cyclic nucleotide. Receptor proteins that bind cyclic nucleotides and distinguish them one from another in a very discrete fashion must surely be recognizing in the same portion of the molecule that the antibody is, so I can't understand the antibody recognizing a cyclic nucleotide which is bound to a receptor protein. If the cyclic nucleotide were bound to a phosphodiesterase which would recognize the phosphate group, it would leave exposed a portion of the molecule that the antibody might recognize. My speculation would be that you might very well be looking at the fixation of cyclic AMP in an inactivated membrane bound phosphodiesterase. Have you considered this, and if so, does that influence your thinking?

Wedner - St. Louis: We considered it. Let me make two points. The protein kinase and an antibody do not look at exactly the same parts of the cAMP molecule. You can show this by two maneuvers. If you do anything to the 2'O position of cAMP you inhibit binding to the protein kinase. If you succinate the 2'O position of cAMP, you increase binding to the antibody by about 100 fold, so they are looking in essence at different or somewhat different parts of the molecule. The obvious experiment, which we are getting ready to do is to look at purified protein kinases, first to bind the cAMP to the protein kinase and then see if we can precipitate it with our antibody. The second point is I do not know the turnover number of phosphodiesterase. I would suspect that the turnover number is very high. If that is true, there is almost no cAMP on a phosphodiesterase. It is taken on, converted to 5' AMP and split off. I would suspect that most of what is on is in the form of 5' AMP rather than cAMP. Our antibody does not recognize 5' AMP at least if you do titration curve it is about 4 logs less sensitive for 5' AMP. I can't believe it is on a phosphodiesterase. I think it is on a protein kinase, but I have not been able to prove that yet.

Schwartz - Bethesda: Were your compartmentalization studies done on Ficoll-Hypaque purified cell populations? I am bothered by the contrast between how rigorous the biochemistry is and how little attention is paid to the fact that the cell population that you have used is quite heterogeneous. Ficoll-Hypaque purified populations consist of monocytes, B lymphocytes, T lymphocytes and are usually also contaminated with a few percent granulocytes. What is the possibility that some of these compartments, which you feel are in different parts of one cell, are coming from different cells?

Wedner - St. Louis: I think you can make the argument, but what we are looking at the way we purify our cells is 95 to 98% lymphocytes. There are less than 3 platelets per lymphocyte. We have looked at the platelets and what we are seeing is not in the platelets. There could be subpopulations of lymphocytes which contain the different compartments but it seems that when you take this whole population and break it up that the most logical thing to conclude is that you are looking at the various compartments in all the cells. The biochemistry follows the immunohistochemistry. You do not see one cell which was a given pattern all of the cells, or most of the cells that are not destroyed in the process of making the slides, show these patterns and it is on the basis that we make the assumption what we are looking at are functional compartments within a single cell.

Parker - St. Louis: I think there is a point that might well warrant clarification. The phosphorylation experiments that Dr. Wedner was describing involved phosphorylating the intact cell, separating out the nuclei and then looking at the residue which would be plasma membrane and cytoplasmic proteins. Were the experiments you were describing with a single cell?

Hadden - New York: Yes, and the proteins analyzed were those isolated from nuclei.

Parker - St. Louis: Then you looked at the nuclei, where Wedner was looking at presumably the cytoplasmic and plasma membrane fractions.

Hadden - New York: Mr. Chairman, if I might match a question with a question, we have seen effects of PHA on cAMP levels in relationship to time and dose response, but we do not consistently see increases. Something is going on overtime and we see a relationship to dose response not only at high concentrations of mitogen but in an optimal mitogenic concentration range. So that while I do not believe that cAMP is involved with the induction of proliferation, I do believe that something is going on with cAMP. The question is, of course, what? Your studies with the membrane localization would suggest that PHA is patching and carrying with it whatever enzyme is binding the cyclic nucleotide. Do you think that the localization process could be patching and somehow cAMP is involved with the patching process; and do you think on the other hand, that the high dose mitogen inhibition that one sees might be attributed to increases in AMP?

Parker - St. Louis: Well, that is a very complicated question. However, if you continue to look at the cAMP localization, it does not go on and cap. So it is not a simple matter of accumulation of proteins with cAMP moving with them. Rather it seems to be associated with the cross-linking of a local accumulation but then the cAMP does not seem to move with it.

Lipsky - Bethesda: I am wondering if you have entertained the possibility that the long term irreversible effects of ouabain may in fact not be on the lymphocyte per se but rather on perhaps a monocyte? The reason I ask this is, of course, that primary effects noted are decrease in stimulatory capacity of these cells in the MLR which Greineder and Rosenthal have shown that stimulating cell in the MLR is a monocyte. Also the decrease in antigen responsiveness also may be a monocyte dependent phenomenon since they function as antigen preventing cells.

Kaplan - Ottawa: Thank you for raising that question. Pamela Wright, in unpublished work showed that when she depleted the lymphocytes of macrophage and in such a way as to reduce practically to zero the mixed lymphocyte reaction and then added back macrophages and she restored the reaction. If the macrophages had been treated for 48 hours with ouabain, there was no effect whatever on them. Indeed our chairman, Alan Rosenthal, has reported both at the last meeting and the previous one, that macrophage function and presentation and concentration of immunologically relevant antigen by them is resistant to ouabain treatment. So we do not think it is on the macrophage.

SECTION V

COUPLING OF CYTOPLASMIC AND NUCLEAR EVENTS TO MEMBRANE SIGNALS

Chairman

HERBERT L. COOPER

National Institutes of Health
Bethesda, Maryland

Co-Chairman

MICHAEL R. QUASTEL

The Soroka Medical Center
Ben Gurion University of the Negev
Beer-Sheba, Israel

COUPLING OF NUCLEAR AND CYTOPLASMIC EVENTS TO MEMBRANE SIGNALS

Herbert L. Cooper
National Institutes of Health, National Cancer Institute,
Laboratory of Pathophysiology, Bethesda, Maryland 20014

CHAIRMAN'S INTRODUCTION

The accumulated research of nearly ten years has revealed a large number of structural and biochemical changes that occur during lymphocyte activation and blastogenesis. We can safely assume that all of these events are direct or indirect consequences of the initiating mitogenic signal: the combination of a surface receptor with a mitogen, or the modification of surface components by other means to produce the activating surface configuration. We can also assume that future work will provide us with still more additions to the list of observed phenomena, and will refine our understanding of already known events. What we must now consider is how these many events are inter-related so as to produce, on the one hand, the smooth transition from small resting lymphocyte to large lymphoblast and thence to mitosis; and on the other, the evolution from a quiescent cell to one expressing a variety of differentiated cell functions such as production of immunoglobulins and/or effector molecules. In particular, how does the surface binding event act as an initiator for the host of well-ordered biochemical events we call blastogenesis?

The answer to this question must be, for the present, highly speculative. In biology, it is much easier to prove that two phenomena occur than to prove that one causes the other. What can we say, then, about 30 or 40 phenomena?

One helpful factor which has become evident during the past five years is that the ability of lymphocytes to shift from G_0 to enter the mitotic cycle in response to an external signal is not biologically unique. Indeed, in virtually every experimental system that has been used, biochemical events resembling those in mitogen-treated lymphocytes have been described (Table I). Details differ, related to tissue and possibly species specificity, but the broad outline is remarkably similar. We can use information from these and other experimental systems to supplement data derived from lymphocytes in formulating our speculations.

In attempting to provide a causal framework to relate all of the events which occur following mitogen treatment we must keep in mind the following observations.

TABLE I
EXPERIMENTAL SYSTEMS SHOWING $G_0 \rightarrow G_1$ TRANSITION IN RESPONSE TO EXTERNAL SIGNALS

Cell Type	Signal
Lymphocyte	Lectins, Ag, etc.
Density-inhibited fibroblasts	Serum (factors) Virus (SV-40, polyoma).
Liver	Partial hepatectomy Hepatotoxins
Uterus	Estrogens
Mammary Gland	Estrogen Prolactin Insulin
Kidney	Contralateral nephrectomy
Salivary gland	Isoproterenol

Cells in the population may not all be activated to reach the same end-point. Some may be stimulated to enter DNA synthesis, others to produce effector molecules, or both. Indeed, it is seldom considered that the response of some cells to mitogen may be death, which may play a significant role in the response of the remaining cells. Therefore, the many events we can measure may not all be going on inside every cell. Some of the events may have no normal physiological role, but may be related to in vitro conditions and the use of lectins as mitogens.

With these points in mind, let us look at the array of events which has been described by workers in this field. I have arranged these in the form of a flow diagram, supplying a speculative framework to interrelate them. (Table II). The first observation is that this is clearly a complicated and confusing system. That is not a trivial point. It means that any simplistic scheme to explain lymphocyte transformation is likely to unhelpful, or only of limited value. Such simple schemes are useful in clarifying segments of the larger question, but should not blind us to that larger question.

I have placed in heavy outlines those phenomena for which there is at least one report in print which has not been entirely contradicted. In light outlines are events which, from studies with other systems, we presume to occur, but which have not been directly demonstrated in lymphocytes. If I have omitted the reader's particular favorite phenomenon I apologize for the oversight, and also for any outright inaccuracies. Where evidence exists, or the presumptive case for causality seems overwhelming, I have connected two events by a heavy arrow. Where a causal relationship is based on informed speculation or outrageous guesswork, I have placed a thin arrow. It will be observed that there are few heavy arrows, which demonstrates the prematurity of this effort. But this exercise may serve to make clear the directions in which future work must lie in order to convert the light outlines and arrows to more solid ones. Space will not permit me to discuss the various phenomena and relationships shown here, each one of which represents a large part of the research career of some investigator and could be expanded in detail to furnish a complicated chart of its own. I will merely consider what the chart shows about the subject of this session: the coupling of membrane events with cytoplasmic and nuclear ones. On the left are various membrane-related events that have been described and which seem to follow directly from the mitogen binding event. The remainder of the chart shows nuclear and cytoplasmic events. Careful examination shows only three connections between these two groups of events. One is from the vague catch-all of "membrane modification" by a tortuous path to activate membrane transport sites. This direct connection is really a confession of ignorance that we must hope will soon be remedied. What are the membrane modifications involved? What characterizes the inactive state of the membrane transport sites (these include sites for active transport, facilitated diffusion, and possibly, passive diffusion)? How does the membrane modification alter the inactive site and so activate it?

A second connection leads from increased pinocytosis, through release of lysosomal enzymes to proteolytic events. In fact, the connection with primary membrane events is very shaky here, and it is not all certain that increased pinocytosis is an immediate consequence of mitogen binding. It may be a secondary phenomenon, based upon prior cytoplasmic changes in energy, electrolyte, and phospholipid metabolism.

The only other connection the present state of knowledge leaves us is through the cyclic nucleotides. Previous speakers have dealt with this area, and I will not dwell further on it. What is needed in this area is a more detailed picture of the minute-to-minute changes in the production and levels of these nucleotides during the mitotic cycle in lymphocytes. Similar work with density inhibited fibroblasts by Seifert suggests that a simple assessment, such as "elevation of cGMP and lowering of cAMP" will not suffice to describe the essential changes which take place.

I will conclude by emphasizing that the similarity of findings between lymphocytes and other growth-regulated systems, together with the large amount of information already accumulated with the lymphocyte system, makes lymphocyte transformation an excellent model for the study of cellular regulatory mechanisms in general. The tendency of immunologists to regard lymphocyte transformation as an immunological phenomenon and to overlay the biochemical events with a miasma of immunological complexity, has tended to limit the use of the lymphocyte as a tool for molecular and cellular biologists. Similarly, the importance of the lymphocyte as a model for the study of cell differentiation and the nature of cell-cell interactions has been underestimated because of stress on immunological aspects. It is to be hoped that this tendency, which has been mirrored in the evolution of the Leukocyte Culture Conferences over the past nine years, will at least be modified in the future.

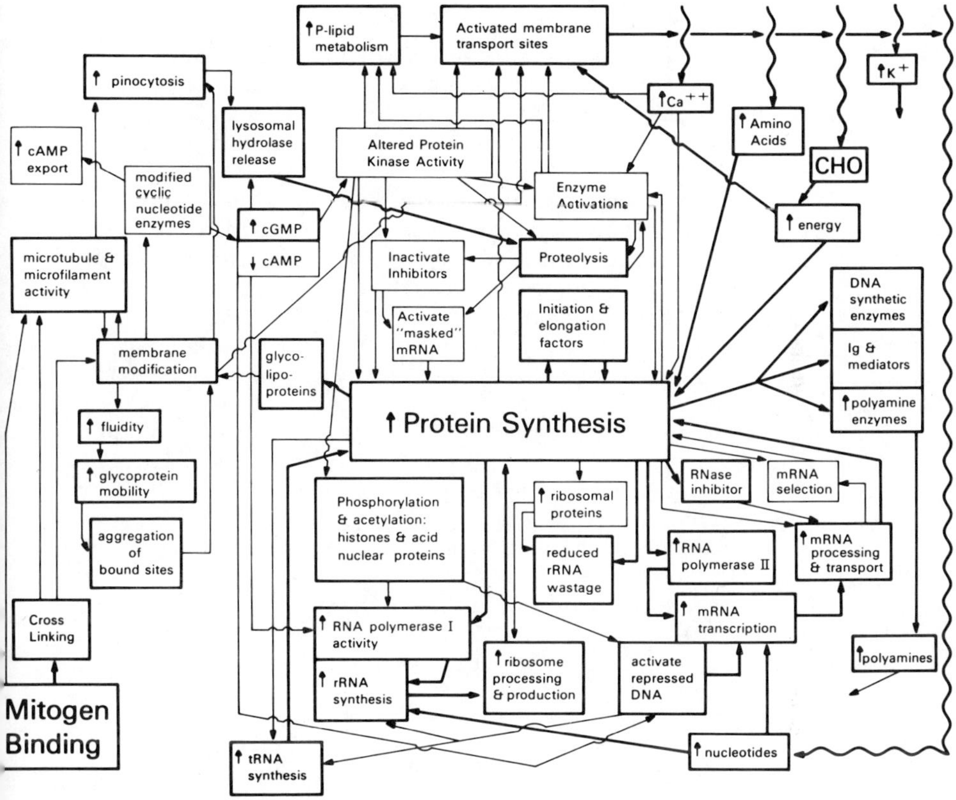

Fig. 1 Flow diagram of known biochemical phenomena associated with lymphocyte activation. Phenomena in heavy outlines have been reported by at least one publication that has not been contradicted. Phenomena in light outlines are inferred from other systems. Heavy arrows indicate a strong probability of causality. Light arrows indicate speculations as to causality.

THE ROLE OF DIVALENT CATIONS IN THE INITIATION OF LYMPHOCYTE ACTIVATION

Vernon C. Maino*, N. Michael Green, and Michael J. Crumpton
National Institute for Medical Research, London, England

It is generally agreed that transformation of lymphocytes by antigens and by nonspecific mitogens (lectins and anti-immunoglobulin sera) is a result of the binding of the mitogen to specific receptors on the cell surface (1-3). Elucidation of the mechanism by which the binding of these effectors to the plasma membrane is translated into lymphocyte activation, is one of the central issues in cellular immunity. Most approaches to this question have been concerned with analyzing the earliest detectable biochemical responses following stimulation. These events include the redistribution of the surface membrane receptors (1,4,5), the enhanced uptake of metabolites (6-9) and ions (9-13), increased cyclic GMP (14) and increased metabolism of phospholipids, especially phosphatidylinositol (15-18). Although the sequence, relative importance and control of these processes are not yet defined, several lines of evidence are consistent with a primary role for Ca^{++}. Since calcium ions have a profound effect on a variety of metabolic and structural functions (19), it has been suggested that intracellular fluctuations of this divalent cation are intrinsically involved in the early events of lymphocyte transformation. Of additional importance is the actively maintained low intracellular concentrations of Ca^{++}. In animal cells the concentration in the cytosol is in the range of 1.0×10^{-6}M to 1.0×10^{-8}M, regardless of the Ca^{++} concentration in the external medium. Increases in this level usually lead to marked changes in cellular activity (19-21). More direct evidence showed that PHA initiated a slow accumulation of Ca^{++} by lymphocytes, and that PHA-stimulated DNA synthesis was inhibited by citrate (22), EDTA (22), or EGTA

Abbreviations used in this paper:

- EDTA: ethylenediamine tetraacetic acid
- EGTA: ethylene glycol bis (β -aminoethyl ether) -N-N'-tetraacetic acid
- LPS: Lipopolysaccharide
- PHA: *Phaseolus vulgaris* phytohemagglutinin
- PI: Phosphatidylinositol

---

*Vernon C. Maino, Ph.D., Allergy & Clinical Immunology Division, National Jewish Hospital and Research Center, 3800 East Colfax Avenue, Denver, Colorado 80206

(23). The inhibition by citrate or EGTA was reversed by Ca^{++}, but Zn^{++} and Fe^{++} were also required for reversal of inhibition by EDTA. It was also shown that binding of PHA was not prevented by the low concentrations (1.4 mM) of EGTA and EDTA that were used for inhibition (24). Furthermore, amino acid uptake was inhibited by EGTA and this inhibition was reversed by Ca^{++} but not Mg^{++} (24). The importance of Ca^{++} on lymphocyte activation was further demonstrated by experiments which show that EGTA does not inhibit transformation when added 16 hours after mitogenic stimulation, and DNA synthesis continues unimpeded in its presence for several days (13). This is approximately the same length of time necessary for mitogens to be bound by surface receptors to irreversibly activate lymphocytes (25,26).

These experiments demonstrated that Ca^{++} entered the cell and that external Ca^{++} was essential for PHA-induced RNA and DNA synthesis, but it was not clear that an influx of Ca^{++} was the only, or even the main, immediate consequence of the binding of PHA.

The recent characterization of an ionophore, A23187 (27,27a) which is specific for divalent metal ions has enabled us to examine more directly the role of Ca^{++} in the initiation of lymphocyte transformation (28). The ionophore is a monobasic carboxylic acid (molecular weight: 531), two molecules of which form a lipid soluble complex with the cation, and thereby facilitate the diffusion of the cation across biological membranes. Thus the ionophore catalyzes transport by: (a) enveloping the ion at a membrane interphase with a consequent dehydration of the ion; (b) diffusing across the membrane as a cation complex; (c) releasing the ion which concomitantly rehydrates at the opposite interphase; and (d) diffusing back uncomplexed to the original interphase to complete the catalytic cycle (29). Ionophores are thus useful models for illustrating dynamic properties of membrane-bound carriers as well as probes for determining the molecular basis of ion discrimination in biological systems.

In the present study we show that A23187 initiates a similar series of events of lymphocyte differentiation to those caused by PHA. We have studied its effect at micromolar concentrations on the morphology of pig lymphocytes, and on the incorporation of (6-^{3}H)-thymidine into DNA.

MATERIALS AND METHODS

Pig lymphocytes were obtained from mesenteric lymph node as previously described (30); about 15% of the cells were stained by a fluorescein-conjugated rabbit antiserum to pig immunoglobulin, and on the basis of these criteria were judged to be B-lymphocytes (M.J. Hayman, personal communication). Assays for the incorporation of ^{3}H-thymidine were performed as described previously (30), using 1 x 10^6 cells in 1 ml of Eagle's medium (containing 1.8 mM $CaCl_2$, 1 mM $MgCl_2$), supplemented with 20% (v/v) fetal calf serum. A23187 (donated by Dr. R. Hamill, Eli Lilly Research Laboratories) was dissolved in ethanol (1.5 mg/ml) and diluted with 0.5 volumes of 10 mM Tris HCl buffer, pH 7.4 or in dimethylsulfoxide (1 mg/ml). Solutions were stored at 2°C in the absence of light, and were used up to 4 weeks after preparation. Purified PHA was purchased from Wellcome Reagents Ltd. Cultures were incubated at 37° in an atmosphere of air containing 5% (v/v) CO_2 for 45 hours prior to addition of ^{3}H-thymidine (1 μCi), and for a further 5 hours before recovery of DNA. The reported values represent the means of duplicate estimations.

Measurement of phosphatidylinositol turnover was determined as described previously (18). Phospholipids were extracted from cell suspensions with $CHCl_3$-methanol (1:2 v/v) according to the method of Kates (31), and separated by chromatography on Wagner papers as described by Kai and Hawthorne (32), and visualized by exposure to I_2 vapor. Radioactive spots were revealed by exposing the chromatograms to x-ray film for one to two days. $^{32}P_i$-labelled PI was identified as a radioactive spot occupying an identical position with a standard sample of PI. Quantitation of radioactivity incorporated into PI was accomplished by removing the area of the PI spot, and counting in Kinard scintillant (33) in a Tri Carb Liquid Scintillation Spectrometer.

RESULTS

Preliminary experiments showed a marked response to low concentrations of A23187 (0.8 μg/ml per 10^6 cells). After 45 hours of culture, the cells were indistinguishable under the light microscope from cells incubated with PHA, and were agglutinated to a similar extent. Examination of electron micrographs (Fig. 1a,1b) revealed that 80-90% of the cells were enlarged, and showed many of the features characteristic of PHA-transformed lymphocytes (62). Thus, the cytoplasm was enlarged and demonstrated marked endocytotic activity. There were many polysomes and an increased number of mitochondria relative to non-stimulated lymphocytes. The nuclei were also enlarged containing dispersed heterochromatin. The plasma membrane was characterized by large pseudopodium-like extensions. No specific lesions of the plasma membrane itself were identified. The principal differences from PHA-treated cells were a lesser development of the rough endoplasmic reticulum and marked abnormalities of the mitochondria, including large

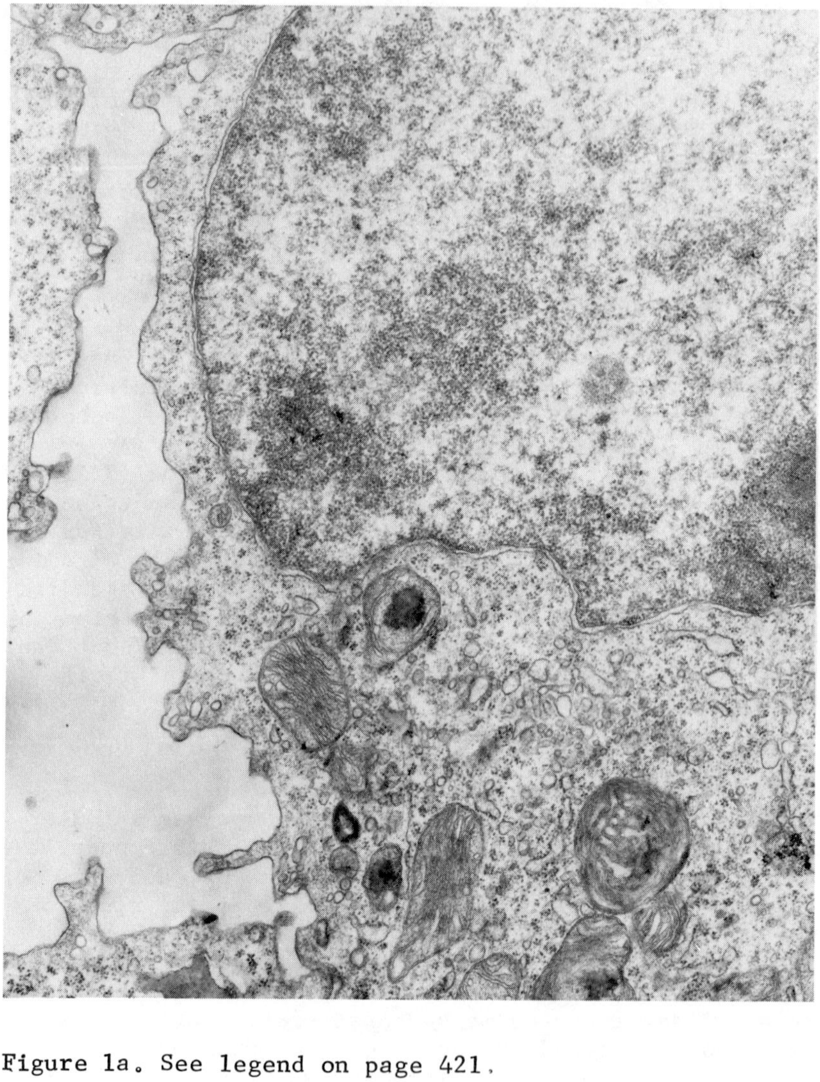

Figure 1a. See legend on page 421.

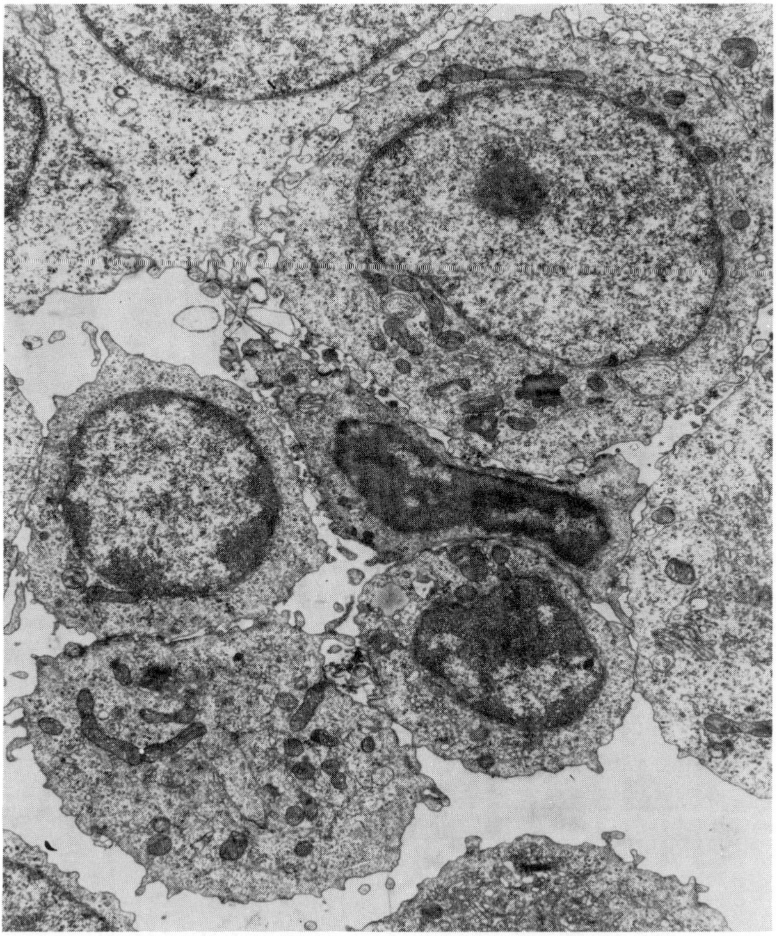

Fig. 1a Typical transformed pig lymphocyte after 45 h in the presence of the ionophore A23187 (0.8 μg/ml). In many cells up to one-third of the mitochondria showed dense granules and/or hypertrophied cristae. In dead cells the proportion of these forms was much higher. Magnification is 30,780.

Fig. 1b Electron micrograph of pig lymphocytes after 45 h in the presence of 0.8 μg/ml of A23187. The greater proportion of lymphocytes (80-90%) resembled blast cells whereas a small proportion (10-15%) were typically unstimulated lymphocytes.

dense granules and hypertrophied cristae in A23187-treated cells. In contrast, with the above enlarged cells, about 10% of the ionophore-treated cells appeared in all respects to resemble normal unstimulated lymphocytes (Fig. 1b). Thus, the possibility that A23187 fails to activate a subpopulation cannot be ruled out at this time. The similarity between the number of B lymphocytes in the total cell population (about 15% as judged using fluorescein-conjugated antiserum to pig immunoglobulin) and the fraction of cells unstimulated by ionophore may be fortuitous but it raises the possibility that the ionophore stimulates only T cells.

To measure quantitatively the stimulus, we followed uptake of ^{3}H-thymidine into DNA after exposure to A23187 for 45 hours. The response to increasing amounts of ionophore is compared with that of PHA in Figure 2. Both PHA and A23187 elicited a sharp increase in thymidine incorporation at critical threshold concentrations. Maximal responses resulted at concentrations of 0.8 - 1.0 μg of A23187 and 2-4 μg of PHA. The dose-response curve for A23187-stimulated cells was much sharper than for PHA at super-optimal concentrations. The absolute amounts of radioactivity varied from one batch of cells to another; however, maximal stimulation by A23187 was usually about one-third to one-fourth of that caused by PHA. Since the morphological evidence suggests 80-90% transformation, it is likely that the lower thymidine incorporation induced by A23187 relative to PHA does not represent a lower proportion of activated cells, but is due to the perturbation of the metabolism of the cell by the penetration of the ionophore into the mitochondria. Thus the lower incorporation ^{3}H-TdR with ionophore is possibly an expression of toxicity and consequent decrease in cell viability and cell number. Further support for a similar mechanism of cellular activation by A23187 and PHA was demonstrated in the experiment shown in Figure 3. The binding of PHA and other mitogens to the surface of lymphocytes is followed by biochemical and morphological changes which appear to follow a predetermined sequence culminating in DNA synthesis which commences after 20 hours and increases to a maximum 48-72 hours after the initial stimulus (1,2). Thus, thymidine incorporation follows a similar time course with the two stimuli.

If both agents act by facilitating entry of Ca^{++} then it should be possible to obtain a response by combining subthreshold amounts of each. The results in Figure 2 showed this effect when the response to A23187 was measured in the presence of 0.3 μg/ml PHA, which by itself had little effect (^{3}H TdR incorporated: 3600 cpm/10^6 cells). This result suggests that largely overlapping populations of cells were

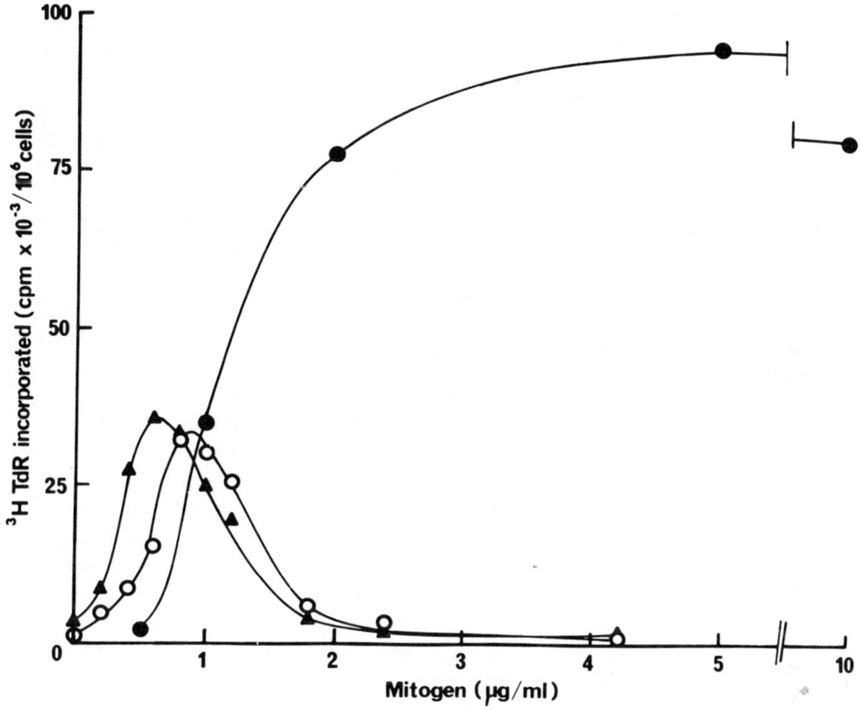

Fig. 2 Incorporation of ^{3}H-thymidine by pig lymphocytes as a function of concentration of A23187 (o) or of PHA (•). The effect of sub-threshold amounts of PHA (0.3 µg) on the response to A23187 is also shown (▲). The assays were performed as described in Materials and Methods using 1 x 10^6 cells from pig mesenteric lymph node in Eagle's medium (containing 1.8 mM $CaCl_2$, 1 mM $MgCl_2$) supplemented with 20% (v/v) fetal calf serum. The reported values represent the means of duplicate estimations.

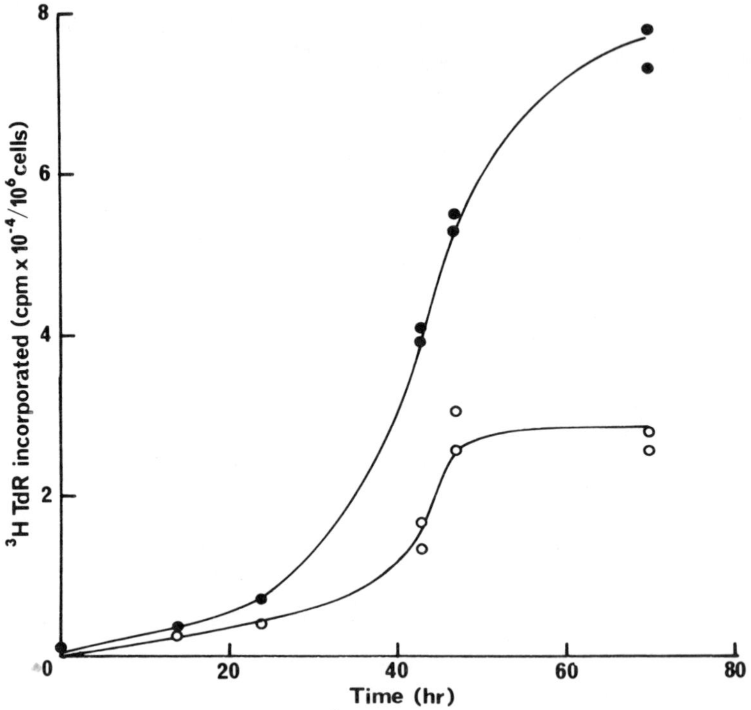

Fig. 3 Time course of ^{3}H-thymidine incorporation by pig lymphocytes in the presence of A23187 (○) and PHA (●). Assays were performed as described in Materials and Methods using amounts of A23187 (0.8 μg/ml) and PHA (2 μg/ml) giving maximal stimulation.

involved in the response to both agents. The converse experiment showing PHA response in the presence of A23187 was complicated by inhibitory effects of the ionophore which will be discussed later.

The critical role of Ca^{++} in both responses was confirmed by lowering its concentration in the medium. Initially the lymphocyte activation response was measured in a basic Ca^{++} free medium supplemented with 10% (v/v) fetal calf serum, instead of the usual level of 20%, to give a final Ca^{++} concentration of 0.3 mM. This lowering of calcium concentration barely affected the response to PHA, but it significantly enhanced the response to the ionophore (Fig. 4). Further lowering of Ca^{++} was effected by addition of sodium citrate at concentrations between 1 mM and 10 mM. The results in Figure 4 show that there was a sharp drop in the response to A23187 when $[Ca^{++}]$ fell below 10^{-4}M. This value approximates the dissociation constant of the complex between Ca^{++} and A23187 (34). The response to PHA was diminished when a slightly lower Ca^{++} concentration was reached (half maximal response at 2×10^{-5}M) in agreement with the results of Alford (23) and Whitney and Sutherland (24). Thus it appears that A23187, like PHA, requires Ca^{++} in the external medium in order to be effective. Addition of 1 mM excess of Ca^{++} over citrate (5 mM and 10 mM) restored the response both to PHA and to A23187 to the level of the controls. Addition of other divalent cations for which A23187 also has an affinity, should inhibit the response if A23187 acts by transporting Ca^{++} into the cell. Figure 5 shows the results of the response of pig lymphocytes to the addition of increasing amounts of Zn^{++} for both PHA and A23187 stimulation. Mn^{++} and Zn^{++} both inhibited the response to ionophore more than that to PHA. With 1×10^{-5}M Zn^{++} the response to PHA was 75% of the control value without Zn^{++}, whereas the response to A23187 was less than 10% of the control. Mn^{++} produced a similar effect, although it was more inhibitory for the PHA response than Zn^{++}. These results are thus consistent with the role of A23187 in divalent cation transport, and a specific role for Ca^{++} in lymphocyte activation.

The inhibitory effects of A23187 on the response to PHA is shown in Figure 6, together with a curve showing the decrease in viability of the cells. Although the two curves show a rough correspondence, it is clear that there is initially a marked inhibition without significant loss of viability. However, above the normal threshold concentration for stimulation by A23187, the activating effect is superimposed upon the declining incorporation of thymidine. It is difficult to explain this peak of activation if a single population of cells,

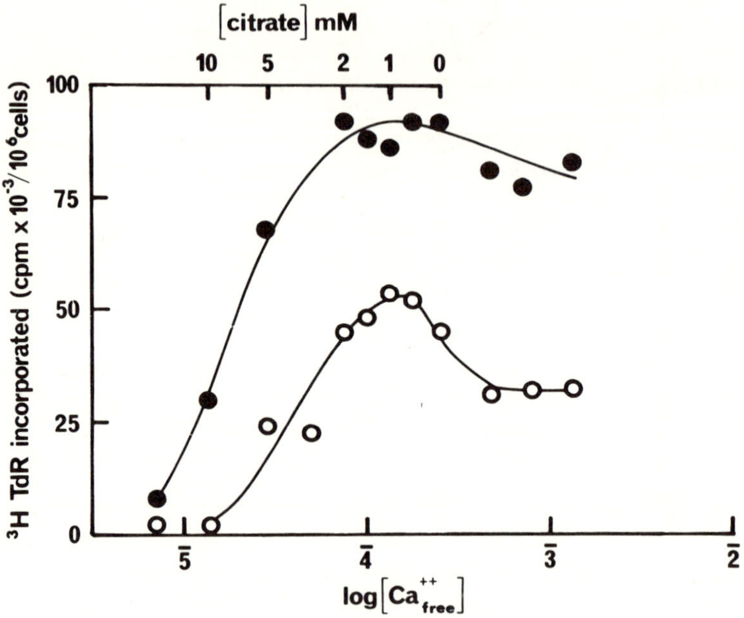

Fig. 4 Effect of concentration of Ca^{++} on the response of lymphocytes to PHA (●) or A23187 (o). The experiments were carried out in fetal calf serum ($[Ca^{++}]$= 3.2 mM, measured spectrophotometrically using murexide (36)) diluted to 1:10 with Ca^{++} free Eagle's medium. Ca^{++} concentration was controlled by addition of Ca^{++} or sodium citrate. In those samples in which Ca^{++} was added it was assumed that 60% was free, the remainder being bound to PO_4 (1 mM, K=1.8 mM) or serum albumin (0.2 mM in Ca^{++} binding sites, K=1 mM). In those samples in which Ca^{++} was complexed by citrate the concentration of bound divalent cation was calculated from the total $[Ca^{++} + Mg^{++}]$ and the concentration of citrate and the dissociation constant K=0.5 mM, which is approximately the same for both ions (34).

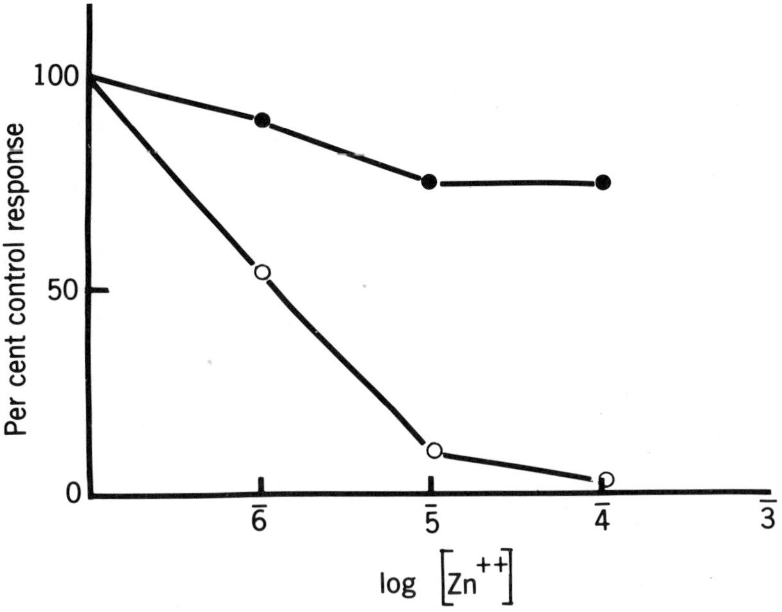

Fig. 5 Effect of concentration of Zn^{++} on the response of lymphocytes to PHA (●) or A23187 (○). The control response represents the stimulation (^{3}H-thymidine incorporation) attained for both mitogens without the addition of Zn^{++}. The assays were performed as described in Materials and Methods in 1 ml of Eagle's medium supplemented with 10% (v/v) fetal calf serum.

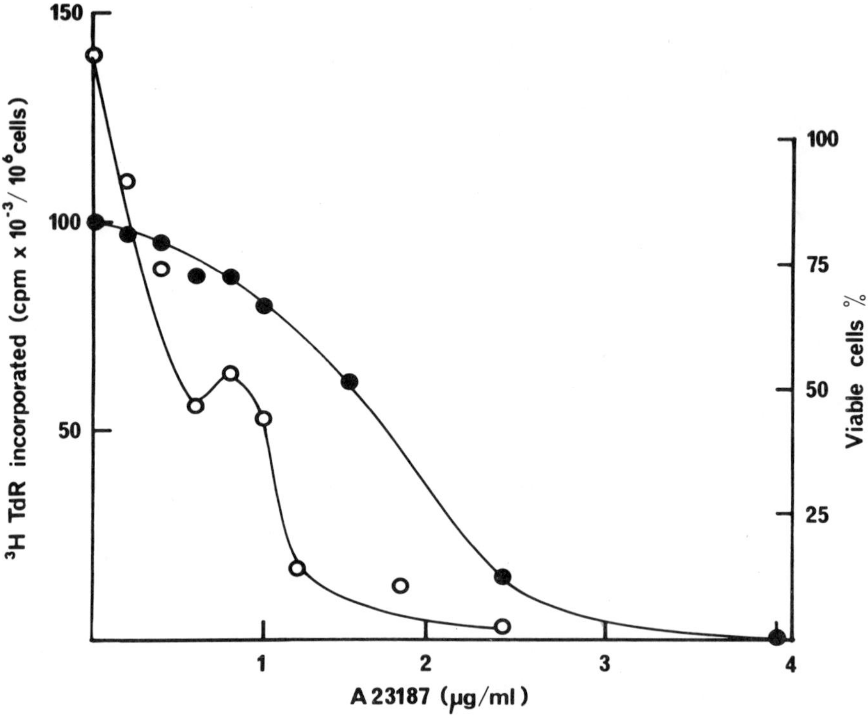

Fig. 6 Effect of A23187 on incorporation of ^{3}H-thymidine into pig lymphocytes stimulated by 2 µg PHA/10^6 cells (○) and on the viability of the cells (●) PHA and A23187 were both added at the beginning of the experiment. The viability after 45 h of incubation with A23187 was measured as the percentage of cells which excluded Trypan blue.

already fully stimulated by PHA is involved, unless A23187 has some effect in addition to facilitating the influx of Ca^{++}. An alternative explanation is that A23187 can stimulate a subpopulation of lymphocytes which are not affected by PHA. This would be consistent with the high proportion of cells which become lymphoblastic under the influence of the ionophore.

It is probable that much of the toxic effect of the ionophore can be explained in terms of its effects on the mitochondria, the morphology of which is grossly disturbed (Fig. 1a). It is well recognized that mitochondria have a large capacity for taking up Ca^{++}, especially in the presence of phosphate and other permeant anions (35), and it is also known that A23187 stimulates this uptake and diverts the energy from both respiration and from exogenous ATP into Ca^{++} transport (36). The excess of internal OH^- exchanges for phosphate and hydroxyapatite precipitates in the mitochondrial matrix. As long as this process continues no energy is available for oxidative phosphorylation. These effects would explain the dense granules which accumulate in the mitochondrial matrix (Fig. 1a) and the decreased incorporation of thymidine which were observed in A23187-stimulated cells (Fig. 2). Since inhibitors of glycolysis and of respiration are known to block transformation (37), it is not surprising that A23187 is also inhibitory. Mouse spleen and rat thymus cells were more sensitive than those of pig to the toxic effects of the ionophore, only 20% surviving after 40 hours at 0.8 µg/ml.

Taken in conjunction with the Ca^{++} transporting role of the ionophore and the known requirement of Ca^{++} for PHA-induced transformation, the above data suggest a simple minimal hypothesis that all events follow from an initial influx of Ca^{++} into the cell. Although we have not looked at a great number of cellular and biochemical events unique to lymphocyte activation, we have established at least by morphological and DNA synthetic criteria that increased Ca^{++} transport (A23187-mediated) is transformatory. In addition, glycolytic activity of rat thymus cells was stimulated by A23187-mediated Ca^{++} transport (M.J. Weidemann and M.J. Crumpton, unpublished observations).

One question that may now come to mind is where in the sequence of events of lymphocyte activation does Ca^{++} influx play its role. We have suggested that the increased Ca^{++} transport is a necessary and sufficient stimulus to elicit the transformatory response and that this response progresses temporally in an identical fashion to mitogen-stimulated lymphocytes. Thus, are there other unique biochemical events which occur during lymphocyte activation which precede, and possibly are necessary for increased Ca^{++} permeability? One early biochemical event in lymphocyte transformation that we have examined in some detail is the enhancement of phosphatidylinositol (PI) turnover (18). Fisher and Mueller (15) have previously reported that PHA rapidly stimulates the selective turnover of $^{32}P_i$ and ^{3}H-myoinositol into PI of human peripheral blood lymphocytes. The enhanced incorporation of P_i into phosphatidylinositol exhibits a number of features which would be expected to be characteristic of an early event in cellular activation. Thus, it is specific, is detected within a few minutes of the stimulus, precedes other biochemical responses such as nucleic acid and protein synthesis, and is apparently localized within the plasma membrane (V.C. Maino, unpublished observations). Previous studies in our laboratory have demonstrated this biochemical response to be specific for mitogenic proteins (18). Furthermore, these results indicated that enhanced PI turnover depends on cross-linkage of surface receptors, and thus is closely related to lymphocyte activation. The experiment shown in Figure 7 demonstrates that the stimulation of turnover of phosphatidylinositol by PHA is dependent upon external Ca^{++}. By lowering the external Ca^{++} concentration (which does not effect PHA binding) with citrate, the enhancement of PI turnover was diminished dramatically. The threshold Ca^{++} concentration for optimal stimulation of $^{32}P_i$ incorporation into PI was 0.1 mM. The inhibition of PI turnover was completely reversed by the addition of 0.2 mM Ca^{++}. Thus, it appears that Ca^{++} is necessary for stimulation of PI turnover by mitogens. The initial transport of Ca^{++} therefore probably precedes this early biochemical event. It is possible that Ca^{++} may directly stimulate the enzyme(s) involved in PI turnover (38), which would explain the immediate response. However, efforts to stimulate PI turnover with A23187 met with only limited success. A23187 at a concentration optimal for blast cell formation and thymidine incorporation stimulated $^{32}P_i$ incorporation into PI only 41% above background, whereas PHA elicited a 600% increase in PI turnover within one hour of stimulation. Furthermore, increasing concentrations of A23187 inhibited the PHA-enhanced $^{32}P_i$ incorporation into PI (Fig. 8). With optimal concentrations of both PHA and A23187 for lymphocyte activation, the PI turnover response was approximately 50% of the response with PHA alone. Thus, the exact role of enhanced PI turnover for lymphocyte activation and its relationship with Ca^{++} transport is unclear at this time. However, it appears that Ca^{++} is necessary for the response. Further insight into the relationship of PI turnover with Ca^{++} uptake will most likely be facilitated when the pathway of PI metabolism, the subcellular locations of

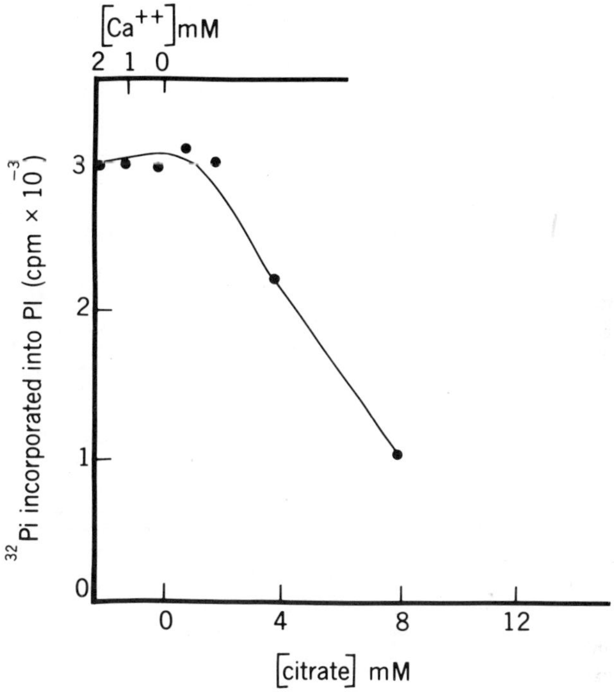

Fig. 7 Effect of concentration of Ca^{++} on stimulation of PI turnover by PHA. 50 μg PHA was added to 5 x 10^7 pig lymphocytes in 5 ml of Ca^{++} and PO_4 free Eagle's medium. The initial medium was 0.1 mM with respect to $[Ca^{++}]$; as a result 0 on the figure represents 0.1 mM Ca^{++}. Incorporation of $^{32}P_i$ into PI was assayed as described in Materials and Methods after one h incubation at 37°C. Values represent the means of duplicate determinations.

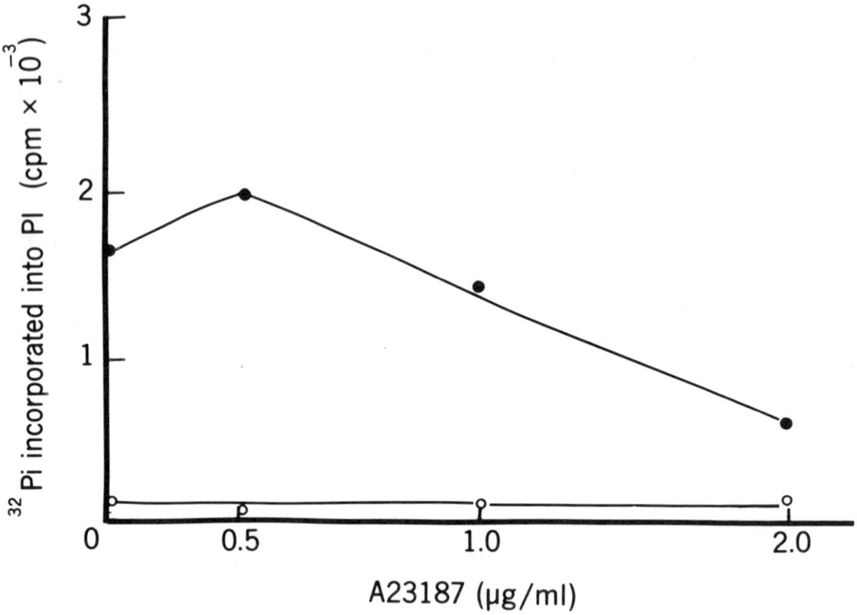

Fig. 8 Effect of A23187 on incorporation of $^{32}P_i$ into PI in pig lymphocytes stimulated by PHA (●) and in lymphocytes with no additions (○). PHA (10 μg/ml) and A23187 were both added at the initiation of the experiment. $^{32}P_i$ incorporation was measured after one hour incubation at 37°C.

the individual steps in the pathway, and the actual step that is affected by PHA are determined.

One would like to know whether Ca^{++} transport as a necessary and sufficient stimulus for thymus-derived lymphocytes is a general mechanism for all cellular differentiating systems. Certainly many intracellular processes are controlled by the concentration of Ca^{++}. At least three of these may be involved prior to DNA synthesis in activated lymphocytes. Activation of phosphorylase kinase (39) is important for the initiation of glycolysis (37); the micro-filament-microtubule system is required for endocytosis (40) and its inactivation blocks transformation (41,42). Thirdly, and possibly most significantly, the level of cyclic GMP in some cells has been shown to be controlled by Ca^{++} (43), and its concentration in lymphocytes rises immediately in response to PHA (14). The secondary effects of cyclic GMP are not yet defined in lymphocytes, but in other tissues it has been implicated in several hormonal responses as a "second messenger". As a number of cellular processes require Ca^{++} for activation (19,21), it is possible that Ca^{++} has a primary role in the stimulation of a variety of differentiating systems. The data we have presented indicate that at least thymus-derived lymphocytes (T cells) from pig lymph nodes are activated by increased Ca^{++} transport (cell preparations from pig lymph nodes are comprised of approximately 85% T cells). The question of whether bone marrow derived lymphocytes (B cells) could be activated by ionophore-mediated stimulation was also explored. Preliminary experiments have utilized mouse spleen cells; however, we have found under our conditions that the ionophore was toxic and failed to mount a mitogenic response at all concentrations tested (Fig. 9). Since A23187 failed to transform 10-15% of the pig lymph node population (the proportion of B cells in this tissue), the possibility exists that B cells are not activated by the same mechanism. However, recently it has been shown that EGTA will inhibit sheep red cell plaques and that this inhibition can be reversed by the addition of Ca^{++} (44). In addition, recent work in our laboratory with mouse spleen cells has revealed a dependency for Ca^{++} for the LPS-induced mitogenic response. The results in Figure 10 show that by lowering the external Ca^{++} with citrate the mitogenic effects of both PHA and LPS were markedly diminished. The addition of 1 mM Ca^{++} completely restored the response to both LPS and PHA. Also both mitogens appeared to have a similar optimal threshold Ca^{++} concentration of about 0.2 mM. Since PHA is thought to be a specific mitogen for T cells and LPS is a specific mitogen for B cells, both populations would appear to have a similar requirement for

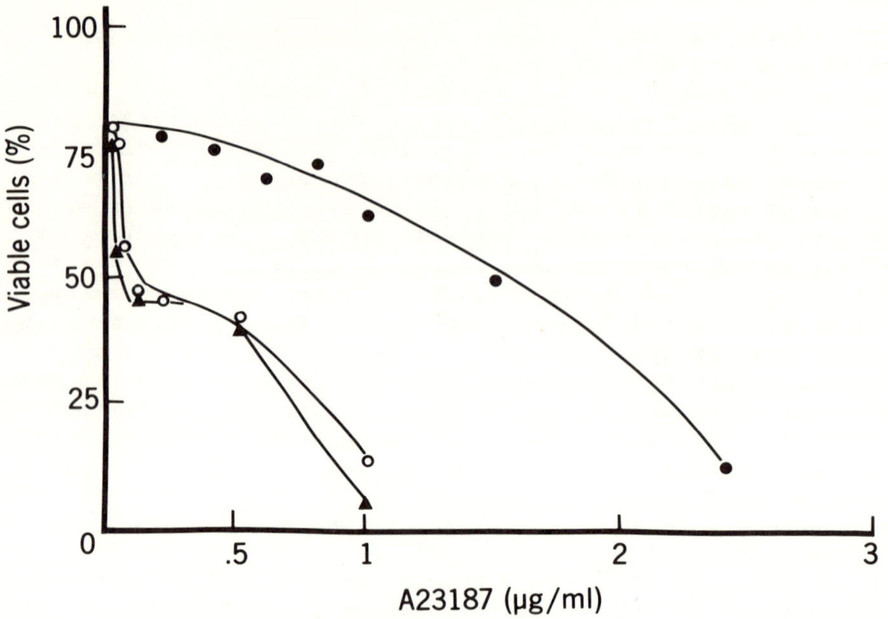

Fig. 9 Effect of A23187 on the viability of pig lymphocytes (●) and mouse spleen lymphocytes in Eagle's BHK medium (○) and RPMI 1640 (▲). The experimental procedure is given under Figure 6.

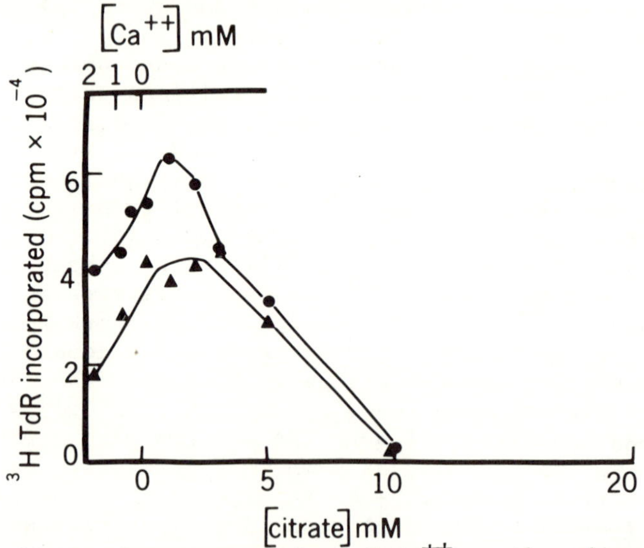

Fig. 10 Effect of concentration of Ca^{++} on the mitogenic response of mouse spleen lymphocytes to 1 μg/10^6 cells of PHA (●) or 10 μg/10^6 cells of LPS (▲). Experimental procedure is described under Figure 4.

Ca^{++} for activation. However, efforts to stimulate mouse B cells with A23187 and Ca^{++} to produce antibody to sheep red blood cells (plaque formation) were not successful (M.J. Crumpton, unpublished observations). Also it was discovered that as low as 0.15 μg of A23187 /10^6 cells depress the LPS mitogenic response by about 60% without any decrease in total number of cells or increase in dead cells compared with the control without LPS (Fig. 11). Furthermore, A23187 (0.3 μg) depressed the LPS induced antibody response by 70%. The results suggest that A23187 either (a) switches off LPS-stimulated cells, or (b) A23187 is cytotoxic for B cells. Although the B cell response appears to require Ca^{++} for activation (LPS-mediated) it is not yet clear whether Ca^{++} alone is a sufficient stimulus in this system as determined by A23187-mediated activation. Part of the answer may be due to differences in sensitivity to the inhibitory effects of ionophore between pig and mouse lymphocytes. Efforts are presently underway to define more clearly the role of Ca^{++} in the activation of B cells.

DISCUSSION

With the exception of the inhibitory properties of the ionophore, the effects of A23187 and PHA are similar. The evidence included: (1) the major morphological changes accompanying transformation; (2) the time course of the response; (3) the enhancement of the response by subthreshold concentrations of PHA, and (4) the inhibition of the response at low Ca^{++} concentration. Assuming the influx of Ca^{++} is sufficient to initiate lymphocyte transformation, and this is the primary consequence of PHA binding, two interesting questions arise: (1) How do mitogens bound to surface glycoprotein receptors influence the permeability of the membrane to Ca^{++} and (2) how does Ca^{++} exert its effect inside the cell?

In answer to the first question, one can postulate two basic mechanisms for Ca^{++} transport across the membrane. Ca^{++} uptake is mainly an energy-independent process, and therefore, its transport is considered to be passive. One concept of Ca^{++} permeability in lymphocytes has been proposed by Whitney and Sutherland (45), which involves the existence of a specific carrier. This proposal is based on saturation kinetics of Ca^{++} uptake and competitive inhibition of Ca^{++} transport by Mn^{++}. Furthermore, it was demonstrated that PHA increased the rate of Ca^{++} influx by decreasing the km for the process, rather than by increasing the Vmax. One candidate that has a possible carrier function for Ca^{++} is phosphatidylinositol. This minor phospholipid in cellular membranes has been shown to undergo increased turnover in different systems when stimulated by

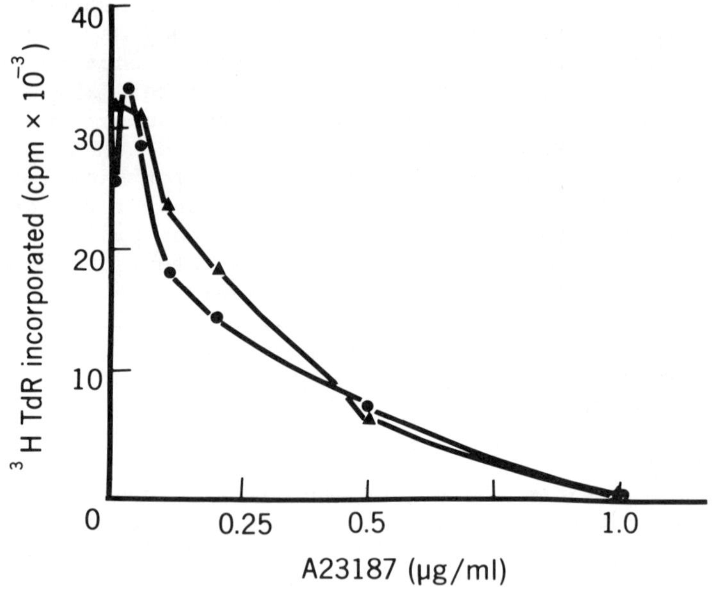

Fig. 11 Effect of A23187 on incorporation of ^{3}H-thymidine by mouse spleen cells stimulated by 2 μg/10^6 cells PHA (▲) or 10 μg/10^6 cells LPS (●). Incorporation of ^{3}H-thymidine by non-stimulated lymphocytes corresponded to background uptake levels when less than 0.02 μg/ml A23187 was added. With concentrations greater than 0.02 μg/ml incorporation was below background.

proper exogenous stimuli (15,18,38,47). Buckley and Hawthorne (46) have furthermore provided evidence that the membrane phosphoinositides may be involved in the regulation of intracellular Ca^{++} in pig erythrocytes. As the amount of phosphoinositide in the membrane was increased, the amount of Ca^{++} bound was also elevated in a 1:1 molar ratio. From the data we have presented it appears that Ca^{++} is necessary for, and possibly activates PI turnover in PHA-stimulated lymphocytes. We have also suggested that PI turnover is an essential event in lymphocyte activation by mitogens (18). However, we were not able to demonstrate significant enhancement of PI turnover with A23187-mediated Ca^{++} transport. Furthermore, A23187 at optimal concentrations for lymphocyte activation inhibited PHA-enhanced PI turnover. Since increased PI turnover is thought to be specific for the mitogenic response, and requires Ca^{++} for activity, the failure of A23187 to stimulate this early biochemical event is puzzling. One explanation may be that Ca^{++} complexed with A23187 in the membrane may not be able to stimulate PI turnover. Alternatively, Ca^{++} uptake and PI turnover are closely linked, and A23187-mediated Ca^{++} transport may bypass the phospholipid turnover requirement for transformation. Lapetina and Michell (47) recently reported a particulate enzyme from rat cerebral cortex catalyzing the formation of cyclic inositol 1, 2 monophosphate (CIP) and inositol 1-phosphate (I 1-P) from PI. This enzyme has been localized in various membrane fractions and can be stimulated by deoxycholate and Ca^{++} to produce increased amounts of CIP and I 1-P from PI. These authors suggested that these breakdown products, especially CIP, might have a "second messenger"-like role in stimulated systems. However, to date there is no direct evidence to support such a role for these products.

As an alternative to a carrier-mediated mechanism for Ca^{++} transport we might propose the existence of polar channels permitting the passage of Ca^{++} into the cell. One of the most striking common features of the action of a variety of mitogens is the requirement for multivalency. This is also true for the release of histamines from mast cells induced by antibodies to IgE, a response which can also be stimulated by A23187 in the presence of Ca^{++} (48). If the receptor proteins span the lipid regions of the membrane then it is plausible to suggest that clusters of receptors might contain polar channels permitting the passage of Ca^{++} into the cell. These channels might be specific for Ca^{++} or they might permit the diffusion of other metabolites. It is well known that lectins do stimulate uptake of other ions, amino acids and nucleotides (6-13), although increased transport of these metabolites may be a secondary consequence of Ca^{++} influx. The argument for a carrier-mediated mechanism for Ca^{++} transport is based upon saturation kinetics for Ca^{++} uptake and competitive inhibition of Ca^{++} transport by other divalent cations. However, these arguments do not necessarily rule out the existence of divalent cation pores which may be formed upon cross-linking of specific membrane receptor proteins by mitogenic stimulants. Such pores would have several ionic sites within the channel that transgresses the membrane. These sites may be occupied by cations and thus become a saturating system. Because of their mutual repulsion positive cations would remain far apart. Thus the number of diffusion sites could greatly exceed the number of ions in the membrane (49), since similarly charged ions must remain a certain distance apart (i.e. several ionic sites along the pore). Such a situation has been found in studies of ionic diffusion across artificial bilayer membranes containing a pore-forming polypeptide, the antibiotic alamethicin (50). In this system, monomers of the alamethicin-cation complex remain on the surface of the bilayer and are non-conducting. Application of a potential drives these monomers into the hydrophilic region where aggregation of six alamethicin molecules produces a conducting channel through the bilayer (51).

Mitogenic stimulants could stimulate Ca^{++} uptake in an analogous fashion by aggregating surface glycoprotein receptor molecules to produce a polar channel for cations. Although both the carrier-mediated and pore-mediated ion transport mechanisms are possible, the PHA-induced pore formation has the attraction of being a simple economical initial step in the transformation process. It is also consistent with the multivalency requirement for lymphocyte activation.

With regard as to how Ca^{++} exerts its effect inside the cell, we can only speculate at this time. However, a number of recent studies with cyclic nucleotides have provided some useful information related to this question. The control of the concentration of cyclic GMP may be of particular interest since it has also been shown that a rise in intracellular concentration of the nucleotide is an immediate consequence of the interaction of mitogens with lymphocytes (14). Moreover, exogenously added cyclic GMP will stimulate RNA and DNA synthesis in mouse spleen lymphocytes (52). The further effects of cyclic GMP in lymphocytes are not known, although in other cells this nucleotide has been implicated in a number of hormonal responses and is clearly an important intermediary in control mechanisms (53). With regard to cellular activation, the proposed regulatory role of cyclic GMP can probably be best understood in context of what is known about cyclic AMP.

Changes in the intracellular concentrations of cyclic AMP are associated with such diverse processes as the regulation of enzyme biosynthesis in bacteria, the transmission of nerve impulses and the inhibition of cell proliferation. However, it is unlikely that bidirectional processes are controlled solely by changes in concentration of cyclic AMP. Goldberg has proposed that cyclic GMP is an antagonist of cyclic AMP in bidirectional processes (54). A feature inherent to this concept is that dual opposing controls can enter into a mutual interaction resulting in synthesis. For example, it has been shown that increased concentrations of cyclic AMP are correlated with contact inhibition of the growth of mouse fibroblasts (55,56). High concentrations of exogenously added cyclic AMP inhibit proliferation and uptake of metabolites. However, when contact inhibited fibroblasts were exposed to insulin they continued to divide with a consequent increase in intracellular cyclic GMP. This increase was also dependent upon the concentration of insulin used (54,57). In addition high concentrations of exogenous cyclic GMP added to mouse fibroblasts were able to overcome the pleiotropic effects of cyclic AMP.

It has also been found that insulin and the cholinergic drug carbamylcholine stimulated intracellular concentrations of cyclic GMP in fat cells and liver slices from rats (57). Carbamylcholine also caused significant elevations in cyclic GMP in human peripheral lymphocytes and rat spleen lymphocytes. However, no such effects could be demonstrated with insulin which is in agreement with the lack of insulin receptors on lymphocytes (63). Additional studies with inhibitors of microtubule assembly suggested that cyclic nucleotides may influence some membrane transport processes by their effects on Colcemid and vinblastine-sensitive structures (58).

Thus if we accept that cyclic AMP and cyclic GMP are in fact playing a vital role in the initiation of the differentiation process, then the understanding of how their intracellular concentrations are regulated in response to external stimuli is of paramount importance. It is intriguing to speculate at this point that the transport of Ca^{++} might directly control the concentrations of these two cyclic nucleotides. The fact that the functions and the metabolism of cyclic AMP and Ca^{++} are closely related has been pointed out in a review by Rasmussen et al. (59). Furthermore, it has recently been shown that Ca^{++} activates AMP phosphodiesterase (60) and cyclic GMP formation (43). Thus the mechanism for PHA-activation of lymphocytes could be explained merely on the basis of increased intracellular concentration of Ca^{++} resulting in increased relative levels of cyclic GMP (guanyl cyclase) and decreased levels of cyclic AMP (phosphodiesterase). In addition, Ca^{++} may regulate other intracellular processes characteristic of lymphocyte transformation. These include activation of phophorylase kinase (39) required for the initiation of glycolysis and the microfilament-microtubule apparatus necessary for endocytosis (40). The sequence of these events relative to changes in cyclic nucleotide concentrations is not yet understood, but it is suggested that these processes may be a secondary response to altered levels of cyclic AMP and cyclic GMP (61).

It is thus apparent that Ca^{++} has a primary role in the signal for initiation of lymphocyte transformation. The mechanism by which Ca^{++} enters the cell as a result of the binding of effector molecules to specific receptors on the cell surface and the process by which Ca^{++} mediates the pleiotropic effects unique to lymphocyte transformation are currently under active investigation.

ACKNOWLEDGMENT

We thank Miss Judy Auger for excellent technical assistance and Dr. J.A. Armstrong for the electron micrographs. A23187 was donated by Dr. R. Hamill, Eli Lilly Research Laboratories, Indianapolis.

REFERENCES

1. Greaves, M. and Janossy, G., Transplant. Rev. 11: 87, 1972.
2. Allan, D. and Crumpton, M.J., Exp. Cell Res. 78: 271, 1973.
3. Fanger, M.W., Hart, D.A., Wells, J.V. and Nisonoff, A., J. Immunol. 105: 1484, 1970.
4. Taylor, R.B., Duffus, W.P.H., Raff, M.C. and de Petris, S., Nature New Biol. 233: 225, 1971.
5. Loor, F., Eur. J. Immunol. 3: 112, 1973.
6. Peters, J.H. and Hausen, P., Eur. J. Biochem. 19: 502, 1971.
7. Peters, J.H. and Hausen, P., Eur. J. Biochem. 19: 509, 1971.
8. van den Berg, J.J. and Betel, I., Febs. Lett. 29: 149, 1973.
9. Averdunk, R., Hoppe-Seyler's Z. Physiol. Chem. 353: 79, 1972.
10. Quastel, M.R. and Kaplan, J.G., Exp. Cell. Res. 63: 230, 1970.
11. Allwood, G., Asherson, G.L., Davey, M.J. and Goodford, P.J., Immunology 21: 509, 1971.

12. Whitney, R.B. and Sutherland, R.M., Cell. Immunol. 5: 137, 1972.
13. Whitney, R.B. and Sutherland, R.M., J. Cell. Physiol. 82: 9, 1973.
14. Hadden, J.W., Hadden, E.M., Haddox, M.K. and Goldberg, N.D., Proc. Nat. Acad. Sci., U.S.A. 69: 3024, 1972.
15. Fisher, D.B. and Mueller, G.C., Biochim. Biophys. Acta 248: 434, 1971.
16. Resch, K., Gelfand, E.W., Hansen, K. and Ferber, E., Eur. J. Immunol. 2: 598, 1972.
17. Ferber, E. and Resch, K., Biochim. Biophys. Acta 296: 335, 1973.
18. Maino, V.C., Hayman, M.J. and Crumpton, M.J., Biochem. J., accepted for publication.
19. Rasmussen, H., Science 170: 404, 1970.
20. Borle, A.B., Fed. Proc. 32: 1944, 1973.
21. Rubin, R.P., Pharmacol. Rev. 22: 389, 1970.
22. Alford, R.H., J. Immunol. 104: 698, 1940.
23. Whitney, R.B. and Sutherland, R.M., J. Cell. Physiol. 80: 329, 1972.
24. Whitney, R.B. and Sutherland, R.M., Biochim. Biophys. Acta 298: 790, 1973.
25. Lindahl-Kiessling, K., Exp. Cell Res. 70: 17, 1971.
26. Novogrodsky, A. and Katchalski, E., Biochim. Biophys. Acta 228: 559, 1971.
27. Hamill, R.L., Gorman, M., Gale, R.M., Higgens, C.E. and Hoehn, M.M., 12th Intersci. Conf. on Antimicrobial Agents and Chemotherapy, Abstr. No. 55 and 56, Amer. Soc. Microbiol., Detroit, 1972.
27a. Chaney, M.D., Demario, P.U., Jones, N.D. and Occolowitz, J.L., J. Amer. Chem. Soc. 96: 1932, 1974.
28. Maino, V.C., Green, N.M. and Crumpton, M.J., Nature 251: 324, 1974.
29. Pressman, B.C., Fed. Proc. 32: 1698, 1972.
30. Allan, D., Auger, J. and Crumpton, M.J., Exp. Cell Res. 66: 362, 1971.
31. Kates, M., in Lab. Techniques in Biochem. and Molecular Biol. (Work, T.S. and Work, E., eds.) North-Holland Publ. Co. Vol. 3: 267, 1972.
32. Kai, L. and Hawthorne, J.N., Biochem. J. 98: 62, 1966.
33. Kinard, F.E., Rev. Scient. Instrum. 28: 293, 1957.
34. Caswell, A.H. and Pressman, B.C., Biochem. Biophys. Res. Commun. 49: 292, 1972.
35. Lehninger, A.L., Carafoli, E. and Rossi, C.S., Advan. Enzymol. 29: 259, 1967.
36. Wong, D.J., Wilkinson, J.R., Hamill, R.L. and Horng, J., Arch. Biochem. Biophys. 156: 578, 1973.

37. Roos, D. and Loos, J.A., Biochim. Biophys. Acta 222: 565, 1970.
38. Lapetina, E.G. and Michell, R.H., Febs. Lett. 31: 1, 1973.
39. Walsh, D.A., Perkins, J.P., Boostrom, L.P., Ho, E.S. and Krebs, E.G., J. Biol. Chem. 246: 1968, 1971.
40. Huxley, H.E., Nature 243: 445, 1973.
41. Yoshinaga, M., Yoshinaga, A. and Waksman, B., Proc. Nat. Acad. Sci., U.S.A. 69: 3251, 1972.
42. Edelman, G.M., Yahura, I. and Warg, J.C., Proc. Nat. Acad. Sci., U.S.A. 70: 442, 1973.
43. Schultz, G., Hardman, J.G., Schultz, K., Baird, C.E. and Sutherland, E.W., Proc. Nat. Acad. Sci., U.S.A. 70: 3889, 1973.
44. Diamanstein, T. and Oldenwald, M.V., Immunology 27: 531, 1974.
45. Whitney, R.B. and Sutherland, R.M., Proc. of Seventh Leukocyte Culture Conference, Daguillard, F., ed., (Academic Press, New York and London, 1973) p. 63.
46. Buckley, J.T. and Hawthorne, J.N., J. Biol. Chem. 247: 7218, 1972.
47. Lapetina, E.G. and Michell, R.H., Biochem. J. 131: 433, 1973.
48. Foreman, J.C., Mongar, J.L. and Gomperts, B.D., Nature 245: 249, 1973.
49. Roy, G., J. Memb. Biol. 17: 199, 1974.
50. Cherry, R.J., Chapman, D. and Graham, D.E., J. Memb. Biol. 7: 325, 1972.
51. Meuller, P. and Rudin, D.P., Nature 217: 713, 1968.
52. Weinstein, Y., Chambers, D.A., Bourne, H.R. and Melmon, K.L., Nature 251: 352, 1974.
53. Kolata, G.B., Science 182: 149, 1973.
54. Goldberg, N., Haddox, M.K., Hartle, D.K. and Hadden, J.W. in Proc. of the Fifth Intl. Cong. of Pharmacol., San Francisco (Karger, Basel, 1973) 5: 146, 1972.
55. Hsie, A.W. and Puck, T.T., Proc. Nat. Acad. Sci., U.S.A. 68: 358, 1971.
56. Prasad, R.N. and Hsie, A.W., Nature New Biol. 233: 141, 1971.
57. Illiano, G., Tell, G.P.E., Siegel, M.I. and Cuatrecasas, P., Proc. Nat. Acad. Sci., U.S.A. 70: 2443, 1973.
58. Kram, R. and Tomkins, G.M., Proc. Nat. Acad. Sci., U.S.A. 70: 1659, 1973.
59. Rasmussen, H., Goodman, D.B.P. and Tenenhouse, A., CRC Crit. Rev. Biochem. 1: 95, 1972.
60. Teo, T.S. and Wang, J.H., J. Biol. Chem. 248: 5950, 1973.
61. Hershko, A., Mamont, P., Shields, R. and Tomkins, G.M.,

Nature New Biol. 232: 206, 1971.
62. Biberfeld, P., Acta Path. Microbiol. Scand. Suppl. 223: 1, 1971.
63. Krug, V., Krug, F. and Cuatrecasas, P., Proc. Nat. Acad. Sci. 69: 2604, 1972.

MICROTUBULES, CYCLIC GMP AND CONTROL OF CELL SURFACE TOPOGRAPHY

Janet M. Oliver
University of Connecticut, School of Medicine, Department of Physiology, Farmington, Connecticut 06032

It has become apparent that the properties of animal cell surfaces are highly correlated with the growth characteristics, metabolic activity and morphology of the whole cell. For example, when cells undergo malignant transformation, we know that changes in membrane properties (increased membrane fluidity (1) and increased mobility of membrane proteins (2)) occur in parallel with loss of contact inhibition of growth and increased agglutinability by lectins (3). Experimental manipulation of the cell surface can also modify the behavior of normal cells. Thus binding of lectins or antibodies to receptors in lymphocyte plasma membranes promotes changes in the topographical distribution of the receptor proteins (4) and also causes dramatic changes in intracellular metabolism and cellular proliferation (5). Similarly, brief exposure of 3T3 fibroblasts to trypsin increases the mobility of membrane proteins (2) and also promotes cell division.

Clearly, it is important to understand the molecular mechanisms that control cell surface topography. Recent studies in our laboratory and others have shown that characteristic changes in surface organization are induced by exposure of cells to exogenous agents (phagocytic particles, lectins and antibodies) and that the surface changes are in turn regulated by structures inside the cell. We shall review studies with polymorphonuclear leukocytes (PMN), fibroblasts and lymphocytes that implicate the cytoplasmic microtubules (MT) as important regulators of surface topography. We will also show that cyclic nucleotides, which are well established as growth regulating substance can modulate the distribution of proteins in cell surfaces. Our data show that 3'5' cyclic GMP (cGMP) antagonizes the stimulation of surface capping of Concanavalin A (Con A) that normally follows MT disruption by colchicine in several cell types. The effect of this nucleotide on Con A capping in PMN from beige (Chediak-Higashi) mice will be described in some detail. Chediak-Higashi (CH) cells show a high degree of capping in the absence of colchicine and this capping is inhibited by cGMP. We propose that MT assembly may be defective in CH cells and that in general, cGMP may promote MT polymerization and thus modify surface topography.

The 'Fluid Mosaic' Membrane

A useful starting point for discussion of membrane organization is Singer and Nicolson's "fluid mosaic" membrane model (6). According to this familiar scheme, biological membranes are composed of a bilayer of lipid in which proteins are embedded. The lipid is fluid so that proteins are free to diffuse and thus to achieve an essentially homogeneous distribution over the cell surface. Frye and Edidin (7) in their now classic cell fusion experiments have presented direct evidence for random mobility of proteins at least under some conditions.

However, there are also some cases in which the movement of surface proteins is clearly not random. A brief review of published studies by Berlin and co-workers, in which the fate of membrane proteins of rabbit peritoneal exudate PMN was followed during phagocytosis, will establish this point.

Effects of Phagocytosis on Surface Topography

During phagocytosis cells internalize extracellular particles by enclosing them with plasma membrane that pinches off to form an intracellular vesicle. This process provides a convenient method to sample the cell surface and to determine whether a particular membrane protein is distributed in a homogeneous or heterogeneous manner. A protein that is distributed homogeneously would be expected to be internalized at random while a protein that is concentrated either in the phagocytic vesicles or in the residual surface membrane must either have a heterogeneous distribution prior to phagocytosis or show segregative movement during phagocytosis. Using this approach we have found that transport carriers for purine nucleosides, purine bases and amino acids are selectively preserved on the cell surface while the glycoprotein receptors for two plant lectins, Con A and Ricinus Communis agglutinin (RCA) are selectively removed from the plasma membrane of phagocytizing cells.

For the transport studies, Tsan and Berlin (8) prepared monolayers of PMN on glass coverslips by incubating cell suspensions for 30 minutes at 37° in a humid chamber (9). After the cells had settled onto the glass, fresh medium containing polystyrene latex beads was added and the cells were allowed to ingest these phagocytic particles. Excess latex beads were then removed by rinsing the coverslips and the initial rate of transport of several radioactive purine compounds and amino acids was measured. Transport substrates were added at concentrations near the Km for transport and incubation times of 45 seconds or less were employed to ensure that transport activity was measured during the initial linear period of uptake. Incubations were terminated by rapid rinsing of the monolayers through four changes of ice-cold medium and the coverslips were digested and radioactivity determined by liquid scintillation counting.

The experimental results (Table I) show that the kinetics of uptake of adenine and lysine are identical before and after removal of a high proportion of the plasma membrane by phagocytosis. It was also found that when lysine transport is blocked by low concentrations of the irreversible inhibitor, p-chloromercuribenzoate sulfonate (PCMBS) before addition of latex beads, then there is no recovery of transport during phagocytosis. Thus, it is unlikely that new carriers are inserted into the membrane to compensate for carriers lost by internalization. This experiment shows that transport carriers are preserved on the outer surface of the cell during phagocytosis. This result is not consistent with the fluid mosaic model which predicts random mobility and by extension random internalization of proteins during phagocytosis.

In contrast with transport carriers, the density of membrane receptors for Con A and RCA in PMN decreases after phagocytosis, indicating selective removal of these glycoproteins (10). PMN were incubated in suspension in the presence of bovine serum albumin (BSA)-coated oil droplets (11) for various times. These phagocytic particles are particularly convenient since they are readily ingested by PMN, excess particles can be removed by washing, and phagocytic vesicles can be isolated from homogenates by flotation. In addition the extent of phagocytosis can be quantified simply by dissolving a dye, Oil Red O, in the paraffin oil used to prepare the phagocytic emulsion and measuring the amount of dye that can be extracted from washed cell pellets spectrophotometrically. After phagocytosis the cells were cooled to 4° and incubated for 30 minutes with ^{125}I-labelled lectins. Excess lectin was removed by washing and the plasma membranes were isolated by sucrose gradient centrifugation of cell homogenates. All steps subsequent to labelling were at 4° since lectin-receptor complexes are readily removed from the cell surface by endocytosis at room temperature and above (see below).

It was found that for both Con A and RCA, phagocytosis was accompanied by a parallel decrease in the specific activity (μg lectin per 100 μg membrane protein) of ligand binding (Fig. 1). That is, receptors for Con A and RCA appear to be selectively removed from the membrane during phagocytosis. As described below, the intrinsic distribution of lectin receptors on cell surfaces is random. Thus, the inclusion of the receptors in phagocytic areas of the membrane indicates

TABLE I
EFFECT OF PHAGOCYTOSIS ON MEMBRANE TRANSPORT IN THE RABBIT ALVEOLAR MACROPHAGE AND POLYMORPHONUCLEAR LEUKOCYTE

Transport Systems	Rate of Transport (pmoles per 45 sec)		P≠
	Control	Phagocytosis	
Alveolar macrophages*			
Lysine	113 ± 27 (6)	126 ± 21 (6)	> 0.2
Adenosine	35 ± 14 (4)	34 ± 14	> .25
Polymorphonuclear Leukocytes+			
Adenosine	11.4 ± 2.1 (3)	10.5 ± .14 (3)	> .2
Adenine	6.0 ± 1.5 (7)	6.3 ± 1.3	> .2

* 0.3 million cells

\+ 2.0 million cells

≠ P values determined from paired differences

From Tsan and Berlin (8)

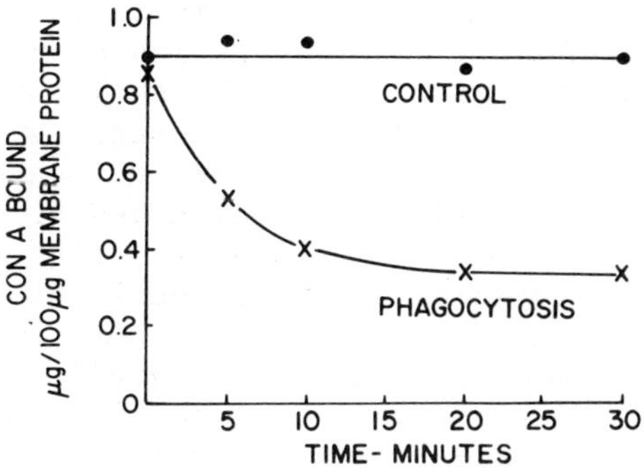

Fig. 1 Effect of phagocytosis on specific activity of CABS of isolated membranes. 2×10^8 PMN were incubated without (control) and with (phagocytosis) emulsion for the times indicated. Cells were then washed, resuspended in modified Hanks' medium containing 7 μg/ml of ^{125}I-Con A and incubated at 4° for 30 minutes. Membranes were isolated by sucrose gradient centrifugation of cell homogenates and the specific activity of bound Con A was determined. In control cells the specific activity is unchanged during incubation at 37°. In contrast, when cells are incubated with emulsion the specific activity of lectin binding decreases in parallel with the extent of phagocytosis.

From Oliver, Ukena and Berlin (10).

segregative movement of proteins during ingestion of oil droplets.

Further analysis showed that the population of lectin receptors that remains on the cell surface after phagocytosis is different from the population that is internalized. As noted, if cells are incubated with lectins at 37° the ligand-receptor complexes are removed from the cell surface by endocytosis. Cells were incubated for 30 minutes at 37° with unlabelled Con A at 50 μg/ml and residual surface-bound Con A was removed by post-incubation of the washed cells with 50 mM α-methylmannose. They were then labelled at 4° with ^{125}I-Con A, and the plasma membranes subsequently isolated and counted. The specific activity of these membranes was very low compared with membranes from cells labelled directly with ^{125}I-Con A (compare Table II, lines 1 and 3), indicating essentially complete removal of Con A receptors after binding of Con A at 37°. Similarly incubation of cells at 37° with RCA induces removal of essentially all of the RCA receptors (compare Table II, lines 7 and 9). On the other hand, if cells are preincubated with RCA at 37°, then cooled to 4° and labelled with ^{125}I-Con A, a decrease of about 50% in specific activity of Con A binding is observed (Table II, line 4). This decrease persists after removal of surface-bound RCA with galactose (Table II, line 5) but it is not observed when cells are incubated with RCA at 4°, when no internalization of lectin-receptor complexes can occur (Table II, line 6). In the reverse experiment, preincubation with unlabelled Con A at 37° reduces the subsequent binding of ^{125}I-RCA at 4° by about 60% (Table II, lines 10 and 11), while incubation with Con A at 4° has no effect on ^{125}I-RCA binding (Table II, line 12). Thus, there are at least three populations of lectin receptors on PMN, one specific for Con A, one specific for RCA and one that binds with (and is internalized at 37° by) both lectins. The joint receptor may represent either a single glycoprotein which binds both Con A and RCA or it may represent separate but closely associated proteins.

After phagocytosis, the amount of ^{125}I-Con A or ^{125}I-RCA that binds at 4° is reduced by about 60%. However, the remaining Con A binding sites (CABS) can be removed only by Con A and not by RCA and vice versa (Table III, lines 2 and 4, 11 and 13). This indicates either that the joint receptor is selectively removed during phagocytosis, leaving behind only receptors specific for a single lectin or that two proteins with separate binding specificities on untreated cells have become disassociated after phagocytosis so that lectin binding with one protein no longer promotes internalization of the second protein. In either case there is clearly a change in

TABLE II

EFFECT OF PREINCUBATION OF CELLS WITH LECTINS AT 37° ON SUBSEQUENT LABELLING AT 4°

Line	Incubations - 37° 1	Incubations - 37° 2	Incubations - 4°	μg Labelled Con A per 100 μg membrane protein	Number of Experiments
(1)	MH	none	---	0.89	15
(2)	Con A	none	---	0.03	4
(3)	Con A	αMM	---	0.17	2
(4)	RCA	none	---	0.45	4
(5)	RCA	gal	---	0.04	2
(6)	---	---	RCA	1.00	2
				μg Labelled RCA	
(7)	MH	none	---	6.99	6
(8)	RCA	none	---	0.39	4
(9)	RCA	gal	---	0.41	2
(10)	Con A	none	---	4.13	4
(11)	Con A	MM	---	4.55	2
(12)	---	---	Con A	6.20	2

2.5×10^8 PMN were incubated for two 30' periods at 37° in 25 ml modified Hanks' medium (MH) in the presence or absence of lectins (50 μg/ml Con A; 30 μg/ml RCA). They were washed and incubated further in 40 ml MH with or without competing sugars (50 mM α-methylmannose; 50 mM D-galactose). Cells were collected, washed, suspended in 10 ml MH and incubated 30' at 4° with ^{125}I-lectin (7.5 μg/ml Con A; 10 μg/ml RCA). Approximately 400 to 500 μg of membranes were subsequently collected and counted. The results of Tables II and III are given for a single experiment. Repeat experiments gave the same qualitative relationships but differed in absolute value.

From Oliver, Ukena and Berlin (10)

TABLE III

EFFECT OF COLCHICINE AND PHAGOCYTOSIS ON LECTIN BINDING

Line	Incubations - 37°			μg Labelled Con A per 100 μg membrane protein	Number of Experiments
	1	2	3		
(1)	none	none	---	0.89	15
(2)	Emulsion	none	---	0.34	15
(3)	Emulsion	Con A	---	0.04	4
(4)	Emulsion	RCA	---	0.36	2
(5)	Colchicine	none	---	0.80	15
(6)	none	RCA	---	0.45	2
(7)	Colchicine	RCA	---	0.45	2
(8)	Colchicine	Emulsion	---	0.45	15
(9)	Colchicine	Emulsion	RCA	0.32	2
				μg Labelled RCA	
(10)	none	none	---	6.99	6
(11)	Emulsion	none	---	2.22	6
(12)	Emulsion	RCA	---	0.43	2
(13)	Emulsion	Con A	---	2.61	4
(14)	Colchicine	none	---	6.85	6
(15)	none	Con A	---	4.13	4
(16)	Colchicine	Con A	---	3.62	4
(17)	Colchicine	Emulsion	---	4.06	6
(18)	Colchicine	Emulsion	Con A	2.90	4

As Table II except cells were preincubated with or without colchicine and with or without emulsion in 100 ml phagocytosis medium during the first two 30' incubations.

From Oliver, Ukena and Berlin (10).

the distribution and nature of lectin receptors after phagocytosis that is inconsistent with the concept that surface proteins move at random.

Willinger and Frankel (12) have recently obtained confirmatory evidence from experiments in which rabbit exudate PMN were prelabelled with radioactive iodine and then allowed to phagocytize on oil-BSA emulsion. Isolated phagolysosomes were enriched for radioactivity twenty-fold indicating selective internalization of iodinated membrane proteins.

Microtubules as Determinants of Protein Segregation during Phagocytosis

Further experiments showed that intracellular microtubules are particularly important in regulating the segregative movement of membrane proteins during phagocytosis. Colchicine is a plant alkaloid that at low doses binds with and disrupts microtubules without affecting any other known cellular property. (At high concentrations, it also inhibits nucleoside (13) and sugar (14) transport).

This drug was found to abolish both the selective preservation of transport sites on cell surfaces and the removal of lectin receptors from membranes during phagocytosis. When monolayers of PMN are incubated with 10^{-6}M colchicine followed by phagocytosis of latex beads (15) or oil droplets (R.D. Berlin, J.S. Fera and J.M. Oliver, unpublished observations) then the rate of transport of adenine or lysine into these cells no longer remains constant but instead decreases as a function of the extent of phagocytosis (Table IV). Vinblastine, another plant alkaloid that disrupts MT, also permits random internalization of carriers but lumicolchicine, a photochemically-derived analogue of colchicine that does not interfere with MT integrity, does not affect the normal preservation of transport carriers in the membrane.

Similarly, colchicine but not lumicolchicine abolishes the selective internalization of lectin receptors during phagocytosis (Fig. 2). After drug treatment there is very little difference in specific activity of ^{125}I-Con A or ^{125}I-RCA binding between untreated cells and cells that were incubated with oil droplets. In addition, the residual surface membrane of colchicine-treated postphagocytic cells contains Con A receptors that can be internalized by incubation at 37° with RCA and vice versa (Table III, lines 8 and 9, 17 and 18). Thus, the essentially complete removal of joint RCA-Con A receptors noted in untreated cells no longer occurs after MT disruption by colchicine.

These data indicate that MT are required for directed movement of certain surface proteins in PMN during phagocytosis.

TABLE IV

EFFECT OF PHAGOCYTOSIS BY POLYMORPHONUCLEAR LEUKOCYTES ON MEMBRANE TRANSPORT OF NONELECTROLYTES WITH OR WITHOUT COLCHICINE PRETREATMENT

Transport System	Per cent depression relative to nonphagocytizing monolayers *	
	No colchicine	Colchicine
Adenine	2.5 ± 2.3 (14) P ≠ > 0.9	42.8 ± 5.0 (14) P <.001
Lysine	6.3 ± 3.9 (5) P > 0.8	43.5 ± 6.3 (5) P <.001

* Values are per cent depression ± standard error of the mean. Number of observations in parentheses.

≠ P values refer to significance of depression from control after phagocytosis.

From Ukena and Berlin (15)

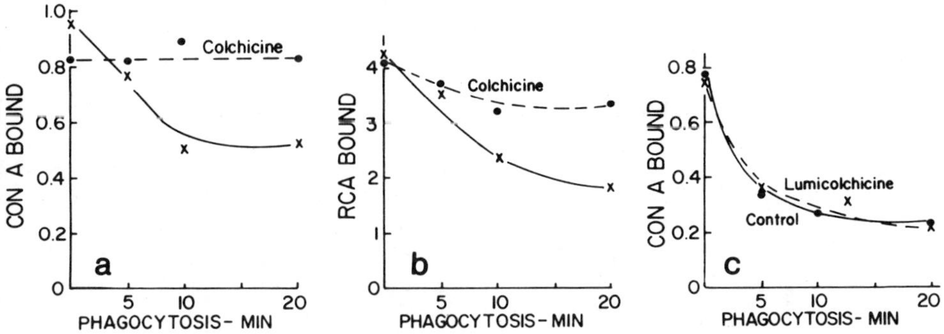

Fig. 2 Effect of colchicine alkaloids on changes in membrane lectin binding sites produced by phagocytosis. Cells were preincubated with 5 μM alkaloid (dashed line) or modified Hanks' medium alone (solid line) for 30 minutes before addition of emulsion and during phagocytosis. (a) Effect of colchicine on CABS labelled with 7.5 μg/ml of Con A. (b) Effect of colchicine on RCA-binding sites. Labelled with 20 μg/ml of RCA. (c) Effect of lumicolchicine on CABS. Labelled with 6.3 μg/ml of Con A. Colchicine abolishes the decrease in specific activity of lectin binding that normally accompanies phagocytosis. This indicates that selective removal of lectin receptors depends on an intact MT system. Lumicolchicine, which does not affect MT integrity has no effect on the segregation of CABS during phagocytosis.

From Oliver, Ukena and Berlin (10).

Disassembly of the MT releases subcellular constraints on movement so that proteins now show random mobility that is reflected by random internalization during phagocytosis. MT can perhaps be viewed as transducers, permitting the cells to phagocytize while preserving essential function such as substrate transport in the proper topographical position (16). Some additional studies of protein and lipid segregation during phagocytosis are described in this publication by Berlin.

We emphasize that the redistribution of membrane proteins during phagocytosis occurs under physiological conditions. Lectin binding at 4° is used simply to titrate changes in receptor distribution and not to induce them. Similarly, in the transport system the only added compounds are transport substrates that are unlikely to have any independent effects on the number or activity of membrane carriers. In Willinger and Frankel's study it is also unlikely that iodination of membrane proteins would directly affect their mobility in the membrane.

The Distribution and Mobility of Con A Receptors on Cell Surfaces

Factors that regulate surface topography have also been studied in systems when ligands are used directly to induce changes in distribution of membrane receptors. In particular, the redistribution of CABS induced by Con A binding to cell surfaces has been studied in our laboratory and others using a range of cell types and a number of different drug treatments. We will review evidence showing that disruption of MT with colchicine promotes redistribution of Con A-receptor complexes into patches and caps in some cells. In addition, we will show that cGMP can modulate this patching and capping, perhaps by stimulating MT assembly.

A variety of techniques are available to study the interaction of Con A (and other lectins and antibodies) with cell surfaces. Ferritin-conjugated Con A can be visualized by electron microscopy on isolated membranes (17) and in thin sections of cells (18); Bernhard and co-workers (19,20,21) have used a Con A-peroxidase technique for labelling thin sections; and hemocyanin labelled Con A has been visualized in three-dimensional replicas of intact cells by Smith and Revel (22) and by Berlin, Karnovsky and co-workers (2,23). In addition, fluorescence-labelled Con A has been extensively used in dynamic studies of receptor movement on cell surfaces (24-28).

These published reports establish an enormous diversity in the distribution of bound Con A on cell surfaces depending on the cell type, the labelling conditions and the presence of drugs. Thus erythrocytes (22) and normal 3T3 mouse fibroblasts (2) show an essentially random distribution of hemocyanin - Con A on surface replicas under most conditions. In contrast to 3T3 cells, SV40 virus transformed 3T3 cells (SV3T3) incubated with Con A at 37° show a patchy distribution of bound ligand, with clumps that are absent from pseudopodia and other peripheral areas of the cell (2,23). Other cells, for example lymphocytes incubated with low (less than 5 μg/ml) concentrations of Con A (4) and SV3T3 fibroblasts (23) will cap only after incubation at 37° with colchicine. We have observed an intermediate distribution of Con A on PMN from human, mink, rabbit and mouse peripheral blood, and rabbit peritoneal exudate. Some caps are seen on direct incubation with fluorescein-conjugated Con A (F-Con A) and the degree of capping increases after colchicine. As noted above, rabbit exudate PMN show very extensive pinocytosis of F-Con A without prior cap formation, while mouse peripheral blood PMN show relatively little internalization except via endocytosis of Con A caps. Monolayers of human peripheral blood PMN as well as 3T3 and SV3T3 fibroblasts also show pinocytic internalization of F-Con A.

A few general properties apply to all cell types. First patch and cap formation appear to result from receptor movement after ligand binding. Cells labelled with lectins at 4° or after fixation invariably show a random distribution of bound ligand (2). Thus the inherent distribution of receptors is random and the surface heterogeneity seen on some cells is due to movement of ligand-receptor complexes.

Metabolic inhibitors such as sodium azide (29) and 2-deoxy-D-glucose (23) also abolish Con A capping, indicating that capping is an active process that is dependent on metabolic energy.

The mechanism of cap formation and the role of MT is not well understood. Edelman and co-workers (4) propose that an intact MT system may serve to anchor Con A-receptor complexes in the membrane. Dissassembly of microtubules releases these restraints on receptor movement so that the ligand-receptor complexes can diffuse in the membrane. According to this model capping results from formation of multiple cross links between the tetravalent Con A and mobile receptors. A similar mechanism was suggested earlier by Berlin to explain the decreased agglutinability of PMN (30) and fibroblasts (3) after colchicine treatment. An alternative proposal is that capping requires whole cell and/or membrane movements (31,32). According to this model colchicine would increase capping by increasing cell motility and surface ruffling.

Con A Capping on Normal and Beige (Chediak-Higashi) Mouse PMN

Some recent studies (28) of the distribution of fluorescein-conjugated Con A (F-Con A) on peripheral blood PMN from two species of mouse, the C57/6J black mouse and the mutant C57/6J beige mouse, have provided some new insights into the mechanism of cap formation and the role of MT. We studied the beige mouse because it seemed likely that the pathological characteristics of its disease could be rationalized as secondary to a defect in MT function. The beige mouse is a homologue of the Chediak-Higashi (CH) syndrome of man (33-37). This is a genetic disease characterized by oculocutaneous pigment dilution, the presence of giant granules in all granule-containing cells and a high incidence of infection due to defective bactericidal activity of leukocytes. Published studies have established that the defect is in the ability of lysosomes to fuse with phagocytic vesicles. CH-PMN phagocytize normally, have a normal content of lysosomal enzymes but discharge their enzymes into internalized vesicles very slowly. The same defect in PMN can be induced in normal PMN by incubation with colchicine to disrupt MT: colchicine-treated cells will phagocytize normally but lysosomal degranulation is severely reduced (38,39). As noted above, incubation of normal PMN with colchicine also causes a large increase in Con A capping, apparently due to release of restraints on receptor mobility. We reasoned that if the same defect in MT function affects degranulation and the diffusion of lectin receptors, then CH cells, which show decreased degranulation, should show increased cap formation after Con A binding. This hypothesis was tested by measuring the ability of normal and CH-PMN to form caps with Con A with and without colchicine.

Purified PMN were allowed to form monolayers on glass coverslips at 37°C. The monolayers were labelled by incubation with F-Con A at 5 μg/ml for 10 minutes, then rinsed thoroughly, fixed for 10 minutes with 2% paraformaldehyde, wet mounted and examined with a Zeiss fluorescence microscope using an FITC interference filter, 500 dichroic mirror, 53 barrier filter and 100 x neofluar objective. Epillumination enabled examination of cells by phase contrast and fluorescence simultaneously. Cells were photographed using Kodak-Tri-X Pan fast black and white film. One hundred cells on each slide were examined at random and assigned to one of three categories according to the observed distribution of fluorescence.

1) Random - showing a diffuse ring of fluorescence
2) Capped - with fluorescence appearing as a polar shell or a more concentrated knob or protrusion

3) Patch - clumped aggregates beneath the periphery of the cell or in larger clusters in the perinuclear region of the cell.

Examples of each labelling pattern are shown in Figure 3. Treatment of labelled cells with 50 mM α-methylmannoside removed labelling of the first two patterns (diffuse or capped). However, aggregates were not removed indicating that the third or patchy pattern corresponds to endocytized Con A that is not accessible to the sugar. Development of the observed patterns appears to follow the sequence 1 to 2 to 3 or 1 to 3 directly (i.e., endocytosis without an intermediate cap stage).

Table V is a summary of the patterns of fluorescence distribution observed on normal and CH-PMN. There is a striking increase in the proportion of cells showing Con A caps and a corresponding decrease in randomly labelled cells among the CH mouse (bg/bg) PMN compared with PMN from normal homozygous (+/+) or heterozygous (+/bg) mice (Line 1). This difference is illustrated in Figures 3a and 3b. Preincubation of monolayers with 1 μM colchicine for 30 minutes greatly enhanced cap formation on normal PMN but had little effect on the extreme degree of capping already present in CH-PMN (Lines 1 and 2). After colchicine treatment the degree of capping in normal PMN was in fact very similar to untreated CH-PMN (compare Figure 3b with 3c). Lumicolchicine (Line 3), had no effect on Con A capping. Thus CH-PMN behave like colchicine-treated normal PMN, indicating that MT function is very likely defective in these cells.

The Effects of Cyclic Nucleotides on Con A Capping on Mouse PMN

A number of published reports have implicated cyclic nucleotides in the control of MT assembly. The results obtained are somewhat controversial, but at least in leukocytes it seems likely that cGMP may promote MT assembly and that cAMP may either inhibit assembly or else have little direct effect on MT. For example, disruption of MT with colchicine inhibits lysosomal enzyme release from PMN that normally follows exposure to zymosan granules or to the C_5a component of complement, and cAMP and agents that elevate intracellular cAMP levels mimic this effect of colchicine (38,39). On the other hand, the MT-stabilizing agent deuterium oxide (D_2O) is reported to enhance lysosomal enzyme release from zymosan or C_5a treated PMN, and cGMP and agents that stimulate cGMP synthesis mimic this effect of D_2O. Similarly, colchicine and cAMP both inhibit histamine release from most cells and D_2O and cGMP enhance histamine release (40). Other studies have shown that binding of phagocytic particles and various ligands

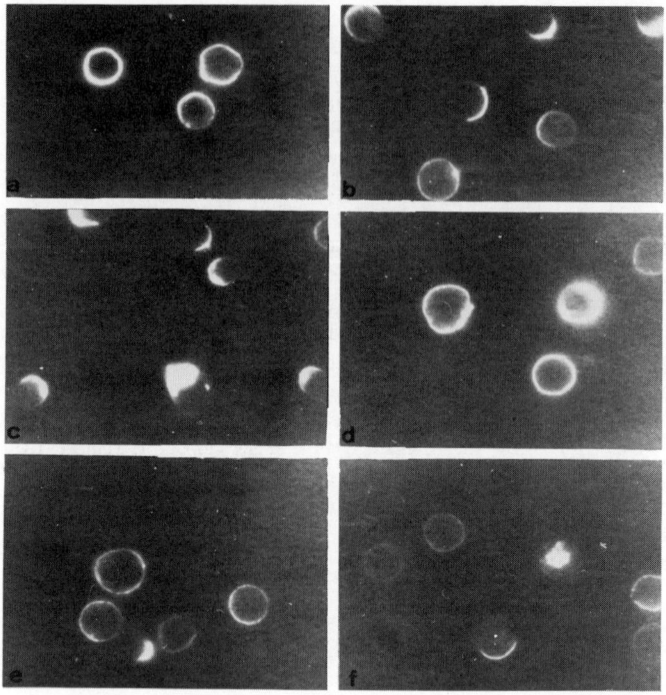

Fig. 3 The distribution of FITC - Con A on peripheral blood PMN from normal and beige (Chediak-Higashi) mice.

a) Normal PMN: the diffuse labelling pattern predominates.

b) CH-PMN: most cells are capped.

c) Normal PMN incubated with 1 μM colchicine: the degree of capping is similar to CH-PMN without colchicine. In one cell the cap has begun to be internalized and fluorescence is seen in a clumpy aggregate.

d) CH-PMN incubated with 0.5×10^{-4}M cGMP: the labelling pattern is diffuse as in the normal PMN.

e) Normal PMN incubated with both cGMP and colchicine: cGMP antagonizes the increased cap formation seen with colchicine alone.

f) CH-PMN incubated with both cGMP and colchicine: most of the label is distributed randomly. One cell shows a central cluster of fluorescence that resembles a central cap. However, this fluorescence cannot be removed by α-methylmannoside and so it most likely represents internalized Con A.

TABLE V

DISTRIBUTION OF FITC - CON A ON PMN FROM NORMAL AND BEIGE (CHEDIAK-HIGASHI) MICE

Line	Preincubation	Postincubation	Normal Per Cent Random	Normal Per Cent Capped	Normal Per Cent Patchy	Chediak-Higashi Per Cent Random	Chediak-Higashi Per Cent Capped	Chediak-Higashi Per Cent Patchy
1	Buffer	None	62	11	27	33	44	23
2	Colchicine (10^{-6}M)	"	23	55	22	32	50	19
3	Lumicolchicine (10^{-6}M)	"	73	11	16	-	-	-
4	c'GMP ($0.5x10^{-4}$M)	"	70	8	22	76	10	14
5	Carbachol (10^{-5}M)	"	87	2	11	72	8	20
6	PMA (10 ng/ml)	"	77	6	17	69	6	25
7	Guanine ($0.5x10^{-4}$M)	"	-	-	-	25	46	29
8	Guanosine ($0.5x10^{-4}$M)	"	-	-	-	30	46	24
9	GMP ($0.5x10^{-4}$M)	"	71	9	20	35	47	18
10	c'AMP (10^{-3}M) + theophylline (10^{-3}M)	"	67	8	25	41	38	24
11	c'GMP ($0.5x10^{-4}$M) + colchicine (10^{-6}M)	"	72	12	16	73	5	22

TABLE V Continued

Line	Preincubation	Postincubation	Normal Per Cent Random	Normal Per Cent Capped	Normal Per Cent Patchy	Chediak-Higashi Per Cent Random	Chediak-Higashi Per Cent Capped	Chediak-Higashi Per Cent Patchy
12	Carbachol (10^{-5}M) + colchicine (10^{-6}M)	None	76	10	14	-	-	-
13	PMA (10 ng/ml) + colchicine (10^{-6}M)	"	75	14	11	-	-	-
14	Buffer	Buffer	70	10	20	-	-	-
15	Buffer	Colchicine (10^{-6}M)	43	38	19	-	-	-
16	Buffer	c'GMP (0.5×10^{-4}M) + colchicine (10^{-6}M)	65	15	20	-	-	-
17	c'GMP (0.5×10^{-4}M)	Colchicine (10^{-6}M)	45	33	22	-	-	-

Cell monolayers were preincubated at 37° in modified Hanks' medium containing 5 mM glucose plus the drugs indicated. After 30 minutes, fresh glucose-free medium containing FITC-Con A (5 µg/ml) plus the same concentration of drugs was added and incubation was continued for 15 minutes. Cells were rinsed and either fixed immediately or postincubated for 20 minutes at 37° with medium containing drugs before fixation. Results were obtained by examining 100 cells at random and assigning each cell to one of the 3 labelling categories (see text) according to the distribution of fluorescence.

All experiments were done using blood from one normal and one CH mouse (3-5 weeks old) and with the same solutions of drugs and FITC-Con A. The data in Lines 1-13 are results from one representative experiment and those in lines 14-etc. from a separate experiment. Some variation was observed between blood obtained from different animals but qualitatively, all effects were the same. The range of capping in absence of drugs varied between 8% and 19% in 11 normal mice and between 36% and 58% in 11 CH mice. In all experiments, about 20% of cells showed a patchy labelling with FITC - Con A.

to animal cells promotes a rapid rise in cGMP levels in the target cells. For example, R.B. Zurier (personal communication) has observed a transient rise in cGMP in human PMN after exposure to zymosan; Hadden and co-workers (41) have demonstrated a rapid increase in cGMP in lymphocytes after incubation with Con A; Cuatrecasas et al. (42) have shown increases in fat cell cGMP after insulin binding; and Rudland et al. (43) found that cGMP levels in 3T3 fibroblasts rise after exposure to fibroblast growth factor (FGF) and phytohemagglutinin (PHA). We (R.B. Zurier and J.M. Oliver, unpublished observations) have also found an increase in cGMP in human PMN after incubation with Con A.

These increases in cGMP are correlated with increases in the numbers of MT that can be seen in thin sections of cells by electron microscopy. Goldstein et al. (44) have observed increased numbers of MT in human PMN after binding of C_5a. Further increases in MT were observed when the cells were also exposed to phorbol myristate acetate (PMA) a cocarcinogen that elevates intracellular cGMP (45). Similarly, Reaven and Axline (46) found a high density of MT and microfilaments adjacent to part of the membrane of macrophages that were in contact with a glass surface or with phagocytic particles. These structures were absent from resting areas of the cell surface.

Thus, data obtained in several different laboratories indicate that MT assembly is induced by binding of phagocytic particles and other ligands with cell surfaces. Increased cGMP synthesis is induced by the same events at the surface and exposure of cells to drugs that stimulate cGMP formation is associated with increases in polymerized MT. These results suggest that the increased cGMP levels resulting from surface events may induce the associated MT polymerization.

We proposed that if elevation of cGMP does in fact stimulate MT assembly, then addition of exogenous cGMP may inhibit Con A capping on CH-PMN.

The effect of cGMP on Con A capping in normal and CH-PMN is listed in Table V. Preincubation of cell monolayers for 30 minutes with cGMP (Line 4) as well as with carbamylcholine (carbachol) (Line 5) and PMA (Line 6) at concentrations that elevate cGMP levels in PMN (39,47) caused a marked reduction in capping in CH cells and had little effect on normal PMN. In fact the degree of capping on CH-PMN after cGMP was virtually identical to the untreated normal PMN (compare Figure 3a with 3d). This effect is chemically specific for the cyclic guanine nucleotide since the potential products of its degradation, guanine (Line 7), guanosine (Line 8) and 5'-GMP (Line 9) did not affect capping. In addition, cAMP had no significant effect on cap formation on either CH or normal cells.

When normal cells were preincubated with cGMP (Line 11), carbachol (Line 12) or PMA (Line 13) simultaneously with colchicine the characteristic stimulation of capping by colchicine (of Line 2) did not occur (Figures 3e and 3f). Thus, elevation of cGMP antagonized the extensive capping of Con A that normally occurred on both untreated CH-PMN as well as on colchicine-treated normal PMN.

Finally, the effect of cGMP is readily reversible. Cap formation on normal PMN was enhanced by incubation with colchicine both before (Line 2) and after (Line 15) labelling with F-Con A and was antagonized by the simultaneous presence of cGMP in both cases (Lines 11 and 16). However, cells incubated with cGMP before labelling and with colchicine after labelling (Line 17) showed a similar increase in capping to cells exposed to colchicine alone.

Effects of cGMP on Con A Capping in other Cells

It seems likely that cGMP inhibition of Con A capping may be a general effect. Experiments with human PMN show that normally only about 12% of cells cap during a 4 minute labelling with 5 μg/ml F-Con A. Preincubation with 1 μM colchicine increases the proportion of capped cells to 66%. cGMP (0.5 mM), Carbachol (10^{-5}M) and PMA (10 ng/ml) had no direct effect on capping but they reduced the colchicine stimulation to 41%, 38% and 22% respectively.

We have also observed changes in the distribution of F-Con A on SV3T3 fibroblasts after incubation with PMA and carbachol. As noted above, surface replica studies with hemocyanin - Con A showed that on SV3T3 cells Con A - receptor complexes move into small patches that are drawn away from the pseudepodia and other peripheral areas of the cell during incubation with lectin at 37° (2,23). In contrast, 3T3 cells show a random distribution of hemocyanin-Con A except after brief trypsinization which induces surface clustering. Colchicine (1 μM) induces cap formation of an SV3T3 cell but does not alter the random distribution of ligand-receptor complexes seen on 3T3 cells. These studies suggested that lectin-receptor complexes are more mobile in the transformed cells and that colchicine releases the remaining restraints on mobility, allowing cap formation to occur.

Fluorescence labelling of fibroblasts under similar conditions as in the surface replica studies (preincubation of cells on monolayers with drugs for 30 minutes followed by 5 minutes labelling with F-Con A (25 μg/ml) and 10 minute post-incubation with drugs) gave very similar results. 3T3 cells show an essentially uniform distribution of label on the surface with some patchiness due to internalization of bound ligand. Colchicine (1 μM) rounds up the cells somewhat but does not alter the labelling pattern. Similarly, cGMP, carbachol and PMA have no effect on the distribution of fluorescence although PMA has a striking effect on morphology causing the cells to spread and adhere tightly to the glass surface. These results are summarized in Table VI, Column 1. In contrast, SV3T3 cells show a grainy but uniform distribution of fluorescence and colchicine causes the cells to round up and develop a significant proportion of caps. cGMP alone is ineffective at antagonizing cap formation, perhaps due to the cell's low permeability, but carbachol and PMA both antagonize Con A capping on SV3T3 cells (Table VI, Column 2). PMA is the most effective drug and is also known to be the most active in elevating cGMP levels in animal cells. It also antagonizes the morphological effects of colchicine.

Table VI (Column 3) also summarizes some preliminary data obtained with primary fibroblasts isolated from beige (CH) mouse embryos. These cells are very large, contain many giant granules and do not show contact inhibition of growth. In contrast with 3T3 and SV3T3 they cap spontaneously with Con A, and carbachol and PMA effectively reverse this capping. Thus, the membrane defect seen with PMN from these mice is also expressed in other cell types.

Mechanism of Cap Formation

These observations provide some new information about the mechanism of cap formation. As noted, two hypotheses have been put forward to explain how capping occurs. One proposes that movement of ligand-receptor complexes in the plane of the membrane is the only requirement for capping, the second proposes that whole cell or membrane movements are required. According to the first theory, MT disruption with colchicine promotes capping by allowing free diffusion of membrane proteins. According to the second proposal colchicine stimulates capping via an increase in cell movement and surface ruffling activity of the whole membrane.

Our data lend support to the first proposal. Cyclic GMP and agents that elevate intracellular cGMP are known to stimulate both random and chemotactic movement of PMA (47) and lymphocytes (48). However, they inhibit capping, at least in PMN. Thus, it seems most likely that Con A capping is due to movement of individual ligand-receptor complexes in the plane of the membrane and that cell or membrane movement, though it may influence the final location of the cap, is not essential for cap formation per se.

TABLE VI

DISTRIBUTION OF FITC - CON A ON 3T3, SV3T3 AND BEIGE (CHEDIAK-HIGASHI) MOUSE EMBRYO FIBROBLASTS

Preincubation	3T3 Percent		SV3T3 Percent		CH Percent	
	Dispersed	Capped	Dispersed	Capped	Dispersed	Capped
Buffer	96	4	85	15	52	48
Colchicine (1 μM)	94	6	66	34	52	48
PMA (10 ng/ml)	98	2	89	11	89	11
Carbachol (10^{-5}M)	95	5	81	19	81	19
PMA (10 ng/ml) plus Colchicine (1 μM)	95	5	92	8	75	25

Fibroblasts grown on glass coverslips were preincubated with the drugs indicated for 30 minutes. The preincubation medium was then replaced with fresh medium (phosphate-buffered saline) containing drugs and F-Con A (25 μg/ml). Labelling was for 5 minutes, after which the cells were postincubated with drugs alone for 10 minutes, then fixed for 10 minutes with 2% formaldehyde, wet mounted and examined as described in the text.

Implications

We have established that events extrinsic to the membrane are involved in the redistribution of membrane proteins that accompanies phagocytosis in PMN and in the redistribution of lectin-receptor complexes that follows Con A binding in several cell types. MT appear to be involved in both systems. They are required for the directed movement of proteins into or out of phagocytic areas of the membrane and disruption of MT by colchicine abolishes this process of segregation. On the other hand, MT appear to immobilize lectin-receptor complexes on cell surfaces and MT disruption by colchicine permits movement of these complexes and cap formation of some cells.

We propose that the degree of MT assembly in cells is highly variable, depending on the presence of extracellular stimuli and the degree of the associated cGMP response. Thus, ligand binding and phagocytosis both stimulate MT assembly and this assembly is dependent on a prior elevation of cGMP levels that follows interaction of ligands and particles with the membrane. In CH cells, it seems likely that the cGMP response is lost, perhaps due to some sort of uncoupling between surface receptors on the outer membrane and guanyl cyclase which is probably associated with the inner membrane (43). Thus, CH cells cap extensively with Con A and also show poor lysosomal degranulation.

Incubation of normal PMN with colchicine results in a cell that behaves very much like the CH-PMN. We suppose that the depolymerization of MT by colchicine antagonizes the MT stabilization due to physiological elevation of cGMP. Alternatively, colchicine may uncouple the linkage between membrane receptors and guanyl cyclase. In either case, exogenous cGMP can overcome the effects of colchicine. We assume that both drugs act on the same MT polymer $\rightleftharpoons$ dimer equilibrium, with colchicine favoring the dimer and cGMP the polymer. The effects of cGMP may, of course, be indirect involving a phosphorylation of a regulatory protein for example.

A similar mechanism may operate in fibroblasts. SV3T3 cells, that normally form patches with Con A and cap after colchicine, may show a suboptimal cGMP response so that MT are normally polymerized to a moderate extent and can be extensively depolymerized by addition of colchicine. Normal 3T3 cells which do not cap with colchicine, may show a greater cGMP response after lectin binding, so that MT polymerization is extensive and cannot be antagonized by colchicine. In support of this postulate, Rudland and co-workers (43,49) have shown that 3T3 cells show large changes in cGMP depending on the growth conditions and the presence of exogenous ligands.

SV3T3 cells on the other hand, appear to have lost the ability to modulate cGMP. The resting levels are quite high but they do not alter under a range of different physiological conditions.

Finally, it seems likely that a critical amount of Con A must bind before cGMP synthesis is stimulated and MT assembly promoted. It is possible that binding of small amounts (less than 5 μg/ml) of Con A to lymphocytes is insufficient to elevate cGMP so that Con A capping occurs as described by Edelman (4). Higher doses of Con A stimulate cGMP synthesis, MT assembly occurs and Con A capping is inhibited.

We should note that in some cells dibutyryl-3',5'-cyclic AMP (dbcAMP) plus theophylline also inhibits Con A capping. For example, Ukena et al. (50) have reported that dbcAMP inhibits patching of Con A on SV3T3 cells and also antagonizes Con A capping due to colchicine. Puck and co-workers (51) have shown that dbcAMP restores normal growth properties to transformed cells. They proposed that this response is due to MT stabilization. However, Yin and Berlin (52) have found that adenine and adenosine can mimic the morphological effects of dbcAMP on SV3T3 cells and like dbcAMP, can antagonize the morphological changes due to colchicine. These effects are seen without any increases in intracellular cAMP levels. It seems possible, therefore, that inhibition of Con A capping by cAMP (which implies MT assembly) may occur indirectly, perhaps via an effect on cGMP degradation or synthesis. According to Gillespie (53) the direct effect of cAMP is inhibition of MT assembly and this is consistent with the colchicine-like effects of cAMP on secretory processes in leukocytes reviewed above.

Experiments are in progress in our laboratory to test whether or not cells that cap readily with Con A (CH-PMN and lymphocytes incubated with low doses of Con A) show lower levels of cGMP and fewer assembled MT in thin sections than cells that do not cap with Con A (normal PMN and lymphocytes incubated with high doses of Con A). In addition, we will measure the effects of Con A binding on guanyl cyclase activity in isolated membranes from these cells to find if a critical concentration of Con A is needed to activate this enzyme and if activation is absent in CH cells.

We have also begun to study the mechanism of membrane-MT interaction and the role of cGMP in MT polymerization using purified tubulin and membranes (54). So far we have been unable to demonstrate a direct effect of cGMP on MT assembly. This may indicate that the cAMP effect on MT assembly is mediated indirectly, perhaps via a phosphorylation reaction catalyzed by a cGMP-dependent protein kinase.

REFERENCES

1. Inbar, M., Shinitzky, M. and Sachs, L., FEBS Letts. 38: 268, 1974.
2. Rosenblith, J.Z., Ukena, T.E., Yin, H.H., Berlin, R.D. and Karnovsky, M.J., Proc. Nat. Acad. Sci. USA 70: 1625, 1973.
3. Yin, H.H., Ukena, T.E. and Berlin, R.D., Science 178: 867, 1972.
4. Edelman, G.M., Yahara, I. and Wang, J.L., Proc. Nat. Acad. Sci. USA 70: 1442, 1973.
5. Hirschhorn, R., in Cyclic AMP, Cell Growth and the Immune Response, eds. W. Braun, L.M. Lichtenstein and C.W. Parker, Springer-Verlag, New York, 1974, p. 45.
6. Singer, S.J. and Nicolson, G.L., Science 175: 720, 1972.
7. Frye, L.D. and Edidin, M., J. Cell Sci. 4: 319, 1970.
8. Tsan, M.F. and Berlin, R.D., J. Exp. Med. 134: 1016, 1971.
9. Hawkins, R.A. and Berlin, R.D., Biochim. Biophys. Acta 173: 324, 1969.
10. Oliver, J.M., Ukena, T.E. and Berlin, R.D., Proc. Nat. Acad. Sci. USA 71: 394, 1974.
11. Stossel, T.P., Pollard, T.D., Mason, R.J. and Vaughan, M., J. Clin. Invest. 50: 1745, 1971.
12. Willinger, M.D. and Frankel, F.R., J. Cell Biol. 63: 373a, 1974.
13. Berlin, R.D., J. Biol. Chem. 248: 4724, 1973.
14. Mizel, S.B. and Wilson, L., Biochemistry 11: 2573, 1972.
15. Ukena, T.E. and Berlin, R.D., J. Exp. Med. 136: 1, 1972.
16. Berlin, R.D., Oliver, J.M., Ukena, T.E. and Yin, H.H., Nature 247: 45, 1974.
17. Nicolson, G.L. and Singer, S.J., Proc. Nat. Acad. Sci. USA 68: 942, 1971.
18. de Petris, S., Raff, M.C. and Mallucci, L., Nature 244: 275, 1973.
19. Bernhard, W. and Avrameas, S., Exp. Cell. Res. 64: 232, 1971.
20. Martinez-Palomo, A., Wicker, R. and Bernhard, W., Int. J. Cancer 9: 676, 1972.
21. Bretton, R., Wicker, R. and Bernhard, W., Int. J. Cancer 10: 397, 1972.
22. Smith, S.B. and Revel, J.-P., Develop. Biol. 27: 434, 1972.
23. Ukena, T.E., Borysenko, J.Z., Karnovsky, M.J. and Berlin, R.D., J. Cell. Biol. 61: 70, 1974.
24. Yahara, I. and Edelman, G.M., Nature 236: 152, 1973.

25. Shoham, J. and Sachs, L., Proc. Nat. Acad. Sci. USA 69: 2479, 1972.
26. Comoglio, P.M. and Guglielmone, R., FEBS. Letts. 27: 256, 1972.
27. Ryan, G.B., Borysenko, J.Z. and Karnovsky, M.J., J. Cell. Biol. 62: 351, 1974.
28. Oliver, J.M., Zurier, R.B. and Berlin, R.D., submitted for publication.
29. Yahara, I. and Edelman, G.M., Proc. Nat. Acad. Sci. USA, 69: 608, 1972.
30. Berlin, R.D. and Ukena, T.E., Nature 238: 120, 1972.
31. Edidin, M. and Weiss, A., Proc. Nat. Acad. Sci. USA 69: 2456, 1972.
32. Unanue, E.R. and Karnovsky, M.J., J. Exp. Med., in press.
33. Gallin, J.I., Bujak, J.S., Patten, E. and Wolff, S.M., Blood 43: 201, 1974.
34. Bennett, J.M., Blume, R.S. and Wolff, S.M., J. Lab. Clin. Med. 73: 235, 1969.
35. Blume, R.S. and Wolff, S.M., Medicine 51: 247, 1972.
36. Davis, W.C., Spicer, S.S., Greene, W.B. and Padgett, G.A., Lab. Invest. 24: 303, 1971.
37. Wolff, S.M., Dale, D.C., Clark, R.A., Root, R.K. and Kimball, H.R., Annals Int. Med. 76: 293, 1972.
38. Zurier, R.B., Hoffstein, S. and Weissmann, G., J. Cell. Biol. 58: 27, 1973.
39. Zurier, R.B., Weissmann, G., Hoffstein, S., Kammerman, S. and Tai, H.H., J. Clin. Invest. 53: 297, 1974.
40. Bourne, H.R., Melmon, K.L. and Lichtenstein, L.M., Science 173: 743, 1971.
41. Hadden, J.W., Hadden, E.M., Haddox, M.K. and Goldberg, N.D., Proc. Nat. Acad. Sci. USA 69: 3024, 1972.
42. Illiano, G., Tell, G.P.E., Siegel, M.I. and Cuatrecasas, P., Proc. Nat. Acad. Sci. USA 70: 2443, 1973.
43. Rudland, P.S., Gospodarowicz, D. and Seifert, W., Nature 250: 741, 1974.
44. Goldstein, I., Hoffstein, S., Gallin, J. and Weissman, G., Proc. Nat. Acad. Sci. USA 70: 2916, 1973.
45. Goldstein, I.M., Hoffstein, S.T. and Weissmann, G., Annals N.Y. Acad. Sci., in press.
46. Reaven, G.P. and Axline, S.G., J. Cell. Biol. 59: 12, 1973.
47. Estensen, R.D., Hill, H.R., Quie, P.G., Hogan, N. and Goldberg, N.D., Nature 245: 458, 1973.
48. Schreiner, G.F. and Unanue, E.R., J. Immunol., in press.
49. Rudland, P.S., Seeley, M. and Seifert, W., Nature 251: 417, 1974.

50. Ukena, T.E., Borysenko, J.Z., Black, P.R., Karnovsky, M.J. and Berlin, R.D., submitted for publication.
51. Puck, T.T. and Jones, C., in Cyclic AMP, Cell Growth and the Immune Response, eds. W. Braun, L.M. Lichtenstein and C.W. Parker, Springer-Verlag, New York, 1974, p. 338.
52. Yin, H.H. and Berlin, R.D., J. Cell Sci., in press.
53. Gillespie, E., in Cyclic AMP, Cell Growth and the Immune Response, eds. W. Braun, L.M. Lichtenstein and C.W. Parker, Springer-Verlag, New York, 1974, p. 317.
54. Becker, J.S., Oliver, J.M. and Berlin, R.D., submitted for publication.

RECEPTOR-CYTOPLASMIC INTERACTIONS AND LYMPHOCYTE ACTIVATION[1]

John L. Wang, Gary R. Gunther, Ichiro Yahara,
Bruce A. Cunningham and Gerald M. Edelman
The Rockefeller University, New York, New York 10021

SUMMARY

The binding of the mitogenic lectin, concanavalin A (Con A), to lymphoid cells can inhibit receptor movement at the cell surface as assayed by patch and cap formation. Conditions that dissociate cellular microtubules, such as low temperature and treatment with colchicine and related drugs, reverse the effect of Con A on receptor mobility. In addition, colchicine inhibits the mitogenic stimulation of lymphocytes by Con A as measured by the incorporation of ^{3}H-thymidine and the appearance of blast cells. Analysis of the activation of human lymphocytes by Con A suggests that the inhibitory activity of colchicine could not be accounted for by diminution in cell viability or by arrest of mitosis at metaphase in stimulated cells. Furthermore, the inhibition of ^{3}H-thymidine incorporation is not due to blockage of thymidine transport or DNA synthesis inasmuch as addition of colchicine has no effect on cells in the S phase of the cell cycle. Kinetic data indicate that colchicine blocks stimulation early in the sequence of events following addition of the mitogen and that the time of inhibition may be correlated with the kinetics of cellular commitment to lectin activation.

INTRODUCTION

The binding of antigens, antibodies, and lectins to lymphoid cells induces a variety of surface and metabolic changes including the stimulation of mitosis. The interactions of these reagents with lymphocytes provide an excellent system for the analysis of receptor-ligand interactions, mitogenic stimulation and growth control. Although the detailed mechanism of lymphocyte stimulation is not known, it is assumed that the surface bound mitogens are responsible for the

[1]This work was supported by U.S. Public Health Service Grants from the National Institutes of Health and by grants from The National Science Foundation. J.L. Wang was supported by a Damon Runyon Cancer Research Fellowship.

[2]Abbreviations used in this paper: Con A, concanavalin A; αMM, α-methyl-D-mannoside.

initiation of cell activation (1,2). No direct evidence has been provided, however, for the presence of a molecule or a system of molecules that links surface receptors to cytoplasmic or nuclear components.

Recent studies on the distribution and mobility of lymphocyte surface receptors and the modulation of receptor movement by lectin reagents suggest that surface receptors may have a common anchorage system within the cell (3-6). It has been proposed that this system involves reversible and cooperative interactions among receptors, microfilaments and microtubules. Such a system may also serve the function of receiving and transducing a variety of external signals to the appropriate intracellular messengers.

In this paper, we discuss experiments designed to correlate the effects of various reagents that modulate both receptor mobility and the stimulation of mitosis in lymphocytes. Our studies have shown that binding of the mitogen concanavalin A (Con A) to lymphocytes can restrict receptor movement at the cell surface. Colchicine and related drugs, which dissociate cellular microtubules, can inhibit the effect of Con A on receptor mobility as well as the stimulation of mitosis in lymphocytes. On the basis of the present studies and those of others, we have formulated an hypothesis to account for the effects of colchicine on the modulation of receptor movement and cell stimulation by Con A.

MATERIALS AND METHODS

The preparation and use of mitogens and antibody reagents have been described elsewhere (3,7,8,9). Human lymphocytes used in the experiments were isolated from the blood of normal donors by density-gradient centrifugation (10). These lymphocytes were cultured in RPMI 1640 medium containing 20% heat-inactivated fetal calf serum. The concentration of Con A used in human lymphocyte stimulation experiments was 20 μg/ml. Murine lymphocytes were from NCS mice (Rockefeller University, New York), and athymic Nu/Nu mice bred on a Balb/C background. These lymphocytes were cultured in medium described by Mishell and Dutton (11). Rabbit peripheral lymphocytes were isolated by the method of Sell and Gell (12) and were cultured in Eagle's Minimum Essential Medium containing 10% fetal bovine serum (11).

Mitogenic responses of lymphocytes were assayed by the incorporation of ^{3}H-thymidine and by the determination of percentage of cells undergoing blast transformation. In order to collect lymphocytes stimulated by Con A at the G_1/S boundary, hydroxyurea (0.5 mM) was added 21 hours after the addition of the mitogen. After waiting another 20 hours, the cells were washed in medium containing 20 μg/ml Con A and finally resuspended in an equal volume of the same medium. The release of the hydroxyurea block at the G_1/S boundary has been shown to be immediate using this washing procedure (7,13). The details of these procedures are described elsewhere (7,14).

RESULTS AND DISCUSSION

Inhibition of receptor mobility by Con A:

The assay used to reveal the effects of various reagents on receptor mobility is that developed by Taylor et al. (15) (Fig. 1). It has been shown that most receptors appear to be distributed diffusely and randomly on the lymphocyte surface. In the presence of multivalent ligands such as antibodies to immunoglobulins, however, the receptors to which they bind can undergo patch formation and subsequent cap formation. Patch formation depends upon the local diffusion of the receptors in the plane of the membrane. Cap formation results when these cross-linked receptor complexes move toward one pole of the cell. This process depends upon cellular metabolism and it appears that microfilaments may be involved although global cell movement or translocation is not required.

Con A strongly affects the ability of cell receptors to form patches and caps (3). If Con A is added at concentrations of 100 μg/ml to lymphocytes at 21° C before treatment with anti-Ig both patch formation and cap formation are inhibited (Table I). This effect depends on the dose of Con A and is reversed by addition of a competitive inhibitor of Con A such as αMM. The inhibition by Con A of receptor movement appears not to be restricted to the surface Ig molecules on B lymphocytes because lectin-binding also inhibits cap formation by a variety of other receptors on both T and B cells: 1) antigen and antibody complexes, 2) the θ antigen of thymocytes, 3) the Con A glycoprotein receptors and 4) H-2 molecules on lymphocytes and fibroblasts.

Several experiments suggest that the ability of the lectin to cross-link receptors may be necessary to induce the modulation (16). Succinyl-Con A is a dimeric derivative of the tetravalent native lectin which competes with Con A for the cell surface receptors. Unlike native Con A, this modified lectin neither forms patches and caps nor does it modulate the mobility of other receptors. If cell-bound dimeric succinyl-Con A is treated with divalent antibodies directed against Con A, however, inhibition of cap formation is again observed (Table I). F_{ab}' fragments of the antibodies to Con A show little or no effect. Thus, the effective

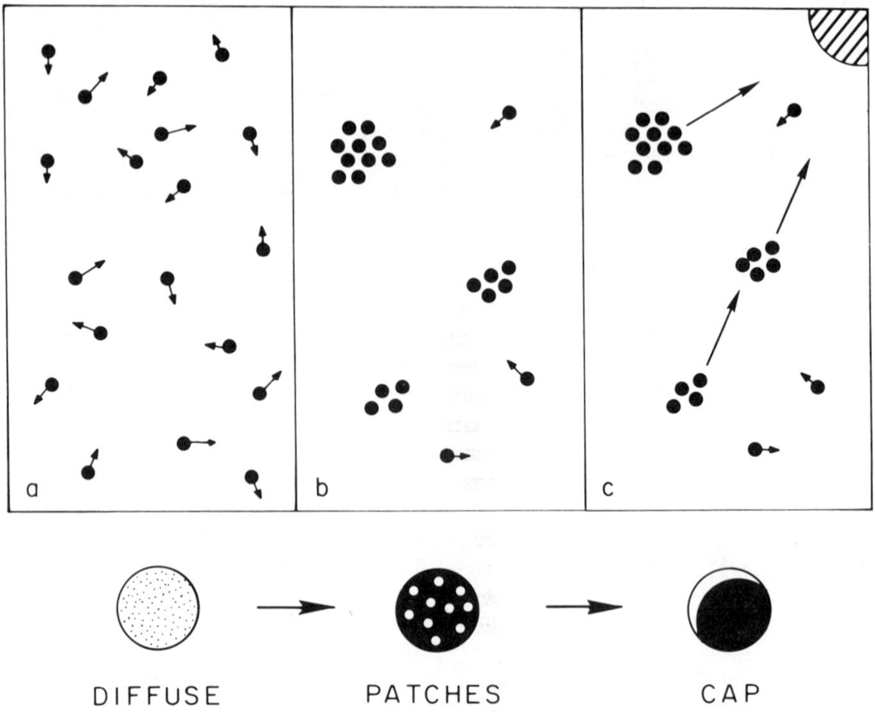

Fig. 1 A schematic diagram of the redistribution of surface immunoglobulin molecules on lymphocytes after binding with anti-immunoglobulin antibodies. The initial random distribution of receptors results in a diffuse labelling pattern. Cross-linking of receptors by the bound ligand leads to cluster or patch formation and coalescence of patches at one pole of the cell results in cap formation.

TABLE I
MODULATION OF RECEPTOR MOBILITY BY CON A AND ANTI-MITOTIC DRUGS[a]

Treatment	% Cap-forming cells
Control	87
Con A	4
Succinyl-Con A[b]	86
Succinyl-Con A + anti-Con A[c]	53
Succinyl-Con A + F_{ab}' anti-Con A	80
Con A + colchicine[d]	30
Con A + vinblastine	51
Con A + vincristine	15
Con A + podophyllotoxin	10
Con A + lumicolchicine	1
Con A + strychnine	1
Con A, 4°, 30 min; 37°, 30 min[e]	30

[a]In order to test for the inhibition by Con A of immunoglobulin receptor cap formation, the percent of cap forming cells obtained with fl-anti-Ig (100 μg/ml) was measured in the presence of Con A (100 μg/ml).

[b]The concentration of succinyl-Con A was 50 μg/ml.

[c]The concentration of succinyl-Con A was 50 μg/ml; the concentration of the antibody reagents was 100 μg/ml.

[d]The concentrations of the various drugs were as follows: colchicine, vinblastine and vincristine, 10^{-4} M; lumicolchicine, 5 x 10^{-4} M; podophyllotoxin and strychnine, 10^{-3} M.

[e]Cells were incubated with Con A at 4° for 30 minutes, washed to remove unbound Con A, and then reincubated at 37° for 30 minutes before cap formation was determined.

valence and possibly the charge of the lectin may play a key role in inducing the inhibition of cell receptor mobility.

Temperature and drug effects on the modulation of receptor mobility by Con A:

A striking feature of the modulation effect is the contrasting results obtained at 4° and 37°. If Con A is applied at 4°, and the cell is washed and returned to 37°, Con A receptors themselves patch and cap (16). In addition, Con A, under these conditions, does not inhibit the movement of any of the other receptors of the cell. This suggested the possibility that proteins capable of dissociation at low temperature such as cytoplasmic microtubules may be involved. Accordingly, a number of reagents known to affect microtubules were employed and were found to reverse partially the inhibition of receptor mobility by Con A (4,17). Colchicine, vinblastine, vincristine, and podophyllotoxin partially reverse the effect of Con A and allow formation of both Con A and anti-Ig caps (Table I). Colchicine and vinblastine are effective at concentrations as low as 10^{-6} M. If the incubation time of the cells is prolonged, the effect with 10^{-6} M colchicine reaches the level obtained with 10^{-4} M drug in about six hours. Neither lumicolchicine, a photo-inactivated derivative of colchicine (18), nor strychnine, which affects erythrocyte membranes (19), showed any effect in the system and the effects of colchicine and related drugs on receptor mobility do not appear to be due to non-specific interactions of these drugs with the lymphocyte membrane. Furthermore, the effects of these drugs are not due to their interaction with Con A. Colchicine does not bind to Con A nor does it cause changes in the aggregation of its subunits. In addition, colchicine does not inhibit the binding of Con A to saccharides or to the cell surface (4).

The results and a variety of additional data (4-6, 20,21) have led to the formulation of a model (4-6) for the modulation of the mobility of cell surface proteins. This model (Fig. 2) assumes that some of these proteins can penetrate the lipid bilayer and through direct or indirect interactions with submembranous proteins, exist in free or anchored states. In view of the effects of cytochalasin B and colchicine on cap formation and its inhibition (5,17,22-24) it is assumed that the basic elements in this system of anchorage are the microfilaments and microtubule-like proteins. Cross-linkage of lectin receptors is assumed to influence the association-dissociation equilibrium of these macromolecular assemblies, shifting their interaction with other receptors or the inner lamella of the bilayer. This provides a system for communi-

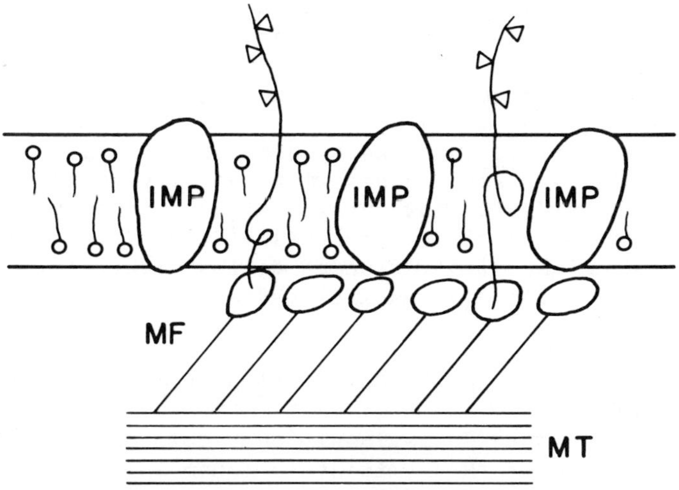

Fig. 2 A hypothetical model for modulation of cell surface receptors by submembranous structures. The model assumes that microfilaments (MF) interact with receptors, possibly *via* a myosin- or spectrin-like structure and that the microfilaments interact in turn with microtubules (MT). The interactions among the various components are assumed to involve reversible association-dissociation reactions. Intramembranous particles (IMP) are not involved in these equilibria.

cating receptor states through the membrane as well as for altering receptor states via intracellular events. Therefore, according to this model, at least four quasi-independent sets of equilibria are possible: 1) cross-linkage of glycoproteins or other receptors with various ligands, 2) interaction of receptors with microfilaments, 3) interaction of microfilaments with formed microtubules, and 4) interaction of tubulin subunits to form the microtubules.

This model implies that the cytoskeleton plays a primary role in modulating receptor behavior and conversely, provides a basis for considering how outside signals interacting with the cell surface may alter the cytoskeleton. To test whether such a system might also influence cell activation, the effect of colchicine on the Con A-stimulated mitosis of lymphocytes has been examined.

Inhibition by colchicine of Con A-stimulated ^{3}H-thymidine incorporation:

It has been observed previously that colchicine inhibits the incorporation of ^{3}H-thymidine in mouse splenic lymphocytes stimulated by Con A (4) and human peripheral blood lymphocytes activated by PHA (25). The degree of inhibition produced by various concentrations of colchicine in the stimulation of both types of lymphocytes by Con A is shown in Fig. 3. At concentrations of 10^{-7} M or above, colchicine reduced the response to Con A by about 70-90% as measured by the incorporation of ^{3}H-thymidine between 48 to 50 hours after the start of the culture. Only a small percent of the lymphocytes present at this time are daughter cells (14,26), so the inhibition by colchicine of the mitogenic response is not due to a simple blockage of the first generation cells at their metaphase.

In the present studies, an inhibitory effect of colchicine has also been observed in three other systems: (a) stimulation of rabbit peripheral lymphocytes by Con A, (b) stimulation of mouse splenic lymphocytes by sodium metaperiodate, and (c) stimulation of athymic nude mice spleen cells by lipopolysaccharide. Because human lymphocytes stimulated by Con A retained a high level of viability under a variety of culture conditions, this system was chosen for further analysis in order to determine the sites of the inhibitory activity of colchicine. In addition, because of the sharp dose dependence between 10^{-7} and 10^{-8} M (Fig. 3), subsequent experiments were performed using 10^{-6} M colchicine so that slight variations in the concentration of the drug would not lead to large changes in the extent of inhibition.

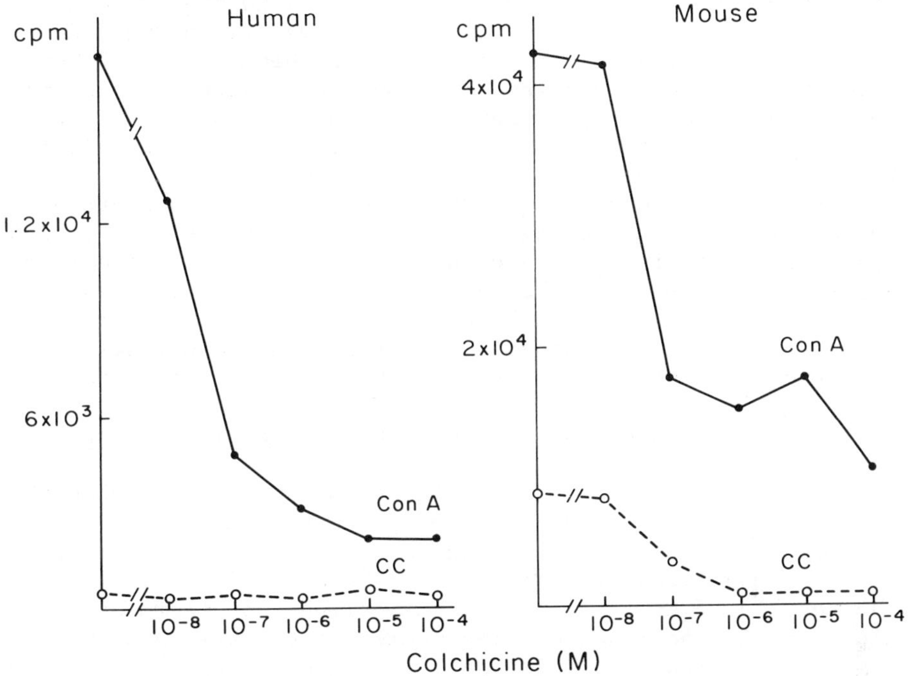

Fig. 3 Effect of the concentration of colchicine on ^{3}H-thymidine incorporation in human and mouse lymphocytes stimulated by Con A. Human lymphocytes were stimulated by 20 μg/ml Con A and mouse lymphocytes were stimulated by 3 μg/ml Con A. Colchicine was present from the beginning of the experiment. The cultures were pulsed with 6 μCi of ^{3}H-thymidine between 48-50 hours. •, incorporation by cultures containing Con A, o, incorporation by cultures in the absence of Con A.

Kinetics of DNA synthesis:

To test whether the effect of colchicine is due to a shift in the onset of DNA synthesis or to a decrease in the final level of synthesis, the kinetics of thymidine incorporation were compared in Con A-stimulated cultures in the presence and absence of colchicine (Fig. 4). It was found that DNA synthesis in cultures without colchicine began about 30 hours after initial Con A binding. In contrast, cultures containing colchicine showed only a low level of DNA synthesis as late as 51 hours. After this time, the level of DNA synthesis in cultures containing colchicine also began to increase, attaining at 70 hours about 30% of the level of cultures containing no colchicine. Lumicolchicine did not have any effect on DNA synthesis in cells stimulated by Con A (Table II). These data suggest that the inhibition by colchicine of mitogenic stimulation of lymphocytes may be partly due to a delaying effect on the onset of DNA synthesis, possibly via the effects of the drug on a process mediated by the cellular microtubules.

Inhibition by colchicine of Con A-stimulated blast transformation:

The effect of colchicine on the stimulation of lymphocytes by Con A was manifested not only by a depressed level of ^{3}H-thymidine incorporation but also by a decrease in the percentage of lymphocytes transformed into blast cells. Fig. 5 shows the results of a representative experiment to determine the kinetics of appearance of blast cells in the presence and absence of colchicine. The data indicate that although some blast cells were observed in cultures containing colchicine, the percentage of these cells was significantly lower than in parallel cultures without the drug. Moreover, it was found that the number of viable small lymphocytes in cultures with colchicine was comparable or slightly higher than the corresponding number in cultures containing no colchicine.

Effect of colchicine on blast cells from stimulated cultures:

To test whether the depressed level of blast cells in colchicine-containing cultures was due to selective killing of these blast cells, lymphocytes stimulated by Con A were synchronized using 0.5 mM hydroxyurea, which has been shown from preliminary experiments to be adequate for collecting cells at the G_1/S boundary as observed by Lohrmann et al. (13). Colchicine was added to half of the cultures six hours before the removal of hydroxyurea. After release of the hydroxyurea block, the number of viable small lymphocytes and blast cells as well as the level of thymidine incorporation was determined. It was found that for at least 30 hours of exposure

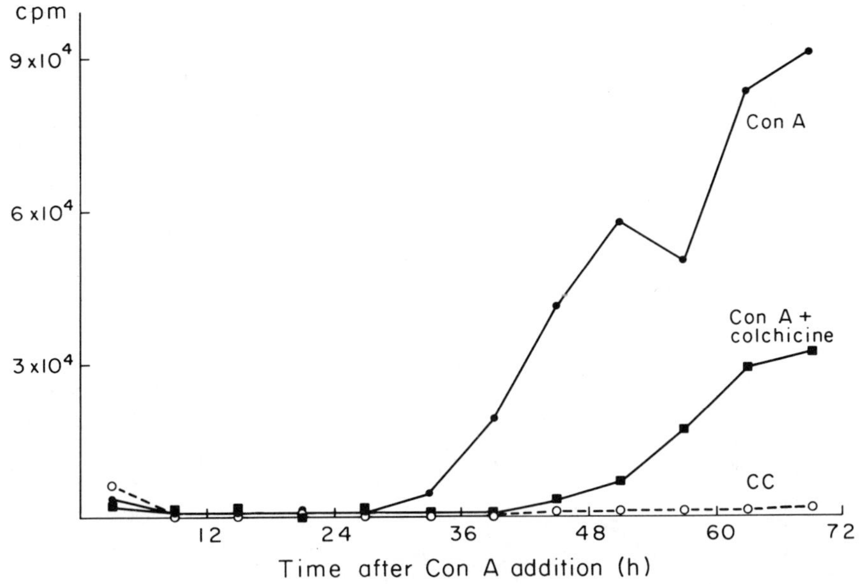

Fig. 4 Effect of colchicine on the kinetics of ^{3}H-thymidine incorporation in human lymphocyte cultures stimulated by Con A. Colchicine was present from the beginning of the experiment. At various times, parallel cultures were pulsed with 6 μCi of ^{3}H-thymidine for 6 hours. Data points are plotted at times corresponding to 3 hours after the start of each pulse. ●, cultures containing Con A (20 μg/ml); ■, cultures containing Con A (20 μg/ml) + colchicine (10^{-6} M); o, cell control.

TABLE II

THE EFFECT OF LUMICOLCHICINE ON MITOGENIC STIMULATION BY CON A

^{3}H-thymidine pulse (h)	Incorporation (cpm) Cell Control	Con A	Con A + Lumicolchicine (10^{-6} M)
24-30	700	800	1,100
30-36	700	4,500	10,100
36-42	700	19,000	31,900
42-48	1,200	41,400	43,200
48-54	1,300	57,700	49,200

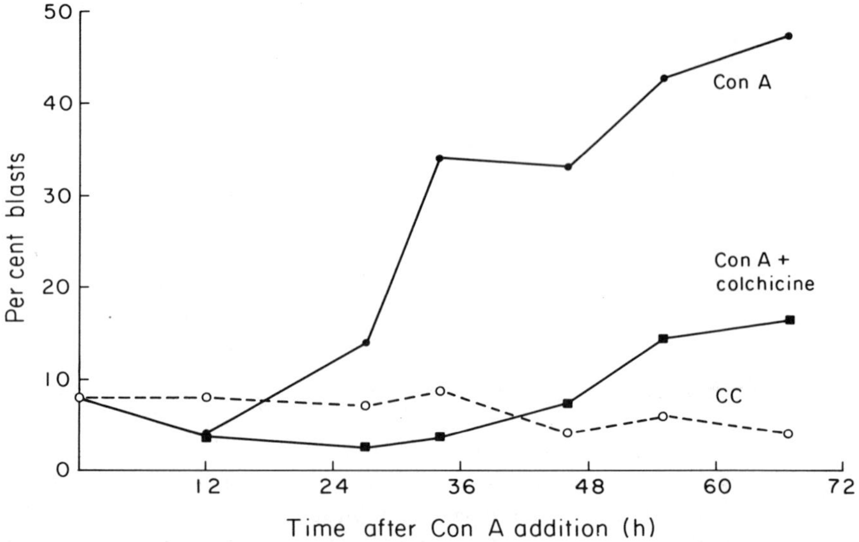

Fig. 5 Effect of colchicine on the kinetics of appearance of blast cells in human lymphocyte cultures stimulated by Con A. Colchicine was present from the beginning of the experiment. At the indicated times, aliquots of cultures were removed and the total number of viable small lymphocytes and blast cells were counted. Data are expressed as the percentage of blast cells present at various times after the addition of Con A. ●, cultures containing Con A (20 μg/ml); ■ , cultures containing Con A (20 μg/ml) + colchicine (10^{-6} M); o, cell control.

to colchicine, there was no appreciable difference in the numbers of blast cells and small lymphocytes (Fig. 6). Colchicine does not, therefore, selectively exert its toxic effects on blast cells. These results suggest that the difference in the number of blast cells in cultures with and without colchicine (Fig. 5) represents a true inhibition by the drug of the transformation of a small resting lymphocyte into a blast cell.

In addition, the data on ^{3}H-thymidine incorporation by blast cells formed in the absence of colchicine but arrested at the G_1/S boundary showed that colchicine had no effect on the level of incorporation (Table III). Under these conditions, therefore, colchicine appears to affect neither ^{3}H-thymidine transport across the membrane nor the process of DNA synthesis itself. This suggests that the depressed level of ^{3}H-thymidine incorporation seen in cultures containing colchicine in Figures 3 and 4 may represent an inhibitory effect of the drug on lymphocytes prior to their transformation into blast cells.

Effects of colchicine added at various times:

In order to define further the period during cell stimulation that is sensitive to colchicine, experiments were performed in which the drug was introduced into the cultures at various times after Con A stimulation (Fig. 7). If colchicine is added at the start of the culture, the incorporation of ^{3}H-thymidine as measured at 48 hours is inhibited by about 90% compared to controls. The extent of inhibition decreases the later colchicine is added. No inhibition is observed if the drug is added 30 hours after the addition of the mitogen.

As previously observed (14), the incorporation of ^{3}H-thymidine at 48 hours in mouse splenic lymphocytes stimulated by Con A is inhibited in a fashion similar to that seen in Fig. 7 when α-methyl-D-mannoside, a competitive inhibitor for Con A, is added to cultures at various times after the addition of the lectin. It was shown that in the saccharide inhibition, the rising level of ^{3}H-thymidine incorporation is proportional to the number of cells committed to lectin stimulation. The similarity in the kinetics of inhibition of stimulation by α-methyl-D-mannoside and by colchicine suggests that the inhibitory effect of the drug might be temporally correlated with cellular commitment.

Our findings on the effects of colchicine on the modulation of receptor mobility and on the stimulation of lymphoid cells have several implications for the hypotheses on receptor-cytoplasmic interactions and mitogenesis in lymphocytes. The data on the inhibition by Con A of cell surface receptor

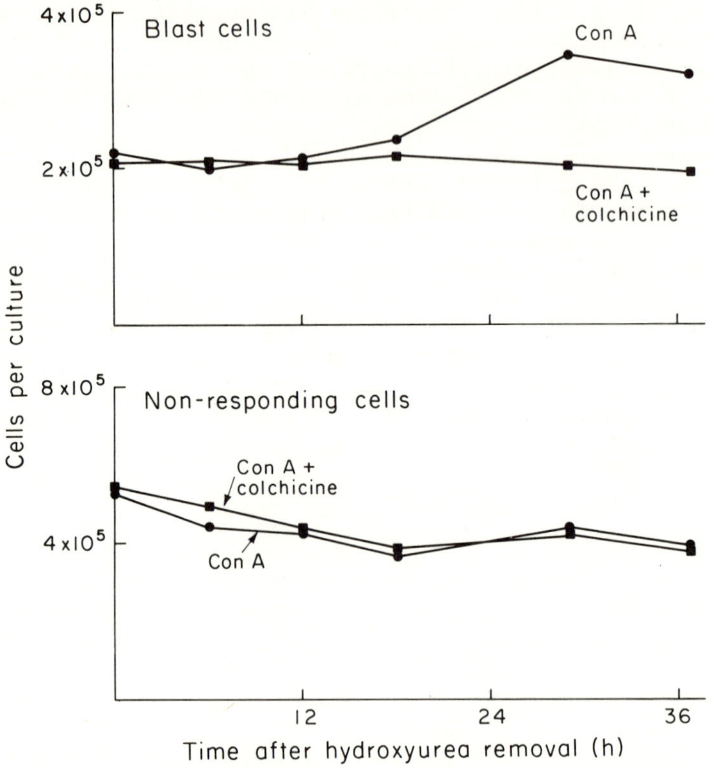

Fig. 6 Effect of colchicine on blast cells collected at the G_1/S boundary using 0.5 mM hydroxyurea in human lymphocyte cultures stimulated by Con A. Colchicine was added to half of the cultures six hours before the removal of hydroxyurea. The data are expressed as the total number of viable small lymphocytes and blast cells at various times after the removal of hydroxyurea. ●, cultures containing Con A (20 μg/ml); ■, cultures containing Con A (20 μg/ml) + colchicine (10^{-6} M).

TABLE III

EFFECT OF COLCHICINE ON ^{3}H-THYMIDINE INCORPORATION BY STIMULATED HUMAN LYMPHOCYTES[a]

^{3}H-thymidine pulse (h after release from hydroxyurea)	Incorporation (cpm)	
	Con A	Con A + Colchicine (10^{-6} M)
1-2	6,300	5,100
5-6	17,300	18,000
8-9	11,900	12,300
12-13	14,200	13,700
16-17	17,300	14,100

[a]Blast cells were collected at the G_1/S boundary using 0.5 mM hydroxyurea in human lymphocyte cultures stimulated by Con A. Colchicine was added to half of the cultures six hours before the removal of hydroxyurea. Cultures with and without colchicine were pulsed with 6 μCi of ^{3}H-thymidine at various times after the removal of hydroxyurea.

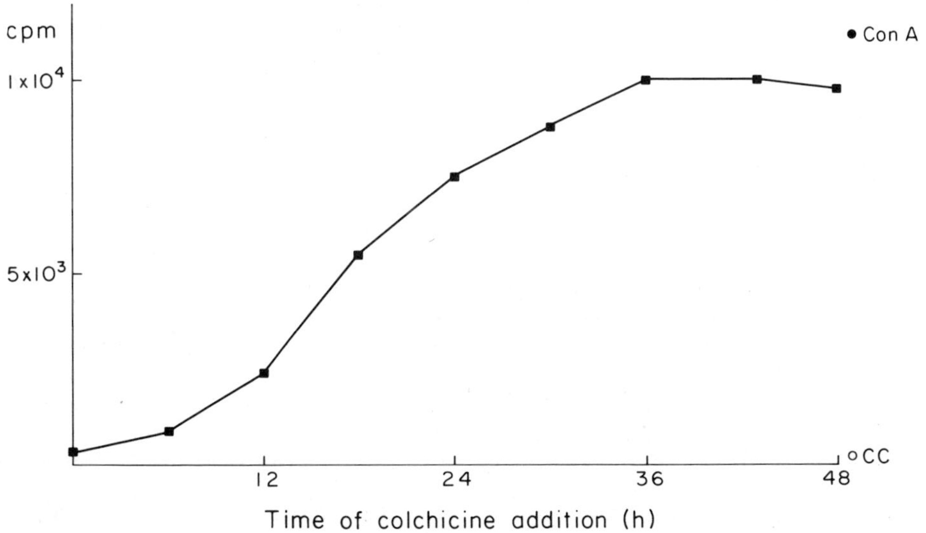

Fig. 7 Effect of colchicine, added at different times after the start of the culture, on the incorporation of ^{3}H-thymidine in human lymphocytes stimulated by Con A. Cultures containing Con A (20 μg/ml) were made 10^{-6} M in colchicine at the indicated time points. The cultures were continued until 48 hours at which time they were pulsed with 6 μCi of ^{3}H-thymidine for two hours and harvested for analysis.

movement and its reversal by colchicine suggests that a colchicine-binding protein system such as the cellular microtubules may mediate the inhibition process (Fig. 2). Inasmuch as microtubules have only been identified in the cytoplasm, this system may represent an example of receptor-cytoplasmic interactions. The inhibition by colchicine of the stimulation of lymphocytes by Con A suggests that mitogenic activation could also be mediated by the same receptor-cytoplasmic interactions.

Alternatively, the effect of colchicine on lymphocyte stimulation could be due to the control exerted by colchicine-sensitive proteins on the transport of various ions and metabolites. In view of the recent findings that Ca^{++} ion uptake is required for lymphocyte stimulation (27,28) and that the Ca^{++}-carrying ionophore A-23187 can stimulate human lymphocytes (J.L. Wang and G.M. Edelman, unpublished observations), it is possible that an intact microtubule assembly may be necessary for Ca^{++} ion transport. Colchicine may exert its inhibitory effects by dissociating the microtubules which regulate this required Ca^{++} ion transport in a fashion analogous to the regulation of other Ca^{++}-sensitive processes such as insulin-release by pancreatic islet cells (29) and granulocyte chemotaxis (30). The demonstration of such a formal analogy between the stimulus-secretion coupling system and lymphocyte mitogenesis promises to contribute to our understanding of the mechanism of lymphoid cell stimulation.

REFERENCES

1. Greaves, M.F. and Bauminger, S., Nature (New Biol.), 235: 67, 1972.
2. Andersson, J., Edelman, G.M., Möller, G. and Sjöberg, O., Eur. J. Immunol. 2: 233, 1972.
3. Yahara, I. and Edelman, G.M., Proc. Nat. Acad. Sci. 69: 608, 1972.
4. Edelman, G.M., Yahara, I. and Wang, J.L., Proc. Nat. Acad. Sci. 70: 1442, 1973.
5. Yahara, I. and Edelman, G.M., Exp. Cell Res., in press, 1974.
6. Edelman, G.M. and Wang, J.L., NIH International Workshop on Cell Surfaces and Malignancy, in press, 1974.
7. Wang, J.L., Gunther, G.R. and Edelman, G.M., manuscript submitted.
8. Cunningham, B.A., Wang, J.L., Pflumm, M.N. and Edelman, G.M., Biochemistry 11: 3233, 1972.
9. McClain, D.A., Wang, J.L. and Edelman, G.M., Cell. Immunol., in press, 1974.

10. Boyum, A., Scand. J. Clin. Lab. Invest. Suppl. 21: 97, 1968.
11. Mishell, R.I. and Dutton, R.W., J. Exp. Med. 126: 423, 1967.
12. Sell, S. and Gell, P.G.H., J. Exp. Med. 122: 423, 1965.
13. Lohrmann, H.-P., Graw, C.M. and Graw, R.G., J. Exp. Med. 139: 1037, 1974.
14. Gunther, G.R., Wang, J.L. and Edelman, G.M., J. Cell Biol. 62: 366, 1974.
15. Taylor, R.B., Duffus, P.H., Raff, M.C. and de Petris, S., Nature (New Biol.) 233: 225, 1971.
16. Gunther, G.R., Wang, J.L., Yahara, I., Cunningham, B.A. and Edelman, G.M., Proc. Nat. Acad. Sci. 70: 1012, 1973.
17. Yahara, I. and Edelman, G.M., Nature 246: 152, 1973.
18. Wilson, L. and Friedkin, M., Biochemistry 6: 3126, 1967.
19. Jacob, H., Amsden, T. and White, J., Proc. Nat. Acad. Sci. U.S. 69: 471, 1972.
20. Yahara, I. and Edelman, G.M., Exp. Cell Res. 81: 143, 1973.
21. Rutishauser, U., Yahara, I. and Edelman, G.M., Proc. Nat. Acad. Sci. 71: 1149, 1974.
22. Raff, M.C. and de Petris, S., Fed. Proc. 32: 48, 1973.
23. de Petris, S. and Raff, M.C., Ciba Foundation Symp. 14: 27, 1973.
24. Yahara, I. and Edelman, G.M., Ann. N.Y. Acad. Sci., in press, 1974.
25. Medrano, E., Piras, R. and Mordoh, J., Exp. Cell Res. 86: 295, 1974.
26. Sasaki, M.S. and Norman, A., Nature 210: 913, 1966.
27. Alford, R.H., J. Immunol. 104: 698, 1970.
28. Allwood, G., Ashirson, G.L., Davey, M.J. and Goodford, P.J., Immunol. 21: 509, 1971.
29. Lacy, P.E., Howell, S.L., Yang, D.A. and Fink, C.J., Nature 219: 1177, 1968.
30. Gallin, J.I. and Rosenthal, A.S., J. Cell Biol. 62: 594, 1974.

LYMPHOCYTE JUNCTIONS: PERMEABILITY AND STRUCTURE

Gilberto M. de Oliveira-Castro
Marcello A. Barcinski and Ionice F. Gaziri

Instituto de Biofisica, Universidade Federal do Rio de Janeiro, Ilha do Fundao, Rio de Janeiro, Brasil

Intercellular communication in most excitable cells is achieved by the release of neurotransmitters. In a few instances, like in the giant motor synapses of the crayfish there is direct electrotonic spread of excitation by low resistance pathways (1). These junctions are not only permeable to current flow but allow direct cell-to-cell flow of molecules and ions.

The study of junctional permeability in non-excitable cells has provided interesting information concerning the intracellular flow of substances and its functional role (2-4). Two complementary approaches have been used for the analysis of epithelial cell coupling: (1) measurements of junctional conductance by means of detecting current flowing from the first to the second of two adjacent cells, (2) by comparing the diffusion of a tracer substance to the neighbouring cell and to the extracellular space.

The electrophysiologic technique uses an ionic current injected in one of two adjacent cells with a simultaneous recording of transmembrane potentials both in the injected (cell 1) and in the neighbour cell (cell 2). The changes induced in membrane potentials provide information regarding the junctional conductance. When this conductance is considerably higher than that of the external membrane and both cells have comparable surfaces in contact with the external bathing solution, the ratio of potential changes in both cells is an indication of junctional permeability (coupling ratio V_2/V_1). If junctional conductance is of the same order of magnitude than that of the external membrane the coupling ratio will approach zero (for details see 2,5-7).

The junctional permeability can also be probed by injecting a marker of low binding capacity to the cell constituents and following its diffusion to adjacent cells. In most epi-

---

This work was supported by the "Banco Nacional de Desenvolvimento Economico (FUNTEC-241)","Conselho Nacional de Pesquisas (CNPq)" and "Conselho de Ensino para Graduados (CEPG) da UFRJ."

thelial junctions that have been probed, permeability to small ions as well as to molecules as large as 10,000 daltons, has been found.

The functional role of this communicating system is not well understood. The interconnected cells form a compartment of finite volume separated from the exterior by diffusion barriers of low permeability - the non-junctional membranes.

The structural basis of the high permeability junction is not definitively established, although considerable evidence suggests the gap junction as containing the coupling pathways. This junctional architecture is a specialized structure of close membrane apposition where an extracellular space of 20 to 40 Angstroms is seen. In freeze-fracture, subunits of the junctional complex appear as a closely packed hexagonal array of membrane particles. The gap junctions have been found both between cells of organized tissues and in tissue culture where electrotonic coupling was described (8). In some non-coupling cells these junctions have not been observed, even after a careful scrutiny (9). A very interesting finding has just been reported (10) where fusion of human cell with a non-coupling mouse cell line not only corrects the junctional permeability defect but simultaneously induces the appearance of gap-junctions. There is a strong possibility that a genetic factor of the human cells corrects the junctional deficiency inducing the formation of gap-junction, establishing as a consequence intercellular communication.

Immunological phenomena, both from an antigenic or cellular viewpoint, reveal very important interactive events with different types of cells participating in different operative systems. This includes for example T-cell "helper" and "suppressor" functions for B-cells involved in antibody production, certain lymphocyte interactions in pure T-cell reaction as well as in cell mediated immunity phenomena that appear to involve interactions between lymphocytes and other types of cells. Some of the basic mechanisms involved in all the described types of cell interactions and cooperations are still obscure, thus deserving further studies.

One of the basic tools to study lymphocyte stimulation is phytohemagglutinin (PHA). Although the mechanism of stimulation is still obscure, one of the earliest detectable changes is the formation of high permeability junctions. In this paper we will discuss some membrane properties of stimulated human lymphocytes, the formation of the permeable junctions and the control mechanism of interlymphocyte communication, as revealed by the electrophysiologic approach.

METHODS

Human lymphocytes were obtained by Ficoll-Hypaque separation and cultured for different periods in RPMI medium. PHA-stimulated clumps were submitted to microelectrode analysis with a standard electrophysiologic set up (for details see 11). Intracellular potentials were recorded in neighbouring cells. The ratio of the voltage-drops recorded in two adjacent cells is expressed as coupling coefficient.

To test the effect of intracellular Ca^{2+} application a micropipette was filled with 95 mM $Ca(OH)_2$, 100 mM EGTA, 100 mM TRIS and calcium was driven by rectangular pulses of 10^{-8} A and 50 ms duration, at a rate of 5 to 10 pulses/s.

This method has several advantages when compared with the $CaCl_2$ - technique described last year in this conference (12): a) injection can be done in one cell (cell 2) and junctional conductance can be simultaneously measured through two adjacent cells (cells 1 and 3), b) the calcium microelectrode is not used for recording, because this procedure can induce error due to its high resistance, c) the calcium-delivery is regular and transference number (0.2) is known (Reuben, in publication).

Cells cultured for 1, 24 or 65 hours with or without PHA were fixed and the ultrastructure of their junctions analyzed. The best fixation was obtained by slight modification of the method of Hirsch and Fedorko (13). Immediately before use, two volumes of 2.5% glutaraldehyde in 0.1 M cacodylate buffer (pH 7.4) were added to one volume of 1% osmium tetroxide, at 0°C. The fixative was gently added to the lymphocytes after removal of the culture medium and replaced after one hour with Earle's salt solution. The cell suspension was then stained with 0.25% uranyl acetate in 0.1 M acetate buffer (pH 6.3). Constant temperature (0°C) was maintained throughout the fixation and staining procedure. To preserve the original cell density and avoid distortions of the clumps, centrifugation was never used. After embedding in EPON mixture the cells were sectioned with an Ultratom III (LKB Products). Thin sections showing silver interference colors were stained with 0.4% lead citrate and examined with an EM6-B electron microscope (A.E.I.). Transjunctional conductance was detected after one hour of PHA treatment. Earlier measurements did not yield reliable results due mainly to frequent cell disaggregation consequent to impalement.

Membrane constants:

A summary of the electophysiologic findings in human lymphocytes is given in Table I. In our previous report,

TABLE I
MEMBRANE PARAMETERS OF HUMAN LYMPHOCYTES AFTER PHA-STIMULATION

Parameters	Range	Mean
Membrane potentials	5-25 mV	12 mV
Input resistances	2.1-8.2 MΩ	4.9 MΩ
Transfer resistances	1.2-3.9 MΩ	2.1 MΩ
Coupling coefficients	0.16-0.60	0.39
No. of cells per clump	3-11	-
Time after PHA-stimulation	20-28 hours	-

TABLE II
COUPLING COEFFICIENT AFTER DIFFERENT PERIODS OF STIMULATION

Time in culture (hours)	Percent of coupled cell pairs	V_2/V_1*		Number of pairs tested
		$\bar{X} \pm$ S.D.	Range	
1	14.3 %	0.18 ±0.08	0.09-0.30	14
24	45.5 %	0.38 ±0.06	0.17-0.45	11
48	75 %	0.42 ±0.05	0.29-0.51	16
65	80 %	0.41 ±0.07	0.30-0.56	10

* Coupling coefficient

due to a biased interpretation of membrane injury, the values below 10 mV and those which decreased as a function of time were considered as unreliable and thus excluded from the final analysis. However, if we consider all the impalements on which a stability of the membrane potential of at least 5 min is observed, several measurements ranging from 5 to 10 mV (inside negative) will be included, and so the final mean value is lowered from the original 18 mV to 12 mV.

Hulser and Webb (14) in non-stimulated bovine lymphocytes have obtained a relationship that extrapolates to zero membrane potential plotting membrane potentials against electrode tip-potentials. The possibility exists that the differences of their results and ours depend on the effect of PHA stimulation (as the findings of Taki (15) seem to suggest), on the use of different species or on the different techniques to prepare electrode-holders and to record the electric characteristics of lymphocytes.

The total input resistances shown on Table I were obtained by the ratio of voltage-drops to the current intensities that produced them (V_1/i_1). Transfer resistances represent the voltage-drops recorded in an adjacent cell divided by the current intensity (V_2/i_1).

The cell clumps that provided these data had different shapes, making detailed analysis of the spread of current and precise estimation of the specific membrane resistance impossible.

Genesis of permeable junctions:

Table II summarizes our findings relating coupling coefficients to time after PHA-stimulation. After one hour of stimulation, about 14% of the cell pairs that were examined showed a significant coupling ratio. Both the percentage of coupling pairs and the junctional permeability (as revealed by the coefficient V_2/V_1) increased up to 48 hours. The mean coupling coefficient at 48 hours is significantly different from the mean at 1 hour ($p<0.005$).

Hulser and Peters (16) manipulating non-stimulated bovine lymphocytes into contact, reported that 15 min after the addition of PHA to their preparation a state of intercellular permeability could already be detected. From our data, we can only say that junctional permeability occurs as early as 1 hour after stimulation and that the percentage of electrotonically coupled cell pairs increases up to 48 hours. A concomitant increase in the coupling coefficient was observed.

Structure of the junctions:

In control cultures, incubated without PHA, no cell contacts were detected even after 24 hours of culture (Fig. 1A). An increase in the area of close membrane contacts could be clearly seen with the increase in time of exposure to PHA from 1 to 48 hours. A few cell contacts were seen within one hour of stimulation, they were very frequent after 24 hours (Fig. 1B) and after 48 hours the cultures presented a tissue-like structure (Fig. 1C).

The structure was heptalaminar with a thickness at the junctions of 200-210 Å (Fig. 2). A gap of 20-40 Å between the two other electron-dense lines could be seen in thin sections that were perpendicular to the junctions. In a few instances the median line of the junctional complex appeared as electron-dense regions, with regular spacing of 90-95 Å.

The only type of specialized cell junctions observed in our preparation were very similar to the gap-junctions previously described.

Uncoupling of the junctions:

The electrical measurements demonstrate that small current carrying ions like K^+ or Cl^- can easily cross lymphocyte junctions. The upper limit of the size of molecules that can pass between these coupled cells remains to be determined.

Loewenstein (17) proposed that Ca^{2+} ions can play an important role in sealing the communicating junctional pathways. Most cells maintain a low level of cytoplasmic calcium by means of an outward pumping system (18,19). In epithelial cells electrotonic uncoupling results after calcium injection or penetration from the outside medium (20). Based on previous observations we analyzed the effect of increasing Ca^{2+} concentration in lymphocyte cytoplasm during continuous monitoring of transjunctional permeability.

The result of one of these experiments is shown in Fig. 3, where the progressive increase in V_1 and the simultaneous decrease in V_3, after calcium application in cell 2, can only be explained by sealing of the permeable pathways from cell 1 to 3 (junctions 1 to 2 and 2 to 3). We have demonstrated that this uncoupling is not due to current flow (21) as it is the case in the salivary gland of Chironomus (22). Control experiments, with KCl filled microelectrodes revealed that the coupling coefficients measured after the injection of a train of depolarizing or hyperpolarizing pulses remains constant. Fig. 4 is from one of these experiments where a coupling ratio of 0.56 is maintained throughout the recording period. Note the fact that transmembrane potentials of the

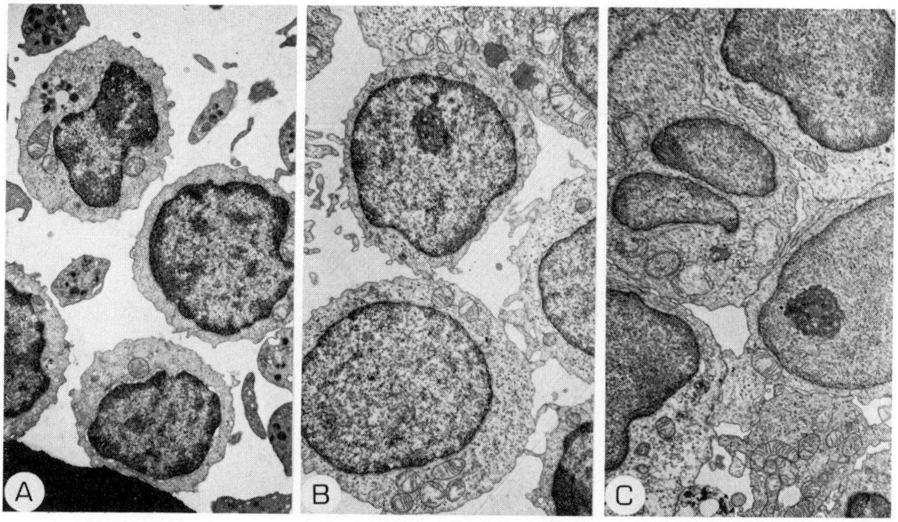

Fig. 1 Lymphocyte cultures after different periods of stimulation by PHA. Magnification X 6,000

A - Control culture after 24 hours of incubation without PHA. Note that no cell contacts are seen.

B - Culture after 24 hours of incubation with PHA. Several contacts between the stimulated cells are present.

C - Culture after 65 hours of incubation with PHA. All the cells are in contact and the culture has a tissue-like appearance.

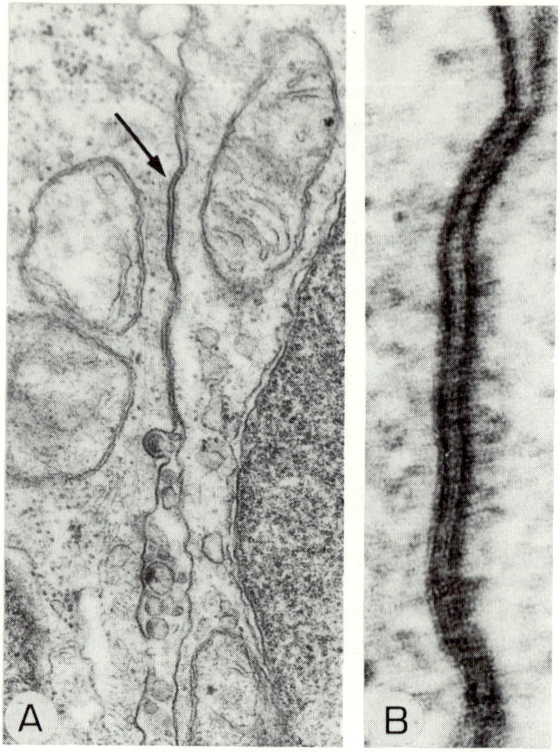

Fig. 2 Junctions between stimulated lymphocytes after 65 hours of incubation with PHA.
A - Portion of the cell junction (X 25,000)
B - Higher magnification of Fig. 2A (arrow). Note the appearance of the junctional complex (X 200,000)

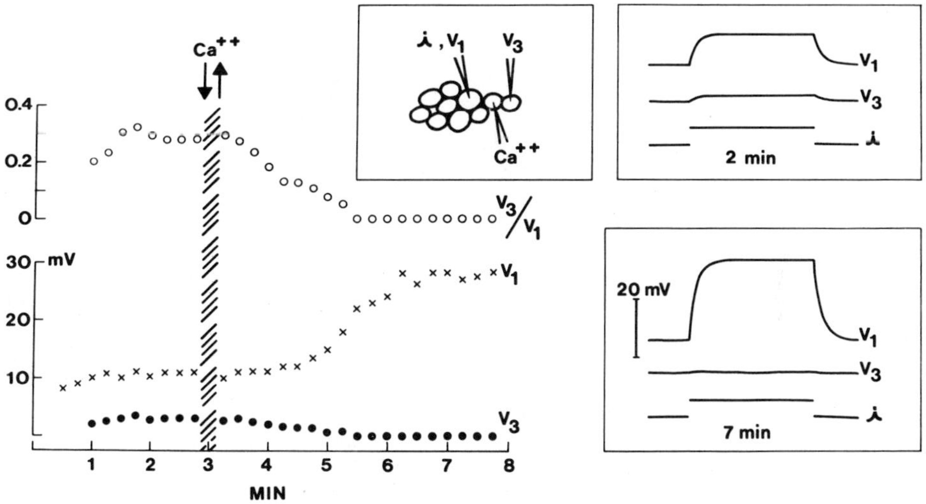

Fig. 3 Effect of calcium injection on junctional conductance. Outward pulses of 50 ms, 2.3×10^{-8} A at a rate of 10 pulses/s during 10 s were applied with Ca^{2+} filled electrode in Cell 2. Note progressive uncoupling with simultaneous increase of V_1 and decrease of V_3. Insets: Samples of oscilloscope records (retouched) of the voltage drops (V_1 V_3) and current (i) before (2 min) and after (7 min) calcium injection. Current pulse duration 50 ms and amplitude 10^{-8}A. Schematic drawing of the cell clump with position of the electrodes (see text for details).

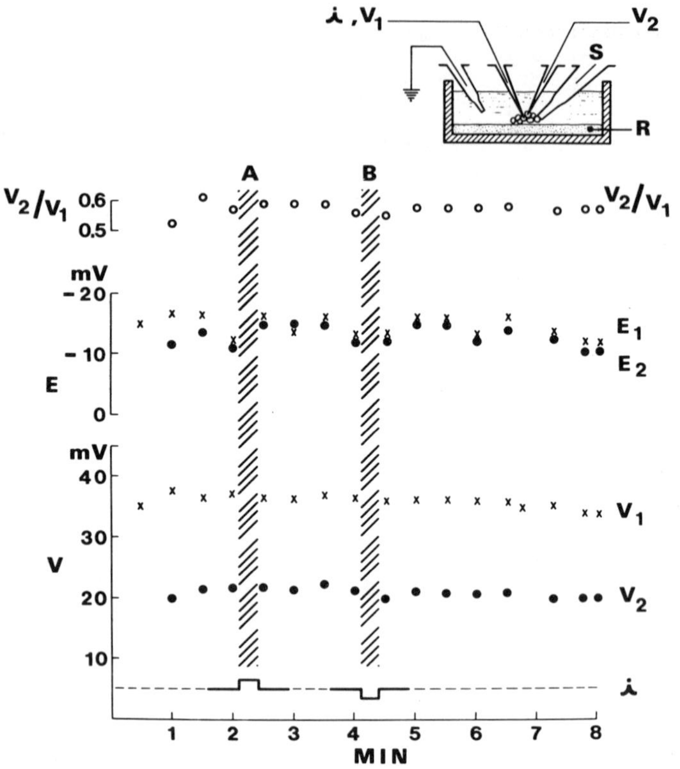

Fig. 4 Effect of current flow on junctional conductance. Transmembrane potentials (E_1, E_2) and voltage drops (V_1, V_2) resulting from current injection through KCl-filled microelectrode in Cell 1. The upper plot represents the coupling coefficient (V_2/V_1). Current pulses (i) applied at A (inward current) and B (outward current) did not affect intercellular conductance (pulses of 50 ms, 10^{-8}A at a rate of 5 pulses/s during 10 s). Inset: Scheme of the experimental set-up showing a lymphocyte clump resting on a neutral resin (R) and immobilized by the suction pipette (S). Microelectrodes (V_1 and V_2) and reference pipette (connected to ground) are also shown.

adjacent cells have a very close value and their fluctuations are simultaneously observed in cells 1 and 2, another indication of intercellular communication. Immediately after inward or outward current injection junctional conductance is not altered.

The cell system that has been more thoroughly analyzed for intercellular communication is the insect salivary gland, where calcium concentration is probably a very important factor controlling junctional permeability (20,23). The most direct evidence of calcium ion involvement in junctional uncoupling is its release from some intracellular storage site during the cyanide-induced uncoupling as detected with the specific calcium indicator aequorin (4). In these experiments a transient rise in aequorin's light output due to inhibition of the energy metabolism on which Ca^{2+} removal from the cytoplasm depends, is described. In the squid giant axon a sustained release of calcium has also been reported in similar condition (24).

The mechanism by which Ca^{2+} can produce this uncoupling is still obscure. In the salivary gland Mg^{2+}, Ba^{2+}, Sr^{2+} and Mn^{2+} are less effective but can also produce sealing of the junctional permeability (20,25). The action of divalent cations on this structure is probably complex and specific, since Co^{++} which would bind in a similar way at a non-specific site does cross the junctions, at least in the crayfish septate axon (26).

The basic difference between the physiology of the lymphocyte-junctions we have described here and other junctions is the fact that transjunctional current flow has no effect on junctional conductance. In Chironomus salivary glands, depolarizing pulses depress junctional conductance (22), but hyperpolarizing currents can restore the high conductance of previously uncoupled cells (27). The retinula cells of the lateral eye of the horseshoe crab can be uncoupled when either depolarizing or hyperpolarizing currents are passed (28). These results illustrate the differences of behavior in transjunctional current when submitted to current flow. It seems impossible to attempt a single explanation for the different situations described above, but it suggests that in this respect the lymphocyte junction is hardly comparable with the other structures.

Many ultrastructural studies have tried to identify the junctional membrane differentiation responsible for the low resistance pathways. The gap-junction is the most likely candidate for this role as stated in the introduction. Our results correlating the existence of gap-junction-like structure with cell coupling are in accordance with previous reports (9,10,29,30). We have not been able to obtain good results with lanthanum impregnation of the extracellular gap and have not used freeze-fracture, both steps being necessary for a definitive demonstration of the junctions we suggest here. The parallel increase of the contact areas and in the coupling coefficient is perhaps the most important of our findings and suggests that the electrophysiologic approach may produce some relevant information related to the mechanism of lymphocyte stimulation.

Junctional coupling allows some speculation concerning its functional role. The possibility of establishing a common intracellular pool for small inorganic ions may be very important in transport-epithelia or in excitable cells. In other cell communities the possibility of using this communication system to control metabolites, chemical signals involved in the control of cell activity and gene function is very exciting. Since the electrophysiologic approach has provided the information described here about the properties of this direct information system in immuno-competent cells, we are stimulated to obtain more data about its detailed mechanism and after that it will be safer to speculate how the different and complex cell interaction systems involved in immunology uses this signaling system.

REFERENCES

1. Furshpan, E.J. and Potter, D.D., J. Physiol. 145: 289, 1959.
2. Loewenstein, W.R., Ann. N.Y. Acd. Sci. 137: 441, 1966.
3. Furshpan, E.J. and Potter, D.D., in Current Topics in Developmental Biology, edited by A.A. Moscona, Academic Press, New York, Vol. 3, p. 95, 1968.
4. Loewenstein, W.R., Fed. Proc. 32: 60, 1973.
5. Bennet, M.V.L., Fed. Proc. 32: 65, 1973.
6. Socolar, S.J., Expt. Eye Res. 15: 693, 1973.
7. Hulser, D.F., in Methods in Cell Biology, edited by D.M. Prescot, Academic Press, New York, Vol. 8, p. 289, 1974.
8. McNutt, N.S. and Weinstein, R.S., Prog. Biophys. Mol. Biol. 26: 45, 1973.
9. Gilula, N.B., Reeves, O.R. and Steinbach, A., Nature 235: 262, 1972.
10. Azarnia, R, Larsen, W.J. and Loewenstein, W.R., Proc. Nat. Acad. Sci. USA 71: 880, 1974.
11. Oliveira-Castro, G.M., Barcinski, M.A. and Cukierman, S., J. Immunol. 111: 1616, 1973.

12. Oliveira-Castro, G.M., Barcinski, M.A. and Cukierman, S., in Lymphocyte recognition and effector mechanisms, edited by K. Lindahl-Kiessling and D. Osoba, Academic Press, New York, p. 535, 1974.
13. Hirsch, J.G. and Fedorko, M.E., J. Cell Biol. 38: 615, 1968.
14. Hulser, D.F. and Webb, D.J., Biophysic 10: 273, 1973.
15. Taki, M., Mie Med. J. 19: 245, 1970.
16. Hulser, D.F. and Peters, J.H., Exptl. Cell Res. 74: 319, 1972.
17. Loewenstein, W.R., J. Colloid. Interface Sci. 25: 34, 1967.
18. Schatzmann, J.J., Experientia 22: 364, 1966.
19. Baker, P.F., in Calcium and cellular function, edited by A.W. Cuthbert, St. Martin's Press, New York, p. 96, 1970.
20. Oliveira-Castro, G.M. and Loewenstein, W.R., J. Membr. Biol. 5: 51, 1971.
21. Oliveira-Castro, G.M. and Barcinski, M.A., Biochim. Biophys. Acta 352: 338, 1974.
22. Socolar, S.J. and Politoff, A.L., Science 172: 492, 1970.
23. Loewenstein, W.R., Nakas, M. and Socolar, S.J., J. Gen. Physiol. 50: 1865, 1967.
24. Blaustein, M.P. and Hodgkin, A.L., J. Physiol. 200: 497, 1969.
25. Loewenstein, W.R., Arch. Intern. Med. 129: 299, 1972.
26. Politoff, A., Papas, G.D. and Bennet, M.V.L., J. Cell Biol. 55: 204a, 1972.
27. Rose, B., Science 169: 607, 1970.
28. Smith, T.G., Baymann, F. and Fuortes, M.G.F., Science 147: 1446, 1965.
29. Pinto da Silva, P. and Gilula, N.B., Exptl. Cell Res. 71: 393, 1972.
30. Hulser, D.F. and Dempsey, A., Z. Naturforsch. C. 28: 603, 1973.

THE RELATIONSHIP BETWEEN "EARLY EVENTS" AND DNA SYNTHESIS IN MITOGEN STIMULATED LYMPHOCYTES

Ido Betel and Kor. J. van den Berg
Radiobiological Institute TNO, Rijswijk (Z.H.)
The Netherlands

INTRODUCTION

One of the key problems in immunobiology is to determine how the interaction of lymphocyte surface receptors with antigens is linked to the metabolic processes in the cell leading to cellular activation and expression of specific immunological functions. Elucidation of this activation mechanism is essential to our understanding of the regulation of differentiation and growth of the lymphocyte as well as other cellular systems.

It is generally accepted that the study of non-specific or polyclonal activation of lymphocytes by mitogens is relevant to the problem of antigenic activation, since apart from the specificity of the antigenic response, lymphocytes activated by mitogens exhibit very much the same properties as antigenically stimulated cells. The recent observations by Coutinho and Möller (1973) indicating that "T independent" antigens may be essentially B-cell mitogens causes the distinction between mitogens and antigens to disappear.

As far as morphological transformation (blast formation) and induction of DNA synthesis are concerned, mitogens show a certain specificity. In the mouse, PHA and Concanavalin A are specific stimulators of thymus derived cells (T-cells) while E. coli lipopolysaccharide is a B-cell mitogen. Pokeweed mitogen (PWM) seems to stimulate both cell populations. (Greaves and Janossy, 1972; Andersson et al. 1972). Binding of the mitogen to the cell surface is a prerequisite for cell activation but seems not to be sufficient, since B-cells bind T-cell mitogens to the same extent as T-cells and vice versa (Stobo et al., 1972). Very shortly after addition of the mitogen to the lymphocytes a number of biochemical processes at the membrane level is stimulated. Large increases have been found in the transport of amino acids, nucleosides, sugars and small ions. Rapid changes have been observed in the acetate incorporation in membrane lipids (Resch and Ferber, 1972) and in the turnover of the phosphate group of membrane phosphatidyl inositol. (Fisher and Mueller, 1968). The relation between these "early events" and the processes leading to DNA synthesis and morphologic transformation are still completely obscure.

It is not known whether any of these early events can occur in a cell population which is, with respect to DNA-synthesis, non-responsive to the mitogen (e.g. T-cells plus LPS) or whether DNA synthesis can be induced without being preceded by some of the early metabolic changes at the membrane level. Experiments designed to elucidate this question could indicate which metabolic changes may be essential for subsequent DNA synthesis and might also detect biochemical differences in the reaction pattern of different cell populations after mitogenic or antigenic stimulation, and contribute to our understanding of the coupling between ligand-binding and intracellular metabolic processes.

METHODS AND MATERIALS

Cell suspensions were prepared from the spleens of 6 - 10 week old CBA mice, congenitally thymusless nu/nu mice and neonatally thymectomized CBA mice as described before (Betel et al. 1974). Corticosteroid resistant thymocytes were obtained from CBA mice by i.p. injection of 2.5 mg hydrocortisone acetate 48 h before removal of the thymus. Preparation of cell suspensions and culture conditions for thymidine incorporation have been described (Betel et al. 1974). α-Amino-isobutyric acid (AIB) transport was measured as described by van den Berg and Betel (1974). Uridine uptake as by Betel et al. (1974). Acetate incorporation and ^{32}P incorporation in phospholipid as described by Betel et al. (1974b).

RESULTS AND DISCUSSION

AIB transport

The results obtained with this parameter have been published (van den Berg and Betel, 1974) and are summarized in Table 1. The results show that Con A does not induce an increased uptake in nude spleen cells as seen in CBA spleen cells. LPS induces a delayed increase in AIB uptake in CBA as well as nude spleen cell suspensions. It was furthermore demonstrated (van den Berg and Betel, 1974) that LPS did not induce an increase in AIB uptake in rat thymocytes which are also unresponsive to LPS with respect to DNA synthesis.

Uridine uptake and incorporation

Results are summarized in Table 2 and show the same patterns as for AIB uptake. Increased uridine uptake and incorporation was found only in association with subsequent DNA synthesis.

TABLE 1
α-AMINOISOBUTYRIC ACID TRANSPORT IN STIMULATED SPLEEN CELLS

Cells	incubation time	AIB transport (cpm) with		
		-	LPS	Con A
CBA spleen	4 h	2627	3446	4350
	24 h	3043	5637	9214
Nu/nu spleen	4 h	1806	2170	1524
	24 h	2234	4550	2110

TABLE 2
URIDINE INCORPORATION IN STIMULATED LYMPHOCYTES

Cells	incubation time	uridine incorporation (cpm*) with			
		-	LPS	Con A	PHA
CBA spleen	24 h	9270	13751	32860	n.d.
Nu/nu spleen	24 h	8718	12990	7461	7984

* In this experiment the sum of uridine incorporation in RNA and soluble intracellular uridine was determined.

Acetate incorporation in lipids

Results are summarized in Table 3 and follow the same rule as previous two parameters.

Turnover of phosphatidyl-inositol phosphate

Here an interesting exception was found. An increase in the turnover of the phosphate group of phosphatidyl inositol was only observed in T-cells when stimulated with T-cell mitogens. LPS had no effect on this parameter in CBA as well as nude spleen cell suspensions (Table 4).

To show that the difference resides in the cells and is not due to the intrinsic properties of the mitogens the effect of PWM was studied as well.

Table 5 shows that corticosteroid resistant thymocytes can be stimulated with PWM and the increase in phospholipid labelling was confined to phosphatidyl-inositol. On the other hand a B-cell suspension obtained from neonatally thymectomized and ALS treated mice showed a strongly decreased response to PHA (DNA and phospholipid); the response to PWM was completely absent in the PI turnover, while thymidine incorporation was stimulated.

Our results indicated that the specificity of the mitogens to stimulate different cell populations is not restricted to DNA synthesis. The specificity is already expressed in the relatively early membrane effects such as AIB transport, uridine uptake and acetate incorporation in lipids.

Since binding properties and receptor redistribution seem to be very similar for a certain mitogen irrespective whether it is bound to B or T cells, it may be speculated that there must be a fundamental difference either in the receptor or in the system that transmits the signal from receptor to the cell interior between T and B cells. On the other hand, our results, while showing a correlation do not prove the existence of a causal relationship between an increase in these parameters and DNA synthesis. For uridine transport an essential role is excluded, since most culture media contain very little uridine and specific inhibition of uridine uptake by persantine has no effect whatsoever on subsequent RNA and DNA synthesis (Peters and Hausen, 1971). Concerning the increased transport of amino acids, we have indications that this increase *per se* is not sufficient to stimulate lymphocytes into DNA synthesis and another action of the mitogens is required as well (K.J. van den Berg, 1974). The increased acetate incorporation is associated with an increased synthesis of poly-unsaturated fatty acids. This change increases the fluidity of isolated cell membranes

TABLE 3

ACETATE INCORPORATION IN LIPIDS OF STIMULATED SPLEEN CELLS

Cells	incubation time	acetate incorporation (cpm) with			
		-	LPS	Con A	PHA
CBA spleen	24 h	2230	5961	8099	12632
Nu/nu spleen	24 h	1156	1946	1012	1143

TABLE 4

^{32}P INCORPORATION IN PHOSPHOLIPID

Cells	mitogen	incubation time	total phospholipid	phosphatidyl inositol
CBA spleen	-	24 h	272	40
	PHA	-	1445	540
	LPS	-	424	80
Nu/nu spleen	-	24 h	402	95
	PHA	-	439	105
	LPS	-	340	75

TABLE 5

EFFECT OF PWM ON PI-PHOSPHATE TURNOVER AND THYMIDINE INCORPORATION IN B AND T CELLS

Cells	^{32}P incorporation*		^{14}C-thymidine incorporation	
	-PWM	+PWM	-PWM	+PWM
Spleen cells from neonatally Tx, ALS treated mice	117	81	1069	6715
Corticosteroid resistant thymocytes	27	105	74	1760

*^{32}P-incorporation was determined after 24 h incubation.

(Ferber et al. 1974). Direct evidence for rapid fluidity changes in intact mouse lymphocytes upon addition of PHA and ConA comes from electron paramagnetic resonance studies with "spin labelled" cells (Barnett et al. 1974). It would be extremely interesting to see whether these changes are correlated with specificity of the mitogens.

Increased turnover of the phosphate group of phosphatidyl inositol is a very general phenomenon in tissues stimulated in many different ways (Lapetina and Michell, 1973) but the significance and the exact mechanism are largely unknown. The effect is clearly unrelated to the change in amino acid or uridine transport induced by mitogens, since this occurs in both B and T cells, while an increase in PI-phosphate turnover is confined to T-cells.

Fisher and Mueller (1971) have shown that gamma-hexachlorocyclohexane (gammexane); a structural analogue of inositol; inhibits the initial increase of PI-phosphate turnover and partly inhibits thymidine incorporation in PHA stimulated human peripheral lymphocytes, suggesting a relationship between these two parameters. However, CBA spleen cells stimulated with PHA or LPS showed no difference in the sensitivity of DNA-synthesis for this inhibitor (Table 6). It is therefore doubtful whether the inhibition in PHA stimulated cells can be ascribed to inhibition of PI-phosphate turnover. Our results do indicate that an early increased turnover rate of PI-phosphate is not required for mitogenic activation of B-cells. Whether or not it is an obligatory step in the activation of DNA synthesis in T-cells still cannot be decided from these experiments.

TABLE 6

EFFECT OF GAMMEXANE ON THYMIDINE INCORPORATION

Gammexane conc. (μM)	Thymidine incorporation (percent of control)		
	nu/nu cells	CBA cells	
	+LPS	+LPS	+Con A
0	100	100	100
1	132	136	113
10	105	84	107
25	66	58	69
50	23	22	31

REFERENCES

1. Andersson, J., Sjöberg, O. and Möller, G., Transplant. Rev. 11: 131, 1972.
2. Barnett, R.E., Scott, R.E., Furcht, L.F. and Kersey, J.H., Nature 249: 465, 1974.
3. Betel, I., Martijnse, J. and van den Berg, K.J., Cell. Immunol., in press, 1974a.
4. Betel, I., van den Berg, K.J., Martijnse, J. and van den Berg, M.D., in Proc. 8th Leucocyte Culture Conf. (Eds. K. Lindahl-Kiessling and D. Osoba). Academic Press. New York, 1974, p. 639.
5. Coutinho, A. and Möller, G., Nature 245: 12, 1973.
6. Ferber, E., Reilly, C.E., De Pasquale, G. and Resch, K., in Proc. 8th Leucocyte Culture Conf. (Eds. K. Lindahl-Kiessling and D. Osoba). Academic Press, 1974, p. 529.
7. Fisher, D.B. and Mueller, G.C., Proc. Natl. Acad. Sci. USA, 60: 1396, 1968.
8. Fisher, D.B. and Mueller, G.C., Biochem. Pharmacol. 20: 2515, 1971.
9. Greaves, M. and Janossy, G., Transplant. Rev. 11: 131, 1972.
10. Lapetina, E.G. and Michell, R.H., FEBS Letters, 31: 1, 1973.
11. Peters, J.H. and Hausen, P., Eur. J. Biochem. 19: 502, 1972.
12. Resch, K. and Ferber, E., Eur. J. Biochem. 27: 153, 1972.
13. Stobo, J.D., Rosenthal, A.S. and Paul, W.E., J. Immunol. 103: 1, 1972.
14. van den Berg, K.J., Thesis, University of Leyden, 1974.
15. van den Berg, K.J. and Betel, I., Cell. Immunol. 10: 319, 1974.

DISCUSSION FOLLOWING SYMPOSIUM
COUPLING OF CYTOPLASMIC AND NUCLEAR EVENTS TO MEMBRANE SIGNALS

Oppenheim - Bethesda: Can you pulse with the ionophore for varying periods of time and get similar effects or does it have to be present continuously?

Maino - Denver: We have not tried pulse exposure, but those kind of experiments may be difficult to do since the ionophore is lipophylic to begin with, so it quickly intergrates itself into membrane.

Oppenheim - Bethesda: Which brings me to a point which is that perhaps there is a limited duration during which it can actually work, which may account for the difference in the degree of thymidine uptake in comparison with PHA. Since PHA can continuously stimulate it produces multiple divisions, you can therefore accumulate more cells in mitosis and perhaps you're not getting as many generations of cells in response to the ionophore.

Maino - Denver: That is quite possible.

Loos - Amsterdam: Roos and Loos have studied the carbohydrate metabolism during phytohemagglutinin stimulation. We have found that in the initial phase of stimulation, glycolysis is enough to furnish you the energy you need and that later on the Krebs cycle is involved. So in relation to that finding, I wonder whether you have studied whether the calcium deposition in the mitochondria is time dependent because it might be that in the initial phase of cell activation, the Krebs cycle acitivty is not necessary.

Maino - Denver: In fact, we have found that in rat lymphocyte that the ionophore will stimulate glycolysis.

Kaplan - Ottawa: I hope this remark is not going to be interpreted as "sticking up for my own ion against yours"! The minimal hypothesis which you put forward, namely that everything depends on an initial calcium influx is fairly maximal. It seems to me the minimal hypothesis would be that the ionophore attacking the membrane of the lymphocyte is itself a mitogen and that as a result of this we have the indispensable calcium influx which we know already to be required from the work of others.

Lucas - Tucson: I would like to point out that this confirms some of our earlier studies with phosphotidylinositol. We found that even with PHA we get this phenomenon with phosphate incorporation into PI in only those cells stimulated to increase DNA synthesis and not in cells that similarly do bind PHA. There is a problem of what is going on metabolically relative to calcium transport whether you induce with PHA or ionophore. I think the ionophore is short-circuiting a system. PHA has probably also short-circuited a system. We are looking at the same end-point, DNA synthesis, but we cannot say that there are uniform events in between stimulation and end-point.

Maino - Denver: That is a good point.

Unidentified Questioner: Have you actually measured total calcium in the cell by stimulation?

Maino - Denver: My coworkers in England apparently have, and it correlates very well with PHA and ^{45}Ca uptake.

Furcht - Minneapolis: Some people have suggested that thymidine incorporation in lymphocyte to mitogen is contact dependent. Berlin's group has shown that colchicine interferes with the Con A mediated agglutination of fibroblast and neutrophils. I was wondering whether you have looked at this in your system to see if there is cell agglutination with colchicine and Con A.

Cunningham - New York: We see no difference in the number of agglutinated cells in cultures that have been incubated with Con A or Con A plus colchicine in the mitogenic assay.

Resch - Heidelberg: What is the real indication that intracellular structures like microtubule or microfilaments are involved in the inhibition of cell proliferation by colchicine? I think it is also evident that colchicine can have a direct membrane effect on for example, aggregation of intramembranous particles.

Cunningham - New York: To answer the question more specifically, we have looked at intramembranous particles under a variety of these conditions, and we do not see any difference in those. The biggest problem with assessment of microtubules is that nobody has seen directly into the cell surface in lymphocytes, so you have to postulate a more elaborate system. I think the inclusion of microfilaments in our case depends on the fact that we at least see inhibition by cytoclasin of cap formation.

MacDermott - Washington: One of the interesting things that we have found in our data in looking at subpopulations of lymphocytes is that B cells are and other cells as well have a markedly delayed proliferative peak. Time wise as compared to T cells and your slide in which you show that the colchicine plus Con A causes this exact same kind of delay, raises a number of points. Maybe you could comment on at least one possibility and that is that the colchicine plus the Con A might be inhibiting one population of lymphocytes.

Cunningham - New York: Yes, that is true.

Davies - London: Have you examined whether your cyclic GMP will also overcome the basic defect of degranulation in your granulocytes from Chediak-Higashi mice?

Oliver - Framington: We have not succeeded in doing that, our problem is simply that so far there are limited numbers of cells. We now have a patient and may be able to study this.

Parker - St. Louis: GTP is well known to influence microtubular aggregation. Has anyone ever shown directly in vitro that cyclic GMP influences microtubular aggregation under conditions in which the formation of GTP in that system has been excluded?

Oliver - Farmington: No one to my knowledge has actually shown a direct effect of cyclic GMP.

Parker - St. Louis: I would be very surprised if a lot of people haven't looked. I cannot quote any specific data but the model systems have been around now for some time. I am sure that the question has been looked at.

SUMMARY OF SYMPOSIUM
LYMPHOCYTE ACTIVATION: THE COUPLING OF MEMBRANE SIGNALS TO NUCLEAR AND CYTOPLASMIC EVENTS

M. R. Quastel
Ben Gurion University of the Negev, Soroka Medical Center, Beer Sheba, Israel.

The activation of the small lymphocyte, characterized by a very complex series of changes in the cell membrane, cytoplasm and nucleus (1), has proven to be a fertile source of research as well as occupation to lymphomaniacs. This cell system is particularly useful for studying how immunological and other agents acting at the level of the cell membrane lead to the activation of this repressed cell.

The induction of these changes is almost certainly brought about as a result of an interaction of the activating agent with cell membrane components, although the concurrent participation of other cells such as macrophages may be obligatory. A major biological problem is how events at the cell surface are translated into the 'mitogenic message' which is transferred into the interior of the cell and the nuclear apparatus. However, activation of the lymphocyte may take place along pathways other than those leading to mitosis. With respect to the interpretation of experimental observations, it must also be realized that, although a number of rapid cellular changes occur after administration of a mitogen, it is difficult to establish which are necessary, which sufficient, which non-contributory and which are secondary to the process of activation.

Following the introduction of mitogens to the medium of lymphocytes cultured in vitro, early membrane changes include rapid uptakes of Ca^{++}(2), amino acids (3,4), K^+ (5,7), nucleotides (8) and sugars (9). Rapid alterations in the intracellular levels of the cyclic nucleotides have been described (10,11). Endocytosis is stimulated (12) and early changes in the metabolism of membrane phospholipids have been reported (13-17). Membrane fluidity may change soon after activation (18). The dynamics of patch and cap formation on lymphocytes show that the plasma membrane of these cells acts as a two dimensional fluid, and raise the possibility that the redistribution of receptors and membrane proteins may be important in lymphocyte triggering (19). Rapid activation of cell membrane Na^+,K^+, ATPase has been proposed because of the enhanced cation transport and the early increase in binding of H^3-ouabain to the lymphocyte surface after treatment with PHA and leukoagglutinin (20). Rapid nuclear changes include enhanced histone acetylation and increased P^{32} turnover in nucleoprotein (21,22). There is increased RNA polymerase activity and accelerated synthesis of ribosomal RNA.

Does the complex series of events in lymphocyte blastogenesis involve a common pathway for different mitogenic agents <u>after</u> the initial activation of the cell membrane has taken place? Although this question cannot at present be answered, and must be qualified by the existence of various lymphocyte subpopulations with different properties, a simple sequential scheme may be considered:

Binding of mitogenic agent	→	Modification of membrane function and structure	↔	Cytoplasmic changes and 'messengers'	↔	Various nuclear and cytoplasmic events of activation

Which are the current candidates for the transmission of the mitogenic message from the membrane to the interior of the cell? Several phenomena have been reported which are not mutually exclusive. These involve the monovalent and bivalent cations, the cyclic nucleotides, and the microtubular assembly.

1. <u>Cations</u>

(a) <u>Calcium ions</u>

The induction of lymphocyte transformation by the ionophore A 23187, reported in this meeting by Maino, Green and Crumpton (23) is of considerable interest. The uptake of Ca^{++} into the lymphocyte is known to be accelerated very soon after the administration of PHA(2), and chelating agents are effective inhibitors of transformation (24,25). The ionophore is thought to facilitate the movement of bivalent cations across the cell membrane (26,27). By showing that the ionophore will initiate transformation, and that the effect is dependent on the level of Ca^{++} in the extracellular medium, a case is put forward for the possibility that the increased flux, or the level of intracellular Ca^{++}, is a necessary and possibly sufficient condition for lymphocyte activation.

Ca^{++} is known to be essential for certain reactions associated with activation, such as the stimulation of phosphatidylinositol turnover by PHA (13). The cleavage of phosphatidylinositol in the presence of lymphocyte membrane components was also shown to be Ca^{++} dependent (28). If increased intracellular Ca^{++} is indeed a critical link in the stimulation process, one can only speculate on its mechanism. Maino and coworkers (23) suggest that the following cellular functions may be affected: (i) control of the levels of cyclic GMP and cyclic AMP, (ii) the microtubular/microfilament system for endocytosis, and (iii) phosphorylase kinase required for the initiation of glycolysis. Might its action perhaps follow passage of Ca^{++} and other metabolites through nuclear pores (29)? Nevertheless, experimental evidence for a scalar increase of Ca^{++} in activated lymphocytes has not yet been obtained, and the role of other cation changes in the cell, following administration of the ionophore, must be clarified.

In the discussion of this paper, Dr. Oppenheim brought up the question whether the time required for the ionophore to cause its effect differed from that of PHA. Dr. Kirschner wondered whether the older findings of lymphocyte stimulation by mercury or zinc might be clarified by the results of the ionophore experiments. Dr. Kaplan pointed out that the minimum hypothesis would be that the ionophore itself is a mitogen, and that as a result of its activating effect, the indispensable increase in Ca^{++} flux takes place. Dr. Lucas suggested that the ionophore might 'short-circuit' the membrane process initiated by the mitogenic lectins.

(b) Potassium

Following the finding that ouabain, a specific inhibitor of the Na^+,K^+, ATPase and of monovalent cation membrane transport, reproducibly and reversibly inhibited PHA-induced lymphocyte transformation (30,31), it was shown that PHA induced a stimulation of K^{42} transport into the cell (5-7), but that it did not affect the rate of cation loss from the cell. The kinetics of the response led to a hypothesis and model whereby PHA caused expression of active K^+ sites on preexisting cryptic membrane Na^+,K^+, ATPase, by a conformational change involving membrane glycoproteins. A resultant alteration of the K^+ level in the cell was proposed to be a necessary (though not sufficient) condition for lymphocyte activation (32,33). Direct evidence for the rapid development of K^+ binding sites on the cell surface of activated lymphocytes is the increased V_{max} but not K_m for K^{42} uptake by PHA-stimulated cells, and indirect evidence is provided by the observation of increased H^3-ouabain binding to the lymphocyte surface within minutes of the administration of PHA or leukoagglutinin (20,34).

This view would postulate that the K^+ level, or perhaps the K^+/Na^+ ratio is an essential aspect of the signal to the nuclear apparatus. Analogous phenomena in other biological systems might be the roles of Na^+ and K^+ levels in the control of chromosomal puffing in Chironomus thummi salivary gland cells, in the control of fertilization (38) and in numerous metabolic systems (39). Activation of the Na^+, K^+, ATPase and K^+ transport may also be associated with the levels of the cyclic nucleotides (40) and with the fluidity of the membrane lipids (41).

2. Cyclic Nucleotides

The recent observation that cyclic GMP is rapidly increased in activated lymphocytes (11) has led to the suggestion that this nucleotide provides the signal for the induction of proliferation, and that cyclic AMP may act as a negative signal in the control of the onset and rate of activation (42,43). It has also been found that cyclic GMP, dibutiryl cyclic GMP and 8 bromo-cyclic GMP have mitogenic properties when administered to lymphocytes in vitro (44). Moreover, fluidization of membrane lipids has been found to be associated with an elevation of the level of cyclic GMP (45). On the other hand, cyclic GMP was also found to increase in polymorphonucleocytes and 3T3 fibroblasts following Con-A and PHA treatment respectively (46), suggesting that the effect is not unique to lymphocyte activation. A cyclic AMP and Ca^{++} mechanism for the control of proliferation of thymocytes has been proposed (47).

3. Microtubules

Wang and his colleagues (48) report that Con-A prevents receptor movement, as measured by patch and cap formation on both B and T cells, and that colchicine inhibits the mitogenic stimulation of lymphocytes by Con-A. As the inhibitory action of colchicine only occurred early in the sequence of events following the administration of the mitogen, it was inferred that the cytoskeleton plays an important role in mediating the signals produced by lymphocyte membrane phenomena to the cell interior.

In the discussion of this paper, Dr. Heidelbach suggested the possibility that colchicine might have also a direct membrane effect. Another discussant (unidentified) confirmed the inhibition of the Con-A and PHA response by colchicine, but noted that the drug had little effect on mitogen induced stimulation of Ca^{++} uptake. With regard to the delay of proliferation by colchicine, Dr. MacDermont suggested the possibility that the effect might not be simply a delay, but could also represent the inhibition of a subpopulation of lymphocytes (? T cells) so that B cells, which can have a delayed peak of proliferation, may then be observed.

In another aspect of this problem, Oliver (46) examined the role of microtubules in cell surface topology and the phagocytic function of polymorphonucleocytes. She reported that cyclic GMP antagonizes the stimulation of surface capping of Con-A that normally follows microtubule disruption by colchicine.

If a unifying theme could cover these diverse observations, it must include the probability that cellular events during the initiation of activation take place both in parallel and in sequence. We may postulate that after the initial membrane interaction with the stimulating agent has taken place, activation is initiated by the coming together of previously inactive enzyme subunits present in the lymphocyte membrane, which are potentially capable of forming active enzymatic complexes (49). From the results of the colchicine experiments, it would appear that such functional complexes should be 'ordered', and that the activated state would be brought about by the formation of an ordered matrix of active enzyme systems on or in the membrane, possibly following the change of fluidity in the lipid layer. Such a matrix could be visualized as analogous to the series of ordered respiratory enzymes of the mitochondrion, and would include enzyme systems needed for the transport of cations and other metabolites, and for the formation of the cyclic nucleotides. If the microtubular assembly is essential to the coming together of the enzyme subunits in an ordered way, perhaps the mechanism of action of colchicine is to prevent the development of the required ordered matrix mediated by the microtubular assembly.

Clearly, much is yet to be learned of the mechanism of signal transmission from membrane to the nucleus and cytoplasm, for which no clear process is yet apparent.

REFERENCES

1. Cooper, H.L., Proceedings of the Ninth Leukocyte Culture Conference (ed. A.S. Rosenthal), Academic Press, 1975, in press.
2. Whitney, R.B. and Sutherland, R.M., J. Cell. Physiol., 82: 9, 1973.
3. Mendelsohn, J., Skinner, Sr. A. and Kornfeld, S., J. Clin. Invest., 50: 818, 1971.
4. van den Berg, K.J. and Betel, I., Exp. Cell Res., 66: 257, 1971.
5. Averdunk, R., Hoppe-Seyler's Physiol. Chem., 353: 79, 1972.
6. Quastel, M.R. and Kaplan, J.G., Exp. Cell Res., 63: 230, 1970.
7. Quastel, M.R., Dow, D. and Kaplan, J.G., Proceedings of the Fifth Leukocyte Culture Conference (ed. J. Harris), Academic Press, pp. 97-123, 1970.
8. Peters, J.H. and Hansen, P., Eur. J. Biochem., 19: 502, 1971.
9. Peters, J.H. and Hansen, P., Eur. J. Biochem., 19: 509, 1971.
10. Smith, J.W., Steiner, A.L., Newberry, W.M., Jr. and Parker, C.W., J. Clin. Invest., 50: 432, 1971.

11. Hadden, J.W., Hadden, E.M., Haddox, M.K. and Goldberg, N.D., Proc. Nat. Acad. Sci., 69: 3024, 1972.
12. Hirschhorn, R., Brittinger, G., Hirschhorn, S. and Weissman, G., J. Cell Biol., 37: 215, 1968.
13. Fisher, D.B. and Mueller, G.C., Proc. Nat. Acad. Sci., 60: 1396, 1968.
14. Kay, J.E., Nature, 219: 172, 1968.
15. Resch, K., Ferber, E., Odenthal, J. and Fischer, H., Eur. J. Immunol., 1: 162, 1971.
16. Resch, K. and Ferber, E., Eur. J. Biochem., 27: 153, 1972.
17. Resch, K., Gelfand, E.W., Hansen, K. and Ferber, E., Eur. J. Immunol., 2: 598, 1972.
18. Kersey, J.H. and Barnett, R.E., quoted by Hadden, J.W. and van den Berg, K.J., 2nd International Conference of Immunology, in press.
19. Raff, M.C. and de Petris, S., Fed. Proc., 32: 48, 1973.
20. Quastel, M.R. and Kaplan, J.G., submitted for publication, 1975.
21. Kleinsmith, L.J., Allfrey, V.G. and Mirsky, A.E., Proc. Nat. Acad. Sci., 55: 1182, 1966.
22. Pogo, B.C.G., Allfrey, V.G. and Mirsky, A.E., Proc. Nat. Acad. Sci., 55: 805, 1966.
23. Maino, V.C., Green, N.M. and Crumpton, M.J., Proceedings of the Ninth Leukocyte Culture Concerence (ed. A.S. Rosenthal), in press, 1975; and Nature, 251: 325, 1974.
24. Alford, R.H., J. Immunol., 104: 698, 1970.
25. Whitney, R.B. and Sutherland, R.M., J. Cell Physiol., 80: 329, 1972.
26. Maino, V.C., Green, N.M. and Crumpton, M.J. Nature, 251: 324, 1974.
27. Pressman, B.C., Fed. Proc., 32: 1698, 1972.
28. Allan, D. and Michell, R.H., Biochem. J., 142: 599, 1974.
29. Maul, G.G., Maul, H.M., Scogna, J.E., Lieberman, M.W., Stein, G.S., Hsu, B.Y. and Bordun, T.W., J. Cell Biol., 55: 433, 1972.
30. Quastel, M.R. and Kaplan, J.G., Nature, 219: 198, 1968.
31. Quastel, M.R. and Kaplan, J.G., Exp. Cell Res., 62: 407, 1970.
32. Quastel, M.R., Wright, P. and Kaplan, J.G., Proceedings of the Sixth Leukocyte Culture Conference (ed. M.R. Schwartz), Academic Press, pp 185-214, 1972.
33. Kaplan, J.G. and Quastel, M.R., Proceedings of the Ninth Leukocyte Culture Conference (ed. A.S. Rosenthal), Academic Press, in press, 1975.
34. Quastel, M.R., Milthorp, P., Kaplan, J.G. and Vogelfanger, I.J., Proceedings of the Eighth Leukocyte Culture Conference (ed. Lindahl-Kiessling, K.), Academic Press, pp 493-

500, 1974.
35. Kroeger, H. and Trösch, W., J. Cell Physiol., 83: 19, 1973.
36. Kroeger, H. and Müller, G., Exp. Cell Res., 82: 89, 1973.
37. Kroeger, H., Trösch, W. and Müller, G., Exp. Cell Res., 80: 329, 1973.
38. Tupper, J.T., Develop. Biol., 38: 332, 1974.
39. Lubin, M., Nature, 213: 451, 1967.
40. Borasio, P.G. and Vassale, M., Am. J. Physiol., 226: 1232, 1974.
41. Grisham, C.M. and Barnett, R.E., Biochemistry, 12: 2635, 1973.
42. Rudland, P.S., Seeley, M. and Seifert, W., Nature, 251: 417, 1974.
43. Goldberg, N.D., O'Dea, R.F. and Haddox, M.K., in Advances in Cyclic Nucleotide Research, Vol. 3 (ed. P. Greengard and G.A. Robinson, Rover, New York), pp 155-223, 1973.
44. Weinstein, Y., Chambers, D.A., Bourne, H.R. and Melmon, K.L., Nature, 251: 352, 1974.
45. Barnett, R.E., Scott, R.E., Furch, L.T. and Kersey, J.H., Nature, 249: 465, 1974.
46. Oliver, J.M., Proceedings of the Ninth Leukocyte Culture Conference (ed. A.S. Rosenthal), Academic Press, in press, 1975.
47. Whitfield, J.F., Rixon, R.H., MacManus, J.P. and Balk, S.D., In vitro, 8: 257, 1973.
48. Wang, J.L., Gunther, G.G., Yahara, I., Cunningham, B.A. and Edelman, G.M., Proceedings of the Ninth Leukocyte Culture Conference (ed. A.S. Rosenthal), Academic Press, in press, 1975.
49. Novogrodsky, A. Proceedings of the Ninth Leukocyte Culture Conference (ed. A.S. Rosenthal), Academic Press, in press, 1975.

SECTION VI

CELLULAR AND MOLECULAR BASIS OF CELL RECOGNITION AND INTERACTION IN NONLYMPHOCYTE SYSTEMS

Chairman

VICTOR GINSBURG

National Institutes of Health
Bethesda, Maryland

Co-Chairman

LUIS GLASER

Washington University School of Medicine
St. Louis, Missouri

STUDIES ON THE CELL SURFACE RECEPTORS FOR IgE

Henry Metzger
National Institute of Arthritis, Metabolism and Digestive Diseases, National Institutes of Health, Bethesda, Maryland.

IgE bound to the surface of mast cells or basophils acts as an antigen receptor. In the presence of antigen such cells rapidly release histamine as well as other vasoactive substances. This system has many similarities to the antigen-induced antibody mediated reactions of lymphocytes. For example aggregation of the immunoglobulin appears to be a critical early event and influx of Ca^{++} may serve as one of the necessary early sequelae.

Although alternative mechanisms can be imagined it seems likely that the IgE mediated stimulation occurs via a change induced in that component with which the IgE interacts: the IgE receptor. We have therefore explored a system in which the IgE cell membrane interaction can be studied rigorously and from which it may ultimately be possible to isolate and characterize such receptors. We have been working with rat basophilic leukemia (RBL-1) cells which can be adopted to cell culture and which retain on their surface membranes highly specific binding sites for IgE (1). The nature of the interaction has been studied in detail (2). Mouse IgE but not human IgE or a variety of other proteins bind to the cells. Binding is via a simple reversible reaction with a $K_A \geq 10^{10}$ M^{-1}. In a limited number of studies normal mast cells showed similar IgE binding properties. The effective number of binding sites on the RBL-1 cells varies depending upon the growth conditions (3).

Recently we have obtained cell free preparations which continue to show specific IgE binding similar to that seen with intact cells (4).

It remains to be proven that the IgE binding component on the RBL-1 cells is a "functional" receptor. So far we have been unable to detect histamine release from these cells via an IgE mediated mechanism. However, studies with a Ca^{++} dependent inophore suggest that the 'block' may occur at some later stage in the release phenomenon (5).

REFERENCES

1. Kulczycki, A., Jr., Isersky, C. and Metzger, H.: The interaction of IgE with rat basophilic leukemia cells. I. Evidence for specific binding of IgE. J. Exp. Med.

139: 600-616, 1974.
2. Kulczycki, A., Jr., and Metzger, H.: The interaction of IgE with rat basophilic leukemia cells. II. Quantitative aspects of the binding reaction. J. Exp. Med. 140: 1974 (in press).
3. Isersky, C., Metzger, H. and Buell, D.: Cell cycle associated changes in receptors for IgE during growth and differentiation of a rat basophilic leukemia cell line. Submitted for publication, 1974.
4. Carson, D., Kulczycki, A., Jr. and Metzger, H.: The interaction of IgE with rat basophilic leukemia cells. III. Release of intact receptors on cell-free particles. J. Immunol. 114: 1975 (in press).
5. Siraganian, R., Kulczycki, A., Jr., Mendoza, G. and Metzger, H.: Manuscript in preparation.

CELL INTERACTION IN THE PACKAGING OF LYSOSOMAL ENZYMES

Elizabeth F. Neufeld
National Institute of Arthritis, Metabolism, and Digestive Diseases, Bethesda, Maryland.

Intercellular cooperation in degradative processes was first observed in the study of fibroblasts cultured from the skin of patients with inherited mucopolysaccharide storage disorders. The abnormal mucopolysaccharide catabolism of such fibroblasts could be "corrected" by mixing cells of patients known to be genetically different, such as Hurler and Hunter, with each other or with normal cells (1). The process was found to be mediated by diffusible macromolecules, or "corrective factors". The process of correction is now understood to be the transfer of hydrolytic enzymes to deficient cells (e.g., of iduronidase to Hurler cells) so that catabolism of mucopolysaccharide is restored to normal (2).

Correction requires uptake of the enzyme from the medium as well as catalytic activity within the lysosomes of the capturing cell. The uptake can occur selectively and efficiently, presumably by recognition of a feature on the enzyme by some receptor on the cell membrane (3). The recognition marker for uptake of enzyme by fibroblasts is different from that required for uptake of plasma proteins by hepatocytes (see lecture by G. Ashwell).

Secretion and selective recapture are probably essential for sequestering hydrolases within lysosomes of connective tissue cells. This is suggested by study of mucolipidoses II and III, two genetic disorders in which several hydrolytic enzymes are inappropriately localized extracellularly rather than intralysosomally (2,4). A defect in the recognition marker on the hydrolases accounts for their extracellular accumulation (5).

REFERENCES

1. Neufeld, E.F. and Fratantoni, J.C., Science 169: 141-146, 1970.
2. Neufeld, E.F., The Biochemical Basis for Mucopolysaccharidoses and Mucolipidoses, in Progress in Medical Genetics, eds. A.G. Steinberg and A.G. Bearn, 10: 81-101, 1974.
3. Hickman, S., Shapiro, L.J. and Neufeld, E.F., Biochem. Biophys. Res. Commun. 57: 55-61, 1974.
4. Neufeld, E.F., Ann. Rev. Biochem., in press, 1975.

5. Hickman, S. and Neufeld, E.F., Biochem. Biophys. Res. Commun. 49: 992-999, 1972.

STUDIES ON THE HEPATIC RECOGNITION AND REMOVAL OF CIRCULATING GLYCOPROTEINS

Gilbert Ashwell and Anatol G. Morell
National Institute of Arthritis, Metabolism, and Digestive Diseases, National Institutes of Health, Bethesda, Maryland, and Albert Einstein College of Medicine, New York, New York.

Studies on the chemical and enzymatic modification of the carbohydrate moiety of serum glycoproteins have provided evidence to indicate that the terminal, nonreducing sugars play a significant role in regulating the serum survival time of circulating glycoproteins. In essence, the hypothesis has been advanced that the presence of sialic acid residues is critical for continued viability in the circulation. Upon treatment with neuraminidase, galactose is exposed as the newly terminal sugar and constitutes a recognition signal for the prompt hepatic removal of the sialic acid-deficient molecule. This concept has been described in a more detailed review which has recently appeared in press (Ashwell and Morell, 1974).

Subsequently, the hepatic binding protein has been isolated in water-soluble form after purification by affinity chromatography (Hudgin et al., 1974). This material, which is free from lipid, has been identified as a glycoprotein in which 10% of the dry weight consists of sialic acid, galactose, mannose, and glucosamine in a molar ratio of 1:1:2:2. Here, as with hepatic plasma membranes, the integrity of the terminal sialic acid residues and the presence of calcium were shown to be absolute requirements for binding. Utilizing this material, it became possible to examine critically a previously postulated role for galactosyl transferase in the hepatic recognition and binding of asialoglycoproteins (Aronson et al., 1973). Under optimally determined conditions, the purified binding protein contained no detectable glycosyltransferase activity for sialic acid, galactose, N-acetylglucosamine or fucose despite a 200-fold increase in the specific binding capacity for asialoglycoproteins (Hudgin and Ashwell, 1974).

Further investigation of the properties of this material by Stockert et al. (1974) has revealed that the purified binding protein possesses the property of agglutinating human and rabbit erythrocytes. The titer is highest for blood group A cells and is readily inhibited by N-acetylgalactosamine. The agglutination phenomenon parallels closely the binding properties of the purified protein in that it requires the addition of calcium and the presence of intact sialic acid residues. Agglutination is blocked in the presence of added asialoglycoproteins and is reversed by the removal of calcium. In short, the hepatic binding protein appears to qualify as the first recognized lectin of mammalian origin.

REFERENCES

1. Aronson, N.N., Tau, L.Y. and Peters, B.P., Biochem. Biophys. Res. Commun. 53: 112-118, 1973.
2. Ashwell, G. and Morell, A.G., Advances Enzymol. 41: 99-128, 1974.
3. Hudgin, R.L. and Ashwell, G., J. Biol. Chem. (in press), 1974.
4. Hudgin, R.L., Pricer, W.E., Ashwell, G., Stockert, R.J., and Morell, A.G., J. Biol. Chem. 249: 5536-5543, 1974.
5. Stockert, R.J., Morell, A.G., and Scheinberg, I.H., Science (in press), 1974.

INTERCELLULAR RECOGNITION AND COMMUNICATION BY CELL SURFACE GLYCOSYLTRANSFERASES AND OLIGOSACCHARIDES[1]

Stephen Roth
Department of Biology, The Johns Hopkins University, Baltimore, Maryland 21218.

Molecular complementarity at the cell surface was suggested years ago as an explanation for many morphogenetic and communicative events in embryogenesis and physiology (1-3). More recently, the complementarity was thought to result from protein-carbohydrate interactions (4,5), at least between cultured fibroblasts.

In the last three years, it has become reasonably clear that some cultured cells contain glycosyltransferases and glycosyl acceptors on their surfaces (6-8). Evidence for this conclusion comes from several different experimental techniques:

1. Intact cells can transfer sugars from their nucleotide derivatives to either endogenous (cell surface) or exogenous (soluble) acceptors.
2. This transfer occurs in the absence of significant degradation of the sugar-nucleotides and in the absence of uptake of the sugar-nucleotide, sugar-phosphate or free sugar. Competitive and direct inhibitors of transport and hydrolysis have been used to establish these points.
3. Untransformed cells require intercellular contact for maximum endogenous glycosylation. Malignant cells do not.
4. Untransformed cells can glycosylate acceptors that are covalently bound to glass coverslips while transformed cells do this poorly.

A wide variety of biological functions have been suggested for these surface components (9) but the best documented is that they act in a lock-and-key fashion to account for some degree of cell recognition.

1. Adhesive specificity between embryonic retinal cells is perturbed by glycosyl acceptors for which there are cell surface transferases (10).
2. Fibroblast adhesion is inhibited by acceptors that are recognized by intact cells and by isolated plasma membranes (11).

---

[1]The work from the author's laboratory reported here was supported by research grants from the National Institute of Child Health and Human Development and The American Cancer Society.

3. Fibroblasts recognize and adhere to some Sepharose beads that have been derivatized with galactose (12) and can transfer monosaccharides to coverslips derivatized with glycolipids (13).

4. Blood platelets seem to adhere to collagen through an interaction between the platelet surface glucosyltransferase and collagen galactosides (14-16).

REFERENCES

1. Weiss, P., Yale J. Biol. Med. 19: 235, 1947.
2. Tyler, A., Growth, 10 (Symposium 6): 7, 1946.
3. Sperry, R., in Organogenesis, Holt, Rinehart and Winston, 1965.
4. Cox, R. and Gesner, B., Proc. Nat. Acad. Sci. 54: 1571, 1965.
5. Gesner, B.M. and Ginsburg, V., Proc. Nat. Acad. Sci. 52: 750, 1964.
6. Roth, S. and White, D., Proc. Nat. Acad. Sci. 69: 485, 1972.
7. Bosmann, H.B., Biochem. Biophys. Res. Comm. 48: 523, 1972.
8. Warren, L., Fuhrer, J.P. and Buck, C.A., in Membrane Mediated Information, American Elsevier, 1974.
9. Roth, S., Quart. Rev. Biol. 48: 541, 1973.
10. Roth, S., McGuire, E.J. and Roseman, S., J. Cell Biol. 51: 536, 1971.
11. Lloyd, C. and Cook, G., J. Cell Sci. 15: 575, 1974.
12. Chipowski, S., Lee, Y.C. and Roseman, S., Proc. Nat. Acad. Sci. 70: 536, 1973.
13. Yogeeswaran, G.A., Laine, R. and Hakomori, S., Biochem. Biophys. Res. Comm. 59: 591, 1974.
14. Barber, A.J. and Jamieson, G.A., Biochim. Biophys. Acta 252: 533, 1971.
15. Bosmann, H.B., Biochim. Biophys. Acta 279: 456, 1972.
16. Jamieson, G.A., in Miami Winter Symposium No. 7, Academic Press, 1974.

SECTION VII

THE ROLE OF THE MACROPHAGE IN THE INITIATION AND REGULATION OF THE IMMUNE RESPONSE

Chairman

ZANVIL A. COHN

The Rockefeller University
New York, New York

Co-Chairman

JOOST J. OPPENHEIM

National Institutes of Health
Bethesda, Maryland

THE SAGA OF THE MACROPHAGE

Zanvil A. Cohn, M.D.
The Rockefeller University, New York, New York

The macrophage is a most difficult cell. Unlike the bone marrow derived lymphocyte which at least superficially has a single, well defined role in life, studies on macrophages have suggested a plethora of roles in the in vitro immune response, the extent of which are largely governed either by the industry and imagination of the investigator or the biosynthetic capabilities of its endoplasmic reticulum.

Although difficult in one sense it is most cooperative in another. In keeping with the Judeo-Christian ethic it helps the less fortunate in its micro-environment and tolerates its peers. It schleps, it ingests, it drinks, it digests, it suffers from dyspepsia and regurgitates upon command. It secretes, it binds, it stimulates, it inhibits, it is even informative - in fact its mini RNA defies the established tenets of molecular and cell biology. It produces super-antigens of Bunyonesque adjuventicity, is a processor of no mean talent - processing is a term near and dear to the heart of many which I can't really define for you and a term which should be dropped from our armamentarium and replaced with more objective expressions. It has a sixth sense and can recognize the good from the bad, the self from the non-self, the benign from the malignant, the B from the T and the C' from the Fc. It fuses, it fragments, it transfers. It ruffles its membranes in a most provocative fashion but replicates with only the greatest of difficulty. It kills, it activates, it modulates, it differentiates. When activated its anger knows no bounds and as you all know it is easily M.I.F.'ed. It flows with the greatest of ease, it balks, it indurates, it is a healer of wounds.

It is a perfect delight for those who culture it in glass and plastic vessels. It spreads, it adheres with a vise-like grip. It is long-lived and like the elderly is satisfied with the simplest of fares - it is even called upon to accept plastic beads as a main course. It does not require the foul smelling reducing agents so important for its more fastidious colleagues. It is in this in vitro environment, removed from its nearest neighbors, that it performs with such gusto and flagrant exhibitionism.

Yet, it is a most difficult cell which in its natural habitat of spleen and lymph node, has elusive and secretive habits. It defies the investigator to follow its tracks after adoptive transfer. It shields itself in a forest of stellate, dendritic cells some of which may trap the proverbial antigen. Its precursors are rather protean and lack unique sensitivities to ionizing irradiation and the mature forms are radioresistant. Therefore, we cannot easily obliterate it and study the resulting defect. In short, we know little about the in vivo role of macrophages in controlling the immune response.

THE ROLE OF MACROPHAGES IN T LYMPHOCYTE ANTIGEN RECOGNITION

A.S. Rosenthal, J.T. Blake, J.J. Ellner
D.K. Greineder and P.E. Lipsky
Section on Biologic Structure, Laboratory of Clinical Investigation, National Institute of Allergy and Infectious Diseases, National Institutes of Health, Bethesda, Maryland 20014.

INTRODUCTION

One view of the mechanism of antigen recognition by thymus-derived lymphocytes holds that the signal for lymphocyte activation comes as a direct result of the interaction of molecules of antigen with a specific lymphocyte receptor. Such direct antigen binding by the T lymphocyte results in the sequence of cellular events leading to RNA, protein and DNA synthesis in a manner analogous to the activation of lymphocytes by mitogens. However, a variety of observations have made this view of the mechanism of antigen-specific signal recognition by T lymphocytes less likely. These include: 1) failure to identify or define an antigen-specific receptor on the T cell comparable to surface membrane immunoglobulin on the B cell, 2) the lack of easily demonstrable direct antigen binding to T lymphocytes and 3) the observation that while obviously necessary, the immuno-specific lymphocyte is not itself sufficient for expression of in vitro correlates of in vivo established cell-mediated immunity. With respect to this latter issue, it is now recognized that the minimal immunogenic unit requires an accessory cell, the macrophage, whose precise role in the antigen recognition process is only now being realized.

This paper will review data concerning the function of the macrophage in antigen recognition by the T lymphocyte. In addition, we will propose a mechanism of antigen-specific lymphocyte triggering which is based on studies from our own laboratory on activation of T lymphocyte proliferation using inbred guinea pigs. This model proposes that antigen recognition involves an initial metabolic dependent uptake of antigen by the macrophage. The immunologically active antigenic moiety escapes macrophage degradation and is presented to the lymphocyte during cell-cell interaction. Lymphocyte activation will follow antigen presentation by the macrophage only if 1) both macrophage and lymphocyte share histocompatibility linked cellular interaction sites and appropriate immune response (Ir) gene products; and 2) the lymphocyte has a receptor specific for the antigen bound to the macrophage. Thus macrophage-lymphocyte interaction is envisaged as playing a central role in the initiation, regulation, and genetic control of the immune response.

THE MACROPHAGE REQUIREMENT FOR IN VITRO ANTIGEN RECOGNITION

Lymphocytes from immunized animals proliferate in vitro when exposed to specific antigens. This is most simply assessed by measurement of ^{3}H-thymidine incorporation into new DNA after 48-96 hours of culture. A number of laboratories have demonstrated that this response was depressed by removing glass adherent cells and enhanced on reintroduction of macrophages (1,2,3,4,5). Thus the possibility was raised that, in this assay, the macrophage might serve as an accessory cell by supporting viability and/or by presenting antigen to the lymphocyte in some highly favorable configuration.

Studies from this laboratory (4,5,6) have shown that the initial antigen binding cell in populations of T lymphocyte enriched lymphoid cells is an adherent, phagocytic radioresistant cell which appears morphologically to be a macrophage. Furthermore, these studies indicated that antigen recognition involved an initial obligatory macrophage uptake of antigen antecedent to interaction of the immunogenic moiety with the T lymphocyte. Macrophages bearing antigen separated from the lymphocytes by a 0.45 micron filter do not cooperate in initiation of lymphocyte proliferation indicating that close contact between antigen-bearing macrophage and lymphocytes is required (1). This latter experiment is of some importance since it counters the possibility that macrophages merely serve as a passive vehicle for antigen carryover.

The importance of the macrophage as an antigen binding cell is further supported by the difficulty in demonstrating direct antigen binding T lymphocytes in the guinea pig (reviewed ref. 7). This contrasts with the B lymphocyte which has, on its surface membrane, easily detectable immunoglobulins with specific antigen binding capacity. In a series of studies in the guinea pig using dinitrophenylated hapten-protein conjugates, Davie and coworkers have characterized the specificity and avidity of the B cell receptors in both non-immune and previously immunized animals. These receptors exhibit hapten but not carrier specificity. When the inhibition of T lymphocyte proliferation and antigen binding to B lymphocytes were studied in the same population, hapten on a heterologous carrier blocked antigen binding to B cells as measured autoradiographically but did not block T lymphocyte proliferation. Again, antigen binding to T lymphocytes was not seen despite its ability to induce antigen-dependent proliferation. While the authors recognize that T lymphocytes must have some antigen-specific receptor, be it immunoglobulin in levels too low for detection or some unique receptor molecule, the data would suggest that for functional expression, direct interaction of antigen with such a receptor either does not occur or is so weak that multivalency must be acquired on a macrophage matrix.

In the mouse, macrophages also maintain lymphocyte viability in Mishell-Dutton cultures (8) and in the mixed leukocyte reaction (9). This trephocytic function can be replaced by reducing compounds such as 2-mercaptoethanol. However, in the guinea pig, 2-mercaptoethanol does not replace the macrophage requirement for antigen recognition (unpublished observation).

UPTAKE AND FATE OF MACROPHAGE-ASSOCIATED ANTIGEN

Macrophage (MΦ) uptake of all soluble proteins proceeds by fluid phase macro- and micro- pinocytosis. However, in the case of some proteins including PPD, KLH and DNP-GPA, there is extensive antecedent membrane binding (10,11,12). This binding step results in marked increase in the uptake of these proteins by macrophages as compared with others such as horseradish peroxidase which are taken up exclusively by fluid phase pinocytosis (13). Following uptake and pinocytosis, there is a period of rapid catabolism during which proteins are degraded to amino acid residues and released from the cell (14). However, in the case of a number of autologous and heterologous proteins including mouse serum albumin and keyhole limpet hemocyanin (15,16), 10-25% of the initially cell-associated protein resists catabolism and is stored in a stable form (approximately one-third of it on the MΦ surface).

The corresponding immunogenicity of MΦ-associated antigen has been studied most extensively by transfer of cells to prime syngeneic recipient mice with subsequent assay of antibody production (11,16). These studies indicated that the stable surface compartment was of major immunologic significance since its removal with trypsin or covering with Fab' fragments of antibody to antigen decreased the eventual level of antibody formation following cell transfer. The quantitative stability of surface MΦ-associated antigen was emphasized by the capacity of antigen-bearing macrophages to transfer full immunogenicity despite preliminary in vitro culture (17).

Recent studies in our laboratory have utilized the immune T lymphocyte proliferative response to MΦ-associated antigen as an in vitro assay of immunogenicity in order to examine the uptake and handling of DNP-GPA by guinea pig macrophages

(12). This hapten-protein conjugate is extensively bound to the macrophage prior to internalization. During the period of rapid catabolism, 80% of the initially cell-bound material is released from the macrophage while 20% of the stable, residual material remains surface-associated (Fig. 1). Preliminary culture of macrophages bearing DNP-GPA for up to 3 days results in progressive fall-off in ^{3}H-thymidine incorporation when such macrophages are mixed with immune T lymphocytes. This fall-off does not reflect either the extent of rapid catabolism nor the eventual qualitative and quantitative stability of residual antigen. Moreover, trypsinization did not further diminish the immunogenicity of pulsed, cultured macrophages.

These findings suggest that during culture there is a progressive decrease in the ability of antigen associated with macrophages to induce proliferative responses in immune T lymphocytes. This might be due either to a loss of macrophage functional efficiency or intracellular redistribution of antigen into some inaccessible pool. The lack of an effect of trypsinization on immunogenicity indicates that surface trypsin-sensitive antigen is not the sole source of immunologically active material, nor is it the critical pool in the case of guinea pig T lymphocyte activation. The mechanism by which intracellular MΦ-associated antigen becomes available to the guinea pig T lymphocytes is unclear. One attractive possibility is that during macrophage-lymphocyte interaction there is directed return of antigen to the cell surface by a process similar to exocytosis.

MORPHOLOGIC CORRELATES OF MACROPHAGE-LYMPHOCYTE INTERACTION

The association of lymphocytes with macrophages has been observed both in vivo and in vitro and in the presence and absence of relevant antigen or specific immunizing procedures (18). The physiologic significance of these associations is only now being understood.

Recent studies from our laboratory have described some of the characteristics of the binding of allogeneic and syngeneic guinea pig lymphocytes to macrophages which occurs in the absence of antigen. Binding occurs rapidly (maximal binding between 30-60 minutes) and is dependent on the number of lymphocytes present during the assay. Some of the features of this phenomenon are summarized in Table I. While it is clear that the binding event requires divalent cations and active macrophage but not lymphocyte metabolism, it remains to be established whether the trypsin sensitivity of the macrophage component represents an indirect effect of trypsin on the cell surface or actual removal of the macrophage recep-

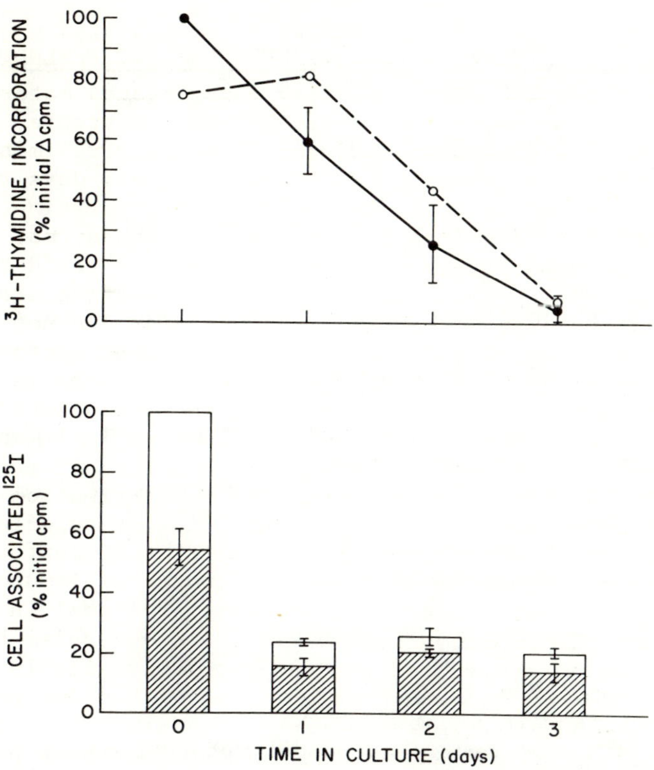

Fig. 1 Comparison of the quantitative aspects of the uptake and handling of DNP-GPA by guinea pig macrophages with resultant immunogenicity.

Lower Panel - Macrophages were pulsed with ^{125}I-DNP-GPA for one hour at 37°, washed and cultured. At the appropriate intervals, aliquots of cells were washed again and trypsin-resistant (hatched bars) and sensitive (unhatched bars) components determined.

Upper Panel - In parallel experiments, groups of macrophages pulsed with 1.0 μg/ml of DNP-GPA under the above conditions were washed, cultured and rewashed with trypsinization of cell aliquots. 50% control (solid line) or trypsin treated (hatched line) macrophages were then added back to peritoneal exudate lymphocytes from DNP-GPA immune animals with determination of ^{3}H-thymidine incorporation at 72 hours.

TABLE I
CHARACTERISTICS OF ANTIGEN-INDEPENDENT AND DEPENDENT MACROPHAGE-LYMPHOCYTE INTERACTION

Antigen-Independent Phase (1 Hour)

1. Reversible
2. Requires metabolically active macrophages but not lymphocytes
3. Requires Ca^{++}
4. Trypsinization of macrophage but not lymphocyte abolishes binding
5. Is not blocked by excess immunoglobulin
6. Does not require serum
7. Both T and B lymphocytes bind in proportion to their frequency in the population
8. Both allogeneic and syngeneic lymphocytes bind to macrophages
9. Not blocked by alloantisera
10. Blocked by cytochalasin-B

Antigen-Dependent Phase (24 Hours)

1. "Irreversible"
2. Does not require serum
3. Requires active macrophage and lymphocyte metabolism
4. Only syngeneic lymphocytes and macrophages interact (except in MLR)
5. Not caused primarily by mediators released by activated lymphocytes
6. Blocked by alloantisera
7. Carrier or conjugate but not hapten specific
8. Not reversed by cytochalasin B or mitomycin
9. Does not occur if antigen-independent phase is first blocked, e.g. by cytochalasin.

tor for lymphocytes. The chemical nature of the interacting molecules is not known although we can state that it does not involve immunoglobulin either on the lymphocyte or macrophage and that the macrophage receptor site does not distinguish between T and B lymphocytes since they bind in proportion to their frequency in the population exposed to macrophages.

More recently, we have extended these observations of macrophage-lymphocyte interaction to examine the effects of antigen and specifically immune lymphocytes on the degree of binding observed (19). We found that the presence of antigen on macrophages significantly affects the course of macrophage-lymphocyte interaction (Table II). While binding in the presence and absence of antigen at one hour of culture was indistinguishable, prolongation of culture revealed a difference between the number of lymphocytes from immunized animals bound to macrophages in the presence of that antigen as contrasted with the control macrophages not exposed to antigen (Fig. 2). These differences were manifest as early as 6-8 hours of culture and were maximal by 18-20 hours of incubation. The degree of binding at 20 hours was dependent on concentration of antigen and was carrier as opposed to hapten specific. The latter observation indicates that the enhanced binding was not due to cytophilic antibody-antigen complexes. Moreover, when antigen-exposed and control macrophages were incubated with lymphocytes in the same chamber, binding was seen predominantly on those macrophages bearing antigen suggesting that binding was not solely the result of a soluble product of lymphocyte activation. To obtain further evidence as to the functional significance of binding, cultures were pulsed at various times with ^{3}H-thymidine and incorporation into lymphocyte DNA assessed autoradiographically. Although antigen-dependent lymphocyte binding is observed as early as six hours, bound lymphocytes were noted to incorporate significant ^{3}H-thymidine only at 48 and 72 hours (Fig. 2). These studies suggest that antigen-dependent binding is an early event in the detection of the antigenic signal.

The relationship of antigen-independent and -dependent physical interactions between macrophages and lymphocytes and the cellular interaction structures hypothesized on the basis of functional assays of T-B collaboration in the mouse by Katz and Benacerraf (20) and of macrophage-lymphocyte interaction in the guinea pig by Shevach et al. (21) is not clear. One interpretation is that morphologically defined physical interactions represent a critical determinant of functional cooperative interactions. Suffice it to say that

TABLE II
EFFECT OF SPECIFIC ANTIGEN ON THE MACROPHAGE BINDING OF LYMPH NODE LYMPHOCYTES FROM PPD* IMMUNE GUINEA PIGS

Macrophage Pulse	Lymph Node Lymphocytes Bound Per 100 Macrophages	
	1 Hour	20 Hours
0	72.3 ± 5.5	22.7 ± 2.0
PPD (100 μg)	75.1 ± 5.6	88.2 ± 3.2

*Tuberculin purified protein derivative

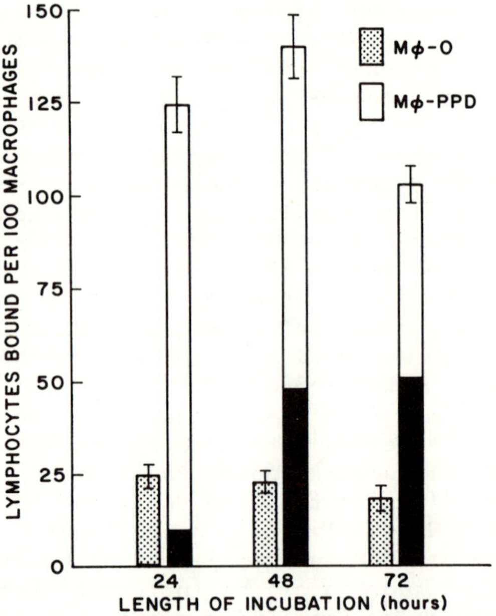

Fig. 2 The binding of LNL from PPD immune animals to syngeneic macrophages during prolonged incubation. The macrophages had previously been exposed to either 100 μg/ml of PPD in PBS (macrophage-PPD) or to PBS alone (macrophage-0) for 60 minutes at 37°C and washed four times prior to use. Each point represents the mean ± standard error of the mean of three replicate experiments. Shaded areas of the bars indicate the number of macrophage-bound LNL which were detected autoradiographically to be incorporating tritiated thymidine.

the definition of such terms as "cellular interaction structures" based solely on functional assays of cellular cooperation is fraught with significant uncertainty.

GENETIC REGULATION OF FUNCTIONAL MACROPHAGE-LYMPHOCYTE INTERACTION

In the guinea pig, activation of T lymphocyte DNA synthesis by macrophages bearing complex protein antigens requires the cooperation of syngeneic macrophages and lymphocytes (22). The failure of allogeneic antigen-bearing macrophages to effectively initiate proliferation of T lymphocytes immune to that antigen was not the result of the presence of an inhibitor of blastogenesis released in mixtures of allogeneic cells or differing kinetics of activation. It is of interest in this regard that we have also found that while antigen-independent binding of lymphocytes by macrophages occurs equally in allogeneic and syngeneic combinations, antigen-dependent binding is observed only in syngeneic combinations of macrophages and lymphocytes (19). In other experiments, functional cooperation between macrophages bearing antigen and lymphocytes immune to that antigen was blocked by specific alloantisera when directed against determinants present on both macrophage and lymphocyte. In preliminary experiments, antigen-dependent binding of lymphocytes to macrophages was also blocked by specific alloantisera. One objection which can be raised to studies of the effect not only of alloantisera but of inhibitors in general is that in the proliferation assay, the response, i.e. ^{3}H-thymidine incorporation, is separated from the initial signal (addition of antigen-bearing macrophages) by periods in excess of 48-72 hours. The MIF assay has the distinct advantage that it can be assessed as early as 18 hours following the mixing of antigen-bearing macrophages and lymphocytes. Therefore, Ben-Sasson and colleagues (25) obviated these objections to some extent by confirming that syngeneic macrophages are required for production of macrophage inhibitory factor (MIF) and by further showing MIF production induced by antigen-bearing macrophages was inhibited by specific alloantisera.

A number of laboratories have suggested that the main function of the products of immune response (Ir) genes (reviewed 23) is in antigen recognition. Since functional and physical macrophage-lymphocyte interactions are controlled by genes linked to the main histocompatibility complex, we began studies exploring this relationship using antigens under genetic control such as the random copolymer of L-glutamic-acid-L-tyrosine (strain 13 responder; strain 2 non-responder) and L-glutamic-acid-L-lysine (strain 2 responder; strain 13 non-responder) and F_1 (non-responder X responder) animals which respond to both synthetic copolymers (24). Parental macrophages pulsed with an antigen, the response to which is controlled by an Ir gene lacking in that parent, will not stimulate F_1 (responder) T lymphocyte DNA synthesis. By contrast, parental macrophages pulsed with antigens controlled by Ir genes which they possess will initiate F_1 T lymphocyte proliferation.

MODEL BINDING FOR THE CELLULAR AND MOLECULAR EVENTS IN ANTIGEN RECOGNITION BY T LYMPHOCYTES

Based on studies from our own as well as other laboratories, we would propose a model for antigen initiated T lymphocyte secretory and proliferative behavior. Recognition of antigen is a cooperative event between the immunospecific T lymphocyte and an antigen-bearing macrophage. Detection of the antigenic signal requires: 1) an antigen specific receptor on the lymphocyte; 2) a functionally defined, genetically determined cellular interaction structure on both macrophage and lymphocyte linked to a serologically defined component of the major histocompatibility complex; 3) antigen bound to metabolically intact macrophages in a form resistant to proteolytic treatment and located in a cellular site whose display on the macrophage surface is restricted; 4) an auxiliary antigen recognition molecule presumably an Ir gene product on either or both macrophage and lymphocyte and 5) a macrophage associated nonspecific membrane trigger site or secretory product.

For the purposes of this discussion, knowledge of the specific molecular character of the antigen T receptor, be it immunoglobulin or otherwise, is not required. Moreover, the reader should note that the data dictate that for initiation of lymphocyte proliferation, antigen must not be presented in an unrestricted fashion. Thus, either H-linked cellular interaction regulates its presentation or the Ir gene product modulates the accessibility of antigen to the lymphocyte receptor. Such modulation could occur by either enzymatic or conformational changes in either the antigen itself or in the antigen presentation process.

Viewed dynamically, macrophage-associated antigen initiation of T lymphocyte proliferation proceeds via an initial antigen-independent phase of binding of the immune lymphocytes to macrophages. Initial macrophage-lymphocyte interaction is followed by a period of antigen-dependent lymphocyte binding to the macrophage and DNA synthesis if and only if 1) both macrophage and lymphocyte share identity of membrane linked H-associated determinants and 2) the macrophage bears the appropriate antigen for which the lymphocyte has an immunospecific receptor. Thus, antigen-independent binding serves to approximate or focus lymphocytes on the macrophage without conveying immune specificity; a subsequent antigen and histocompatibility dependent physical and functional phase of interaction conveys specificity and an "auxiliary receptor" or Ir gene product regulates genetically controlled immune responses.

Support for a model, in which the Ir gene product is not the antigen specific receptor is found in the studies of Hämmerling and McDevitt (26) who find equivalent numbers of antigen binding T and B lymphocytes in non-responder as well as responder mice and in the studies of Kapp et al. (27), using the terpolymer L-glutamic-acid-L-alanine-L-tyrosine in which immunized non-responder mice develop suppressor rather than helper T cells.

In addition, Shevach et al. (21) have recently shown that functional cooperation between macrophages bearing antigen and lymphocytes requires the presence on both of products of the classic Ir gene as well as products of separate genes linked to the serologically definable histocompatibility complex. These studies used inbred strain 2, strain 13, and phenotypically characterized outbred guinea pigs and showed that macrophages bearing genetically controlled antigens do not initiate proliferation in lymphocytes from immunized animals unless both possess identical Ir and serologically defined histocompatibility products. Schwartz and Paul (28) used peritoneal exudate lymphocytes from congenic mice strains in an in vitro proliferation assay and found that histocompatibility linked constraints on macrophage-associated antigen presentation to T lymphocytes parallel those in the guinea pig. Similarly, Erb and Feldmann have shown that syngeneic macrophages are more efficient than allogeneic macrophages in generation of T helper cells.

The cellular events in antigen recognition which we have proposed are summarized in Table III and shown diagrammatically in Figure 3. The use of "equations" should not be interpreted by the reader as indicating any absolute correctness of a given event in the series but rather that it represents a convenient shorthand or framework for discussion, experimental design and future modification.

To what extent are these events either sufficient or necessary for signal transduction during the recognition process? Recent observations have stimulated us to consider the possibility that the macrophage may have a specialized region of surface membrane or secrete a product which acts as a non-

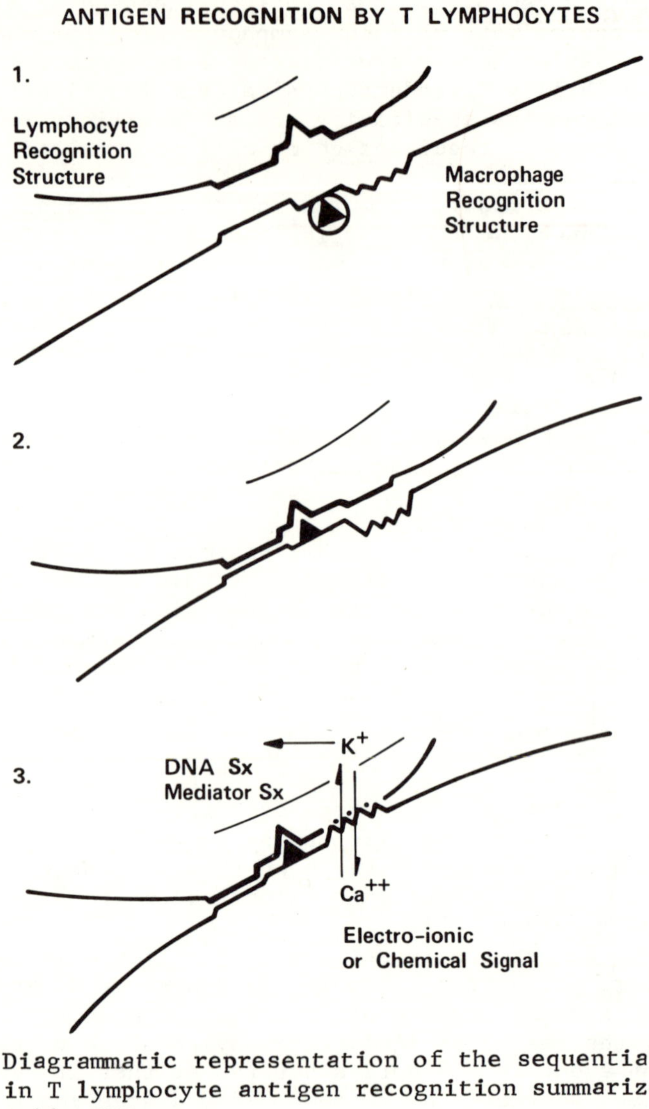

Fig. 3 Diagrammatic representation of the sequential events in T lymphocyte antigen recognition summarized in Table III.

TABLE III

MACROPHAGE-LYMPHOCYTE INTERACTION IN ANTIGEN RECOGNITION

$$M\Phi + Ag + L \xrightarrow{1} (M\Phi \cdot Ag) + L \overset{2}{\rightleftharpoons} (M\Phi \cdot Ag \cdot L)_A \xrightarrow{3} (M\Phi \cdot Ag \cdot L)_B \xrightarrow{4} (M\Phi \cdot Ag \cdot 2L)_C$$

MΦ = macrophage
L = lymphocyte
Ag = antigen (soluble or cell bound)

<u>Step #1.</u> Metabolic dependent uptake of soluble protein antigens by the macrophage into an immunologically relevant cell compartment.

<u>Step #2.</u> Antigen-independent, reversible macrophage-lymphocyte interaction between allogeneic or syngeneic lymphocytes and macrophages. This step may serve as alloantigen presentation step in MLC between strain 2 and 13 guinea pigs or aldehyde modified surface activation of lymphocyte proliferation. Blocked by cytochalasin-B.

<u>Step #3.</u> Antigen-dependent "irreversible" macrophage-lymphocyte interaction. Stabilization is blocked specifically by alloantisera against H-linked gene products and presumably regulated by an Ir gene product.

<u>Step #4.</u> Hypothetical trigger step. Transduction of signal may involve chemical or electrotonic events at cell surface which modify or perturb the lymphocyte surface as does mitogen. In this model antigen serves only to select which clone of lymphocytes is expanded; antigen interaction with receptor need not be the triggering event.

specific trigger for initiation of T lymphocyte activation. Modification of the membrane of living macrophages by selectively generating aldehydes on sialic acid with $NaIO_4$ (30) or on galactose with neuraminidase-galactose oxidase treatment (31) yields a population of macrophages which are capable of initiating allogeneic and syngeneic T lymphocyte proliferation non-specifically (unpublished observation). Equivalent numbers of lymphocytes similarly modified, mitomycin-C treated, and mixed with untreated lymphocytes do not induce lymphocyte proliferation. It is also of interest that macrophages pulsed with the non-specific mitogen phytohemagglutinin, PHA, activate allogeneic lymphocyte proliferation to the same extent as syngeneic macrophages further suggesting that triggering of lymphocyte proliferation by macrophage bound signals is not always constrained by H-linked genetic factors (22).

Other data indicating a central role for macrophages in antigen recognition by T lymphocytes are the finding that in the guinea pig, the macrophage is the principal cell responsible for stimulating the mixed leukocyte reaction between strain 2 and 13 guinea pigs (32). Thus, while the macrophage cannot present soluble protein antigens to allogeneic lymphocytes (maximal ^{3}H-thymidine incorporation at 48-72 hours), paradoxically the interaction between the allogeneic macrophage and lymphocyte would appear to be sufficient to trigger lymphocyte proliferation (maximal ^{3}H-TdR incorporation at 144 hours) in a mixed leukocyte culture. Since we have already shown that both allogeneic and syngeneic lymphocytes physically bind to macrophages under normal circumstances (19), one is tempted to speculate that this interaction site may be closely associated with an antigen recognized by allogeneic lymphocytes during the MLR.

It would appear that at least in the guinea pig, initiation of cell proliferation by either soluble antigenic proteins, cell-bound alloantigen, or by aldehydes on carbohydrates appears to result morphologically in extended stabilization of the binding of the lymphocyte to the macrophage membrane. Is the induction of lymphocyte proliferation therefore simply the consequence of stabilization of the lymphocyte on the macrophage membrane, albeit in some ordered relationship? Is this hypothetical non-specific triggering event, enzymatic, ionic, or electrical? This latter possibility must be considered in view of our recent demonstration that the macrophage is an electrically active cell exhibiting spontaneous hyperpolarizing currents across its plasma membrane (33).

The basic similarity between this model and that of Katz and Benacerraf (20) for T-B collaboration stresses the central role of cellular interaction in a variety of immune phenomena. The existence of other macrophage dependent pathways for T lymphocyte antigen recognition which do not require direct physical interaction but instead are mediated by soluble macrophage or lymphocyte products has not been excluded and indeed seems highly likely to occur.

Nonetheless, we consider this model a reasonable alternative recognition system to that classically envisaged where interaction between antigen and a receptor on the surface membrane is a sufficient antigenic signal. Perhaps its greatest value is that it is subject to direct experimental verification at both a morphological and functional level.

REFERENCES

1. Cline, M.J. and Swett, V.C., J. Exp. Med. 128: 1309, 1968.
2. Oppenheim, J.J., Leventhal, B.G. and Hersh, E.M., J. Immunol. 101: 262, 1968.
3. Seeger, R.C. and Oppenheim, J.J., J. Exp. Med. 132: 44, 1970.
4. Rosenstreich, D.L. and Rosenthal, A.S., J. Immunol. 110: 934, 1973.
5. Waldron, J.A., Horn, R.G. and Rosenthal, A.S., J. Immunol. 111: 58, 1973.
6. Rosenstreich, D.L. and Rosenthal, A.S., J. Immunol. 112: 1085, 1974.
7. Davie, J.M. and Paul, W.E., in Contemporary Topics in Immunobiology, edited by M.D. Cooper and N.C. Warner, Vol. 3, p. 171, 1974.
8. Chen, C. and Hirsch, J.G., J. Exp. Med. 136: 604, 1972.
9. Bevan, M.J., Epstein, R. and Cohn, M., J. Exp. Med. 139: 1025, 1974.
10. Waldron, J.A.,Jr., Horn, R.G. and Rosenthal, A.S., J. Immunol. 112: 746, 1974.
11. Unanue, E.R. and Cerottini, J.C., J. Exp. Med. 131: 711, 1970.
12. Ellner, J.J. and Rosenthal, A.S., J. Immunol., in press.
13. Steinman, R.M. and Cohn, Z.A., J. Cell Biol. 55: 186, 1972.
14. Gordon, S. and Cohn, Z.A., Int. Rev. Cytol. 36: 171, 1973.
15. Unanue, E.R., Cerottini, J.C. and Bedford, M., Nature New Biol. 222: 1193, 1969.
16. Schmidtke, J.R. and Unanue, E.R., J. Immunol. 107: 331, 1971.

17. Unanue, E.R. and Askonas, B.A., J. Exp. Med. 127: 915, 1968.
18. Lipsky, P.E. and Rosenthal, A.S., J. Exp. Med. 138: 900, 1973.
19. Lipsky, P.E. and Rosenthal, A.S., J. Exp. Med., in press.
20. Katz, D.H. and Benacerraf, B., Transpl. Rev., in press.
21. Shevach, E.M., Lee, L. and Ben-Sasson, S.Z., this volume.
22. Rosenthal, A.S. and Shevach, E.M., J. Exp. Med. 138: 1194, 1973.
23. Benacerraf, B. and Katz, D.H., in Adv. in Cancer Res., in press.
24. Shevach, E.M. and Rosenthal, A.S., J. Exp. Med. 138: 1213, 1973.
25. Ben-Sasson, S.Z., Shevach, E.M., Green, I. and Paul, W.E., J. Exp. Med. 140: 383, 1974.
26. Hämmerling, G. and McDevitt, H.O., J. Immunol. 112: 1734, 1974.
27. Kapp, J.A., Pierce, C.W. and Benacerraf, B., J. Exp. Med. 140: 172, 1974.
28. Schwartz, R. and Paul, W.E., manuscript in preparation.
29. Erb, P. and Feldmann, M., unpublished observations.
30. Novogrodsky, A. and Katchalski, E., PNAS 69: 3207, 1972.
31. Novogrodsky, A. and Katchalski, E., PNAS 70: 1824, 1973.
32. Greineder, D.K. and Rosenthal, A.S., J. Immunol., in press.
33. Gallin, E.K., Wiederhold, M.L., Lipsky, P.E. and Rosenthal, A.S., submitted for publication.

SECRETION BY MACROPHAGES OF TWO MOLECULES MODULATING CELL PROLIFERATION

Emil R. Unanue, Jesus Calderon and Jeanne-Marie Kiely
Department of Pathology, Harvard Medical School, Boston, Massachusetts 02115

It is well accepted that macrophages play an important role in immune induction as well as in cell-mediated immune reactions. In immune induction, macrophages have been thought to act as an antigen focusing cell favoring T and B cell interactions (1). In cell-mediated immunities, phagocytes form the major component of the inflammatory reaction - their concentration and activation appear to curb the growth of bacteria, viruses or tumor cells. The question has arisen whether phagocytes may carry out some of their diverse roles by secreting molecules that moderate the activities of surrounding cells. The purpose of this paper is to report on two contrasting biological activities found in cultures of peritoneal exudate cells rich in macrophages. Both activities are found in cultures of macrophages markedly depleted of T lymphocytes by treatment with anti-θ antibodies and complement. The fluids are generated by cultures of peritoneal macrophages in regular minimal essential medium with 5% fetal calf sera. Figure 1 depicts the general procedure.

Inhibition of Cell Proliferation

Culture fluids from peritoneal exudate cells rich in macrophages markedly inhibit the spontaneous proliferation of EL-4 leukemia cells, P815 mastocytoma and 3T3 transformed or non-transformed fibroblasts; also the cultures inhibit the proliferation of spleen lymphocytes induced by Concanavalin A or bacterial endotoxin as well as antibody formation in vitro by Mishell-Dutton method. Most of these results have just been published in an article in Proc. Natl. Acad. Science (2). Figures 2 and 3 are examples of the effects of the culture fluids on EL-4 cells. The following results have been obtained:

1. The material responsible for the inhibition is dialyzable (Fig. 2). By the use of filters of differing sizes, the material is thought to have a molecular size of 1400 daltons or less.

This work was supported by National Institutes of Health Grant AI 10091 and NCI Grant 14723.

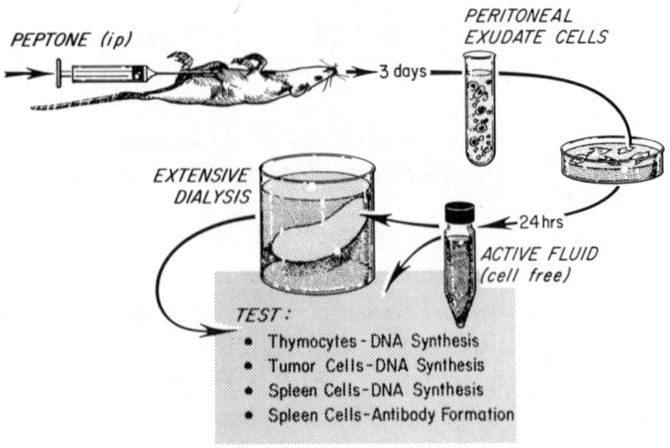

Fig. 1 The graph depicts the procedure used in most experiments. Mice are injected i.p. with peptone and their peritoneal cells harvested and cultured. The culture fluids are tested prior to or after dialysis.

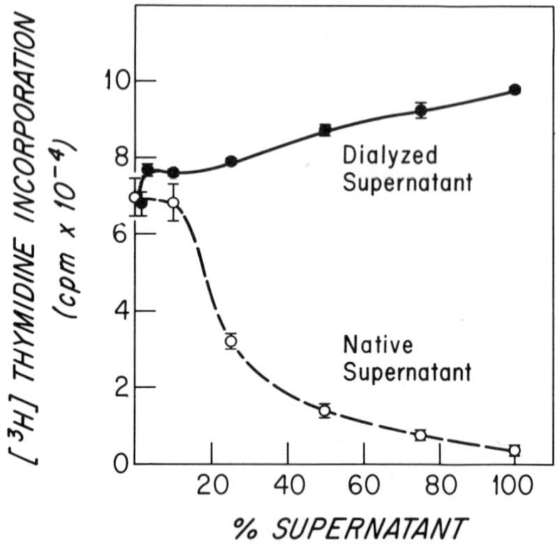

Fig. 2 EL-4 leukemia cells were incubated with macrophage culture fluid for 8 hrs, the last 4 in the presence of tritium labeled thymidine. Culture conditions: 5×10^5 EL-4 cells, 1 ml or RPMI 1640 medium with 5% fetal calf serum. % supernatant refers to concentration (v/v) of undialyzed or dialyzed macrophage fluid.

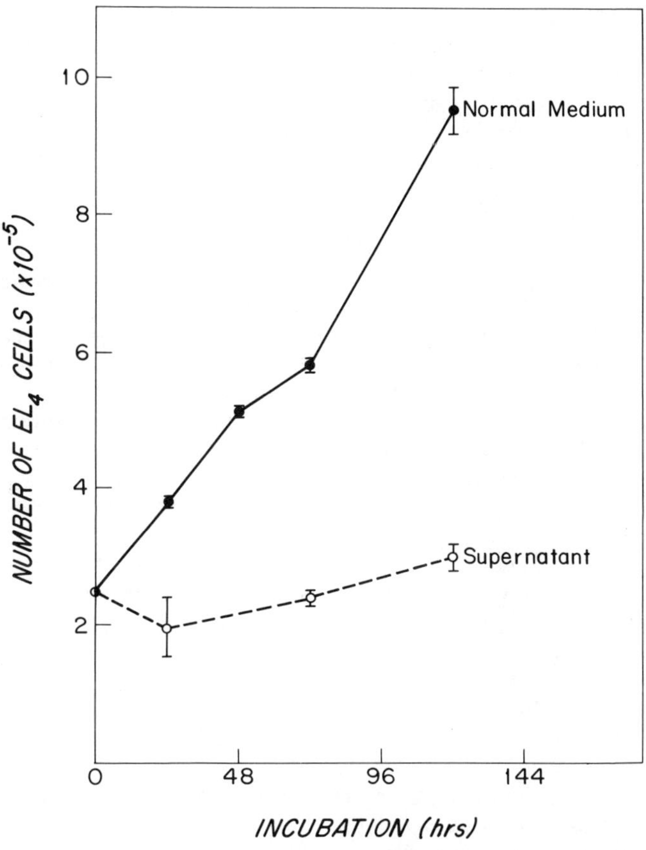

Fig. 3 In this experiment, the EL-4 leukemia cells were grown for the indicated times in normal medium or in the macrophage culture fluids.

2. The inhibitory material resists tryptic digestion and treatment with phosphodiesterase; it resists boiling and freeze-thawing.

3. The effects on cells is cytostatic. Cells cultured in the inhibitor will recover their normal growth activity if cultured in fresh medium.

4. The inhibitor is synthesized by the macrophages and is not simply released from cells dying in the culture. The experiment in Figure 4 substantiates this point (3).

5. The inhibitor is synthesized in equal amounts by peritoneal macrophages from unstimulated, peptone stimulated, or *Listeria monocytogenes* infected mice. Spleen macrophages synthesize it in small amounts.

6. The inhibitor acts on differing strains of mice and also acts on cells from other species. Murine inhibitor can act on human lymphocytes and vice versa.

7. The inhibitor can be absorbed by the target cells and removed from solution.

8. Inhibiting activity has not been found in cultures of fibroblasts or EL-4 leukemia cells. Spleen "non-adherent" cells secrete it in small amounts probably because of a small number of residual macrophages.

Stimulation of Lymphocyte Proliferation

The experiments so far have been done only with peritoneal exudate cells activated by peptone. One experiment has employed cultures of human spleen macrophages. In most experiments, the effects of the stimulatory molecule are best seen following removal of the inhibitory molecule. The following results have been obtained:

1. The cultures stimulate proliferation of thymocytes. This proliferation is seen best after removal of the inhibitor by dialysis (Fig. 5). The stimulatory molecule enhances the PHA response of thymocytes (3).

2. The stimulatory activity allows spleen cells from nude mice in vitro to make an antibody response both of the IgM and IgG class.

3. The stimulatory molecule enhances the response of hapten carrier primed cells. Indeed, spleen cells from mice primed to Fluorescein hemocyanin (FITC-KLH) will develop 3 to 4 times more PFC, both of IgM and IgG class. This response to FITC on a heterologous carrier is also enhanced but much less.

4. Finally, anti-θ-treated spleen cells from FITC-KLH primed mice will make a peritoneal anti-FITC response when cultured in the macrophage culture fluids.

5. The stimulatory substance crosses strains and species.

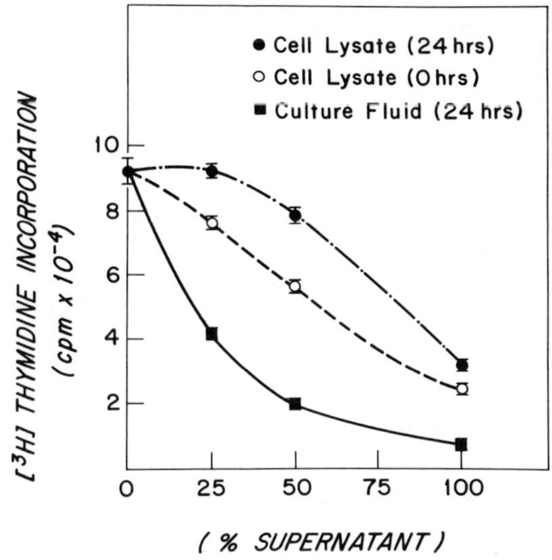

Fig. 4 This figure shows that the inhibitor is synthesized and released by the macrophage monolayer in culture. Macrophages were planted and immediately lysed by freeze-thawing. Another set was cultured 24 hrs at which time the fluids were obtained as well as a lysate from the cells. Each was tested for inhibitor on EL-4 cells, as in Figure 2. Inhibitor is formed in the initial cell lysate but it is released in larger amount after 24 hrs of culture.

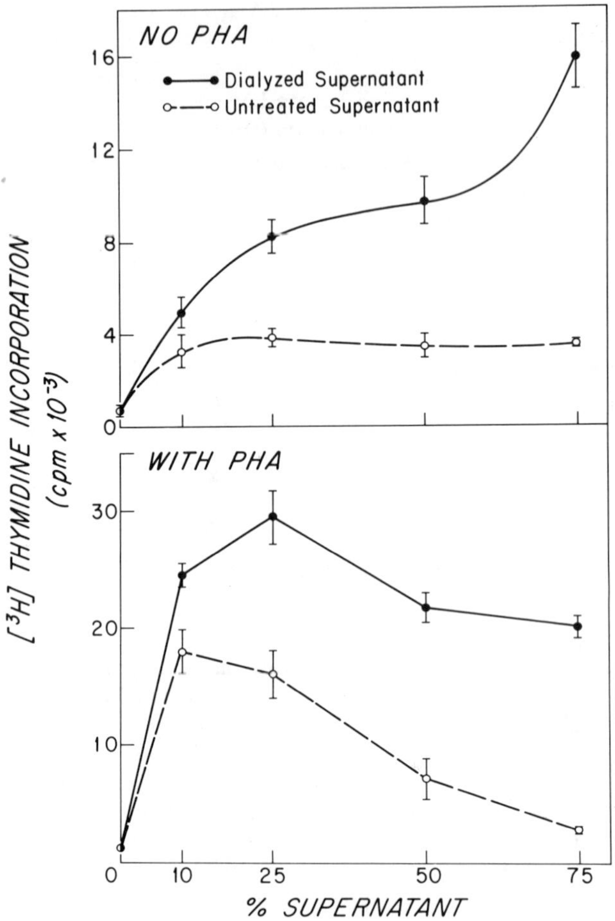

Fig. 5 Thymocytes were incubated with dialyzed or untreated cultures from 72 hrs, with or without PHA, the last 12 hrs in the presence of titrated thymidine. Culture conditions: 5×10^6 thymocytes in 1 ml of RPMI 1640 medium with 5% fetal calf serum; PHA: 1 μg per ml of PHA-P from Difco Labs.

6. The stimulatory substance can be generated from cultures of peritoneal exudate cells depleted of T cells and from peritoneal cells obtained from x-irradiated mice. It has not been obtained from cultures of nonadherent spleen cells.

In summary, the stimulatory molecule appears to act on both T and B cells. Indeed, it stimulates proliferation of thymocytes and enhances the response of carrier primed T cells. It also allows B cell preparations to make a significant response probably as a result of proliferation and differentiation. Unfortunately, we have no idea at present on the requirement for its production by macrophages, that is to say whether its production reflects, as we suspect, the state of activation of the macrophage. Similar stimulatory substances were previously reported by Gery et al. (4), Schrader (5) and Wood (6).

In summary it would appear that phagocytes can secrete at least two kinds of molecules that have important modulatory effects on surrounding cells. The stimulatory molecule could play an important role in responses of lymphocytes and may be responsible for some of the helper effects previously reported. The role of the small molecular weight inhibitor is not clear, since it should readily diffuse in tissue fluids. It may be that the end result inhibition or stimulation depends on several factors: number of macrophages, their state of activation, their relationship to the target cells (for example if forming a granuloma), the nature of the target cells, etc. The presence of these potentially important molecules again focus on the macrophages as an important regulatory cell in immunity.

REFERENCES

1. Unanue, E.R., Adv. Immunol. 15: 95-154, 1972.
2. Calderon, J., Williams, R.T. and Unanue, E.R., Proc. Natl. Acad. Sci. USA, in press, 1974.
3. Calderon, J. and Unanue, E.R., Nature, in press, 1975.
4. Gery, I., Gershon, R.K. and Waksman, B.H., J. Exp. Med. 136: 128-142, 1972.
5. Schrader, J.W., J. Exp. Med. 138: 1466-1488, 1973.
6. Wood, D.D. and Gaul, S.L., J. Immunol. 113: 925-933, 1974.

GVH INDUCED IMMUNOSUPPRESSION: REGULATORY FUNCTION OF MACROPHAGES IN THE HUMORAL IMMUNE RESPONSE[1]

Robert Elie[2] and Wayne S. Lapp[3]
Department of Physiology, McGill University, Montreal, Canada

Three cell types, thymus derived (T), bone marrow derived (B) and adherent or accessory (A) cells are involved in the induction of antibody synthesis to most antigens. It has been well documented that B cells are the precursors of the antibody forming cells and in order to be activated they require the cooperation of T and A cells (1-3). The nature of the T cell helper function and the tole of A cells have not been clearly defined. From studies dealing with the mechanism of the graft-versus host (GVH) induced immunosuppression, further information has been obtained on the role of T and A cells. It has been demonstrated that GVH induced immunosuppression is due, at least in part, to a defective T cell helper function since the immune response can be restored by a nonspecific mediator released by T cells in the absence of antigen (4,5). In this paper we demonstrate that the A cell population, which is necessary for the induction of the immune response, can also inhibit the immune response. The results suggest that the A cell population exerts its influence by regulating T cell helper function.

MATERIALS AND METHODS

GVH reactions were induced in 8 to 12 week old C57BL/6 X A F_1 (B6AF_1) mice by an intravenous injection into each mouse of 75 x 10^6 spleen and lymph node cells obtained from adult A strain mice. Four to five days later one half of the group was used as spleen cell donors for in vitro studies, the other half received 5 x 10^8 sheep red blood cells (SRBC) in order to confirm that immunosuppression was induced.

Single cell suspensions were prepared from various lymphoid organs by gently tamping them through a 50 mesh stainless steel screen (4,5).

Restoration experiments were performed using a modified

---

[1]This work was supported by grants from the Medical Research Council and the National Cancer Institute of Canada.

[2]MRC of Canada Fellow.

[3]MRC of Canada Scholar.

Marbrook culture chamber (MMCC) (14). The modification consisted of two concentric culture compartments separated by a 0.45 μm cell impermeable Millipore filter. The experimental protocol consisted of culturing 10 x 10^6 responding spleen cells from 4-5 day GVH animals (GVH-SC) plus 5 x 10^6 SRBC in the outer compartment (OC) of the MMCC and 3 x 10^6 restoring cells in the inner compartment (IC). CMRL 1066 medium supplemented with 10-15% fetal calf serum (FCS) was used. The cell cultures were incubated at 37°C in a 5% CO_2 humidified atmosphere for 4 days. The number of direct PFC per culture vessel was determined by a modification of the Cunningham and Szenberg technique (6,7). Thymus derived cells were killed with an AKR anti-theta C3H serum plus complement as described previously (8).

In vitro experiments designed to investigate the immunosuppressive action of A cells on the PFC response to SRBC were performed using the following protocol. A cells and non-adherent (NA) cells were prepared from spleen obtained from normal and GVH mice. The spleen cells were suspended in Eagle's medium supplemented with 10% FCS at a concentration of 10 x 10^6 cells per ml. The cells were incubated in plastic Petri dishes for 1.5 hours at 37°C after which the NA cells were removed by washing 2X. The A cells were scraped from the plastic Petri dishes with a rubber policeman. Both NA and A cell populations were washed once and suspended in CMRL 1066 medium supplemented with 10-15% FCS. Increasing numbers of A cells were added to 5 x 10^6 normal NA cells and 2.5 x 10^6 SRBC and cultured in a standard Marbrook chamber under the conditions described above.

RESULTS

Restoration of the in vitro PFC response by GVH lymphoid cells.

It has been previously reported that the in vitro PFC response of GVH-SC can be restored across a cell impermeable membrane by normal thymus (TC) lymph node (LNC) and spleen cells (SC) but not by normal bone marrow cells (BMC). The results in Table 1 show that TC and LNC taken from GVH mice at different times post GVH induction restored the PFC response of GVH-SC as well as normal TC and LNC. Table 1 also demonstrates that, within a short time after the incubation of the GVH reaction, SC from GVH mice lost their ability to restore the PFC response of GVH-SC. However, when A cells were removed from day 5 GVH-SC and the non-adherent cell fraction (GVH-NA cells) were placed in the IC of the MMCC it was observed that the GVH-NA cells were able to restore the in vitro PFC response of GVH-SC as well as normal SC, LNC

TABLE I

RESTORATION OF THE PFC RESPONSE OF GVH-SC BY GVH LYMPHOID CELLS ACROSS A CELL IMPERMEABLE MEMBRANE [a]

Days post GVH Induction for obtaining restoring cells	PFC response of GVH-SC (% of normal)		
	Source of restoring cells		
	Thymus	Lymph nodes	Spleen
0[b]	65	69	80
2	27	29	45
5	60	80	0
7	54	95	3
10	68	70	0
17	79	106	0

[a]The results of 3 experiments were pooled and are presented as the mean percent PFC response of normal spleen cells.

The GVH control PFC response was 14.0% of the normal response.

[b]Restoring cells taken from normal non-GVH mice.

and GVH-LNC (Table II). It was further demonstrated that treatment of GVH-NA cells with an anti-theta serum and complement abrogated their restoring ability (Table II).

Functional properties of A cells in the spleens of GVH mice.

The previous results suggest that A cells taken from the spleens of GVH mice exert a suppressive effect on T cell helper function. Experiments were therefore designed to determine whether the suppressive effect was due to qualitative or quantitative changes in the A cell population. The first series of experiments tested for a correlation between the number of A cells and the in vivo PFC response of GVH mice. GVH reactions were induced in a large number of animals on day 0. At different times post GVH induction a minimum of three mice were sacrificed to determine the proportion of A cells in the spleen and a minimum of two mice were injected with SRBC and the number of PFC was determined four days later. Similar determinations were performed at the same time on normal animals. The results presented in Table III show that the total number of cells and the proportion of A cells in the spleen increased with time and reached a maximum value by day 10-11 post GVH induction followed by a decrease at day 15. Table II also demonstrates an inverse relationship between the number of A cells and the total number of PFC per spleen.

Experiments were then designed to determine whether a qualitative change had also occurred in the A cell population of GVH spleens. The results shown in Table IV demonstrate that A cells from GVH-SC could substitute for normal cells and cooperate with normal NA cells in the development of the in vitro PFC response to SRBC. The results also show that regardless of the source of A cells (normal SC or GVH-SC), the in vitro PFC response was related to the ratio of A to NA cells in culture (Table IV). The maximum PFC response was obtained when the mixture contained 1 A to 10 NA cells. When the ratio of A to NA cells exceeded that level the PFC response was inhibited. These results suggest that the suppressive effect exerted by A cells on T cell helper function in GVH spleens is due mostly to a quantitative rather than a qualitative change in the A cell population.

DISCUSSION

Previous reports on in vivo (8,9) and in vitro (4,5) restoration experiments have shown that the GVH induced immunosuppression is due, at least in part, to the deficiency of a non-specific T cell factor which is essential for the T cell helper function. It has been demonstrated that the in vitro

TABLE II

RESTORATION OF THE IN VITRO PFC RESPONSE OF GVH-SC BY GVH-SC DEPLETED OF A-CELLS

Responding Cells[a]	Restoring Cells[b]	Mean PFC ± SE per culture vessel	% of Normal
NSC		2228 ± 226	100
GVH-SC	-	68 ± 42	3
GVH-SC	NLNC	649 ± 53	29
GVH-SC	NSC	648 ± 58	29
GVH-SC	GVH-LNC	773 ± 217	60
GVH-SC	GVH-SC	136 ± 14	6
GVH-SC	GVH-NAC	569 ± 48	25
GVH-SC	GVH-NAC+Anti θ	61 ± 40	3

[a]10×10^6 responding spleen cells were cultured in the OC with 5×10^6 SRBC.

[b]3×10^6 restoring cells were cultured in the IC which was separated from the OC by a cell impermeable membrane.

TABLE III

CORRELATION BETWEEN THE NUMBER OF ADHERENT CELLS AND THE TOTAL PFC RESPONSE OF GVH-SPLEENS

Origin of spleens	Mean number of cells per spleen ± SE($\times 10^6$)	% A Cells	Mean PFC ± SE per spleen($\times 10^3$)
Normal B6AF$_1$	53.2 ± 2.6	4.8	129.4 ± 14.5
Day 2 GVH	73 ± 7.5	10	60.9 ± 1.5
Day 3 GVH	85 ± 5	14.5	16.5 ± 1.1
Day 5 GVH	91 ± 2.2	21.7	0
Day 7 GVH	106.3 ± 3.4	28.5	0
Day 10 GVH	111 ± 5.5	33	0
Day 11 GVH	136.3 ± 18.3	21	0
Day 15 GVH	109.5 ± 2.5	15	0

TABLE IV

EFFECT OF VARIOUS NUMBERS OF A CELLS TAKEN FROM NORMAL OR GVH SPLEENS ON THE PFC RESPONSE OF NORMAL NA CELLS

Responding Cells	Source of A cells[a]	Ratio of A to NA cells	$\overline{X}$ PFC ± SE
NSC	-		880 ± 106
GVH-SC	-		287 ± 37
NSC-NA	-		220 ± 80
"	GVH-SC	1:50	197 ± 30
"	"	1:25	272 ± 93
"	"	1:10	690 ± 51
"	"	1: 5	442 ± 21
"	"	1: 1	119 ± 50
"	NSC	1:50	178 ± 36
"	"	1:25	313 ± 98
"	"	1:10	1200 ± 62
"	"	1: 5	343 ± 119
"	"	1: 1	141 ± 51

[a]Adherent cells were taken from normal and day 5 GVH spleens.

PFC response of 5 day GVH-SC can be restored by a soluble factor released by normal thymus, spleen and lymph node T cells in the absence of antigenic stimulation (4). The results shown in Table I demonstrate that TC and LNC taken from GVH mice at different times post GVH induction restored the PFC response of GVH-SC as well as normal TC and LNC. Table I also shows that within a short period of time post GVH induction, splenic TC lost their restoring ability. However, when the A cell fraction was placed in the IC of the MMCC, it restored the PFC response of GVH-SC (Table II). The restoring ability of the GVH-NA cells was abrogated by a treatment with an anti-theta C3H serum and complement (Table II). These results suggest that T cells in GVH spleens were capable of producing and releasing the T cell factor which is essential for the T cell helper function, but their helper function was suppressed by the A cell population. A correlation was demonstrated between the in vivo PFC response and the number of AC in the spleen of GVH mice. The total number of PFC per spleen decreased with time post GVH induction, while the number of A cells showed a marked increase. However, the A cells from GVH spleens did not appear to be solely suppressor cells but served to function as regulatory cells in the inductive phase of the immune response since it was shown that GVH-A cells could replace normal A cells and cooperate with normal NA cells to produce an anti-SRBC response. The maximum PFC response was obtained at the ratio of one A cell to ten NA cells for both normal and GVH-AC. When the number of A cells to NA cells exceeded the 1 to 10 ratios, the PFC response was markedly suppressed. These results seem to indicate that the excessive number of A cells was responsible for the depressed humoral immune response observed in GVH animals. Results reported by others seem to confirm this speculation. It has been shown by Hoffmann and Dutton (10) that an excessive concentration of A cell product suppressed the immune response of mouse spleen cells. Waldman and Gottlieb (11) have recently reported that high ratios of macrophages to lymphocytes inhibited PHA transformation of rat T cells. They suggested that the suppressive effect was elicited by a low molecular weight, heat stable factor released into the culture fluid by viable macrophages, and was not released as a consequence of cell death. Therefore, it appears that small numbers of A cells stimulate T cells whereas large numbers inhibit T cell function.

Although present results suggest that suppression is due to a quantitative change in the A cell population; a qualitative change cannot be ruled out. Two sets of results presented here suggest that the A cell population had undergone a qualitative change with an increase in the number of suppressor cells. Firstly, GVH-A cells were not as efficient as normal A cells in reconstituting the PFC response of the normal NA cell population. Secondly, spleens from day 3 and day 15 GVH mice contained approximately the same percentage of splenic A cells, however day 15 GVH mice were total immunosuppressed, whereas day 3 GVH mice produced a significant PFC response. Experiments by others have shown that splenic and peritoneal A cells taken from GVH mice 7 or more days post GVH induction released an inhibitory factor which suppressed the immune response and normal T cell function (12, 13, 14). Scott has obtained similar results with A cells taken from mice injected with Corynebacterium parvum (14).

In summary, these results suggest that at least one type of A cell in the GVH spleen is not simply an inhibitory cell but rather a regulatory cell and most likely functions to regulate normal T cell helper function. It is proposed that this regulatory effect is exerted via a feedback control mechanism, which is common to many biological control systems, in which the same A cell and/or its product may both stimulate and suppress T cell function.

ACKNOWLEDGMENT

We thank Mrs. H.C. Lee, Mrs. A. Lee Foon, and Miss R. Siegrist for their careful technical assistance.

REFERENCES

1. Miller, J.F.A.P. and Mitchell, G.F., Transplant. Rev. 1: 3, 1969.
2. Claman, M.N. and Chaperon, E.A., Transplant. Rev. 1: 92, 1969.
3. Unanue, E.R., Adv. Immunol., 15: 95, 1972.
4. Elie,R., Abrahams, R., Parthenais, E., and Lapp, W.S., Proc. 8th Leuc. Cult. Conf., Acad. Press, 175, 1974.
5. Parthenais, E., Elie, R., and Lapp, W.S., Cell Immunol. 13: 164, 1974.
6. Cunningham, A.J. and Szenberg, A., Immunology 14: 599, 1968.
7. Kongshavn, P.A.L., and Lapp, W.S., Immunology 22: 227, 1972.
8. Lapp, W.S., Wechsler, A., and Kongshavn, P.A.L., Cell. Immunol., 11: 419, 1974.
9. Grushka, M., and Lapp, W.S., Transplantation, 17: 157, 1974.
10. Hoffmann, M., and Dutton, R.W., Science, 172: 1047, 1971.
11. Waldman, S.R., and Gottlieb, A.A., Cell. Immunol., 9: 142, 1973.
12. Sjöberg, O., Clin. Expl. Immunol., 12: 365, 1972.
13. Nelson, D.S., Nature, 246: 306, 1973.
14. Scott, M.T., Cell. Immunol., 5: 469, 1972.

A NOVEL ADHERENT CELL IN MOUSE LYMPHOID ORGANS*

Ralph M. Steinman+ and Zanvil A. Cohn
The Rockefeller University, New York, New York 10021

Cellular immunologists have long been separating mouse spleen cell suspensions into two populations: one which adheres to glass and plastic surfaces, and another which does not (1,2,3). The former are thought to serve an essential accessory function in the generation of immune responses by the nonadherent cells. More specifically, many believe that the adherent cells are macrophages which process antigens for effective lymphocyte triggering.

Several years ago, we became skeptical of the ability of macrophages to process antigens by any of the currently proposed mechanisms (4). We decided to look at the adherent cells of mouse spleen and to our surprise, found them to be a heterogeneous population of nucleated and nonnucleated cells, as well as cell debris. Mononuclear phagocytes were infrequent and in fact, the most prominent well spread adherent cell had morphological features that were quite distinct from typical macrophages (Fig. 1). Because of their elongate and/or stellate configurations, and because they formed and retracted many tiny cell processes in the living state, we decided to call them "dendritic" cells (5). In choosing this term, we did not mean to imply that they are identical to other "dendritic" cells (dendritic macrophages, dendritic reticular cells) that have been postulated to exist in situ (e.g. 6,7).

Additional cytological features of dendritic cells included the large irregularly shaped nucleus with its peripheral refractile band of heterochromatin and small nucleolus, and many spherical phase dense mitochondria. In contrast to macrophages, dendritic cells lacked surface ruffles, phagocytic inclusions, pinocytic vesicles and numerous lysosomes (Fig. 1).

In mouse spleen, it appears that all cells capable of exhibiting the appearance of dendritic cells will adhere to a glass or plastic surface in a single plating under appropriate conditions (5). They can then be counted and are found to

---

*Supported by grants AI07012 & AI01831 from the U.S. Public Health Service.

+Leukemia Society of America Scholar.

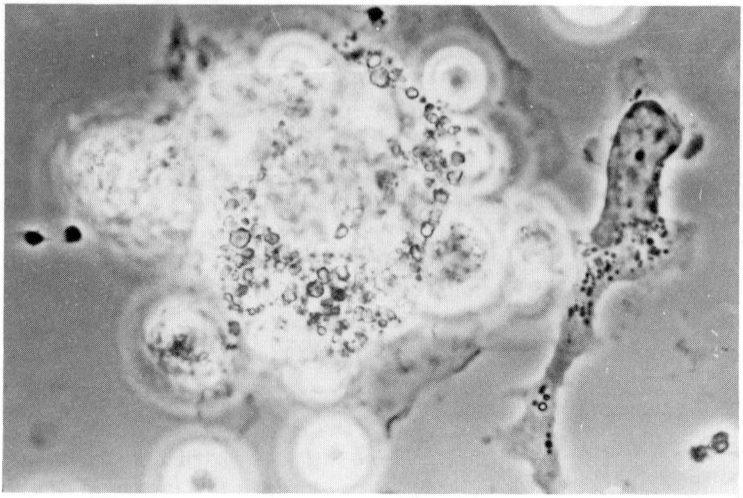

.g. 1 A typical dendritic cell (left) has an extended shape, many phase dense cytoplasmic granules (mitochondria), and an irregularly shaped nucleus with a peripheral band of heterochromatin. The plane of focus for the macrophage (right) is on the numerous perinuclear lysosomes (arrow). These cells were exposed to the pinocytic marker, soluble horseradish peroxidase, which was then localized cytochemically. The macrophage lysosomes were reactive indicating the active endocytic capacity of this cell type, in contrast to the dendritic cell. 3500X.

comprise 1% or less of the total nucleated cells. They are found in even smaller numbers in the adherent populations of cell suspensions from lymph node and Peyer's patch, but are not present in bone marrow, thymus, liver, peritoneal cavity or exudate. Of course these numbers depend on assumptions that we have not proved, viz., that all the dendritic cells in a given organ can be released into single cell suspensions and that they will then adhere to glass. In any case, cells which have the morphological features of adherent, mouse dendritic cells have been found in lymphoid cell suspensions from all other mammals tested, including man. In each case, they lack an essential differentiated feature of mononuclear phagocytes, i.e, the ability to bind and interiorize appropriately opsonized particles. Most of our work has been performed on mouse spleen, since dendritic cells there are easiest to obtain and to characterize.

Following the identification of a morphologically distinct type of adherent cell, our first goal was to prove that it differed significantly from the other well known cell type that adheres to glass - the mononuclear phagocyte. The differences are numerous and impressive (9,10,11) (Table I). In particular, dendritic cells do not actively endocytose a full spectrum of marker materials, both in vitro (9) (Fig. 1), and apparently in vivo (12). Also, on prolonged cultivation in vitro, they come off of the surface to which they initially adhered (most within 12-14 hrs) and then exist in suspension as a nondividing, but nonphagocytic, highly irregularly shaped cell (Fig. 2).

Further tests on dendritic cells isolated in vitro showed that they lacked the functional features of all other cell types either known to exist, or thought to exist, in lymphoid organs (9,10,11). They lack the surface markers of lymphocytes; they do not appear to give rise to hematopoietic colonies; they do not synthesize collagenous-like macromolecules; and they do not bind antigen or immune complexes like the presumptive antigen retaining follicular dendritic cells (6,7,8).

It is clear though that the best proof that dendritic cells are a novel cell type will come with the identification of a distinctive function(s) that this adherent cell performs. We have not established such a function. However, we have been studying some interesting phenomena that involve the multiplication and differentiation of antibody secreting cells in vitro and in vivo. In both dendritic cells, at least by virtue of guilt by association, may play a role.

The cells we work with are a low density fraction of mouse spleen. Empirically, we noted that most splenic dendritic

TABLE I

DENDRITIC CELLS ARE NOT MONONUCLEAR PHAGOCYTES

A. They have a different morphology by phase contrast (fixed and live), electron microscopic, and cytochemical (acid phosphatase, membrane adenosinetriphosphatase, Prussian blue) criteria.

B. They do not actively endocytose pinocytic (soluble HRP, colloidal thorium dioxide and carbon) or phagocytic (latex, heat-killed S. albus, HRP-antiHRP aggregates, antibody and/or complement coated RBC) markers.

C. They do not persist as glass adherent cells in vitro, and do not differentiate into mononuclear phagocytes.

D. They do not require collagenase treatment for bulk release into single cell suspensions.

E. They are found primarily in the white pulp nodules of mouse spleen.

F. They are not detected in the adherent cell population of mice given relatively low doses of ionizing irradiation and corticosteroids.

G. They have a more rapid turnover (4-5X) in situ in mouse spleen.

H. They are detected in adherent cell populations of mouse spleen and lymph node only after the first 2-3 weeks of life.

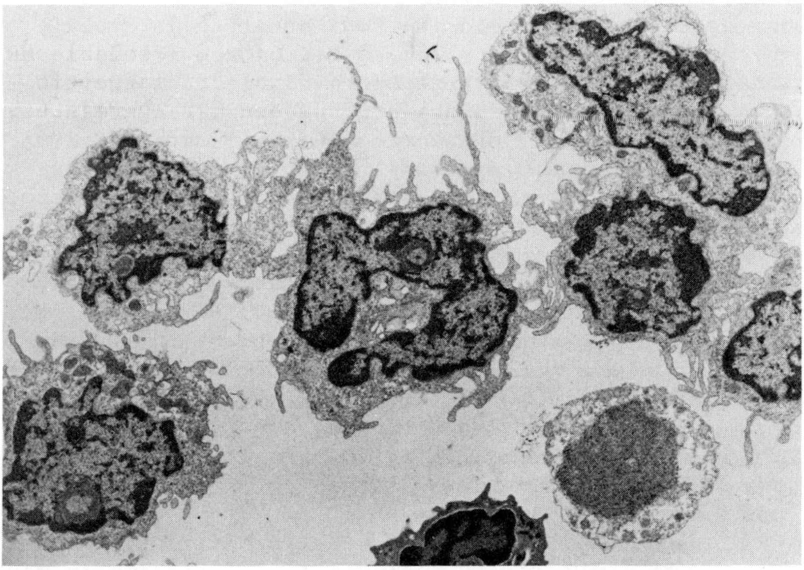

Fig. 2 A highly enriched population of adherent dendritic cells was cultivated for 2 days. Dendritic cells quickly (12-14 hrs) disappear from the glass surface but can then be found in the culture supernatant in substantial numbers. The nonadherent cells are shown in this micrograph. They have the same cytologic features as they possessed initially: the nucleus is large, irregularly shaped, and has a peripheral band of heterochromatin; the cytoplasm is "dendritic" in shape and contains relatively few organelles. 4800X.

cells, but only 5-10% of the total nucleated cell population, float to the top of BSA columns (p=1.082) following application of suitable centrifugal forces (9). When the low density cells are shaken at 37°C, and at appropriate cell concentrations, aggregates with an average of 10 cells quickly form (Fig. 3). These clusters can in turn be separated from the nonaggregated cells by velocity sedimentation in Ficoll. The nonaggregated cells account for 70-85% of the low density population, lack the ability to form significant numbers of clusters, and are severely depleted of dendritic cells.

Morphologic examination of the cluster fraction, first at the light level, showed many cells with the irregularly shaped nuclei and chromatin pattern of dendritic cells. The other cells were predominantly medium sized lymphocytes. They had circular shapes and their nuclei contained considerable nuclear heterochromatin. They lack both diffuse and granular, cytochemical, peroxidatic activity, as would occur if they were of an erythroid or myeloid nature. In thin sections, it again appears that the main cell types in the clusters are dendritic cells and medium sized lymphocytes (Fig. 4). Other cell types could be recognized less frequently in clusters, e.g., large lymphocytes, normoblasts, plasma cells, and mononuclear phagocytes.

Since other workers have called attention to aggregates of mouse spleen cells (13,14) it is important to mention that the low density clusters that we are studying are not the only aggregates that can develop in splenic cell suspensions. For example, the high density (80-90% of the total cells) fraction will form cell aggregates under the same conditions used in the low density population. These clusters are very different from the low density ones that we are studying. In particular, they consist largely of phagocytic and erythroid cells.

The fate of the low density aggregates on continued cultivation in vitro proved extremely interesting. Considering the morphological changes first, we found that within a day, numerous large lymphocytes appeared and within another day, the clusters consisted largely of immature and mature plasma cells, the cytoplasma of which stained readily with a fluoresceinated anti-I_g. In the electron microscope, most of the circular cells had the characteristic nuclear morphology and cytoplasmic rough endoplasmic reticulum of mature and immature plasma cells (Fig. 5). Dendritic cells were easy to recognize, although their surface had become even more irregular and dendritic. They were present in roughly similar numbers to that seen initially in the clusters - as determined by counting cells with appropriate nuclear cytological fea-

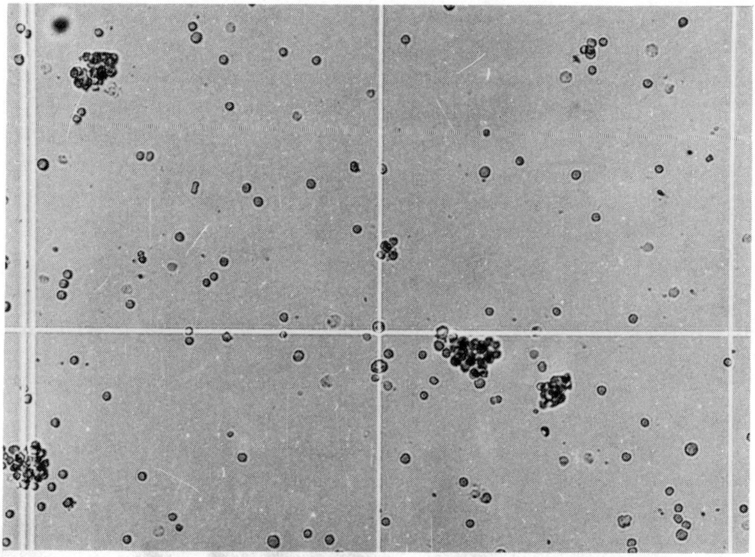

Fig. 3 A preparation of low density mouse spleen cells that has been agitated at 37°C for 45 min. is seen to contain several cell aggregates. 500X.

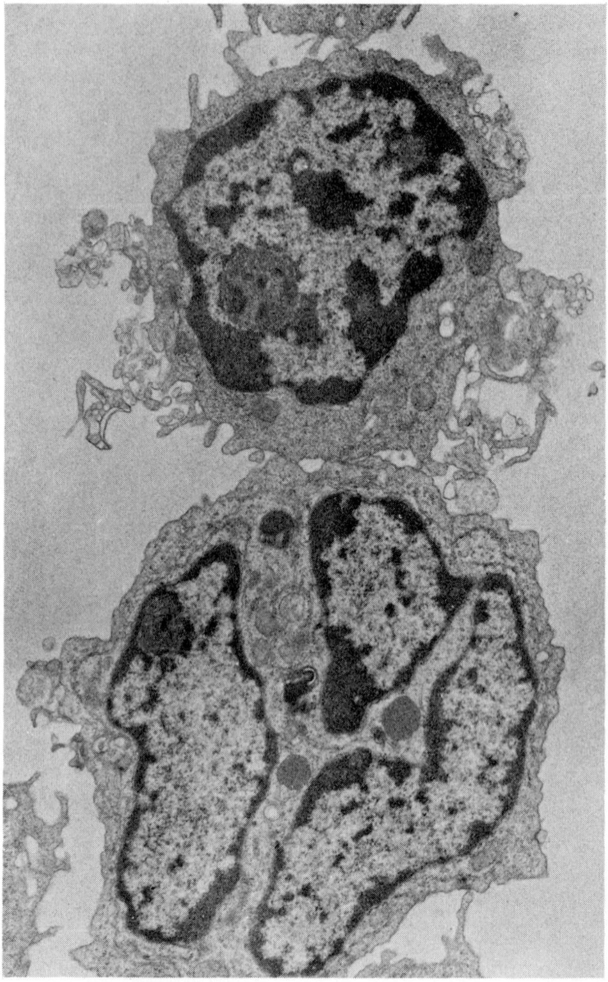

Fig. 4 Cells from the cluster fraction of low density spleen cells are predominantly of two types. Dendritic cells (bottom) have large irregularly shaped nuclei, a smooth surface initially, and relatively few cytoplasmic organelles, viz. there are scattered mitochondria, lysosomes, and slips of rough endoplasmic reticulum. The "medium-sized" lymphocytes have nuclei which are circular in shape and contain more prominent nucleoli and chromatin masses. The cytoplasm is dense and contains many particles, presumably ribosomes. 13,300X.

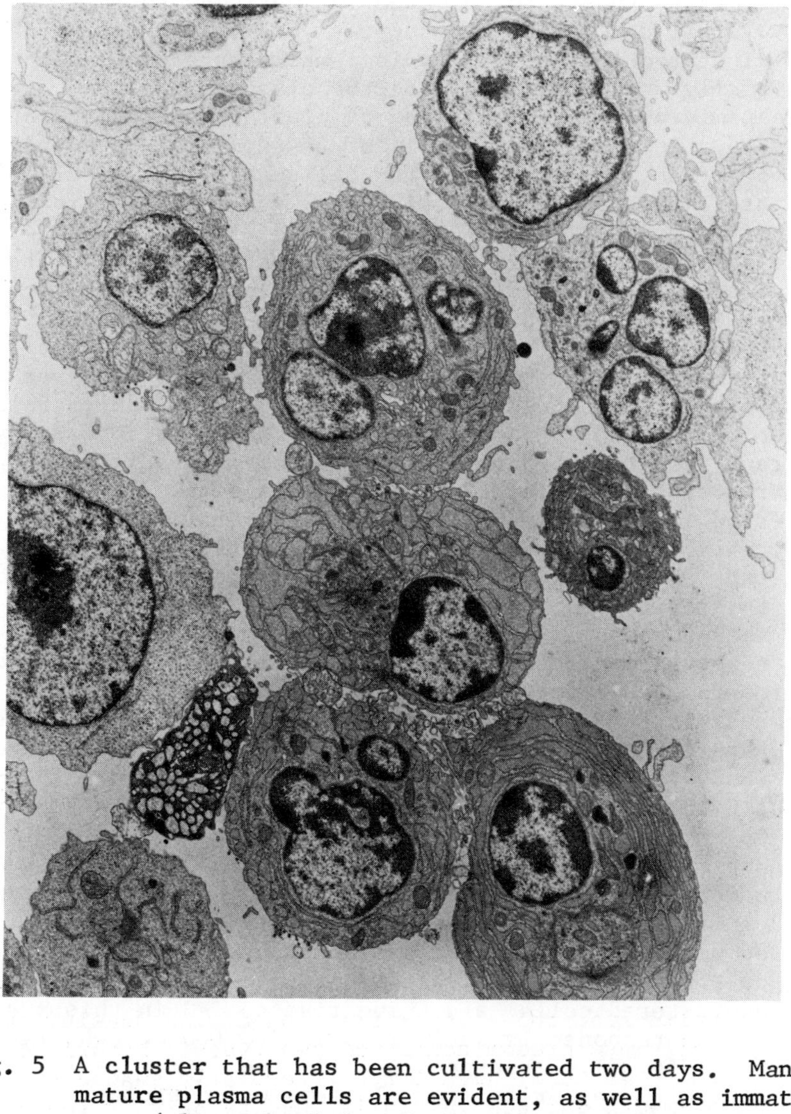

Fig. 5 A cluster that has been cultivated two days. Many mature plasma cells are evident, as well as immature ones (along the left of the micrograph). The latter have less developed rough endoplasmic reticulum, and their nuclei are more vesicular or blastoid in appearance. A dendritic cell (arrow) exhibits many cytoplasmic processes that contain few organelles, and an irregularly shaped nucleus. 4400X.

tures in 1 μ plastic sections. Dendritic cell processes were often closely opposed to the maturing antibody secreting elements. Small and medium lymphocytes were infrequent in the 2 day cluster cultures, and typical macrophages were seen only occasionally and usually outside the clusters.

Light microscopic examination of clusters cultured continuously in a low dose (0.1 μc/ml) of ^{3}H-thymidine revealed that more than 70% of the cluster cells were radiolabeled 2 days later. However, almost all cells with nuclear morphology of dendritic cells were not labeled.

The proliferative activity of the clusters was quantitated with pulses of ^{3}H-thymidine at varying time intervals. Thymidine uptake increased exponentially beginning immediately and lasting about 36 hrs (Fig. 6). With respect to the initial uptake, thymidine incorporation increased some 3-4 fold per culture and some 12-16 fold per viable cell. By 72 hrs, proliferation had fallen off considerably. These kinetics are quite different from that induced by antigen or nonspecific mitogens in vitro, since proliferation only begins after a considerable lag period in the latter instances. The kinetics make it likely that the proliferation and differentiation of plasma cells that we are following have been triggered in the animal prior to sacrifice.

We therefore administered known antigens to mice 2 days prior to obtaining a splenic cluster population. Most of our experiments have involved the primary response to intravenously administered sheep red blood cells. Clusters obtained two days after giving the antigen contained a low incidence of antibody forming cells, as detected by a plaque assay in gel. With culturing, specific antibody forming cells appeared (Fig. 6) and in large numbers, even in the absence of further exogenous antigen. The addition of sheep red cells to the culture does prolong the phase of rapid exponential expansion of PFC numbers, so that the peak responses are 2-4 times that obtained without the addition of exogenous antigen (Fig. 7). Other cell fractions, e.g., high density cells, or low density nonclustered cells, also give rise to PFC in this system, but the response is smaller by one to two orders of magnitude (Fig. 7). We suspect even in the former cell fractions, that PFC may arise from clusters which escaped the initial separation procedures, or which formed via dendritic cell precursors which apparently exist in spleen (10).

The quantitative data obtained following administration of sheep cells in situ are impressive in two respects. First, the degree of enrichment of specific antibody secreting cells (3-5% of the total viable cells) is large. Even higher enrichments can be obtained (6-20%) if the cluster cultures are

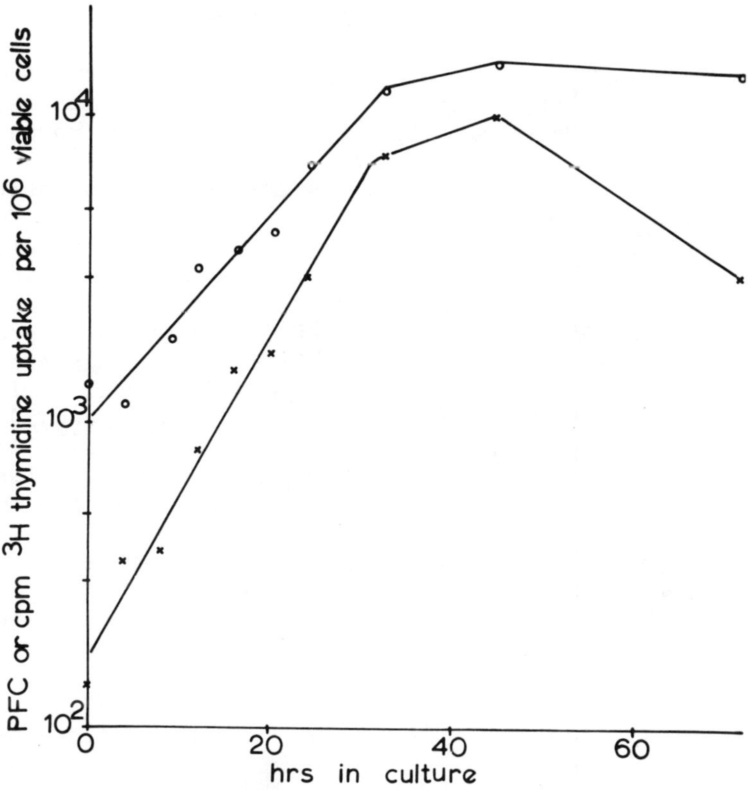

Fig. 6 Uptake of ^{3}H-thymidine (1 hr pulses; o-------o) and formation of direct plaque forming cells (PFC; x-------x) in cluster cultures made from animals primed 2 days previously with sheep red blood cells i.v. In the absence of additional red cells in vitro, thymidine uptake and PFC expand exponentially for 1½ days, plateau, and then begin to subside.

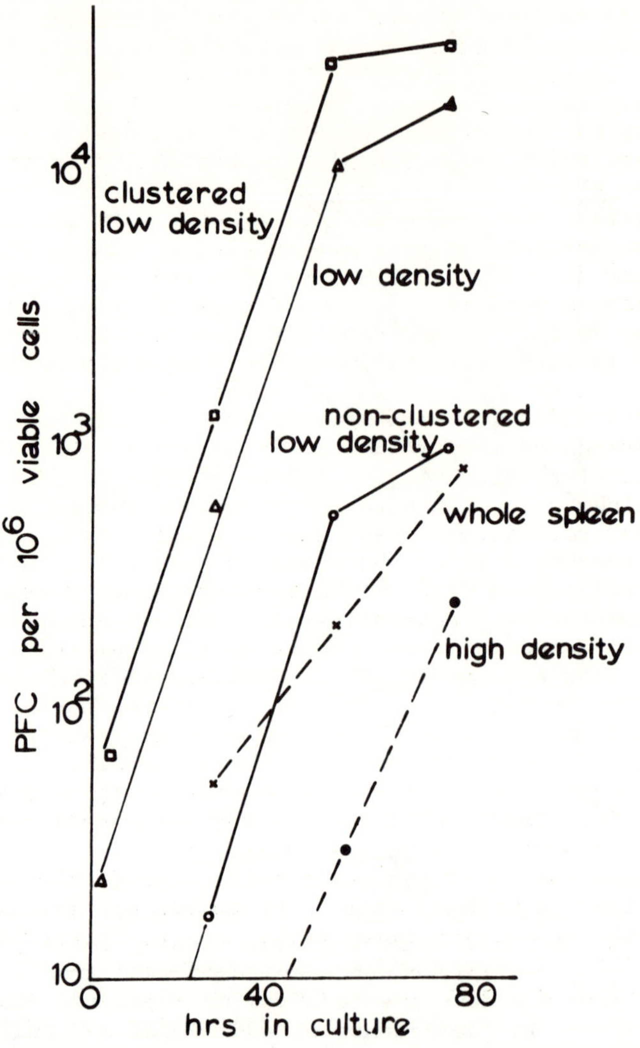

Fig. 7 Formation of direct PFC in various cell fractions taken from mice primed 2 days previously with sheep red blood cells i.v. Sheep cells have been added to the culture as well. The capacity to form PFC is greatly enriched in the cluster fraction.

begun 3 rather than 2 days after antigen administration in situ. Also the numbers are probably an underestimate. In many instances, the plaques contain more than one cell in their centers, because it is difficult to dissociate the clusters completely prior to assay. Moreover, the hemolysis in gel assay may not detect all those specific antibody secreting cells that exist, e.g., Cunningham has estimated that a liquid monolayer plaque assay is 3 times more sensitive than the gel system (15). The second feature about these quantitative data is the suggestion that the entire plaque forming capacity of a spleen may at some stage involve clusters, i.e., an association between dendritic cells and proliferating antibody secreting cells. Thus 4 days after giving sheep red cells i.v., or in vitro, to unprimed mice, the yield of PFC is of the order of $10^3/10^6$ cells. Clusters, representing about 1/30 - 1/50th of the total spleen, can be obtained 2 days after antigen administration in situ, and in the subsequent 2 days in vitro give rise to 3-5 x 10^4 PFC/10^6, i.e., a 30-50 fold enrichment over the total splenic response.

Similar results, at least qualitatively, have been obtained with another antigen, horseradish peroxidate (HRP). Here animals are primed with HRP in complete Freunds adjuvant several months before boosting with soluble enzyme intravenously. Two days later, clusters and other cell fractions are made from spleen. Antibody forming cells are identified cytochemically following the binding of exogenous HRP to cytoplasmic anti-HRP in fixed preparations. Anti-HRP cells are rarely found at the start of the experiment, but within 2 days in vitro, many of the clusters contain positive cells. Few specific plasma cells form in the other cell fractions, i.e., low density nonclustered cells, and high density cells.

The association of dendritic cells with developing plasma cells occurs in vivo as well. The spleen of germ free mice was studied at varying times after an i.p. dose of 2 x 10^8 sheep cells. At the start of the experiment, there is little morphological evidence of an antigen response, i.e., there are few germinal centers, large lymphocytes, or plasma cells. A sequence of morphological changes then ensues that has been well described for a number of antigens in many species (16, 17). By day two, the periarteriolar sheaths contain many blasts (?plasmablasts) and by day three, many aggregates of immature plasma cells begin accumulating at the border of red and white pulp. In close association with these proliferating and maturing plasma cell precursors are cells with the same cytological features as the dendritic cells identified in vitro (Fig. 8). The nuclei are irregular in shape and have a small nucleolus and a peripheral band of heterchromatin.

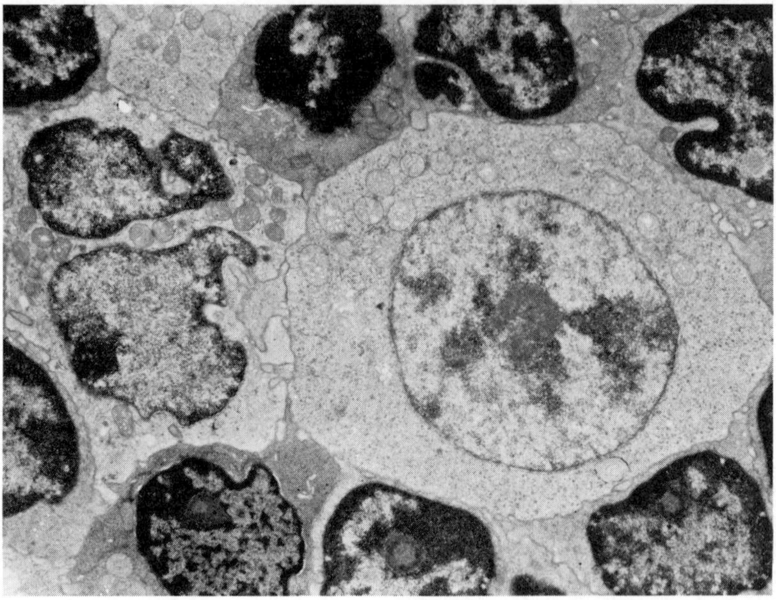

Fig. 8 A micrograph from the periarteriolar region of germ-free, splenic white pulp 2 days after administration of sheep red blood cells i.p. Many lymphoblasts (Lb) are present and contain large swollen nuclei, large nucleoli, and abundant cytoplasmic polysomes. These blasts (? plasmablasts) are frequently in contact with cells (De) having the same cytological features as the dendritic cells being studied in vitro. 8,500X.

TABLE II
PROPERTIES OF LOW DENSITY MOUSE SPLEEN CELL FRACTIONS

Property	Clusters	Non-clusters
Dendritic cells/total cells in fraction *	.16	.03
% of cells killed with		
Rabbit complement	7	6
Rabbit + anti kappa	14	44
Rabbit + anti theta	9	28
% of cells labeling with a 1 hr pulse of ^{3}H-thymidine	22	7

* The number of dendritic cells cannot be counted directly in the clusters, and is determined from the difference between the total number of dendritic cells in the low density population minus those in the non-cluster fraction.

The cytoplasm has relatively few organelles and is clearly very irregular in outline. Adjacent lymphocytes often insert cytoplasmic fingers or interdigitations into the body of the dendritic cell. In contrast, typical mononuclear phagocytes are much less frequent in these zones of a developing plasma cell response. Only after plasma cells have matured and move into the marginal zone and red pulp, are macrophages - which normally occupy these areas - to be found in significant numbers.

In conclusion, dendritic cells are a novel adherent cell type that can be distinguished from mononuclear phagocytes. Their function is not known. However, both in vivo and in vitro, they are associated with the proliferation and maturation of antibody forming cells. The meaning of this association is unclear and will probably require that we learn to effectively separate dendritic cells from the plasma cell precursors. It seems likely though that dendritic cells will have to be reckoned with in analyzing the complex events which must be involved in the generation of a plasma cell response.

REFERENCES

1. Mosier, D.E., Science 158: 1573, 1967.
2. Mosier, D.E. and Coppelson, L.W., Proc. Nat. Acad. Sci. U.S. 61: 542, 1968.
3. Pierce, C.W., J. Exp. Med. 130: 345, 1969.
4. Steinman, R.M. and Cohn, Z.A., J. Cell Biol. 55: 186, 1972.
5. Steinman, R.M. and Cohn, Z.A., J. Exp. Med. 137: 1142, 1973.
6. White, R.G., French, V.I. and Stark, J.M., J. Med. Microbiol. 3: 65, 1970.
7. Nossal, G.J.V., Abbot, A., Mitchell, J. and Lummus, Z., J. Exp. Med. 127: 277, 1968.
8. Hanna, M.G., Jr., and Szakal, A.K., J. Immunol. 101: 949, 1968.
9. Steinman, R.M. and Cohn, Z.A., J. Exp. Med. 139: 380, 1974.
10. Steinman, R.M., Lustig, D.S. and Cohn, Z.A., J. Exp. Med. 139: 1431, 1974.
11. Steinman, R.M. and Cohn, Z.A., in Mononuclear Phagocytes, 2nd edition, ed. R. van Furth, Blackwell Scientific, London, 1974.
12. Steinman, R.M., Adams, J. and Cohn, Z.A., Submitted for publication.
13. Mosier, D.E., J. Exp. Med. 129: 351, 1969.

14. Pierce, C.W. and Benacerraf, B., Science 166: 1002, 1969.
15. Cunningham, A.J., Nature 207: 1106, 1965.
16. Thorbecke, G.J. and Keuning, F.J., J. Infect. Dis. 98: 157, 1956.
17. Largevoort, H.L., Lab. Invest. 12: 106, 1963.

THE MACROPHAGE AS SECRETORY CELL[1]

Saimon Gordon[2], Jay C. Unkeless[3] and Zanvil A. Cohn
The Rockefeller University, New York, New York

INTRODUCTION

Macrophages play a central role in host defense against infection and in the inflammatory response. These functions depend, in part, on their widespread tissue distribution, specialized phagocytic properties and their ability to become metabolically activated in the course of infection and to gain an enhanced microbicidal capacity (1,2). The macrophages of the body originate from common bone marrow precursors which migrate through the blood as monocytes and enter tissues such as lungs (alveolar macrophages), liver (Kupffer cells) and spleen to constitute the mononuclear phagocyte system.

It has been possible to elucidate some of the cellular properties which underlie macrophage function by studying purified cell populations under controlled conditions of cultivation. Macrophages bear plasma membrane receptors which participate in a variety of cellular recognition reactions and contain an elaborate vacuolar apparatus involved in intracellular digestive function. A possible extracellular mode of macrophage function has, however, received little attention.

This report will describe recent studies in our laboratory which show that cultivated macrophages release high levels of enzymes into their culture medium and that the level of enzyme secretion can vary in a striking manner with the functional state of the cell. These products include lysozyme (3) and a variety of specific neutral proteases including plasminogen activators (4,5), collagenase (6) and elastase (7). These studies indicate that the mononuclear phagocyte system serves as a major source of secretion products which enable macrophages to exert a selective effect on their extracellular environment.

---

[1]Supported in part by grants (AI07012 and AI01831) from the U.S. Public Health Service, N.I.H. grant CA 08290-09 and The Rockefeller Foundation.
[2]Leukemia Society of America Inc. Scholar.
[3]Present address: Center for Cancer Research, Massachusetts Institute of Technology, Cambridge, Massachusetts 02139.

LYSOZYME

A. Introduction

Lysozyme (muramidase) is a widely distributed enzyme which hydrolyses N-acetyl muramic β 1, 4 N-acetyl glucosamine linkages in the bacterial cell wall, with resultant lysis of highly susceptible organisms like Micrococcus lysodeikticus (8). Although a great deal is known about its structure and enzymology, the cellular sources of lysozyme in body fluids have not been clearly identified. High concentrations are found in leukocytes, especially polymorphonuclear leukocytes (PMN) and BCG stimulated rabbit alveolar macrophages (9), and large amounts accumulate in the serum and urine of patients and animals bearing monocytic leukemia (10,11). Lysozyme is found in both the azurophil and specific granules of the PMN (12,13) and the rabbit alveolar macrophage is able to release a large fraction of its intracellular lysozyme into the medium during phagocytosis (14) and cultivation in vitro (15). The mechanism of lysozyme release from these leukocytes is, however, unknown.

B. Lysozyme secretion by mononuclear phagocytes

We have studied lysozyme production by three types of cultivated macrophage - unstimulated mouse peritoneal cells, stimulated mouse peritoneal macrophages harvested 4 days after i.p. injection of thioglycollate medium, and human blood monocytes. As shown in Fig. 1, there was a substantial net increase in total lysozyme present in each case, up to 20-fold in 3 days. The intracellular content remained relatively constant whereas extracellular lysozyme continued to increase daily, representing more than 90% of total production per day. These findings made it likely that lysozyme was actively secreted by the cells. It was also remarkable that these cells, which differ strikingly with regard to stimulation and maturity, all showed essentially the same pattern of lysozyme production and release.

The lysozyme secreted by mouse macrophages was shown in further studies to resemble authentic lysozyme in antigenicity, its size of 14,000 mw and in its highly cationic charge. Biosynthesis and secretion in vitro were confirmed by identifying ^{14}C labelled lysozyme in culture medium obtained from mouse macrophages which had been previously incubated in radioactive amino acids.

Human PMN containing substantial amounts of lysozyme did not produce lysozyme in vitro as these cells died rapidly in culture. Lysozyme secretion in vitro was found to be a specific macrophage trait absent in a variety of lymphoid, epithelioid and fibroblast cell lines.

C. Regulation of lysozyme secretion

Macrophages continued to secrete lysozyme at a remarkably

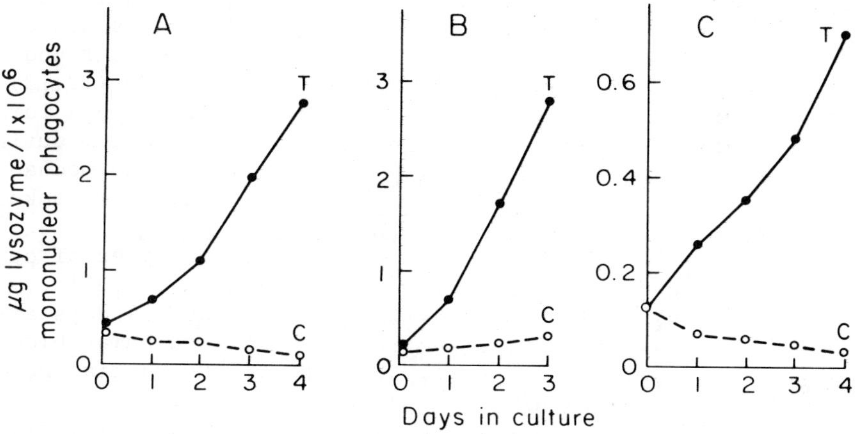

Fig. 1 Lysozyme production by mononuclear phagocytes in culture (A) Unstimulated mouse peritoneal macrophages, (B) Thioglycollate stimulated mouse peritoneal macrophages, (C) Human blood monocytes. C, cell lysate; T, total cell lysate and cumulative medium content.

constant rate over many days in culture. Lysozyme production - 0.75 μg/1 x 10^6 cells/day - represented up to 2.5% of total cell protein. We searched for, but failed to find specific inducer molecules in the culture medium, since lysozyme production continued at a high level even in serum free media. Macrophages apparently failed to secrete lysozyme only after total inhibition of protein synthesis, for example by treatment with cycloheximide.

We know nothing about the site of production or secretory pathway of the enzyme. Indirect evidence suggests, however, that lysozyme secretion is quite distinct from "classic" lysosomal function and the digestive tract of the macrophage. The first indication of such a difference was the observation that maximal lysozyme production and secretion were observed at far lower serum concentrations, 5% fetal bovine serum, than induction of acid hydrolases (16). Second, a variety of typical acid hydrolases, such as cathepsin D and N-acetyl β glucosaminidase, showed neither net synthesis nor significant release into the culture medium, less than twenty per cent of total activity in 3 days, under conditions favoring lysozyme secretion. Finally, a massive phagocytic load of latex particles or erythrocytes had no effect on the content or distribution of lysozyme in mouse macrophages.

All these studies suggested that lysozyme is a cell-specific, bulk secretion product of cultivated mononuclear phagocytes, and that its production in culture is regulated in a unique, unknown, fashion.

PLASMINOGEN ACTIVATOR (P.A.)

A. Introduction

The fibrinolysin plasmin is generated from plasminogen, its serum zymogen, by tissue proteases of restricted specificity which function as plasminogen activators. Avian and mammalian fibroblasts release high levels of plasminogen activators following transformation by oncogenic viruses (17-20), but the significance of this enzyme for abnormal growth remains unclear, since it is also present in a wide variety of normal cells (21), including PMN (22) and renal cells (23). Since fibrin deposition and lysis play an integral role in tissue injury and inflammation, it was also of interest to examine purified macrophage populations for production of fibrinolytic activity. Mononuclear phagocytes are a rich source of acid proteases, such as cathepsin D, which are thought to play a role in intracellular digestion (24), but proteases capable of acting extracellularly, at neutral pH, had not been characterized.

B. <u>Production and secretion of P.A. by macrophages</u>

Full experimental details are given elsewhere (4,5). Fibrinolytic activity was measured by solubilisation of a thin film of [^{125}I] fibrin, in the presence of plasminogen, by intact cells, cell fractions or conditioned serum-free medium. The sensitivity of the assay could be enhanced by destroying inhibitors of plasmin, such as α_2 macroglobulin, by acid treatment of the fetal bovine serum.

The fibrinolytic activity of intact macrophages on [^{125}I] fibrin-coated dishes is illustrated in Fig. 2. Thioglycollate stimulated macrophages showed rapid fibrinolysis; virtually all of this activity was plasminogen-dependent since it was not observed when the serum was depleted of plasminogen. In contrast to the high fibrinolytic activity of thioglycollate-stimulated cultures, that of the unstimulated cells was very low and barely exceeded the background level in cell-free control dishes. Dose response studies showed that the activity per cell of thioglycollate stimulated macrophages was at least 100-fold higher.

Further studies showed that conditioned medium and cell lysates from thioglycollate-stimulated macrophages contained considerable levels of plasminogen activator activity, whereas the same preparations from unstimulated macrophages showed no comparable activity (Table 1). The specific activity of the plasminogen activator is 60 times higher in CM from stimulated macrophages than in cell lysates, suggesting that this enzyme is actively secreted by the cells. Although lysozyme production and secretion were similar to those of plasminogen activator, lysozyme was however secreted by both stimulated and unstimulated cells.

The time course of plasminogen activator formation in cultures of stimulated macrophages is presented in Fig. 3A. Also shown for comparison is the production of lysozyme in the same cultures (3B). These cells, which had been transferred to serum-free lactalbumin hydrolysate medium after 24 hr in culture, synthesized and released P.A. over a period of at least 3 days. The intracellular activity remained relatively constant, whereas the total activity in the medium continued to increase and greatly exceeded that present in the cells. This fact provides additional evidence for secretion of the enzyme. The kinetics of PA secretion were similar to those of lysozyme secretion by the same cells.

C. <u>Properties of macrophage plasminogen activator</u>

Since the enzyme remained active in the presence of sodium dodecyl sulfate (SDS) it was possible to identify directly the molecular species of P.A. released by macrophages (Fig. 4A). A major peak of P.A. was found at a mol. wt. of 48,000,

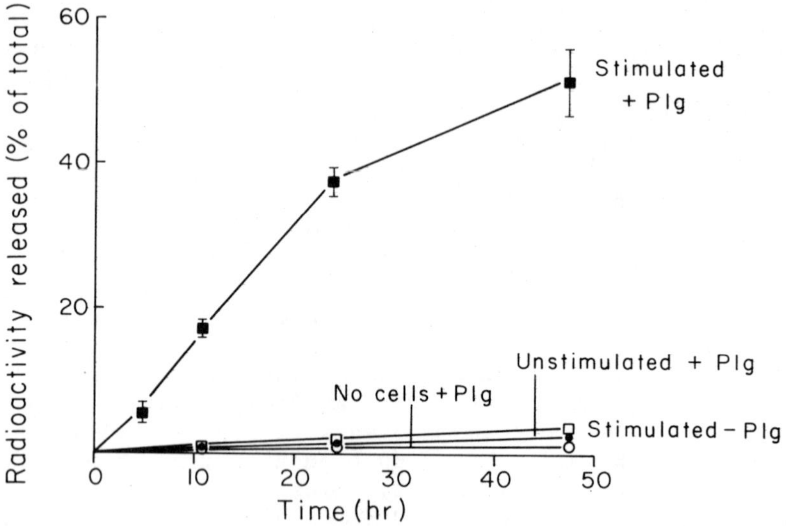

Fig. 2 Fibrinolysis by thioglycollate-stimulated and unstimulated macrophages. 1.5×10^6 stimulated or 4×10^6 unstimulated macrophages were plated on 35 mm [^{125}I] fibrin coated dishes in MEM containing 5% of FBS and 60 μg/ml STI and incubated for 24 hr. After washing, MEM containing 5% of human serum or 5% of plasminogen depleted human serum (total volumed 3 ml) was added and 0.3 ml samples were removed as indicated (■ - ■). Stimulated cells, human serum; (□ - □) Stimulated cells, plasminogen depleted serum; (● - ●) unstimulated cells, human serum; (O - O) control dishes without cells, human serum.

TABLE 1

Specific activities of plasminogen activator and lysozyme

	Total macrophage No.	Plasminogen-dependent fibrinolysis	Total protein	Plasminogen activator sp act	Lysozyme	Lysozyme sp act
		U/dish	mg	U/mg	μg/dish	μg lysozyme/mg protein
Stimulated macrophages	4×10^6					
Cell lysate (24 h)		17	0.60	28	0.43	0.70
Cell lysate (72 h)		55	0.75	74	1.3	1.7
Conditioned medium (24-72 h)		770	0.17	4,500	11	62
Unstimulated macrophages	12×10^6					
Cell lysate (24 h)		0	0.70	0	0.40	0.57
Cell lysate (72 h)		5.0	0.40	12.0	0.85	2.1
Conditioned medium (24-72 h)		0.60	0.22	3.0	11	50

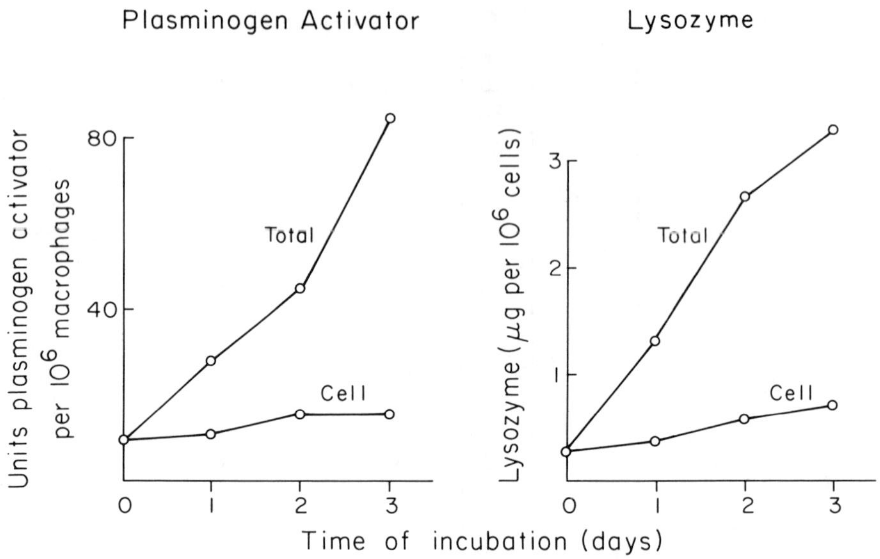

Fig. 3 Production and secretion of plasminogen activator and lysozyme. 4 x 10^6 thioglycollate stimulated macrophages were cultured for 24 hr in MEM containing 5% of FBS, washed three times and incubated in MEM containing 0.2% of lactalbumin hydrolysate. Cell lysates and conditioned medium were prepared from replicate dishes daily for 3 days subsequently.

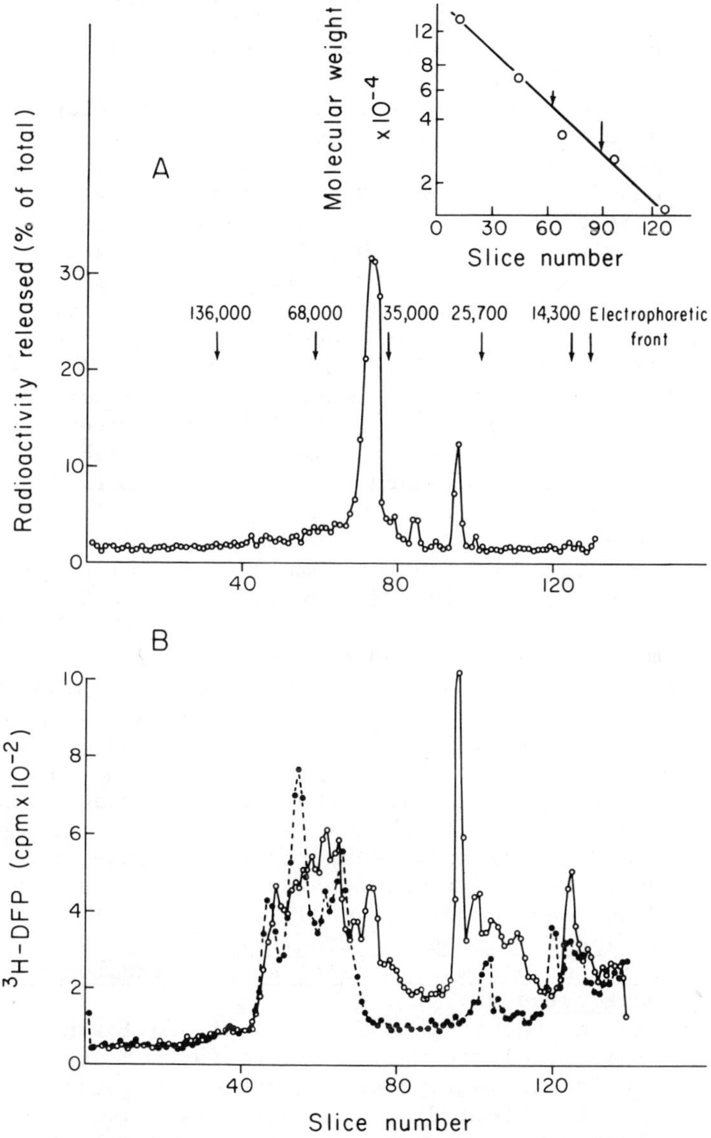

Fig. 4 Correlation of plasminogen activator and ^{3}H-DFP labelling in conditioned medium from stimulated and unstimulated macrophages (A) Plasminogen activator activity. (B) ^{3}H-DFP labelling of stimulated macrophage-conditioned medium, (O - O); unstimulated macrophage-conditioned medium (● - ●).

and a smaller peak at 28,000. All of the fibrinolysis was strictly dependent on the presence of plasminogen. The mol. wt. of the major macrophage activator (48,000) was very close to that of the plasminogen activators released by transformed mouse fibroblasts.

The striking differences in P.A. activity of thioglycollate-stimulated and unstimulated cells prompted further attempts to extend the comparison of macrophage products in the respective conditioned media. Since the macrophage P.A. was readily inhibited by di-isopropylfluorophosphate (DFP) samples of both cultures were incubated with ^{3}H-DFP and analysed by SDS-polyacrylamide gel electrophoresis to indicate the spectrum of serine enzymes present. As shown in Fig. 4B, the radioactivity profile of ^{3}H-DFP labelled proteins contained multiple peaks in both samples, but the two peaks in the medium from thioglycollate-stimulated cells which corresponded with plasminogen activator activity (4A) were absent from unstimulated macrophage C.M. The main activator peak represented 0.15 μg of P.A. released by 2 x 10^{7} cells in 48 hr, during which time this culture produced 40 μg of lysozyme.

In further studies nitrophenyl-p-guanidino benzoate (NPGB), which forms a stable acyl-enzyme complex with serine proteases of trypsin-like specificity (25), was used to inhibit the binding of ^{3}H-DFP by macrophage enzymes. The ^{3}H-DFP labelling of several peaks in medium from stimulated cultures was inhibited by NPGB (Fig. 5A); consistent with their sensitivity to NPGB, the labelling of the P.A. peaks was particularly decreased. In contrast, (Fig. 5B), the labelling of all DFP reactive species released by unstimulated cultures was completely resistant to NPGB. Thus the plasminogen activators produced by stimulated macrophages were identified as trypsin-like enzymes and it was shown that the serine enzymes released by the two kinds of macrophages differed both in specificity and in electrophoretic mobility.

D. Regulation of macrophage fibrinolysis

(i) Induction of plasminogen activator by endotoxin stimulation and phagocytosis.

The striking enhancement in fibrinolysis following stimulation with thioglycollate medium led to further studies on its regulation in the macrophage. A variety of other intraperitoneal irritants including endotoxin was tested, but macrophages harvested from such animals showed only a partial increase, less than 10% of the fibrinolytic activity of the thioglycollate stimulated cells. A possible role for phagocytosis was also examined by feeding formalin-treated sheep erythocytes or latex particles to populations of unstimulated macrophages, but such treatment again induced only a small

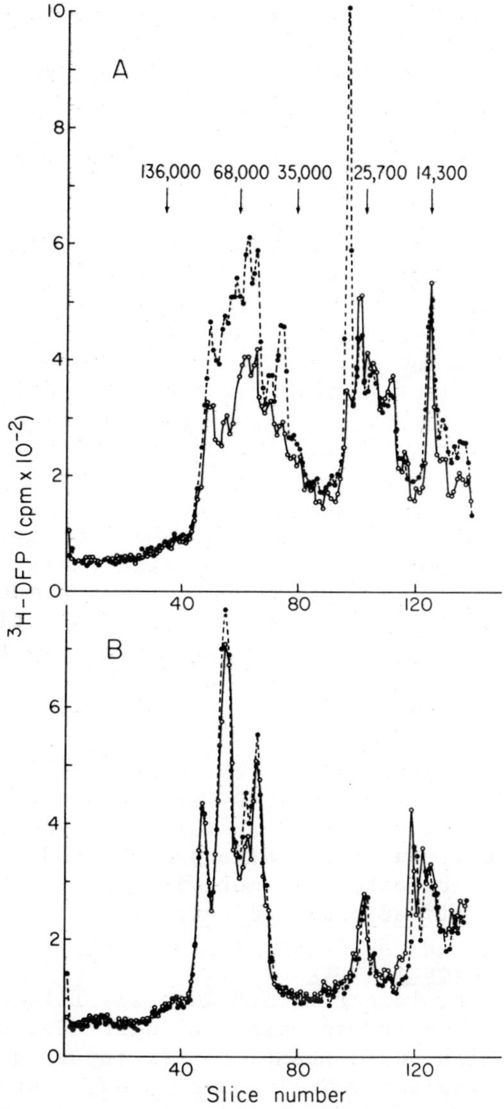

Fig. 5 Inhibition by NPGB of ^{3}H-DFP labelling of proteins in conditioned medium from stimulated and unstimulated macrophages. (A) stimulated macrophage conditioned medium (● - ●) without NPGB; with 8 x 10^{-5}M NPGB (O - O).

fraction, less than 10%, of thioglycollate cell activity. However, when partially stimulated macrophages, obtained 4 days after intraperitoneal injection of endotoxin, were allowed to phagocytose latex particles *in vitro*, a potent stimulatory effect on fibrinolytic activity was achieved. Fig. 6 illustrates an experiment which shows that combined treatment with endotoxin and latex dramatically stimulated plasminogen activator levels in macrophage cell lysates and C.M., in comparison with each treatment alone. Stimulation after combined treatment was apparent within one day of phagocytosis and P.A. levels increased 5-fold in both cell lysate and medium during further cultivation. Such cells secreted 70 to 85% of total P.A. produced per day.

We concluded from these studies that combined treatment with endotoxin and latex had a synergistic effect on the induction of macrophage P.A. As illustrated in Table II, this induction process was selective for the P.A. since such treatments had no comparable effect on secretion of lysozyme or the production of an intracellular acid protease, cathepsin D.

(ii) Priming the macrophage

The macrophage P.A. could therefore be induced by a 2-stage process which involved a priming reaction, after which the partially stimulated cells produced little enzyme, and a phagocytic trigger which further stimulated production and secretion. It now became possible to design experiments to study the contribution of each step independently.

Fig. 7 illustrates an experiment which shows that macrophage fibrinolysis was stimulated by latex after a variety of intraperitoneal stimuli, except for thioglycollate cells which already displayed maximal activity without phagocytosis. The priming efficiency of these stimuli diminished in the order thioglycollate, endotoxin, mineral oil, BCG antigen and fetal calf serum. Priming was therefore not unique to endotoxin and was correlated with the intensity of the inflammatory stimulus.

(iii) The phagocytic trigger

The role of particle nature and load in triggering endotoxin primed macrophages was next examined. Table III shows that other particles could stimulate fibrinolysis as effectively as latex, including *M. lysodeikticus*, heat aggregated γ globulin and immune complexes of horse radish peroxidase (HRP)-anti HRP. These materials were all rapidly ingested by the endotoxin primed macrophages and the degree of stimulation of fibrinolysis, 2-6 fold, was proportional to particle dose. We concluded that the triggering reaction was associated with a general aspect of the phagocytic process.

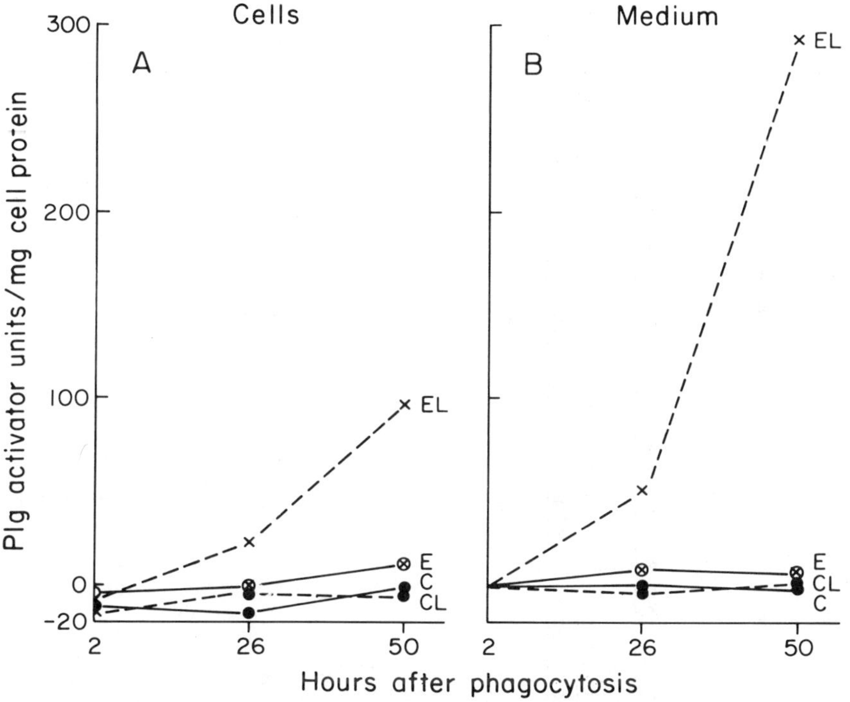

Fig. 6 The effect of endotoxin stimulation and latex phagocytosis on macrophage plasminogen activator in cells (A) and conditioned medium (B). 4 x 10^6 macrophages were cultivated in Dulbecco's medium and 10% FBS for 1 day. The cells were then fed latex and placed in Dulbecco's medium + 0.05% LH either immediately or 26 hrs after phagocytosis. Plasminogen dependent fibrinolysis was defined as the difference in radioactivity released in the presence and absence of plasminogen. Results for plasminogen activator in medium show secretion between 2-26 and 26-50 hrs respectively. (●---●) control: (●---●) control fed latex; (x---x) endotoxin; (x---x) endotoxin fed latex.

TABLE II

SELECTIVE INDUCTION AND SECRETION OF MACROPHAGE ENZYMES

Treatment	Plasminogen Activator U/mg	Plasminogen Activator % secreted	Lysozyme μg/mg	Lysozyme % secreted	Cathepsin D U/mg	Cathepsin D % secreted
Unstimulated	- 3	-	56	95	.38	< 20
" + Latex	15	-	31	92	.44	11
Endotoxin	55	100	80	98	.39	< 20
" + Latex	250	87	39	95	.43	11
Thioglycollate	800	83	37	94	.52	25

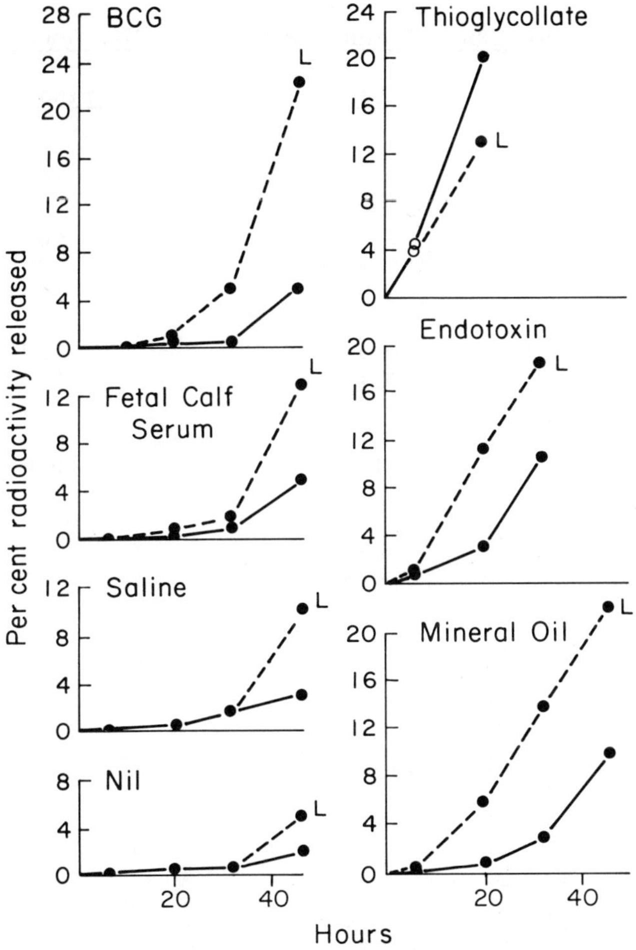

Fig. 7 Macrophage priming and fibrinolysis _in situ_. Number of macrophages plated: Thioglycollate 6 x 10^5, others 4-6 x 10^6 (●---●) no latex; (●---●) fed latex after 24 hrs cultivation then washed and placed in AT-FBS for assay.

TABLE III

PHAGOCYTIC TRIGGERS FOR ENDOTOXIN PRIMED MACROPHAGES

Particle	Load	Fibrinolysis % Control
Nil	-	100
Latex	5×10^8	400
	1.5×10^8	330
	0.5×10^8	250
M. lysodeikticus	0.5 mg	310
	0.17 mg	210
	0.06 mg	130
Ht. Agg. γ glob.	150 μg	650
	15 μg	350
Peroxidase - Antibody complexes	30 μg	540
	3 μg	240

Although the initial trigger was not specific for a particular substance the ability to induce a sustained fibrinolytic response did, however, depend in a striking manner on the persistence of the phagocytized particle within the cell. Fig. 8 illustrates an experiment in which endotoxin primed macrophages were fed either M. lysodeikticus, a particle which is rapidly digested, or latex, which is non-digestible. Both types of particle triggered macrophage fibrinolysis within 8 hrs, but the subsequent time course of induced fibrinolysis was quite different. Fibrinolysis increased and persisted after ingestion of latex, but the activity of cells fed M. lysodeikticus diminished within a day and approached that of unfed cells after further cultivation. Similar experiments on non-radioactive plates confirmed that P.A. was secreted only transiently by endotoxin stimulated macrophages after ingestion of M. lysodeikticus, whereas the latex laden macrophages continued to release high levels of enzyme for at least 9 days. The induction of P.A. in the macrophage thus contrasts clearly with that of intracellular acid hydrolases like acid phosphatase which is only stimulated by particles of a digestible nature (26).

COLLAGENASE, ELASTASE AND OTHER NEUTRAL PROTEASES

Experiments performed in collaboration with Z. Werb of the Strangeways Laboratory, Cambridge, show that thioglycollate stimulated mouse peritoneal macrophages also secrete a variety of other neutral proteases which are low or absent in unstimulated cell cultures unless the cells have been allowed to ingest latex particles. These include a collagenase (6) which cleaves reconstituted collagen fibrils in a specific manner, an elastase which solubilises elastin fibrils (7) and a neutral azocaseinase. These enzymes have been partially characterized and compared with related enzymes present in PMN and other mammalian cells. The macrophage collagenase is a metalloprotease inhibited by chelating agents such as EDTA or o-phenanthrolene, as well as by α_2 macroglobulin, but it is insensitive to DFP. The elastase is sensitive to EDTA, unlike the plasminogen activator, as well as to DFP. The mouse macrophage elastase does not readily cleave synthetic substrates like acetyl-$(ala)_4$-chloromethylketone, unlike porcine pancreatic elastase, and is insensitive to soybean trypsin inhibitor, unlike the human PMN elastase.

The production and release of collagenase by oil induced guinea-pig peritoneal macrophages treated with endotoxin or lymphokines in vitro has also been described recently by other investigators (27,28).

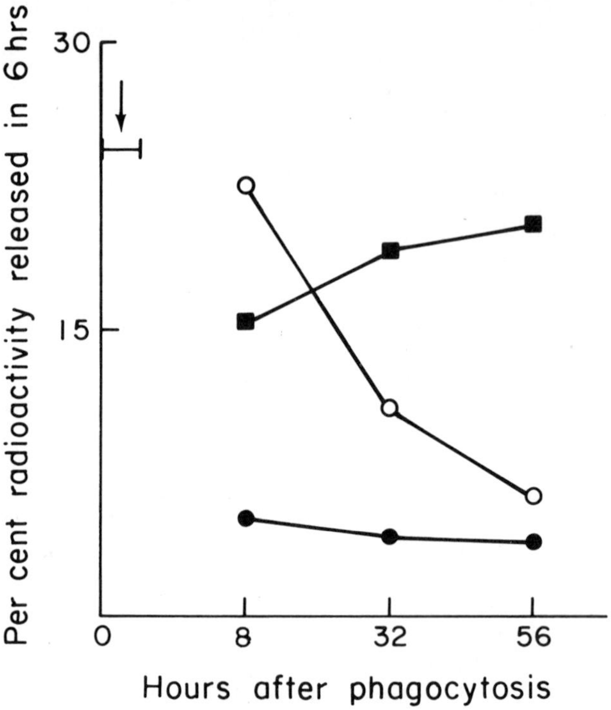

Fig. 8 Fibrinolysis by endotoxin primed macrophages after phagocytosis of latex or M. lysodeikticus. 5 x 10^6 macrophages were cultured on ^{125}I labelled fibrin dishes for 24 hrs in the presence of STI. After phagocytosis (arrow) the cultures were maintained in Dulbecco's medium containing 10% FCS and STI until assay in AT-FCS at 2, 26 and 50 hrs. (●---●) endotoxin primed macrophages; (O---O) fed M. lysodeikticus; (■---■) fed latex.

OTHER MACROPHAGE SECRETION PRODUCTS

In addition to the enzymes discussed, a great variety of products has been detected in macrophage conditioned medium (Table IV). Among these are other enzymes like esterases and acid hydrolases, components of the complement system and factors which have been implicated in hemopoietic colony stimulation (CSF) and perhaps lymphocyte function. The evidence that these are indeed secreted by macrophages is good for products like C_2, C_4, pyrogen and CSF, but it is inconclusive for other products, which require further characterization.

Another illustration of the diversity of macrophage secretion is given in Fig. 9, which shows the SDS polyacrylamide gel patterns of CM obtained from stimulated and unstimulated mouse macrophage cultures after incubation in a mixture of radioactive amino acids (4).

The gels stained for protein and the corresponding autoradiogram shows protein species common to both types of macrophage, for example lysozyme (4); but other proteins are unique to each type of CM. The heterogeneity of these labelled proteins and of the DFP-reactive serine enzymes (Fig. 4) indicate the extent and variety of synthetic and secretory activity in the macrophages, and their striking modification by thioglycollate stimulation.

DISCUSSION

These studies have demonstrated that macrophages accumulate and release a great variety of products in culture. In view of their considerable biosynthetic capacity, extended life span and enhanced activity in the course of many diseases, we conclude that the mononuclear phagocyte system represents a secretory organ of considerable biological significance, and one which displays novel mechanisms for regulating its secretory activity. Our studies have shown that the hydrolases of cultured macrophages can be differentiated into at least three classes which differ in their control mechanisms. The plasminogen activator and other neutral proteases are inducible enzymes associated with macrophage stimulation and markedly influenced by phagocytosis, whereas lysozyme secretion is a continuous process, independent of cell stimulation and unaffected by phagocytosis. The secretion of PA and lysozyme is unaffected under conditions where there are considerable increases in intracellular acid hydrolases. These differences in enzyme regulation suggest distinct functions for each class of enzyme.

Mouse macrophages provide a simple in vitro system to measure enzyme secretion and its control. High levels of plas-

TABLE IV
MACROPHAGE SECRETION PRODUCTS

ENZYMES	Ref.
Lysozyme	3, 15
Neutral proteases:	
Plasminogen activator	4, 5
Collagenase	6, 27
Elastase	7
Azocaseinase	7
Esterases	29, 4
Acid hydrolases	30
PROTEINS	
Complement	
C_2	31, 32
C_4	33, 31, 32
(C_5)	34
Ss	35
Pyrogen	36
Interferon	37
FACTORS	
Macrophage CSF, Granulocyte CSF	38, 39, 40, 41
Lymphocyte	42, 43, 44, 45
Plasmacytoma	46

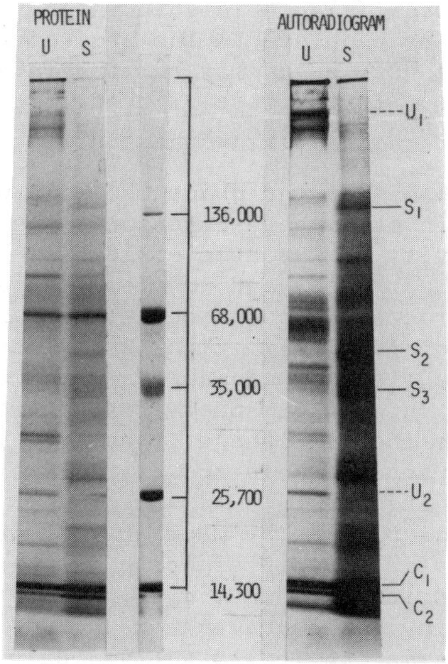

Fig. 9 Incorporation of [^{14}C] amino acids into products secreted by stimulated and unstimulated macrophages. [^{14}C] labelled conditioned medium from 1 x 10^7 unstimulated or 3 x 10^6 thioglycollate stimulated macrophages was electrophoresed on a 8-15% polyacrylamide slab gel after concentration and dialysis.

minogen activator can be induced in mouse macrophages by a single intraperitoneal injection of thioglycollate medium or, in two stages, namely by endotoxin injection in vivo followed by latex phagocytosis in vitro. Thioglycollate treatment induces higher levels of activity and latex phagocytosis does not further enhance secretion or fibrinolysis. A likely explanation for this difference is that phagocytosis of thioglycollate medium takes place in the peritoneal cavity and that some of the constituents may be undegradable and persist in the macrophages. In the case of thioglycollate treatment both stimuli are thus delivered simultaneously and result in a maximal effect.

Much remains to be learned about the role of various secretory products in macrophage function. The extracellular microbicidal activity of lysozyme could be enhanced by synergistic action with antibody and complement (47). The formation of neutral proteases could account for many aspects of macrophage function in inflammation. These enzymes attack fibrin, collagen and elastin, all abundant extracellular proteins, in a highly specific manner. By controlled release of these proteases the macrophage is able to limit excessive fibrin deposition and fibrosis and help restore normal function after tissue injury.

The formation of plasmin by macrophages could not only initiate fibrinolysis, but interaction of plasmin with the enzymatic cascades involved in coagulation, complement and kinin formation provide several pathways for generating peptides that are pharmacologically active, particularly in their effects on vascular permeability (48,49). In addition plasmin may also affect the properties of other cells in the immediate environment (50).

The requirement of phagocytosis for secretion of the P.A. and the fall in fibrinolytic activity and secretion after digestion of the particles may serve to localize closely the generation of plasmin by macrophages. The continued generation of plasmin, collagenase and elastase associated with intracellular persistence of the phagocytosed particles could serve to perpetuate chronic granulomatous inflammation and contribute significantly to tissue injury.

The concept of macrophage activation was developed to explain the non-specific enhancement of macrophage bactericidal capacity observed after infection by BCG and other agents (51). Very little is known about the cellular properties which are associated with macrophage activation and the absence of a suitable biochemical marker has been a particular handicap in the study of this process. Our studies show that the inudction of neutral proteases can accompany macrophage activation and that such macrophage secretory products could play a role in some of the effector functions of the activated macrophage.

SUMMARY

Pure cultures of unstimulated mouse peritoneal macrophages, thioglycollate stimulated mouse peritoneal macrophages and human blood monocytes synthesize and secrete substantial amounts of lysozyme _in vitro_. The rate of lysozyme production and secretion is remarkably constant under a variety of culture conditions. It is independent of cell stimulation and unaffected by massive phagocytosis of latex particles.

Thioglycollate stimulated macrophages produce, accumulate and continuously release high levels of plasminogen activator in culture whereas unstimulated macrophages do not. The major macrophage plasminogen activator is a trypsin-like serine protease of mol. wt. 48,000. The secretion products released by stimulated and unstimulated macrophages have been compared by SDS-polyacrylamide gel electrophoresis after chemical labelling with ^{3}H-DFP or biosynthetic labelling with ^{14}C amino acids. These procedures show that some proteins are found in both cultures, whereas others are uniquely secreted by each type of macrophage. The serine enzymes released by the two kinds of macrophage differ in specificity and electrophoretic mobility.

Endotoxin stimulated macrophages contain and release relatively low levels of plasminogen activator, but these primed cells can be triggered to produce and secrete high levels of enzyme by phagocytosis of latex. Under conditions where the plasminogen activator is induced and secreted there are no effects on the production and/or release of lysozyme or intracellular acid hydrolases. Discovery of a 2-stage procedure for inducing macrophage P.A. made it possible to study the role of cell priming and phagocytosis separately. Although endotoxin was the most effective priming agent, other intraperitoneal irritants like mineral oil and BCG antigen also stimulate macrophage fibrinolytic activity after phagocytosis.

Latex and a variety of other particles such as _M. lysodeikticus_, aggregated γ globulin and immune complexes showed dose-dependent stimulation of fibrinolysis by endotoxin primed macrophages. Although the initial phagocytic trigger was not specific for the substance employed, the ability to induce a sustained response depended on the persistence of the phagocytized particle within the cell. Fibrinolysis and secretion of plasminogen activator continued at high levels for at least 9 days after uptake of latex, a non-digestible particle whereas P.A. was secreted only transiently after ingestion of rapidly digested *M. lysodeikticus*.

Thioglycollate stimulated macrophages also release a specific collagenase, an elastase and neutral azocaseinase, all of which are low or absent from medium conditioned by unstimulated macrophages.

The production and release of a variety of enzymes and other biologically active molecules provides a mechanism by which the macrophage can modify the extracellular milieu of the organism at both the local and systemic level.

REFERENCES

1. Gordon, S. and Cohn, Z.A., Int. Rev. Cytol. 36: 171, 1973.
2. Nelson, D.S., Critical Rev. Microbiol. 1: 353, 1972.
3. Gordon, S., Todd, J. and Cohn, Z.A., J. Exp. Med. 139: 1228, 1974.
4. Unkeless, J.C., Gordon, S. and Reich, E., J. Exp. Med. 139: 834, 1974.
5. Gordon, S., Unkeless, J.C. and Cohn, Z.A., J. Exp. Med. 140: 995, 1974.
6. Werb, Z. and Gordon, S., manuscript in preparation.
7. Werb, Z. and Gordon, S., manuscript in preparation.
8. Imoto, T., Johnson, C.N., North, A.C.T., Phillips, D.C. and Rupley, J.A., in "The Enzymes" (ed. by P.D. Boyer, Academic Press, New York, 3rd ed.), p. 666, 1972.
9. Cohn, Z.A., in "Symp. Int. Soc. of Cell Biol." (ed. by K.B. Warren), 7: 101, 1968.
10. Osserman, E.F. and Lawlor, D.F., J. Exp. Med. 124: 921, 1966.
11. Riblet, R.J. and Herzenberg, L.A., Science (Washington, D.C.) 168: 1595, 1970.
12. Baggiolini, M., Hirsch, J.G. and de Duve, C., J. Cell Biol. 45: 586, 1970.
13. Bretz, U. and Baggiolini, M., J. Cell Biol. 63: 251, 1974.
14. Cohn, Z.A. and Wiener, E., J. Exp. Med. 118: 1009, 1963.
15. Heise, E.R. and Myrvik, Q.N., J. Reticuloendothelial Soc. 4: 510, 1967.
16. Cohn, Z.A. and Benson, B., J. Exp. Med. 121: 835, 1965.
17. Unkeless, J.C., Tobia, A., Ossowski, L., Quigley, J.P., Rifkin, D.B. and Reich, E., J. Exp. Med. 137: 85, 1973.
18. Ossowski, L., Unkeless, J.C., Tobia, A., Quigley, J.P., Rifkin, D.B. and Reich, E., J. Exp. Med. 137: 112, 1973.
19. Quigley, J.P., Ossowski, L. and Reich, E., J. Biol. Chem. 249: 4306, 1974.
20. Unkeless, J., Danø, K., Kellerman, G.M. and Reich, E., J. Biol. Chem. 249: 4295, 1974.

21. Astrup, T., Fed. Proc. 25: 42, 1966.
22. Wunschmann-Henderson, B., Horwitz, D.L. and Astrup, T., Proc. Soc. Exp. Biol. and Med. 141: 634, 1972.
23. Ali, S.Y. and Lack, C.H., Biochem. J. 96: 63, 1965.
24. Dingle, J.T., Poole, A.R., Lazarus, G.S. and Barrett, A.J., J. Exp. Med. 137: 1124, 1973.
25. Fasco, M.J. and Fenton, J.W., Arch. Biochem. Biophys. 159: 802, 1973.
26. Axline, S.G. and Cohn, Z.A., J. Exp. Med. 131: 1239, 1970.
27. Wahl, L.M., Wahl, S.M., Mergenhagen, S.E. and Martin, G.R., Proc. Natl. Acad. Sci. U.S.A. 71: 3598, 1974.
28. Wahl, L.M., Wahl, S.M., Martin, G.R., and Mergenhagen, S.E., Fed. Proc. 33: 618, 1974.
29. Wiener, E. and Levanon, D., Science 159: 217, 1968.
30. Davies, P., Page, R.C. and Allison, A.C., J. Exp. Med. 139: 1262, 1974.
31. Wyatt, H.V., Colten, H.R., and Borsos, T., J. Immunol. 108: 1609, 1972.
32. Colten, H.R., J. Clin. Invest. 51: 725, 1972.
33. Littleton, C., Kessler, D. and Burkholder, P.M., Immunol. 18: 693, 1970.
34. Levy, N.L., Snyderman, R., Ladda, R.L. and Lieberman, R., Proc. Nat. Acad. Sci. U.S.A. 70: 3125, 1973.
35. Saunders, D. and Edidin, M., J. Immunol. 112: 2210, 1974.
36. Bodel, P., J. Exp. Med. 140: 954, 1974.
37. Smith, T.J. and Wagner, R.R., J. Exp. Med. 125: 559, 1967.
38. Golde, D.W. and Cline, M.J., J. Clin. Invest. 51: 2981, 1972.
39. Chervenick, P.A. and LoBuglio, A.F., Science 178: 164, 1972.
40. Moore, M.A.S., Williams, N. and Metcalf, D., J. Natl. Cancer Inst. 50: 591, 1973.
41. Eaves, A.C. and Bruce, W.R., Cell Tissue Kinet. 7: 19, 1974.
42. Bach, F.H., Alter, B.G., Solliday, S., Zoschke, D.C. and Janis, M., Cel. Immunol. 1: 219, 1970.
43. Hoffmann, M. and Dutton, R.W., Science 172: 1047, 1971.
44. Gery, I. and Waksman, B.H., J. Exp. Med. 136: 143, 1972.
45. Wood, D.D. and Gaul, S.L., J. Immunol. 143: 925, 1974.
46. Namba, Y. and Hanaoka, M., J. Immunol. 109: 1193, 1972.
47. Inoue, K., in Research in "Immunochemistry and Immunobiology" (ed. by J.B.G. Kwapinski, University Park Press, Baltimore), 6: 177, 1972.
48. Ratnoff, O.D., Adv. Immun. 10: 145, 1969.
49. Miller, R.L., Reichgott, M.J. and Melmon, K.L., J. Infect. Dis. 128: Suppl. S144, 1973.

50. Ossowski, L., Quigley, J.P., Kellerman, G.M. and Reich, E., J. Exp. Med. 138: 1056, 1973.
51. Mackaness, G.B., in "Mononuclear Phagocytes" (ed. by R.v. Furth), Blackwell, Oxford , p. 461, 1970.

DISCUSSION FOLLOWING SYMPOSIUM THE ROLE OF THE MACROPHAGE IN THE INITIATION AND REGULATION OF THE IMMUNE REPONSE

Unanue - Boston: We have discussed this many times, the data are very pretty and the hypothesis fits in very nicely with the data of that particular experimental model. But there are a series of other data in the mouse in which macrophages simply do not have those restrictions. This suggests that other pathways of interaction exist. There are also data that macrophages may secrete modulating materials which have no genetic restrictions and which do not require physical interaction between the lymphocyte and the macrophage. I think it is a mistake for all of us to simply depend on a particular experimental system.

Rosenthal - Bethesda: I would agree with that. I am certainly not trying to present the "universal field theory" of immunology. There must surely be other pathways operative for cell triggering. I must say, however, that data will be presented later in the conference that genetic restriction between macrophage and T lymphocyte also exists in the mouse as well as the guinea pig in terms of how T cells recognize antigen.

Cooper - Bethesda: I am interested in the early nonspecific phase of interaction. Have you ever done the following reciprocal experiments? Following the fall off of the nonspecific interaction, have you ever removed the non-bound lymphocytes and placed them in a fresh macrophage monolayer and vice versa? What happens?

Rosenthal - Bethesda: Yes, Peter Lipsky has done those experiments. Lymphocytes not bound to the monolayer will bind just as well as fresh lymphocytes to a second fresh macrophage monolayer. Conversely "used" macrophages bind just as well to fresh lymphocytes.

Greaves - London: Is it not possible that the B cell will have much less stringent requirements for this kind of interaction than the T cell?

Rosenthal - Bethesda: I think that is perhaps part of what Unanue was trying to make and I do not think Emil and I have that great a difference in the way we think about the system. It may well be that the recognition of the antigen by B cells is less stringent in its requirements for the configuration of presentation and the nature of the components that are necessary to be recognized.

Paul - Bethesda: I think you alluded earlier in your talk to the possibility that a nonadherent precursor might exist. I wonder if you would be willing to say a word or two about that?

Steinman - New York: The finding is that if animals are lethally irradiated, we don't find dendritic cells in the adherent population. If these are then reconstituted with F_1 bone marrow and nonadherent spleen cells, then adherent dendritic cells with the histocompatibility marker of the F_1 donor repopulate the spleen.

Korn - Bethesda: Can you comment on the presence of Fc receptor on your cell?

Steinman - New York: Well we have tried a number of approaches to demonstrate the Fc receptor and simply can't find it. We have also tried a number of markers such as aggregates of peroxidase and antiperoxidase as well as red cells oxidized in various ways.

Mosier - Bethesda: As much as I would like to believe these observations for reasons of personal bias, there are two particular observations that give me a lot of trouble. One is that you say that the dendritic cells you describe are not present until two or three weeks of age. Both in our laboratory and the laboratory of Dr. Ed Gault, functional murine adherent cells can be prepared from spleens of mice only one week old. The second problem is that you just mentioned you do not obtain active dendritic cells from irradiated mice. However, adherent cells capable of supporting immune response are radio-resistant and can be obtained at various times following very high doses of radiation.

Steinman - New York: Although we don't detect dendritic cells by these morphological criteria in the adherent cell population in neonatal mice, it does not mean that they are not there. Contrasting them with typical macrophages; if you collagenase treat your spleens, you can isolate considerable numbers of typical phagocytes but you can't say that they were not there in absence of collagenase. In addition, I think there is some evidence that if you use homogeneous populations of peritoneal macrophages that you can obtain some of the adherent cells functions that you get from adherent cell in mouse spleen; that doesn't exclude a role for these adherent dendritic cells in immune phenomenon.

Breck - San Francisco: You mentioned that the antigen is not interiorized in B cells and I wondered how do these cells differ from the dendritic macrophages that Nossal talks about in the lymph nodes? Also have you followed these adherent in long cultures, that is, could these particular cells turn into the typical dendritic macrophages?

Steinman - New York: I think the problem we have in relating our work to Nossal's "dendritic" macrophage is that no good in situ morphological studies exist of antigen retention and its association with distinct cell types. What you usually see are autoradiographs of some cell processes with grains associated with them; you really do not know where those processes are coming from. In fact some people would argue that antigen retention is on B lymphocytes via a combination of Fc and complement receptors.

Oppenheim - Bethesda: The conspicuous absence of lymphocyte supernatants as an activator for macrophage synthesis and release of enzymes is very provocative. I wonder if you are saving that for the future or whether it does not work.

Gordon - Montreal: We have not looked at that extensively. In the in vivo situation induction could involve the participation of other cell types and serum factors as well.

Havemann - Marburg: Does your plasminogen activator or your neutral proteases activate or change membrane receptors in lymphocytes?

Gordon - Montreal: Yes. I think that it is an interesting possibility that these enzymes might also serve to modify cell surfaces rather than just serum substrate, but we have no information on that.

Nelson - Sydney: To what extent do neutrophils contribute to the amount of lysozyme and other enzymes released?

Gordon - Montreal: No neutrophils are present in the system we use and we have gone to considerable trouble to show that. I do not want to give you the impression that neutrophils don't contain or release their enzymes during the very short period during which they remain viable in culture. They do, but the biosynthetic apparatus is lacking in these cells, so the mechanism of release is quite likely to be different from that of macrophage.

Oppenheim - Bethesda: Let me ask a question in reference to the small dialyzable inhibitor. Are you sure that in fact it is not merely inhibiting the thymidine incorporation without actually interfering with cell growth and proliferation?

Unanue - Boston: Yes, that was a very important consideration. That is why I showed the first slide where we do not measure incorporation of tritiated thymidine but just the number of cells in culture at different times. We also did mitotic indices and as you saw mitosis was turned off with this inhibitor. Also you turn off the PFC response when the undialyzed macrophage culture or fluid is present. So I do not think you simply have cold thymidine competing with radioactive thymidine.

Nelson - Syndey: I do not want to steal what is left of my own very feeble thunder this afternoon, but I would like to say that this sort of thing, production of stimulators and inhibitors, seems to be a very general phenomenon in macrophages not only of mice but also in rats and guinea pigs. At least in our hands there are terrible technical problems and minor variations in things like the medium and the serum which will produce very marked changes in the detailed findings, although not so much in the principals. In rats, if I could cite Dr. Keller's work, his factor is a macromolecular produced only by activated but not by "normal" macrophages and susceptible to freezing and thawing. There exists similar difference among workers asking related questions all along the line.

SECTION VIII

GENETIC CONTROL OF IMMUNOCOMPETENT CELL INTERACTIONS

Chairman

HUGH O. MCDEVITT

Stanford University School of Medicine
Stanford, California

Co-Chairman

DAVID H. SACHS

National Institutes of Health
Bethesda, Maryland

GENETIC CONTROL OF IMMUNOCOMPETENT CELL INTERACTIONS

Hugh O. McDevitt
Division of Immunology, Department of Medicine, School of Medicine, Stanford University, Stanford, California 94305

Despite a wealth of experimental data, the precise mechanisms by which immunocompetent cells -- thymus derived lymphocytes (T cells), bone marrow derived precursors of antibody producing cells (B cells), and macrophages -- interact to generate cell mediated immunity, humoral antibody responses, or allogeneic reactions between cells, remain a mystery. One possible approach in unraveling this complex problem may lie in the utilization of the histocompatibility linked specific immune response genes (_Ir_ genes). Following the initial demonstration of linkage between genetic control of specific immune responses and the major histocompatibility complex (MHC) in the mouse (1) it rapidly became apparent that a great many _Ir_ genes were linked to the species' major histocompatibility complex (2). Although the mechanism of action of these genes remains a puzzle (for reviews see 2,3,4), the close association between specific _Ir_ genes and the species' major histocompatibility complex has permitted new approaches to the problem of genetic regulation of cell-cell interaction. The initial approach to this problem was made by Shevach et al. (5) who showed that alloantisera directed against determinants controlled by the MHC of the guinea pig can inhibit the activation of T lymphocytes. These experiments were conducted with F_1 hybrids between strains which differed in their capacity to respond to synthetic antigens. The responder or nonresponder status of these animals is linked to the MHC against which the alloantisera are directed. Alloantisera against the responder MHC inhibited T lymphocyte activation by antigens to which only the responder strain is capable of making a response. Alloantisera against the nonresponder histocompatibility type had no effect. The same results were obtained in the reciprocal direction, using other antigens in which the responder and nonresponder histocompatibility types are reversed. These initial results suggested that the alloantisera were directed either against an _Ir_ gene product which functioned as the T lymphocyte antigen receptor, or which was situated in close proximity on the cell surface to the T lymphocyte antigen receptor.

Further studies have extended these observations and suggested that some gene product coded for by the major histocompatibility complex may be involved in macrophage - T lymphocyte interaction in this type of in vitro system. In studies by Shevach and Rosenthal (6), it was shown that antigen bearing macrophages from a responder, but not a nonresponder, parental line were able to activate (responder x nonresponder) F_1 T lymphocytes (the F_1 T lymphocytes themselves being responders). Alloantisera against the MHC shared by the parental macrophages and the F_1 T cells blocked antigen induced lymphocyte stimulation. When the alloantisera were directed against determinants found only on the F_1 T lymphocytes, they only inhibited recognition of antigen, the response to which was controlled by an Ir gene linked to that histocompatibility type. These results suggested that antigen activation of T lymphocytes required more than one cell type, involved more than a simple antigen receptor on the surface of T lymphocytes, and involved gene products in the major histocompatibility complex. Elegant studies by Bluestein (7) showed that divalent $(Fab)_2$ fragments of alloantibodies would mediate this suppressive effect, but monovalent Fab fragments were unable to do so. These data suggest that the gene product of the major histocompatibility complex which was the target of this suppressive effect was not the antigen receptor itself, although it might be intimately associated with it on the cell surface.

A series of in vivo experiments by Katz and his colleagues (8) using a somewhat similar approach, tested the interaction between (responder x nonresponder)F_1 T cells with either nonresponder or responder parental B cells. They demonstrated that successful interaction between T and B cells occurred in response to antigens under Ir gene control only when both the T and B cells shared the responder histocompatibility type. These results suggested either that the Ir genes were expressed in B cells, or that some part of the major histocompatibility complex was responsible for efficient interaction between T cells and B cells in the generation of an in vivo immune response. Similar, although not identical, results were obtained by Benacerraf et al. (9) in an in vitro system. Thus, the evidence from a number of different in vitro experimental systems indicates that genes in the major histocompatibility complex play a major role in the interaction of immunocompetent cells in the generation of an in vitro immune response, both for T lymphocyte antigen induced stimulation and for the generation of the humoral antibody response.

Recent advances in our understanding of the genetic fine structure of the major histocompatibility complex (H-2) in the mouse also appear to have a bearing on this problem. Following detailed mapping of the Ir-1 locus within the H-2 complex (10), it rapidly became apparent that genes in this region of the

H-2 complex, now designated as the I region, controlled several immunologic functions. These include control of the capacity to respond to certain antigens (Ir genes) (10), stimulation in the mixed lymphocyte culture reaction in vitro (11) and the graft versus host reaction in vivo (Lad genes), and quite possibly the expression of cell surface structures responsible for effective interaction between T cells and B cells (8). In addition to these properties, recent experiments in a number of laboratories have shown that the I region determines a complex series of alloantigens which have now been termed the I region associated antigens (Ia antigens) (for a review see 12). Thus, at least four functions have been mapped in the I region, and there now appears to be a strong likelihood that this region plays a major role in the genetic regulation of immunocompetent cell interactions. A number of studies have shown that the Ia antigens are not present on all tissues, but are definitely present on B lymphocytes and on some subpopulation of T lymphocytes (12,13) as well as on sperm, epidermal cells and macrophages (14). The presence or absence of Ia antigens on subpopulations of B and T lymphocytes is currently a matter of great interest and bears directly on the problem of syngeneic cell interactions as well as allogeneic cell interactions in vitro (15).

Our current understanding of the distribution and density of Ia antigens on subpopulations of immunocompetent cells is incomplete and requires much further study before a definite pattern can be established. Initial attempts to correlate Ia antigens with specific immunocompetent cell functions (15-18) are also incomplete. It seems clear that anti Ia antisera, in contrast to antisera directed against products of the H-2K or H-2D regions are capable of inhibiting in vitro antibody responses to sheep red cells (17,18). The demonstration by Lonai (15) that elimination of Ia positive cells by anti-Ia and complement blocks MLC stimulation by purified T or B cell subpopulations, while similar treatment has no effect on the ability of T cells to respond in the MLC reaction or to mediate cytotoxicity in the cell mediated lympholysis reaction suggests that in allogeneic cellular interactions Ia antigens may be the antigens responsible for stimulation in the mixed mixed lymphocyte culture reaction but that both the responding cell in the MLC reaction and the cytotoxic cell either lack Ia antigens or have them in an extremely low density on the cell surface. These experiments must be pursued further, and verified by more sensitive techniques before it can be stated with confidence that MLC responder cells and cytotoxic T cells lack Ia antigens. Similarly, analytical experiments will have to be carried out to determine whether the suppressive effect of anti-Ia sera on in vitro antibody responses is directed at T cells, B cells or both.

While it seems clear from a number of laboratories that the majority (80-90%) of B cells are Ia positive, the situation with respect to T cells is much less clear. Early experiments (13a) indicated that only a small subpopulation of T lymphocytes is Ia positive. An additional puzzling finding is the demonstration by Segal (19) that anti-Ia sera plus complement markedly suppress spleen cell responsiveness to phytohemagglutinin and concanvalin-A. This latter finding is reproducible and occurs with anti-Ia antisera which kill only 5-10% of nylon wool purified T cells isolated from normal spleen cell populations. One possible interpretation of these experiments is that T cell-T cell interaction is required for responsiveness to lymphocyte mitogens. In addition, the finding that cytotoxic T cells appear to be Ia negative brings up the interesting speculation that Ia negative T cells may serve cytotoxic and perhaps suppressor T cell functions, while Ia positive T cells function as helper T cells in initiating humoral antibody responses. Clearly much more extensive experiments with isolated and purified cell populations, using much more sensitive techniques for Ia detection on these cell populations, will be required before these complex interrelationships can be sorted out.

If one may be permitted a sweeping generalization, it appears that the Ia antigens may be responsible for efficient cellular interaction in syngeneic combinations and for the mixed lymphocyte culture reaction and the graft versus host reaction in allogeneic combinations (12,20). This generalization is based on the fragmentary evidence cited above, and on a survey of the strain distribution of Ia specificities in comparison with the strain distribution of Ir genotypes and the strain distribution of MLC reactivity. There appears to be a close correlation between Ia differences among congenic strains and the occurrence of a mixed lymphocyte culture reaction between these strains. No obvious correlation has yet emerged between the distribution of Ia specificities and particular _Ir_ genes (12).

A very large number of questions about this type of genetic regulation of immunocompetent cell interactions remains to be answered. First, are _I_ region antigens involved in all types of immunocompetent cell interactions? Second, are the Ia antigens solely responsible for cell interaction, or do they have pleiomorphic effects including specific immune response gene action? Third, does the recent evidence by Dickler and Sachs that the Ia antigens may function as the Fc receptor on B lymphocytes (21) also imply that Ia antigens serve a similar function on T cells? Fourth, are Ia antigens expressed on all types of immunocompetent T cells or only on certain types of competent T cell populations (see above)? Fifth, what is the significance of the fact that the Ia antigens appear to be an extremely complex set of alloantigens coded for by at least three subregions of the I region of the H-2 complex? Does this indicate that these molecules are responsible for several different types of immunocompetent cell interactions and are specialized for interaction between macrophages and T cells, macrophages and B cells, T and B cells, or different subtypes of T and B cells. These are but a few of the questions concerning genetic regulation of immunocompetent cell interactions which have been studied by in vitro techniques which remain before us. In addition, we must keep in mind the fact that there are many other alloantigenic systems represented on the surface of lymphocytes which are not linked to the major histocompatibility complex, but which may have equally important effects on immunocompetent cell interactions. While the availability of congenic resistant strains bearing recombinant H-2 chromosomes has permitted relatively rapid progress in our understanding of the effect of genes in the H-2 complex on immunocompetent cell interaction, we must not lose sight of the fact that other genes (for example, the M locus (22) and the Ly antigens), may have equally important or more important effects. Hopefully, many of these questions will be resolved in the near future.

REFERENCES

1. McDevitt, H.O. and Tyan, M.L., J. Exp. Med. 128: 1, 1968.
2. Benacerraf, B. and McDevitt, H.O., Science 175: 273, 1972.
3. McDevitt, H.O. and Benacerraf, B., Adv. Immunology II: 31, 1969.
4. McDevitt, H.O. and Landy, M., Genetic Control of Immune Responsiveness, Academic Press, 1972.
5. Shevach, E.M., Paul, W.E. and Green, I., J. Exp. Med. 136: 1207, 1972.
6. Shevach, E.M. and Rosenthal, A.S., J. Exp. Med. 138: 1213, 1973.
7. Bluestein, H.G., J. Exp. Med. 140: 481, 1974.
8. Katz, D.H., Hamaoka, T., Dorf, M.E., Maurer, P.H. and Benacerraf, B., J. Exp. Med. 138: 734, 1973.
9. Benacerraf, B., Kapp, J.A., Pierce, C.W. and Katz, D.H., J. Exp. Med., in press, 1974.
10. McDevitt, H.O., Deak, B.D., Shreffler, D.C., Klein, J., Stimpfling, J.H. and Snell, G.D., J. Exp. Med. 135: 1259, 1972.

11. Bach, F.H., Widmer, M.B., Bach, M.L. and Klein, J., J. Exp. Med. 136: 1430, 1972.
12. Shreffler, D.C. and David, C.S., Adv. Immunol., in press, 1974.
13. Frelinger, J.A., Niederhuber, J.E., David, C.S. and Shreffler, D.C., J. Exp. Med. 140: 1273, 1974.
14. Hämmerling, G.J., Mauve, G., Goldberg, E. and McDevitt, H.O., Immunogenetics, in press, 1974.
15. Lonai, P., Proc. Ninth Leukocyte Culture Conf., in press, 1974.
16. Nabholz, M., Vives, J., Young, H.M., Meo, T., Miggiano, V., Rijnbeck, A. and Shreffler, D.C., Eur. J. Immunol., in press, 1974.
17. Pierce, C.W., Kapp, J.A., Solliday, S.M., Dorf, M.E. and Benacerraf, B., J. Exp. Med. 140: 921, 1974.
18. Frelinger, J.A., Niederhuber, J.E. and Shreffler, D.C., Science, in press, 1975.
19. Segal, S., Unpublished observations.
20. McDevitt, H.O., Bechtol, K.B., Hämmerling, G.J., Lonai, P. and Delovitch, T.L., in The Immune System, Genes, Receptors, Signals, eds. E.E. Sercarz, A.T. Williamson and C.F. Fox, 1974, p. 597.
21. Dickler, K.B. and Sachs, D.H., J. Exp. Med. 140: 779, 1974.
22. Abbasi, K., Demant, P., Festenstein, H., Holmes, J., Huber, B. and Rychlikova, M., Transp. Proc. 5: 1327, 1973.

GENETIC CONTROL OF MACROPHAGE-T LYMPHOCYTE INTERACTION

Ethan M. Shevach, Linda Lee, and Shlomo Z. Ben-Sasson
Laboratory of Immunology, NIAID, National Institutes of Health, Bethesda, Maryland 20014

HISTOCOMPATIBILITY REQUIREMENT FOR THE ACTIVATION OF IMMUNE T LYMPHOCYTE PROLIFERATION BY MACROPHAGE-ASSOCIATED ANTIGEN

Studies on genetic control of the immune response have raised the possibility that products of the immune response (Ir) genes play a role in antigen recognition by the T lymphocyte. Thus, alloantisera can specifically block the activation of T lymphocytes by antigens, the response to which is linked to the presence of histocompatibility specificities against which the alloantisera are directed (1). Because of the critical role of the macrophage in the process of T cell activation, particularly as measured by lymphocyte proliferation, we investigated the importance of histocompatibility determinants in the interaction of antigen pulsed macrophages with immune T lymphocytes (2,3).

The initial studies were performed with purified protein derivative of tuberculin (PPD), an antigen not known to be under unigenic control. When macrophages were obtained from strain 2 animals, pulsed with PPD, washed to remove unbound antigen, and then added to immune strain 2 peritoneal exudate lymphocytes (PELs) significant stimulation of T cell proliferation was observed (ΔCPM 179,000); however, when these same macrophages were added to immune strain 13 PELs little DNA synthesis was seen (ΔCPM 6,100). Similar results were observed when antigen pulsed strain 13 macrophages were mixed with immune strain 13 or strain 2 PELs. Combinations of parental and F_1 cells did lead to significant T cell proliferation although in some cases this activation was less efficient than the syngeneic combinations.

These studies clearly established that the efficient interaction of antigen-pulsed macrophages with immune T lymphocytes required identity between macrophage and T lymphocyte at some portion of the major histocompatibility complex. We next turned to the question whether macrophages obtained from a parental animal that lacks a given Ir gene can activate F_1 T lymphocyte proliferation to an antigen the response to which is controlled by that gene. The results of a typical experiment are shown in Table I. F_1, strain 2, or strain 13 macrophages were pulsed with the 2,4-dinitrophenyl derivative of a copolymer of L-glutamic acid, L-lysine (DNP-GL, an antigen

TABLE I

ACTIVATION OF F_1 LYMPHOCYTE PROLIFERATION--EFFECT OF ALLOANTISERA

ANTIGEN	F_1 MΦ	STRAIN 2 MΦ	STRAIN 13 MΦ
	Normal Guinea Pig Serum		
DNP-GL	14,539*	23,402	2,992
GT	11,447	4,816	28,186
PPD	28,117	21,919	28,048
PHA	34,710	55,021	67,538
	13 Anti-2 Serum		
DNP-GL	0	1,840	69
GT	13,878	0	22,335
PPD	19,333	2,167	31,648
PHA	50,208	61,249	75,052
	2 Anti-13 Serum		
DNP-GL	16,517	29,368	0
GT	1,071	1,368	0
PPD	18,538	26,113	119
PHA	60,098	78,963	77,449

*Results are expressed as Δ cpm/culture; each value is the mean of three determinations.

the response to which is controlled by a 2-linked Ir gene), a copolymer of L-glutamic acid, L-tyrosine (GT, an antigen the response to which is controlled by a 13-linked Ir gene), PPD, and the mitogen phytohemagglutinin (PHA). After a 60 minute incubation at 37°C, the macrophages were washed and added to immune F_1 PELs and cultured for three days in the presence of normal guinea pig serum (NGPS), anti-2, or anti-13 sera. Stimulation of F_1 T cell proliferation was observed when the F_1 cells were cultured with pulsed F_1 macrophages in the presence of NGPS. However, when the F_1 cells were mixed with strain 2 macrophages that had been pulsed with GT, much less stimulation (ΔCPM 4,816) was seen than when strain 13 (ΔCPM 28,186) or F_1 (ΔCPM 11,447) macrophages were used. Similarly, when strain 13 macrophages were cultured with F_1 cells in NGPS, the response to the DNP-GL pulsed macrophages was much less (ΔCPM 2,922) than that produced by F_1 (ΔCPM 14,539) or strain 2 macrophages (ΔCPM 23,402) pulsed with DNP-GL. Thus, parental macrophages pulsed with an antigen whose Ir gene they lack were capable of initiating F_1 T-cell proliferation. However, the magnitude of stimulation was approximately 1/10 that seen when macrophages from either the responder parent or the F_1 were used.

We hypothesized that one possible explanation for the inefficient activation of F_1 T cell proliferation by the non-responder macrophage was that during macrophage-T lymphocyte interaction the two cells came into close functional contact in areas of shared histocompatibility. Furthermore, it is likely that the antigen recognition sites controlled by the Ir genes on the surface of the T lymphocyte are physically related to the sites of macrophage-T lymphocyte interaction and both are linked to the serologically determined histocompatibility antigen. Thus, DNP-GL pulsed strain 13 macrophages activate F_1 cells poorly because the Ir gene controlled antigen recognition sites on the T lymphocyte are physically related to the strain 2 macrophage-T cell interaction sites, while the main contacts between the cells are at the strain 13 macrophage interaction sites. Thus, DNP-GL although brought to the general proximity of the T lymphocyte is, relatively speaking, unavailable to the Ir gene controlled DNP-GL antigen recognition sites.

These studies also allow us to understand more clearly the site of alloantiserum induced inhibition of T cell proliferation. When parental macrophages were pulsed with any antigen and added to F_1 T cells, an alloantiserum directed against the parental histocompatibility (H) antigen reacted with both the lymphocyte and the macrophage, inhibited macrophage-lymphocyte interaction, and abolished antigen induced lymphocyte transformation. However, when F_1 cells were stimulated by DNP-GL pulsed strain 13 macrophages, anti-2 serum completely inhibited the response (Table I). In this situation the anti-2 serum can only act on the lymphocyte and must in some manner interfere with the Ir gene controlled DNP-GL antigen recognition site.

HISTOCOMPATIBILITY REQUIREMENT FOR OTHER IN VITRO ASSAYS OF CELL MEDIATED IMMUNITY

A large number of studies have demonstrated that T lymphocytes mediate a number of immune phenomena including lymphokine production, specific cytotoxicity, mixed lymphocyte responsiveness, and helper function in antibody production, as well as the ability to proliferate in vitro in response to antigen (4). Previous studies from our laboratory have demonstrated that a second major T lymphocyte function, the production of migration inhibition factor (MIF) can also be specifically blocked by alloantisera (5). Thus, when (2x13) F_1 animals were immunized with DNP-GL and with the 2,4-dinitrophenyl derivative of guinea pig albumen (DNP-GPA, an antigen the response to which is controlled mainly by a 13-linked Ir gene) and MIF production was assayed by an indirect procedure in which the migrating cell population lacked the H antigen against which the alloantiserum was directed, anti-2 serum blocked MIF production by F_1 cells in response to DNP-GL, but not DNP-GPA, while anti-13 serum had the opposite effect.

In all of these experiments antigen was presented to the immune lymphocytes in the form of macrophage-associated antigen. Although rigorous studies have not been performed to evaluate the absolute requirement for the macrophage for antigen presentation in the induction of the production of MIF, these results indicate that macrophage-associated antigen is an efficient stimulus for MIF production. It was therefore of interest to evaluate the histocompatibility requirement for macrophage-T lymphocyte interaction in the activation of MIF production, an in vitro assay of cell mediated immunity that does not require T cell proliferation.

Strain 2 and strain 13 macrophages obtained from non-immune donors were pulsed with 100 μg/ml of DNP-GL, washed four times and then added to a population of strain 2 peritoneal exudate cells (PEC) obtained from a DNP-GL immune animal. When strain 2 macrophages were used, marked inhibition of migration was seen at final concentrations of antigen pulsed macrophages of 5% and 20%; slight, but significant inhibition of migration was still observed at a final concentration of 1% antigen pulsed macrophages (Fig. 1A). When strain 13 DNP-GL pulsed macrophages were used significant inhibition of migration was observed only at macrophage concentrations of 20%; even at this concentration strain 2 macrophages were much more efficient activators of MIF production (91% vs 32% inhibition).

One difficulty that arises in the interpretation of these studies is that some of the antigen presented on allogeneic macrophages may be transferred or carried over to the syngeneic macrophages in the strain 2 PEC population. In order to evaluate this possibility, we pulsed strain 2 and strain 13 macrophages with DNP-GL at final concentrations ranging from 1-1000 µg/ml. After a 60 minute incubation at 37°C, the antigen-pulsed macrophages were washed four times and mixed at a final concentration of 10% with strain 2 PEC obtained from a DNP-GL immune donor. Strain 2 macrophages pulsed with any amount of DNP-GL were capable of activating a significant amount of MIF production (Fig. 1B); on the other hand, significant MIF production was only seen when strain 13 macrophages were pulsed with 1000 µg/ml of DNP-GL.

These results demonstrated that the histocompatibility requirements observed for the activation of DNA synthesis by macrophage-associated antigen were also applicable to the activation of MIF production by macrophage-associated antigen. Allogeneic macrophages may be capable of activating the production of a small amount of MIF when they are used at high concentrations or when pulsed with very large amounts of antigen. However in these situations it is possible that all the stimulation observed is secondary to antigen that is carried over to the syngeneic macrophages in the immune PEC population.

We next turned to the evaluation of the histocompatibility requirement for effective macrophage-T lymphocyte interaction in the stimulation of early protein synthesis. This assay does not require T cell proliferation and is probably a good in vitro measure of cell mediated immunity. Thus, Levy and Rosenberg have demonstrated that the ability of tissues to show activity in this in vitro assay was correlated with the development of delayed hypersensitivity in guinea pigs (6); cells from immune animals responded in vitro at times when in vivo antibody synthesis was not yet detectable. Cells from animals immunized so as to elicit only antibody production and not cellular immunity did not respond to antigen in vitro. More recently, it has been demonstrated that in the mouse the response to PHA and alloantigens as measured in the protein synthesis assay was a T cell response as it could be abolished by treatment with anti-theta antibody or adult thymectomy (7).

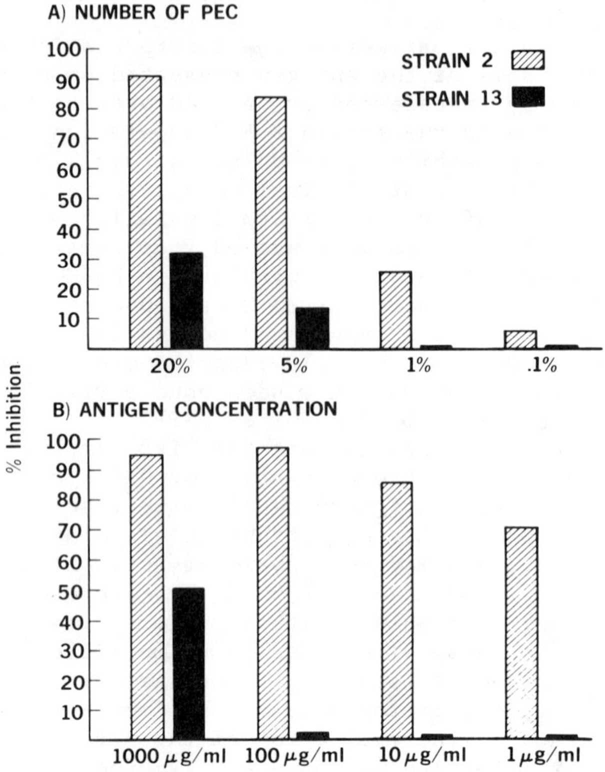

Fig. 1 Histocompatibility requirement for macrophage-T interaction for the production of MIF. A) Strain 2 and strain 13 macrophages were pulsed with 100 μg/ml of DNP-GL and mixed with immune strain 2 PEC at the concentrations indicated. B) Strain 2 and strain 13 macrophages were pulsed with 1-1000 μg/ml DNP-GL and mixed with immune PEC at a final concentration of 10%. The % inhibition is calculated as follows:

$$\left[1 - \frac{\text{area of migration with antigen}}{\text{area of migration with no antigen}}\right] \times 100$$

Although the absolute requirement for the macrophage in the presentation of antigen for the stimulation of protein synthesis by T cells has not been established, preliminary studies demonstrated that macrophage-associated antigen was very effective in the activation of immune T cells to synthesize protein. In the present studies, strain 2 or strain 13 macrophages obtained from non-immune donors were pulsed with 100 μg/ml of DNP-GL, washed and then added to immune strain 2 PELs. Significant stimulation of protein synthesis was observed at macrophage concentrations of 10% and 20% when strain 2 macrophages were used (Fig. 2A). Strain 13 antigen pulsed macrophages at final concentrations of 10% and 20% were also able to activate significant stimulation of protein synthesis, but at each concentration of macrophages used the percent stimulation was only about half that seen when strain 2 macrophages were used.

Experiments to rule out the possibility of antigen carry over were performed by pulsing strain 2 and strain 13 macrophages with 1-1000 μg/ml of DNP-GL. After these cells were washed, they were added to immune strain 2 PELs at a final concentration of 10%. Significant stimulation of protein synthesis was observed when strain 2 macrophages were pulsed with 10, 100, or 1000 μg/ml of DNP-GL. However, significant stimulation was observed only when strain 13 macrophages were pulsed with 1000 μg/ml or DNP-GL (Fig. 2B).

These results demonstrated that allogeneic macrophages were markedly less effective than syngeneic macrophages in the induction of the early stimulation of protein synthesis. The possibility that allogeneic macrophages were capable of activating some stimulation when used in high concentrations or when pulsed with large amounts of antigen could not be ruled out.

ROLE OF THE GPL-A LOCUS IN MACROPHAGE-T LYMPHOCYTE INTERACTION

We have previously evaluated the ability of macrophages from outbred animals which bore the serologically defined 2 or 13 specificities to activate T cell proliferation in cells derived from the two inbred strains (2). A general correlation existed between the sharing of strain 2 or strain 13 H determinants and the ability of macrophages from outbred guinea pigs to stimulate the immune inbred T cells. Thus, 2^+ macrophages, whether 13^- or 13^+, were superior to 2^- macrophages in the activation of strain 2 lymphocytes. Similarly 13^+ macrophages, whether 2^+ or 2^-, were superior to 13^- macrophages in the stimulation of 13 lymphocytes. These findings suggested that macrophage-T lymphocyte interaction was regulated by the serologically defined 2/13 H antigens themselves

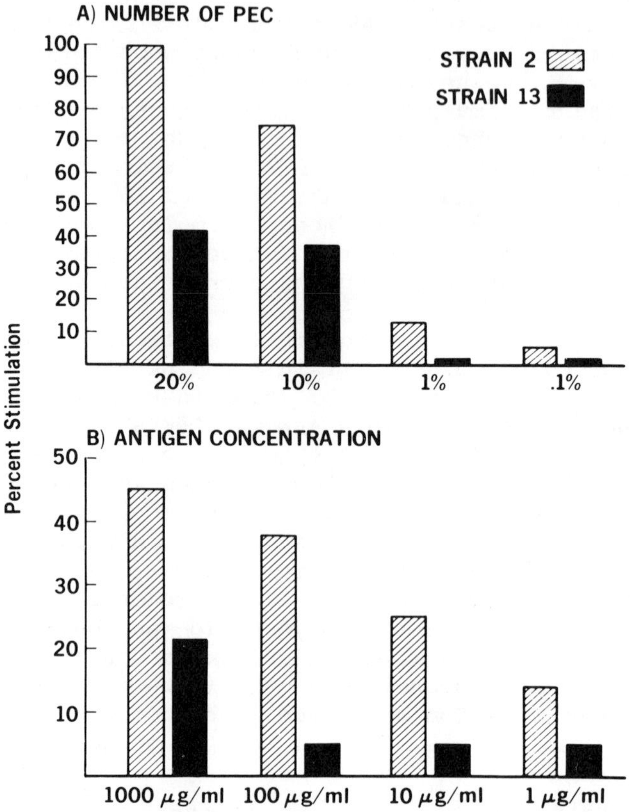

Fig. 2 Histocompatibility requirement for macrophage-T interaction for the stimulation of protein synthesis. A) Strain 2 and strain 13 macrophages were pulsed with 100 μg/ml of DNP-GL and added to immune strain 2 PELs at the final concentrations indicated. B) Strain 2 and strain 13 macrophages were pulsed with 1-1000 μg/ml of DNP-GL and added to immune strain 2 PELs at a final concentration of 10%. The percent stimulation was calculated as follows:

$$\left[\frac{\text{CPM with antigen}}{\text{CPM with no antigen}} - 1\right] \times 100$$

or by the products of gene(s) closely linked to the H antigen gene.

Over the past several years, de Weck and coworkers have independently described a system of guinea pig H antigens which differ from the 2/13 antigens (8,9). They have identified at least one locus with 4-5 well defined alleles; this locus has been termed the GPL-A locus, and the alleles B, C, D, E etc. Strain 2 and strain 13 animals both bear the B antigen. Breeding studies have demonstrated that the GPL-A locus is linked to the 2/13 locus and both are probably products of the guinea pig major histocompatibility complex (MHC) (10). More recent studies have clarified the relationship between the antigens controlled by the GPL-A locus and the 2/13 locus. The antigens controlled by the GPL-A locus have a wide tissue distribution and are present both on lymphoid and non-lymphoid cells. In contrast, the 2/13 antigens are restricted in distribution to lymphoid cells and probably macrophages and have an increased density on the B lymphocyte compared to the T lymphocyte (11). More detailed studies of these antigens using lactoperoxidase catalyzed surface iodination have revealed that the B antigen has a molecular weight of $\sim$ 45,000 daltons with an associated $\sim$ 11,000 molecular weight component. The 2/13 antigens have molecular weights in the range of $\sim$ 25,000-30,000 daltons with no associated low molecular weight component (12). In both tissue distribution and molecular size the 2/13 antigens resemble the immune response associated (Ia) antigens of the murine MHC, while the B antigen resembles the D or K antigens. Indeed, the low molecular weight peak observed in association with the B antigen may be the guinea pig analog of the low molecular polypeptide chain found in the HL-A antigens of humans which has been shown to be B_2-microglobulin (13).

We have compared the ability of macrophages from outbred animals which have been typed for both the 2/13 specificities and the GPL-A antigens in their ability to activate proliferation in T cells derived from inbred strain 2 and strain 13 animals. Strain 13 macrophages pulsed with GT, DNP-GPA, PPD or PHA activated significant DNA synthesis when mixed with immune strain 13 PELs (Fig. 3A). Very little stimulation of DNA synthesis was observed when GT, DNP-GPA or PPD pulsed strain 2 macrophages were mixed with strain 13 lymphocytes; strain 2 macrophages pulsed with PHA activated strain 13 cells to the same extent as syngeneic macrophages. Both groups of macrophages which were obtained from $2^{+}13^{+}D^{+}B^{-}$ animals activated T cell proliferation in the strain 13 cells comparable to that produced by the inbred strain 13 macrophages.

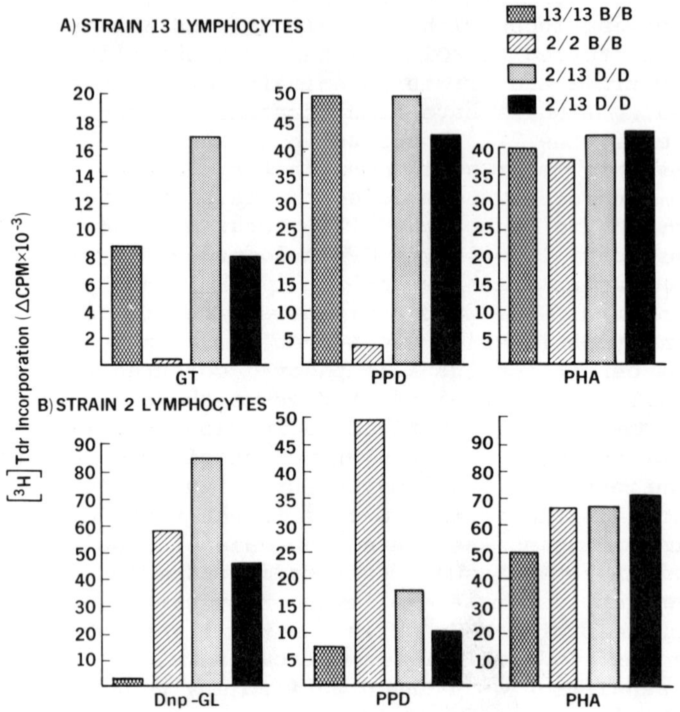

Fig. 3 Role of the GPL-A antigens in macrophage-T cell interaction. A) 13^+B^+, 2^+B^+, $2^+/13^+D^+$, and $2^+/13^+D^+$macrophages were pulsed with GT 200 μg/ml, PPD 100 μg/ml, and PHA 10 μg and then added to immune strain 13 lymphocytes. B) The same macrophages were pulsed with DNP-GL 100 μg/ml, PPD, or PHA and then mixed with strain 2 lymphocytes. The amount of radioactivity incorporated into cellular DNA was determined as previously described (2).

We next evaluated the ability of these same populations of macrophages to activate strain 2 lymphocyte proliferation (Fig. 3B). When strain 2 macrophages were pulsed with DNP-GL, PPD or PHA and mixed with strain 2 lymphocytes marked stimulation of DNA synthesis was observed; strain 13 DNP-GL or PPD pulsed macrophages activated very little DNA synthesis when mixed with strain 2 cells. On the other hand, both groups of outbred macrophages which were obtained from 2^+D^+ animals (which were also responders to DNP-GL) were capable of initiating significant DNA synthesis by strain 2 T cells when they were pulsed with DNP-GL, but were only slightly better than strain 13 macrophages in activating strain 2 lymphocytes when pulsed with PPD.

The results of these and other studies performed using different combinations of outbred macrophages with either inbred or outbred T lymphocytes have demonstrated that it is the antigens controlled by the 2/13 region and not the antigens controlled by the GPL-A locus that are the critical determinants of effective macrophage-T lymphocyte interaction as measured by the activation of T cell proliferation. It should be noted that outbred 13^+ macrophages when pulsed with GT, DNP-GPA (the response to both is controlled by Ir genes linked to the 13 alloantigen gene), or with PPD (response not under unigenic control) are capable of activating both inbred and outbred 13^+ lymphocytes to the same extent as inbred strain 13 antigen pulsed macrophages. Outbred 2^+ macrophages when pulsed with DNP-GL (the donor is always a DNP-GL responder) are capable of activating immune strain 2 T cell proliferation; however, outbred 2^+ macrophages when pulsed with PPD, irrespective of their GPL-A type, are on the whole very poor activators of strain 2 T cell proliferation. The explanation for this result is not known but it is likely that most 2^+ outbred animals differ from inbred strain 2 animals by some of the 2-linked interaction sites necessary for effective macrophage T cell collaboration when PPD is used as the antigen.

Recent studies performed in collaboration with Dr. H. Rühl have demonstrated that anti-B serum is capable of inhibiting the in vitro proliferative response of B^+ lymphocytes. In contrast to the specific inhibition of linked Ir gene product function produced by anti-2 or anti-13 serum, the inhibition produced by anti-B serum is general in nature (Table II). Thus, anti-B serum produced a 60-90% inhibition of the response to the antigens DNP-GL and PPD and to the mitogen PHA. The inhibition was not due to non-specific toxic effects of the serum as this same lot of serum had no effect on the pro-

TABLE II
INHIBITION OF LYMPHOCYTE PROLIFERATION BY ANTI-B SERUM

ANTIGEN	2/2 B/B Lymphocytes		13/13 D/D Lymphocytes	
	SERUM			
	NGPS	ANTI-B	NGPS	ANTI-B
0	1,197*	2,353	1,215	819
DNP-GL	41,588	9,848	---	---
PPD	86,617	9,942	73,866	103,781
PHA	57,991	8,585	53,987	60,765

*Results are expressed as cpm/culture; each value is the mean of three determinations.

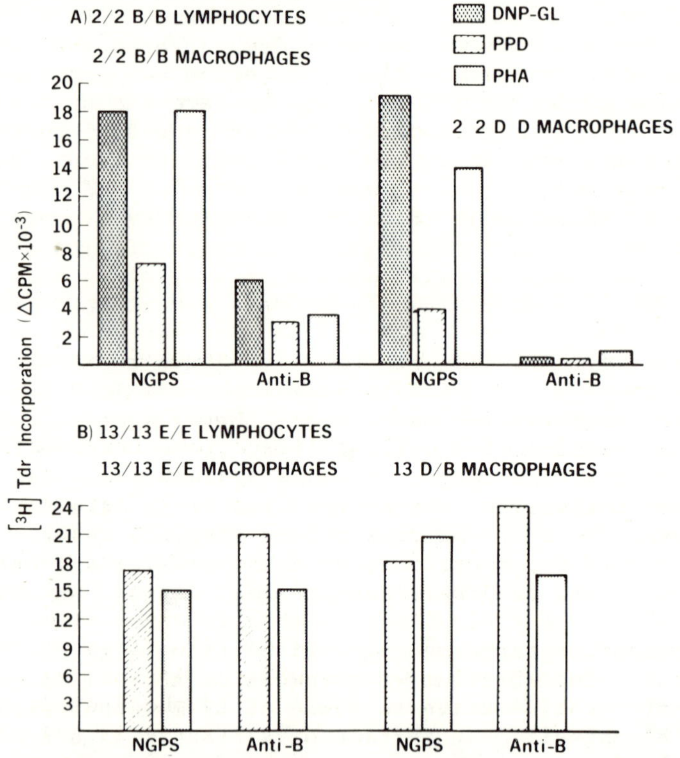

Fig. 4 The cellular site for the inhibition of lymphocyte transformation with anti-B serum. A) 2^+B^+ macrophages and 2^+D^+ macrophages were pulsed with DNP-GL 100 μg/ml, PPD 100 μg/ml, or PHA 10 μg/ml, mixed with inbred strain 2 lymphocytes, and cultured for three days in NGPS or anti-B serum. B) 13^+E^+ and $13^+D^+B^+$ macrophages were pulsed with PPD or PHA, mixed with 13^+E^+ lymphocytes and cultured for three days in NGPS or anti-B serum.

liferative response of D^+B^- T cells.

Since macrophage-T lymphocyte interaction is independent of specificities coded for by the GPL-A locus, we were able to avaluate the target (macrophage or T cell) for the inhibitory effect of anti-B serum. Strain 2 (2^+B^+) PELs were cultured with antigen pulsed syngeneic or allogeneic (2^+D^+) macrophages in the presence of NGPS or anti-B serum. Significant stimulation of DNA synthesis was observed when the lymphocytes were cultured with either set of macrophages in NGPS (Fig. 4A). Marked inhibition of T cell proliferation was observed when the cells were cultured in anti-B serum independent of the H type of the macrophage donor. Indeed, the inhibition appeared to be somewhat greater when the 2^+D^+ macrophages were used.

In order to rule out the possibility that anti-B serum might inhibit T cell proliferation by also acting on the macrophage, T lymphocytes were obtained from a 13^+E^+ outbred animal (Fig. 4B). Syngeneic (13^+E^+) and allogeneic (13^+B^+) macrophages were pulsed with PPD or PHA and mixed with the immune T lymphocytes. No significant difference was observed when the cells were cultured in either NGPS or anti-B serum. Thus, the inhibitory effect of anti-B serum on T cell proliferation was mediated by blocking determinants present only on the T lymphocyte.

ROLE OF THE Ir GENE PRODUCT IN MACROPHAGE-T LYMPHOCYTE INTERACTION

We have demonstrated that the 2/13 antigens or cell surface antigens coded for by closely linked genes appear to be responsible for effective macrophage-T lymphocyte interaction for the activation of T cell proliferation. Previous studies have shown that the target antigens against which the T cell inhibitory activity of the anti-2 or anti-13 sera were directed were the 2/13 antigens rather than the specific Ir gene products (14). Thus, in inbred and backcross animals the gene controlling responsiveness to the polymer, L-glutamic acid, L-alanine (GA) is linked to the strain 2 alloantigen gene. However, outbred animals can be identified which are GA^+, but 2^-. Anti-2 serum was capable of inhibiting the in vitro proliferative response to GA of lymphocytes derived from GA^+2^+ animals, but the serum had little if any effect on the GA response of T cells from GA^+2^- animals. It would therefore be of interest to evaluate the capacity of GA pulsed macrophages derived from a GA^+2^- animal to activate immune strain 2 T cell proliferation. In similar fashion, the capacity of GA pulsed strain 2 macrophages to activate GA^+2^- T cells can be studied.

When strain 2 macrophages were pulsed with DNP-GL, GA, PPD, or PHA, washed and mixed with immune strain 2 lymphocytes, significant stimulation to all the antigens tested and the mitogen was observed (Table III). Antigen pulsed $GA^{+}2^{-}$ macrophages were incapable of activating strain 2 lymphocytes to any of the antigens tested. On the other hand, GA pulsed macrophages obtained from a $GA^{+}2^{+}$ animal were capable of activating strain 2 T cell proliferation. This experiment demonstrated that the Ir gene product alone was not the cell surface structure mediating macrophage-T cell interaction. Similar results were observed when $GA^{+}2^{-}$ lymphocytes were tested (Table IV). Significant stimulation in response to GA was only seen when $GA^{+}2^{-}$ macrophages were used; neither GA pulsed strain 2 nor strain 13 macrophages could activate significant DNA synthesis in $GA^{+}2^{-}$ T cells. It should be noted that $GA^{+}2^{-}$ animals are uniformly 13^{+} and that GT and DNP-GPA pulsed strain 13 macrophages were capable of activating $GA^{+}2^{-}13^{+}$ T cells.

It is also of interest to evaluate the role of the alloantigen in macrophage-T lymphocyte interaction. A number of outbred animals have been identified in which relationships between Ir genes and alloantigens exist which differ from those found in the inbred animals (Table V). Strain 13 animals respond to GT and to low dose immunization with DNP-GPA; strain 2 animals respond to neither of these antigens. Backcross analysis demonstrated that both the GT and the DNP-GPA genes were linked to the 13 alloantigen gene. A number of different patterns of response were identified in outbred animals. Group I resembles the inbred 13 and responds to both GT and DNP-GPA as does group III which resembles the $(2 \times 13)F_1$. Group IV outbreds are $2^{-}13^{+}$ but respond only to DNP-GPA and not GT; conversely, group V outbreds respond to GT and not DNP-GPA.

We have used macrophages from the $GT^{-}DNP\text{-}GPA^{+}13^{+}$ animals to evaluate the role of the alloantigen in macrophage-T cell interaction. When strain 13 lymphocytes were stimulated by GT or DNP-GPA pulsed syngeneic macrophages significant stimulation of DNA synthesis was observed (Table VI). Strain 2 macrophages were much less effective in the activation of strain 13 lymphocytes to any of the antigens tested. Macrophages from the $GT^{-}DNP\text{-}GPA^{+}13^{+}$ animal were capable of stimulating strain 13 cells when pulsed with DNP-GPA but were markedly less effective in activating strain 13 cells when pulsed with GT; indeed, in this experiment they were no more effective than strain 2 macrophages. Antigen pulsed macrophages from the $GT^{+}DNP\text{-}GPA^{+}13^{+}$ outbred animals were capable of activating strain 13 lymphocytes to all antigens tested.

In order to evaluate whether the failure of the GT^{-} macrophage to activate strain 13 T cell proliferation to GT was

TABLE III
ACTIVATION OF STRAIN 2 LYMPHOCYTES
BY MACROPHAGE-ASSOCIATED ANTIGEN

ANTIGEN	MACROPHAGE TYPE			
	STRAIN 2	$GL^-GA^+2^-13^+$	$GL^+GA^+2^+13^-$	$GL^-GA^-2^-13^+$
GL	19,833*	0	15,219	1,894
GA	2,217	0	2,048	448
PPD	71,709	8,266	23,721	8,987
PHA	53,401	45,291	32,835	40,214

*Results are expressed as Δ cpm/culture; each value is the mean of three determinations.

TABLE IV
ACTIVATION OF LYMPHOCYTES FROM $GL^-GA^+2^-13^+$ ANIMALS
BY MACROPHAGE-ASSOCIATED ANTIGEN

ANTIGEN	MACROPHAGE TYPE		
	$GL^-GA^+2^-13^+$	STRAIN 2	STRAIN 13
GA	13,818*	998	1,146
DNP-GPA	7,764	4,730	10,650
GT	3,996	0	6,195
PPD	46,496	19,693	18,338
PHA	29,486	27,975	29,201

*Results are expressed as Δ cpm/culture; each value is the mean of three determinations.

TABLE V
RELATIONSHIP BETWEEN HISTOCOMPATIBILITY TYPE AND RESPONSIVENESS TO GT AND DNP-GPA (low dose)

STRAIN	PHENOTYPE	GT RESPONSE	DNP-GPA RESPONSE
Inbred 2	$2^+/2^+$	-	-
Inbred 13	$13^+/13^+$	+	+
$(2x13)F_1$	$2^+/13^+$	+	+
$(2x13)F_1x2$	$2^+/2^+$	-	-
	$2^+/13^+$	+	+
Outbred I	$2^+/13^-$	-	-
Outbred II	$2^-/13^+$	+	+
Outbred III	$2^+/13^+$	+	+
Outbred IV	$2^-/13^+$	-	+
Outbred V	$2^-/13^+$	+	-

TABLE VI
ACTIVATION OF STRAIN 13 LYMPHOCYTES BY MACROPHAGE-ASSOCIATED ANTIGEN

ANTIGEN	MACROPHAGE TYPE			
	STRAIN 13	STRAIN 2	$GT^-DNP-GPA^+13^+$	$GT^+DNP-GPA^+13^+$
DNP-GPA	22,588*	5,080	19,411	18,558
GT	37,384	6,721	4,235	28,229
PPD	88,330	8,738	75,596	89,163
PHA	79,642	66,415	70,926	58,411

*Results are expressed as Δ cpm/culture; each value is the mean of three determinations.

specific for the 13 alloantigen haplotype, we next studied the ability of macrophages obtained from a $2^{+}GT^{-}DNP\text{-}GPA^{+}13^{+}$ outbred animal to activate lymphocytes from a $2^{+}GT^{+}DNP\text{-}GPA^{+}13^{+}$ outbred animal. Strain 13 macrophages were capable of activating DNA synthesis in these outbred cells when they were pulsed with either DNP-GPA or GT. Macrophages from strain 2 animals could only activate $2^{+}13^{+}$ lymphocytes when they were pulsed with GA, the response to which is controlled by a 2-linked Ir gene, but were markedly less effective in activating $2^{+}13^{+}$ lymphocytes when pulsed with DNP-GPA or GT. Both sets of macrophages obtained from the $2^{+}/GT^{-}DNP\text{-}GPA^{+}13^{+}$ animals behaved similarly. They were capable of activating the $2^{+}13^{+}$ lymphocytes for the 2-linked GA response and were very effective activators when pulsed with DNP-GPA. However, no stimilation was seen when the GT^{-} macrophages were used to activate for GT.

In a small number of experiments performed thus far macrophages from the $GT^{+}DNP\text{-}GPA^{-}13^{+}$ animals were ineffective in activating 13^{+} cells for DNP-GPA, but were very effective in activating T cell proliferation when pulsed with GT. Although the results of these studies would imply that both the alloantigen and the Ir gene product are necessary for effective macrophage-T lymphocyte interaction, the interpretation of these experiments is somewhat more difficult in that absorption studies have demonstrated that the lymphocytes from either the $GT^{-}DNP\text{-}GPA^{+}13^{+}$ animals, or the $GT^{+}DNP\text{-}GPA^{-}13^{+}$ animals bear less of the "13" alloantigen. Thus, lymph node and spleen cells from $GT^{+}DNP\text{-}GPA^{+}13^{+}$ animals readily removed all the cytotoxic activity for inbred strain 13 cells from the 2 anti-13 serum, while following absorption with $GT^{-}DNP\text{-}GPA^{+}13^{+}$ cells or $GT^{+}DNP\text{-}GPA^{-}13^{+}$ cells significant cytotoxic activity for inbred 13 cells was retained. It would therefore appear that the macrophages obtained from these non-responder animals were also deficient in some part of the 13 alloantigen complex.

COMMENTS

In these studies we have demonstrated that the activation of T cells to proliferate, produce MIF, and synthesize protein requires identity between antigen-pulsed macrophage and T lymphocyte at some portion of the MHC. The one major aspect of T cell function that has not been examined is the histocompatibility requirement for the induction of the T helper cell. Studies in the mouse have demonstrated that allogeneic antigen-pulsed macrophages were capable of activating both a primary and secondary antibody response in vitro. In-

TABLE VII

ACTIVATION OF LYMPHOCYTES FROM $2^+/GT^+DNP\text{-}GPA^+13^+$ OUTBRED ANIMALS BY MACROPHAGE-ASSOCIATED ANTIGEN

ANTIGEN	MACROPHAGE TYPE				
	$2^+/GT^+DNP\text{-}GPA^+13^+$	STRAIN 13	STRAIN 2	$2^+/GT^-DNP\text{-}GPA^+13^+$	$2^+/GT^-DNP\text{-}GPA^+13^+$
GA	8,594*	700	5,559	8,323	6,993
DNP-GPA	15,936	12,389	8,899	15,114	10,297
GT	9,171	6,093	0	0	757
PPD	28,047	29,579	23,777	48,698	34,881
PHA	59,382	40,173	47,872	36,809	42,359

*Results are expressed as Δ cpm/culture; each value is the mean of three determinations.

deed, macrophages from the genetic non-responder mouse when pulsed with the terpolymer L-glutamic acid, L-alanine, L-tyrosine (GAT) were equivalent in their ability to activate non-adherent responder spleen cells in the primary in vitro antibody response (15). In the present report we have ruled out the possibility that the difference between the guinea pig and mouse studies is the requirement of the former for T cell proliferation; MIF production and early protein synthesis do not require T cell proliferation and both have the same histocompatibility requirement as the proliferating T cell. Although T helper function may be unique and not require antigen presented on histocompatible macrophages, other studies using somewhat different in vitro techniques have demonstrated a histocompatibility requirement for optimal macrophage-T cell interactions in the induction of T helper activity (16).

In certain respects the histocompatibility requirements for effective macrophage-T lymphocyte interaction in the guinea pig resemble the histocompatibility requirements for effective T-B collaboration in the mouse. Thus, in the mouse effective collaboration occurred only when the interacting T + B cells shared gene identity in the MHC. Mixtures of F_1 and parental T and B lymphocytes cooperated effectively in response to antigens not under the control of individual Ir genes in one of the two parental strains. However, when the primed carrier T cells were derived from F_1 (responder x nonresponder) animals, it was found that F_1 T cells could provide helper function for B cells of responder, but not of the non-responder parent (17). The relative ineffectiveness of the antigen-pulsed non-responder macrophage in the activation of F_1 (responder x non-responder) T cell proliferation appears to be an entirely analogous situation. Indeed, the fundamental explanation for both of these experimental observations is probably the same.

We have demonstrated that the cell surface structures responsible for macrophage-T lymphocyte interaction are the products of the region controlling the 2/13 antigens or the products of closely linked genes. The GPL-A antigens which appear to be analogous to the murine D or K antigens both in tissue distribution and molecular size play little if any role in macrophage T cell collaboration. Here again, these observations are analogous to the studies in the mouse where mapping experiments have demonstrated that it is the products of the I region that are responsible for T-B collaboration although the role of K region gene products has not been fully excluded (18). The genes in the I region governing the cell surface antigens mediating cell-cell interactions have been termed by Katz and Benacerraf cell interaction or <u>CI</u> genes

(19). We would also suggest that it is likely that the products of the CI genes in the guinea pig are the serologically demonstrable 2/13 alloantigens.

Which structures on the macrophage are responsible for macrophage-T cell interaction - the products of the Ir genes or the products of the CI genes? A number of different hypothetical possibilities for the relationship between antigen and macrophage surface structures are shown in figure 5. However, before commenting on proposed mechanisms of macrophage antigen presentation, it is necessary to make a number of comments about the T lymphocyte antigen recognition structures. A number of recent experimental observations suggest that the Ir gene product is not the primary antigen binding receptor of the T lymphocyte. Thus, unprimed responder and non-responder mice bear equivalent numbers of antigen-binding T cells to Ir gene controlled antigens (20). The administration of GAT to non-responder mice in most instances fails to lead to the development of helper T cell function, but rather a population of specific suppressor T cells is induced (21). These two observations suggest that the non-responder can recognize antigen. One must therefore postulate that the Ir gene products are probably non-clonally distributed on T cells and are capable of interacting with antigen with a low degree of energy and specificity and acting in an auxiliary fashion with the prime antigen-binding receptor of the T lymphocyte. The antigen recognition complex on the T cell would then consist of a receptor, an Ir gene product and a CI gene product. A similar model has been suggested by Katz and Benacerraf (19). The genetic origin and chemical nature of the prime T cell antigen binding receptor are at present unknown.

Our studies have demonstrated that identity of Ir gene products alone was not sufficient for effective macrophage-T collaboration (Fig. 5B). These results were derived from experiments in the GA system where $GA^{+}2^{-}$ macrophages failed to activate $GA^{+}2^{+}$ lymphocytes and vice versa. One caveat that must be mentioned in reaching this conclusion is that the identity of the GA genes in $GA^{+}2^{+}$ and $GA^{+}2^{-}$ animals has not been proven. Studies performed in the GT system suggest that identity of alloantigen or CI alone is also not sufficient for effective macrophage-T cell collaboration (Fig. 5A). Macrophages from a GT^{-} animal though 13^{+} were incapable of activating strain 13 T cell proliferation in response to GT. It would therefore appear likely that both the CI gene products and the Ir gene products are expressed on the macrophage surface and both probably acting in concert are required for macrophage-T interaction. What is the relationship of the antigen to CI-Ir complex? In figure 5C antigen is randomly dis-

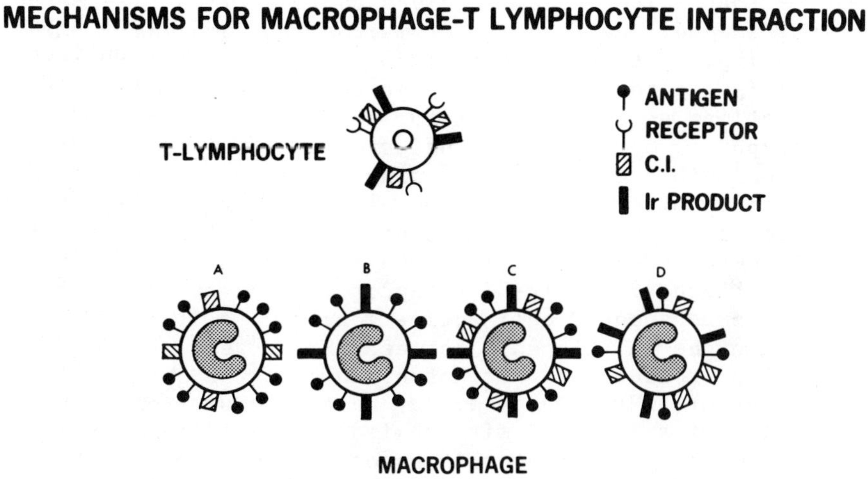

Fig. 5 A) The macrophage bears only the cell interaction gene product; antigen is randomly distributed. B) The macrophage bears only the Ir gene product; antigen is randomly distributed. C) The macrophage bears both the Ir and the C.I. gene products and antigen is randomly distributed. D) Same as C, except immunologically relevant antigen is intimately associated with the Ir-CI complex.

tributed on the surface of the macrophage, but the only antigen that is immunologically relevant is that which is in close proximity to the Ir-CI complex. Alternatively (Fig. 5D), the only site of immunologically relevant antigen on the macrophage is that which is intimately associated with the Ir-CI complex. Indeed, one function of macrophage "processing" of antigen might be to present antigen to the Ir-CI complex. A more intriguing possibility is that the Ir gene product itself functions as the receptor for antigen on the surface of the macrophage. Our earlier studies on the failure of alloantisera to inhibit macrophage binding of antigen would not rule out this possibility, as the Ir gene product would most likely be serologically silent (2). The proposed Ir-CI antigen complex on the surface of the macrophage would then function to present antigen to the Ir-CI receptor complex of the T lymphocyte with subsequent T cell activation.

Although the relationship between the Ir gene products and the CI gene products are not known, some of the observation made in the GT, DNP-GPA system may best be explained by the association of a given CI with an individual Ir gene product. Thus, the GT Ir gene product is associated with a specific 13 CI, while the DNP-GPA Ir gene product is associated with a different 13 CI. In either case, the non-responder animal is deficient in both the Ir gene product and its associated CI. This hypothesis would explain why the $GT^{-}13^{+}$ and $DNP\text{-}GPA^{-}13^{+}$ cells are incapable of removing all the cytotoxic activity from the anti-13 serum.

CONCLUSION

Some of the models we have proposed for macrophage-T interaction should be readily testable by the appropriate experiments. It is likely that the interaction between antigen-pulsed macrophage and immune T lymphocytes is mediated by the same mechanisms that govern the interaction of all immunocompetent cells.

REFERENCES

1. Shevach, E.M., Paul, W.E. and Green, I., J. Exp. Med. 136: 1207, 1972.
2. Rosenthal, A.S. and Shevach, E.M., J. Exp. Med. 138: 1194, 1973.
3. Shevach, E.M. and Rosenthal, A.S., J. Exp. Med. 138: 1213, 1974.
4. Katz, D.H. and Benacerraf, B., Adv. Immunol. 15: 1, 1972.
5. Ben-Sasson, S.Z., Shevach, E.M., Green, I. and Paul, W.E., J. Exp. Med. 140: 383, 1974.

6. Rosenberg, S.A. and Levy, R., J. Immunol. 108: 1080, 1972.
7. Gorczynski, R.M. and Rittenberg, M.B., J. Immunol. 112: 47, 1974.
8. Sato, W. and deWeck, A.L., Z. Immunitactsforsch. 149: 49, 1972.
9. Geczy, A.F. and deWeck, A.L., Eur. J. Immunol. 4: 483, 1974.
10. Shevach, E.M., Geczy, A.F. and deWeck, A.L., manuscript in preparation.
11. Shevach, E.M., Finkelman, F., Ben-Sasson, S.Z. and Green, I., manuscript in preparation.
12. Finkelman, F., Shevach, E.M., Vitetta, E., Green, I. and Paul, W.E., J. Exp. Med., in press.
13. Grey, H.M., Kubo, R.T., Colon, S.M., Poulik, M.O., Cresswell, P., Springer, T., Turner, M. and Strominger, J.L., J. Exp. Med. 138: 1608, 1973.
14. Shevach, E.M., Green, I. and Paul, W.E., J. Exp. Med. 139: 679, 1974.
15. Kapp, J.A., Pierce, C.W. and Benacerraf, B., J. Exp. Med. 138: 1121, 1973.
16. Erb, P. and Feldmann, M., unpublished observations.
17. Katz, D.H., Hamaoka, T., Dorf, M.E. and Benacerraf, B., J. Exp. Med. 138: 74, 1973.
18. Katz, D.H., Dorf, M.E. and Benacerraf, B., J. Exp. Med. 140: 290 1974.
19. Katz, D.H. and Benacerraf, B., Trans. Reviews 22: in press, 1974.
20. Hämmerling, G. and McDevitt, H.O., J. Immunol. 112: 1734, 1974.
21. Kapp, J.A., Pierce, C.W. and Benacerraf, B., J. Exp. Med. 140: 172, 1974.

ON THE MECHANISM OF ALLOANTISERUM MEDIATED SUPPRESSION OF HISTOCOMPATIBILITY-LINKED Ir GENE CONTROLLED IMMUNE RESPONSES OF GUINEA PIGS*

Harry G. Bluestein
University of California, San Diego, School of Medicine, La Jolla, California 92037

The importance of genetic factors in determining immune responses has been appreciated since the infancy of immunology as a scientific discipline. It is only in the past decade, however, that the use of antigens having limited structural heterogeneity has permitted the identification of unique genetic loci controlling the ability of an animal to respond immunologically to a specific antigen (1). Since the identification of the PLL gene in guinea pigs, an autosomal dominant gene controlling immune responsiveness to the synthetic polypeptide poly-L-lysine (2), a large number of specific immune response (Ir) genes have been identified in several mammalian species (3). It has become clear during the course of these studies that Ir genes are not randomly distributed within the genome. Rather, they form part of highly organized linkage groups which include loci controlling the expression of other lymphocyte membrane antigens in addition to multiple Ir gene loci. In mice, for example, the ability to make antibodies to each of three branched multichained synthetic polypeptide antigens (T,G)-A--L, (H,G)-A--L, and (Phe, G)-A--L is controlled by alleles at the Ir-1 locus. Ir-1 is linked to the major histocompatibility locus of mice, H-2 (4). The development of congenic and recombinant inbred strains of mice have permitted a sophisticated degree of genetic mapping of the chromosomal regions containing the H-2 loci. At least three distinct Ir gene loci have been identified: Ir-1, Ir-IgG and Ir-IgA, each of which can be located between, and clearly separated from, the D and K ends of the H-2 locus (5,6).

In guinea pigs, because of the relatively inefficient reproductive rate, sophisticated genetic studies have not yet been possible. The knowledge of the major histocompatibility (H) region, therefore, is very rudimentary. However, by taking advantage of the fact that random bred guinea pig strains express Ir gene function, linkage of Ir gene loci

---

*Supported by U.S.P.H.S. Research Grant Al 10931 from the National Institute of Allergy and Infectious Diseases.

has been demonstrated. The PLL gene and the GA gene, controlling responses to a random copolymer of L-glutamic acid and L-alanine, are inherited as linked but separable loci which behave as alleles or pseudoalleles to the GT gene, which controls immune responsiveness to a random linear copolymer of L-glutamic acid and L-tyrosine (7). The PLL, GA and GT guinea pig Ir genes are also linked to the major histocompatibility loci (8). The coinheritance of the phenotypic expression of those genes; i.e. the observation that an animal's ability to respond immunologically to an antigen was associated with the susceptibility of its lymphocytes to lysis by specific antihistocompatibility sera, provided the evidence of that linkage (9,10). Histocompatibility-linked Ir genes have now been identified in several species and thus their existence appears to be a general phenomenon (11).

The demonstration of linkage between histocompatibility and Ir loci is intriguing. Ir genes are identified functionally and their importance for antigen recognition well established. However, there is little known about the molecular nature of the Ir gene product itself. In contrast, histocompatibility antigens have been isolated and characterized physicochemically but until recently their biologic function was totally mysterious. Direct evidence for the involvement of histocompatibility antigens in the development of specific immune responses, however, has been provided by the demonstrations that efficient macrophage-T cell and T cell-B cell cooperation requires that the interacting cell types share a common histocompatibility antigen (12-14). Furthermore, stimulation of Ir gene controlled immune responses in guinea pig lymphocytes in vitro requires that the macrophage and T cell share the histocompatibility antigen to which the Ir gene is linked (15). These studies have suggested, therefore, that histocompatibility antigens may have an important immunologic function, that of facilitating the cooperative interaction between immunocompetent cells.

The involvement of histocompatibility antigens in the development of specific immune responses has also been suggested by the demonstration that histocompatibility-linked Ir gene controlled immune responses can be specifically suppressed in vitro by antisera containing antibodies directed at the linked histocompatibility antigens (16-18). In guinea pigs anti-strain 2 alloantisera prepared by immunizing inbred strain 13 guinea pigs with the lymphoid cells from inbred strain 2 animals can specifically inhibit responses to the synthetic polypeptide antigens DNP-GL and GA, both of which are controlled by strain 2 histocompatibility-linked Ir genes. Conversely, strain 2 anti-strain 13 alloantisera inhibit responses to GT, a strain 13 associated immune response. Antigen stimulated T cell proliferation, measured as tritiated thymidine (^{3}H-TdR) incorporation, has been the in vitro response most thoroughly studied. Recently, Ben-Sasson et al. (19) have shown similar effects of alloantisera on MIF production.

The specificity of the suppressive effects of the anti-histocompatibility sera for immune responses controlled by linked Ir genes is most clearly demonstrated using responding cells from F_1 animals. A representative experiment is shown in Table 1. Hybrid (2 x 13)F_1 guinea pigs possess the full complement of both strain 2 and strain 13 linked Ir genes. After immunization with both DNP-GL and GT their lymphoid cells respond to either of those antigens in vitro with increased ^{3}H-TdR incorporation. The addition of a strain 13 anti-strain 2 alloantiserum to such cultures profoundly suppresses DNP-GL stimulated tritiated thymidine incorporation without appreciably affecting GT stimulation. GT stimulation, on the other hand, is suppressed by a strain 2 anti-strain 13 alloantiserum at concentrations that do not appreciably affect DNP-GL stimulation. Each of the lymphoid cells from the F_1 animals possess both strain 2 and strain 13 lymphocyte membrane antigens. The exquisite specificity of the alloantiserum mediated suppression for immune responses controlled by linked Ir genes indicates a very close physical relationship between the site of action of the Ir gene product and the lymphocyte membrane antigen at which the alloantiserum is directed.

IDENTIFICATION OF THE TARGET ANTIGEN

The mechanisms through which alloantisera suppress specific immune responses has been an area of considerable interest because it promises to provide some insights into the process of antigen recognition by T cells. An initial area of inquiry has been characterization of the lymphocyte membrane antigen that serves as the target for the alloantisera. Antibodies directed at the guinea pig histocompatibility antigens are known to be present in the alloantisera. It is also possible that these sera contain antibodies directed at Ir gene controlled lymphocyte membrane antigens. Since the alloantisera used to suppress a specific immune response were produced in animals that were genetic non-responders to that antigen, their lymphoid cells might lack the specific antigen receptor or other Ir gene product. The immunizing lymphocytes, on the other hand, were obtained from genetic responders who must have those membrane markers on their cells. The anti-Ir gene product antibodies that might be present in such

TABLE I

THE SPECIFICITY OF ALLOANTISERUM-MEDIATED SUPPRESSION OF HISTOCOMPATIBILITY-ASSOCIATED IMMUNE RESPONSES IN VITRO[a]

Antigen	Antiserum[c]	^{3}H-TdR Incorporation[d] cpm
__[b]	NGPS	614
DNP-GL	NGPS	12,768
DNP-GL	13 Anti-2	679
DNP-GL	2 Anti-13	10,947
GT	NGPS	4,990
GT	13 Anti-2	4,323
GT	2 Anti-13	897

a) Lymphocytes were obtained from (2 x 13)F_1 guinea pigs immunized to DNP-GL and GT.
b) __ = No antigen added.
c) All sera added at 1/150 final dilution. NGPS = normal guinea pig serum; 13 Anti-2 = strain 2 anti-strain 13.
d) Data are expressed as the mean cpm of triplicate cultures.

TABLE II

REQUIREMENT FOR STRAIN 2 HISTOCOMPATIBILITY ANTIGENS FOR THE EFFECTIVE SUPPRESSION OF GA STIMULATED ^{3}H-TdR INCORPORATION BY ANTI-STRAIN 2 ALLOANTISERA

H-Type [a]	No. of Animals	% Suppression[b] GA mean	GA (range)	GT mean	GT (range)
Strain 2+	8	60	(30-88)	14	(0-51)
Strain 2-	5	10	(0-21)	10	(0-25)

a) The presence of strain 2 histocompatibility antigens was determined by the susceptibility of the animal's lymphocytes to anti-strain 2 serum mediated cytotoxicity using the ^{51}Cr-release assay.

b) $$\text{\% Suppression} = 100 \times \frac{\text{Antigen Stimulated } ^3\text{H-TdR Incorporation in 13 anti-2 treated cultures}}{\text{Antigen Stimulated } ^3\text{H-TdR Incorporation in NGPS treated cultures.}}$$

alloantisera could be responsible for its in vitro immunosuppressive effects.

The lymphocytotoxic activity of the alloantisera, as measured in a ^{51}Cr-release assay, is due to antibodies directed at the major histocompatibility antigens. Analysis of backcross progeny derived from matings between (2 x 13)F_1 guinea pigs and strain 13 animals has shown the co-inheritance of the susceptibility to lymphocytotoxicity by anti-strain 2 alloantisera and the ability to accept a transplantable strain 2 histocompatibility antigen bearing leukemia (20). Furthermore, absorption of the anti-strain 2 alloantisera by non-lymphoid tumor cells from strain 2 animals completely removes its cytotoxic capacity (21). We, therefore, compared the cytotoxic capacity of an alloantiserum with its ability to suppress a histocompatibility associated immune response in vitro (18). The curves obtained when plotting the cytotoxic capacity of an alloantiserum and its capacity to suppress antigen stimulation in vitro as a function of decreasing antiserum concentration are strikingly parallel. These results suggest that the same antibodies are responsible for both properties and thereby implicate the histocompatibility antigens as the target for the suppressive actions of the alloantisera. Further support for this interpretation is provided by the observation by Shevach et al. (22) that absorption of the anti-strain 2 sera by strain 2 histocompatibility antigen bearing lymphoid cells depleted the cytotoxic capacity of the alloantisera in parallel with depleting its ability to suppress strain 2 histocompatibility-linked Ir gene controlled immune responses in vitro.

The GA gene is linked to the strain 2 histocompatibility locus and to the PLL gene. In earlier studies of immune responses to GA in random bred guinea pigs, we identified a small number of animals that responded to GA but not to DNP-PLL (7). Those animals do not have strain 2 histocompatibility antigens on their lymphocytes. We have made use of those animals, who presumably resulted from crossovers between the GA and PLL Ir genes, to provide a more direct demonstration that anti-histocompatibility and not anti-Ir gene product antibodies are responsible for the suppression of specific immune responses in vitro (17,18). Lymphocytes from random bred Hartley guinea pigs that had responded to both GA and GT were characterized for the presence of strain 2 and strain 13 histocompatibility antigens. The effect of anti-strain 2 alloantisera on the proliferative response of those cells to both GA and GT was assessed in vitro. As shown in Table II suppression of GA stimulated responses by the anti-2 serum was observed in cultures of lymphoid cells bearing strain 2 histocompatibility antigens. The specificity of suppression for the strain 2 histocompatibility-associated immune response is verified in the random bred animals by the lack of significant suppression of GT responses. The anti-strain 2 alloantisera had no significant effect on either the GA or the GT stimulated tritiated thymidine incorporation in the lymphocytes from responder guinea pigs that did not have strain 2 histocompatibility antigens on their lymphocytes. Thus, the effective suppression of Ir gene controlled immune responses requires the presence of the linked histocompatibility antigens. Since those animals lacking strain 2 histocompatibility antigens were genetic responders, the lack of suppression of their in vitro immune responses by the anti-strain 2 sera demonstrates that anti-receptor or anti-Ir gene product antibodies are not responsible for the alloantiserum mediated immunosuppression.

The target antigen through which guinea pig alloantisera mediate their suppressive function has also been the subject of studies by Shevach and co-workers (22). They have demonstrated that cells from normal or DNP-GL immunized strain 2 guinea pigs were equally effective in removing the inhibitory effects of the anti-strain 2 serum on DNP-GL stimulated ^{3}H-TdR incorporation in vitro. They concluded, therefore, that it is unlikely that the suppression of T cell proliferation is due to the presence of antibodies specific for clonally expressed, Ir gene controlled antigenic determinants on T lymphocytes. In another report (23), they have confirmed the observation that the anti-strain 2 serum inhibits the in vitro response to GA of sensitized lymphocytes that have strain 2 histocompatibility antigens, but does not suppress the response of GA responder lymphocytes that lack those antigens. In addition, they showed that an antiserum prepared in strain 13 guines pigs against cells from a random-bred GA responder lacking strain 2 histocompatibility antigens did not inhibit the GA response of cells from a $(2 \times 13)F_1$ animal; while an antiserum prepared against GA responder cells possessing strain 2 histocompatibility antigens was capable of specifically inhibiting GA responses. These studies provide additional evidence that the alloantiserum mediated suppression occurs via antibodies directed at histocompatibility antigens rather than via antibodies specific for Ir gene determined specificities.

MECHANISM OF SUPPRESSION

The mechanisms by which antibodies directed at histocompatibility antigens suppress Ir gene product function in vitro have not been adequately defined. Several models have been proposed, including (a) a steric hindrance model in which antigen access to its receptor is physically blocked by the antibody molecules attached to the very closely associated histocompatibility antigens; (b) the alloantisera may suppress stimulation by masking the histocompatibility antigens, thereby preventing optimum macrophage-T cell interaction and perhaps preventing presentation of the antigen to its receptor; or (c) rather than preventing antigenic stimuli from reaching the appropriate sites on the lymphocyte membrane, the antibodies attached to the histocompatibility antigens may interfere with the ability of the lymphocytes to respond to those stimuli. In an attempt to begin to distinguish among those alternatives, we have tested for the suppression of antigen stimulation in vitro by immunoglobulin fragments derived from the suppressive alloantisera (24).

Fab, $F(ab)'_2$, and Fc fragments were prepared from the immunoglobulin fraction of a strain 13 anti-strain 2 alloantiserum. The Fab and Fc fragments were obtained by chromatographic separation of a papain digest of the IgG fraction on DEAE cellulose. $F(ab)'_2$ fragments were prepared by pepsin digestion of the IgG. None of the fragments were cytotoxic for strain 2 histocompatibility antigen-bearing lymphocytes. The Fab and $F(ab)'_2$ fragments, however, retained their antigenic specificity as demonstrated by their ability to inhibit the cytotoxicity of the intact anti-strain 2 serum when lymphocytes from a (2 x 13)F_1 guinea pig were used as target cells. The cytotoxicity of the anti-2 serum was unaffected by the addition of Fc fragments. None of the immunoglobulin fragments inhibited the cytotoxicity of an anti-strain 13 serum for the same target cells, thus demonstrating the specificity of the Fab and $F(ab)'_2$ fragments for strain 2 histocompatibility antigens.

The efficiency of binding of divalent antibody fragments to its antigen is considerably greater than the binding of monovalent fragments of the same antibody. In an attempt to define the difference in relative binding to strain 2 histocompatibility antigens by the Fab and $F(ab)'_2$ fragments prepared from anti-strain 2 sera, the effect of each of the fragments on cytotoxicity mediated by the intact antiserum was measured as a function of increasing competitor fragment concentration. As expected, there was considerable difference in their effectiveness. Essentially complete inhibition of the alloantiserum mediated cytotoxicity was achieved with the $F(ab)'_2$ fragments at concentrations as low as 0.06 mg/ml. A concentration of 0.45 mg/ml of the Fab fragments was required to achieve the same degree of competitive inhibition. The divalent fragments, therefore, appear to bind to the histocompatibility antigens with approximately tenfold greater efficiency than do the monovalent Fab fragments.

We have tested the effectiveness of the immunoglobulin fragments as suppressors of antigen stimulation in vitro. The $F(ab)'_2$, Fab and Fc fragments were added to cultures of lymphocytes obtained from (2 x 13)F_1 guinea pigs that had been immunized to DNP-GL and GT, and the effect on antigen stimulated ^{3}H-TdR incorporation was measured. A representative experiment is shown in Table III. The $F(ab)'_2$ fragments completely suppress the proliferative response to DNP-GL but not to GT. The effect of the $F(ab)'_2$ fragments, therefore, like that of its parent intact antibody, is specific for immune responses controlled by strain 2 histocompatibility-linked Ir genes. Unlike the divalent molecule, the monovalent Fab fragments did not suppress the response of the lymphocytes to DNP-GL. The Fc fragments also had no significant effect. Thus, although both the Fab and $F(ab)'_2$ fragments derived from the anti-strain 2 alloantisera have antigen combining sites specific for strain 2 histocompatibility antigens, only the $F(ab)'_2$ fragments retain the capacity to suppress the histocompatibility-linked Ir gene controlled immune response in vitro.

It was necessary to determine whether the differences observed between the Fab and $F(ab)'_2$ fragments are due solely to the more avid binding of the divalent fragment to its target antigen. Toward that end we tried to measure the magnitude of the difference between the highest non-suppressive concentration of Fab and the lowest effective concentration of $F(ab)'_2$ fragments. For practical reasons the highest concentration of Fab that could be achieved was 0.45 mg/ml. There was no significant suppression of DNP-GL stimulated responses at that concentration of monovalent fragment. Dilution studies with $F(ab)'_2$, however, have shown that the proliferative response of the lymphocytes to DNP-GL is exquisitely sensitive to the suppressive effects of the divalent fragments. Virtually complete suppression of the DNP-GL stimulated responses were observed at concentrations of $F(ab)'_2$ fragments as low as 60 μg/ml (Table IV). At 4.8 μg/ml the divalent fragments were still consistently effective, producing 23% mean suppression. Thus, there is a 100-fold difference between the lowest documented suppressive concentration of $F(ab)'_2$ and the highest Fab concentration that we tested. Obviously this is a minimum estimate since higher concentrations of Fab, had they been achieved, might also have been ineffective. In contrast, as we have shown, there is only a tenfold difference between the concentrations

TABLE III

THE EFFECT OF Fab, Fc, AND $F(ab)'_2$ FRAGMENTS DERIVED FROM ANTISTRAIN 2 SERA ON ANTIGEN-STIMULATED ^{3}H-TdR INCORPORATION IN VITRO[a]

Antigen	Ig Fragment	^{3}H-TdR Incorporation[c]
—	—	606
DNP-GL	—	8,996
DNP-GL	$F(ab)'_2$ (0.12)[b]	671
DNP-GL	Fab (0.45)	7,830
DNP-GL	Fc (0.32)	8,787
GT	—	3,904
GT	$F(ab)'_2$ (0.12)	4,275
GT	F(ab) (0.45)	3,222
GT	Fc (0.32)	3,815

a) Lymphocytes were obtained from individual (2 x 13)F_1 guinea pigs immunized to both DNP-GL and GT.

b) The quantities in parentheses indicate the final in vitro concentration in mg/ml of each fragment.

c) The data are expressed as the mean cpm of triplicate cultures.

TABLE IV

SUPPRESSION OF DNP-GL STIMULATED ^{3}H-TdR INCORPORATION BY ANTISTRAIN 2 SERUM AND $F(ab)'_2$ FRAGMENTS[a]

Antiserum[b]		$F(ab)'_2$ Fragments[b]	
Dilution	% Suppression	% Suppression	Concentration µg/ml
1/75	100	95	60
1/375	86	76	24
1/1875	25	23	4.8

a) Lymphocytes were obtained from (2 x 13)F_1 guinea pigs immunized to both DNP-GL and GT.

b) Strain 13 anti-strain 2 serum was added to the cultures at the final dilutions indicated. $F(ab)'_2$ fragments derived from the same alloantiserum were added at the final in vitro concentrations indicated.

of Fab and $F(ab)'_2$ needed to compete equally with the intact antibody for binding to the histocompatibility antigen in the cytotoxicity assay. These results indicate that it is the divalent character of the $F(ab)'_2$ fragments, not their increased efficiency in combining with histocompatibility antigen, that is required for the effective suppression of antigen stimulation in vitro.

We have compared the intact anti-histocompatibility sera and their $F(ab)'_2$ fragments for their ability to suppress in vitro responses (Table IV). The whole antiserum produced an average of 25% suppression of DNP-GL stimulated ^{3}H-TdR incorporation at a 1/1875 dilution. With an immunoglobulin concentration of 20 mg/ml in the guinea pig alloantiserum and assuming a molecular weight of 150,000 for the guinea pig IgG, that dilution represents an IgG concentration of 7×10^{-5} μmoles/ml. The $F(ab)'_2$ fragments produce similar suppression (mean 23%) at a concentration of 4.8×10^{-5} μmoles/ml. Thus, the divalent $F(ab)'_2$ fragments **suppress** as effectively as the intact antibody. The elimination of the Fc portion of the alloantibodies has not altered their effectiveness in suppressing antigen stimulation.

The demonstration that the $F(ab)'_2$ fragments are as effective as the intact antibody molecules suggests that steric hindrance is an unlikely explanation for the mechanism of alloantiserum mediated suppression. The Fc piece of the antibody is the portion that would be most likely to produce the steric inhibition and yet its removal has no effect on the efficiency of the molecule in suppressing antigen stimulation. The inability of the Fab fragment to suppress antigen stimulation suggests that the mechanism of the suppression is not simply a masking of histocompatibility sites, and, thereby, preventing efficient macrophage-T cell interaction. The monovalent fragments, as shown by their inhibition of antibody-mediated cytotoxicity, effectively mask the histocompatibility antigens but they do not suppress antigen stimulation. These experiments, then, suggest that the alloantiserum does not prevent the antigenic stimulus from gaining access to the lymphocyte and its antigen receptor.

The possibility that the alloantisera suppress by preventing the interaction of sensitized lymphocytes with the antigenic stimulus has also been examined by studying the effects of delayed addition of the alloantiserum to the antigen stimulated cultures. The cultures were initiated by adding DNP-GL to lymphocytes from strain 2 guinea pigs that had been immunized to that antigen. Anti-strain 2 alloantiserum was added to the cultures at various times after their initiation. As shown in Figure 1, effective suppression of DNP-GL stimulated

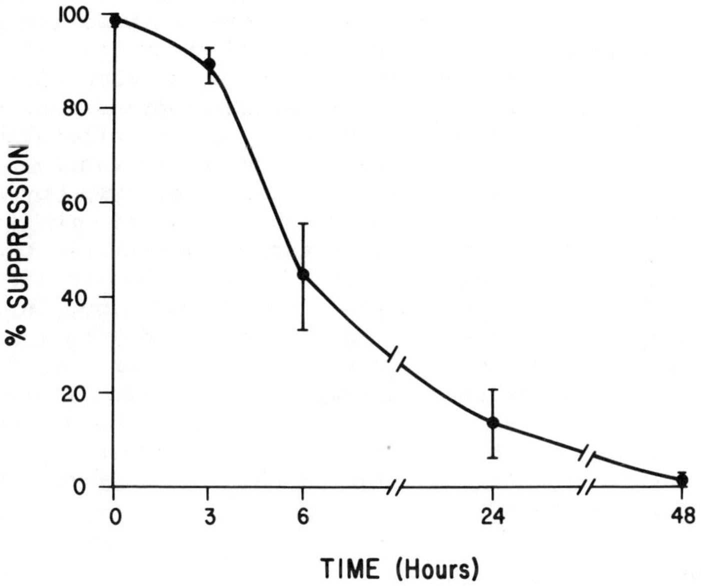

Fig. 1 The effect of delayed addition of strain 13 anti-strain 2 alloantiserum on the suppression of DNP-GL stimulated ^{3}H-TdR incorporation in vitro. Each point represents the mean % suppression observed in four experiments, each using lymphocytes from individual strain 2 guinea pigs immunized to DNP-GL. The brackets indicate $\pm$ the SE.

tritiated thymidine incorporation does not require that the alloantiserum be present at the time of the initiation of the cultures. Essentially complete suppression (greater than 90%) is noted when addition of the anti-2 serum is delayed up to 3 hours. Adding the alloantiserum 6 hours after initiation of the cultures still resulted in a significant suppression although, as indicated by the wider standard error brackets, the degree of suppression observed is quite variable. After a 24 hour delay however, there is only slight suppression, and by 48 hours after initiation of the cultures the alloantiserum had no effect on DNP-GL stimulated ^{3}H-TdR incorporation. It seems likely that the interaction of sensitized lymphocytes with the antigenic stimulus is well underway within the first few hours in culture. It has been clearly demonstrated that there is significant uptake of antigen by macrophages after exposures as brief as five minutes (25). In addition, in vitro lymphocyte proliferation has been observed when lymphoid populations have been exposed to antigen for only ten minutes (26). The suppressive effects of the alloantisera when they were added several hours after initiation of the cultures indicate, therefore, that the antisera can exert their effects after antigenic stimuli have already reached the lymphocyte membrane.

CONCLUSION

The lymphocytes of hybrid F_1 animals bear histocompatibility antigens from both parental strains; yet, a given antihistocompatibility serum suppresses only those immune responses controlled by Ir genes linked to that histocompatibility specificity. This suggests that the genetic linkage of histocompatibility and Ir loci has its counterpart in the phenotypic expression of the products of those genes on the lymphocyte membrane. These two membrane markers appear to be physically linked. The nature and distribution of the Ir gene product, however, has not yet been defined. If, in fact, it is the T cell antigen receptor, it might be expected to be clonally expressed. The Ir gene product - histocompatibility antigen complex would, therefore, be found on some cells and not others. Shevach et al. (23) have proposed an interesting model for this situation. They have drawn an analogy with the constant and variable region genes of immunoglobulin heavy chains that are separate but genetically linked. Together they produce a single polypeptide chain. They likened the Ir gene to the variable gene and the histocompatibility marker to the constant region and postulated that all T cells would have the 2/13 antigens while an individual Ir gene product would have clonal distribution.

The hypothesis that the Ir gene product is the antigen receptor has not yet been experimentally verified. The possibility that Ir genes code for lymphocyte membrane molecules that are not themselves receptors remains a viable alternative. The physical interaction of the Ir gene product with other membrane markers such as antigen receptors may alter their function, and thereby influence specific antigen recognition. Clonal expression of the Ir gene product, then, would not be required, but its physical association with histocompatibility antigen on the membrane remains necessary. Each T cell may have arrayed on its surface histocompatibility antigens and associated Ir gene products physically ordered in a sequence mimicking the relationship of their respective genetic loci. We have shown that the anti-strain 2 serum is more efficient in suppressing DNP-GL than it is GA stimulated ^{3}H-TdR incorporation although both responses are controlled by strain 2 histocompatibility linked Ir genes (24). Our studies of the relationships among those genes within the genome have shown that the PLL gene (which controls the response to DNP-GL) is much more closely linked to strain 2 histocompatibility loci than is the GA gene (7). Those observations document further similarities between the functional relationship of histocompatibility antigens to the development of Ir gene-controlled immune responses and the genetic relationship of the histocompatibility locus to Ir gene loci. This suggests that the major histocompatibility region of the chromosome is copied as a polycistronic message, and, after translation, expressed "en bloc" on the lymphocyte surface.

Any hypothetical model attempting to explain the mechanism of alloantiserum mediated suppression of Ir gene-controlled responses must take into account both the requirement for divalent antihistocompatibility antibody and, also, the effective suppression with alloantiserum after the interaction of T cell with antigen has occurred. There are at least two models that may satisfy the requirements: (a) The attachment of the alloantibody to its T cell membrane target antigen may induce a conformational change in the physically linked Ir gene product; i.e. an allosteric interaction, that leads to the release of the antigen from its receptor. In support of this model, studies of PHA stimulation have shown that the continued presence of the mitogen on the lymphocyte membrane is required for optimal stimulation (27). By analogy it is likely that a similar requirement exists for the continued presence of antigen. Further support for this model can be derived from our observation that the apparent affinity of the GA receptor for its antigen differs depending on whether or not the responder lymphocytes have strain 2 histocompatibility antigen on their membrane (17). One interpretation of that study is the suggestion that the receptor conformation may be altered by the presence of the histocompatibility antigen.

An alternative model proposes that (b) The attachment of the antibody to the histocompatibility antigen may lead to a dynamic alteration of a localized region of the lymphocyte membrane, in effect "depolarizing" it, thereby interrupting the transmission of the signal being generated by the antigen in its receptor. This model fits especially well with the requirement for divalent alloantibody, since it has been shown that divalent antibody is required for the redistribution of lymphocyte membrane markers (28-30). At the present time there is no compelling reason to favor one of these models over the other. The end result of both mechanisms, however, is the same. The alloantisera suppress antigen stimulated T cell proliferation by a dynamic interaction with the lymphocyte membrane that switches off the so-called "second messenger".

REFERENCES

1. McDevitt, H.O. and Benacerraf, B., Adv. Immunol. 11: 31, 1969.
2. Levine, B.B., Ojeda, A. and Benacerraf, B., J. Exp. Med. 118: 953, 1963.
3. Gasser, D.L. and Silvers, W.K., Adv. Immunol. 18: 1, 1974.
4. McDevitt, H.O., Bechtol, K.B., Grumet, F.C., Mitchell, G.F. and Wegmann, T.G., Prog. Immunol. 1: 495, 1971.
5. McDevitt, H.O., Deak, B.D., Shreffler, D.C., Klein, J., Stimpfling, J.H. and Snell, G.D., J. Exp. Med. 135: 1259, 1972.
6. Lieberman, R., Paul, W.E., Humphrey, W., Jr. and Stimpfling, J.H., J. Exp. Med. 136: 1231, 1972.
7. Bluestein, H.G., Green, I. and Benacerraf, B., J. Exp. Med. 134: 471, 1971.
8. Benacerraf, B., Bluestein, H.G., Green, I. and Ellman, L., Prog. Immunol. 1: 485, 1971.
9. Ellman, L., Green, I., Martin, W.J. and Benacerraf, B., Proc. Nat. Acad. Sci. U.S. 66: 332, 1970.
10. Bluestein, H.G., Ellman, L., Green, I. and Benacerraf, B., J. Exp. Med. 134: 1529, 1971.
11. Benacerraf, B. and McDevitt, H.O., Science 175: 273, 1972.
12. Rosenthal, A.S. and Shevach, E.M., J. Exp. Med. 138: 1194, 1973.
13. Kindred, B. and Shreffler, D.C., J. Immunol. 109: 940, 1972.

14. Katz, D.H., Hamaoka, T., and Benacerraf, B., J. Exp. Med. 137: 1405, 1973.
15. Shevach, E.M. and Rosenthal, A.S., J. Exp. Med. 138: 1213, 1973.
16. Shevach, E.M., Paul, W.E. and Green, I., J. Exp. Med. 139: 679, 1974.
17. Bluestein, H.G., Fed. Proc. 32: 985 (abs), 1974.
18. Bluestein, H.G., J. Immunol. 113: 410, 1974.
19. Ben-Sasson, S.Z., Shevach, E., Green, I. and Paul, W.E., J. Exp. Med. 140: 383, 1974.
20. Ellman, L., Green, I. and Benacerraf, B., J. Immunol. 107: 382, 1971.
21. Martin, W.J., Ellman, L., Green, I. and Benacerraf, B., J. Exp. Med., 132: 1259, 1970.
22. Shevach, E.M., Paul, W.E. and Green, I., J. Exp. Med. 139: 661, 1974.
23. Shevach, E.M., Green, I. and Paul, W.E., J. Exp. Med. 139: 679, 1974.
24. Bluestein, H.G., J. Exp. Med. 140: 481, 1974.
25. Rosenstreich, D.L. and Rosenthal, A.S., J. Immunol. 110: 934, 1973.
26. Kasakura, S., J. Immunol. 103: 1078, 1969.
27. Mendelsohn, J., Skinner, A. and Kornfeld, S., J. Clin. Invest. 50: 818, 1971.
28. Taylor, R.B., Duffus, P.H., Raff, M.C. and de Petris, S., Nature New Biol. 233: 225, 1971.
29. Unanue, E.R. and Karnovsky, M., Transplant. Rev. 14: 184, 1973.
30. Kourilsky, F.M., Silvestre, D., Neuport-Sautes, C., Loosfeld, Y. and Dausset, J., Eur. J. Immunol. 2: 249, 1972.

IN VITRO STUDIES OF THE CELLULAR INTERACTIONS IN AN ANTIBODY RESPONSE CONTROLLED BY AN IMMUNE RESPONSE (Ir) GENE(S)

Judith A. Kapp, Carl W. Pierce and Baruj Benacerraf
Department of Pathology, Harvard Medical School
Boston, Massachusetts 02115

The study of genetic control of immune responses has been greatly advanced by the observations that inbred strains of guinea pigs and mice develop different degrees of immunity when injected with antigens of limited diversity such as: 1) synthetic polymers of limited numbers of amino acids; 2) antigens differing only slightly from host proteins; and 3) antigens which differ substantially from host proteins but which are administered in low doses (1,2). The development of high levels of circulating antibody and/or cellular immunity to these antigens is controlled by individual autosomal dominant immune response (Ir) genes. Many of these genes are linked to those genes controlling the expression of histocompatibility antigens (3,4). In mice the position of several of these Ir genes has been localized within the I region of the H-2 gene complex between the K and S regions (5).

The antibody response by inbred strains of mice to the random terpolymer of L-glutamic acid^{60}-L-alanine^{30}-L-tyrosine^{10} (GAT) is controlled by an H-2 linked Ir gene. Animals bearing the H-2^{a,b,d,f,j,k,r,u,v} haplotypes behave as responder mice, synthesizing specific antibody after injection of GAT. Mice bearing the H-2^{n,p,q,s} haplotypes are nonresponder mice, synthesizing no detectable antibody after injection of GAT. However, both responder and nonresponder mice produce GAT-specific antibody when immunized with GAT complexed to methylated bovine serum albumin (GAT-MBSA) (6,7). The latter observation demonstrates the nonresponder mice, like responder mice, have precursors of antibody-producing cells (B cells) capable of synthesizing GAT-specific antibody if the appropriate thymus-derived cell (T cell) helper function is provided.

This investigation was supported by U.S. Public Health Service Grants AI-09929 and AI-09897 from the National Institute of Allergy and Infectious Diseases.

C.W. Pierce is the recipient of a Research Career Development Award from the National Institute of Allergy and Infectious Diseases.

We have been studying the development of antibody responses by spleen cells from responder and nonresponder mice to soluble GAT, GAT-MBSA and macrophage-bound GAT (GAT-MΦ) in vitro. Cells producing antibody specific for GAT are enumerated in a hemolytic plaque assay using GAT coupled to sheep erythrocytes (GAT-SRBC) as indicator cells (8). In Table I, the GAT-specific IgG PFC responses developed by spleen cells of several strains of mice after incubation with each of these three forms of GAT are depicted. Cultures were incubated with 1 to 10 μg soluble GAT, GAT-MBSA containing approximately 5 μg GAT or syngeneic macrophages bearing 1 to 3 nanograms of GAT. All cultures developed GAT-specific IgG PFC responses after incubation with GAT-MBSA while only responder mice developed GAT-specific responses when incubated with soluble GAT or macrophage-bound GAT. No direct (IgM) PFC specific for GAT have been detected in cultures of spleen cells from any of the mice tested.

The cellular requirements for the development of GAT-specific PFC responses in cultures of spleen cells from responder and nonresponder mice have also been investigated (9). Macrophages are required for the development of primary GAT-specific PFC responses in vitro by non-adherent spleen cells (lymphoid cells) from responder mice stimulated with GAT-or GAT-MBSA and by non-adherent spleen cells from nonresponder mice stimulated with GAT-MBSA. Macrophages from responder and nonresponder mice supported equally the development of PFC responses by allogeneic non-adherent spleen cells in cultures stimulated with SRBC and GAT-MBSA. However, macrophages from responder mice did not confer the ability to develop GAT-specific PFC responses to non-adherent nonresponder spleen cells in cultures stimulated with GAT. Thus, the responder status to GAT is determined by the genotype of the non-adherent cells, not by the genotype of the macrophages. Not only did nonresponder macrophages support the development of GAT-specific PFC responses by non-adherent responder lymphoid cells in cultures stimulated with soluble GAT but, like responder macrophages, they stimulated comparable responses when the only GAT in the cultures was provided in nanogram amounts bound to the macrophages. Thus, macrophage function with regard to GAT is the same in cultures of spleen cells from responder and nonresponder mice and the macrophage is not responsible for the failure of nonresponder spleen cells to develop PFC responses to soluble GAT.

These results, demonstrating that macrophages need not be

TABLE I

DEVELOPMENT OF GAT-SPECIFIC IgG PFC RESPONSES IN VITRO

Strain	H-2	GAT-Specific IgG PFC/culture[a] GAT	GAT-MBSA	GAT-MΦ[b]
C57BL/6	b	480	400	465
B10.S	s	<10	420	<10
DBA/1	q	<10	420	<10
(C57BL/6 x DBA/1)F_1	b/q	420	540	675

[a]Representative experiment for each strain - Day 5 PFC Response.

[b]GAT-bound to syngeneic MΦ(1 to 3 nanograms GAT/7 x 10^4 MΦ/culture).

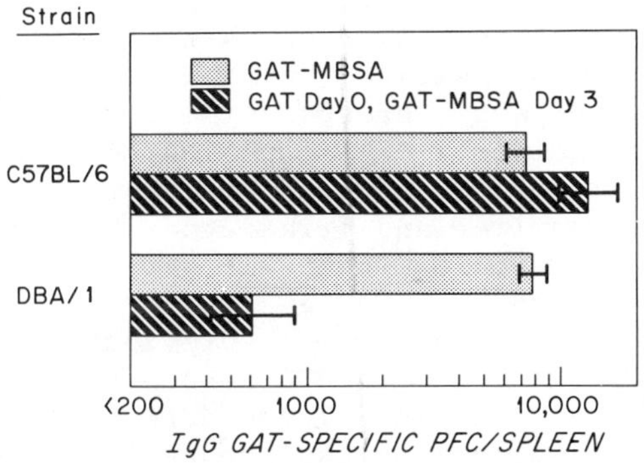

Fig. 1 Effect of GAT on responses to GAT-MBSA by responder and nonresponder mice in vivo. Normal C57BL/6 (H-2^b) and DBA/1 (H-2^q) mice were injected with 10 μg GAT in Maalox-pertussis on day 0. Three days later, these mice and normal mice were injected with GAT-MBSA containing 10 μg GAT in Maalox-pertussis i.p. Seven days later the mice were sacrificed and their spleens examined for GAT-specific IgG PFC. The graph shows the geometric means of the PFC responses of 8 to 12 mice/group, the brackets indicate the standard error of the mean. Reproduced with permission of Rockefeller Institute Press from Kapp, J.A., Pierce, C.W., Schlossman, S. and Benacerraf, B., J. Exp. Med. 140: 172, 1974.

histocompatible with the lymphoid cells to support the development of a PFC response in vitro, are in apparent contradiction with the results of Rosenthal and Shevach (10,11). They have demonstrated that, in guinea pigs, antigen-bearing macrophages and sensitized T cells must be syngeneic for successful development of antigen-induced DNA synthetic responses. These differences in requirements for syngenicity could reflect differences in which lymphoid cell response (T or B) is being measured and its sensitivity to activation or that in the strains of mice that we have used no restriction in macrophage-lymphoid cell interaction is observed. Resolution of these differences awaits development of tissue culture systems for the study of antibody responses by guinea pig lymphoid cells and antigen-specific DNA synthetic responses by mouse T cells.

Treatment of spleen cells from responder mice with anti-theta (θ) serum and complement (C) prior to culture initiation abolishes the development of PFC responses to SRBC, GAT and GAT-MBSA. Similar treatment of spleen cells from nonresponder mice abolishes the development of PFC responses to SRBC and GAT-MBSA (9). It has been shown that antigen-primed T cell helper function is radio-resistant whereas unprimed T cells and B cells are radiosensitive (12). We therefore examined GAT-primed responder and nonresponder mice for the development of radioresistant helper T cell function (9). Spleen cells from responder mice which had been primed with GAT in Maalox-pertussis 30 days earlier and irradiated just prior to culture initiation reconstituted GAT-specific PFC responses by B cells from responder mice to both GAT and GAT-MBSA. Spleen cells from X-irradiated nonresponder mice previously primed with GAT failed to restore GAT-specific PFC responses to GAT or GAT-MBSA by nonresponder B cells. These data suggest that the defect in genetic nonresponder spleen cells is the failure of their T cells, after interaction with GAT, to provide helper function for the initiation of the B cell response.

In addition, we have found that injection of nonresponder mice with GAT prior to immunization with GAT-MBSA depresses the development of the GAT-specific response in vivo and also that incubation of spleen cells from nonresponder mice with GAT together with GAT-MBSA decreases the anti-GAT PFC response developed in vitro (13). As shown in Figure 1, 10 μg of GAT in Maalox and pertussis suppressed the GAT-specific response by nonresponder (DBA/1) mice immunized with GAT-MBSA, whereas the same dose of GAT in Maalox and pertussis does not depress, and usually increases the GAT-specific PFC response developed by responder (C57BL/6) mice injected with GAT-MBSA. Although not shown, injection of Maalox-pertussis alone on day 1 caused no decrease in GAT-specific PFC responses to GAT-MBSA in either the responder or nonresponder strains of mice. Furthermore, injection of 10 μg GAT in Mallox and pertussis on day 1 did not decrease the primary immune response to SRBC injected 3 days later in either of the strains of mice.

Similarly, the immune response in vitro to GAT-MBSA by spleen cells from all nonresponder mice is suppressed by incubation with 5 μg soluble GAT together with GAT-MBSA (Figure 2). The same concentration of soluble GAT did not suppress the immune response of responder mice to GAT-MBSA.

That the B cells of nonresponder mice are more sensitive to suppression by soluble GAT than are responder B cells is shown in Figure 3. Using (responder X nonresponder) F_1 spleen cells from mice primed with GAT-MBSA and irradiated just prior to culture initiation, we can see that the immune response to GAT-MBSA by nonresponder (B10.S) B cells, but not by responder (C57BL/6) B cells is suppressed by soluble GAT.

To determine whether the unresponsiveness induced in vivo in nonresponder mice injected with 10 μg GAT is due to B cell tolerance, we have investigated the immunocompetence of T and B cells from spleens of nonresponder mice previously rendered unresponsive by injection of GAT (14). Normal nonresponder B cells were prepared by treating normal spleen cells with anti-θ serum and C. Helper T cells were provided by spleen cells from GAT-MBSA primed, X-irradiated nonresponder mice. B and T cells from GAT primed nonresponder mice were fractionated from spleen cell suspensions by passage over a Sephadex G-200 column to which purified rabbit anti-mouse FAB antibodies were bound (14). Cells which were not retained by the column are T cells and other nonimmunoglobulin-bearing cells (15). The cells bound to the column (B cells) were eluted by competitive inhibition using media containing 10% normal mouse serum.

The results of a representative experiment are shown in Table II. Groups A through D are controls showing that the anti-θ and C-treated normal spleen cells, irradiated GAT-MBSA-primed spleen cells, and column-purified T and B cells did not develop GAT-specific PFC responses to GAT-MBSA when cultured separately. The B and T cell sources used for reconstitution are shown to be competent in Group E. Column purified T and B cells cultured together developed a PFC response to SRBC (Group F) but failed to develop a significant GAT-specific PFC response to GAT-MBSA (Group G). Column-purified T cells from GAT-primed mice were unable to provide helper function to normal B cells (Group H). However, B cells from GAT-primed mice responded very well to GAT-MBSA when cultured with carrier-primed helper T cells (Group I). This experiment demonstrates that unresponsiveness induced by injection of

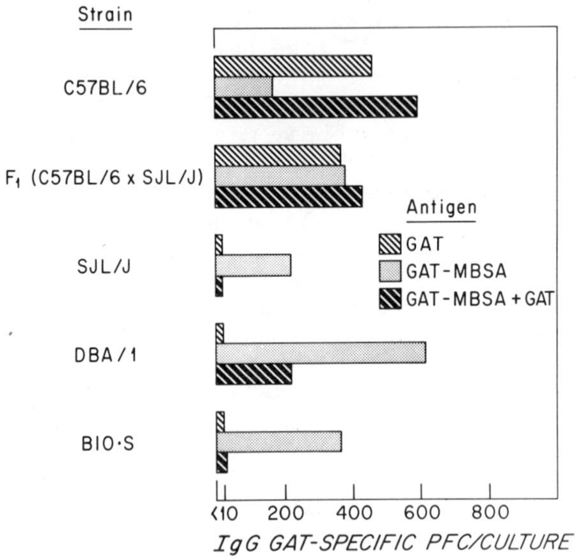

Fig. 2 Effect of GAT on responses to GAT-MBSA by spleen cells from responder and nonresponder mice in vitro. 10^7 spleen cells were cultured with 10 μg GAT, or GAT-MBSA containing approximately 5 μg GAT, or both. PFC responses determined on day 5. Reproduced with permission of Rockefeller Institute Press from Kapp, J.A., Pierce, C.W., Schlossman, S. and Benacerraf, B., J. Exp. Med. 140:172, 1974.

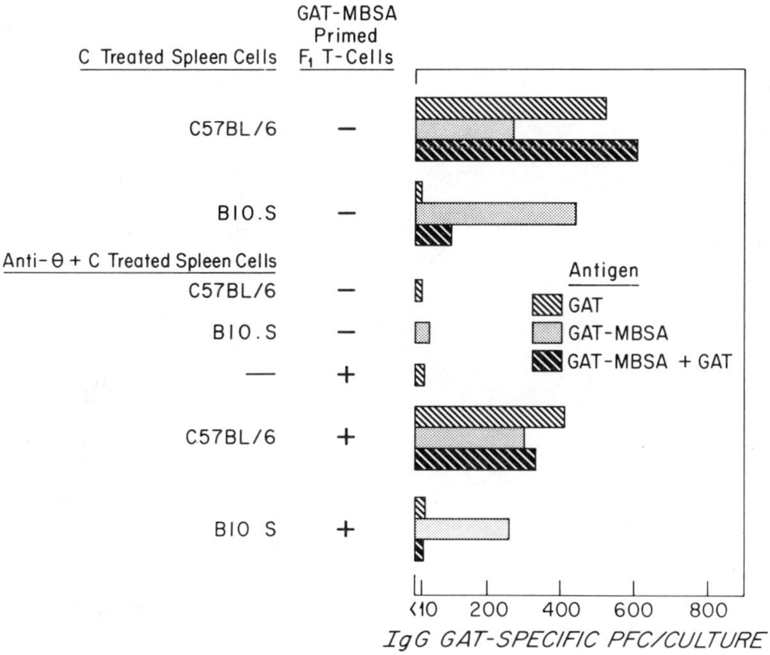

Fig. 3 Effect of GAT on responses to GAT-MBSA in cultures of responder and nonresponder B cells in the presence of GAT-MBSA primed irradiated F_1 spleen cells. 10 μg soluble GAT, or GAT-MBSA containing approximately 5 μg GAT, or both were added to cultures of 10^7 normal C57BL/6 or B10.S spleen cells or to cultures containing 10^7 F_1 T cells plus 10^7 B cells from responder or nonresponder mice. Reproduced with permission of Rockefeller Institute Press from Kapp, J.A., Pierce, C.W., Schlossman, S. and Benacerraf, B., J. Exp. Med. 140:172, 1974.

TABLE II

CHARACTERIZATION OF COLUMN-PURIFIED B AND T CELLS FROM GAT-PRIMED DBA/1 MICE

Group	DBA/1 Spleen Cells: 10^7 Normal Treated with Anti-θ and C	DBA/1 Spleen Cells: 10^7 X-Irradiated GAT-MBSA Primed[b]	GAT-Primed Column-Purified[c] $5x10^6$ B	GAT-Primed Column-Purified[c] $5x10^6$ T	Antigen	IgG PFC/Culture[a] Experiment 1 SRBC	Experiment 1 GAT	Experiment 2 SRBC	Experiment 2 GAT
A	+	-	-	-	GAT-MBSA	-	0	-	0
B	-	+	-	-	GAT-MBSA	-	35	-	0
C	-	-	+	-	GAT-MBSA	-	0	-	0
D	-	-	-	+	GAT-MBSA	-	0	-	0
E	+	+	-	-	GAT-MBSA	-	1,080	-	1,375
F	-	-	+	+	SRBC	585	-	1,000	-
G	-	-	+	+	GAT-MBSA	-	50	-	40
H	+	-	-	+	GAT-MBSA	-	15	-	0
I	-	+	+	-	GAT-MBSA	-	1,255	-	785

[a]Cultures harvested after incubation with 10^7 SRBC or 5 μg GAT as GAT-MBSA for 5 days.

[b]Mice primed with GAT-MBSA containing 10 μg GAT 60 days before irradiation (800R) and sacrifice.

[c]Mice primed with 10 μg GAT in Maalox 3 days before sacrifice.
Column-purified splenic B cells: Experiment 1 - 85% Ig bearing
Experiment 2 - 96% Ig bearing

Column-purified splenic T cells: Experiment 1 - 7% Ig bearing
Experiment 2 - 0% Ig bearing

10 μg GAT in vivo is not due to B cell tolerance but rather to a failure of the T cells to provide helper function for the B cell response.

Column-purified T and B cells from GAT-primed DBA/1 mice were also examined for their ability to suppress the immune response by spleen cells from normal DBA/1 mice to GAT-MBSA (14). The results (Table III) demonstrated that spleen cells from GAT-primed DBA/1 failed to respond to GAT-MBSA in vitro (Group C and E) and that they suppress the immune response of spleen cells from normal DBA/1 mice (compare Groups B and D). Suppressor cell activity was present in the column fraction which contained T cells (Group F), and was not present in the fraction containing B cells (Group G). The identification of suppressor cells as T cells has been confirmed by the observation that the suppression caused by GAT-primed DBA/1 spleen cells can be inhibited by treatment of the spleen cells with anti-θ serum and C (14).

Data presented in the experiments thus far have shown that the generation of suppressor cells in nonresponder mice depends upon previous exposure to GAT. The findings that mice injected with GAT fail to respond to subsequent challenge with GAT-MBSA, but respond normally to a subsequent injection of SRBC (13) and that cultures containing GAT-induced suppressor cells are nevertheless able to respond to SRBC (Table II) suggests, but does not prove, that the suppression is antigen-specific. Since the demonstration of suppression involves exposure to GAT-MBSA, it is possible that GAT-MBSA may be required to induce active suppressors in cultured spleen cells from mice previously primed with GAT. Therefore, the specificity of the suppressive activity must be examined by the addition of both GAT-MBSA and SRBC to cultures containing normal and GAT-primed spleen cells. Such an experiment is presented in Table IV. All spleen cell cultures respond to SRBC and GAT-primed spleen cells did not inhibit the normal spleen cell response to SRBC (Group F). Addition of GAT-MBSA to cultures of normal spleen cells stimulated by SRBC did not inhibit the development of anti-SRBC PFC responses (Group G) nor was this response inhibited if GAT-primed spleen cells were present (Group H). Therefore, the suppressor cells elicited by GAT inhibit only GAT-specific B cell responses.

Yet to be resolved is the site of action of suppressor T cells in the GAT system. In other systems, suppressor T cells have been shown to exert their effects at the levels of the carrier-specific helper T cells (16) or on T cells which mediate delayed hypersensitivity reactions (17). The results of our experiments do not permit us to distinguish whether

TABLE III

SUPPRESSIVE ACTIVITY OF COLUMN PURIFIED T CELLS FROM GAT-PRIMED DBA/1 MICE

Group	DBA/1 Spleen Cells: 10^7 Normal	DBA/1 Spleen Cells: 10^7 GAT-Primed[b]	GAT-Primed Column-Purified[c]: 5×10^6 B	GAT-Primed Column-Purified[c]: 5×10^6 T	Antigen	IgG GAT-Specific PFC/Culture[a]: Expt. 1	IgG GAT-Specific PFC/Culture[a]: Expt. 2
A	+	---	---	---	GAT	0	0
B	+	---	---	---	GAT-MBSA	515	600
C	---	+	---	---	GAT-MBSA	15	35
D	+	+	---	---	GAT-MBSA	105	10
E	---	---	+	+	GAT-MBSA	50	40
F	+	---	---	+	GAT-MBSA	0	35
G	+	---	+	---	GAT-MBSA	935	725

[a]Cultures harvested after incubation with 5 μg GAT or 5 μg GAT-MBSA for 5 days.

[b]Mice primed with 10 μg GAT in Maalox 3 days before sacrifice.

[c]Column-purified cells from the same cell suspension as described in (b) and as described in Table IV.

Reproduced with permission of Rockefeller Institute Press from Kapp, J.A., Pierce, C.W. and Benacerraf, B., J. Exp. Med. 140:648, 1974.

TABLE IV
ANTIGEN SPECIFICITY OF SUPPRESSOR T CELLS

	DBA/1 Spleen Cells			IgG PFC/ Culture[a]	
Group	10^7 Normal	10^7 GAT-Primed[b]	Antigen	SRBC	GAT
A	+	---	GAT-MBSA	---	525
B	---	+	GAT-MBSA	---	10
C	+	+	GAT-MBSA	---	0
D	+	---	SRBC	250	---
E	---	+	SRBC	530	---
F	+	+	SRBC	640	---
G	+	---	SRBC and GAT-MBSA	470	---
H	+	+	SRBC and GAT-MBSA	635	---

[a]Cultures harvested after 5 days incubation with 10^7 SRBC or GAT-MBSA containing 5 μg GAT or both.

[b]Mice primed with 10 μg GAT in Maalox 11 days before sacrifice and culture initiation.

Reproduced with permission of Rockefeller Institute Press from Kapp, J.A., Pierce, C.W. and Benacerraf, B., J. Exp. Med. 140:648, 1974.

suppressor T cells generated by GAT in DBA/1 mice act on GAT-specific B cells or on carrier-specific helper T cells which, by binding GAT-MBSA, could be susceptible to specific suppression. This issue may be resolved by investigating whether the suppressor T cells generated by GAT can also suppress the antibody response to the carrier in nonresponder mice injected with GAT bound to an immunogenic carrier.

Since these latter data demonstrate that nonresponder mice do have T cells which recognize GAT and that after exposure to GAT these T cells develop suppressor activity, we were prompted to ask whether there was any set of circumstances in which GAT-specific helper T cell function could be demonstrated in nonresponder mice (18). To examine this question, nonresponder DBA/1 mice were immunized with GAT, GAT-MBSA or MBSA in Maalox-pertussis or 2×10^6 DBA/1 macrophages bearing 15 to 20 nanograms of GAT. After 30 days spleen cells from these mice were examined for radioresistant, GAT-specific helper T cell function. The results from several experiments have been summarized as the number of positive responses/total number of experiments in Table V. Responses were judged to be positive if the number of GAT-specific PFC observed in the experimental groups was greater than or equal to 20% of the response in cultures of the same number of normal spleen cells incubated with GAT-MBSA. Although not shown, neither B cells nor T cells developed GAT-specific PFC responses when incubated separately with GAT-MBSA. As we have shown before, spleen cells from X-irradiated GAT-primed DBA/1 mice provide no helper cell function for DBA/1 or F_1 B cells incubated with GAT, GAT-MBSA or macrophage-bound GAT. Spleen cells from MBSA-primed DBA/1 did cooperate with DBA/1 and F_1 B cells in cultures stimulated with GAT-MBSA but not in those stimulated with GAT or GAT-MΦ. This demonstrates 1) that parental T cells can cooperate with F_1 B cells and 2) that under these conditions, the immune response is antigen specific and not the result of an allogeneic stimulation. GAT-MBSA-primed DBA/1 T cells did provide helper function for DBA/1 B cells when incubated with GAT-MBSA but not when incubated with GAT or GAT-MΦ. In contrast, GAT-MBSA-primed DBA/1 T cells provided helper cell functions for the response by F_1 B cells stimulated with GAT-MBSA or GAT-MΦ and for the response to GAT in half of the experiments. This observation suggests that nonresponder mice immunized with GAT-MBSA developed GAT-specific T cells which provided helper cell function to F_1 B cells in the absence of MBSA in the cultures.

To determine whether the development of GAT-specific helper cells in nonresponder mice requires that GAT be presented as a complex to an immunogenic carrier or whether it only depends

TABLE V

ANTIGEN SPECIFICITY OF DBA/1 HELPER T CELLS

DBA/1 T Cells Primed To[a]	B Cell Source	Antigen in Culture GAT	GAT-MBSA	GAT-MΦ
GAT	DBA/1	0/2[b]	0/2	0/2
GAT	F_1	0/3	0/3	0/2
MBSA	DBA/1	0/2	2/2	0/2
MBSA	F_1	0/4	5/5	0/4
GAT-MBSA	DBA/1	0/4	7/7	0/6
GAT/MBSA	F_1	4/7	8/8	6/7
GAT-MΦ	DBA/1	0/1	3/3	0/2
GAT-MΦ	F_1	1/2	3/3	2/2

[a]Mice primed 30 to 60 days prior to X-irradiation (800R) and sacrifice.

[b]Number of positive responses/total number of experiments.

upon presentation of GAT in a bound form, nonresponder mice were injected with 2 x 10^6 syngeneic macrophages bearing 15 to 20 nanograms of GAT. As shown in the last column of Table V, GAT-specific helper cells were demonstrable in GAT-MΦ-primed, irradiated DBA/1 spleen cells. These spleen cells cooperated with DBA/1 B cells in response to GAT-MBSA but not GAT or GAT-MΦ and with F_1 B cells in response to all three forms of GAT. Thus, it is not necessary that GAT be complexed to an immunogenic carrier in order to elicit GAT-specific helper cell functions. We can draw several conclusions from these data:

1) There is a difference between responder and nonresponder T cell function after exposure to GAT. Injection of GAT results in the development of GAT-specific helper T function in responder mice and GAT-specific suppressor T cell function in nonresponder mice.

2) GAT-specific helper cell function can be demonstrated in nonresponder mice if they are immunized with GAT bound to MBSA or to macrophages.

3) There is apparently a difference between GAT-specific responder and nonresponder B cells. This is demonstrated by the greater sensitivity of nonresponder B cells to inhibition by soluble GAT in vitro (Figure 3) and by the fact that nonresponder B cells are less responsive to stimulation by macrophage-bound GAT (Table I), even though demonstrable GAT-specific helper T cells can be detected in nonresponder mice after injection of macrophage-bound GAT (Table V). However, we must note that the failure of nonresponder B cells to be stimulated by macrophage-bound GAT is not absolute. In the presence of GAT-primed X-irradiated F_1 spleen cells, nonresponder B cells can develop GAT-specific PFC response when stimulated with macrophage-bound GAT (19).

These observations raise important questions concerning the mechanism of Ir gene regulation of the immune response. The first relevant issue is whether the generation of suppressor T cells in nonresponder mice is unique to the GAT system or may be generalized to other systems and species. The next question is what has the data revealed about Ir gene control of the immune response? Since we have demonstrated that nonresponder mice have GAT-specific T cells that are easily stimulated by GAT to become suppressor T cells and that by the appropriate experiments one can demonstrate GAT-specific helper T cells in nonresponder mice immunized with GAT bound to MBSA or to macrophages, we must conclude that the differences between responder and nonresponder mice is not in the absolute capacity to generate helper T cells specific for GAT or in the capacity of responder and nonresponder B cells to produce GAT specific antibody.

It appears that both responder and nonresponder T cells recognize GAT but some event distal to that recognition event governs the development of helper or suppressor activity. It is, therefore, possible that Ir genes behave as regulator genes determining the function of T and B cells after interaction with antigen.

ACKNOWLEDGEMENTS

The authors thank Ms. Fern De La Croix for her excellent technical assistance in these studies.

REFERENCES

1. Benacerraf, B. and McDevitt, H.O., Science (Wash. DC) 175: 273, 1972.
2. McDevitt, H.O. and Benacerraf, B., Adv. Immunol. 11: 31, 1969.
3. Martin, J.W., Maurer, P.H. and Benacerraf, B., J. Immunol. 107: 715, 1971.
4. Merryman, C.F. and Maurer, P.H., J. Immunol. 108: 135, 1972.
5. Dunham, E.K., Dorf, M.E., Shreffler, D.C. and Benacerraf, B., J. Immunol. 111: 1621, 1973.
6. Dunham, E.K., Unanue, E.R. and Benacerraf, B., J. Exp. Med. 136: 403, 1972.
7. Gershon, R.K., Maurer, P.H. and Merryman, C.F., Proc. Nat. Acad. Sci. USA 70: 250, 1973.
8. Kapp, J.A., Pierce, C.W. and Benacerraf, B., J. Exp. Med. 138: 1107, 1973.
9. Kapp, J.A., Pierce, C.W. and Benacerraf, B., J. Exp. Med. 138: 1121, 1973.
10. Rosenthal, A.S. and Shevach, E.M., J. Exp. Med. 138: 1194, 1973.
11. Shevach, E.M. and Rosenthal, A.S., J. Exp. Med. 138: 1213, 1973.
12. Katz, D. and Benacerraf, B., Adv. Immunol. 15: 1, 1972.
13. Kapp, J.A., Pierce, C.W. and Benacerraf, B., J. Exp. Med. 140: 172, 1974.
14. Kapp, J.A., Pierce, C.W., Schlossman, S. and Benacerraf, B., J. Exp. Med. 140: 648, 1974.
15. Schlossman, S.F. and Hudson, L., J. Immunol. 110: 313, 1973.
16. Basten, A., in Immunological Tolerance: Mechanism and Potential Therapeutic Applications. eds. B. Benacerraf and D. Katz, Academic Press, New York, in press.

17. Claman, H.N., Phanuphak, P. and Moorehead, J.W., in Immunological Tolerance, Mechanism and Potential Therapeutic Applications. eds. B. Benacerraf and D. Katz, Academic Press, New York, in press.

18. Kapp, J.A., Pierce, C.W. and Benacerraf, B., manuscript in preparation.

19. Benacerraf, B., Kapp, J.A., Pierce, C.W. and Katz, D., J. Exp. Med. 140: 185, 1974.

GENETIC CONTROL OF THE STIMULATOR AND EFFECTOR FUNCTION IN ALLOGENEIC LYMPHOCYTE INTERACTION: THE EXPRESSION OF *I* REGION GENE PRODUCTS ON T AND B LYMPHOCYTES[1,2]

Peter Lonai[3]
Division of Immunology, Department of Medicine, School of Medicine, Stanford University, Stanford, California.

The *I* region is a small genetic region within the MHC (H-2) of the mouse, situated between the *H-2K* and *Ss* loci. Three *I*-region subdivisions, designated *Ir-1A*, *Ir-1B* and *I-C* have been derived (1-4). Gene(s) localized in this chromosomal segment control at least five immunological phenomena. *Ir* genes which are dominant and autosomal, regulate antigen specific responsiveness both in humoral and cell mediated immunity (5-8). Gene(s) controlling cell interaction structures permit specific cooperative interaction between T and B lymphocytes and/or macrophages (9-12). The *Lad* gene products function as lymphocyte activating determinants. They are the most immunogenic, but are not the only antigens which can stimulate MLR, GVHR and some graft rejection reactions (13-15). The *H-2I* locus specifies classical transplantation antigens (16). *Ia* loci code for cell membrane alloantigens, and are detected by alloantisera (1, 16). Recent studies by Dickler and Sachs (18) suggest that the receptors for the Fc part of aggregated immunoglobulins and antigen-antibody complexes are situated in close proximity to Ia antigens on the cell membrane, and may in fact be the *Ia* gene product.

---

[1]This research was supported by USPHS research grant AI 07757 and the National Cancer Institute #1 CB 43941.

[2]Abbreviations used in this paper: MHC - major histocompatibility complex; MLR - mixed lymphocyte reaction; GVHR - graft versus host reaction; CML - cell mediated lympholysis; MEM - minimal essential medium (Eagle); FCS - foetal calf serum; 2-ME - 2-mercathoethanol; PHA - phytohaemagglutinin; and LPS - *E.coli* lipopolysaccharide. For abbreviations used for the genetic nomenclature of the H-2 complex, see references 3 and 4.

[3]Recipient of Senior Dernham Fellowship D-195 from the American Cancer Society, California Division.

It is not known whether these experimentally separable phenomena are, in fact, distinct. Their cellular-molecular mechanism is not known. It is not clear whether these phenomena are controlled by separate genes (like, Ir, Lad, lymphocyte cooperation, etc.) or are different phenotypic manifestations of the same gene. The importance of understanding these relationships is apparent from the fact that the H-2 linked regulation of immune responsiveness, the lymphocyte-lymphocyte, lymphocyte-macrophage cooperation, and the stimulation of lymphocytes by antigen are very basic immunological phenomena, and their detailed understanding holds the promise of a more comprehensive view of immunology and probably of cellular interactions in general.

In this report, the role of I-region gene products in the MLR and CML *in vitro* immune reactions is examined. Areas studied include: 1) The expression of I-region gene products on T and B lymphocytes as studied by the MLR; 2) The role of Ia bearing cells in the MLR and CML; 3) The nature of the target cell in the CML; and 4) The nature of the CML killer cell.

MATERIALS AND METHODS

Mouse Strains. The congenic and congenic recombinant mouse strains used and their F_1 hybrids are included in Table 1. Most of the mice have been bred in our colony. Some of the animals, and most of the breeding nuclei, were gifts from Drs. Shreffler (Dept. of Human Genetics, The University of Michigan Medical School, Ann Arbor, Michigan), Klein (Dept. of Microbiology, The University of Texas Southwestern Medical School, Dallas, Texas), and Stimpfling (The McLaughlin Research Institute, Columbus Hospital, Great Falls, Montana). F_1 hybrid mice were specially bred for these studies for two reasons: 1) To avoid the influence of genes not linked to H-2; or 2) To examine the effects of restricted regions within the H-2 complex. The animals were sacrificed at two to five months of age. In each experiment, either female or male mice were used.

Culture Conditions. MLR cultures were prepared as described previously by Lonai and McDevitt (19), and the CML cultures as described by Nabholz (20,21). Briefly, stimulation by irradiated cells in the MLR was monitored by ^{3}H-thymidine uptake. Triplicate cultures of both syngeneic (e.g. haplotype k against k), and allogeneic (e.g. haplotype k against b) combinations were established. A stimulation index was calculated as follows: $\text{S.I.} : \frac{\text{mean cpm allogeneic}}{\text{mean cpm syngeneic}}$.

P values were determined by the t test. Cell mediated killing was determined by the release of ^{51}Cr from radiolabeled PHA and/or LPS stimulated target cells. The percent specific release was obtained by subtracting the spontaneous release from that obtained in the presence of stimulated lymphocytes (21). Three parallel cultures were run for each point.

Preparation and Characterization of Cell Suspensions. Thymocyte suspensions were prepared from animals previously injected i.p. with india ink to facilitate the removal of the parathymic lymph nodes. Isolated T cell suspensions from lymph nodes or spleen were separated by the nylon wool filtration method (22). T cell preparations were tested before each experiment with anti-θ serum plus complement, and only suspensions exhibiting 90% or greater θ-positivity were used. B cell suspensions were made by treating the original cell suspension with anti-θ serum and complement. Only B-cell suspensions containing 10% or less θ positive cells were used.

Serological Reagents. The following alloantisera were used: anti-H-2^k (C3H.SW, hereafter referred to as CSW, anti-C3H/DiSn); anti-H-$2K^k$ ((A.TL x B10.A(5R))F_1 anti-B10.A); anti-Ia^k ((A.TH x B10.S)F_1 anti-(A.TL x B10.S)F_1, serum #007)(23) and (A.TH anti-A.TL, serum #013); anti-Ia^s (A.TL anti-A.TH, serum #014); and anti-θ (($A^{\theta AKR}$ x AKR.B)F_1 anti-A thymoma) and (AKR/J anti-C3H)(25)). Serum #007 plus complement killed 30-40% of lymph node cells and 50-60% of spleen cells in the dye exclusion assay (25). No significant killing was observed with this antiserum with thymus or nylon wool purified T cells. Antisera #013 and #014 gave 40-50% lysis with lymph node targets, 60-70% lysis with spleen, and not more than 10% killing (generally not significant) with nylon wool purified T cells. Rabbit complement obtained from selected rabbits, and adsorbed on mouse liver powder and mouse (BALB/C) spleen cells, was used. Each batch of complement was tested for toxicity in the dye exclusion assay and in culture.

Blocking and Elimination Experiments. The ability of antisera to block the CML reaction was determined by adding predetermined dilutions of antiserum, specific for the radiolabeled target cells or killer cells, at the initiation of the CML test. In elimination experiments conducted in both the MLR and CML assays, the appropriate cells were incubated with an excess of antiserum and complement and washed three times before being introduced into the cultures. Equal numbers of viable cells from treated or untreated suspensions were added to the cultures, unless indicated otherwise.

RESULTS

1. The Expression of I-Region Gene Products on T and B Cells in the MLR. When recombinant mouse strains differing only in the *I*-region became available, it was expected that alloantisera produced between such animals might react with the *Ir* gene product and would be specific for T lymphocytes. However, most of the early studies with such antisera showed preponderant B cell specificity, and some authors claimed that these antibodies were directed exclusively against B lymphocytes (23,26). Because of numerous, if not unequivocal, evidence for T cell expression of the *Ir* gene product (6,27-30), we chose an independent, non-serological test to investigate the expression of *I*-region gene products on T lymphocytes.

Our working hypothesis is the following: If *I*-region genes are not expressed on T cells, then an MLR directed against T-stimulator cells in a combination differing only in the *I*-region should be negative. On the other hand, a positive response in such a combination would imply that *I*-region gene products are expressed on T lymphocytes. Accordingly, a large number of strain combinations describing different subregions of the *H-2* complex were tested in the MLR. Part of this study has already been published (19,30).

Table I demonstrates 15 selected combinations tested in the MLR. Combinations 1 and 2 differ for the whole *H-2* complex. Combinations 3-6 analyze the MLR between strains differing only in the *H-2K* and *H-2D* regions. Combinations 7 and 8 examine *I* and *S* region differences. Combinations 10 to 15 describe different subregions of the *I*-region.

We will analyze Table I according to these groups. Combinations 1 and 2 demonstrate that all four cell types used in these reactions (lymph node cells, thymocytes, nylon wool purified T cells, and isolated B cells) can stimulate when incompatibility exists for the entire *H-2* complex. T cells seem to be weaker immunogens than B cells or nonfractionated lymph node lymphocytes. In combination 2, isolated lymph node T lymphocytes were used as responder cells. This experiment was performed to demonstrate that the ^{3}H-thymidine incorporation observed with T-stimulator cells is not due to the recognition of non-irradiated B cells in the responder fraction by irradiated T cells in the stimulator fraction ("backstimulation" (see 31,32)). Similar experiments were performed with other combinations with the same result (combinations 7-10 and 12, Table I).

Combinations 3 and 4 differ only in the *H-2K* region while 5 and 6 differ in the *H-2D* region. These experiments were performed to determine whether *K* and *D* gene products express

Lad determinants. No consistent results were obtained. Combinations 4 and 6 gave significant stimulation values, but 3 and 5 were negative in repeated experiments. These data suggest that the K and D products are marginally stimulatory in the MLR, and support the findings of Meo et al. (14,33).

Combinations 7 and 8 differ in the I- and S regions. It can be seen that in these combinations both T and B cells stimulate and that T cells seem to be slightly more immunogenic. Since it has been shown that the Ss locus does not code for lymphocyte activating determinants (13, and our unpublished results), these experiments can be interpreted as a demonstration of the expression of I-region gene products on both T and B cells.

The MLR activity of determinants controlled by different subregions of the I-region is demonstrated in the remaining combinations of Table I. Three points emerge from the data: 1) There are combinations (9 and 10) in which both T and B cells are almost equally stimulatory (comparable to that observed in combinations which differ for the whole I-region). Stimulation in combination 9 is attributed to disparity in the Ir-1A subregion since minimal stimulation was observed in another combination (4) which included the same H-2K region incompatibility. These data demonstrate that Ir-1A, and possibly Ir-1B, controls determinants expressed on both T and B cells. 2) In some cases no stimulation is observed with either T or B lymphocytes (11 and 15). Failure to stimulate might reflect identity in Lad gene products, or genetic non-responsiveness. Alternatively, in combination 11, the Ir-1B subregion may not specify Lad determinants. It is interesting to note that neither of these combinations differ in Ia specificities (see last column of Table I). 3) Several combinations demonstrate a sizable difference between the stimulation obtained with T and B cells (12-14). In combination 12, B cells stimulate far better than T cells (incompatibility in Ir-1B, I-C, and S regions) in agreement with experiments of Fathman et al. (34), while in combinations 13 and 14 (incompatibility in the I-C and S regions) the reverse is true. In fact, in the latter two combinations, it is questionable whether B cells stimulate at all. This low or negligible response to one of the two main lymphocyte subpopulations in the MLR may mean that the same determinant is present in unequal frequency on the membrane of the two cell types, or that certain I-region gene products are preferentially expressed on either T or B lymphocytes.

The available data permit a limited comparison between Ia and Lad determinants. The last column of Table I includes the Ia specificities determined by each haplotype according

TABLE I

THE EXPRESSION OF Lad GENES ON T AND B CELLS AS STUDIED BY THE MLR

COMBINATION	RESPONDER	STIMULATOR	H-2-MAP K	I-A	I-B	1-C	S	D	STIMULATION INDEX	P	Ia SPECIFICITIES
1	B10.BR-L*	B10.S-L	k	k	k	k	k	k	11.05	0.001	1,2,3,7
	B10.BR-L*	B10.S-Thy**	s	s	s	s	s	s	2.28	0.001	(4,5,9)
2	CSW-T***	C3H-T	b	b	b	b	b	b	29.18	0.001	3,8,9
	CSW-T***	C3H-B****	k	k	k	k	k	k	64.18	0.001	(1,2),3(7)
3	A.TL-L	A.AL-L	s	k	k	k	k	d	1.35 ± 0.6+	N.S.++	---
			k	k	k	k	k	d			
4	(A.TL x B10/Sn)F_1-L	B10.A(4R)-L	s/b	k/b	k/b	k/b	k/b	d /b	3.11 ± 0.77	0.02	---
			k	k	b	b	b	b			
5	A.TH-L	A.SW-L	s	s	s	s	s	d	1.66 ± 1.03	N.S.	---
			s	s	s	s	s	s			
6	(A.AL x C3H.Q)F_1L	C3H-L	k/q	k/q	k/q	k/q	k/q	d/q	1.86	0.01	---
			k	k	k	k	k	k			
7	A.TH-T	A.TH-T	s	s	s	s	s	d	12.17	0.001	4,5,9
	A.TH-T	A.TH-B	s	k	k	k	k	d	8.80	0.001	(1,2,3,7)
8	B10.AQR-T	B10.T(6R)-T	q	k	k	d	d	d	17.06	0.001	1,2,3,6,7
	B10.AQR-T	B10.T(6R)-B	q	q	q	q	q	d	3.63	0.02	3,(5,9,10)
9	(B10.HTT x B10/Sn)F_1-T	B10.A(4R)-T	s/b	s/b	s/b	k/b	k/b	d/b	20.7	0.001	3,4,7,8,9
	(B10.HTT x B10/Sn)F_1-T	B10.A(4R)-B	k	k	b	b	b	b	12.4	0.001	(1,2),3
10	(A.TH x B10.HTT)F_1-T	A.TL-T	s/s	s/s	s/s	s/k	s/k	d/d	21.93	0.001	4,5,7,9
		A.TL-B	s	k	k	k	k	d	31.47	0.001	(1,2,3),7

11	B10.A(4R) x B10.A(5R)F_1-L	B10.A(2R)-L	k/b	k/b	b/b	b/d	b/d	b/d	1.05	N.S.	1,2,3,6,7,8,9
	B10.A(4R) x B10.A(5R)F_1-L	B10.A(2R)-Thy	k	k	k	d	d	d	1.01	N.S.	1,2,3,6,7
12	B10.A(4R)-T	B10.A(2R)-T	k	k	b	b	b	b	2.93	0.01	1,2,3
	B10.A(4R)-T	B10.A(2R)-B	k	k	k	d	d	b	15.65	0.001	1,2,3(6,7)
13	(B10.HTT x A/J)F_1-L	A.TH-L	s/k	s/k	s/k	k/d	k/d	d/d	1.38	0.05	1,2,3,4,6,7,9
	(B10.HTT x A/J)F_1-L	A.TH-Thy	s	s	s	s	s	d	8.11	0.001	4,(5),9
	(B10.HTT x A/J)F_1-L	A.TH-T							4.64	0.001	
	(B10.HTT x A/J)F_1-L	A.TH-B							1.20	N.S.	
14	B10.S(7R)-L	B10.HTT-T	s	s	s	s	s	k	2.23	0.01	4,5,9
	B10.S(7R)-L	B10.HTT-B	s	s	s	k	k	d	0.97	N.S.	4,(7),9
15	A/J-L	A.AL-T	k	k	k	d	d	d	0.95	N.S.	1,2,3,6,7
	A/J-L	A.AL-B	k	k	k	k	k	d	1.18	N.S.	1,2,3,7

FOOTNOTES:
* L = lymph node cells
** Thy = thymocytes
*** T = nylon wool purified T cells
**** B = anti-theta + complement purified B cells
+ N.S. = non-significant
++ standard error of 3-5 experiments

to David and Schreffler (1,2,16). Specificities which are known to be different between the responder and stimulator haplotype are enclosed in brackets. Two observations suggest that Ia and Lad determinants may be identical. First, whenever Ia incompatibility exists, a positive MLR is obtained. Second, in combinations where no Ia disparity exists (11 and 15), no MLR is observed. In contradiction to this interpretation is the fact that, in some combinations, B cells, which are known to express Ia specificities, do not stimulate. For example, in combination 14, responder and stimulator cells differ in Ia.7, yet no stimulation with B cells is observed. Delovitch and McDevitt (36) have demonstrated that this determinant is present on B cells. This problem may not be resolved until the cellular distribution of other Ia and Lad products has been examined, but it may suggest that different lymphocyte populations express different Ia specificities, which are recognized as Lad products in the MLR.

In summary, the data suggest that the H-2 complex codes for lymphocyte activating surface structures on both T and B cells. Such determinants can be found in at least four distinct genetic regions: H-2K, Ir-1A, I-C, and H-2D. Some of these determinants are expressed preferentially on T or B cells. Further studies will be required to determine whether other regions in the H-2 gene complex control Lad determinants, and to discern whether Ia and Lad are governed by the same or different genes.

2. The Role of Ia Bearing Cells in MLR and CML. The possible role of Ia bearing cells in MLR and CML was studied by selectively eliminating this cell population with anti-Ia serum and complement prior to initiation of the cultures. Utilizing this approach one can examine the function of the cell carrying the Ia determinant. However, such studies yield no information on the possible role of the Ia determinant itself. Direct blocking experiments, in which the appropriate Ia anti-serum is added to the culture medium were not feasible since the nonpurified or ammonsulfate fractionated Ia sera in our hands were non-specifically toxic under culture conditions.

In the experiments to be described, an anti-Ia^k (A.TH anti-A.TL) antiserum was used. This antiserum reacts with cells expressing Ia specificities of haplotype k (e.g. haplotypes k, a and t1), but not with cells expressing Ia^s determinants (haplotypes s and t2). Table II shows the effect of anti-Ia^k sera on the stimulator cells in the MLR. Experiments 1 and 2 demonstrate that the elimination of Ia reactive cells (35-45%) from the stimulating lymph node cell suspensions decreases the MLR. This effect is due to complement

TABLE II

THE EFFECT OF ANTI-Ia SERA ON THE MLR. TREATMENT OF THE STIMULATOR CELL.

EXP.	RESPONDER	STIMULATOR	TREATMENT	CPM SYNGENEIC	± S.E.	CPM ALLOGENEIC	± S.E.	STIMULATION INDEX	P
1	A.TH-L	A.TL-L*	---	11161	1500	68009	13487	6.09	0.001
	A.TH-L	A.TL-L	C' only			67150	6459	6.01	0.001
	A.TH-L	A.TL-L	anti-Ia^k only			87408	9569	7.83	0.001
	A.TH-L	A.TL-L	anti-Ia^k + C'			38076	7830	3.41	0.01
2	B10.S-L	B10.A-L	C' only	1185	49	17901	1015	15.11	0.001
	B10.S-L	B10.A-L	anti-Ia^k only			19193	3410	16.20	0.001
	B10.S-L	B10.A-L	anti-Ia^k + C'			4836	1300	4.08	0.01
3	B10.S-L	B10.A-Thy	medium only	2811	544	8373	879	2.98	0.001
	B10.S-L	B10.A-Thy	anti-Ia^k only			9325	317	3.32	0.001
	B10.S-L	B10.A-Thy	anti-Ia^k + C'			4069	634	1.45	0.05
4	B10.S-L	B10.BR-T	C' only	392	99	36906	1812	94.15	0.001
	B10.S-L	B10.BR-T	anti-Ia^k only			30497	2189	77.80	0.001
	B10.S-L	B10.BR-T	anti-Ia^k + C'			2160	1095	5.51	0.05
5	A.TL-L	A.TH-L	---	13661	590	81230	4189	5.95	0.001
	A.TL-L	A.TH-L	C' only			91583	3283	6.70	0.001
	A.TL-L	A.TH-L	anti-Ia^k + C'			80831	7317	5.92	0.001

* Abbreviations L, T, Thy - see footnote to Table I.

mediated cytotoxicity, since neither anti-Ia nor complement alone influenced the reaction. In experiments 3 and 4, the effect of anti-Ia treatment on stimulator T-cells was studied. Treatment of both thymocytes and isolated T cells with anti-Ia serum and complement resulted in a significant decrease in MLR stimulation. Experiment 5 is included as a specificity control and shows that treatment of A.TH stimulator cells with the A.TH anti-A.TL serum had no effect on the MLR.

In the previous section, we have demonstrated the expression of _I_-region gene products on both T and B cells. The present results verify that T cells do stimulate, and further suggest that this stimulatory population expresses Ia determinants.

In the next group of experiments, the anti-Ia sensitivity of the responder-cells was studied. It can be seen from Table III that the responder function is not sensitive to anti-Ia treatment. Moreover, anti-Ia treatment of aliquots from the same cell suspensions which exhibited impaired stimulator function, had intact responder capacity in the MLR (in experiments 1 and 2 of Table III and in experiments 1 and 2 of Table II the same cells were used in the reciprocal combination). This phenomenon was observed in repeated experiments.

The results obtained with anti-Ia treatment of cells participating in the MLR were confirmed in subsequent experiments using CML cultures (Table IV). Treatment of responder cells with anti-Ia serum and complement did not influence the development of killer cells (experiment 1, Table IV). Treatment of nylon wool purified T-stimulator cells, however, completely abolished the ensuing target cell killing (experiment 2, Table IV).

In summary, it has been demonstrated that Ia bearing T lymphocytes express determinants which function in allogeneic lymphocyte stimulation. The responding cell in these reactions -- known to be a T-lymphocyte (36,37,32) -- does not appear to be sensitive to anti-Ia treatment.

3. The Nature of the Target in the CML Reaction. PHA or LPS blast cells labelled with ^{51}Cr were cultured with _in vitro_ stimulated killer cells and alloantisera specific for the target cells were added. In some experiments, antisera specific for the responder cell were added to study their effect on killer cells. No complement was added, and during the short term incubation (3-4 hours) of the CML-test cultures, no toxic effect due to mouse serum was observed.

Experiment 1 (Table V) demonstrates that anti-H-2, anti-H-2K, and anti-Ia sera, all specific for the target cell, inhibit lytic activity. The effect of the anti-H-2K and anti-

TABLE III

THE EFFECT OF ANTI-Ia SERA ON THE MLR. TREATMENT OF THE RESPONDER CELL.

EXP.	RESPONDER	STIMULATOR	TREATMENT	CPM SYNGENEIC	± S.E.	CPM ALLOGENEIC	± S.E.	STIMULATION INDEX	P
1	A.TL-L	A.TH-L*	---	20625	3917	194246	3220	9.42	0.001
	A.TL-L	A.TH-L	C' only			198821	12450	9.64	0.001
	A.TL-L	A.TH-L	anti-Ia^k only			196251	11930	9.52	0.001
	A.TL-L	A.TH-L	anti-Ia^k + C'			205187	17280	9.95	0.001
2	B10.A-L	B10.S-L	medium only	1197	143	18207	315	15.21	0.001
	B10.A-L	B10.S-L	C' only			15595	565	13.03	0.001
	B10.A-L	B10.S-L	anti-Ia^k only			12777	734	10.67	0.001
	B10.A-L	B10.S-L	anti-Ia^k + C'			23204	773	19.39	0.001

* For abbreviations - see footnote to Table I.

TABLE IV

THE EFFECT OF ANTI-Ia^k SERUM ON THE CML. TREATMENT OF PRECURSOR CELLS.

EXP.	RESPONDER	STIMULATOR	TARGET	TREATMENT	SPECIFIC ^{51}Cr RELEASE AT KILLER/TARGET RATIO 5/1	10/1	20/1
1	B10.BR-L*+	B10.S-L	B10.S-PHA	Medium only	15.89 ± 2.83	20.49 ± 0.72	30.97 ± 4.25
	B10.BR-L	B10.S-L	B10.S-PHA	C' only	16.49 ± 2.92	24.18 ± 1.45	29.45 ± 1.18
	B10.BR-L	B10.S-L	B10.S-PHA	Anti-Ia^k only	21.59 ± 2.68	27.78 ± 1.63	34.46 ± 2.02
	B10.BR-L	B10.S-L	B10.S-PHA	Anti-Ia^k + C'	22.55 ± 1.22	27.42 ± 0.64	32.34 ± 1.63
2	B10.S-L	B10.BR-T+	B10.BR-PHA	Medium only	44.13 ± 1.3	49.82 ± 0.92	54.76 ± 3.2
	B10.S-L	B10.BR-T	B10.BR-PHA	Anti-Ia^k + C'	-1.49	0.40	-0.89

* Abbreviations L,T - see footnote in Table I.

+ In experiments 1 and 2 the responder or the stimulator cell, respectively, was treated at day 0, before the cultures were set up.

TABLE V

THE EFFECT OF ALLOANTISERA ON THE CML. BLOCKING OF THE KILLER OR TARGET CELLS.

EXP.	RESPONDER	STIMULATOR	TARGET	TREATMENT	KILLER/TARGET RATIO	SPEC. ^{51}Cr RELEASE	± S.E.	P*
1	B10.S	B10.BR	B10.BR-PHA	---------	10/1	40.04	2.85	
	B10.S	B10.BR	B10.BR-PHA	norm. mouse se.	10/1	37.78	2.72	N.S.**
	B10.S	B10.BR	B10.BR-PHA	anti-H-2^k	10/1	7.91	0.86	0.001
	B10.S	B10.BR	B10.BR-PHA	anti-H-$2K^k$	10/1	25.60	1.56	0.001
	B10.S	B10.BR	B10.BR-PHA	anti-Ia^k	10/1	31.39	1.29	0.01
2	B10.BR	B10.S	B10.S-PHA	---------	10/1	42.55	0.73	
	B10.BR	B10.S	B10.S-PHA	norm. mouse se.	10/1	47.14	3.35	N.S.
	B10.BR	B10.S	B10.S-PHA	anti-H-2^k	10/1	31.37	2.62	0.001
	B10.BR	B10.S	B10.S-PHA	anti-Ia^k	10/1	44.01	1.50	N.S.
3	B10.S	B10.BR	B10.BR-PHA	norm. mouse se.	10/1	19.83	2.22	
	B10.S	B10.BR	B10.BR-PHA	anti-Ia^k	10/1	11.90	1.81	0.01
	B10.S	B10.BR	B10.BR-LPS	norm. mouse se.	20/1	38.09	0.58	
	B10.S	B10.BR	B10.BR-LPS	anti-Ia^k	20/1	23.11	3.92	0.001
	B10.S	B10.BR	A.TL-PHA	norm. mouse se.	10/1	11.56	0.75	
	B10.S	B10.BR	A.TL-PHA	anti-Ia^k	10/1	6.51	1.71	0.001
	B10.S	B10.BR	A.TL-LPS	norm. mouse se.	20/1	10.16	2.08	
	B10.S	B10.BR	A.TL-LPS	anti-Ia^k	20/1	1.79	0.52	0.001

* t test, P values compared to control in each experiment.

** N.S. = not significant.

Ia sera (both restricted in specificity) was significantly weaker than that of the anti-H-2 serum (directed against the whole complex). In contrast, killer cell activity was not altered by anti-Ia serum, and was only slightly depressed by the anti-H-2 serum. (Experiment 2, Table V).

It has been shown that the H-2K and H-2D antigens are the major target antigens of the CML reaction (20,38). Since the anti-Ia serum partially blocked this response (Experiment 1) further experiments were conducted to determine whether Ia antigens might also serve as a target in cell mediated lysis. In experiment 3 (Table V) B10.S killer cells sensitized against B10.BR stimulator cells (sensitized against $H\text{-}2K^k$ and Ia^k and $H\text{-}2D^k$ determinants) were tested against B10.BR (expressing $H\text{-}2K^k$, Ia^k and $H\text{-}2D^k$ determinants) or A.TL (expressing $H\text{-}2K^s$, Ia^k and $H\text{-}2D^d$ determinants) PHA or LPS target cells. The anti-Ia^k serum had a weak but significant blocking effect with B10.BR target cells, and a stronger inhibitory effect with A.TL target cells. These experiments suggest that target cell killing may be directed against Ia antigens present on both T(PHA) and B(LPS) target lymphocytes.

4. The Nature of the Killer Cell. Killer cells were prepared by *in vitro* sensitization of lymphocytes against irradiated stimulator cells. The presence of surface antigens on these allo-sensitized cells was examined utilizing anti-Ia, anti-H-2K, and anti-θ alloimmune sera. Immunoglobulin bearing cells were detected by immunofluorescent staining. It was found that the majority (90% or more) of the sensitized cells carried responder type (H-2K) histocompatibility antigens. The preparations contained varying amounts of θ, Ia and Ig positive cells. The Ia-type of the cells in all of the experiments was that of the responder, i.e.: in case of B10.BR, B10.A or A/J responder cells, it reacted with anti-Ia^k, and not with anti-Ia^s, while if B10.S or A.SW were the responder cells, the resulting stimulated cells reacted with anti-Ia^s and not with anti-Ia^k. Double labeling immunofluorescence experiments to determine presence and amount of Ia^+, Ig^+ versus Ia^+, θ^+-cells is currently under investigation.

The expression of Ia determinants on the killer cells was subsequently examined. B10.A spleen cells were sensitized against B10.S spleen cells *in vitro* (Table VI). The sensitized cells were then subjected to anti-Ia and complement treatment (38% trypan blue positive cells). The treated cells and appropriate controls were introduced into the CML test, and their killer capacity against B10.A PHA-transformed target cells was determined. It can be seen that the anti-Ia treatment had no effect on the killer capacity of the sensitized cells, suggesting that the killer cell is not an Ia

TABLE VI

EFFECT OF ANTI-Ia AND COMPLEMENT ON THE KILLER ACTIVITY OF B10.A CELLS SENSITIZED AGAINST B10.S SPLEEN CELLS.

EXP.	TREATMENT	% STAINED CELLS (TRYPAN BLUE) AFTER TREATMENT	SPECIFIC ^{51}Cr RELEASE AT KILLER/TARGET RATIOS 5/1	10/1	20/1
1	medium only	11	18.34 ± 1.59	25.02 ± 3.34	34.88 ± 1.59
2	anti-Ia only	18	13.63 ± 0.50	23.34 ± 0.49	28.10 ± 0.37
3	C' only	17	12.80 ± 3.43	19.59 ± 0.59	26.50 ± 0.72
4	anti-Ia + C' uncorrected*	38	12.87 ± 0.66	17.65 ± 0.68	26.78 ± 0.62
5	anti-Ia + C' corrected for viable cells	38	15.64 ± 3.47	23.39 ± 0.10	30.97 ± 0.74

* In this experimental point the killer cells were added to the target cells on the basis of total cell number counted after the last wash following treatment (20% positive cells). All others were corrected for viable cell numbers.

positive cell.

DISCUSSION

The strategy employed in this study was to use Lad and Ia determinants as cell markers, and to study their presence on cells of more or less known function. It is apparent that in this approach the method of cell separation is crucial. For example, purification of T cells on nylon wool columns yields cell populations with approximately 10% B cell contamination. However, this contamination appears to be negligible, since the number of B cells is less than that required for significant stimulation in both MLR and CML assays (unpublished results). Furthermore, in combinations where only T cells stimulate, or where T cells stimulate much stronger than B cells, possible B cell contamination should have no effect. "Backstimulation", in which irradiated T stimulator cells may recognize responder B cells and induce them to proliferate, is also unlikely, since the same results were obtained when purified T cells were reacted with purified T cells, and experiments designed to create conditions for "backstimulation" exhibited comparatively much weaker, if any response, than that observed between allogeneic I-region different T lymphocytes.

A second problem concerning cell separation techniques involves possible selection for certain lymphocyte subpopulations. For example, it has been suggested that nylon wool purification may exclude Ia positive T cells as well as B lymphocytes (39). In order to control for this possibility, we have performed several experiments with Sepharose bound anti-immunoglobulin columns to eliminate B cells, and obtained results comparable to those observed with the nylon wool technique (unpublished).

The lymphocyte subpopulations examined in this report are characterized according to their function and Ia or θ-positivity in Table VII. Several points are striking: 1) Responder T lymphocytes in the MLR and killer T lymphocytes in the CML are insensitive to treatment with anti-Ia serum plus complement. Either cells which participate in these responses do not express Ia products, or they express such determinants in sublytic density. We are unable to distinguish between these two possibilities with the available data. If the second alternative is correct, it would necessitate postulating the existence of two Ia-positive T cell populations, one sensitive to anti-Ia plus complement, the other not. The existence of at least one Ia-positive T cell population, however, was clearly demonstrated. Frelinger et al. have demonstrated the expression of Ia determinants on a fraction of T

TABLE VII

CHARACTERIZATION OF LYMPHOCYTE SUBPOPULATIONS BY THEIR FUNCTION AND Ia POSITIVITY

	Ia	θ
A. Cell mediated immune reactions		
1. Sensitization phase (MLR)		
Responder cell	-	+
Stimulator cells - T	+	+
Stimulator cells - B	+	-
2. Effector phase (CML)		
Sensitized T cells		
Killer	-	+
Helper (?)	+	+
Target lymphoblasts		
PHA (T)	+	+
LPS (B)	+	-
B. Migration characteristics		
(Sensitized T cells)		
Spleen seeking	+	+
Lymph node seeking	-	+

cells (39). Our results support their findings: 1) Approximately 70% of spleen lymphocytes are killed with our anti-Ia^k serum (#013), implying both B and T cell killing. 2) Our MLR experiments on one hand demonstrate that T cells express Lad determinants controlled by the I-region, and on the other suggest that the Ia and Lad loci of the I-region may be identical. 3) Treatment of T stimulator cells with anti-Ia serum plus complement inhibited the MLR response. In order to distinguish whether the responder T cells express Ia determinants or not, it may be necessary to resort to direct fluorescent staining methods, such as that used by Wofsy et al. (40). For the purpose of discussion, we will assume that responder and killer lymphocytes are Ia-negative T cells.

I-region products expressed on both T and B lymphocytes function as lymphocyte stimulating determinants. Treatment with anti-Ia sera plus complement inhibited the stimulation by either cell type. These results suggest that cells which express Ia determinants can induce a proliferative response, but shed no light on the possible identity of Ia and Lad products. Comparisons between Ia incompatible and Lad incompatible combinations, and between Ia compatible and Lad compatible combinations, suggested possible identity of Ia and Lad. However, the observation that B lymphocytes of certain haplotypes, known to express specificity Ia.7 (35) do not stimulate in the MLR either argues against such a possibility, or suggests the existence of different Ia.7 specificities on T and B cells. Further study will be required to resolve this problem.

The target for cell mediated lysis can be an Ia-positive cell. Previous studies have demonstrated that the products of the H-2K and H-2D loci can serve as target antigens for CML (38,20). In this report, blocking experiments in which anti-Ia sera were incubated with the target cells in the absence of complement, suggested that killer activity might also be directed against the Ia gene product.

The Ia positive T cells observed in these experiments appear to have a passive role in the allogeneic lymphocyte interactions studied. These cells can stimulate the MLR and CML, and may appear in sensitized lymphocyte populations, but do not function as responder or killer cells. If we try to fit the Ia positive and Ia negative T cells into the two main functional T cell classes, it seems to be tempting to assume that the Ia positive cells may function as helper cells, while the Ia negative ones as suppressor cells.

A hypothetical scheme describing cellular interactions and lymphocyte differentiation is presented in Figure 1. According to this scheme, Ia negative responder T cells recognize

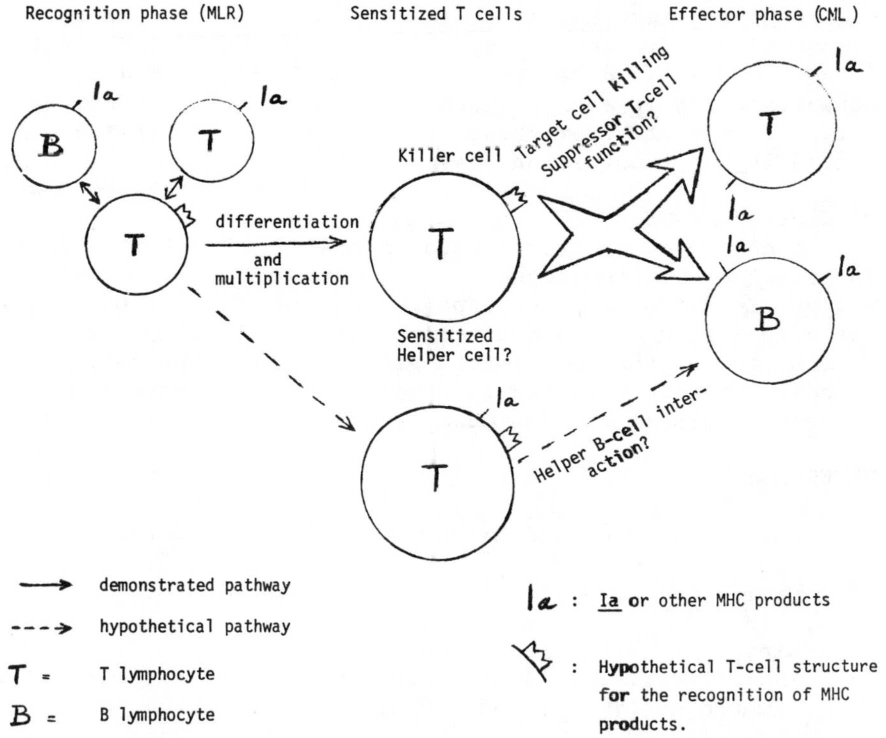

Figure 1. Hypothetical Scheme for Lymphocyte Interactions.

determinants present on Ia-positive, allogeneic T and/or B lymphocytes (recognition phase). The nature of the receptor involved in this recognition process is not known. It may be the Ir gene product, or it may be the product of some other H-2 or non H-2 gene. The responder cells proliferate and differentiate into killer cells capable of lysing T and B lymphocytes (effector phase). The sensitized killer T cells do not express Ia determinants, and their activity is directed against the Ia and H-2 gene products. The receptor involved in the effector phase may be identical to that involved in the recognition phase.

ACKNOWLEDGEMENTS

The continuous critical encouragement of Dr. H.O. McDevitt is gratefully acknowledged. The contribution of Dr. D.B. Murphy, in discussing the experiments and ideas as well as in the preparation of the manuscript, was invaluable. The fourth part of the experiments presented here was performed in collaboration with Dr. T.L. Delovitch. The excellent technical help of Mr. Ken Odom is gratefully acknowledged.

REFERENCES

1. Shreffler, D.C. and David, G.S., Adv. in Immunology, in press.
2. David, C.S. and Shreffler, D.C., Transplantation 18: 313, 1974.
3. Klein, J., Bach, F.H., Festenstein, H., McDevitt, H.O., Shreffler, D.C., Snell, G.D. and Stimpfling, J.H., Immunogenetics 1: 186, 1974.
4. Shreffler, D.C., David, C.S., Götze, D., Klein, J., McDevitt, H.O. and Sachs, D., Immunogenetics 1: 189, 1974.
5. McDevitt, H.O., Deak, B.D., Shreffler, D.C., Klein, J., Stimpfling, J.H. and Snell, G.D., J. Exp. Med. 135: 1259, 1972.
6. Benacerraf, B. and McDevitt, H.O., Science 175: 273, 1972.
7. Lieberman, R., Paul, W.E., Humphrey, W. Jr. and Stimpfling, J.H., J. Exp. Med. 136: 1231, 1972.
8. Melchers, I., Rajewsky, K. and Shreffler, D.C., Eur. J. Immunol. 3: 756, 1973.
9. Kindred, B. and Shreffler, D.C., J. Immunol. 109: 940 1972.
10. Katz, D.H., Hamaoka, T., Dorf, M.E. and Benacerraf, B., Proc. Nat. Acad. Sci., U.S.A. 70: 2626, 1973.
11. Katz, D.H., Hamaoka, T. and Benacerraf, B., J. Exp. Med.

137: 1405, 1973.
12. Shevach, E.M. and Rosenthal, A.S., J. Exp. Med. 138: 1213, 1973.
13. Bach, F.H., Widmer, M.I., Bach, M.L. and Klein, J., J. Exp. Med. 136: 1430, 1972.
14. Meo, T., David, C.S., Nabholz, M., Miggiano, V. and Shreffler, D.C., Transplant. Proc. 5: 377, 1973.
15. Klein, J. and Park, J.M., J. Exp. Med. 137: 1213, 1973.
16. Klein, J., Hauptfeld, M. and Hauptfeld, V., Immunogenetics 1: 45, 1974.
17. David, C.S., Shreffler, D.C. and Frelinger, J.A., Proc. Nat. Acad. Sci., U.S.A. 70: 2509, 1973.
18. Dickler, H.B. and Sachs, D.H., J. Exp. Med. 140: 779, 1974.
19. Lonai, P. and McDevitt, H.O., J. Exp. Med. 140: 1317, 1974.
20. Nabholz, M., Vives, J., Young, H.M., Meo, T., Miggiano, V., Rijnbeck, A. and Shreffler, D.C., Eur. J. Immunol., in press, 1974.
21. Lightbody, J.F., Bernoco, D., Miggiano, V. and Ceppellini, R., G. Batteriol. Virol. Immunol. 64: 243, 1971.
22. Julius, M.H., Simpson, E. and Herzenberg, L.A., Eur. J. Immunol. 3: 645, 1973.
23. Hämmerling, G.J., Deak, B.D., Mauve, G., Hämmerling, U. and McDevitt, H.O., Immunogenetics 1: 68, 1974.
24. Reif, A.E. and Allen, J.M.V., Nature 209: 521, 1966.
25. Boyse, E.A., Old, L.J. and Chouroulinkov, I., Meth. Miol. Res. 10: 39, 1964.
26. Sachs, D.H. and Cone, J.L., J. Exp. Med. 138: 1289, 1973.
27. Lonai, P. and McDevitt, H.O., J. Exp. Med. 140: 977, 1974.
28. Bechtol, K.B., Wegman, I.G., Freed, J.H., Grumet, F.C., Chesebro, B.W., Herzenberg, L.A. and McDevitt, H.O., Cellular Immunol. 13: 264, 1974.
29. McDevitt, H.O., Bechtol, K.B., Freed, J.H., Hämmerling, G.J. and Lonai, P., Ann. Immunol. (Inst. Pasteur) 125c: 175, 1974.
30. McDevitt, H.O., Bechtol, K.B., Hämmerling, G.J., Lonai, P. and Delovitch, T.L., in "The Immune System, Genes, Receptors, Signals" (ed. by Sercarz, E.E., Williamson, A.R. and Fox, C.F.), p. 597, 1974.
31. Harrison, M.R. and Paul, W.E., J. Exp. Med. 138: 1602, 1973.
32. von Boehmer, H., J. Immunol. 112: 70, 1974.
33. Meo, T., Vives, G., Rijnbeck, A.M., Miggiano, V.G., Nabholz, M. and Shreffler, D.C., Transpl. Proc. 5: 1339, 1973.
34. Fathman, C.G., Handwerger, B.S. and Sachs, D.H., J. Exp.

Med. 140: 835, 1974.
35. Delovitch, T.L. and McDevitt, H.O., Immunogenetics, in press, 1974.
36. Potworowsky, E.F. and Nairn, R.C., Nature 213: 597, 1967.
37. Salerno, A., Courcon, F. and Grabar, P., Ann. Inst. Pasteur (Paris) 112: 38, 1967.
38. Alter, B.J., Schendel, D.J., Bach, M.L., Bach, F.M., Klein, J. and Stimpfling, J.H., J. Exp. Med. 137: 1303, 1973.
39. Frelinger, J.A., Niederhuber, J., David, C.S. and Shreffler, D.C., J. Exp. Med. 140: in press, 1974.
40. Wofsy, L., Baker, P.C., Thompson, K., Goodman, J., Kimura, J. and Henry, C., J. Exp. Med. 140: 523, 1974.

DISCUSSION FOLLOWING SYMPOSIUM
GENETIC CONTROL OF IMMUNOCOMPETENT CELL INTERACTIONS

McDevitt - Stanford: I would interpret your data to say that you can separate an alloantigen function from an Ir function.

Bluestein - La Jolla: That is right.

McDevitt - Stanford: Why does that lead you to postulate three molecular products: an Ir product, an interaction product and a receptor product?

Bluestein - La Jolla: Nothing in my studies suggests that the Ir gene product isn't the receptor. That is other people's work. I am drawing on their data.

Dickler - Bethesda: In the mouse the Ia antigens are very polymorphic. You would have only the 2 and 13 in your outbreds. Is it possible that there are other alloantigens besides 2 and 13?

Bluestein - La Jolla: There is a whole other system of alloantigens in the guinea pig which has been developed in Dr. de Weck's laboratory in Berne and over the past year collaborative studies between the two labs along with Dr. Andrew Dempsey in Berne have really characterized these antigens. The so-called "BCD 'series" of guinea pig alloantigens appear not to be involved in macrophage-T cell collaboration. It is the 2-13 like antigens that are involved. There may be other Ia antigens or other antigens like 2-13 antigens which we will probably be able to identify serologically.

Shevach - Bethesda: In the outbreds there is an absolute correlation between 2 antigen positivity and response to DNP-GL. In other words we at the moment don't find outbred animals that have the 2 antigens and do not respond to DNP-D-GL and vice versa. There is thus a lack of polymorphism in the guinea pig and we can use outbred animals to get these results.

Paul - Bethesda: One of the critical experiments you presented was the demonstration that a given dose of the monovalent antibody fragment was equivalent to that of the divalent antibody fragment in its capacity to block cytotoxicity. One critical difference exists however between antibody mediated cytotoxicity and antigen dependent activation of cells and that is the time required for the two activities. In the latter, presumably antigen and receptor must be together for a long period of time; in the former antibody and antigenic site contact is short. Moreover, one would anticipate that the dissociation rate of univalent antibody would be very much more rapid than divalent, so what would be very interesting to know is if you extend the period of time for your cytotoxicity tests to a very much longer period, will any difference develop in the blocking activity of these two agents?

Bluestein - La Jolla: I have not done that experiment. I think that is a very good point.

de Weck - Berne: The sera you are working with, which are 2 anti-13 and 13 anti-2, are by no means monospecific. We will show some data this afternoon which indicate that you have on the one hand antisera against what may be Ir gene products and on the other hand antisera against histocompatibility antigens which may be the equivalent of the K and D region antigens of mice. It seems to us that antisera directed against histocompatibility antigens in mouse also do inhibit response to some antigens, so that I think the interpretation of the mechanism of inhibition is still somewhat open to discussion.

Bluestein - La Jolla: I think first of all that there are clearly at least two different types of suppression that one can see with antihistocompatibility sera, and one is a general suppression of lymphocyte function. You can suppress mitogenic responses and you can suppress non-specifically with antihistocompatibility sera. On the other hand there is the very specific Ir gene-linked kind of suppression that we are talking about which does not have that broad specificity. I think that is very true that there are polyvalent antisera but the absorption experiments from Dr. Shevach's lab, absorbing with the strain 2 positive cells which have everything but the Ir gene-related antigens, that is a B cell line from strain 2 animals, show that such cells very effectively absorb out this suppressive activity.

Janeway - Bethesda: I do not think you specified how you did your cytotoxic assays and my concern is that it appeared that you are only killing some of your cells at "maximum" killing. Is it possible that your fragments are going on to totally different cells and might have very different affinities on cells that are not even being picked up in the cytotoxic assay and that those other cells are the ones you are measuring in your inhibition assay?

Bluestein - La Jolla: I think that that is very possible, certainly what we are measuring in the cytotoxicity assay, the chromium release cytotoxicity assay, is essentially a B cell that is being killed. What we are suppressing in vitro is a T cell.

Herzenberg - Stanford: If you can't see helper function in the presence of suppressors, how do you know that helpers aren't generated in the animals in which you generate suppressors?

Kapp - Boston: I am not sure that every time you have suppressor cells you do not see helper cell function. Certainly in this data, in a primary immune response that is true; but, we do have cases in which suppressor cells do not suppress an immune response mediated by GAT primed T cells.

Herzenberg - Stanford: The question was really, are you sure whether or not you do have helpers generated at the same time as suppressors?

Kapp - Boston: Well if they are generated at the same time, and you measured them with an F_1, B cell, then they are equally susceptible to the suppressor. This does not say that they are not there, just that we do not have those data.

Simpson - Middlesex: I would like to ask about some of the controls that you have used and about the nature of your stimulating cell both before irradiation and afterward. I ask this because in doing similar sorts of experiments we have found that in a variety of strain combinations, some of which certainly overlap your Ia specificities, we are unable to get stimulation with peripheral, nylon column purified T cells. One can get what looks to be an MLR if one uses improperly irradiated cells. I wonder whether in fact you have confirmed the effect of your irradiation in every experiment by using PHA stimulation of your irradiated cells to make quite sure of your effects?

Lonai - Stanford: No, we have not.

PANEL DISCUSSION

Janeway - Bethesda: I have a question for Dr. Kapp. You said that if you put GAT on macrophages from nonresponders, they would stimulate spleen cells from responders. If you take B cells from responders and nonresponders and T cells from F_1's primed with mBSA-GAT you showed that both types of B cells will stimulate with mBSA-GAT, but that the only responders stimulate with GAT or with mBSA-GAT plus GAT. Would you interpret that?

Kapp - Boston: Under these culture conditions B cells from responders and nonresponders seem to be different in their ability to be suppressed, or probably tolerized, by those doses of soluble GAT. We initially thought that was what we would find in vivo, but with the low dose that it takes (we can induce suppressor in vivo with 1 to 10 micrograms of GAT) that is not the case.

Janeway - Bethesda: But in this case there is no source of nonresponder T cells so that the suppressive effect must be on the B cell.

Kapp - Boston: That is correct. I would say that their difference exists in both places.

Janeway - Bethesda: It seems to me that it cannot be a cooperative difference. It must be a difference in the B cell receptors for GAT.

Kapp - Boston: Correct.

Janeway - Bethesda: Have you tried it with a second marker linked with the GAT so that you could examine B cells which have a different receptor, say by using something like DNP-GAT to see if in this instance the free GAT no longer blocks the response?

Kapp - Boston: No, but it's a good idea.

Feldmann - London: I would like to ask Peter Loani a question. Were your experiments really based on the assumption that nylon wool purified populations do not have macrophages? Have you really checked how many macrophages there are on these suspensions and how active macrophages may be inducing MLRs and cytotoxic responses?

Lonai - Stanford: If you add macrophages to these cells you will not regain the response.

Feldmann - London: And you get no effect at all?

Lonai - Stanford: No, there was no significant return of

the response.

Siegel - New York: I may be missing something about the techniques which I would appreciate if Dr. Shevach or others would answer. We know that in the presence of soluble antigen, if we want lymphocytes to transform, they need cooperation with syngeneic macrophages. Now since the lymphocytes are not familiar with the intention of the investigator, and if they would be transformed as we heard yesterday from the paper of Greineder and Rosenthal by allogencic macrophages alone, how could we know that they were not transforming due to allogeneic macrophages in response to soluble antigen. Allogeneic macrophages are present in both cases.

Shevach - Bethesda: In all of these cultures I avoided the mixed lymphocyte reaction for the most part. In every one of the cultures where we mix allogeneic macrophages with lymphocytes there is a mixed lymphocyte culture. It is much less pronounced than the data that Greineder presented yesterday, because in the guinea pig the maximum MLR is seen at six days and assays that I am doing are all done at three days. Dr. Rosenthal and I did the appropriate controls to rule out the possibility that the MLR inhibits antigen induced proliferation. I think that is effectively ruled out.

Wedner - St. Louis: A question to Peter. If I understood you right you stated that PHA blasts are Ia positive. On which line of evidence are you basing this conclusion?

Lonai - Stanford: The only evidence is that the response against them was suppressed by anti-Ia.

Thorsby - Oslo: Do pure anti-H2D and H2K antisera exist now? If so have these antisera been used in inhibition studies of MLC reactions? The reason I ask this question is that in some human studies we have obtained evidence that seems to indicate that anti-HLA antibodies seem preferentially to inhibit a responder cell in MLC and not the stimulating cell. If we use the same antibodies against effect cells in CML tests we cannot find an inhibition. Have you done similar studies in the mouse?

Lonai - Stanford: I have not used them in the MLR. In the CML, anti-H2K suppresses the target cell and not the responder and anti whole H-2 had a small but significant effect on the responder.

Simpson - Middlesex: We have another discrepancy in data because I find that splenic macrophages that are defined by adhering to plastic and lack of response to LPS and PHA are extraordinarily good stimulators for both MLC and cytotoxicity.

Lonai - Stanford: I have not worked with splenic macrophages.

Handwerger - Minneapolis: In the data in which you show T cells being a stimulator and their response being removed by anti-Ia plus complement treating of the stimulator population, what was the source of those T cells, thymocytes or nylon wool?

Lonai - Stanford: I think that slide showed nylon wool purified T cells. But we also did it with thymocyte cells.

Wedner - St. Louis: I would have a similar type of question for Dr. Lonai. You have demonstrated very nicely that by treating your stimulator cells with anti-Ia plus complement diminish the capacity to stimulate. Now if one goes along with the data which are known, anti-Ia kills easily B lymphocytes but not T lymphocytes. In my mind there is some kind of contradiction. If you remove the B cells by anti-Ia plus complement you decrease stimulating capacity. Yet, on the other hand you have a normal stimulating capacity by nylon purified T lymphocytes. How do you explain this type of result?

Sachs - Bethesda: A related question is why was complement necessary if you can't see any killing of the T cells with anti-Ia plus complement?

Lonai - Stanford: I honestly can say that I do not know. I could come up with several suggestions but they have no proof.

Fathman - Bethesda: I would like to respond to Dr. de Weck's question, if I may. In blocking experiments that Dr. Schwartz and I have carried out, and which we will talk about later this afternoon, we have been able to effectively block MLC stimulators with antisera directed toward that I region whereas we diminish but do not obliterate the responder population with antisera directed toward the entire MHC in mice.

Sachs - Bethesda: Both Dr. Shevach and Dr. Bluestein use the same example of outbred guinea pigs which, if I am correct, are 2 negative but GA positive. In one case anti-2 does not eliminate GA responsiveness and in the other case lymphocytes from such animals don't collaborate with 2 positive macrophage. You have both interpreted your data to indicate that there are separable cell surface products. Now if there were really only one product but it was polymorphic, and the GA gene in the outbred were different than the GA gene of the strain 2 animal, then these experiments would not be conclusive.

Shevach - Bethesda: That is the major worry, and Dr. Bluestein actually had some data a number of years ago that suggested that that might be the case. You can then go back and ask the same thing in the mouse. Is the TG-AL gene in one strain of mouse the same as the TG-AL gene in another strain of mouse? It could be that complicated.

Bluestein - LaJolla: Yes, I think that is a worry and the data I have presented in the past have looked at the response to GA with decreasing antigen dose. We found that GA positive strain 2 positive cells lost their ability to respond to GA much sooner than the GA positive strain 2 negative cells. That is, it appeared to be a difference in the affinity of the GA receptor. That does not necessarily mean that the GA genes are different. By interactions between the histocompatibility antigens and perhaps the antigen receptor, the conformation may be changed just a little bit while the basic molecule may not be changed at all.

Sachs - Bethesda: But if they were different the only relevant antiserum would be the one made against that particular outbred guinea pig.

Bluestein - La Jolla: So far neither Ethan nor myself have been able to block GA responses with antisera made against those GA positive strain 2 negative cells.

SUMMARY OF SYMPOSIUM
GENETIC CONTROL OF IMMUNOCOMPETENT CELL INTERACTIONS

David H. Sachs
National Institutes of Health, Bethesda, Maryland

It is clear from the papers presented in this session and from the lively discussion which they generated, that despite many recent advances, we do not yet have all of the information we will need to discern the mechanism of genetic control of immunocompetent cell interactions. Three of the papers presented (Drs. Shevach, Bluestein and Kapp) relied heavily on the use of antigens to which genetic control of responsiveness has been described. Such antigens have played, and will undoubtedly continue to play, a major role in determining the mechanism of this control. One might ask why one would study responses to such unusual molecules rather than to ordinary antigens if one were seeking to understand general mechanisms of cell interactions. The reasoning lies in the fact that, just as in diseases related to inherited metabolic deficiencies, the study of abnormal responses can provide insight into normal mechanism in a biologic system.

For most common antigens all strains of mice are capable of mounting an adequate immune response. It is only when one tests rather restricted antigens, that one detects deficient responses in some strains. Unless there is a deficient response to a particular antigen in at least one strain of mice there will be no phenotype permitting the definition of an Ir gene. This is because genetic mapping studies require a discernible phenotypic marker, the segregation of which can be followed in progeny studies. However, there is no reason to expect that the pathways by which immune responses to the more common antigens are mediated should differ mechanistically from those mediating responses to these restricted antigens. It seems more reasonable to expect that deficient responders have a defect in some component of the normal pathway of immune response and that it is the lack of alternative pathways for response to unusual antigens that permits the definition of an Ir gene for that antigen. Thus, in studying the mechanism of Ir gene control one can hope to unravel the complex interactions which are involved in the normal control of immunocompetent cell interactions.

It is undoubtedly more than fortuitous that Ir genes of a variety of species have now been found closely linked to the major histocompatibility complex (MHC). Since the histocompatibility antigens are cell-surface molecules and since cooperation between different cell types in the immune response is presumably mediated via cell surface components, the notion that molecules determined by this genetic region are involved in cell cooperation has gained wide acceptance. The recently described Ia antigens, which are cell surface molecules determined by genes in the same region within the MHC as that implicated in Ir control, have received special attention in this regard because of their more restricted tissue distribution. The hypothetical models for cellular interactions which have been presented during this session and in other sessions of this meeting will be useful if they serve to integrate diverse experimental data and to suggest relevant new experiments to further elucidate the mechanism of control of these interactions.

The final paper in this session (Dr. Lonai) assessed the relative roles of products of the major histocompatibility locus in allogeneic interactions between immunocompetent cells. The clear requirement for different cell types in such interactions lends support to the concept that allogeneic reactions may be an analog of physiologic interactions between different immunocompetent syngeneic cells in regulation of the immune response. Such studies may thus also help to elucidate the mechanisms of immune regulation. In addition, they may shed light on the functional significance and the very reason for existence of the major histocompatibility complex.

SECTION IX

REGULATION OF THE IMMUNE RESPONSE: CELLULAR MECHANISMS

Chairman

WILLIAM E. PAUL

National Institutes of Health
Bethesda, Maryland

Co-Chairman

CARL W. PIERCE

Harvard Medical School
Boston, Massachusetts

REGULATION OF INTERACTION OF IMMUNOCOMPETENT CELLS

William E. Paul
Laboratory of Immunology, National Institute of Allergy and Infectious Diseases, National Institutes of Health, Bethesda, Maryland 20014.

One of the most interesting and important problems of contemporary immunology is to delineate the mechanisms by which antigen-specific activation of lymphocytes is achieved and by which such activation is regulated. A large body of work performed over the last several years has suggested that, in many cases, specific interaction of antigen with cell surface receptors on lymphocytes is insufficient to activate such cells to mediate the full range of specific immune responses. In particular, the activation of proliferative responses of thymus-dependent (T) lymphocytes appears to be most efficient if antigen is presented on the surface of syngeneic macrophages; similarly, the capacity of "T-dependent" antigens to stimulate antibody production by thymus-independent (B) lymphocytes requires (or is enhanced by) the activity of specific T lymphocytes.

The symposium to be presented today will deal primarily with the latter problem. We will consider the mechanisms by which T lymphocytes and their products aid in (or are required for) the activation of specific B lymphocytes. As a prelude to this discussion, I will briefly review some of the principal theories put forth to explain T-B cell collaboration and point out some of the problems which must be faced in choosing among them and in fully understanding their implications.

Perhaps the earliest proposal concerning the mechanism of T-B collaboration was the "antigen-focusing" theory (here referred to as the direct antigen-focusing theory)(Table I). This was suggested by Mitchison, Rajewsky and Taylor (1) in their attempts to explain the phenomenon of cell collaboration in hapten-carrier systems. It had previsouly been observed, both in vivo and in transfer experiments, that anamnestic anti-hapten antibody responses were most efficient if the secondary challenge employed the hapten-carrier conjugate used for primary immunization. Carrier specificity in the elicitation of secondary anti-hapten antibody responses appears to reflect a paucity (?absence) of functional hapten specific T lymphocytes (2). Therefore, T lymphocyte recognition of a hapten-carrier complex depends principally upon the presence of T lymphocytes bearing receptors specific for the carrier. Furthermore, for an optimal anti-hapten response to occur,

TABLE 1

DIRECT ANTIGEN-FOCUSING THEORY

(Mitchison et al.)

1. Carrier effect in collaboration in hapten carrier systems is explained by T lymphocyte binding antigen through carrier specific receptors and concentrating or "focusing" antigen on its surface so that the haptenic determinants of the hapten-carrier complex can be bound by B lymphocytes with hapten-specific receptors.

2. Objection: T lymphocytes appear to bind antigen less efficiently (or in smaller quantities) than do B lymphocytes making the T lymphocyte an unlikely antigen-concentrating cell.

Table 2

INDIRECT ANTIGEN-FOCUSING THEORY

(Feldmann et al.)

1. T lymphocytes postulated to bind antigen through receptor of "IgT" class and to release complexes of antigen and IgT.

2. IgT-antigen complexes appear to be highly cytophilic for macrophages and when complexes bound to macrophages are encountered by specific B lymphocytes, the latter cells become activated.

3. Free (non-macrophage bound) IgT-antigen complexes are highly tolerogenic for B lymphocytes explaining both specific tolerance and antigenic competition.

4. Objections: Difficulty in demonstrating IgT and proving it is a T cell product; difficulty in explaining genetic constraints in lymphocyte collaboration.

the hapten and the carrier to which the animal had previously been primed must be presented as a single molecular complex. A straight forward explanation for this phenomenon is that the hapten-carrier complex is initially bound by carrier specific receptors on the T cell surface and that this binding serves to increase the local antigen concentration to the extent that B lymphocytes could bind sufficient antigen to their surface receptors to be triggered. This theory in its simplest form requires the T lymphocyte to be a much more efficient binder of antigen than the B lymphocyte. The observations that specific B lymphocytes are extremely efficient in binding antigen and that T lymphocyte binding of antigen is relatively difficult to demonstrate (3), suggests that the concept that the role of the T lymphocyte is simply to concentrate antigen on its surface is an oversimplification.

Partly because of this difficulty, Feldmann and his colleagues developed an alternative thesis, which may be termed the indirect antigen-focusing theory (4) (Table 2). They proposed that T lymphocytes bore surface immunoglobulin, most probably of a unique class (IgT), that this immunoglobulin functioned as an antigen-binding receptor and, after binding antigen, the resulting antigen-IgT complex was shed from the surface of the T cell. These complexes appeared to be highly cytophilic for macrophages (5) and the encounter of specific B lymphocytes with antigen-IgT complexes bound to macrophages was proposed to result in the activation of B lymphocytes. Furthermore, Feldmann and his colleagues provided evidence that when macrophages were absent from an in vitro system, shed complexes of IgT and antigen were highly tolerogenic for B lymphocytes (6), presumably because the direct binding of such complexes to the receptors of B lymphocytes provides a tolerogenic signal whereas the same complexes, presented on the surface of macrophages, provides an activation signal. Finally, Schrader and Feldmann pointed out that such a thesis could also explain the phenomenon of antigenic competition (7) in that complexes of IgT and antigen, resulting from the activation of one set of specific T lymphocytes, might saturate all macrophage sites which bind antigen-IgT complexes. When a second antigen was added to the system, complexes of IgT and the new antigen would be unable to bind to the already saturated macrophage binding sites but would, instead, interact directly with receptors of B lymphocytes rendering the latter unresponsive. Hence, an ongoing response to one antigen would inhibit a response to a second antigen and, indeed, lead to B lymphocyte tolerance to that antigen.

The Feldmann hypothesis is a most attractive one in that it quite easily accounts for a variety of the phenomena observed in immune responses. Certain objections, however, have been raised which indicate that the situation is by no means quite so straight forward. Firstly, the issue of the existence of IgT remains unresolved. The demonstration of immunoglobulin on T lymphocyte surfaces is still highly controversial (8,9), particularly so in view of data demonstrating that peripheralized T lymphocytes can passively acquire small amounts of IgM from the serum (10). Moreover, some recent studies of antigen-binding by thymocytes and T lymphocytes have indicated that anti-immunoglobulin causes little or no inhibition of such binding while anti-histocompatibility sera may cause very marked inhibition of binding (11). Similarly, many in vitro functions of T lymphocytes are blocked poorly, if at all, by anti-immunoglobulin antibodies, but efficiently by sera directed at antigens encoded in the MHC (12).

It has been pointed out, moreover, both by Feldmann and others, that the factor produced by T lymphocyte-rich cell populations and identified as IgT may not necessarily be a synthetic product of the T lymphocyte, as other cell types are present in the responding cell populations.

An alternative proposal to explain the mechanism of T-B collaboration comes from immunogenetic studies of such collaboration and from the apparent inability of the IgT hypothesis to adequately deal with the genetic constraints of lymphocyte cooperation. Early in the study of cellular collaboration in the immune response of mice to sheep erythrocytes and to other T-dependent antigens, it was observed that if bone marrow cells and thymocytes were obtained from semi-allogeneic sources, collaboration was very inefficient compared to that obtained if these cells were derived from syngeneic donors (13). It was then shown that the failure of such collaboration was related to disparity in the MHC of the donor mice (14), but an analysis of this phenomenon was seriously hampered by the fact that mixtures of histoincompatible lymphocytes often led to B lymphocyte stimulation mediated by "allogeneic effects." Katz and his colleagues developed a system, based on irradiation of helper cell populations, which ablated the capacity of these cells to mediate allogeneic effects without abolishing their specific helper activity (15). They could then test, either in vivo or in vitro, the capacity of primed helper cells to collaborate with B lymphocytes of various genetic types in the mediation of specific antibody responses. Their results indicated that helper cells and B cells derived from congenic mice with distinct histocompatibility regions failed to collaborate or collaborated inefficiently unless certain critical genetic regions within the MHC were held in common (Table 3). In particular, similarity at the

TABLE 3

MHC CONTROL OF T-B COLLABORATION
(Katz et al.)

1. Allogeneic thymocytes and bone marrow cells collaborate poorly in SRBC response.

2. Genetic analysis of collaboration between irradiated carrier primed helper cells and hapten specific B cells reveals that concordance in the I region of the mouse H-2 complex is required for efficient collaboration.

3. In MHC-linked Ir gene controlled systems, helper cells obtained from F_1 crosses of responder by non-responder parents collaborate with B lymphocytes from the responder parent much more efficiently than with B lymphocytes from the non-responder parent.

4. Problems: Biochemical and genetic mechanisms that allow self-recognition in highly polymorphic systems are required.

TABLE 4

MACROPHAGE T-LYMPHOCYTE INTERACTION
(Rosenthal and Shevach)

1. Most efficient if macrophages and lymphocytes are obtained from syngeneic or semi-syngeneic donors.

2. When parental macrophages are used to present antigen to F_1 lymphocytes, alloantisera directed at antigen shared by both cell types completely blocks T lymphocyte activation.

3. In MHC-linked Ir gene controlled systems, macrophages from responder parent are much more efficient than macrophages from non-responder parent in presenting antigen to F_1 lymphocytes.

4. Macrophage-T lymphocyte interaction appears to be controlled by an MHC encoded structure closely related to the Ir gene product.

I region of the murine MHC seemed particularly crucial for cell collaboration (16). In a recent study by Janeway et al., similarity limited to the I-A subregion of the I region was sufficient to allow efficient collaboration between primed helper cells and B lymphocytes (17).

A close parallel to these results is found in the studies of the genetic constraints of macrophage-T lymphocyte interaction, in which sharing of certain alloantigens and of Ir genes appears to be critical for T cell activation (18-20) (Table 4). These experiments suggest that products of the major histocompatibility complex, particularly of genes closely linked to immune response genes, play critical roles in the mediation of specific cellular collaboration.

A recent exciting development in the study of T-B collaboration is the report, from several groups, that soluble factors isolated from thymocytes and T lymphocytes may mediate specific interaction, at least in certain circumstances (Table 5). Thus, Taussig and his colleagues (21) have isolated a factor from "educated thymocytes" which will aid bone marrow cells to make an antibody response, in a transfer system, to the branched chain polymer (T,G)-A--L. This factor appears to be antigen-specific in that it can be removed on a (T,G)-A--L immunoadsorbent; it differs from IgT in that it fails to be removed by anti-immunoglobulin immunoadsorbents but it is removed by adsorbents bearing antibodies directed at MHC antigens encoded within the K end of the H-2 complex (K or I region antigens) (22). Tada and his associates have described a suppressive factor which is isolated from the thymus of immunized mice; this factor also appears to be antigen-specific, non-Ig in nature and to bear determinants encoded within the MHC (23).

Although the demonstration of specific factors which play important roles in cellular interactions provides a major extension of the studies of genetic regulation of cellular collaboration, it also raises certain difficulties which will require resolution. Thus, although irradiated specific helper cells fail to collaborate with DNP-specific precursors across an I region MHC barrier, soluble factors from C3H ($K^kI^kS^kD^k$) "educated" thymocytes collaborate with bone marrow cells from C3H.SW ($K^bI^bS^bD^b$) mice in response to (T,G)-A--L. This difference in capacity to collaborate across an I region barrier may be a function of the use of different antigens, different strains of mice, different collaborating cell types or the use of cells versus soluble factors. It is clear, however, that an understanding of the molecular mechanisms of collaboration will require a detailed exploration of these apparently disparate results.

TABLE 5

SOLUBLE ANTIGEN-SPECIFIC COLLABORATING AND SUPPRESSING FACTORS IN B CELL ACTIVATION
(Taussig et al.; Tada et al.)

1. Factors isolated from "educated" T cells can aid bone marrow cells in responses to certain antigens; factors isolated from thymuses of immunized animals can suppress responses in certain circumstances.

2. Factors appear to be specific; they can be removed by antigen on immunoadsorbents.

3. Factors are not removed by anti-Ig immunoadsorbents but are removed by antisera directed at MHC products.

4. In one genetically controlled system, "helper" factor obtained from non-responder ($H\text{-}2^k$) T cells collaborates with responder ($H\text{-}2^b$) bone marrow cells in initiating response.

5. Problem: The chemical nature of antigen-binding moiety and the degree of its specificity must be defined.

It is still too early to attempt a complete synthesis of the molecular and cellular basis of cell collaboration but we must soon deal with the apparent discrepancy between factors which appear to involve immunoglobulins (i.e., IgT) and those which are principally based on MHC controlled products. Unfortunately, there are major technical differences between the experiments demonstrating "IgT" factors and those yielding MHC type factors. It would obviously be critical to widen the models used in each case so that a picture of the generality of the two types of factors can be achieved. A simple resolution is that both types of mechanisms exist and are utilized to different extents in distinct systems. Nonetheless, before accepting such a compromise proposal, serious efforts to test each hypothesis in detail will be needed.

Both the indirect antigen-focusing theory and the thesis that MHC products are critical to cell collaboration have wide ranging implications. Indeed, the concept that cellular collaboration is based upon recognition of MHC antigens provides a powerful tool for the analysis of an enormous variety of phenomena in the regulation of cell interactions. This theory requires the delineation of genetic and biochemical mechanisms which allow both for a high degree of polymorphism and, at the same time, for rather precise recognition of identity (or similarity) of certain interaction structures. Moreover, implicit in the concept that specific MHC encoded cooperating factors exist is the requirement for specific MHC antigen-binding structures (?receptors). The preliminary evidence that anti-immunoglobulin antibodies do not remove the factors bearing MHC antigens indicates that these factors differ from conventional immunoglobulin molecules. The understanding of the chemical and genetic basis of this type of antigen recognition is among the most challenging and exciting of modern biological problems.

REFERENCES

1. Mitchison, N.A., Rajewsky, K. and Taylor, R.B., in "Developmental Aspects of Antibody Formation and Structure" (ed. by J. Sterzl and I. Riha), Academic Press, New York, p. 547, 1970.
2. Paul, W.E., Transplant. Rev. 5: 130, 1970.
3. Davie, J.M. and Paul, W.E., in "Contemporary Topics in Immunobiology" Vol. 3 (ed. by M.D. Cooper and N.L. Warner) Plenum Press, New York, p. 171, 1973.
4. Feldmann, M. and Basten, A., J. Exp. Med. 136: 49, 1972.
5. Feldmann, M., J. Exp. Med. 136: 737, 1972.

6. Feldmann, M., Nature New Biology 242: 82, 1973.
7. Schrader, J.W. and Feldmann, M., Eur. J. Immunol. 3: 711, 1973.
8. Marchalonis, J.J., Cone, R.E. and Atwell, J.L., J. Exp. Med. 135: 956, 1972.
9. Vitetta, E.S., Bianco, C., Nussenzweig, V. and Uhr, J.W., J. Exp. Med. 136: 81, 1972.
10. Hunt, S.V. and Williams, A.F., J. Exp. Med. 139: 479, 1974.
11. Hämmerling, G.J. and McDevitt, H.O., J. Immunol. 112: 1734, 1974.
12. Shevach, E.M., Paul, W.E. and Green, I., J. Exp. Med. 139: 661, 1973.
13. Mitchell, G.F. and Miller, J.F.A.P., J. Exp. Med. 128: 821, 1968.
14. Kindred, B. and Shreffler, D.C., J. Immunol. 109: 940, 1972.
15. Katz, D.H., Hamaoka, T. and Benacerraf, B., J. Exp. Med. 137: 1405, 1973.
16. Katz, D.H., Graves, M., Dorf, M.E., DiMuzio, H. and Benacerraf, B., J. Exp. Med., in press.
17. Janeway, C.A., Jr., Werblin, T., Lieberman, R. and Paul, W.E., unpublished observations.
18. Rosenthal, A.S. and Shevach, E.M., J. Exp. Med. 138: 1194, 1973.
19. Shevach, E.M. and Rosenthal, A.S., J. Exp. Med. 138: 1213, 1973.
20. Schwartz, R.H. and Paul, W.E., manuscript in preparation.
21. Taussig, M.J., Nature 248: 234, 1974.
22. Taussig, M.J. and Munro, A.J., Nature 251, 63, 1974.
23. Tada, T., Proceedings of the IX Leucocyte Culture Conference, in press.

EVIDENCE FOR THE CONTROL OF LYMPHOCYTE INTERACTIONS BY GENE PRODUCTS OF THE I REGION OF THE H-2 COMPLEX[1]

David H. Katz and Dieter Armerding[2]
Department of Pathology, Harvard Medical School, Boston, Massachusetts 02115

Delineation of the complex events involved in requisite cooperative interactions between thymus-derived (T) and bone marrow-derived (B) lymphocytes in the development of antibody responses to all but a selected class of antigens has been the subject of intensive analysis in many laboratories (reviewed in Ref. 1). Several hypotheses have been developed over the past years to explain the mechanism of T-B cell cooperation. These have included direct or indirect antigen presentation by T cells and the active regulation of B cell responses by either direct membrane interaction with T cells or by the interaction with soluble mediators produced and released by T cells (1). The latter hypothesis stems largely from: 1) the demonstration that T cells activated by means other than the constituent determinants of the antigen employed, such as by allogeneic cell interactions, were highly effective in providing helper function for antigen-specific B cells (2); and 2) the identification of soluble products from culture supernatants of activated T cells capable of acting directly on B cells in the development of antibody responses (3-11). More recently, it has been demonstrated that the most optimal cell interactions occur between T and B lymphocytes bearing identical histocompatibility antigens, thus adding an additional condition to consider in the mechanism of such cell interactions (12-19).

In this paper, we shall briefly review the data that have demonstrated genetic restrictions of T-B cell interactions and discuss our own recent observations on the isolation and characterization of an active T cell mediator of cooperative responses. The collective implications of the findings to be described draw us to the conclusion that the control of such cell-cell interactions involve product(s) of a gene or genes in the I region of the H-2 complex and may well be mediated

---

[1] Studies supported by Grants AI-10630 and AI-09920 from the National Institutes of Health.

[2] Research Fellow of the Damon Runyon Memorial Fund for Cancer Research, Inc.

by products of _I_ region genes located on the cell surface membranes of immunocompetent lymphocytes.

THE ROLE OF HISTOCOMPATIBILITY GENE PRODUCTS IN CELL-CELL INTERACTIONS

The major points to emphasize in this regard are the following: 1) The participation of histocompatibility gene products in cell-cell interactions reflects a physiological control mechanism utilizing readily accessible structures that are dynamic constituents of the surface membranes of lymphoid cells; and 2) Like other physiologic control mechanisms, it is not an _absolute_ system in the sense of precluding any alternate, albeit less efficient, routes to the same destination.

The evidence for the existence of genetic restrictions in T-B cell interactions can be succinctly summarized as follows: Under conditions in which isogeneic or syngeneic lymphoid cells interact together to develop humoral immune responses, cells lacking certain critical identities in histocompatibility genes fail to successfully interact (12-19). The succinctness of the preceding statement must be elaborated upon to include the very important qualifications that: 1) demonstration of a "failure" to obtain a response in the system employed in our studies has been shown _not_ to reflect a suppression phenomenon (16) and 2) the positive responses obtained are those which are elicited by interactions between antigen-specific T and B cells and not by allogeneic cell interactions. The importance of the latter statement stems from the fact that in our first experiments on this question, we made the totally unexpected observation that a heavily irradiated mouse possesses sufficient numbers of active residual T cells to exert an enhancing allogeneic effect on a small population of adoptively transferred histoincompatible primed B lymphocytes (20). The central problem was, therefore, to design an experimental scheme that specifically circumvented the possible contribution to the results of a complicating allogeneic effect. This was accomplished for in vivo cell transfer studies by using an F_1 hybrid as the recipient of T and B cells from the respective parental strains against which the semiallogeneic host would be genetically incapable of reacting. The intricacies of the experimental protocol employed have been described in detail elsewhere (13,14).

The protocol and data from a representative experiment using congenic-resistant mouse strains demonstrating the involvement of the _H-2_ gene complex in physiologic lymphocyte interactions is shown in Fig. 1. The DNP-primed B cells were derived from congenic-resistant B10.A (_H-2^a_) donor mice and

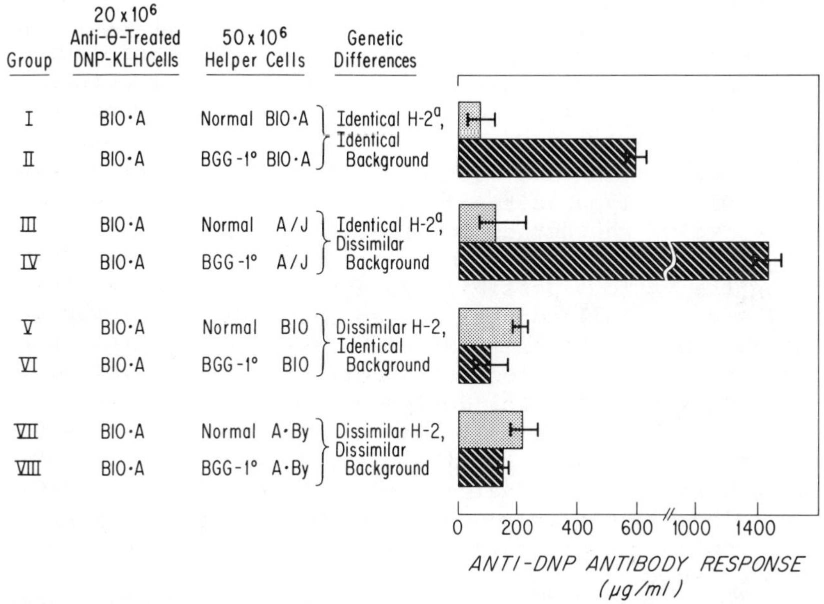

Fig. 1 Failure of physiologic cooperative interactions to occur between T and B lymphocytes differing at the major histocompatibility locus. The scheme followed is described in text. Recipients for all cell combinations were (A x B10)F_1 hybrids. Combinations and strain origins of T and B cells and the relevant genetic differences are indicated. Recipients in groups I-VIII were secondarily challenged with 50 μg of DNP-BGG. Mean serum anti-DNP antibody levels of groups of five mice on day 7 after secondary challenge are illustrated. Horizontal bars represent ranges of the standard errors (Ref. 14).

the relevant genetic similarities and/or differences are listed for each combination (14). Groups I and II demonstrate the intact cooperative functional capacities of the irradiated (in situ) bovine gamma globulin (BGG)-primed T cells and the anti-θ-treated DNP-primed B cells of syngeneic B10.A origin within the environs of (A x B10)F_1 irradiated recipients (Group I). Similarly, BGG-primed T cells derived from A/J donors, which are identical with B10.A at the major $\underline{H-2}^a$ locus but dissimilar with respect to background genotypes are capable of exerting a clear helper effect in cooperating with B cells of B10.A origin. In sharp contrast, T cells from A.By or B10 donors, which are both $\underline{H-2}^b$, fail to cooperatively interact with B lymphocytes from B10.A mice. This is true irrespective of whether or not the genetic background other than H-2 is identical such as in the case of B10 donor cells.

Among the various possible explanations for the failure of physiologic T-B cell cooperation to occur across the major histocompatibility barrier certain possibilities which are trivial in terms of biological significance have been ruled out either by experimental design or by direct experimentation (13,14,16). These possibilities have been described and discussed in detail elsewhere (13,14,16,18) and will not be reiterated here. These findings led us to conclude, therefore, that a relevant gene or genes in the major histocompatibility complex control physiologic cooperative interactions between T and B cells in the immune response. Subsequent studies were designed to determine more precisely the genetic region of H-2 primarily involved.

In our initial studies (13), no cooperation occurred with mixtures of T and B cells from BALB/c ($\underline{H-2}^d$) and A/J ($\underline{H-2}^a$) donors, respectively. Since these particular strains are identical at S and the entire D end of the H-2 complex, these results indicate that gene identities only at S and D are insufficient to permit optimal cooperative interactions. In another experiment we asked the reciprocal question--i.e. whether identities at only K and I are sufficient to allow effective cooperation. This experiment was carried out in vitro using lymphocytes from A/J and B10.BR mice which differ from one another at genes in the S and D regions of the H-2 complex but are identical at K and I regions (17). As shown in Fig. 2, DNP-primed B cells from A/J donors developed effective cooperative responses with KLH-primed T cells from both isogeneic A/J donors and from B10.BR donors. In the reciprocal mixed cell cultures, DNP-primed B cells from B10.BR mice interacted with KLH-primed T cells from A/J as well as isogeneic T cells in response to DNP-KLH. These data do not reflect non-specific allogeneic effects as an explanation for

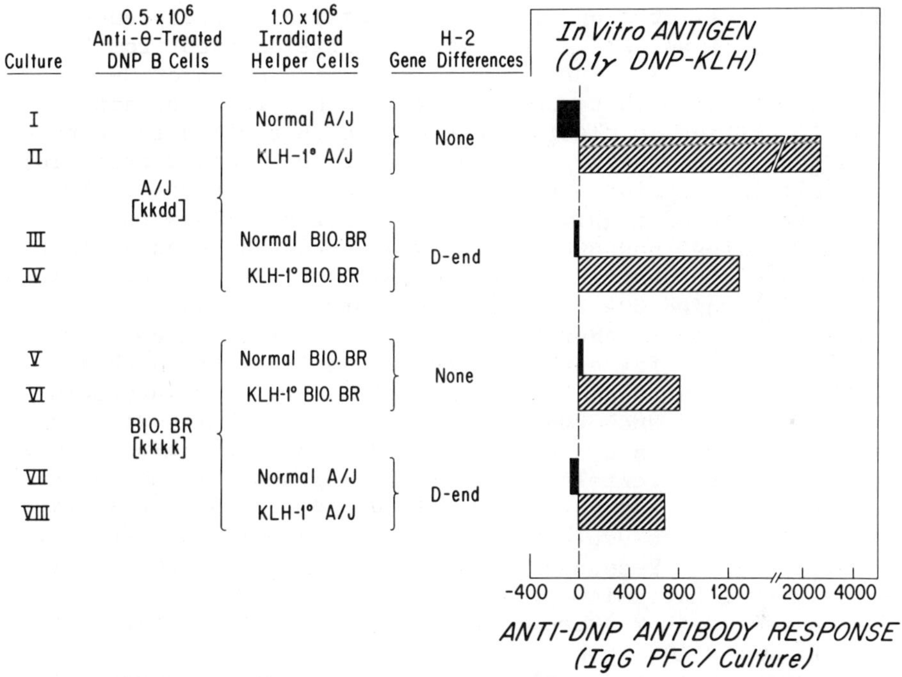

Fig. 2 DNP-ASC-primed spleen cells from A/J and B10.BR donors were depleted of T cells by in vitro treatment with anti-θ serum plus complement and then cultured with either irradiated normal or KLH-primed spleen cells in the combinations indicated. Primed donors were immunized with either DNP-ASC in alum or KLH in CFA six months earlier and then boosted with 25 μg of the respective antigen in saline one month prior to culture. Cells were cultured with either no antigen (not shown) or DNP-KLH. The background responses of non-stimulated cultures have been subtracted from the numbers of DNP-specific PFC developed in cultures containing DNP-KLH--hence, the negative values depicted here. IgG (indirect) DNP-specific PFC responses are shown. Responses in the IgM class (not shown) were parallel (Ref. 17).

successful cooperation between A and B10.BR lymphocytes since, as shown here, reciprocal controls using irradiated normal rather than KLH-primed cells failed to develop secondary responses to DNP-KLH. The development of cooperative responses between A/J and B10.BR which differ for genes in _S_ and _D_ but are identical for _K_ and _I_ region genes indicate that the critical genes involved in T-B cell cooperation exist in the latter regions.

The data described above raise the intriguing possibility that identities at the _I_ gene region either alone or together with _K_ or _D_ end identities are the critical determinants for T-B cell interaction. By taking advantage of our previous demonstration of highly effective cooperation between reciprocal combinations of parental and F_1 hybrid T and B lymphocytes when the carrier molecule employed is one to which both parental strains are genetic responders (13), we next asked the question of whether F_1 carrier-primed T cells can serve as helper cells for either or both parental B cells when: a) the response to the carrier molecule employed is under genetic control such that one parental strain is a responder and the other is a nonresponder, and b) the determinant specificity of the parental B cells being assessed is not under genetic control and bears no relationship to the specificity of the carrier molecule (15). The experimental system utilized an immune response gene controlling responses to the terpolymer L-glutamic acid-L-lysine-L-tyrosine (GLT) to which A strain mice (_H-2_^{a}) are nonresponders whereas BALB/c (_H-2_^{d}) and (BALB/c x A)F_1 hybrids (CAF_1) are responders.

As shown in Fig. 3, BALB/c (Group II) and A/J (Group VIII) DNP-specific B cells are effectively "helped" by _KLH_-specific F_1 T cells in responses to DNP-KLH. Similarly, _GLT_-primed CAF_1 T cells cooperate very well with B cells from BALB/c donors in response to either soluble or macrophage-bound DNP-GLT (Groups IV and VI). In marked contrast, however, these same GLT-specific F_1 T cells fail to serve as helper cells for DNP-specific B cells from A/J donor mice irrespective of whether soluble or macrophage-bound DNP-GLT is employed for secondary challenge (Groups X and XII). These results demonstrate, therefore, that GLT-primed CAF_1 T cells can provide for responder BALB/c, but not for nonresponder A/J, the required stimulus for the anti-DNP responses of DNP-specific B cells of these respective parental strains to the DNP conjugate of GLT (15).

The two preceding experiments permit us to draw the conclusion that the histocompatibility genes involved in controlling optimal T-B cell interactions are located to the left of the _S_ region of _H-2_--i.e. somewhere in the _K_ and/or _I_ re-

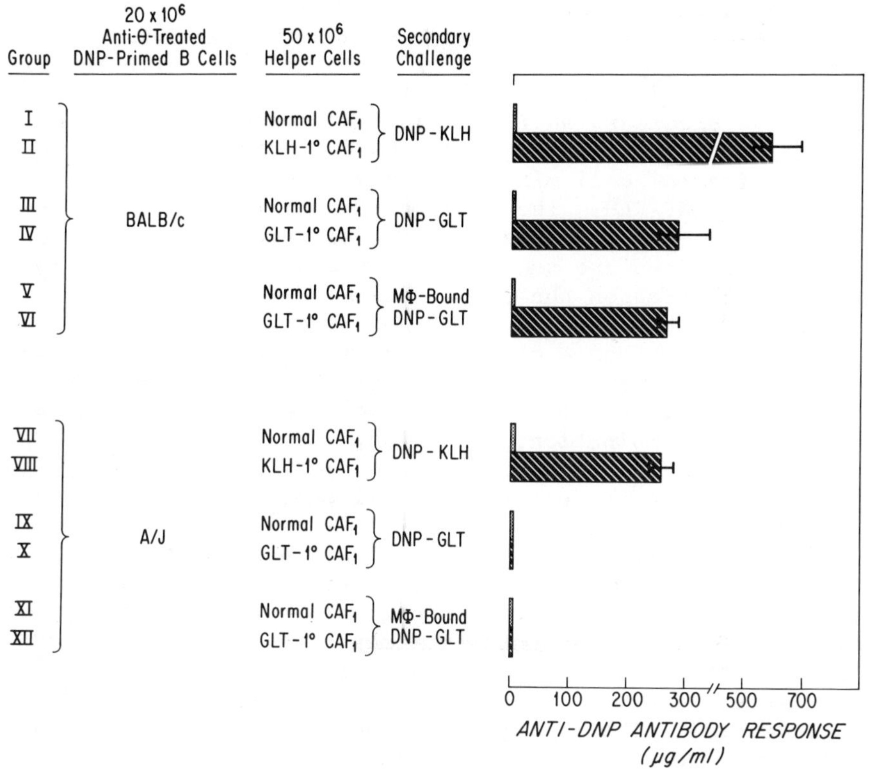

Fig. 3 Involvement of Ir gene in control of T and B lymphocyte interactions. Recipients for all cell combinations were CAF_1 hybrids. The transfer scheme is described in the text. Combinations, strain origins and specificities of T and B cells are indicated. Secondary challenge was performed intraperitoneally with either 20 μg of soluble DNP-KLH or 100 μg of soluble DNP-GLT or intravenously with 10^7 F_1 macrophages (MØ) containing 2.4 μg DNP-GLT per mouse as indicated. Mean serum anti-DNP antibody levels of groups of five mice on day 7 after secondary challenge are illustrated. Horizontal bars represent ranges of the standard errors.

gions(s). The existence of appropriate inbred and recombinant strains of mice differing at known loci of the H-2 complex makes it possible to define more precisely which region(s) and/or subregion(s) of the complex contains the relevant gene or genes. The following experiment was designed to answer this question using mixtures of T and B cells from congenic mice A.TL ($H\text{-}2^{t1}$), A.TH ($H\text{-}2^{t2}$), A.AL ($H\text{-}2^{a1}$), A.SW ($H\text{-}2^{s}$), inbred A/J ($H\text{-}2^{a}$) and (A x A.TH)F_1 ($H\text{-}2^{a/t2}$) in in vitro cooperative secondary anti-DNP antibody responses (19).

The left side of Fig. 4 depicts the protocol and various combinations of cell mixtures analyzed for cooperative responses to DNP-KLH. The gene regions of the H-2 complexes are symbolized and the gene region differences among the various combinations are summarized for convenience. The relevant data are depicted on the right side of Fig. 4. Cultures I-IV demonstrate the capacity of syngeneic mixtures of T and B cells from A.TL and A mice to cooperate in vitro, and reciprocal mixtures of such cells to interact together despite the existence of gene differences in both the K and S regions of H-2 in this combination. Likewise, when gene differences are restricted to only one of these respective regions (K or S), the capacity for effective T-B cell cooperation remains intact as evidenced by the ability of T cells from both A.TL and A donors to interact with B cells from A.AL donors (cultures V and VI). In marked contrast, however, is the inability of either of these primed and functionally intact T cell populations from A.TL or A donors to cooperate with B cells from either A.TH or A.SW mice (cultures VII, VIII, X and XI); in the latter combinations gene differences exist in the I region as indicated. The failure of either A.TH or A.SW DNP-primed B cells to respond in these cultures is not a reflection of an incapacity of these B cells to function since irradiated KLH-primed cells from (A x A.TH)F_1 donors were able to provide helper function for A.TH parental cells (culture IX) and for A.SW cells which differ only at genes in the D region (culture XII). Moreover, the capacity of A.TH primed cells to develop effective syngeneic cooperative interactions was not inhibited or diminished by the addition of irradiated KLH-primed spleen cells from A.TL donors thereby ruling out a possible suppression mechanism, due to e.g. an allogeneic effect, as an explanation for these findings. Thus, in a syngeneic cooperative control culture response, A.TH cells developed 1253 IgG anti-DNP PFC in the absence of A.TL cells and 1486 anti-DNP PFC in the presence of 1.0×10^6 irradiated KLH-primed A.TL cells (19).

The results of this experiment provide compelling evidence for the existence of the gene or genes controlling optimal

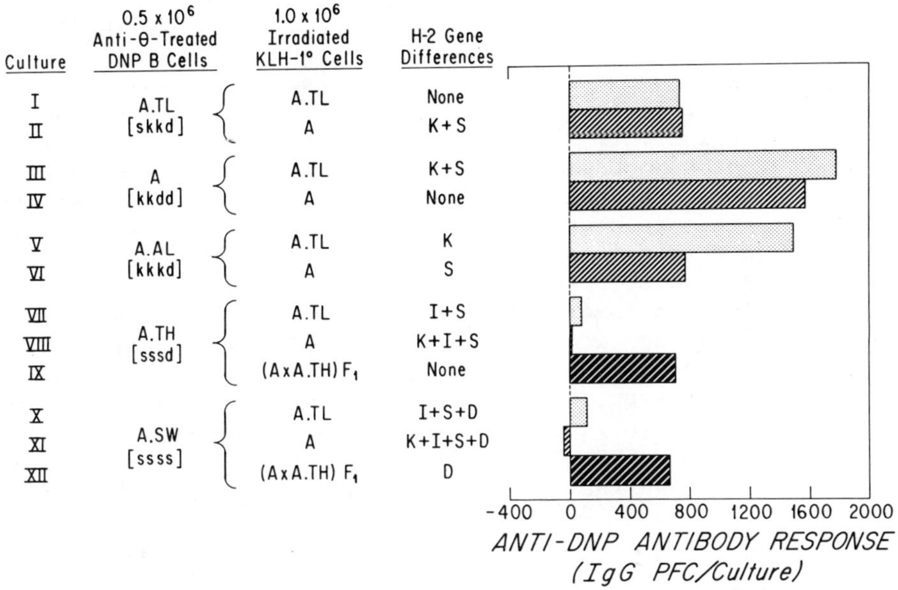

Fig. 4 DNP-ASC-primed spleen cells from A.TL, A, A.AL, A.TH and A.SW mice were depleted of T cells by in vitro treatment with anti-θ serum plus complement and then cultured with irradiated KLH-primed spleen cells from A.TL, A or (A x A.TH)F_1 donors in the combinations indicated. Cells were cultured with either no antigen (not shown) or DNP-KLH. The background responses of non-stimulated cultures have been subtracted from the numbers of DNP-specific PFC developed in cultures containing DNP-KLH (hence, the negative value depicted in culture XI). IgG (indirect) DNP-specific PFC responses are shown. Responses in the IgM class (not shown) were parallel (Ref. 19).

T-B cell cooperative interactions in the designated I region of the H-2 gene complex. This point is particularly emphasized by the inability of A.TL to provide helper T cell function for A.TH or A.SW B cells in which combinations there are gene identities in the K region but differences in I region genes.

The precise genetic mapping of these genes will require experiments using mixtures of T and B cells from strain combinations with differences and identities at one or more subregions of the I region. The I region has been divided into three subregions, Ir-1A, Ir-1B and I-C, on the basis of documented crossovers between definable Ir genes and also Ia serological specificities (21). The data presented here indicate that the genes responsible for control of T-B cell interactions in this system are located in the Ir-1A and/or Ir-1B subregions. This follows from the fact that A↔A.TL and A↔A.AL combinations, all of which are identical for all genes in Ir-1A and Ir-1B, were capable of effectively interacting despite the existence of gene differences in the I-C subregion. However, despite the fact that the I-C subregions of these strains are derived from different paternal haplotypes, they may nevertheless conceivably share genes coded for in the I-C subregion which permit effective T-B cell cooperation. Further definition of these possibilities should be forthcoming in the near future.

BIOLOGICAL, BIOCHEMICAL AND IMMUNOLOGICAL PROPERTIES OF A T CELL PRODUCT ACTIVE IN TRIGGERING B LYMPHOCYTES

A major effort in studies on the mechanism of T-B cell interactions has been focused on the identification of various biologically active substances capable of influencing lymphocyte function in antibody responses to different antigens elicited in in vitro systems. The demonstration by Dutton et al. (3) that supernatants obtained from short-term cultures of histoincompatible mouse spleen cells contained a non-antigen-specific biologically active mediator capable of markedly affecting in vitro antibody responses to thymus-dependent antigens provided evidence for the existence of such a possible mediator. In this section we will briefly review our published observations (11) and then present our most recent data on the biological, biochemical and immunological properties of an active moiety produced in supernatants of short-term (24 hours) in vitro reactions between T cells specifically activated to foreign alloantigens and the corresponding target cell population. We have termed the factor that we have been studying "allogeneic effect factor" (AEF) since its biological action on in vitro antibody responses appears to be identical to the in vivo phenomenon known as the allogeneic effect (2).

A. Biological properties of AEF

The AEF preparations that are the focus of this discussion have been prepared by culturing DBA/2 ($\underline{H\text{-}2}^d$) T cells (which had been activated for 6 days against (C3H x DBA/2)F_1 ($\underline{H\text{-}2}^{k/d}$) target lymphocytes in irradiated DBA/2 hosts) for 24 hours with irradiated (C3H x DBA/2)F_1 target spleen cells (11). The culture supernatants from such allogeneic cell mixtures are biologically active in enhancing in vitro immune responses when such supernatants are obtained as early as 12 hours after culture initiation of allogeneic cell cultures, display peak activity at 24 hours and thereafter become progressively more suppressive. We have prepared numerous AEF from various genetic mixtures of effector and target cell populations and all appear to behave in analogous fashion biologically. The AEF are active as unfractionated (crude) supernatants or, as will be discussed below, as highly purified fractions obtained by gel and/or ion exchange chromatography.

The principal biological activity of AEF that has been studied in depth is the capacity of this material to functionally replace the requirement for helper T cells in in vitro antibody responses. The earlier work of Dutton et al. (3) and Schimpl and Wecker (4,5) made it clear that such factors were active in this regard insofar as in vitro responses to particulate erythrocyte antigens or haptenated erythrocytes were concerned. Our own studies extended these observations to soluble DNP-protein conjugates in which case AEF can reconstitute helper cell function in responses of T cell-depleted primed spleen cells, but the extent to which AEF will either exhibit a reconstituting effect or permit the development of secondary in vitro anti-DNP responses to heterologous DNP-carriers is clearly determined by the form in which the specific DNP determinants are presented to the B cell (11). Thus, as shown in Fig. 5, cultures of DNP-KLH-primed lymphocytes in which the presence of AEF failed to stimulate responses to the heterologous conjugate, DNP-fowl gamma globulin (FγG), added to the culture in free soluble form, could develop very good secondary responses to the same conjugate bound to macrophage surface membranes. Moreover, the response to macrophage-bound DNP-FγG in the presence of AEF was totally abolished by addition of free soluble DNP-FγG to the same culture. It appears, therefore, that the presence of a sufficient quantity of free DNP-carrier molecules might render all or some of the DNP-specific B cells refractory to the action of AEF or, for that matter, the product(s) of carrier-

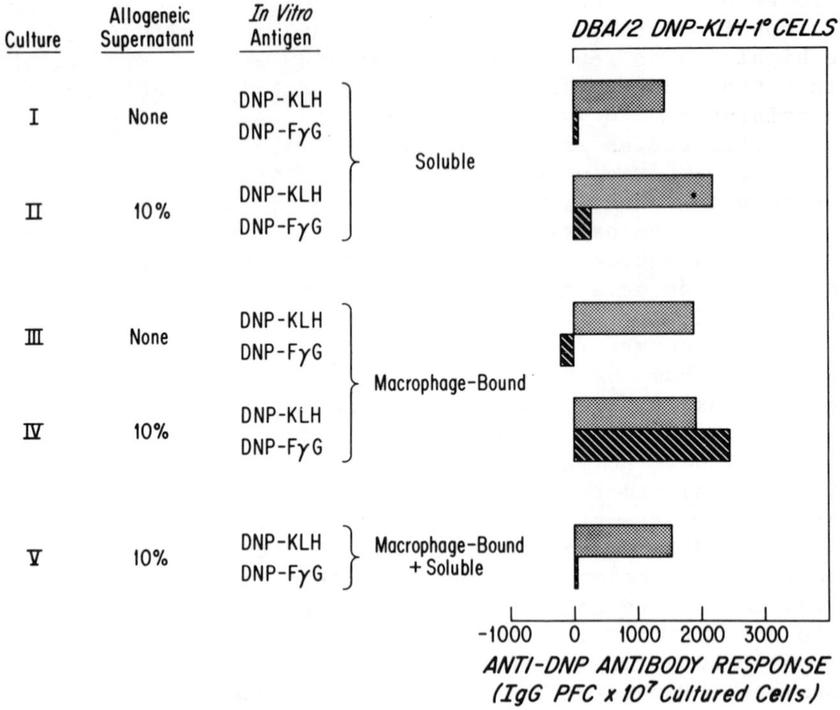

Fig. 5 The capacity of AEF to bypass the requirement for carrier-specific helper T cells in secondary in vitro anti-DNP antibody responses. Spleen cells from DNP-KLH-primed (4 months) DBA/2 were cultured at a cell density of 1.67×10^6 cells/ml with or without antigen in the presence or absence of 10% allogeneic supernatant as indicated. Soluble DNP-KLH or DNP-FγG were added at a concentration of 1 μg/ml (cultures I, II and V). Macrophage-bound antigen was added on 2.5×10^4 macrophages per well (cultures III, IV and V). The IgG anti-DNP responses after four days are shown. Background values for non-antigen stimulated control cultures have been subtracted from each corresponding experimental value for purposes of clarity (Ref. 11).

stimulated T cells if the latter are in any way different from AEF.

The strongest evidence that our active AEF acts directly on B cells stems from the capacity of AEF to stimulate B cells exposed to DNP conjugated to the D-glutamic acid, D-lysine (D-GL) copolymer (11). This compound has been demonstrated to be highly tolerogenic for DNP-specific B cells both in vivo and in vitro under normal circumstances (22-25). However, when administered to appropriately-primed animals during a critical time period after induction of an in vivo allogeneic effect, DNP-D-GL can provide a definite inductive stimulus for primary or secondary anti-DNP antibody responses (1,2,22, 26). Since no demonstrable T cell function specific for the D-GL carrier has been demonstrated, these observations provided the strongest indirect proof that the in vivo allogeneic effect is mediated by a direct interaction on the responding B cells. The capacity of AEF to permit in vitro responses to DNP-D-GL, as shown in Fig. 6, constitutes conclusive evidence, therefore, that the active moiety involved is acting directly on B lymphocytes (11).

B. Biochemical properties of AEF

The experiments performed thus far on the physicochemical features of AEF indicate that the active component(s) consists of protein and/or glycoprotein which is heat - labile (56°C, 1 hour), thereby indicating the importance of tertiary structure to activity and is in the molecular weight range of 30,000 - 45,000 (11,27). Moreover, the active moiety appears to consist of two components associated non-covalently (27).

Initially, crude allogeneic supernatants were fractionated first on Sephadex G-200 yielding multiple peaks of which only one contained active enhancing AEF (11). As shown in Fig. 7, this peak from G-200 was then subjected to chromatography on Sephadex G-100 yielding two fractions of which only one eluting in the range of an insulin marker (m.w. 36,000) was biologically active in triggering B lymphocyte responses to SRBC (11). On this basis we have estimated the m.w. of AEF to be in the range of 30,000 - 45,000.

The active fraction B (Fig. 7) from Sephadex G-100 was further chromatographed on DEAE cellulose using an exponential gradient from 0.01 - 0.15 M sodium phosphate buffer, pH 8.1. Under these conditions, all of the biologically active material eluted at low salt concentration in a second peak from the column indicating that AEF is homogeneous with regard to charge and that it could be either positive or nearly neutral (27). The active fraction of AEF obtained by DEAE-cellulose chromatography was radio-labelled with ^{125}I by the

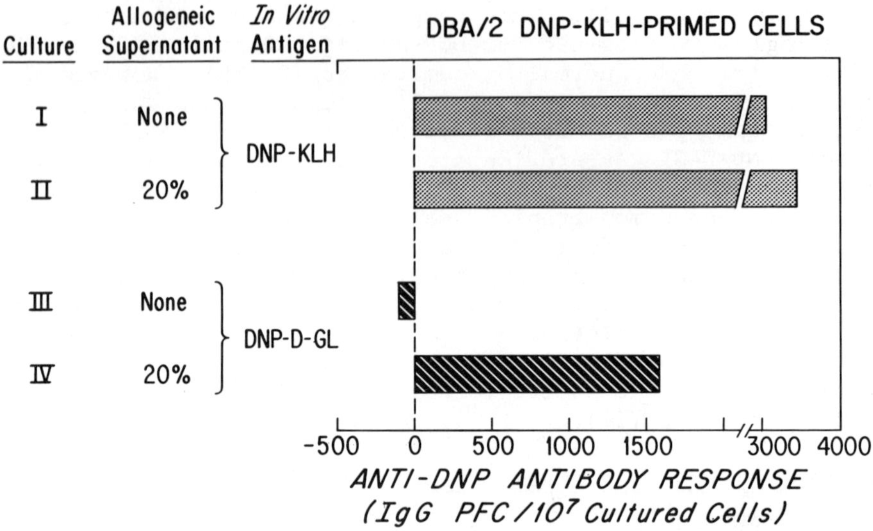

Fig. 6 The capacity of AEF to induce B cell responses to DNP-D-GL. DNP-KLH-primed spleen cells from DBA/2 mice were cultured at a density of 1.67 x 10^6 cells/ml with either DNP-KLH or DNP-D-GL (1 μg/ml) in the presence or absence of AEF (20%) as indicated. The background responses of non-stimulated cultures have been subtracted from the number of DNP-specific PFC developed in cultures containing antigen (hence, the negative value depicted in culture III). IgG (indirect) DNP-specific PFC responses are shown. Responses in the IgM class (not shown) were parallel (Ref. 11).

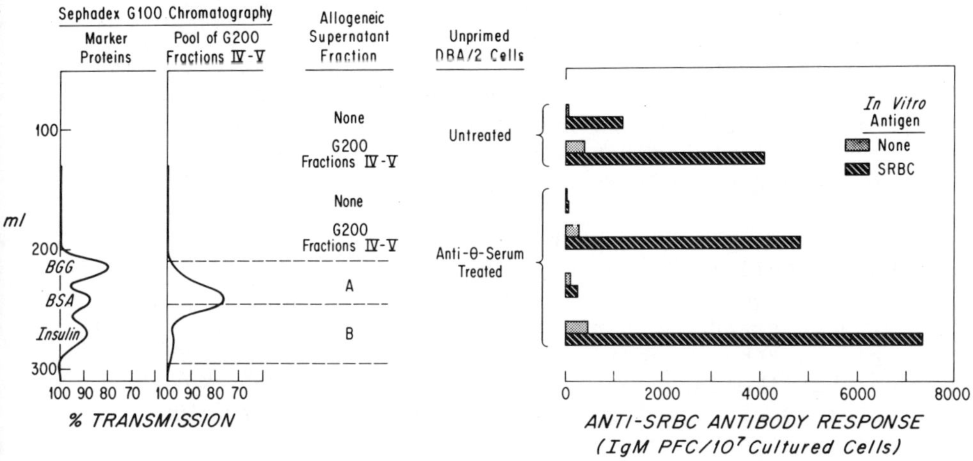

Fig. 7 Comparison of activities of unseparated and Sephadex G-100-fractionated AEF on primary anti-SRBC responses of untreated and anti-θ serum-treated DBA/2 spleen cells. Fractions IV and V from G-200 were pooled and subjected to further chromatography on G-100. The elution patterns of this supernatant and the corresponding markers BGG (m.w. 150,000), BSA (m.w. 65,000) and insulin (m.w. 36,000 in hexameric form) are shown on the far left. The primary IgM anit-SRBC responses of untreated and anti-θ serum-treated DBA/2 cells in the presence of the various fractions indicated (50% concentration in reference to the original unseparated supernatant) are shown (Ref. 11).

chloramine-T method and then subjected to electrophoresis on 10% SDS-polyacrylamide gel. The electrophoretic patterns of both unreduced AEF and AEF following reduction with 2-mercaptoethanol are shown in Fig. 8 (27). The patterns were identical and consisted of a large molecular peak which coincided with molecular weight of 47,000 and a smaller peak in the molecular weight range of 11,500 (extrapolated). The molecular weight determinations were made by comparison with parallel gels containing marker proteins with known molecular weights. It should be pointed out, however, that this by no means accurately establishes the molecular weights since the presence of carbohydrate in varying quantities will affect the migration of such a substance in SDS gel. Indeed, the large molecular weight band stains positively with periodic acid Schiff's reagent indicating that it is probably a glycoprotein (27).

The electrophoretic pattern illustrated in Fig. 8 is remarkably similar to that observed when solubilized H-2 antigens are subjected to SDS-acrylamide electrophoresis (28). We were interested, therefore, in determining the biological activity of the two definable molecular species observed. Since the likelihood of recovering biologically active material by elution from SDS-acrylamide gel is extremely low, we attempted to obtain comparable dissociation of AEF into subfractions by chromatography on Sepharose-6B in 6M guanidine HCl. The elution pattern obtained by such chromatography is depicted on the far left panel of Fig. 9 (27). The corresponding biological activities of the tested fractions (after removal of guanidine HCl by dialysis) are shown on the middle and right panels of Fig. 9. Six peaks (II-VII) of UV-absorbing material were demarcated as indicated by the dotted lines. Two heavier molecular weight fractions eluting after the void volume (fractions II and V) and two lighter molecular weight fractions (VI and VII) were tested for activity on in vitro primary anti-SRBC antibody responses of untreated (middle panel) and anti-θ serum-treated (far right panel) spleen cells from DBA/2 mice.

As shown in Fig. 9, the addition of unfractionated AEF enhanced the response of untreated spleen cells and fully reconstituted the response of spleen cells depleted of T cells by anti-θ serum treatment. Fraction II from the column substantially enhanced the response of untreated spleen cells at the 1:5 concentration (around 3-fold) but exerted only a partial effect on anti-θ-treated cells. The 1:10 concentration of fraction II and 1:5 of fraction V had no significant effect on responses of either the intact or depleted cell population. The lighter molecular weight materials in frac-

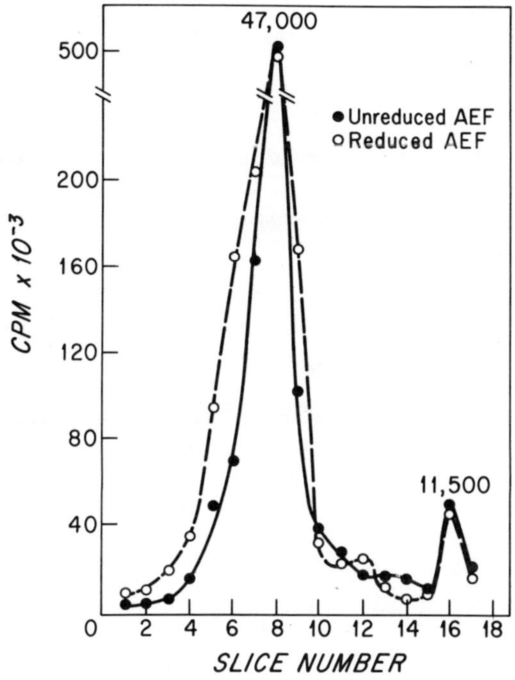

Fig. 8 Electrophoretic pattern of AEF on SDS-polyacrylamide gel. The biologically active fraction of AEF purified on Sephadex G-100 followed by chromatography on DEAE-cellulose was radio-labelled with ^{125}I by the chloramine-T method and then subjected to electrophoresis on 10% SDS-polyacrylamide gel either unreduced or following reduction by 2-mercaptoenthanol. Gels were cut into 3 mm thick slices and counted in a gamma counter. Molecular weights of the major peaks are indicated (Ref. 27).

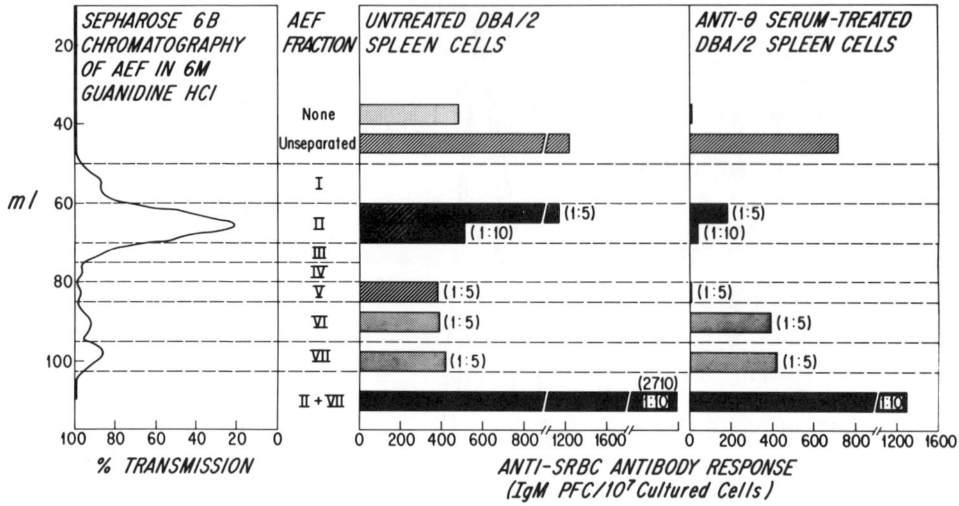

Fig. 9 Activities of AEF fractions obtained by chromatography on Sepharose 6B in guanidine-HCl. Biologically active AEF purified by gel chromatography on Sephadex G-100 was chromatographed on Sepharose-6B in quanidine HCl. The elution patterns of this supernatant are shown on the far left panel. The corresponding biological activities of unseparated AEF, the tested fractions (after removal of guanidine-HCl by dialysis) and combination of fractions II and VII on primary IgM anti-SRBC responses of untreated (middle panel) and anti-θ serum-treated DBA/2 spleen cells (right panel) are shown (Ref. 27).

tions VI and VII exerted no detectable effect on the response of untreated spleen cells but did manifest some activity in reconstituting the responses of the anti-θ-treated cells which was approximately 50% of the activity exhibited by the unfractionated AEF. The striking observation, however, is the effect obtained when fraction II and fraction VII were mixed together prior to addition to the cultures. This mixture exerted a marked enhancing effect on the response of untreated spleen cells (around 5-fold over the normal response) and displayed the highest reconstituting activity on the response of anti-θ-treated cells, which was almost 50% greater than that obtained with unfractionated AEF. It should be noted that: 1) the final concentration of the fraction II and VII components of the mixture was 1:10 (cf. the low activity of the 1:10 concentration of fraction II), and 2) the observed activity of the mixture is substantially greater than the additive effects expected from the biological activity of the individual components themselves (27).

This finding strongly indicates, therefore, that the active moiety of AEF may consist of a bicomponent complex--one heavy and one light--which are associated non-covalently. The variable effects of the individual heavy (fraction II) and light (fraction VI and VII) components on the in vitro responses in Fig. 9 deserves some comment. The activity of fraction II predominantly on the intact spleen cell population suggests that this heavier molecule may exert enhancing effects on T lymphocyte function either directly or via effects on macrophages. The capacity of the lighter material in fraction VII to partially reconstitute responses of anti-θ-treated spleen cells while exerting little or no effects on intact spleen cell responses suggests that this molecule by itself exerts little or no effect on T lymphocyte function but some limited influence on B cells. The optimal influence on B cells, however, clearly requires the concomitant presence of the heavy and light components. It is not possible to know from these data whether the heavy and light molecules must become associated in order to function in the manner observed. Studies currently are underway to determine this point.

C. Immunological properties of AEF

Thus far, we have analyzed the immunological properties of AEF by various immunochemical and functional techniques. The results of these procedures are summarized in Table I. Immunochemical analysis has shown that AEF purified on Sephadex G-100 does not react or cross-react with any heterologous antisera directed against immunoglobulin determinants (27).

In our initial studies on the activity of AEF, we found

TABLE I
IMMUNOLOGICAL PROPERTIES OF AEF (SEPHADEX G-100 Fr. B) DERIVED FROM DBA/2 (H-2^d) T CELLS

a) Reactions in Gel (Ouchterlony) - (Ref. 27)

1) Negative - Rabbit anti-mouse λ
Rabbit anti-mouse κ
Rabbit anti-mouse μ
Rabbit anti-mouse $\gamma 1$
Rabbit anti-mouse $\gamma 2$
Rabbit anti-mouse α
Rabbit anti-mouse IgG

b) Removal by immunoabsorbents

Activity of AEF removed by:

1) Anti-H-2^d - (Ref. 29)

2) Anti-Ia - (Ref. 35)

that although it did not manifest any specificity for antigens against which the in vitro antibody responses were directed, AEF did exhibit some strain-specific properties suggesting a relationship to antigens or gene products coded in the major histocompatibility gene complex (11). Thus, as shown in Fig. 10, purified AEF from supernatants of DBA/2 ($H\text{-}2^d$) activated T cells, although fully reconstituting B cell responses of DBA/2 and BALB/c mice, only slightly reconstituted responses to SRBC of B cells from C57BL/6 ($H\text{-}2^b$) mice. Moreover, preliminary absorption studies with spleen cell populations have demonstrated that spleen cells from normal DBA/2 mice are considerably effective in absorbing the biological enhancing activity from AEF derived from DBA/2 T cells, whereas cells from strains of other _H-2_ haplotypes are not very effective (27).

The aforementioned observations prompted us to explore the relationship of AEF to histocompatibility antigens by functional analysis. Thus, experiments were designed to determine whether immunoadsorbents prepared with antisera reactive with _H-2_ determinants would specifically remove the biologically active moiety of AEF. In the first experiments we found that antisera directed against the entire $H\text{-}2^d$ haplotype would indeed remove the activity of AEF derived from DBA/2 T cells (29). Recently, investigations in several laboratories using anti-lymphoid cell antisera prepared between recombinant mice differing at genes present in the _I_ region of the _H-2_ complex identified a new antigen system, which has been termed _Ia_, coded for by genes in the _I_ region; the _Ia_ antigens have been found to exist predominantly on B cells and macrophages and to varying extents on T cells (21, 30-34). Accordingly, we considered the possibility that gene products in this region may be involved in regulatory cell interactions in immune responses. The experiment presented in Fig. 11 demonstrates that the active enhancing factor(s) in AEF can be removed by an immunoadsorbent prepared with an anti-_Ia_ antiserum indicating that, indeed, the biologically active moieties responsible for T-B cell interactions are probably products of genes in the _I_ region of the _H-2_ gene complex (35). The following antisera were used: 1) B10.A anti-B10--this antiserum contains antibodies reactive with antigens coded by genes in the _I_ region of $H\text{-}2^d$ (Ia.8) but not with antigens coded by genes in either _K_ or _D_ regions of $H\text{-}2^d$; 2) (B6A)F_1 anti-B10.D2--this antiserum contains predominantly antibodies reactive with specificity _H-2.31_ present on cells from animals with the $H\text{-}2^d$ allele; recent analyses have demonstrated the presence also of antibodies reactive with a new _Ia_ specificity present in $H\text{-}2^d$ (36).

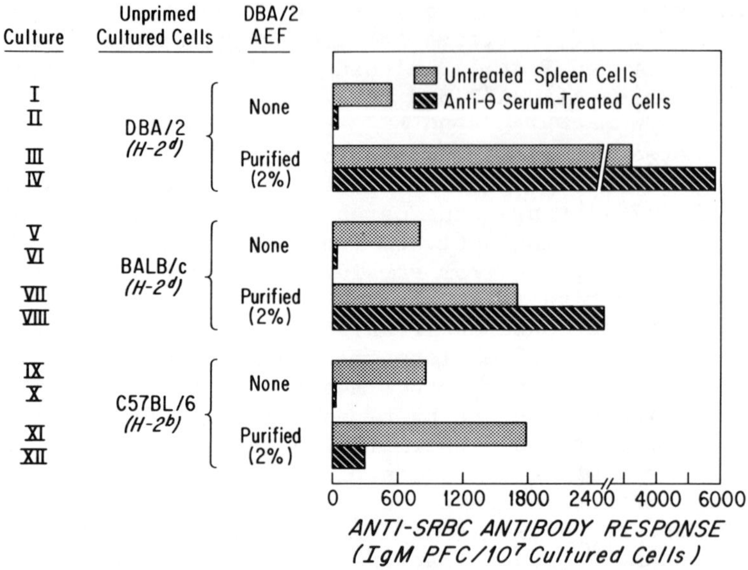

Fig. 10 Reconstitution of helper cell function in primary in vitro anti-SRBC antibody responses of various mouse strains with Sephadex G-200-purified AEF derived from DBA/2 activated T cells. Normal untreated or anti-θ serum-treated spleen cells from DBA/2, BALB/c and C57BL/6 mice were cultured with SRBC in the absence or presence of AEF for 4 days as indicated. The background responses of non-stimulated cultures have been subtracted from the numbers of SRBC-specific PFC developed in cultures containing antigen. IgM PFC responses are shown (Ref. 11).

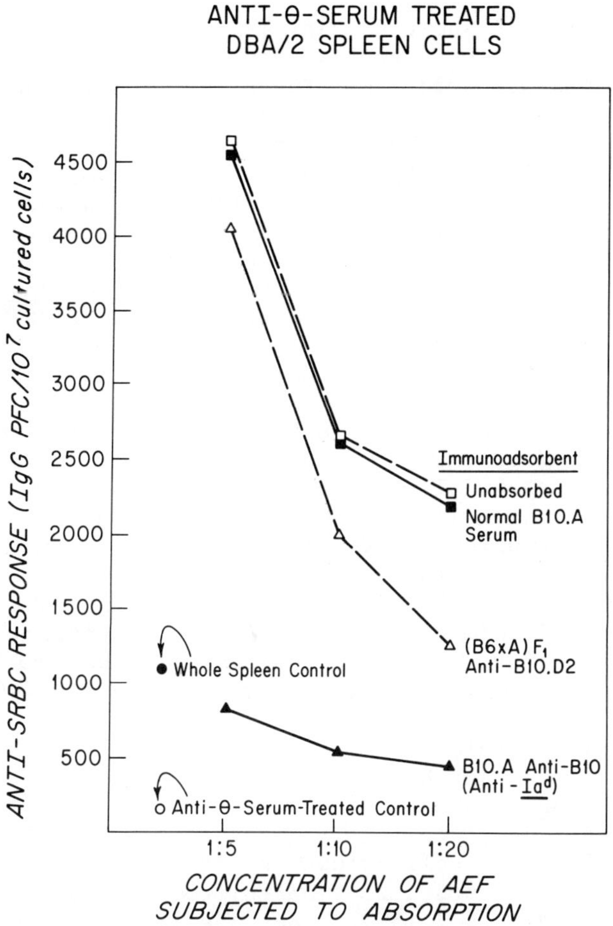

Fig. 11 Removal of biological activity of AEF derived from DBA/2 (H-2^d) T cells by an anti-Ia^d immunoadsorbent. Three different concentrations of AEF were subjected to immunoadsorbents prepared from either normal B10.A serum or from (B6A)F_1 anti-B10.D2 or B10.A anti-B10 (anti-Ia^d) ascites. These AEF were then tested and compared to unabsorbed AEF for activity on the in vitro response to SRBC of anti-θ serum-treated DBA/2 spleen cells. Control responses of whole spleen cells and anti-θ-treated cells in the absence of AEF are shown to the left. The IgG antibody responses are presented and paralleled the IgM response pattern of these same cultures (not shown) (Ref. 35).

In the experiment shown in Fig. 11, three different concentrations of AEF were directly absorbed independently by immunoadsorbents prepared from $(B6A)F_1$ anti-B10.D2 and B10.A anti-B10 alloantisera and by an adsorbent prepared from normal B10.A serum. These AEF were then compared to unabsorbed AEF for biological activity on the in vitro response to SRBC of DBA/2 B lymphocytes. As shown in Fig. 11, cultures of untreated control whole spleen cells developed primary IgG anti-SRBC responses of around 1200 PFC; anti-θ-treatment diminished this response to around 150 PFC. The addition of unabsorbed AEF to such anti-θ serum treated B cells reconstituted and enriched the response markedly and in a dose-related manner at all three concentrations of AEF employed. The AEF subjected to the adsorbent prepared from normal B10.A serum exhibited virtually identical biological activity. The AEF obtained from the immunoadsorbent prepared from $(B6A)F_1$ anti-B10.D2 serum retained essentially normal biological activity at the highest concentration (1:5), but the lowest concentration subjected to absorption (1:20) was around 45% lower in activity than the normal serum control. The AEF subjected to the B10.A anti-B10 (anti-Ia) immunoadsorbent, on the other hand, exhibited markedly diminished (80% or more) activity at all three concentrations indicating substantial reactivity of this antiserum with the biologically active component(s) of AEF (35).

We conclude, therefore, that the biologically active enhancing moieties of AEF bear Ia determinants and therefore are most probably gene products of the I region of the H-2 gene complex. Recent data from other investigators have shown that an antigen-specific T cell product could be specifically adsorbed by immunoadsorbents prepared from antisera directed against the K end of H-2 (37). Since the latter antisera may contain antibodies reactive with specificities of both K and I regions, it is possible that the use of selective anti-Ia sera may yield results consistent with those presented here. Taken collectively, these observations indicate that I region gene products may be intimately involved in the mechanism of cell-cell interactions and responsible for the regulation of immune responses.

CONCLUSIONS

The data and observations presented in this paper have shown that the regulatory interactions between T and B lymphocytes in immune responses are governed by products of genes that appear to be located in the I region of the H-2 gene complex. We originally hypothesized, based on data obtained in this system, that there must exist on the surfaces of lymphocytes (and macrophages) certain molecules coded for by genes in the H-2 complex responsible for permitting effective cell-cell interactions (13-18). These molecules, which have recently been termed by us as cell-interaction or CI molecules (38), probably are present in varying quantities on T and B lymphocytes and macrophages and are envisaged by us as interacting by homology during the cooperative interactions between such cells. The possible mechanisms by which such molecular interactions may occur have been elaborated upon at length elsewhere (38). The genetic mapping data presented here together with the observations that demonstrate the capacity of anti-Ia antiserum to absorb the biologically active molecule(s) from AEF demonstrate that CI molecules are products of genes located in the I region of H-2.

It should be mentioned that the number of specific H-linked Ir genes that have been identified and the specific manner in which they permit immune responses to distinct antigens to take place, particularly at the T cell level, have suggested that they are somehow involved in either the specificity or the function of the T cell antigen receptors and may, therefore, be clonally expressed in this class of lymphocytes.

The implications of our experiments on the F_1-parent cooperative responses to DNP-GLT described above have been discussed by us (14,18,38) to mean either: 1) that Ir genes may also be expressed and non-clonally in B cells of responder animals, and/or 2) that the activation of the Ir gene product determines the activation in turn of the molecules involved in T-B cell interactions coded for by the same haplotype. According to the latter alternative, there would be no requirement for the functional expression of the Ir gene product in the B cell but only a requirement for the Ir gene product or associated I region gene product(s) from the T cell to govern the interaction with the B cell at the histocompatibility "acceptor" site on the B cell surface.

Finally, it should be noted that the Ir-1A and Ir-1B subregions where the CI genes appear to be mapped by our studies (Fig. 4) are precisely the subregions where all known immune response or Ir genes have been mapped (39). However, as discussed at length elsewhere (38), while this suggests an association between Ir and CI gene products in lymphocyte function, it does not imply that these are necessarily products of the same gene(s).

ACKNOWLEDGMENTS

We are most appreciative to all of our colleagues who participated with us in the performance of these experiments. We are particularly indebted to Professor Baruj Benacerraf for constant critical attention to the questions at hand and to Drs. Martin Dorf, Toshiyuki Hamaoka, Zelig Eshhar and David Sachs and Miss Mary Graves and Mr. Henry DiMuzio for their invaluable contributions to the accomplishment of these studies. We also thank Ms. Deborah Maher for excellent secretarial assistance in the preparation of the manuscript.

REFERENCES

1. Katz, D.H. and Benacerraf, B., Adv. Immunol. 15: 1, 1972.
2. Katz, D.H., Transpl. Rev. 12: 141, 1972.
3. Dutton, R.W., Falkoff, R., Hirst, J.A., Hoffmann, M., Kappler, J.W., Kettman, J.R., Lesley, J.F. and Vann, D., Prog. Immunol. p. 355, 1971.
4. Schimpl, A. and Wecker, E., New Biol. 237: 15, 1972.
5. Schimpl, A. and Wecker, E., J. Exp. Med. 137: 547, 1972.
6. Feldmann, M. and Basten, A., J. Exp. Med. 136: 722, 1972.
7. Britton, S., Scand. J. Immunol. 1: 89, 1972.
8. Gorczynski, R.M., Miller, R.G. and Phillips, R.A., J. Immunol. 110: 968, 1972.
9. Vann, D.C. and Galloway, P.C., J. Immunol. 110: 1542, 1972.
10. Watson, J., J. Immunol. 111: 1301, 1973.
11. Armerding, D. and Katz, D.H., J. Exp. Med. 140: 19, 1974.
12. Kindred, B. and Shreffler, D.C., J. Immunol. 109: 940, 1972.
13. Katz, D.H., Hamaoka, T. and Benacerraf, B., J. Exp. Med. 137: 1405, 1973.
14. Katz, D.H., Hamaoka, T., Dorf, M.E. and Benacerraf, B., Proc. Nat. Acad. Sci. 70: 2624, 1973.
15. Katz, D.H., Hamaoka, T., Dorf, M.E., Maurer, P.H. and Benacerraf, B., J. Exp. Med. 138: 734, 1973.
16. Katz, D.H., Hamaoka, T., Dorf, M.E. and Benacerraf, B., J. Immunol. 112: 855, 1974.
17. Katz, D.H., Dorf, M.E. and Benacerraf, B., J. Exp. Med. 140: 290, 1974.
18. Katz, D.H. and Benacerraf, B., in The Immune System: Genes, Receptors, Signals - Proceedings of the 1974 I.C.N.-U.C.L.A. Symposium on Molecular Biology, eds. E.E. Sercarz, A.R. Williamson and C. Fred Fox, Academic Press, New York, 1974, p. 569.
19. Katz, D.H., Graves, M., Dorf, M.E., DiMuzio, H. and Benacerraf, B., J. Exp. Med., in press.
20. Hamaoka, T., Osborne, D.P., Jr. and Katz, D.H., J. Exp. Med. 137: 1393, 1973.
21. Shreffler, D.C. and David, C.S., Adv. Immunol., in press.
22. Katz, D.H., Davie, J.M., Paul, W.E. and Benacerraf, B., J. Exp. Med. 134: 201, 1971.

23. Katz, D.H., Hamaoka, T. and Benacerraf, B., J. Exp. Med. 136: 1404, 1972.
24. Hamaoka, T. and Katz, D.H., J. Exp. Med. 139: 1446, 1974.
25. Nossal, G.J.V., Pike, B.L. and Katz, D.H., J. Exp. Med. 138: 312, 1973.
26. Osborne, D.P., Jr. and Katz, D.H., J. Exp. Med. 137: 991, 1973.
27. Armerding, D. and Katz, D.H., manuscript in preparation.
28. Silver, J. and Hood, L., Nature 279: 765, 1974.
29. Armerding, D., Dorf, M.E. and Katz, D.H., manuscript in preparation.
30. David, C.S., Shreffler, D.C. and Frelinger, J.A., Proc. Nat. Acad. Sci. 70: 2509, 1973.
31. Hauptfeld, V.D., Klein, D. and Klein, J., Science 181: 167, 1973.
32. Sachs, D.H. and Cone, J.L., J. Exp. Med. 138, 1289, 1973.
33. Hämmerling, G.J., Deak, B.D., Mauve, G., Hämmerling, U. and McDevitt, H.W., Immunogenetics 1: 68, 1974.
34. Unanue, E.R., Dorf, M.E., David, C.S. and Benacerraf, B., Proc. Nat. Acad. Sci. U.S.A., in press.
35. Armerding, D., Sachs, D.H. and Katz, D.H., J. Exp. Med. 140: in press.
36. Sachs, D.H., Fathman, C.G., Cone, J.L. and Dickler, H.B., Transplant. Proc., in press.
37. Munro, A.J., Taussig, M.J., Campbell, R., Williams, H. and Lawson, Y., J. Exp. Med., in press.
38. Katz, D.H. and Benacerraf, B., Transpl. Rev. 22: in press.
39. Benacerraf, B. and Katz, D.H., Adv. Cancer Res. 21: in press.

CELL INTERACTIONS IN ANTIBODY PRODUCTION - PROBLEMS OF HETEROGENEITY, DIVERSITY AND REGULATION

Marc Feldmann, Reginald Gorczynski, Peter Erb and Catherine Desaymard

ICRF Tumour Immunology Unit, Department of Zoology, University College London and Department of Experimental Immunobiology, Wellcome Research Laboratories, Beckenham, Kent, England

While there are certain advantages, and some satisfaction, in the formulation of logically consistent theories of the mechanism of the immune response (1,2,3) there are many disadvantages, as often the theory becomes dominant and clouds observations, which may be misinterpreted or ignored if they do not fit the preconceived 'theory' (2). Regrettably, at the present time, immunological 'theories' cannot be of the breadth of theories in physics, but are merely approximations or generalizations drawn from a set of consistent observations. It is not possible to do any better at present, as we do not yet know the full extent of the complexity and heterogeneity of the cellular components of the immune response. Without this information meaningful detailed concepts of the mechanisms of the immune response cannot be formulated on a rational basis. It should be remembered that it is less than 10 years since the dichotomy of T and B lymphocytes became apparent and our working concepts had to be drastically revised and redescribed on the basis of the precise cell type. Because of evidence for heterogeneity of lymphocytes it seems essential to investigate the full extent of the diversity of the components of the lymphoid system - not only of lymphocytes, but also of accessory cells (and of the diverse functions of various immunological mediators), to provide a baseline for

---

<u>List of Abbreviations</u>

TNP KLH - trinitrophenylated Keyhole limpet haemocyanin
DNP POL - dinitrophenylated polymeric flagellin
IgT - Ig derived from T cells
MIF - macrophage migration inhibitory factor
ConA - Concanavalin assay
PHA - phytohaemagglutinin
(TG)AL - copolymer of tyrosine and glutamic acid, attached to side chains of polyalanine on a polylysine backbone
GAT - copolymer of glutamic acid, alanine and tyrosine

studies in depth of the mechanisms of triggering and suppression in antibody production. This article will briefly present some of the problems and progress in this path, and discuss to what extent traditional concepts or dogma still hold.

PROBLEMS

1. <u>Heterogeneity of lymphocytes</u>

A. <u>T cells</u>. This has been recognized for several years, since Raff and Cantor (4) described T_1 and T_2 cells, short lived and long lived recirculating cells. This has been a useful classification, especially as there is evidence for the synergy of these two cell types in the graft-versus-host response (5) and in the cytotoxic responses in vitro (6,7). Antigenic differences between different T cells exist - T_1 cells have more Thy 1 (θ) antigen than T_2, differences have been suggested in Ly markers and in Fc receptors. Needless to say we do not know the functional significance of these antigenic differences, nor the differences in how these cells are stimulated to perform their various functions which are interrelated, since there is increasing evidence of T-T interactions- in graft-versus-host response, in the cytotoxic response, and preliminary evidence in the generation of helper cells (Feldmann, Kilburn and Levy, unpublished data). The increasingly obvious suppressor effect of T cells (reviewed in 8) also chiefly acts on other T cells.

B. <u>B cells</u>. Playfair and Purves (9) reported that B cells in spleen behaved differently from those in bone marrow, suggesting a heterogeneity of B cells. Further evidence for this concept has been obtained by cell fractionation studies, by density (10), size (11) or cell electrophoresis (12). Fig. 1 shows some size separations, on spleen cells of mice primed to TNP KLH, which indicate that in these spleens, IgM cells responding to the thymus dependent antigen (TNPKLH) are different to those responding to the thymus independent antigen (DNPPOL). Similarly the peak of the IgG response is different from the IgM. Whether these separations indicate basic differences in the cell types, maturational changes or more trivial explorations is not yet clear, although it should be stressed that comparable differences have been found by Dr. Ken Shortman using cell electrophoresis (12).

These functional differences are additional to those already known for surface markers - immunoglobulin receptors of difference classes, Fc and C3 receptors, and Ia antigens. How these differences are reflected in different triggering requirements is not known, but it cannot be assumed that they are the same for all B cells, as there is increasing evidence that IgG responses have different requirements to IgM or IgE

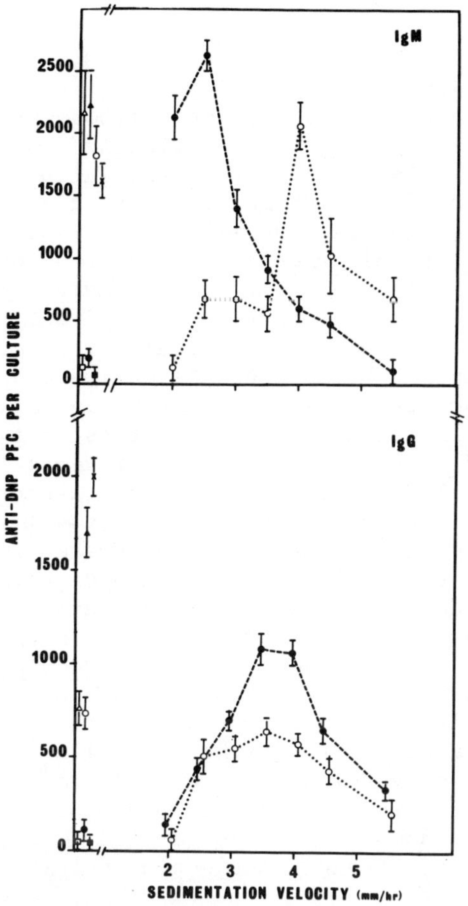

Fig. 1 Sedimentation profile of B cells responding to DNP POL (open circles) or TNP KLH (closed circles) in anti-Brθ treated spleen cells of TNP KLH primed mice. The upper panel of this figure shows the data for IgM-PFC,the lower panel that for IgG-PFC. Data to the left of this figure show the response from unfractionated cells, and in the various control cultures described below:

Open and closed squares: Response of anti Ig treated spleen cells with DNP POL or TNP KLH.

Open and closed triangles: Response of untreated spleen cells with DNP POL or TNP KLH.

Cross: Response of combination of anti-Brθ treated cells and anti-Ig treated cells to TNP KLH.

All figures are arithmetic means (± standard deviation) of 3 cultures per group.

responses (13,14).

2. Cell interactions

The cell interactions involved in antibody production are now known to be highly complex, evolving in 3 semi discrete stages - the generation of helper cells, which then release molecules which interact with macrophages, finally B cells are triggered after interaction with these macrophages.

A. Generation of helper cells. Stimulation of helper cells requires the interaction of T cells with macrophages. Peter Erb has found that this interaction is indirect, and that T cells and macrophages need not physically interact. Supernatants of antigen incubated macrophages induced helper cells, provided the supernatant came from macrophages of the appropriate mouse strain (Table 1). Similarities in the H-2 region were important, and the gene permitting interactions to KLH was mapped in the IrI-A region. These supernatants were specific both for the appropriate antigen and for the mouse strain. Other supernatants have been generated which are not antigen or strain specific which function on polymeric antigens (15). Thus there are 2 possible pathways of T macrophage interaction in the generation of helper cells which seem to result in similar cells (15).

The evidence for T-T interaction in the generation of killer cells in vitro (6,7) prompted a search for such an interaction in the generation of helper cells. Using the oxidized ferrodoxin system developed by Julia Levy, evidence consistent with such interaction has been found (Feldmann, Kilburn and Levy, unpublished observations), though its generality in other, stronger antigenic systems is not yet known.

B. Release of mediators by helper cells. Two basically different types of mediators of cell collaboration have been reported antigen specific, or non specific. In general, experimental protocols have been used which only look at one or the other activity and it seems likely that usually both types of factors participate in B cell triggering. The requirements of the two activities may vary with the form of antigen used (particulate or soluble), but arguments as to which is the 'more important' are as meaningless as discussing which is more important in a motor car - the steering or the engine.

Antigen specific factors, Ig in nature (IgT), were reported by Feldmann et al. (16), by Rieber and Riethmüller (17) and by Tada et al. (18). The Ig specific factors are 7S in size.

Smaller, non-immunoglobulin specific factors have also been reported. Lachmann and Amos (19) described 'antigen

TABLE I

MACROPHAGE REQUIREMENT FOR THE INDUCTION OF T HELPER CELLS

1st CULTURE HELPER CELL INDUCTION		2nd CULTURE COOPERATIVE RESPONSE ANTI DNP RESPONSE		
SOURCE	ANTIGEN (μg/ml)	CELLS ADDED TO 2nd CULTURE	CHALLENGE (μg/ml)	AFC/CULTURE
Spleen	10^{-1}KLH	10^5	10^{-1} TNP KLH	207
T_s	"	"	"	33
T_s + M(CBA)	"	"	"	270
T_s + M(Balb/c)	"	"	"	43
T_s + S/N(CBA)	"	"	"	257
T_s + S/N(Balb/c)	"	"	"	30
NIL	-	-	"	37
"	-	-	NIL	37
"	-	-	DNP POL	523

Helper cells induced from spleen or macrophage depleted spleen cells (T_s)using nylon wool.

Macrophages (M) purified from induced peritoneal exudate by treatment with anti T antiserum.

Both periods of cultures were 4 days. Data pooled from 2 experiments (15).

specific MIF', Munro and Taussig have found a non-immunoglobulin antigen specific mediator which binds to columns of anti H-2 antisera (20). Other workers have found specific cooperative factors but have not characterized them (21,22). The existence of multiple antigen specific T cell factors is puzzling. It is not yet known whether all of these are 'T-B cooperative factors' or whether some, such as that of Munro and Taussig are involved in the generation of helper cells. The resemblance of the latter, obtained from populations of splenic activated T cells, rich in macrophages, with the macrophage factor described by Erb, is striking. Nevertheless it is conceivable that factors generated with antigens with H-2 linked genetic responsiveness patterns may differ from the specific factors generated with other (non H-2 linked) antigens. Why this should be so is not known, but perhaps these two pathways enhance the probability of optimal antigen presentation.

Non-specific factors which augment responses have been described by various workers (23, 24, 25). These are widely assumed to be of importance in cell cooperation, although the assay systems for these factors, in vitro immune responses are well known to be highly susceptible to a variety of agents-nutritional, adjuvants and the like which can modify responses. By their very non-specificity it is difficult to perform adequate controls. To preserve the overall specificity of antibody responses, it is assumed that these substances act at short range, or are rapidly degraded in vivo. Alternately, the problem of specificity is circumvented if it is assumed that these act in conjunction with specific factors.

The cell synthesizing IgT is not known. It could be made by T cells of the helper lineage or by a subpopulation of B cells. Current methods of cell separation have not permitted a solution. It should be stressed, however, that the T cell biosynthesis of <u>any</u> mediator has yet to be proven. MIF in the guinea pig, long assumed to closely reflect delayed sensitivity and hence T cell function, was found to be made by B lymphocytes (26). The macrophage origin of some non-specific factors is suggested by the work of Waldmann and Munro (27). Reports of diverse MW of non-specific factors (25,28) together with some genetic restrictions, suggests a heterogeneity of non-specific factors. Because of the problems of the assay system, it cannot be assumed that all molecules derived from T cell cultures which exhibit non-specific stimulatory activity in vitro are in fact mediators of cooperation in vivo. Conversely, nor can it be assumed that all specific mediators of cooperation in vitro must work in vivo, although the assays of Riethmüller (17), Taussig (20) and Tada (18) are in vivo.

Cultures of T cells release specific and non-specific factors which suppress antibody production (18, 29, 30). The relationship between stimulatory and suppressive factors is of major importance in the regulation of the antibody response. As with specific cooperative factors, both Ig and non Ig factors have been reported. Ig factors were found in the supernatant of activated mouse T cells (29), whereas Tada et al. (18) reported a low MW, non Ig specific suppressor derived from rat thymocytes. It is not known whether the IgT suppressor is the same as helper IgT, acting in a suppressive manner at high concentration (directly on lymphocytes), or whether low concentrations of the non Ig suppressor may be involved in cooperation. Whether the same non-specific factors which stimulate responses can also suppress them is also not clear, although Concanavalin A stimulation induces both activities (30), but PHA does not.

C. Macrophage-B cell interaction. The question of the heterogeneity of help and suppression by activated T cells raises the corollary question of how many cells are involved in the release of these factors. At the moment there is little information, except for the elegant limiting dilution experiments of Hunter (Marrack) and Kappler (31) which suggest that the cells which release non-specific factors are different to those which release specific factors.

It is not yet known why IgT complexed antigen is so highly immunogenic for B cells. Initial speculations, that the form of antigen presentation was 'optimal' in some ill defined way, seem to be correct for B cell precursors of IgM, but only part of the answer for IgG precursors. The concept of 'critical antigen presentation' was proposed to unify the mechanism of B cell triggering with thymus dependent antigens, which are polymers with repeating determinants (reviewed in 32). The antigen presentation concept of cooperation implies a passive role of the macrophage, which merely provides the appropriate surface. Attempts to duplicate macrophage function by conjugation of antigen onto inert beads were successful, supporting the antigen presentation concept, but only for IgM responses (33). Other results consistent with the antigen presentation role of IgT were obtained by autoradiography-labelled IgT bound to the surface of macrophages at 4° in a polar ('cap') distribution (34). Functional studies, using activated T cell supernatants made with different antigens, have also supported the antigen presentation concept of a repeating lattice of determinants, since it was found that the prior exposure to a supernatant (at 4°) inhibited the macrophages' capacity to bind a second supernatant in an immunogenic way (35). This observation is easily explained by the antigen presentation concept as inhibition of lattice formation.

A more active role for macrophages in cooperation was proposed by Waldmann and Munro (27) who suggested that these cells, after the influence of T cells, release non-specific factors. Recently we have found evidence that low concentrations of IgT derived from two lymphoid cell tumors bearing T cell markers influence IgG production in a non-specific way (36). IgG augmentation occurred with thymus dependent and independent antigens, and in preliminary experiments was macrophage dependent, suggesting that IgT may induce the production of a factor required for the generation of IgG plaque forming cells.

It was interesting that high concentrations of IgT (derived from supernatant of cultured lymphomas WEH1122 and EL4) inhibited both IgM and IgG responses, intermediate concentrations blocked IgM only, supporting concentrations blocked IgM only, supporting the greater importance of lattices of antigen for triggering IgM precursors. Experiments with mice undergoing antigenic competition revealed that there was a block of cooperative activity at the level of the macrophage surface (37). Since this block could be induced by T cell supernatants containing, but not by those depleted of IgT, it was concluded that IgT was the blocking agent (35).

Macrophage-T cell interactions have genetic restriction, both in the guinea pig (38) and in the mouse (15). Whether such restrictions also apply for macrophage-B cell interactions is not clear, although some published experiments suggest that this is not the case (39).

3. Genetics of immune responses

A. H-2 linked responses. McDevitt and Benacerraf suggested that non-responder animals in H-2 linked responses had no functioning T cell receptors for that antigen (40). However non-responders to (T-G)A-L were found to have equivalent numbers of antigen binding cells (41) and their thymus and spleen cells can be triggered to produce helper cells in vivo (42) or in vitro (43), as shown in Table 2. These results indicate that the lesion of 'non-responsiveness' in this system is not at the generation of helper cells. Taussig et al. (42) found that responder bone marrow was needed to cooperate with non-responder thymus, suggesting the lesion is at the B cell level.

However, studies of a different H-2 linked antigen, GAT, in vitro by Kapp and Benacerraf (44) did not demonstrate the presence of helper cells, but revealed abundant suppressor

TABLE 2
CBA HELPER CELLS TO (T-G) -- A-L

CELLS	ANTIGEN	CHALLENGE	ANTI DNP RESPONSE
10^5CRT	10^{-1} μ TGAL	NIL	50
"	"	DNP POL	925
"	"	DNP TGAL	740
"	"	DNP KLH + TGAL	70
NIL	-	DNP TGAL	80

Induction of helper cells in CBA (H-2^k) cortisone resistant thymus cells in vitro by culture with 10^{-1} μg (TG)AL for 4 days. 10^5 resulting helper cells were cultured with 1 μg/ml DNPTGAL for 4 days. Results are helper cell dependent IgM plaque forming cells.

cells in non-responders. Thus these early results suggest diverse mechanisms of H-2 linked non-responsiveness.

B. Surface antigens of cooperating cells. Genetic restrictions on cooperating cells have been reported in two instances - in T/B cooperation (44) and in T-macrophage cooperation (15). Genetic mapping studies suggest that similar loci are involved, the K and/or I regions for T/B cooperation, and the I (probably Ir-I-A) region for T/M interaction (to KLH). Since macrophages are involved in the triggering of T lymphocytes, and of B lymphocytes it is conceivable that both genetic restrictions operate by the same underlying mechanism. This conclusion, however, may not be true, if it can be shown that primed T cells (which are radio-resistant) do not need macrophages for stimulation into helper cells, and if it can be shown that T cells tolerant to the histocompatible B cells (or macrophages) still cannot cooperate. It is not impossible that histocompatible B cells are 're-jected', in a subtle way, manifest as incapacity to respond, by histoincompatible T cells. Whatever the eventual mechanism of unresponsiveness of histocompatible T and B cells, these studies (44) have demonstrated the importance of the H-2 region antigens in lymphocyte function.

Mechanisms of lymphocyte triggering and tolerance

This is the gut issue of immunology. Real progress has been slow. Attempts have been made in the last few years to substantiate and elaborate the antigen presentation/epitope density concepts of B cell triggering, first formulated on the basis of work in vitro with DNP-POL and DNP-MON of various conjugation ratios. The need for polymeric structure (and repeating determinants) was pointed out by the MON/POL difference in thymus dependence (32) which has been substantiated in other ways, e.g. hydrolysed LPS is thymus dependent (Coutinho, personal communication), thymus dependent antigens conjugated onto beads are thymus independent (33). No convincing evidence against this concept has yet been present--low MW of thymus independent antigens may still, depending on their chemical nature, mean quite long antigens, capable of interlinking many receptors.

Evidence to support the epitope density concept has been obtained with polysaccharides - levan, dextran (linear and branched) and pneumococcal polysaccharide type III, and with synthetic poly amino acids, the copolymer of D glutamic acid and lysine. In each case 'low' conjugates were immunogenic (and not tolerogenic), and high conjugates were tolerogenic. These results, both in vitro (45) and in vivo (46) stress the importance of the exact manner of 'antigen presentation'.

Although the results so far have been chiefly for IgM responses, preliminary results for IgG responses have also shown the need for a higher epitope density for tolerance than for immunity.

Analogous results have been obtained by James Howard and his colleagues (47) in studying the response to the α1-3 and α1-6 linkage in dextrans. They found that the α1-3 link, present at lower frequency, was much more immunogenic and less tolerogenic than the α1-6 link. These results are of particular importance, as they verify the epitope density phenomenon with natural antigens. Regrettably we do not know at present how a B lymphocyte perceives changes in epitope density. The suggestion that excess conjugation destroys mitogenic sites (2) is wrong, as the weak (if any) 'mitogenicity' of these compounds does not change with conjugation, and the phenomenon occurs naturally with dextrans (i.e. no conjugation).

Other aspects of lymphocyte triggering have been studied by others. Coutinho and Möller (2) proposed that B cell activation depends on a single (non-specific) mitogenic stimulus. B cell receptors for antigen merely concentrate antigen for the mitogenic receptors. B cell tolerance is due to excess activation. This idea cannot be discussed in full here, but this theory (like other theories) does not fit many of the facts - such as the existence of non-mitogenic, non Ig secretion activating thymus independent antigens (33, DNP Ficoll, 47); the existence of non-immunogenic tolerogens (48,49); the phenomenon of epitope density as discussed above; the capacity of antibody to facilitate tolerance (50); that cross linking of antigens onto beads makes them direct B cell stimulators without being mitogenic (33, unpublished data); that cross linking of PHA or ConA (or aggregated PWM) makes them B cell mitogens (51,52); that anti Ig antibody (or anti Ig $F(ab)_2$) but not Fab stimulates proliferation in the rabbit (53), or antibody secretion (54).

However, it is still possible that some thymus independent (TI) antigens, such as LPS, which are strong mitogens, are so because they are mitogenic. But most of the TI antigens do not operate by this pathway for the IgM response. Other B cells (e.g. from lymph node, or precursors of IgG) are unlikely to be triggered by a mitogenic mechanism because these cells respond poorly to mitogens.

Two signal theories have been proposed to explain this discrimination between tolerance and immunity and the mechanism of B cell triggering (1). These cannot be discussed at length, here, but the ever increasing evidence for 'true thymus independent antigens'(32) makes this theory, at best,

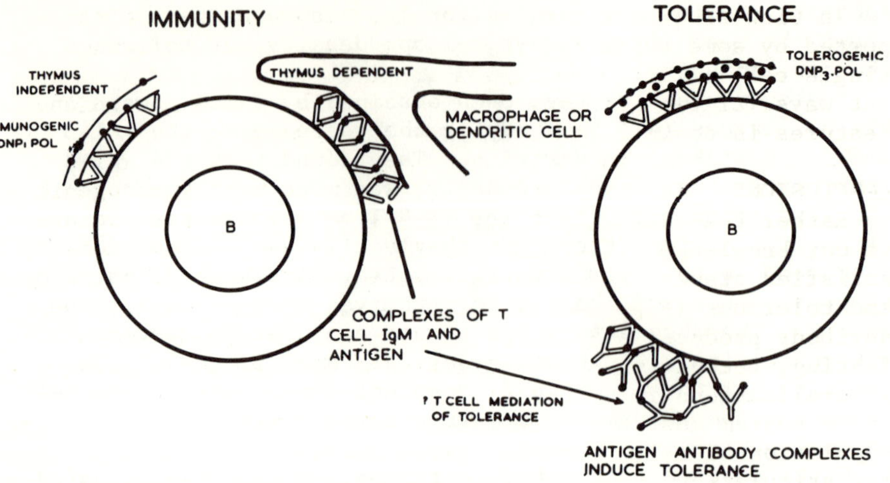

Fig. 2 Scheme of discrimination between immunity and tolerance, based on the concept of epitope density.

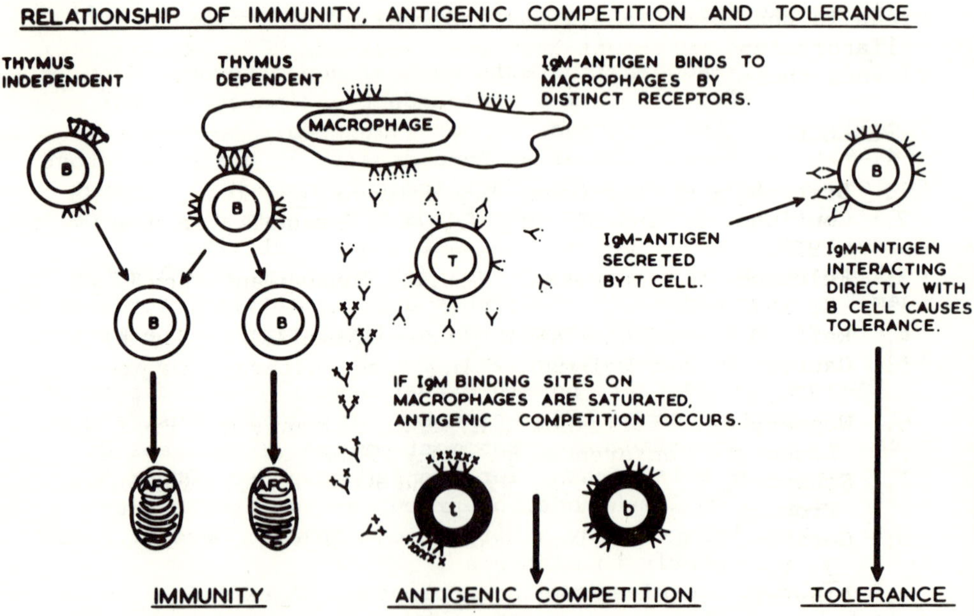

Fig. 3 Role of IgT in regulation of the antibody response - in stimulation, antigenic competition and T cell suppression.

only true for some antigens.

In view of contrasting theories of stimulation, all supported by some evidence, the possibility remains that a single cell or different cells could be activated in different ways (cf two pathways of complement activation, with some features in common, but different early steps).

CONCLUSIONS

Rather than scanning a wide range of problems and gaps in our knowledge, it would have been easier to take a self-satisfied stance, and discuss some generalization of immunity and tolerance (Fig. 2), or the role of IgT in regulation of antibody production (Fig. 3). That would have been self-defeating, because we now know that the data base for these generalizations is incomplete. The heterogeneity of the lymphoid system must have functional significance, and each population of cells must be tested to find its role and rules for stimulation and suppression. The problems discussed above were of lymphocyte biology. What is even more difficult is to place these problems back in their physiological in vivo contexts - of lymphoid architecture and recirculation. A dramatic indication of the depth of our ignorance in these areas is that despite all our 'sophisticated' discussions of collaboration, we do not know precisely where in the lymphoid tissues these interactions take place.

REFERENCES

1. Bretscher, P. and Cohn, M., Science 169: 1042, 1970.
2. Coutinho, A. and Möller, G., Adv. Immunol., in press, 1974.
3. Feldmann, M. and Nossal, G.J.V., Transplant. Rev. 13: 3, 1972.
4. Raff, M.C. and Cantor, H., Prog. Immunol. 1: 83, 1973.
5. Cantor, H. and Weisman, I.L., Prog. Allergy, in press, 1974.
6. Wagner, H., Röllinghoff, M. and Feldmann, M., Proc. 8th Leukocyte Conference, p. 111, 1974.
7. Schendel, D.J., Alter, B.J. and Bach, F.H., Transplant. Proc. 5: 1651, 1973.
8. Gershon, R.K., Current Topics Microbiol. and Immunol. 3: 1, 1973.
9. Playfair, J.H.L. and Purves, E., Nature 231: 149, 1971.
10. L'Age Stehr, J. and Herzenberg, L.A., J. Exp. Med. 131: 1093, 1970.
11. Gorczynski, R.M. and Feldmann, M., Cell. Immunol., in press, 1974.

12. Schlegel, R.A., von Boehmer, H. and Shortman, K., Cell. Immunol., in press, 1974.
13. Cheers, C. and Miller, J.F.A.P., J. Exp. Med. 136: 1661, 1972.
14. Katz, D.H., Hamaoka, T., Newberger, P.E. and Benacerraf, B., in press, 1974.
15. Erb, P. and Feldmann, M., submitted for publication, 1974.
16. Feldmann, M., J. Exp. Med. 136: 737, 1972.
17. Rieber, E.P. and Riethmüller, G., Z. Immun. Forschung 147: 262, 1974.
18. Tada, T., Okamura, K. and Taniguchi, M., Allergology Int. Congress 323, Excerpta Medica, p. 473, 1974.
19. Lachmann, P.J. and Amos, H.E., Immunopathology 6: 65, 1971.
20. Munro, A. and Taussig, M.J., Nature 251: 64, 1974.
21. Yu, K. and Gordon, J., Nature New Biol. 244: 20, 1973.
22. Gisler, R.H., Staber, F., Rude, E. and Dukor, P., Eur. J. Immunol. 3: 650, 1973.
23. Dutton, R.W., Falkoff, R., Hurst, J.A., Hoffman, M., Kappler, J.W., Kettman, J.R., Lesley, J.F. and Vann, D., Prog. Immunol. 1: 355, 1971.
24. Schimpl, A. and Wecker, E., Nature New Biol. 237: 15, 1972.
25. Armerding, D. and Katz, D.H., J. Exp. Med. 140: 19, 1974.
26. Bloom, B., Ann. N.Y. Acad. Sci., in press, 1974.
27. Waldmann, H. and Munro, A., Nature 243: 256, 1974.
28. Watson, J., Epstein, R., Nakoinz, I. and Ralph, P., J. Immunol. 110: 43, 1973.
29. Feldmann, M., Eur. J. Immunol. 4: 667, 1974.
30. Dutton, R.W., J. Exp. Med. 136: 1445, 1972.
Rich, R. and Pierce, C.W., J. Exp. Med. 137: 205, 1973.
31. Marrack, P. (Hunter) and Kappler, J., manuscript submitted, 1974.
32. Feldmann, M., Contemp. Topics Molec. Immunol. 3: 57, 1974.
33. Feldmann, M., Greaves, M.F., Parker, D.C. and Rittenberg, M.B., Eur. J. Immunol. 4: 593, 1974.
34. Cone, R.E., Feldmann, M., Marchalonis, J.J. and Nossal, G.J.V., Immunology 26: 43, 1974.
35. Feldmann, M. and Schrader, J.W., Cell. Immunol. 14: 255, 1974.
36. Feldmann, M., Boylston, A.W. and Hogg, N.M., in press, 1974.
37. Schrader, J.W. and Feldmann, M., Eur. J. Immunol. 3: 711, 1973.

38. Rosenthal, A. and Shevach, E., J. Exp. Med. 138: 1194, 1973.
39. Kapp, J.A., Pierce, C.W. and Benacerraf, B., J. Exp. Med., 138: 1121, 1973.
40. Benacerraf, B. and McDevitt, H.O., Science 175: 273, 1972.
41. Hämmerling, G., Masuda, T. and McDevitt, H.O., J. Exp. Med. 137: 1180, 1973.
42. Taussig, M., Mozes, E. and Isac, R., J. Exp. Med. 140: 301.
43. Feldmann, M., submitted for publication, 1974.
44. Benacerraf, B. and Katz, D.H., Adv. in Cancer Res., in press, 1974.
45. Desaymard, C. and Feldmann, M., submitted for publication, 1974.
46. Desaymard, C. and Howard, J.G., submitted for publication, 1974.
47. Howard, J.G., Courtenay, B.M. and Vicari, G., submitted for publication, 1974.
48. Roelants, G.E. and Goodman, J.W., Nature, 227: 175, 1970. Hamilton, J.A. and Miller, J.F.A.P., Eur. J. Immunol. 3: 457, 1973.
49. Golan, D.T. and Borel, Y., J. Exp. Med., 134: 1046, 1971.
50. Diener, E. and Feldmann, M., Transplant. Rev. 8: 76, 1972.
51. Greaves, M.F. and Bauminger, S., Nat. New Biol. 235: 67, 1972.
52. Andersson, J., Edelman, G.M. and Möller, G., Eur. J. Immunol. 2: 233, 1972.
53. Fanger, M.W., Immunological Communications 1: 315, 1972.
54. Kishimoto, T. and Ishizaka, K., J. Immunol., in press, 1974.

ANTIGEN-SPECIFIC ENHANCING AND SUPPRESSIVE T CELL FACTORS IN THE REGULATION OF ANTIBODY RESPONSE

Tomio Tada
Department of Pathology, School of Medicine
Chiba University, Chiba, Japan

INTRODUCTION

There has recently been a growing awareness that thymus-derived lymphocytes (T cells) have dual regulatory function, stimulatory and suppressive, in certain antibody responses by bone marrow-derived lymphocytes (B cells). Besides the well known enhancing helper effect of T cells, it has been well substantiated that the latter suppressive effect of T cells plays an important role over the wide range of homeostatic regulation of the immune responsiveness (reviewed in 1). The detailed analyses from various laboratories on the regulatory function of T cells are now leading us to the more fundamental understanding of the nature of T-B cell interactions, in general.

An important clue to the further analysis of cell interactions has arisen from the recent discovery of diffusable substances from T cells either released by antigenic stimulation or extractable by physical means (2-16). These include both antigen-specific and non-specific factors, while the joint action of these factors may be of great importance for determining the magnitude and class distribution of antibody responses.

Our previous studies demonstrated two different antigen-specific T cell components that suppress and enhance IgE antibody response in the rat (15,16,21). Both factors were extractable from mechanically disrupted thymocytes or spleen cells obtained from primed animals. The specificity of the factors was warranted by their affinity to the immunizing antigen but not to unrelated protein molecules. However, the physicochemical, as well as immunochemical properties of these two factors differed greatly from each other: The suppressor molecule contained no immunoglobulin determinants as revealed by absorption with various anti-immunoglobulin antibodies, whereas the enhancing factor contained Fab and μ chain determinants. The molecular weight of the suppressor component was less than 100,000, while that of the enhancing factor was between 200,000 and 100,000, closely resembling IgT described by Feldmann and coworkers (8,9). Heterologous anti-thymocyte serum could absorb the suppressive activity, while the enhancing activity was unaffected. These results suggest that there are two distinct antigen-specific T cell factors having opposite functions, and that the balance of these factors and the different sensitivity of B cells at discrete stages of maturation may be of crucial importance for the regulation of antibody response.

A similar antigen-specific suppression by primed T cells was found in IgM and IgG antibody responses in the mouse (17,18). BALB/c J mice immunized with the dinitrophenyl derivative of keyhole limpet hemocyanin (DNP-KLH) with the aid of Bordetella pertussis vaccine (BPV) produced a good sequential production of DNP-specific IgM and IgG antibodies. The passive transfer of T cells from KLH-primed syngeneic donors could suppress IgG and/or IgM antibody responses, dependent upon the time when T cells were transferred. Conversely, BGG-(bovine γ globulin)-primed T cells could suppress anti-DNP antibody response of recipients against DNP-BGG, while they failed to do so if recipients were immunized with DNP-KLH. Furthermore, we were able to demonstrate that the suppression was induced by thymocytes and spleen cells bearing θ-isoantigen. With this well-defined system, some interesting facets of T cell-mediated suppression have been revealed. These are summarized as follows: 1) The suppressive effect of carrier-specific T cells on the hapten-specific antibody response primarily depends on the time of administration of the T cells to the immunized host. 2) IgG antibody response is, in general, more susceptible than IgM antibody response to the suppressive effect of T cells. 3) T cell-dependent process of the antibody response is more severely affected by suppressor T cells than the relatively T cell-independent process. 4) B cells having higher affinity receptor for the hapten are more susceptible than those with low affinity receptor to the suppressive influence of T cells, and thus suppressor T cells cause the arrest of antibody maturation. 5) To elicit strong suppression, haptens must be present on the same carrier molecule by which suppressor T cells have been stimulated. The latter two findings were important as suspects to the presence of antigen-specific factor which mediated this suppressive cell interaction.

These results naturally led us to investigate the subcellular component of suppressor T cells in this well-defined antibody response in mice. The present communication deals with three different approaches involved in investigating the antigen-specific suppressive component, and in determining its nature and activity in the hapten-specific IgM and IgG antibody responses in the mouse. These approaches are: 1) to find out the in vivo inhibitory effect of the T cell product on primary antibody response, 2) to characterize the factor in an in vitro culture system in which primed cells were stimulated with relevant antigen, and 3) to produce heterologous antibody against the T cell factor.

I shall attempt here to demonstrate, using the above three experimental system, the nature and activity of the antigen-specific T cell factor of the mouse in relation to the possible nature of the antigen-binding receptor of T cells. However, many of the data presented here are more preliminary than conclusive mostly due to the lability of the component to various separation and absorption procedures, and more extensive analyses are currently being carried out in our laboratory.

EXPERIMENTS AND RESULTS

1. In vivo activity of the suppressive T cell factor

Donor BALB/c J mice, six weeks old, were immunized with 100 μg of soluble KLH twice at a two-week interval. Animals were killed two weeks after the second injection, and their thymuses and spleens were removed. Single cell suspensions of thymocytes and spleen cells were prepared at a concentration of 1 x 10^9 cells per ml in Eagle's minimal essential medium (MEM) containing 4 mM HEPES. The cells were disrupted by sonication for 5 min, and cell-free extracts were obtained by ultracentrifugation at 40,000 G for 2 hrs. These will be designated as KLH-primed T-ext and S-ext. For some control studies, similar extracts were prepared from thymocytes and spleen cells obtained from normal BALB/c mice and from those immunized with unrelated antigen, BGG or egg albumin (EA).

To test the effect of primed cell extracts, recipient animals were immunized by a single intraperitoneal injection of 100 μg of DNP_{880}-KLH mixed with 10^9 Bordetella pertussis vaccine followed by an intravenous injection of T-ext or S-ext of various origins corresponding to 1 x 10^8 cells. Anti-DNP antibody-producing cells were enumerated of their spleens by the hemolytic plaque method of Cunningham and Szenberg (19) 6 days after the immunization and the transfer of extracts using sheep erythrocytes (SRBC) coupled with dinitrophenylated bovine serum albumin (DNP_{31}-BSA) by $CrCl_3$. The plaque forming cells (PFC) detected without addition of anti-mouse IgG were considered to be IgM producers (direct PFC), and those detected only with anti-IgG to be IgG producers (indirect PFC).

As reported previously (17), normal BALB/c mice immunized with DNP-KLH plus BPV produced a good primary antibody response against the hapten as measured by direct and indirect PFC on day 6. When such animals were injected with KLH-primed T-ext and S-ext, they did not show any significant decrease in the number of IgM antibody-forming cells, whereas the number of IgG antibody forming cells was greatly reduced (Table 1). The somewhat large deviation observed in the number of indirect PFC in the group given KLH-primed T-ext is primarily due to the presence of completely unresponding animals. On the other hand, both T-ext and S-ext from animals immunized with an unrelated antigen (BGG) showed no significant suppressive effect on this antibody response.

A reverse experiment was performed using DNP-BGG instead of DNP-KLH as immunizing antigen. In this case the response was lower than that obtained by immunization with DNP-KLH. However, BGG-primed T-ext and S-ext exerted a significant suppressive effect on the anti-hapten antibody response, provided animals were immunized with DNP-BGG, but not with DNP-KLH.

In order to confirm the specificity of the observed suppression, KLH-primed T-ext was absorbed with an immunoadsorbent composed of homologous (KLH) or heterologous antigen (BGG). The KLH-primed T-ext was incubated with Sepharose 4B particles coupled with KLH or BGG for 6 hrs with occasional shaking, and the absorbed extract was obtained by suction in a small column with a glass filter. The extract was likewise injected into animals that were immunized with DNP-KLH. As depicted in Table 2, the suppressive effect of KLH-primed T-ext was largely absorbed with KLH, while it was not affected by absorption with BGG. Here again, there were some completely refractory animals in the suppressed groups resulting in the large standard error values in the indirect PFC. The differences were statistically significant as can be seen in Table 2. The results clearly indicate that KLH-primed T-ext contains a soluble component having specificity and affinity to KLH, which preferentially suppresses IgG antibody response against hapten coupled to the homologous carrier.

The approximate molecular weight of the suppressor component was studied using fractions separated by gel-filtration with Sephadex G-200. Before applying to the Sephadex column, the sample was dialyzed against an equal quantity of borate buffered saline, and the dialysate was collected. The extract was further dialyzed against a large quantity of saline, and then was applied to a recycling Sephadex G-200 column (2.5 x 90 cm x 2) with a constant upward flow of borate buffered saline at 4°C. Figure 1 shows a representa-

TABLE 1
SUPPRESSION OF HAPTEN-SPECIFIC ANTIBODY RESPONSE BY CARRIER-PRIMED THYMOCYTE OR SPLEEN CELL EXTRACT

Immunizing antigen: DNP-KLH

Extract injected[a]	PFC/spleen on day 6	
	Direct	Indirect
---	7,280 ×÷ 1.24	11,100 ×÷ 1.15
KLH-primed T-ext	7,800 ×÷ 1.19	1,120 ×÷ 4.20
KLH-primed S-ext	4,790 ×÷ 1.37	2,930 ×÷ 1.45
BGG-primed T-ext	9,100 ×÷ 1.78	10,500 ×÷ 2.18
BGG-primed S-ext	9,650 ×÷ 1.79	13,600 ×÷ 1.96

a. Extracts corresponding to 1 x 10^8 cells were injected intravenously into recipients simultaneously with the immunization with 100 μg of DNP-KLH and 10^9 *Bordetella pertussis* vaccine.

TABLE 2
EFFECT OF ABSORPTION WITH ANTIGENS ON SUPPRESSIVE EFFECT OF KLH-PRIMED THYMOCYTE EXTRACT

Extract injected[a]	Absorption with	PFC/spleen on day 6	
		Direct	Indirect
---	---	5,360 ×÷ 2.08	7,700 ×÷ 1.47
KLH-primed T-ext	---	7,370 ×÷ 2.18	699 ×÷ 5.99*
KLH-primed T-ext	KLH	10,300 ×÷ 1.33	3,970 ×÷ 2.27**
KLH-primed T-ext	BGG	5,590 ×÷ 2.71	469 ×÷ 6.40***

*$P < 0.05$ **$P > 0.2$ ***$P < 0.02$

a. KLH-primed T-ext corresponding to 1 x 10^8 cells was absorbed with immunoadsorbent composed of KLH or BGG and injected into recipient on day 0.

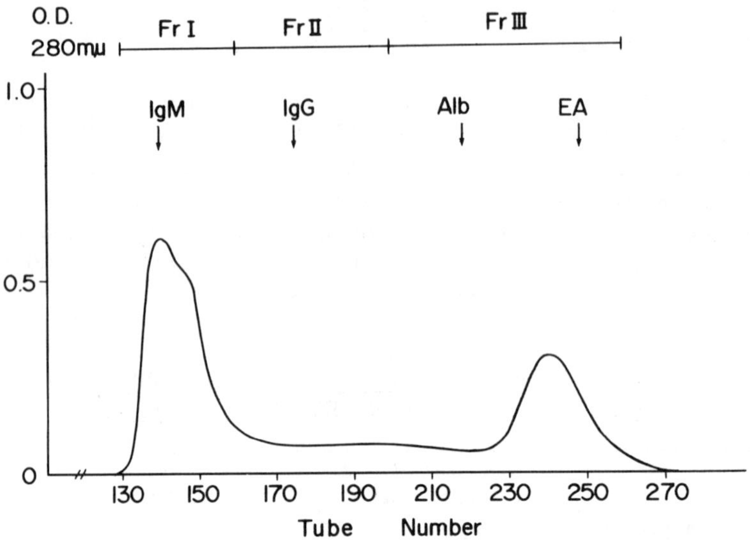

Fig. 1 Elution profile of S-ext from a Sephadex G-200 column. S-ext corresponding to 2 x 10^9 KLH-primed spleen cells was applied to a column of Sephadex G-200 (2.5 x 90 cm x 2). Arrows indicate the position of the marker proteins eluted at the identical condition. Three fractions were separated in the figure, and were tested for the presence of suppressor activity.

tive elution profile of S-ext. A similar elution pattern was obtained with T-ext. The arrows in the figure indicate the position of marker proteins with known molecular weights eluted under the identical condition. Three fractions were separated as indicated in the figure; Fr. I contained proteins eluted with the void volume, Fr. II those corresponding to 7S IgG, and Fr. III corresponded to those having molecular weight less than 100,000. These fractions were concentrated by pressure dialysis, and further tested for the presence of the suppressor activity.

Table 3 shows the distribution of suppressive activity among those fractions of KLH-primed S-ext. A significant suppression in indirect PFC was observed only in Fr. III which had molecular weight range between 100,000 and 30,000 daltons. The results indicate that suppressor component is not a usual immunoglobulin, and are consistent with those found with the rat suppressor molecule (16).

More recently, we found that the above suppressor component failed to suppress the antibody response of histoincompatible recipients. For example, the suppressor factor from C57BL mice did not suppress the response of BALB/c mice, while under the same experimental condition it suppressed the antibody response of syngeneic C57BL recipients. As will be apparent in the later section, the suppressor molecule contains genetic expression of the H-2 complex, and thus it may be able to interact only with histocompatible lymphoid cells.

2. Characterization of the antigen-specific suppressive component in an in vitro culture system

In order to characterize further the nature of the suppressive T cell component, primed spleen cells were cultured in Marbrook culture bottle containing 1 ml of cell suspension in MEM enriched with 10% fetal calf serum in the inner chamber of the bottle, and 10 ml of the same medium outside the chamber separated by a cellophane membrane. 1×10^7 spleen cells obtained from donor mice primed with 100 μg of DNP-KLH + BPV one month before were cultured in the chamber in the presence of 0.1 μg/ml of DNP-KLH at 37°C in the atmosphere containing 90% air and 10% CO_2. This condition was found to be optimal to obtain *in vitro* secondary antibody response against the DNP determinant.

At the beginning of the culture, T-ext or S-ext from various sources was added to the inner chamber of the bottle. Maximal PFC response occurred on day 4 or 5 in varying numbers depending on the culture condition. Table 4 shows the result of one experiment. The addition of KLH-primed T-ext or S-ext significantly reduced the number of indirect

TABLE 3

DISTRIBUTION OF SUPPRESSIVE ACTIVITY OF SPLEEN CELL EXTRACT IN SEPHADEX G-200 FRACTIONS IN IN VIVO ANTIBODY RESPONSE

Fraction added[a]	Marker protein	PFC/spleen on day 6	
		Direct	Indirect
---	---	10,850 $\overset{\times}{\div}$ 1.53	4,160 $\overset{\times}{\div}$ 1.43
Fr. I	IgM	7,680 $\overset{\times}{\div}$ 1.67	7,440 $\overset{\times}{\div}$ 1.65
Fr. II	IgG	11,830 $\overset{\times}{\div}$ 1.70	5,421 $\overset{\times}{\div}$ 2.20
Fr. III	Alb-EA	3,520 $\overset{\times}{\div}$ 3.93	640 $\overset{\times}{\div}$ 4.88*

*P < 0.02

a. Fractions of KLH-primed spleen cell extract corresponding approximately to 1 x 10^8 cells were injected into recipients on day 0.

TABLE 4

SUPPRESSION OF IN VITRO HAPTEN-SPECIFIC PFC RESPONSE BY THYMOCYTE OR SPLEEN CELL EXTRACT FROM ANIMALS IMMUNIZED WITH CARRIER

Extract added[a]	PFC/culture	
	Direct	Indirect
---	138 ± 133	1,397 ± 438
Normal T-ext	117 ± 32	1,347 ± 456
Normal S-ext	98 ± 42	1,175 ± 201
KLH-primed T-ext	319 ± 90	277 ± 201
KLH-primed S-ext	108 ± 77	339 ± 203
EA-primed T-ext	154 ± 35	1,311 ± 578

a. T or S-ext corresponding to 1 x 10^7 cells were added to each culture tube on day 0.

(IgG) PFC, while the number of direct PFC was not significantly altered. On the other hand neither the extracts from normal nor EA-primed thymocytes and spleen cells showed such suppressive activity. The results confirm the specificity observed in the *in vivo* studies.

The dose of the T-ext required for a significant suppression was determined by adding varying amounts of KLH-primed T-ext corresponding to the different number of thymocytes. The extract corresponding to 10^6 primed thymocytes produced significant suppression of indirect PFC response. Neither the greater nor smaller amount of T-ext exerted enhancing effect on the *in vitro* antibody response. Since the suppressive activity of the T-ext varied from one sample to the other, the dose corresponding to 1×10^7 cells was routinely used for the rest of the study.

The molecular weight of the suppressive component was also studied using the fractions obtained by Sephadex G-200 gel-filtration. The fractions of KLH-primed T-ext, as well as the dialysate, corresponding to approximately 10^7 original thymocytes were added to culture tubes in which 10^7 primed spleen cells were cultured in the presence of 0.1 μg/ml of DNP-KLH.

The result is shown in Table 5. Despite that the direct PFC response varied among groups with different fractions, only the Fr. III eluted later than 7S IgG caused a significant suppression of indirect PFC response. Although it was often experienced that the suppressive activity of T-ext was greatly reduced by such separation procedures, only the Fr. III produced variable but significant degree of suppression in several separate experiments. Since the upper limit of the molecular weight of this fraction is about 100,000 daltons, it is concluded that the suppressive component has smaller molecular weight than usual immunoglobulins being consistent with the results of *in vivo* studies.

In order to confirm the specificity of the suppressor component, the crude extract of KLH-primed thymocytes, as well as the suppressor fraction, Fr. III, were absorbed with the immunoadsorbent composed of KLH or unrelated antigen, ascaris extract (Asc). The samples were incubated with the respective immunoadsorbents in small columns for 6 hours at 4°C, and then were eluted with cold saline. Putatively the same amount of the absorbed materials was added to the culture to test the suppressive activity. As shown in Table 6, suppressive activity of KLH-primed T-ext was almost completely absorbed by KLH-immunoadsorbent but not by Asc-immunoadsorbent. It was also found that the suppressor molecule in Fr. III also had an affinity for KLH. Here

TABLE 5
DISTRIBUTION OF SUPPRESSIVE ACTIVITY OF THYMOCYTE EXTRACT IN SEPHADEX G-200 FRACTIONS IN IN VITRO ANTIBODY RESPONSE

Fraction added[a]	PFC/culture	
	Direct	Indirect
---	441 ± 77	1,791 ± 343
dialysate	21 ± 85	1,530 ± 138
Fr. I	586 ± 250	1,476 ± 647
Fr. II	82 ± 68	1,411 ± 610
Fr. III	43 ± 23	315 ± 258*

*P < 0.001

a. Fractions of T-ext corresponding to 1 x 10^7 cells were added to each culture tube on day 0.

TABLE 6
ABSORPTION OF THE SUPPRESSIVE COMPONENT BY ANTIGEN

Suppressor added	Absorbed[a] with	PFC/culture	
		Direct	Indirect
---	---	691 ± 204	3,180 ± 460
KLH-primed T-ext	---	1,472 ± 520	856 ± 273
KLH-primed T-ext	KLH	609 ± 242	2,804 ± 555
KLH-primed T-ext	Asc	1,335 ± 297	750 ± 408
Sephadex Fr. III	---	1,044 ± 188	872 ± 400
Sephadex Fr. III	KLH	628 ± 206	2,549 ± 777

a. KLH-primed T-ext and the Sephadex Fr. III were absorbed with immunoadsorbents composed of KLH or Asc before addition to the culture tubes.

again, there were no meaningful differences in the number of direct PFC between the groups.

The immunochemical nature of the suppressor molecule was studied by absorption with various antibodies in the form of solid immunoadsorbents. KLH-primed T-ext was similarly absorbed with immunoadsorbents passing through columns containing Sepharose 4B with γ globulin fractions of various antibodies. As shown in Table 7, the suppressive activity was not significantly absorbed with anti-heavy chain antibodies, although sometimes the passage through anti-μ columns slightly reduced the suppressive activity. On the other hand, the suppressive activity was greatly reduced by absorption with rabbit anti-mouse thymocyte serum (ATS), C57BL anti-H-2^d (anti-DBA/2 mastocytoma) antisera consistently in repeated experiments. A guinea pig anti-serum which had been raised against the KLH-T cell factor complex (anti-Tr, see below) was also capable of absorbing suppressor activity. In a preliminary absorption experiment using anti-mouse light chain antiserum, the activity was mostly preserved after absorption.

However, it has been experienced repeatedly that the suppressor activity is unstable to the absorption and separation procedures, and that in some occasions the activity disappeared even after keeping the material for a few days in the refrigerator. It was also noted that the suppressor activity was easily visualized when the spleen cells had been primed only once 4 weeks before harvesting, whereas the same suppressor material produced only a slight suppression of the anti-hapten antibody response of hyper-immunized spleen cells. Since the condition most suitable to emerge suppressor T cells has not been established, the suppressor activity of T-ext varied from one sample to the other. It was only noted that if the cells were taken from animals given third injection of KLH two days before harvesting, the extract showed no suppressive activity but rather enhanced the response. Such variables also have prevented more exact study of the nature of the suppressive component.

Although our experiments are not conclusive as yet and need further confirmation, it is most probable that the presently described antigen-specific suppressor molecule is not a usual immunoglobulin but a unique T cell surface molecule which is closely related to the antigen-binding receptor of T cells and is probably linked to H-2 antigens. The next experiment has been undertaken to investigate this possibility.

TABLE 7
EFFECT OF ABSORPTION WITH VARIOUS ANTIBODIES ON THE SUPPRESSIVE EFFECT OF KLH-PRIMED T EXTRACT

Extract added	Absorbed with[a]	PFC/culture	
		Direct	Indirect
---	---	892 ± 364	1,652 ± 471
KLH-primed T-ext	---	498 ± 498	313 ± 308
KLH-primed T-ext	Anti-μ	750 ± 623	561 ± 368
KLH-primed T-ext	anti-γ	965 ± 369	288 ± 391
KLH-primed T-ext	ATS	594 ± 275	1,001 ± 241
KLH-primed T-ext	Anti-H-2^d	582 ± 228	1,387 ± 350
KLH-primed T-ext	Anti-Tr[b]	768 ± 281	1,431 ± 569

a. Absorption was carried out by passing through a column of immunoabsorbent composed of one of the antisera.

b. See section 3.

3. An attempt to produce a heterologous antiserum reactive to the T cell component

Since above studies indicated that immune T-ext contained a soluble subcellular component which can specifically bind the immunizing antigen, an attempt was made to produce antibody against the active component. In this study we used A/He (H-2^a) mice to obtain the suppressive component, because this strain has k alleles only in the K and I regions of the H-2 complex. Mice were immunized with 100 μg of KLH twice at a two-week interval, and thymocyte and spleen cell extracts were obtained by the same procedure used in the previous experiment. The extracts were centrifuged at 78,000 G for 2 hours to remove any of the material sedimentable at this centrifugation. They were mixed with a small amount of associated form KLH which had been collected by repeated ultracentrifugations. The mixture was kept overnight at 0°C with shaking, and then was centrifuged at 78,000 G for 2 hrs to recover KLH which might have bound any reactive substance in the T or S extract. The sediments were washed twice with cold saline by ultracentrifugation.

The sediments which were possibly composed of KLH and KLH-reactive component(s) from lymphoid cells were included in Freund's complete adjuvant, and were injected intracutaneously into guinea pigs via the four footpads. Guinea pigs were boosted intramuscularly with the same material 4 to 5 times at two-week intervals. The extracts obtained from more than 300 mice were used to immunize a single guinea pig. The sera were taken two weeks after the last immunization.

The antisera were repeatedly absorbed with a large amount of KLH-immunoadsorbent to remove anti-KLH antibodies present in the antisera. Some antisera having high cytotoxic activity against bone marrow cells were further absorbed with cell suspensions of bone marrow, kidney and liver to render them specific for lymphoid cells. Although they did not react with mouse serum proteins in immunodiffusion, an additional absorption was made by adding small amount of normal A/He serum. Since the antiserum raised against KLH-S-ext complex killed 100% of both thymocytes and spleen cells, it was absorbed with either thymocytes or splenic B cells that had been obtained by the treatment of normal spleen cells with anti-θ and C followed by centrifugation on Ficoll-Hypec. The resulting antisera, which are only cytotoxic for T or B cells, are designated as anti-Tr or anti-Br, since they are theoretically reactive to the determinants associated with antigen-binding receptor of T or B cells.

The anti-Tr which had been obtained by immunization with

KLH-T-ext complex killed 100% of thymocytes in the cytotoxic assay, but killed less than 40% spleen cells under the same condition. The same was also true for the anti-Tr which had been obtained by absorption of anti-KLH-S-ext complex with splenic B cells. Both antisera showed rather weak cytotoxicity on splenic T cells which were purified by passage through a nylon fiber column. On the other hand, anti-Br, which was obtained by absorption with thymocytes, killed about 60% of spleen cells. The specificity of these antisera was further warranted by the inhibition of PFC of immune spleen cells: If the spleen cells from mice immunized with DNP-KLH were treated in vitro with anti-Br and C, the number of DNP-specific PFC was greatly reduced, whereas the treatment with anti-Tr and C had little effect on the number of PFC.

The cytotoxic tests performed on lymphoid cells from various inbred strains of mice having different H-2 genotypes revealed that there were no strain differences in the reactivity with these antisera, which indicates that the antibodies are not directed against alloantigens present on T or B cells. This is not surprising, if we assume that guinea pigs, being heterologous animals, may not be able to recognize alloantigens associated with lymphoid cells, while they are able to respond to common antigenic determinants shared by all mouse strains to which allogeneic mice cannot produce antibody. It was also apparent that the cytotoxic activity was not related with θ and Ly antigens. The detailed description of these antisera will be published elsewhere.

The most important point considered critical to the nature of anti-Tr is whether or not this antiserum, in fact, reacts with receptor-associated antigen on T cells. The first approach was made to study whether the anti-Tr could absorb the suppressive factor or not. Since there were no strain differences in the reactivity with T cells, the suppressive T-ext of BALB/c mice was absorbed with an immunoadsorbent composed of anti-Tr, and was tested in in vitro culture system as described in section 2. It was found that the suppressive activity of KLH-primed T-ext was greatly reduced by single passage through the column of anti-Tr immunoadsorbent (Table 7).

Since this in vitro secondary antibody response involves a cooperation between KLH-primed T cells and DNP-primed B cells, a small amount of anti-Tr was added in culture tubes to see whether the presence of anti-Tr could inhibit this anti-DNP antibody response. In the same experimental system, two different anti-$H\text{-}2^d$ antisera were also tested for their inhibitory activity. As shown in Table 8, the presence of either anti-Tr or anti-H-2 antisera greatly reduced the number of DNP-specific PFC. Thus, it seems that anti-Tr interferes with some processes of T-B cell cooperation, either by blocking the antigen recognition of T cells or by reacting with T cell product(s) required for B cell stimulation.

CONCLUSION AND SUMMARY

Although at the present moment, there are no conclusive data to show the definitive nature of the suppressive component from T cells, the above approaches give a vague picture of the antigen-specific T cell component involved in cooperating events in the antibody response. The suppressive component I described here has specificity and affinity for the carrier molecule by which animals have been primed. The component is unlikely to be a usual immunoglobulin with respect to the immunochemical and physicochemical properties, although the possibility of the presence of V region-like determinants is not completely eliminated. On the other hand, the activity was absorbed with anti-sera reactive for surface determinants of T cells (ATS and anti-Tr) and anti-H-2 antibodies. Thus, it seems that the active component may be closely related to the antigen receptor of T cells which is different from immunoglobulin, and perhaps be linked to some of the expressions of H-2 complex. The component is clearly analogous to that reported by Taussig and Munro (20) with respect to the immunochemical and physicochemical properties, although their antigen-specific factor enhances rather than suppresses the T cell-dependent antibody responses. The apparent discrepancy between these two independent studies should be clarified in further experiment. Moreover, there are some similarities between the suppressive T cell component and the "allogeneic effect factor" which has recently been characterized by Armerding and Katz (7), despite that the latter has no antigen-specificity. Thus, it seems that both the similarity and heterogeneity of these T cell factors warn the further comparative studies in the future.

However, an important point concerning the presently described suppressive T cell component is that it does not seem to be a so-called T cell replacing factor. This has been shown by the fact that the suppressor effect of primed T cell extract was only envisaged when the helper T cells and precursor B cells were coexisting in the culture; it neither enhanced nor suppressed the response of T cell-depleted spleen cells. Also, we have tried to detect the

TABLE 8

SUPPRESSION OF IN VITRO ANTIBODY RESPONSE OF DNP-KLH-PRIMED SPLEEN CELLS BY ANTI-Tr AND ANTI-H-2 ANTISERA

Antiserum added[a]	dose	Anti-DNP PFC/culture	
		Direct	Indirect
	μl		
Normal guinea pig serum	50	669 ± 217	2,543 ± 573
Anti-Tr	50	322 ± 108	734 ± 220
Normal C57BL serum	50	602 ± 198	1,755 ± 343
Anti-H-2^d	50	57 ± 40	102 ± 58
Anti-H-2^d	10	490 ± 253	358 ± 340

a. DNP-KLH-primed spleen cells were cultured in the presence of 0.1 μg of DNP-KLH and respective antiserum.

effect of the KLH-specific suppressor component in the culture in which DNP-specific B cells and EA-specific T cells were stimulated with the mixture of DNP-EA and DNP-KLH. The KLH-primed T-ext neither enhanced nor suppressed the DNP-specific antibody response even though DNP-KLH was present in the culture. It only suppressed the background DNP-specific PFC response which was elicited by DNP-KLH in the culture of spleen cells obtained from mice primed with DNP-EA. The results now suggest that the KLH-specific suppressor molecule does not directly act on B cells, but rather suppresses the emergence of KLH-specific helper effect of T cells. This possibility has been presented by Gershon (1) in several experimental models, and perhaps is relevant to interpret the well-known effect of antigenic competition, while there is again a disparity in the specificity. However, due to the low yield of the factor, as well as to the instability of the activity during the separation and absorption procedures, it is extremely difficult to draw a conclusive picture of our T cell component at the present time, and we are currently conducting some more confirming experiments.

Notwithstanding the above restrictions, we were able to raise a heterologous antiserum which can absorb the suppressive activity of T cell extract. The antiserum was more reactive to thymocytes than to splenic T cells, and interferes with the T cell-dependent antibody response _in vitro_. Since anti-H-2 antisera and ATS produced a similar suppression, it cannot be concluded at the present moment that this antiserum, in fact, reacts with determinants associated with T cell receptor. However, it is tempting to consider that the antigen recognition unit of T cells is a complex molecule containing both alloantigens coded by genes in the H-2 complex and an additional common antigen which is shared by strains of mice, although the structure of receptor site for antigen in this whole unit is still controversial. The presence of such a common determinant reactive to anti-Tr in thymocytes more than in splenic T cells suggests that the polymorphism of the T cell components has some bearing on the diverse cooperative events, e.g. suppression and enhancement via specific and nonspecific pathways, in the immune response.

Although we were not so far able to demonstrate enhancing activity of T cell extracts as had been envisaged in the rat system, some studies were started to explore the enhancing T cell factor from the similar source. Since it was noted that the T-ext or S-ext obtained a few days after the tertiary immunization with KLH had no suppressive activity, but rather enhanced the antibody response of the recipients (see p. 781), we are currently working on the detection and separation of enhancing T cell component. As the specificity of the enhancing and suppressive T cell components in the rat is obviously the same, it is probable that the suppressor and helper molecules may share a structural homology, while the other molecular entity may have considerable heterogeneity. The possibility that the antigen-binding receptor of T cells contains primordial V region moiety still remains to be clarified, inasmuch as the presence of carrier-specific IgMs-like molecule has been reported from various laboratories (8,9,21,22). In view of the multiplicity of T cell factors reported by several investigators, the clarification of the interrelationship between these factors should be a critical point in understanding the regulation of antibody responses as pointed out by Katz and Benacerraf (23).

ACKNOWLEDGMENTS

I am extremely grateful to my collaborators Drs. M. Taniguchi, N. Shinohara, T. Takemori, A. Yano, K. Hayakawa and T. Matsuzawa who participated in the above studies.

REFERENCES

1. Gershon, R.K., in Contemporary Topics in Immunobiology, Vol. 3, Edited by Warner, N.L., Plenum Press, New York, 1974.
2. Dutton, R.W., Falkoff, R., Hirst, J.A., Hoffmann, M., Kappler, J.W., Kettman, J.R., Lesley, J.F. and Vann, D., in Prog. Immunol., Vol. 1, pp. 355, Edited by Amos, B., Academic Press, New York and London, 1972.
3. Schimple, A. and Wecker, E., Nat. New Biol. 237: 15, 1972.
4. Rubin, A.S. and Coons, A.H., J. Immunol. 108: 1597, 1972.
5. Gorczynski, R.M., Miller, R.G. and Phillips, R.A., J. Immunol. 110: 968, 1972.
6. Doria, G., Agarossi, G. and DiPietro, S., J. Immunol. 108: 268, 1972.
7. Armerding, D. and Katz, D.H., J. Exp. Med. 140: 19, 1974.
8. Feldmann, M. and Basten, A., J. Exp. Med. 136: 722, 1974.
9. Feldmann, M., Cone, R.E. and Marchalonis, J.J., Cell Immunol. 9: 1, 1973.
10. Rosenthal, M., Stastny, P. and Ziff, M., J. Immunol. 111: 1119, 1973.

11. Sjöberg, O., Anderson, J. and Möller, G., J. Immunol. 109: 1379, 1972.
12. Watson, J., J. Immunol. 111: 1301, 1973.
13. Kishimoto, T. and Ishizaka, K., J. Immunol. 111: 1194, 1973.
14. Taussig, M.J., Mozes, E. and Isac, R., J. Exp. Med. 140: 301, 1974.
15. Tada, T., Okumura, K. and Taniguchi, M., J. Immunol. 111: 952, 1973.
16. Okumura, K. and Tada, T., J. Immunol. 112: 783, 1974.
17. Tada, T. and Takemori, T., J. Exp. Med. 140: 239, 1974.
18. Takemori, T. and Tada, T., J. Exp. Med. 140: 253, 1974.
19. Cunningham, A.J. and Szenberg, A., Immunology 99: 859, 1968.
20. Taussig, M.J. and Munro, A.J., Nature 251: 63, 1974.
21. Taniguchi, M. and Tada, T., J. Immunol., in press, 1974.
22. Gisler, R.H., Staber, F., Rude, E. and Dukor, P., Eur. J. Immunol., in press, 1974.
23. Katz, D.H. and Benacerraf, B., Transplant. Rev., Vol. 22, in press, 1974.

ANTIGEN-SPECIFIC T CELL FACTOR IN CELL COOPERATION AND GENETIC CONTROL OF THE IMMUNE RESPONSE

Michael J. Taussig and Alan J. Munro
Immunology Division, Department of Pathology, University of Cambridge, Tennis Court Road, Cambridge, England

An important discovery in immunology in recent years has been that genes of the major histocompatibility (H) locus, in addition to coding for the antigens involved in allograft rejection, play a fundamental role in cellular immunological processes. The T cell receptor system, cell cooperation and genetic control of immune responsiveness all seem to be bound up with products of the H complex. The important genes involved are found in the I region of the H-2 complex in the mouse, which includes the H-linked Ir-genes, controlling the response levels to a variety of antigens (1,2), and genes coding for the Ia antigens on the lymphocyte surface (3). What role do these genes and their products play in immune processes at the cellular and molecular level? In order to answer this question, we have studied the properties of an antigen-specific T cell product or 'factor' which is able to replace the requirement for T cells in certain immune responses, notably cooperation with B cells (4). The T cell factor appears to carry a binding site for antigen, and significantly is a product of the left-hand side (K or I regions) of the H-2 complex of the mouse (5,6). The evidence is very suggestive that the T cell factor is the soluble expression of the T cell receptor, and that the latter is therefore, at least in part, an H-2 (K or I) product. The factor has also proved useful in studying the cellular basis of genetic control of immune responses, since it can be used as a direct measure of the functional capacity of T and B cells (7,8).

MATERIALS AND METHODS

Preparation of T cell factor.

The preparation and test of the T cell factor are shown diagrammatically in Figure 1 (a,b). The factor is prepared from mouse thymocytes which have been specifically primed or

---

Abbreviations

Synthetic polypeptide antigens:

(T,G)-A--L - poly(tyr,glu)-polyDLala--polylys.
(Phe,G)-A--L - poly(phe,glu)-polyDLala--polylys.
(H,G)-A--L - poly(his,glu)-polyDLala--polylys.
(Phe,G)-Pro--L - poly(phe,glu)-polypro--polylys.

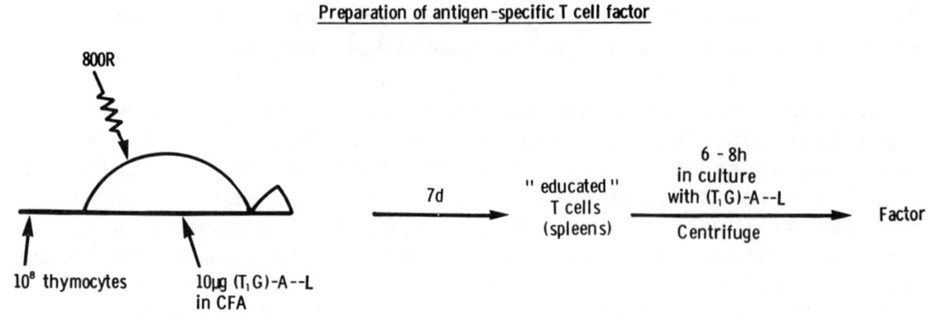

Fig 1a

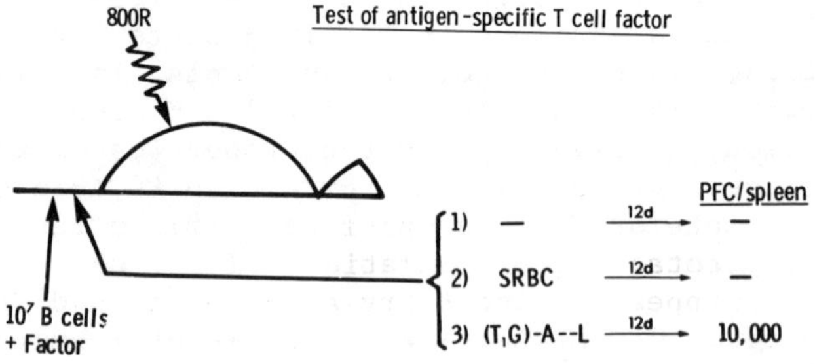

Fig 1b

Fig. 1 a) Preparation and b) Test of Cooperative Activity of T Cell Factor Specific for (T,G)-A--L.

'educated' by antigen. The antigens used to date have been the synthetic polypeptides prepared by Sela and colleagues, chief among which in this work is (T,G)-A--L (9). Education of T cells involves transfer of 10^8 thymocytes into an irradiated (800 r) recipient which is immunized with (T,G)-A--L in CFA. 7 days later the spleens of these animals, containing the educated T cells, are removed and a cell suspension is prepared in minimal Eagle's medium. The suspension is cultured for a period of 6-8 hours together with the priming antigen in a small Petri dish. At the end of this time the cells are removed by centrifugation - the supernatant is used as the source of T cell factor. Note that no factor activity is produced if the antigen is either omitted from the culture or replaced by a non-cross-reacting antigen; treatment of the cells with anti-θ serum also prevents factor production (4). The factor is assayed by transfer together with 10^7 bone marrow (B) cells and soluble antigen into a second irradiated recipient, syngeneic with the bone marrow donor (Fig. 1b). 12 days later the spleens of these recipients are removed and assayed for plaque forming cells. Table I illustrates the levels of direct p.f.c. to (T,G)-A--L obtained in a typical experiment. Approximately 10,000 p.f.c. per spleen are regularly obtained, compared with a background of 1-200. One spleen equivalent of T cell factor produces as high a number of p.f.c. as $5x10^6$ viable educated T cells, or 10^8 normal thymocytes.

A notable - and somewhat inconvenient - feature of the T cell factor is its instability. When stored at 4° the factor activity decays rapidly after 24-36 hours; it has also proved very difficult to retain activity by freezing. For this reason, factor is prepared fresh for each experiment. It remains to be seen whether the instability is inherent in the structure of the factor, or caused by enzymes released during the in vitro incubation, though enzyme inhibitors have little stabilizing effect.

RESULTS

Specificity of T cell factor

The most significant property of the T cell factor is its specificity for antigen - factor produced to (T,G)-A--L will not cooperate in the response to a non-cross-reacting antigen such as sheep red cells (4). Moreover, the (T,G)-A--L-specific factor can be removed by a (T,G)-A--L immunoadsorbent ((T,G)-A--L coupled to CNBr-activated sepharose) (6). This indicates that the factor possesses a specific binding site for antigen. An important question, to which future effort will be directed, is how far this factor binding site resembles the antigen-

TABLE I

ACTIVITY OF A SPECIFIC COOPERATIVE T CELL FACTOR IN VIVO

Cells etc. transferred into irradiated recipients	Antigen	$\log_{10}$ P.F.C./spleen
B cells[a]	(T,G)-A--L[e]	2.122
B cells + normal T cells[b]	(T,G)-A--L	3.516
B cells + T cells educated to (T,G)-A--L[c]	(T,G)-A--L	3.868
B cells + Factor produced by T cells educated to (T,G)-A--L[d]	(T,G)-A--L	3.910
	Standard Error	0.210

a) 10^7 bone marrow cells. b) 10^8 thymocytes. c) 1 spleen equivalent (approx. 5×10^6 cells). d) 1 spleen equivalent. e) 10 µg in solution.

TABLE II

ADSORPTION OF T CELL FACTOR BY ALLOANTISERA ADSORBENTS

Factor	Adsorbent	$\log_{10}$ P.F.C./spleen[a]
-*		2.133
$H\text{-}2^d$ (B10.D2)	-	4.107
	anti-$H\text{-}2^d$	2.220
	anti-$H\text{-}2^k$	4.198
	Standard Error	0.145
-*		2.220
$H\text{-}2^k$ (B10.BR)[b]	-	4.166
	anti-$H\text{-}2^k$	2.112
	anti-$H\text{-}2^d$	4.170
	Standard Error	0.130

*Control - B cells + (T,G)-A--L transferred into irradiated recipients without T cells factor.

a) Direct anti-(T,G)-A--L plaque forming cells (geometric means).

b) Activity measured in B10 (high responder) recipients.

Antisera used: anti-$H\text{-}2^d$ - DBA/2 spleen cells injected into B10.BR.
anti-$H\text{-}2^k$ - CBA injected into B10.D2.

binding site on conventional antibody in terms of size, specificity, affinity, etc. At this time, the only relevant piece of information on this point is that the factor for (T,G)-A--L cross-reacts significantly in terms of activity with the closely related polypeptide, (Phe,G)-A--L, in which tyrosine (T) is replaced by phenylalanine (Phe). In collaboration with Edna Mozes and Ronit Isac (Rehovot), factor for (T,G)-A--L was prepared in C3H.SW mice, high responders to both antigens, and was able to cooperate well with B cells and (Phe,G)-A--L. Antibodies to (T,G)-A--L and (Phe,G)-A--L also cross-react significantly. It remains to be seen if the (T,G)-A--L factor can be absorbed by a (Phe,G)-A--L -coated column, or if it can be separated into specific and cross-reactive fractions. The cross-reactivity is only found between these very closely related antigens - (T,G)-A--L factor does not cross-react with (Phe,G)-Pro--L, in which the poly-alanine (A) side chains are replaced by poly-proline (Pro).

Molecular nature of the T cell factor.

The most obvious possibility is that the factor is conventional immunoglobulin (Ig), but for the following reasons it seems clear that the factor is non-Ig in nature.

a) Attempts to remove the factor by passage over anti-Ig immunoadsorbents have been uniformly negative. The reagents used include anti-Fab, anti-IgM and anti-Fv sera (5,6).

b) The molecular weight of the factor, as determined by Sephadex chromatography is in the region of 50,000 - the factor runs just behind BSA on Sephadex G100 (6).

c) The factor has been positively identified as a product of the H-2 complex by absorption with anti-H-2 sera (5,6).

The experiments which established that the factor is an H-2 product are summarized in Table II. Factor for (T,G)-A--L was prepared in various congenic strains of the B10 series, and tested before and after passage over congenically-raised alloantisera, immobilized as immunoadsorbents. As shown in Table II, factor produced by B10.D2 mice was removed by an anti-H-2^d column, but not by anti-H-2^k; and vice versa, B10.BR (H-2^k) factor was removed by anti-H-2^k but not by anti-H-2^d adsorbents (5,6). This result has now been obtained with other haplotypes, and it appears a general rule that the factor carries alloantigens of the strain in which it is produced.

Further mapping of the factor has been achieved using the H-2^a haplotype, a recombinant which has the K,Ir-1A and Ir-1B loci of H-2^k, and the I-C, Ss and D loci of H-2^d. The factor produced by B10.A (H-2^a) mice is removed completely by anti-H-2^k sera, and not at all by anti-H-2^d, as shown in Table III (6). Further use of antisera raised specifically against the

TABLE III

IDENTIFICATION OF T CELL FACTOR AS A PRODUCT OF THE LEFT-HAND (K) SIDE OF H-2

Factor	Adsorbent	$\log_{10}$ P.F.C./spleen+
-*		1.462
H-2^a (B10.A)	-	2.875
	anti-H-2^k	1.301
	anti-H-2^d	2.740
	Standard Error	0.142

*Control - B cells and (T,G)-A--L transferred into irradiated recipients without T cell factor.
+Measured in B10.D2 recipients; $\log_{10}$ geometric means anti-(T,G)-A--L direct p.f.c.
Antisera used: anti-H-2^k - CBA spleen cells injected into B10.D2.
anti-H-2^d - DBA/2 injected into B10.BR.

TABLE IV

T CELL FACTOR IN GENETIC CONTROL OF RESPONSE TO (T,G)-A--L

Factor	B Cell Origin	
	High Responder	Low Responder
High Responder	+ (6,850)	- (93)
Low Responder	+ (5,788)	- (109)

High responder - C3H.SW Low responder - C3H/HeJ
Direct p.f.c. to (T,G)-A--L (geometric means).

Data from reference 7.

K or D sides of H-2, or rendered specific by absorption, confirmed this result - that the factor for (T,G)-A--L is a product of the left-hand (K) side of H-2, either of the K or I region. Antisera specific for the I region antigens (anti-Ia sera) are now available and it should soon be possible to discover whether the factor is a product of the K or I genes (see Addendum). Note that these experiments can establish only that various H-2 gene products form at least part of the factor; it is possible that the factor is not entirely H-2 in nature and that a product of a non-H-2 gene becomes coupled to an H-2 product for activity.

These observations on the nature of the antigen-specific cooperative factor invite comparison with those on other T cell products which can replace T cells in antibody production. Feldmann, for example, has described a 7S Ig, which is antigen specific, and which may constitute a new Ig class - T cell produced Ig (IgT, IgX) (10). The possibility that this is in reality a B cell product passively absorbed by T cells (11), casts some doubt on whether this molecule is in fact the primary T cell receptor. Nevertheless, it may well constitute an alternative, Ig-mediated cooperation mechanism. On the other hand, Katz and coworkers have described a non-specific T cell product which is secreted by T cells when cultured with foreign alloantigens, and which can trigger B cells in the presence of antigen (12,13). This allogeneic effect factor (AEF) has now been characterized as an H-2 product, and in particular an I-region product (14). It seems quite possible that AEF and the H-2 derived antigen-specific factor described here are closely related molecules, perhaps even identical. It would be important to establish whether AEF has a binding site for the alloantigens against which it is produced, despite its non-specific action on B cells; and conversely, whether the H-derived antigen-specific factor can have a non-specific stimulatory action on B cells if present in sufficiently high concentration.

Antigen-specific T cell factor in genetic control of the immune response.

Recently, in collaboration with Drs. Edna Mozes and Ronit Isac of the Weizmann Institute, we have utilized the properties of the T cell factor to analyze the cellular basis of genetic control in the antibody response to (T,G)-A--L. The strains under study were C3H.SW ($H\text{-}2^b$), which are high responders to (T,G)-A--L; C3H/HeJ($H\text{-}2^k$), low responders to (T,G)-A--L; and SJL ($H\text{-}2^s$), which are low responders to all polypeptides built on the A--L backbone. The main question under study was, in which cell population does the genetic defect in low responders occur? Since (T,G)-A--L is a thymus-dependent antigen, an antigen-specific defect in response could be attributed to either T or B cell lines, or both. Much evidence, most of it circumstantial in nature, has suggested that T cell recognition of (T,G)-A--L is impaired in low responders (1, 15-19). On the other hand, estimates of T and B cell precursor cell numbers by limiting dilution techniques indicated that, in low responders of $H\text{-}2^k$ haplotype, the defect was expressed in B cells (1,20). In order to try to resolve this rather extreme conflict of view, the T cell factor was used to measure directly the responsiveness of T and B cells of high and low responder strains.

In the first experiments, the congenic resistant strains C3H.SW (high responders) and C3H/HeJ (low responders) were used. T cells of both strains were educated to (T,G)-A--L, and T cell factor prepared in vitro. The factors were tested for their ability to cooperate in vivo with B cells of high responder (C3H.SW) origin in the antibody response to (T,G)-A --L. As shown in Table IV, both high and low responder factors were equally capable of cooperating with high responder B cells; moreover, titration of the two factors did not reveal any inter-strain differences (7,8). By this criterion, therefore, no T cell defect for (T,G)-A--L exists in low responders. However, when the B cells of the two strains were compared for the ability to respond to antigen with factor of either strain, only the high responder B cells were able to mount an immune response (Table IV). Controls established that C3H/HeJ B cells responded efficiently to other antigens, for which that strain is a high responder, when combined with appropriate T cell factors of either C3H/HeJ or C3H.SW origin. The conclusion was thus drawn that the cellular basis of low responsiveness to (T,G)-A--L in $H\text{-}2^k$ mice lies in the inability of B cells to respond to antigen and the appropriate T cell signal.

In order to explain these results, it was proposed that for thymus-dependent antibody responses, B cells must receive and respond to a signal or stimulus, put out by T cells, in the form of the antigen-specific T cell factor. How is the factor signal received by B cells, and how is this event controlled by the H-2 complex? The recent discovery of a complex antigenic system, Ia, coded for by genes in the I region of H-2 and expressed mainly on B cells (3), suggests a possible explanation. Ia antigens on the B cell surface may constitute a receptor for the T cell factor; the spectrum of Ia specificities expressed by each strain may thus determine the antigens to which the B cells of that strain can respond. This suggestion goes only part of the way to explaining the high degree of antigen specificity shown in genetically controlled responses. The total interaction of B cell Ig receptor - antigen - T cell factor - B cell Ia receptor may determine whether or not the B cell of a given specificity can be successfully triggered.

The situation in low responders of H-2^s haplotype, such as SJL, differs from that in H-2^k mice. The SJL strain is incapable of antibody responses to any antigen built on the A--L backbone, including (T,G)-A--L. When similar experiments to those described for C3H/HeJ were carried out, it was found that the defect in SJL mice is expressed in both T and B cell populations (21). Educated T cells of SJL origin were incapable of producing T cell factor for (T,G)-A--L; and SJL B cells were unable to respond to T cell factors of high responder animals. It seems that SJL T cells lack the specific receptor capable of recognizing the A--L structure; and that in addition SJL B cells cannot respond to antigen and T cell factor.

DISCUSSION

Although it seems inappropriate at the present time to speculate extensively on the cellular mechanisms involved in cell cooperation, genetic control, etc., the following points emerge from our work so far.

1) <u>The T cell receptor</u>. It seems most unlikely that T cells would release an antigen-specific, nonimmunoglobulin product that was different, in terms of antigen recognition, from the T cell receptor. The specific T cell factor and the T cell receptor must therefore be closely related molecules - at least carrying the same binding site, and possibly being identical molecules. On this basis it would be predicted that the T cell receptor is a product of the K or I regions of the H-2 complex, and this is in good agreement with several recent experiments which show that T cell recognition of antigen is blocked by anti-H-2 sera (22-24).

2) <u>Cell cooperation</u>. The T cell factor is capable of replacing T cells in thymus-dependent immune responses in vivo, but it is obviously difficult to tell whether this is the major mechanism of cell cooperation under normal circumstances. It is possible that so-called 'physiologic' cooperation operates via direct contact between T and B cells, by interaction of cell bound T cell receptor (factor) and a B cell 'acceptor' (2,25). One notable difference between cooperation using T cell factor and experiments involving T cell-B cell interaction, is that the factor operates well across histocompatibility barriers in most allogeneic combinations so far tried,

while efficient cellular interactions apparently require T and B cell identity in at least part of the H-2 complex (probably the I region)(2,25,26). As yet, however, we have not systematically compared T cells and T factors in their cooperation with histoincompatible B cells, and this may well be crucial in discovering the normal mechanism of cooperation under physiological conditions.

Since the T cell factor is able to replace T cells in cooperative responses, it must presumably be able to deliver a stimulatory signal to B cells, and the B cell must be capable of receiving such a signal. This raises the possibility of a B cell receptor for the T cell factor, and, as already mentioned, the Ia antigens are candidates for this role. It has recently been found that the Fc receptor on B cells is blocked by anti-Ia sera (27). The functional significance of the Fc receptor is not clear, and it is possible that it is in reality the receptor for T cell factor and that the latter bears a certain structural similarity to Fc.

3) <u>Genetic control of the antibody response</u>. The significant outcome of the work described here on genetic control is the conclusion that a defect in response may appear to be associated with either the B cell or T cell populations, or both, depending on the antigen and the strain. A certain controversy still surrounds the basis of genetic control of the antibody response to (T,G)-A--L in low responders of the $H\text{-}2^k$ haplotype (e.g. C3H/HeJ). The use of the T cell factor has indicated that T cells of the low responders are functionally as efficient as those of the high responders, but that low responder B cells are unable to respond to the combination of antigen and T cell factor. These results are in obvious contrast with those obtained with allophenic mice, in which the responsiveness of low responder B cells to (T,G)-A--L was also tested directly and was not found to be defective (28,29). At present it is very difficult to reconcile these two sets of results, but they may reflect differences in the two methods of assaying B cell function.

Finally, following on the discovery that various functional T cell products are associated with the major histocompatibility system, the possibility arises that other 'lymphokines' are also products of the same system. Results of experiments in which MIF, SMAF, etc. are absorbed with alloantisera are now eagerly awaited, with the fascinating possibility that the T cell mediators will turn out to be a closely related family of molecules coded for by the major H complex.

ADDENDUM

Since this manuscript was written further results have been obtained on the nature and activity of the T cell factor. These are in summary as follows.

1. Molecular nature of the factor

We have now used antisera specific for the K and I regions to absorb the factor. Clearcut results were obtained using antisera kindly supplied by Drs. Shreffler and David. An antiserum raised specifically against the K locus (anti-K^k, A.QR anti-B10.A) failed to remove the T cell factor produced in $H\text{-}2^k$ mice (B10.BR). However, an antiserum directed against antigens of the I region (anti-I^k, A.TH anti-A.TL) did completely remove this factor. Anti-H-2 sera provided by Dr. Staines, made specific for the I region by absorption with mouse red cells, gave similar results. Taken together with the results already described, we may conclude that the (T,G)-A--L - specific factor is a product of the I region, either of Ir-1A or Ir-1B genes. Since the response to (T,G)-A--L is under the control of genes in the Ir-1A locus, the factor is quite likely coded for by this locus.

2. Histocompatibility requirements for cell cooperation

We have now systematically examined the ability of the T cell factor to cooperate with allogeneic B cells and confirm that there is apparently no requirement for H-2 compatibility between the T cell factor and the B cells. This remains a puzzling difference from results obtained for cooperation between T cells and B cells (25).

3. Genetic control of the immune response

A possible objection to our experiments which tested the responsiveness of high and low responder B cells is that these were not tested in identical recipients - high responder recipients were used for high responder B cells, whereas low responder recipients were used for low responder B cells. In order to rule out a possible host contribution to the result - the possibility of radiation-resistant T cells being important in the recipient and only being available in high responder animals has been suggested - we have now used F1 animals as recipients for both high and low responder B cells. When B10 (high responder) or B10.A (low responder) B cells were tested by transfer together with T cell factor and (T,G)-A--L into irradiated F1 (B10xB10.A) recipients, only the high responder, B10.B cells responded. The result confirms our previous conclusion that low responsiveness of mice of $H\text{-}2^k$ haplotype is the result of a B cell defect in response to (T,G)-A--L and the T cell factor.

REFERENCES

1. McDevitt, H.O. and Landy, M., eds. Genetic Control of Immune Responsiveness. Academic Press, New York, London, 1972.
2. Benacerraf, B. and Katz, D.H., Advances in Cancer Research, in press, 1974.
3. Shreffler, D.C. and David, C.S., Advances in Immunology, in press, 1974.
4. Taussig, M.J., Nature, 248: 234, 1974.
5. Taussig, M.J. and Munro, A.J., Nature, 251: 63, 1974.
6. Munro, A.J., Taussig, M.J., Campbell, R., Williams, H. and Lawson, Y., J. Exp. Med., in press, 1974.
7. Taussig, M.J., Mozes, E. and Isac, R., J. Exp. Med., 140: 301, 1974.
8. Mozes, E., Progress in Immunology II (Proceedings of the Second International Congress of Immunology), in press, 1974.
9. Sela, M. and Fuchs, S. in Handbook of Experimental Immunology (ed. D.H. Weir) vol. 1, ch. 1. Blackwell Scientific Publications, Oxford, 2nd. Edition, 1973.
10. Feldmann, M. and Nossal, G.J.V., Transplantation Reviews, 13: 3, 1972.
11. Hudson, L., Sprent, J., Miller, J.F.A.P. & Playfair, J.H.L., Nature, 251: 60, 1974.
12. Armerding, D. and Katz, D.H., J. Exp. Med., 140: 19, 1974.
13. Katz, D.H. and Benacerraf, B., The Immune System: Genes, Receptors, Signals - Proceedings of the 1974 I.C.N.-U.C.L.A. Symposium on Molecular Biology. C. Fred Fox, ed., Academic Press, New York, 1974.
14. Katz, D.H., Progress in Immunology II (Proceedings of the Second International Congress of Immunology) in press, 1974.
15. McDevitt, H.O., J. Immunol., 100: 485, 1968.
16. Grumet, F.C., J. Exp. Med., 135: 110, 1972.
17. Mitchell, G.F., Grumet, F.C. and McDevitt, H.O., J. Exp. Med., 135: 126, 1972.
18. Ordal, J.C. and Grumet, F.C., J. Exp. Med. 136: 1195, 1972.
19. Hämmerling, G.J., Masuda, T. and McDevitt, H.O., J. Exp. Med., 137: 1180, 1973.
20. Lichtenberg, L., Mozes, E., Shearer, G.M. and Sela, M. Eur. J. Immunol. 4: 430, 1974.
21. Mozes, E., Isac, R. and Taussig, M.J., J. Exp. Med., submitted, 1974.
22. Hämmerling, G.J. and McDevitt, H.O., J. Immunol., 112: 1734, 1974.

23. Shevach, E.M., Paul, W.E. and Green, I., J. Exp. Med., 136: 1207, 1972.
24. Shevach, E.M., Green, I. and Paul, W.E., J. Exp. Med., 139: 679, 1974.
25. Katz, D.H., Hamaoka, T., Dorf, M.E. and Benacerraf, B., Proc. Nat. Acad. Sci. USA, 70: 2624, 1973.
26. Katz, D.H., Hamaoka, T., Dorf, M.E., Maurer, P.H. and Benacerraf, B., J. Exp. Med. 138: 734, 1973.
27. Dickler, H. and Sachs, D., J. Exp. Med., in press, 1974.
28. McDevitt, H.O., Bechtol, K.B., Hämmerling, G.J., Lonai, P. and Delovitch, T., The Immune System: Genes, Receptors, Signals, Proceedings of the I.C.N.-U.C.L.A. Symposium on Molecular Biology. C. Fred Fox, ed. Academic Press, 1974.
29. Bechtol, K.B., Wegmann, T.G., Freed, J.H., Grumet, F.C., Chesebro, B.W., Herzenberg, L.A. and McDevitt, H.O., Cell Immunol., 13: 264, 1974.

TOLERANCE INDUCTION TOWARDS DETERMINANTS OF THE MAJOR HISTOCOMPATIBILITY COMPLEX: EVIDENCE FOR DELETION OF T CELL SUBSETS.

H. von Boehmer, J. Sprent, M. Nabholz and N. Erhardt
Basel Institute for Immunology, 487 Grenzacherstrasse, Basel, Switzerland.

Recently evidence has been accumulating that tolerance to cell bound antigens in certain experimental situations is either dependent on humoral serum blocking factors (1,2,3,4, 5) or on specific suppressor cells (6). Some authors maintain that self tolerance is also mediated by factors present in serum (7). Most authors studying tolerance in chimeric mice produced either by fusion of embryos (5), injection of histo-incompatible cells (2) or x-irradiation and reconsitution with bone marrow (4) have tested the immunological reactions in a microcytotoxicity assay using fibroblasts as target cells. Unfortunately it is not clear whether this assay system reflects recognition of histocompatibility antigens of the MHC and killing by T cells. Furthermore, it is evident that in all experiments some numbers of immunocompetent T cells were present during the induction of chimerism. To avoid these eventually complicating factors we have studied T cell tolerance to serologically as well as to lymphocyte defined determinants of the MHC using the mixed lymphocyte reaction (MLR) and the cell mediated lympholysis (CML) as test systems. Tolerance was studied in bone marrow chimeras produced by injecting lethally x-irradiated F_1 hybrids with relatively high numbers of T cell depleted (anti θ treated) bone marrow cells from both parents. For these mice we use the term 'tetraparental bone marrow chimeras' (TBM). Stem cell differentiation in this situation should resemble that under physiological conditions where potentially self reactive cells are eliminated during ontogeny.

MATERIAL AND METHODS

Mice: CBA/J, DBA/2, F_1 (CBA x DBA/2), C57BL and F_1 (CBA x C57BL) mice were used.

Cells: Thymus, bone marrow and lympnode cells were prepared as described under (8).

Bone Marrow Chimeras: F_1 (CBA x DBA) mice were x-irradiated with 900r and injected intravenously with 7.5 x 10^6 CBA as well as 7.5 x 10^6 DBA T cell depleted bone marrow cells.

Lymphoid cell chimerism was checked at various times after irradiation with anti-H-2 sera produced in either CBA or DBA mice by injecting spleen cells from the opposite strain.

Mixed lymphocyte reaction was carried out as described previously (8).

Cell mediated lympholysis: The assay system used was that described in detail by Nabholz et al. (9) employing PHA stimulated and ^{51}Cr labelled lymphocytes as target cells.

RESULTS

Lymphoid cell chimerism in TBM

Lymphoid cell chimerism in the chimeras was studied from 2 months (Chimera I) after injection up to 10 months (Chimera X). As shown in Table I both anti-CBA as well as anti-DBA serum was highly specific and killed up to 100% of F_1 (CBA x DBA) lymphnode cells. With lymphoid cells from the chimera, however, neither anti-CBA nor anti-DBA serum alone produced more than 71% lysis. By contrast, incubation with both antisera simultaneously produced 100% lysis. Summation of the values obtained with anti-CBA and anti-DBA serum in separate incubation of cells from the same chimera gave figures ranging from 101 - 110% lysis. Thus it is clear that the vast majority of cells in the chimeras were of donor origin and that very few - if any - host cells were present. Furthermore, it appears that the proportions of cells from the two parental populations were quite similar and that chimerism was predictable and stable over a long time period.

Specific unresponsiveness of lymphocytes from TBM to host type MLR determinants

As shown in Table II thymus and lymphnode cells from TBM showed specific unresponsiveness to MLR determinants of either CBA or DBA lymphocytes in a 'one way' mixed lymphocyte reaction, whereas a significant response was obtained when 'third party' C57 stimulator cells were present in the cultures. In these studies 3×10^6 thymus cells or 1×10^6 lymphnode cells were cultured with 5×10^5 Mitomycin C treated stimulator cells. No evidence could be found that chimeric cells showed higher levels of 'background' thymidine incorporation than cells from normal F_1 (CBA x DBA) mice. Chimeric cells, by themselves unresponsive to CBA MLR determinants, failed to suppress the response of normal CBA lymphocytes to DBA stimulators in vitro as shown in Table III: CBA lymphnode cells were equally well stimulated by F_1 (CBA x DBA) cells and by chimeric (CBA + DBA) cells. Thus DBA cells obtained from the chimera were stimulatory for normal CBA cells and chimeric CBA cells did not suppress the response of normal CBA to these MLR determinants.

TABLE I
CHIMERISM IN TETRAPARENTAL BONE MARROW CHIMERAS

CELLS	INCUBATION WITH			
	anti CBA + C'	anti DBA + C'	saline + C'	anti CBA anti DBA + C'
CBA LN	94%	14%	19%	95%
DBA LN	15%	99%	10%	99%
F_1 (CBAxDBA) LN	99%	98%	11%	-
Chimera I LN	71%	30%	13%	99%
Chimera II LN	67%	35%	10%	100%
Chimera III LN	51%	58%	19%	100%
Chimera IV LN	60%	50%	22%	98%
Chimera V TDL	63%	40%	13%	-
Chimera VI LN	45%	50%	16%	97%
Chimera VII TDL	53%	51%	0%	99%
Chimera VIII LN	70%	40%	22%	98%
Chimera IX LN	68%	42%	21%	98%

LN = lymphnode
TDL = thoracic duct lymphocytes
C' = complement

% = dead cells

TABLE II

RESPONDER CELLS	STIMULATOR CELLS				
	F_1 (CBA x DBA) spleen	C57 spleen	CBA spleen	DBA spleen	No Stimulators
Chimera Thymus	2,380 ± 111	9,750 ± 1,403	2,278 ± 589	2,432 ± 132	2,252 ± 551
Chimera LN	11,654 ± 1,399	40,899 ± 1,205	12,737 ± 1,233	11,603 ± 1,329	12,630 ± 1,435
F_1 (CBA x DBA) Thymus	3,822 ± 442	22,751 ± 1,806	3,838 ± 110	3,635 ± 427	3,418 ± 167
F_1 (CBA x DBA) LN	11,709 ± 494	64,804 ± 2,760	11,826 ± 2,622	12,058 ± 1,329	11,070 ± 1,467
CBA Thymus	17,048 ± 1,502	15,608 ± 3,086	3,079 ± 683	26,663 ± 1,739	2,654 ± 819
CBA LN	77,810 ± 5,186	78,777 ± 4,209	12,728 ± 671	95,236 ± 4,869	13,470 ± 1,356

M = Mitomycin C treated. Values represent counts per minute
LN = lymphnode cells

TABLE III
STIMULATORY CAPACITY OF CHIMERIC LYMPHOCYTES FOR NORMAL PARENTAL LYMPH NODE CELLS

STRAIN + ORGAN	CELL NUMBER PER CULTURE	CALCULATED BACKGROUND	OBSERVED ^{3}H-T UPTAKE
Chimera (CBA + DBA) lymph node	1×10^6	-	$12{,}630 \pm 1{,}435$
CBA lymph node	1×10^6	-	$13{,}470 \pm 1{,}356$
Chimera lymph node + CBA lymph node 1:1	1×10^6	13,050	$75{,}253 \pm 7{,}028$
F_1 (CBA x DBA) lymph node	1×10^6	-	$11{,}070 \pm 1{,}467$
F_1 lymph node + CBA lymph node 1:1	1×10^6	12,270	$62{,}243 \pm 822$

In these experiments stimulating cells were not treated with Mitomycin C.

Specific tolerance in chimeric cells to host type H-2 antigens

Spleen cells from chimeric (CBA + DBA) mice were not able to destroy F_1 (CBA x DBA) PHA-stimulated and ^{51}Cr labelled targets when incubated with the targets immediately after preparation. Significant lysis of F_1 targets, however, was observed with CBA or DBA cells cultured with F_1 (CBA x DBA) stimulators for 5 days in vitro, Fig. 1. Thus there was no evidence that cytotoxic lymphocytes directed against host determinants had been generated in vivo.
When chimeric cells were cultured with F_1 (CBA x C57) or F_1 (DBA x C57) stimulators good lysis was obtained on 'third party' C57 targets but not on DBA or CBA targets, Fig. 2. However, CBA and DBA targets were lysed when normal DBA or CBA responders were cultured with the F_1 stimulator cells (Fig. 2b + 2c). Thus no precursors for killer cells against host type antigens could be demonstrated despite the fact that a good response against 'third party' C57 antigens was obtained in the same culture vessel. This also ruled out the possibility that cytotoxic cells were not generated because of the absence of a significant MLR (10).

Failure to demonstrate specific suppressor cells

It was conceivable that the specific unresponsiveness to host type H-2 antigens was maintained by specific suppressor cells. If this was the case one would expect that chimeric cells suppress the response of normal CBA to DBA or of normal DBA to CBA in a specific fashion. This was tested by culturing chimeric cells with normal CBA cells and stimulating this mixture with DBA cells. In control cultures F_1 (CBA x DBA) cells were given to the normal CBA cells. The results illustrated in Fig. 3a and 3b show that chimeric cells cultured at a 1:1 ratio with CBA or DBA cells did not suppress the response of the latter against DBA or CBA respectively. There was also no evidence found that chimeric cells had an effect on responses of normal CBA or DBA cells against 'third party' C57 antigens (not shown here).

Failure to demonstrate specific blocking serum factors in CML

The possibility that serum blocking factors were involved in maintaining a state of tolerance was investigated by testing the ability of serum from the chimera to suppress the generation of cytotoxic cells. Table IV demonstrates that at high concentration both normal mouse serum as well as the chimera serum blocked the production and/or action of killer cells but that this effect was nonspecific.

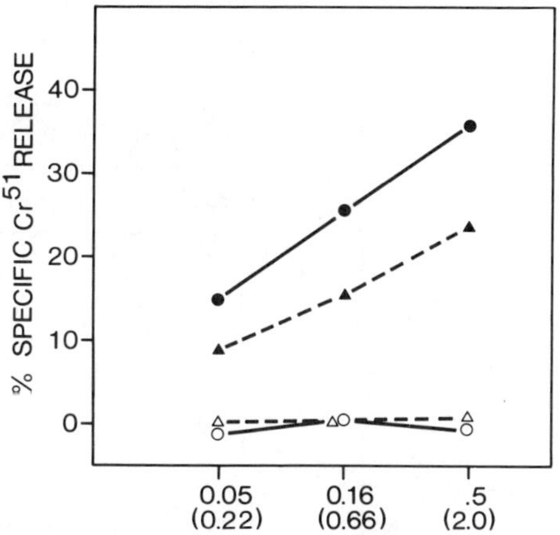

Fig. 1 Killing of F_1 (CBAxDBA) targets.
●-● DBA spleen cells activated in vitro by F_1 (CBA x DBA) Mitomycin-C-treated stimulators.
▲-▲ CBA spleen cells activated in vitro by F_1 (CBA x DBA). o-o spleen cells from chimera I
Δ-Δ spleen cells from chimera II
To each well are added an (unknown) number of killer cells which are the descendants of the (known) number (x 10^6) of responders indicated on the abscissa (see also methods section). In parentheses are given the numbers (10^6) of unstimulated chimera cells/well.

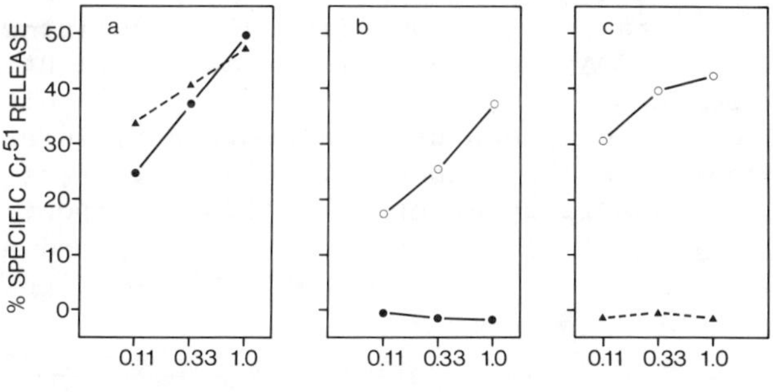

Fig. 2a Killing of C57 targets by chimeric spleen cells activated in vitro by F_1 (DBA x C57) ▲-▲ or by F_1 (CBA x C57) ●-● .

b Killing of CBA targets by DBA spleen cells o-o , or chimeric spleen cells ●-● , activated in vitro by F_1 (CBA x C57) cells.

c Killing of DBA targets by CBA spleen cells o-o, or chimeric spleen cells ▲-▲ , activated in vitro by F_1 (DBA x C57) cells. Numbers on abscissa indicate the number (10^6) of responders/well (see Fig. 1).

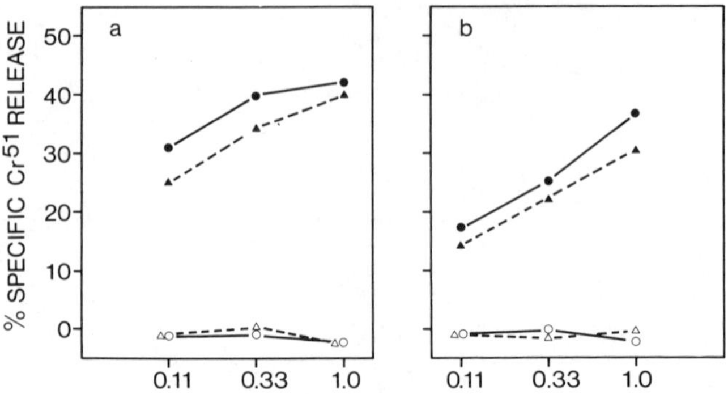

Fig. 3a Killing of CBA targets by DBA spleen cells cultured at 1:1 ratio with chimeric spleen cells (●-●) or F_1 (CBA x DBA) spleen cells (▲-▲) and stimulated by F_1 (CBA x C57) cells. o-o chimeric cells alone, Δ-Δ F_1 (CBA x DBA) cells alone, both stimulated by F_1 (CBA x C57).

Fig. 3b Killing of DBA targets by CBA spleen cells cultured at a 1:1 ratio with chimeric spleen cells ●-● or with F_1 (CBA x DBA) spleen cells ▲-▲ and stimulated by F_1 (DBA x C57) cells. o-o chimeric spleen cells alone, Δ-Δ F_1 (CBA x DBA) spleen cells alone, both stimulated by F_1 (DBA x C57) cells. Numbers on abscissa indicate responders (10^6)/well (see Fig. 1).

TABLE IV

EFFECT OF SERUM ON LYMPHOLYSIS

RESP	No $(\times 10^6)$	STIM	SERUM*	DIL	TARGETS CBA	DBA	C57
CBA	1.0	C57xDBA	-		15±2	56±5	38±1
	.5		-		17±2	59±1	40±3
	.25		-		12±0	23±3	21±2
	1.0		CHIM	1/6	7±1	14±3	12±2
	1.0			1/12	11±1	50±4	36±1
	1.0		NMS	1/6	9±2	15±2	13±1
	1.0			1/12	12±0	60±2	18±1
DBA	1.0	C57xCBA	-		33±4	15±0	30±4
	.5		-		30±4	15±0	26±2
	.25		-		13±1	13±1	16±1
	1.0		CHIM	1/6	17±6	10±1	13±2
	1.0			1/12	49±5	13±0	34±3
	1.0		NMS	1/6	27±3	10±0	19±1
	1.0			1/12	43±1	12±0	27±0
C57	1.0	CBAxDBA	-		58±3	52±1	18±1
	.5		-		62±3	48±3	18±1
	.25		-		41±1	48±7	18±1
	1.0		CHIM	1/6	36±2	39±4	12±1
	1.0			1/12	59±2	49±3	15±1
	1.0		NMS	1/6	41±3	31±4	11±1
	1.0			1/12	49±5	42±4	12±2
	SPONTANEOUS RELEASE				16±1	23±1	24±1

Equal numbers of responders and irradiated stimulator cells from normal mice were cultured in 120 μl medium containing, where indicated, mouse serum at various concentrations in wells of Falcon 3040 microculture plates. On day 3, 50 μl medium was added to each culture. On day 5 50 μl of PHA-stimulated target cells (10^4) were added to each well and cytolytic activity determined as described in Material and Methods. Results are given as ^{51}Cr released as % of the total uptake determined by Zaponin cytolysis.

*Sera: CHIM: pooled serum from four (CBA+DBA) tetraparental bone marrow chimeras.
NMS: normal serum from F_1 (CBA+DBA) mice
Sera were not heat inactivated.

DISCUSSION

The results obtained are not easily comparable with those reported by Hellstrom et al. in chimeric dogs (1), in neonatally induced chimeras (2) and with those of Wegmann et al. (5) and Phillips et al. (6) in tetraparental or allophenic mice. It is conceivable that the different test systems as well as the different experimental animals have contributed to the differing results. From the results reported here it seems clear that tolerance in tetraparental bone marrow chimeras to MLR determinants and to serologically defined antigens of the MHC is not reversible and is not maintained by either suppressor cells or blocking serum factors. Thus the results indicate deletion of functional cytotoxic lymphocytes and their precursors. Conceivably suppressor cells or blocking serum factors may have contributed to the deletion.

TBM chimeras are of potential value in substituting for allophenic mice in certain experimental situations. The ease with which they can be prepared and their predictable and stable chimerism make them especially suitable as experimental models.

SUMMARY

Tetraparental bone marrow chimeras were produced by injecting lethally x-irradiated F_1 hybrids with relatively high numbers of T cell depleted bone marrow cells from both allogeneic parental strains. The mice survived in excellent health and showed a stable, approximately 50:50 (parent: parent) lymphoid cell chimerism lasting at least 10 months after irradiation; regeneration of host type immunocompetent cells was very limited.

Lymphocytes from the chimera showed specific unresponsiveness to MLR determinants of the host. No suppressor cells suppressing the MLR response of normal parental strain cells against host determinants could be demonstrated. Similarly specific tolerance and lack of suppressor cells as well as blocking serum factors could be demonstrated in cell mediated lympholysis.

The result suggests specific deletion of functional T cells reactive to host type MLR and CML determinants.

REFERENCES

1. Hellström, I., Hellström, K.E., Storb, R. and Thomas, E.D., Proc. Natl. Acad. Sci. USA, 66: 65, 1970.
2. Hellström, I. and Hellström, K.E., Nature, 230: 49, 1967.
3. Wright, P.W., Hargreaves, R.E., Bansal, S.C., Bernstein,

I.D., and Hellström, K.E., Proc. Natl. Acad. Sci., 70: 2539, 1973.

4. Hellström, I. and Hellström, K.E., Cell. Immunol., 7: 73, 1973.
5. Wegmann, T.G., Hellström, I. and Hellström, K.E., Proc. Natl. Acad. Sci., 68: 1644, 1971.
6. Phillips, S.M. and Wegmann, T.G., J. Exp. Med., 137: 291, 1973.
7. Wekerle, H., Cohen, I.R. and Feldman, M., Nature, New Biology, 241: 25, 1973.
8. von Boehmer, H., Shortman, K. and Adams, P., J. Exp. Med. 136: 1648, 1972.
9. Nabholz, M., Vives, J., Young, H.M., Meo, T., Miggiano, V., Rijnbeck, A. and Shreffler, D.C., Europ. J. of Immunol., 4: 378, 1974.
10. Alter, B.J., Schendel, D.J., Bach, M.L., Bach, F.H., Klein, J. and Stimpfling, J.H., J. Exp. Med. 37: 1303, 1973.

SUPPRESSION OF LYMPHOCYTE PROLIFERATIVE RESPONSES BY MURINE LYMPHOMA CELLS

Guy D. Bonnard* and Ronald B. Herberman
Laboratory of Immunodiagnosis, National Cancer Institute, Bethesda, Maryland 20014, USA

There has been much interest in the responses of lymphocytes to allogeneic tumor cells *in vitro*. In a few studies the *in vitro* conditions which allowed for maximal proliferative and cytotoxic responses to the major histocompatibility antigens on the tumor cells were examined. Such information obtained from these allogeneic models could then be applied to measure the responses of lymphocytes to tumor associated antigens on syngeneic tumor cells (1). In the experiments to be described, rat thymus cells, which responded well to allogeneic thymus or spleen cells in mixed lymphocyte cultures (MLC), did not respond to tumor cells of the W/Fu rat lymphoma (C58NT)D (NTD cells), as measured by tritiated thymidine incorporation. Furthermore, the responses of thymus cells to alloantigens and mitogens were strongly inhibited by low numbers of NTD cells added in these cultures. Similar results were obtained in the mouse, using spleen cells as responding cells and various mouse lymphoma lines.

MATERIALS AND METHODS

The animals employed were 8-12 week old rats of the inbred strains W/Fu, BN, LEW and F344.

The tumor cells used (NTD cells) were from the ascites of W/Fu rats injected intraperitoneally with the Gross-virus-induced lymphoma (C58NT)D (2), maintained by weekly passages. NTD cells growing in continuous tissue culture were also available, as kindly provided by Dr. R. Nowinski (3). They were maintained and fed daily in medium RPMI 1640 with 20% FBS. NTD ascites cells were also cryopreserved in 10% DMSO in liquid nitrogen vapors, using a 1° C/min planned automated freezing program (GV Planer Ltd) (4).

Preparation of the cell suspensions: After careful dissection, thymuses (and occasionally spleens) were teased apart in BSS. Cell suspensions were passed through a sterile gauze, counted, washed twice and resuspended in medium at a concentration of 5×10^7 cells/ml. Whenever needed, stimulating cells were treated at 10^7 cells/ml with mitomycin-C (Sigma, 25 μg/ml, 30 min), washed 3 times, counted and re-

*Supported by the Foundation of the Swiss Academy of Medicine

suspended in medium at 5×10^7 cells/ml. Fifty μl representing 2.5×10^6 cells of each responding and stimulating cell population were then distributed to duplicate tubes containing one ml medium. NTD cells were used both to test for stimulation of some of the responding cell populations and for their interference with MLC reactions when used as third party. NTD ascites cells were washed twice in BSS. In some experiments NTD ascites cells were treated with mitomycin-C as described above, but nontreated NTD ascites cells were also found adequate, since they did not proliferate in these cultures. NTD tissue culture cells received 5000 R X-irradiation. NTD cells were then resuspended in medium at 4×10^6 cells/ml and 50 μl representing 2×10^5 NTD cells were distributed to tubes already containing the responding cells, the MLC or medium alone.

Culture conditions: Bicarbonate-buffered RPMI 1640 (NIH Media Unit) with 5% FBS (Gibco) and 5×10^{-5}M 2-mercaptoethanol (Schwarz-Mann) was used. In some experiments 25×10^{-3}M Hepes solution (Microbiological Associates) was added. One ml aliquots of medium were distributed to 75 x 12 mm plastic tubes (Falcon). After distribution of the cells, the cultures were put in a 5% CO_2, humidified 37° C incubator. The mitogens phytohemagglutinin (PHA; Burroughs Wellcome; 2 μg/ml) Concanavalin A (Con A; Calbiochem; 2 μg/ml) and pokeweed mitogen (PWM; Gibco; final dilution 1:100) were used in some experiments.

^{3}H-thymidine pulse and cell harvest: At appropriate times 0.2 aliquots were transferred from each tube to 2 wells of a flat-bottom Microtiter plate (Falcon), thus providing 4 replicates for each reaction. The plates were then pulsed with ^{3}H-thymidine (1 μCi in 20 μl RPMI 1640 per well, specific activity 6.7 mCi/mM, Schwarz-Mann). After 4 or 6 hr incubation the cells were harvested using a multiple automated sample harvester and processed for scintillation counting.

The results of the thymus cells stimulations or of the MLC with and without added NTD cells are given as the median counts per minute (cpm) of the quadruplicate samples without corrections. The cpm for thymus cells responses to any antigen are presented together with the cpm of these thymus cells cultured alone. Cpm values for all stimulating cells in these experiments were obtained and they are not reported since they did not usually exceed 100 cpm.

RESULTS

Thymus cells were found to be highly reactive in MLC by the criterions of blast transformation, cellular proliferation and ^{3}H-thymidine incorporation. However, thymus cells failed in this system to generate detectable cytotoxic effector-cells as measured in a ^{51}Cr-release assay. Further, thymus cells were characterized by a low spontaneous proliferation and a very good ability to stimulate in MLC. Thus thymus cells vs. mitomycin-C treated thymus cell interactions were used for most of these experiments and evaluated in terms of ^{3}H-thymidine incorporation. The ability of allogeneic thymus cells to respond to various cells of the W/Fu rat and in particular to NTD cells was studied first (Table I). Whereas thymus cells responded very well to mitomycin-C treated W/Fu thymus cells and spleen cells, they did not respond to various doses of NTDm cells. This could be demonstrated with responding thymus cells of various strains. NTDm cells added to allogeneic or syngeneic thymus cells usually reduced their spontaneous proliferation.

We had previously postulated that atypical lymphoid cells, which had failed to stimulate in MLC, would have in fact inhibited the proliferation of lymphocytes (5). This was studied here by allowing NTDm cells to interfere as a third party with MLC reactions. When 2×10^5 NTDm cells were added to the mixture of 2.5×10^6 thymus cells responding to 2.5×10^6 allogeneic mitomycin-C treated thymus or spleen cells, the MLC responses were completely inhibited (Table II). This was true for MLC harvested at various times after their onset and for MLC with various strain combinations. Whereas 2×10^5 NTDm cells always caused strong inhibition, the effects of lower numbers of NTDm cells were dose-dependent (Table III). When low numbers of NTDm cells were used, differences between NTDm cells in individual experiments became apparent.

NTD cells of various sources were then used and results are shown in Table III. NTD cells, not treated with mitomycin-C, could be used since they did not proliferate *in vitro* and they also inhibited MLC. NTD tissue culture cells, which had undergone several passages *in vitro*, were used after 5000 R X-irradiation and they were found similarly inhibitory. Cryopreserved NTD or NTDm cells retained good viability and also their capacity to inhibit MLC. In no case did any of these cell preparations stimulate significantly allogeneic thymus cells. Intact NTD cells were required for the inhibition, since aliquots of cell suspensions subjected to freeze thawing or which had been incubated at 156°C for 20 min did not inhibit.

The nature of the effects of the NTD cells on the MLC were then studied. Both NTDm and NTD cells had a very poor survival in culture (this was confirmed by studies on their proliferation rates which dropped considerably with the first

TABLE I

^{3}H-THYMIDINE UPTAKE (cpm) BY THYMUS CELLS OF VARIOUS STRAINS CULTURED WITH W/Fu MITOMYCIN-C TREATED THYMUS CELLS (TCm), SPLEEN CELLS (SCm) OR NTD CELLS (NTDm)

Strain	Exp. no.	Incubation time (hr)	W/Fu mitomycin-C treated cells added			
			none	TCm	SCm	NTDm *
BN	1	72	2530	-	14130	256
		96	1702	-	40000	1720
		120	436	-	75157	1111
		144	1457	-	71074	1204
		168	266	-	33541	306
BN	2	96	4065	36884	-	473
	3	96	542	18845	12913	61
F344	2	96	288	7924	-	336
	4	96	1420	18509	22575	32
LEW	2	96	5184	14233	-	145
W/Fu	5	72	488	-	-	104
		96	608	-	-	125
		120	891	-	-	221
		144	3230	-	-	481

* In experiments 1 and 2 doses of NTDm ranging from 5×10^3 to 4×10^5 were tested and no stimulation was seen.

TABLE II

^{3}H-THYMIDINE UPTAKE (cpm) IN RAT MLC WITH AND WITHOUT THE ADDITION OF 2 x 10^{5} NTDm CELLS*

MLC				Cpm in thymus cells				
		Exp.	Incubation		with mitomycin-C treated cells			
R	SM	no.	time (hr)	alone	TCm	TCm+NTDm	SCm	SCm+NTDm
W/Fu vs	BN	3	96	357	33239	117	44203	66
BN vs	W/Fu	3	96	542	18845	48	12913	381
W/Fu vs	F344	4	96	508	17945	32	17490	165
F344 vs	W/Fu	4	96	1420	18509	48	22572	652
W/Fu vs	BN	6	72	400	12300	120	-	-
			96	462	25218	469	-	-
			120	456	43645	163	-	-
			144	252	16157	503	-	-
LEW vs	BN	2	96	5184	47076	932	-	-
F344 vs	BN	2	96	288	14405	1787	-	-

* Abbreviations same as in Table I

TABLE III

INHIBITION OF RAT MLC BY VARIOUS CONCENTRATIONS AND VARIOUS PREPARATIONS OF NTD CELLS

MLC			Exp.	Cpm in thymus cells					
				alone	with mitomycin-C treated stimulating cells				
						with various concentrations of NTD			
R		Sm	no.			$2x10^5$	$5x10^4$	$2.5x10^4$	$5x10^3$
a. NTDm cells were used:									
W/FU	vs	BN	10	224	29529	77	5585	13444	19908
W/Fu	vs	F344	8	740	14831	96	597	3009	6609
b. Untreated NTD cells were used:									
W/Fu	vs	BN	11	337	15563	47	327	4929	22282
F344	vs	W/Fu	8	3983	13210	57	264	447	1988
c. Cryopreserved NTDm cells were used:									
BN	vs	W/Fu	9	1789	20406	187	-	-	-
d. Cryopreserved NTD cells were used:									
BN	vs	W/Fu	9	1789	20406	102	-	-	-
e. 5000 R X-irradiated NTD tissue culture cells were used:									
BN	vs	W/Fu	12	2214	21956	278	2432	4308	6981

48 hr). MLC without NTD cells at 96 hr had 30% surviving responding cells of which approximately 60% were blasts. MLC containing NTD cells had only slightly less surviving cells, but no blasts were seen. Only very small numbers of NTD cells were used and it seemed unlikely that they would deplete the medium of one of its essential components. Indeed, spinning the cells and replacing the culture medium at 24 hr did not restore the responses of MLC which contained NTD cells. Fresh NTDm ascites cells completely inhibited MLC when added 24 hr after initiation of the mixed cultures. In a further experiment, fresh 5000 R X-irradiated NTD tissue culture cells were added at time zero and 24 hr and 48 hr after the onset of MLC. Complete inhibition occurred at time zero and 24 hr, but no inhibition was seen at 48 hr. Thus NTD cells acted on a rather early step of the MLC reaction.

To determine whether the inhibitory effect was on the responding population and whether the inhibition was a more general phenomenon we performed experiments with mitogens. Results shown in Table IV indicated that NTD cells also inhibited the responses of thymus cells to standard doses of the mitogens Con A, PHA and PWM.

We were not able to consistently demonstrate inhibitory activity from the supernatants of 2 x 10^5 NTD ascites cells cultured for 24 hr. However, supernatants of NTD tissue culture cells, grown at concentrations ranging between 0.5 and 2.5 x 10^6 cells/ml for various periods of time, were strongly inhibitory. Some differences between individual supernatants were seen (Table V). The results indicated that a factor inhibiting MLC was actively produced under appropriate culture conditions. They also strongly suggested, but did not prove, that NTD ascites cells acted by the production of this factor within the microenvironment of the MLC cell pellet.

DISCUSSION

In these experiments cells of the W/Fu rat lymphoma (C58NT)D (NTD cells) were found unable to stimulate allogeneic normal thymus cells *in vitro* under all conditions tested. Low numbers of NTD cells also inhibited thymus cell responses to alloantigens and mitogens. The lack of stimulation seemed to be related to this strong inhibition of lymphocyte proliferation, rather than to a lack of MLC and other stimulatory antigens on the NTD cells. We have obtained similar findings when a number of mouse tumors were tested for their ability to stimulate allogeneic or syngeneic spleen cells and to inhibit MLC. Those tumor cells which did not

TABLE IV

^{3}H-THYMIDINE UPTAKE (cpm) BY THYMUS CELLS OF VARIOUS STRAINS CULTURED WITH MITOGENS, WITH AND WITHOUT THE ADDITION OF 2 x 10^5 NTD CELLS*

		Cpm in thymus cells						
Strain	Exp. no.	alone	with ConA	PHA	PWM	ConA and	PHA NTD	PWM cells
W/Fu	3	357	8140	-	13448	88	-	119
BN	3	542	43490	-	16079	313	-	452
W/Fu	7	1250	48919	-	18563	125	-	143
W/Fu	8	740	-	-	12342	-	-	83
F344	8	3983	33755	-	28632	9111	-	124
BN	9a	1789	32553	33625	6427	15278	269	419
	9b	1789	32553	33625	6427	260	112	112

* Various sources of NTD cells. Exp. no. 3 and 8: NTDm cells; no. 7: NTDm cells added at 24 hr of culture; no. 9a: cryopreserved NTDm cells; no. 9b: cryopreserved NTD cells. Incubation time: 96 hr.

TABLE V
INHIBITION OF RAT MLC BY VARIOUS CONCENTRATIONS OF MILLIPORE FILTERED (0.22 μ) SUPERNATANTS OF NTD TISSUE CULTURE CELLS

Cpm in MLC without supernatants	Supernatant code	Cpm in MLC with supernatants at the final dilutions				
		1:10	1:30	1:100	1:300	1:1000
29207	# 1	101	142	332	750	1345
	# 2	1299	1075	1880	4346	9470
	# 3	1526	614	3172	2956	3990
	# 4	5061	30831	25876	42161	43274
	# 5	3702	11682	21836	46922	48244

stimulate would also inhibit lymphocyte proliferation. Inhibitory and non-inhibitory cells could be found within the same tumor system; some sublines of the EL4 lymphoma were inhibitory while others were not. Allogeneic and syngeneic spleen cells would not respond to stimulatory EL4 cells in the presence of small numbers of inhibitory EL4 cells (Bonnard and Herberman, in preparation).

We failed to demonstrate stimulation of normal thymus cells using NTD and thus provide no evidence here that NTD cells carry MLC and/or other stimulatory antigens. However, since NTD cells inhibited the proliferative capacity of the responding thymus cells, the question arises of whether NTD cells might carry MLC or other stimulatory antigens which were not detected due to the inhibition. Clearly a goal of these studies is to learn how to obtain cells devoid of inhibitory activity, which could then be tested in both allogeneic and syngeneic systems for the presence of stimulatory antigens. The spontaneous occurrence of both stimulatory and inhibitory cells in the mouse EL4 system could be an exception and, more generally, methods must be found to overcome the inhibition exerted by some tumor cells. In a previous study from this laboratory (6), spleen cells from W/Fu rats immune to NTD cells responded to NTD cells in vitro. This shows that at least some lymphocytes may proliferate in response to tumor associated antigens on the NTD cells. It suggests that spleen cells of NTD-immune animals may escape the inhibition of proliferative responses due to NTD cells.

Whenever some cells are found not to stimulate in MLC or in syngeneic systems, we feel now that clear evidence must also be brought that the cells do not inhibit the proliferative response to their own antigens. This concept was raised when the atypical blast cells of a patient with a congenital leukaemoid reaction failed to respond or stimulate in MLC. They were thought to have produced an inhibitor of lymphocyte proliferation, because the morphologically normal cells of the mother also did not respond nor stimulate adequately. A virus (7,8) or another product of the atypical blasts was possibly involved and responsible for the distrubed functions in MLC (5). In more recent human studies, cells from thymus derived lymphoblastoid cell lines (T-LCL) have been shown unable to stimulate in MLC (9,10). So far, five T-LCL have been tested as third party for inhibition in MLC (11,12) and inhibition was observed in one instance, in which the line was contaminated with mycoplasma organisms (11). In the case of the 4 other T-LCL it was clear that they did not inhibit and most probably lacked MLC and other stimulatory antigens (12). This, together with studies on inhibitory factors produced by

B-LCL (13) suggests that the inhibition of lymphocyte proliferation in man may not be uniquely associated with thymus derived cells, but more generally with the products of some tumor cells. So far, the NTD cells have not been characterized in regard to their T or B cell origin.

The present findings with NTD cells have many features in common with other recent reports (13,14,15) which have indicated that tumor cells or their products inhibited lymphocyte proliferation. Both the inhibition by tumor cells and the recovery of an inhibitory factor in supernatants may well be observed only under very specific conditions. A possible explanation for the inhibitory action of tumor cells on lymphocytes is derived from many observations showing that viruses interfere with lymphocyte responses *in vitro* (7,8). The NTD cells used in this study are producers of RNA C-type viruses, the role of which awaits further investigation. It is likely, and we are presently investigating the possibility, that quite a number of tumor lines are contaminated with even more common murine viruses. Infections of cultures with mycoplasma may also inhibit lymphocyte proliferative responses (11,16). It was unlikely that our rat NTD (or mouse EL4) ascites cells were contaminated *in vivo* by mycoplasma and we failed to detect such a contamination on repeating testing. Likewise, the corresponding tissue culture lines were not infected.

The present observations show that particular tumor cells, which failed to stimulate allogeneic or snygeneic lymphocytes *in vitro*, should be tested for inhibition of various lymphocyte functions. In particular, they may strongly inhibit proliferation rather than lack the needed antigens for stimulation of lymphocytes. This concept may be an important factor in experimental studies which aim at the sensitization of lymphocytes to tumor associated antigens *in vitro* and, maybe, *in vivo*.

REFERENCES

1. Wagner, H. and Roellinghoff, J. Exp. Med. 138: 1, 1973.
2. Geering, G., Old, L.J. and Boyse, E.A., J. Exp. Med. 124: 753, 1966.
3. Nowinski, R.C. and Peters, E.D., J. Virol. 12: 1104, 1973.
4. Holden, H.T., Oldham, R.K. and Herberman, R.B., Fed. Proc. 33: 806, 1974.
5. Bonnard, G.D., Gautier, E., Haller, R., Lemos, L. and Jeannet, M., unpublished observation, 1972.
6. Glaser, M., Herberman, R.B., Kirchner, H. and Djeu, J.Y., Cancer Res. 34: 2165, 1974.

7. Olson, G.B., Dent, P.B., Rawls, W.E., South, H.A., Montgomergy, J.R., Melnick, J.L. and Good, R.A., J. Exp. Med. 128: 47, 1968.
8. Hayri, P., Rago, D. and Defendi, V., J. Natl. Cancer Inst. 44: 1311, 1970.
9. Han, T. and Minowada, J., Clin. Exp. Immunol. 15: 535, 1973.
10. Royston, I., Graze, P.R. and Pitts, R.B., J. Natl. Cancer Inst. 53: 361, 1974.
11. Lightbody, J. and Royston, I., Personal communication, 1974.
12. Royston, I., Pitts, R.B., Smith, R.W. and Graze, P.R., Transpl. Proc. 7, in press, 1975.
13. Hersch, E.M., McCredie, K.B. and Freireich, E.J., Clin. Exp. Immunol. 17: 463, 1974.
14. Wong, A., Mankovitz, R. and Kennedy, J.D., Int. J. Cancer 13: 530, 1974.
15. Rodey, G.E., Sprader, J.E. and Bortin, M.M., Cancer Res. 34: 1289, 1974.
16. Barile, M.F. and Leventhal, B.G., Nature 219: 752, 1968.

SUPPRESSOR AND HELPER EFFECTS OF SENSITIZED T-CELL SUBPOPULATIONS ON PROLIFERATIVE T-CELL RESPONSES[1]

Jerry A. Bash[2], Helen G. Durkin[3] and Byron H. Waksman
Department of Pathology, Yale University School of Medicine, New Haven, Connecticut

Regulation of the immune response has been shown by many investigators to involve interactions between T-lymphocyte subpopulations (Review in 1). Using DNA-synthesis in response to mitogens or antigens as a measurement of T-lymphocyte responses *in vitro* we have previously shown that the normal rat spleen contains thymus-dependent adherent cells, distinct from macrophages, which suppress responses to high doses of PHA or Con A or to allogeneic cells in culture (2-5). We report here an extension of these studies; antigenic stimulation is shown to activate regulatory cells in both spleen and lymph node which are separable on the bases of adherence and buoyant density into helper and suppressor cell subpopulations.

Inbred DA male rats, 4-6 months of age, were given 5-500 mg of ovalbumin (OA) or 5×10^8-5×10^{10} sheep red blood cells (SRBC) intravenously 3-24 hours before sacrifice. Alternatively, 5-8 week old animals received 200 μg OA in complete Freund's adjuvant into both hind footpads 9 days before sacrifice.

Washed spleen and lymph node cells (LNC) were fractionated as follows. Nonadherent subpopulations of each were obtained by incubation on glass wool columns (2). Cells from the draining inguinal lymph nodes of rats receiving footpad injections of OA in adjuvant were also separated on discontinuous BSA density gradients into bands A to D of increasing density plus a pellet P (6). Purified macrophages were obtained from peritoneal cells of normal or antigen-treated donors (2). Cells were cultured in microtiter plates (7), each well receiving one drop (0.025 ml) of the target cell suspension (5×10^5 cells), a second drop containing varying numbers of test cells or medium, and one drop of PHA and/or OA in 24% normal DA rat serum. Macrophages were included in the first cell suspension at 1 or 2% concentration (0.5 or 1×10^4 cells) in some experiments. Cultures were incubated

[1]Research supported by USPHS Research Grants AI-06112 and AI-06455.

[2]Postdoctoral Trainee, NIH Grant AI-291-08.

[3]Postdoctoral Fellow, Arthritis Foundation.

for 72 hr and pulsed with ^{3}H-thymidine during the final 24 hr. Results are expressed as mean TCA-precipitable counts per minute (cpm) of duplicate experimental cultures minus background. In the tables only peak responses are presented.

Injection of antigen intravenously produced dose-dependent suppression of spleen cell responses to PHA (Figure 1), manifested as early as 3-6 hours after injection and maximal at 12-24 hours (Figure 2). After removal of glass wool adherent cells the response was actually enhanced above the normal level, and addition of peritoneal macrophages failed to restore suppression (Table I). Similar findings were obtained with LNC (Figure 3). Draining LNC of rats sensitized by footpad injections of 200 μg OA in adjuvant showed comparable suppression of the PHA response, also alleviated by passage of the cells through glass wool (Table II).

Whereas the LNC of rats given large intravenous doses of antigen failed to show a proliferative response to antigen added to the cultures, LNC from the footpad-sensitized animals did show such a response. However, the response to OA + PHA was below that seen with either stimulus alone. Removal of adherent cells rendered the responses of normal and sensitized cells essentially equivalent.

Since the cells responsible for the proliferative response to OA as well as cells suppressing PHA responses are both adherent to glass wool (2, 8), it was impossible to distinguish the two populations by this means. Having previously employed discontinuous BSA gradients to separate OA-responsive cells from PHA-responsive cells (8), we entertained the possibility of separating a suppressor cell population as well using this technique. Accordingly, aliquots of B band cells were added to various target populations responsive to PHA. B band cells, poorly responsive to PHA themselves, greatly potentiated the response of band D from sensitized or normal rats as well as that of normal nonadherent spleen or lymph node cells (Table III). A comparable potentiation was not obtained by the addition of macrophages.

In further experiments, bands B and C were added separately to various responding cell populations. In the experiment presented in Table IV, band B again potentiated the PHA response of sensitized D band cells while band C significantly suppressed it. Responses of normal spleen and LNC remained essentially unchanged by addition of either. In a further experiment, increasing numbers of B band cells were added to sensitized D band cells. As seen in Table V, B band cells, rather than potentiating the response, markedly suppressed in this experiment. Of particular interest was the paradoxical decrease in suppressor activity as increasing numbers of B

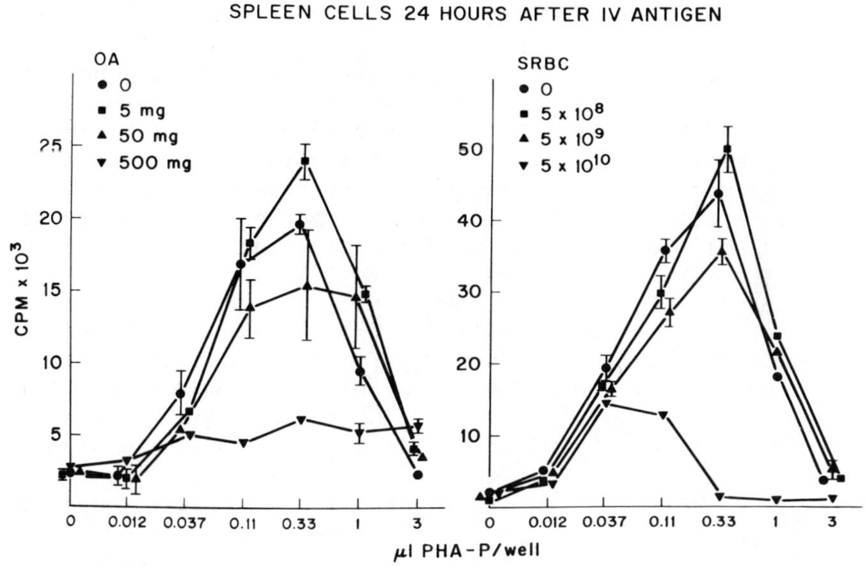

Fig. 1 Response (^{3}H-thymidine uptake) of spleen cells from rats given various intravenous doses of antigen 24 hours before culture with different amounts of PHA.

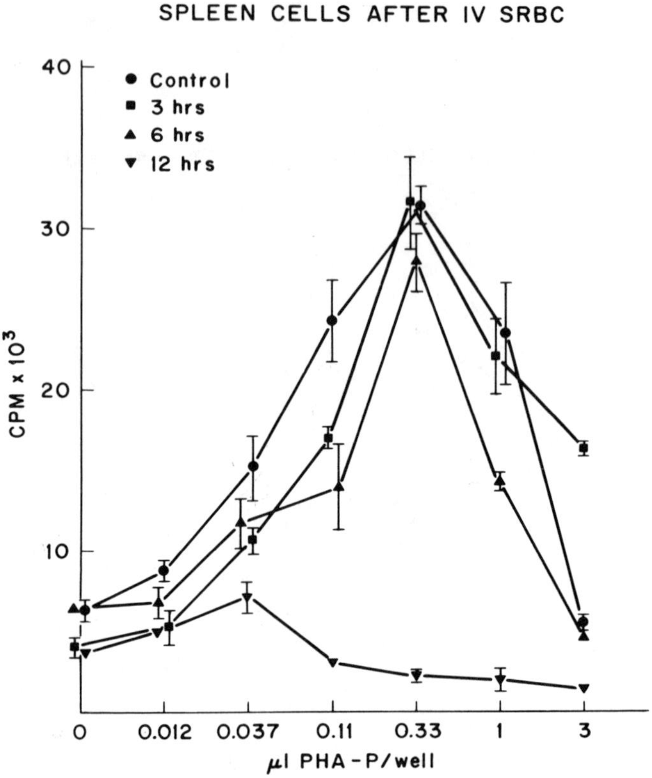

Fig. 2 Response of spleen cells from rats given an intravenous dose of 3 x 10^{10} SRBC at various times prior to culture with PHA.

TABLE I
ABROGATION OF ANTIGEN-INDUCED* SUPPRESSION BY REMOVAL OF ADHERENT CELLS

		^{3}H-T Incorporation (CPM x 10^3) PHA (µl/ml)	
Donor	Spleen Cells	0	4.5
Normal	Whole	6.3	31.5
Suppressed	Whole	5.6	2.7
	Nonadherent	2.9	54.5
	Nonadherent + 1% Macrophages	2.4	43.6

* 3 x 10^{10} SRBC, I.V., 24 hrs.

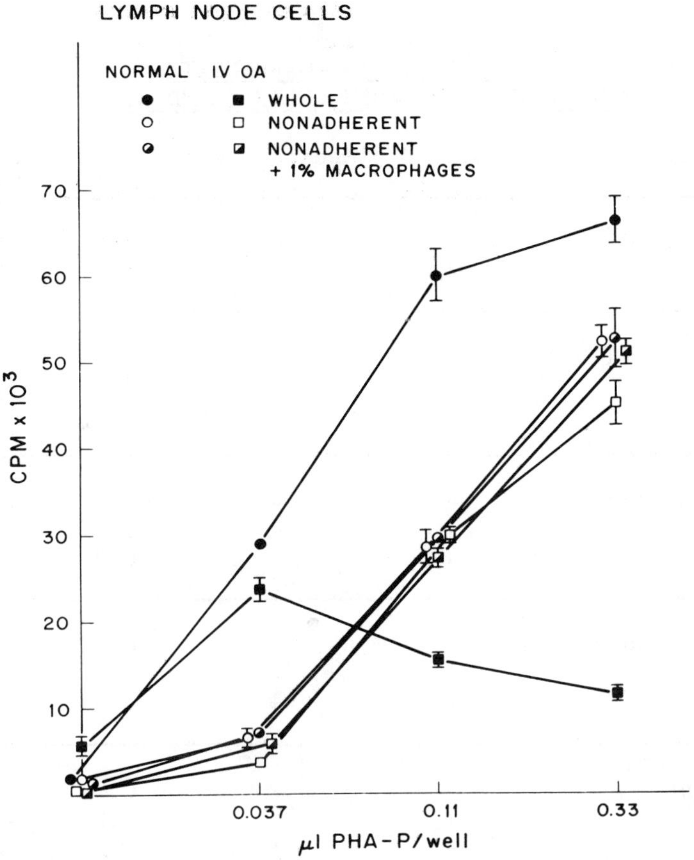

Fig. 3 Response of lymph node cells from normal rats and rats injected intravenously with 100 mg OA 24 hours before culture with PHA. Added macrophages are from normal donors.

TABLE II

SUPPRESSION WITH WHOLE AND NONADHERENT LYMPH NODE CELLS OF RATS AFTER FOOTPAD SENSITIZATION*

Added to Culture:		^{3}H-T Incorporation (CPM x 10^3)$^+$			
OA	PHA	Whole Cells		Nonadherent Cells	
(810 μg/ml)	(4.5 μl/ml)	Normal	OA Sens.	Normal	OA Sens.
-	+	8.3	2.8	7.8	8.2
+	-	<1	5.4	<1	<1
+	+	7.8	<u><1</u>	6.9	8.8

* 200 μg OA + CFA in footpads, 9 days.

$^+$ Δcpm at 48-72 hr.

TABLE III
HELPER ACTIVITY IN LOW-DENSITY
SENSITIZED LYMPH NODE CELLS (BAND B)

Target Cells 5×10^5	Band B 1×10^5	Macrophages 1×10^4	^{3}H-T Incorporation* OA+	PHA++
Sens. LNC, unfractionated	-	-	3.4	5.0
-	+	-	1.7	0.5
Sens. LNC, band D	-	-	1.7	8.7
+	+	-	3.2	99.1
+	-	+	2.0	7.0
Normal LNC, band D	-	-	0.3	18.3
+	+	-	0.4	102.8
+	-	+	0.2	40.0
Normal LNC, Nonadherent	-	-	0.8	14.3
+	+	-	0.1	44.5
+	-	+	0.8	18.0
Normal Spleen, Nonadherent	-	-	0.1	17.8
+	+	-	0.2	141.3
+	-	+	0.2	30.1

* Δcpm at 48-72 hr.
+ 270 μg/ml.
++ 1.5 μl/ml.

TABLE IV

SEPARATION OF HELPER AND SUPPRESSOR ACTIVITY IN SENSITIZED LYMPH NODE CELLS (BANDS B AND C)

Target Cells 5 x 10^5	Band B 1 x 10^5	Band C 1 x 10^5	^{3}H-T Incorporation* OA$^+$	PHA^{++}
Sens. LNC, unfractionated	-	-	6.6	68.7
-	+	-	1.9	0.1
-	-	+	0.2	4.5
Sens. LNC, Band D	-	-	2.8	77.5
+	+	-	8.6	105.3
+	-	+	3.1	7.7
Normal LNC, Nonadherent	-	-	0.1	51.4
+	+	-	16.9	85.6
+	-	+	0.1	84.5
Normal Spleen, Nonadherent	-	-	0.1	65.0
+	+	-	4.2	87.0
+	-	+	1.0	76.3

* Δ cpm at 48-72 hr.

\+ 270 μg/ml.

++ 1.5 μl/ml.

TABLE V

DOSE-DEPENDENT SUPPRESSOR ACTIVITY

Sensitized LNC Band D, x 10^5	Sensitized LNC Band B, x 10^5	^{3}H-T Incorporation* OA$^+$	PHA^{++}
5.0	-	13.2	54.2
5.0	0.5	1.9	0.4
5.0	1.0	2.6	15.5
5.0	2.0	20.5	21.6
-	5.0	19.2	8.8

* Δcpm at 48-72 hr.

\+ 270 μg/ml.

++ 1.5 μl/ml.

band cells were added. This result was reminiscent of the finding of Haskill and Axelrad that the blast cell fraction of spleen cells from mice primed with SRBC suppressed the small lymphocyte PFC response when added in small numbers but failed to suppress in larger numbers (9). As in the present experiments, the large cell subpopulation potentiated the response in some experiments (10) and suppressed it in others (9).

By separating sensitized LNC subpopulations and recombining them in different proportions we are apparently able to tip the balance of the response in either a positive or negative direction. The response of unfractionated LNC may be maintained at a given level by a regulator cell producing an inhibitory factor in amounts subject to feedback control. Increased concentrations of regulator cells would result in a turn-off suppression or an apparent potentiation. Alternatively, two separate subpopulations of regulatory cells may be involved, a suppressor cell population which exerts its effect at low cell concentrations and a minor "helper" cell population which can overcome suppression and produce potentiation at relatively higher cell concentrations. On the molecular level regulation might be mediated through antagonistic lymphokines such as mitogenic factor (11) and an inhibitor of DNA synthesis (12). Since low density bands of LNC also contain increased numbers of activated macrophages a role for these cells in the regulatory interaction must be considered. However, in our experiments, neither the addition of 2% purified macrophages nor selective removal of macrophages with high concentrations of carrageenan (13) affect the level of response to PHA.

REFERENCES

1. Gershon, R.K., in " Contemporary Topic in Immunobiology" volume 3 (ed. by Cooper, M.D. and Warner, N.L.), Plenum Publishing Corporation, New York, p. 1, 1973.
2. Folch, H., Yoshinaga, M. and Waksman, B.H., J. Immunol. 110: 835, 1973.
3. Folch, H. and Waksman, B.H., Cell. Immunol. 9: 12, 1973.
4. Folch, H. and Waksman, B.H., Cell. Immunol. 9: 25, 1973.
5. Folch, H. and Waksman, B.H., J. Immunol. 113: 127, 1974.
6. Colley, D.G., Shih-Wu, A. and Waksman, B.H., J. Exp. Med. 132: 1107, 1970.
7. Williams, M., Cell. Immunol. 9: 435, 1973.
8. Durkin, H.G. and Waksman, B.H., in "Lymphocyte Recognition and Effector Mechanisms" (ed. by Lindahl-Kiessling, K.

and Osoba, D., Academic Press Inc., New York) p. 599, 1973.
9. Haskill, J.S. and Axelrad, M.A., Nature New Biol. 237: 251, 1972.
10. Haskill, J.S. and Axelrad, M.A., Nature New Biol. 231: 219, 1971.
11. Wolstencraft, R.A. and Dumonde, D.C., Immunol. 18: 599, 1970.
12. Smith, R.T., Bauscher, M.S. and Adler, W.H., Amer. J. Path. 60: 495, 1970.
13. Lake, W.W., Bice, D., Schwartz, H.J. and Salvaggio, J., J. Immunol. 107: 1745, 1971.

DISCUSSION FOLLOWING SYMPOSIUM
REGULATION OF THE IMMUNE RESPONSE:
CELLULAR MECHANISMS

Cooper - Bethesda: In considering the heterogeneity of the T cell population, is there evidence to suggest that you are dealing with an initially heterogeneous population or rather a population of uniform cells which differentiate heterogeneously in the course of the response or even possibly a uniform population of cells which produce a host of factors which do different things following the response?

Feldmann - London: At the moment this is exactly one of the problems. There is no evidence as to where the heterogeneity arise. I think some of the heterogeneity arises before antigenic encounter. We don't know how many different factors are made by one cell and how many are made by different cells. We don't know if IgT is made by the same cell that makes suppressor IgT let alone whether these two molecules are the same.

E. Möller - Stockholm: I would like to ask you about the specificity of this factor you call IgT because you believe there is antigen specificity in the factor. We have debated this for years. I don't think there is very good proof of that. This must be a completely nonspecific product which seems to have the same effect as your IgT, or am I confused?

Feldmann - London: You are probably confused. I went through the experiment faster than I would have liked, but the chairman's clock induces very compulsive traits! Essentially the evidence for specificity of IgT is really based on some very old experiments performed in Australia. You have to have the right population of educated T cells and you had to have the right antigen. The way you can show specificity for the released molecule is in fact by showing that the cooperative effect was blocked by excess free carrier. Cooperation was only demonstrated if one had the appropriate hapten carrier and if you had excess free carrier present during the generation of supernatant, the capacity to induce a response was lost.

Oppenheim - Bethesda: Now you come up with a specific supernatant from macrophages which has not been found in most systems and in addition to a nonspecific or at least a genetically controlled one. Could you expand that a tiny bit so we can have a little raft to hold on to?

Feldmann - London: What is really transferring information in this system is antigen plus some component of the macrophage which is involved in triggering. We do not know much about the degree of specificity of the macrophage component to that antigen, and we will presume that they probably bind many others if not all antigens with the same degree of specificity.

Shearer - Bethesda: Do you know for certain whether Edna Mozes has verified that the TG-Pro-L unresponsiveness in the SWR mouse is also attributable to Ir 3 H-2 link or does she know that yet?

Taussig - Cambridge: She was doing that. I have not heard from her yet. I should have pointed that out.

Shearer - Bethesda: It was very interesting that your (T,G)-A--L factor in the presence of (T,G)-A--L antigen gave 19,000 plaques and in the presence of (Phe,G)-A--L antigen gave 16,000 plaques. Now this would speak for a cross reactivity at the factor level. Now we know from early work of McDevitt, however, that the strain distribution for (Phen,G)-A--L and (T,G)-A--L are quite different. So this would suggest in fact that the Ir defect for (T,G)-A--L is at a level other than the factor.

Taussig - Cambridge: I can explain that in rather a round about or complicated way, but superfically that is true. You have to make some rather special argument to get around that.

Feldmann - London: The main difference in the experiments you showed and our own are really in the form of the assay system used plus some minor details as to how the supernatants are generated. I think the major thing is that you use an in vivo while we use an in vitro assay. The major criticism I feel is your use of bone marrow as a source of B cells. I don't know whether you have done purified B cells from bone marrow. I think that the bone marrow contains a really quite different population of cells from the B cells in the spleen. The other major issue is the fact that you are really studying a phenomenon greatly removed in time from your experimental manipulations. I would feel that in view of a number of events which are known to occur between injection of antigen and a generation of antibody at the other end, that it could be acting upon any number of different pathways. The induction of T helper cells is quite complex and macrophage factors could act on subpopulated T cells present in your system.

Taussig - Cambridge: It is quite true. It is a valid criticism to say that we have not cleared the bone marrow of theta positive cells. I personally don't like the idea that the T cell factor is acting on another type of T cell for various reasons. I would agree that there is no reason to suppose that it is not acting by a macrophage as an intermediate cell. It is rather difficult using an in vivo system to detect how exactly each cell is going to recognize one thing and not another because it is very much of a black box type of situation. I also have an equal antipathy towards purely in vitro systems where I think certain manipulation situations may be just unphysiological.

Tomasi - Rochester: Let's assume for a moment that your factoris a complex of either specific or nonspecific between the Ir region peptide and a V region of an immunoglobulin. Have you or Dr. Tada done any experiments using antisera for adsorption which were directed against antigenic determinants on V regions or perhaps could you design an experiment with an idiotypic antisera to see whether that absorbed your factor? Admittedly, it is not a whole immunoglobulin.

Taussig - Cambridge: Yes. We have used circle anti-Fv serum which will probably partially answer that. Those are not active. This is an antibody against the Fv fragment which contains the V domains of both heavy and light chains. The idiotype experiment was a good one and I think that will be a second approach to it.

Paul - Bethesda: May I just embroider a little on Tom's question? Since we would assume that the pool of V regions that is expressed in immunoglobulin would probably not be expressed in a H-2 linked pool of variable regions such existed, one might anticipate that a V region type factor and the Ia factors would be on separate molecules; and, if that were the case, then the material coming through an anti-Ia column and an antigen column if mixed together might function. Have you had an opportunity to try that?

Feldmann - London: No, that is a very good idea! I'll write that down on a piece of paper.

Greaves - London: Could I ask if any of the speakers have

tested their factors for cytophilic capacity and in particular if the T cell produces the factor which the macrophage receives. Also is there any indication that responder and nonresponder macrophages are different?

Feldmann - London: No. We have not done that and I think it would be important to see whether the material is absorbed onto cells.

Katz - Boston: We have put AF onto various cell sources from various strains and find that the strain that absorbs out is the strain that produces it. In fact, something that was rather surprising to us, but really no conclusion can be drawn from it, is that even the target alloantigens strain failed to absorb it out. So there is a requirement for homology on some cell and I don't know which cell because we used whole spleen cell populations.

Dickler - Bethesda: Dr. Taussig has had a chance to test his factor with anti-immunoglobulins. I wonder if Feldmann has had a chance to test his with anti-Ia's?

Feldmann - London: The antisera against the Fv regions do block helper cell induction very efficiently. This blocking activity can be removed by immunoglobulin coupled onto Sepharose. Presumably it is anti-immunoglobulin and is the activity which can block helper cell induction.

McDevitt - Stanford: One thing about these sera is that they are not pure Fv and they do contain anti-Ig as well. This you can show by doing immunodiffusion with them. Is it not possible that your anti-Ig in that case is cross reacting with a H2 product of similar structure? I think there are a lot of problems with these sort of experiments. The thing which was striking to us was that for blocking helper cell induction, anti-μ doesn't work. What heavy chain class is IgT? I think that difficulty could be explained by the different ways the experiment is done. To answer Howard Dickler's question, if you throw in Ia anti-Ia antisera, one of which David Sachs sent me, you block the generation of helper cells. This still does not tell us what cell they react with. For example, the factor which Peter Erb has mapped which comes from macrophages, maps in the Ir-1A region, probably the same region as Dr. Taussig's factor. If one were to stretch the point, one could say it is conceivable that Dr. Taussig's factor is very similar to that and is in fact acting at a much earlier stage in the response. It is conceivable, to make the other point, that a second signal in the generation of many of these responses is one of these related products.

SUMMARY OF SYMPOSIUM
WILL WE RECOGNIZE THE ELEPHANT WHEN WE SEE IT?

Carl W. Pierce
Department of Pathology, Harvard Medical School
Boston, Massachusetts 02115

The immune mechanism of mammals is now appreciated to be an extremely complex, integrated and delicately balanced multi-cellular system whose activation via the stimulation of antigen-reactive precursor cells by specific antigen results in the development of two distinctly different types of immune responses: cellular and humoral immune responses. This Symposium has concentrated on some of the complexities involved in the development and regulation of antibody responses, which are now generally accepted to involve the participation of three functionally distinct types of cells: macrophages; thymus-derived cells (T cells); and, the precursors of the antibody producing cells (B cells). The development of optimal antibody responses by B cells to "thymus-dependent" antigens, such as hapten-protein conjugates, requires macrophages for "antigen presentation" and the cooperative or helper functions of T cells. Antibody responses to "thymus-independent" antigens, such as pneumococcal polysaccharides, appear not to require the helper functions of T cells; the participation of macrophages in these types of responses remains unresolved because of technical difficulties in obtaining absolutely macrophage-free lymphocyte populations for study. It is now appreciated that T cells, in addition to their cooperative or helper functions in B cell responses to antigen, may also have negative regulatory functions and suppress B cell responses to both thymus-dependent and independent antigens. Thus, the T cell appears to be the critical regulator of the expression of B cell responses to antigen. Still to be resolved, however, are the precise mechanisms by which T cells mediate these regulatory influences. The speakers in this Symposium have discussed the evidence for cell-to-cell contacts and soluble products, such as an immunoglobulin or "IgT" molecule or molecules encoded by the H-2 gene complex, and the participation of macrophages in these regulatory activities of T cells.

The rather simplified view of the segment of the immune mechanism involved in antibody responses, as presented above, is, I hope, sufficiently general that it is acceptable to everyone. However, is our current view of the mechanisms involved in the regulation of expression of antibody responses sufficiently complete that we can describe the entire system as an elephant? You are absolutely correct in asking at this point whether the author has taken leave of his senses (which is a distinct possibility) when equating the antibody response with an elephant. However, I am reminded of an ancient story which probably has parallels in all cultures of the world. A group of blind persons encountered a docile elephant while they were out for a stroll. An aggressive sightless one dared touch the beast, and upon encountering a leg and after examining it thoroughly declared "Not to fear, this is only a tree". Another, who had encountered one of the tusks of the elephant, declared "I agree, for I have found a limb of the tree". Another, however, who had examined an ear, declared "I'm confused, for I have obviously found a fan". Still another, after examining the trunk of the elephant, disagreed with the rest and said "You must be mistaken, this is a snake and if we don't leave in a hurry we will all be in trouble". As the group cautiously withdrew from the docile elephant, they engaged in a heated discussion about the nature of what they had just encountered. This discussion was really not unlike the current controversy about the mechanisms involved in the regulation of the antibody response. Each member of the group argued from his experiences with the part of the elephant that he had examined and each was adamant that his perceptions were correct and described the whole of what the group had encountered. Obviously, this group was unable to resolve their conflicts since they were unable to perceive the whole elephant. A moral to this story, if it has a moral, might be that one cannot accurately describe the entirety of a beast (the antibody response?) by examining only a single part while being unable to see the whole beast. Indeed, one may come to erroneous conclusions about the nature of the whole beast from one's limited observations. A corollary to this story would suggest that one should not criticize the observations of others because they disagree with one's own perceptions, since all of us may not be able to perceive the entire elephant at the present time. So much for philosophy, elephants and their relationship to antibody responses and whether our current perception of the antibody response may be correct or incorrect.

In this Symposium, we have heard four distinguished investigators who have described their perceptions of the mechanisms involved in the regulation of the expression of B cell responses to antigen. These perceptions, derived from each investigator's experiments of Nature, although possibly superficially contradictory in certain parts, must be telling us something about the nature of the entire elephant. It is now up to us to put all these pieces of information together and to describe as best we can the current state of the art. The observations described by our speakers can be broken down in several ways: specific vs non-specific T cell factors; immunoglobulin vs H-2 gene complex T cell factors; enhancing vs suppressing T cell factors; and factors which require the participation of macrophages vs factors which appear to act directly on B cells or on other T cells.

Marc Feldmann has described an immunoglobulin or IgT factor produced by T cells or cytophilic for T cells which acts in a specific manner to enhance or suppress B cell responses to antigen. It appears that the membrane of the macrophage which binds the antigen-IgT complex is the critical site for regulation of B cell responses to antigen by IgT. Indeed, there is much to be said for the simplicity and conservative nature of this model. It explains both the helper and suppressor functions of T cells by a single class of molecules and the end result appears to depend on the relative concentrations of IgT on the macrophage membrane. This model also implies that a single subset of T cells and a single factor are responsible for both the "helper" and "suppressor" functions of T cells.

On the other hand, we have heard three elegant and detailed studies about T cell factors which appear to be products of the H-2 gene complex, and not immunoglobulin in nature, and which may be either mediators of cooperative or suppressor functions of T cells. Two of these types of factors appear to be antigen-specific. The factor described by Taussig and Munro enhances B cell responses to antigen, whereas the factor described by Tomio Tada suppresses B cell responses to antigen. A third factor, described by Katz and Armerding, is also a product of the H-2 gene complex, but is not antigen-specific in the usual sense. This factor has been shown to substitute for the helper functions of T cells in B cell responses to antigen. At present, the exact relationship among these three factors and the opposing effects they mediate is unknown.

There are many questions that can be raised about these observations, both about the experimental systems where the observations were made and how these can be integrated into our overall view of the nature of the elephant. It does seem apparent, however, that products of the H-2 gene complex are of crucial importance in the regulation of expression of antibody responses. It is also apparent that many other observations, some of which do not concern H-2 gene complex products, must be integrated into the emerging view of regulation of antibody responses. All of these diverse and seemingly contradictory observations about regulatory mechanisms are most probably relevant, but in our "blindness" we are still unable to integrate them properly and accurately describe the elephant. It is more than likely that a process as complex as the antibody response has multiple pathways for achieving the same end result, just as the complement and coagulation systems have more than one pathway for activation. There is no reason to believe that the antibody response will be a simple one-pathway process. We must remain alert for these alternate pathways and explore them whenever we confront them. We must also continue to attempt to integrate all the diverse observations of the many investigators experimenting in the area of regulation of the antibody response into our models so that when we at last encounter the real elephant we will be able to recognize it.

AUTHOR INDEX

SUBJECT INDEX

D

E

F

G

H

I

J

L

M

A 5
B 6
C 7
D 8
E 9
F 0
G 1
H 2
I 3
J 4